This book is dedicated to
Nancy Tobey
A loving wife for thirty years,
An outstanding mother of three children,
A dedicated elementary teacher,
A helpful friend to so many

Contents

CHAPTER 3

CHAPTER 4

CHAPTER 5

CHAPTER 6

CHAPTER 7

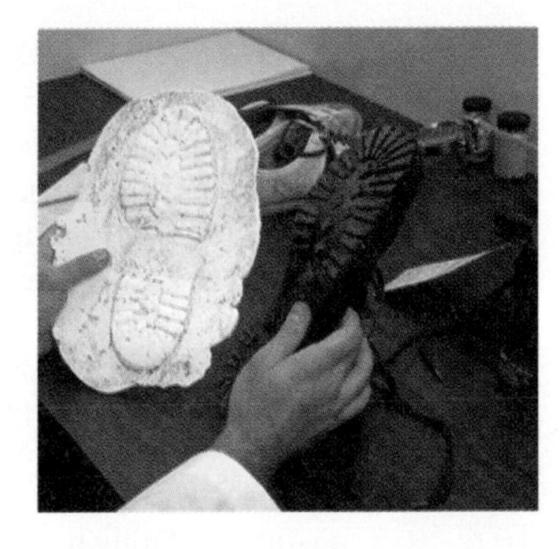

CHAPTER 8

CHAPTER 9

CHAPTER 10

Preface

A Message to the Mathematics Instructor

This text is designed to be used in a variety of class settings: lecture based classes, discussion oriented classes, self-paced classes, mathematics laboratories, and computer or audio-visual supported learning centers. The original manuscript and the first two editions of this book have been class tested by students across the country. The book was developed to help students learn and retain mathematical concepts. Special attention has been given to problem solving in this new edition. This text is written to help students organize the information in any problem solving situation, to reduce anxiety, and to provide a guide which enables the student to proceed with the problem solving process. *Basic College Mathematics,* Third Edition, is the first in a series of three texts that include the following:

Slater/Tobey	*Basic College Mathematics,* Third Edition
Tobey/Slater	*Beginning Algebra,* Fourth Edition
Tobey/Slater	*Intermediate Algebra,* Third Edition

This series is designed to prepare students for any mathematics course that is required for students to reach their future goals.

WHY DID WE CHANGE THE THIRD EDITION?

We have listened to teachers across the country, both those who are enthusiastic Slater/Tobey users as well as teachers who have not used the text. We have responded to their ideas and suggestions and have incorporated their invaluable insight into this edition. We have retained the same Table of Contents as the Second Edition but have extensively revised the exercises and the application problems.

EMPHASIS

DEVELOPING PROBLEM SOLVING ABILITIES

Throughout the country there is a renewed interest in improving the critical thinking, reasoning, and problem-solving skills of students. The interest is evident in government, the business community, and the national associations: AMATYC, NCTM, AMS, NADE, and MAA. Faculty have been encouraged to place greater emphasis on these areas of critical thinking, reasoning, and problem solving. In light of this focus, we have carefully drsigned *Basic College Mathematics* to facilitate this objective. It has been carefully structured to improve the students' ability to reason and solve applied problems. It is written to incorporate modeling and connecting to other disciplines. The student is encouraged to read, write, and communicate mathematical ideas. Where scientific calculators and the Internet will help student problem solving, provision is made to use these technologies. Students are exposed to new ideas to encourage exploration and mathematical study.

INCREASED NUMBER AND QUALITY OF APPLIED PROBLEMS

Other faculty and students have joined the authors in writing many new applied problems increasing the quantity, quality, and diversity of the world problems in the book. Now almost every exercise set in the text has some applied problems. The applications come from other academic disciplines, everyday life, and increased emphasis on global issues beyond the borders of the United States. Students will only engage in rich experiences that encourage thinking, analysis, and intellectual investigation if they face realistic and interesting application problems that attract them in the pursuit of knowledge. Over 50% of the applied problems of the new Third Edition have been changed in order to help students in this area.

PROVIDING A MATHEMATICAL BLUEPRINT FOR PROBLEM SOLVING

Often students in *Basic College Mathematics* course flounder because "they don't know where to begin" in the problem solving task. The new Third Edition has been re-written with a new emphasis on analyzing an applied problem. This unique feature helps students to begin the problem solving process and to plan the steps to be taken along the way; it provides them with an outline to organize their approach to solving problems. Often the hardest part in problem solving is determining where to begin. Once students fill in the blueprint, they can refer back to their plan as they do what is needed to solve the problem. Because of its flexibility, this feature can be used with single-step problems, multi-step problems, applications, and nonroutine problems that require problem solving strategies. Students will not need to use the blueprint to solve every problem. It is available for students who are faced with a problem with which they are not familiar, to alleviate their anxiety, to show them where to begin, and to assist them in the steps of reasoning.

INTERPRETING GRAPHS, CHARTS, AND TABLES IN THE PROBLEM SOLVING PROCESS

When students encounter mathematics in real world settings such as in the daily paper, USA Today, Time, Newsweek, or Sports Illustrated, or similar published materials, they often encounter data represented in a graph, a chart, or a table and are asked to make a reasonable conclusion based on the data presented. This emphasis on graphical interpretation is a trend that is continuing with the expanding technology of our day. The number of mathematical problems based on charts, graphs, and tables has been significantly increased in the Third Edition. Students are asked to make simple interpretations, to solve medium level problems, and to investigate challenging applied problems based on the data shown in a chart, graph, or table.

EXPANDED USE OF PUTTING YOUR SKILLS TO WORK APPLICATIONS

This widely-praised feature in the Second Edition has been expanded with new problems and more in depth student investigation for the new Third Edition. These non-routine application problems challenge the students to synthesize the knowledge they have gained so far and apply it to a totally new area. These unusual problems encourage critical thinking and a variety of approaches of analysis. Each problem is specifically arranged for **Cooperative Learning Activities and Group Investigation** of mathematical problems that pique student interest. Students are given an opportunity to work together to help one another discover mathematical solutions to extended problems. These investigations feature open-ended questions and extrapolation of data to areas beyond what is normally covered in such a course.

INTERNET CONNECTIONS

As an integral part of each Putting Your Skills to Work Problem students are exposed to an interesting application of the Internet and encouraged to continue their investigation. This use of modern technology inspires students to have confidence in their abilities to successfully use mathematics. It helps them to realize that they will not end their use of mathematics at the end of the course. Rather they are inspired to use mathematics and modern technology in the investigation of other disciplines and in their own pursuit of knowledge.

EXAMPLES AND EXERCISES

MASTERING MATHEMATICAL CONCEPTS

The examples and exercises in this text have been carefully chosen to guide students through *Basic College Mathematics*. We have incorporated several different types of exercises and examples to assist your students in retaining the content of this course.

Chapter Pretests
Each chapter opens with a concise pretest to familiarize the student with the learning objectives for that particular chapter.

Practice Problems
These are found throughout the chapter, after the examples, and are designed to provide your students with immediate practice of the skills presented. The complete worked out solution of each Practice Problem is contained in the back of the book in the answer section with yellow page trim on the edge.

To Think About
These critical thinking questions follow many of the examples in the text and also appear in the exercise sets. They extend the concept being taught; providing the opportunity for all students to stretch their minds, to look for patterns, and to make conclusions based on their previous experience.

Exercise Sets
Exercise Sets are paired and graded. This design helps to ease the students into the problems; and the answers provide students with immediate feedback.

Cumulative Review Problems
Almost every exercise set concludes with a section of *Cumulative Review Problems*. These problems review topics previously covered, and are designed to assist students in retaining the material. Many addition applied problems have been added to the Cumulative Review Sections.

Scientific Calculator Problems
Calculator boxes are placed in the margin of the text to alert students to a scientific calculator application. In the exercise section a scientific calculator icon is used to indicate problems that are designed for solving with a calculator.

REVIEWING MATHEMATICAL CONCEPTS

At the end of each chapter we have included problems and tests to provide your students with several different formats to help them review and reinforce the ideas that they have learned. This assists them not only with that specific chapter, but reviews previously covered topics as well.

Chapter Organizers
The concepts and mathematical procedures covered in each chapter is reviewed at the end of the chapter in a unique *Chapter Organizer*. This device has been extremely popular with faculty and students alike. It not only lists concepts and methods, but provides a completely worked out example for each type of problem. Students find that preparing a similar chapter organizer on their own in higher level math courses becomes an invaluable way to master the content of a chapter of material.

Verbal and Writing Skills
Writing exercises provide students with the opportunity to extend a mathematical concept by allowing them to use their own words, to clarify their thinking, and to become familiar with mathematical terms. These exercises have been included at the beginning of exercise sets to set the stage for the practice that follows, or at the end of the practice as a summary.

Review Problems
These problems are grouped by section as a quick refresher at the end of the chapter. These problems can also be used by the student as a quiz of the chapter material.

Tests
Found at the end of the chapter, the *Chapter Test* is a representative review of the material from that particular chapter that simulates an actual testing format. This provides the students with a gauge to their preparedness for the actual examination.

Cumulative Tests
At the end of each chapter is a *Cumulative Test*. One-half of the content of each *Cumulative Test* is based on the math skills learned

in previous chapters. By completing these tests for each chapter, the students build confidence that they have mastered not only the contents of the present chapter but of the previous chapters as well.

SUPPLEMENTS

FOR THE INSTRUCTOR

Annotated Instructor's Edition

- Complete student text
- Teaching Tips in the margin
- Answers appear next to every exercise.
- Answers to all Pretests, Review Problems, Tests, and Cumulative Tests.
- Instructor's Materials are in the front of the book. A thought provoking document prepared by the AMATYC called "Crossroads in Mathematics: Standards for Introductory College Mathematics Before Calculus." will help provide new ideas for teaching Basic College Mathematics.

Instructor's Solutions Manual

Worked solutions to all even-numbered exercises from the text and complete solutions for chapter review problems and chapter tests.

Test Item File: (Instructor's Manual with Tests)

Features nine tests per chapter plus four forms of a final examination, prepared and ready to be photocopied. Of these tests, 3 are multiple choice and 6 are free response. Two of the free response tests are cumulative in nature. The manual also contains 4 forms of a final examination.

TestPro (IBM and Macintosh)

This versatile testing system allows the instructor to easily create up to 99 versions of a customized test. Users may add their own test items and edit existing items in WYSIWYG format. Each objective in the text has at least one multiple choice and free response algorithm. Free upon adoption.

FOR THE STUDENT

Student Solutions Manual

Worked solutions to all odd-numbered exercises from the text and complete solutions for chapter review problems and chapter tests.

MathPro (INTERACTIVE COMPUTER LEARNING AND TUTORING): IBM and Macintosh

This tutorial software has been developed exclusively for Prentice Hall and is text-specific to this Tobey/Slater text. *MathPro* is designed to generate practice exercises based on the exercise sets in the text, and *MathPro* will provide the student with interactive help on each exercise. If a student is having difficulty working any exercise, they can ask to see a step-by-step example or to get step-by-step help in solving the exercise. At the end of each exercise, results are provided and when necessary, remediation is provided by referring the student to a specific section of the text for additional skills practice or review of key concepts. Students can also practice taking a test by accessing the Quiz mode. A record keeping function allows the instructor to track individual or section progress.

INTERACTIVE VIDEO LEARNING AND TUTORING

The Complete Video Series

Every section of this text is explained in detail on a videotape featuring Professor Michael Mayne and Professor John (Biff) Pietro of Riverside Community College. These experienced mathematics faculty members have been successfully teaching telecourses and traditional courses in developmental mathematics for several years. They have prepared the videotapes for several other Prentice-Hall mathematics texts and are considered to be two of the most effective mathematics teachers at the college level in the entire country. They are selected by a group of college students as the best mathematics teachers using a video presentation when compared to seven other mathematics faculty. Their videotapes are filled with humor and variety, and are designed to put the student at ease. The content of the tape follows the exact format of the textbook. Professors Mayne and Pietro work out the solutions to several even numbered problems in each exercise set.

GUIDE FOR STUDENTS

How to Use Basic College Mathematics, Third Edition to enrich your class experience and prepare for tests.

CHAPTER 1

Whole Numbers

Each chapter opens with an application. These applications relate to the material you find in the chapter as well as an extended discovery called *Putting Your Skills to Work* which you will encounter later in the chapter.

Just ten years ago hardly anyone in the United States leased a car for private use. New car leasing ads appear everywhere and growing numbers of Americans are leasing their own private car. But is this really a good decision? Could you figure out the mathematics to determine if leasing or buying would be best for you? Turn to the Putting Your Skills to Work Problem on page 57 and find out.

page 1

Putting Your Skills to Work

BUYING OR LEASING A CAR

Today over $\frac{1}{3}$ of all potential new car buyers decide to lease the car instead of purchasing it. What are the mathematics of buying versus leasing? How do we compare the costs?

Charlie and Sue Downey are considering leasing a new Dodge Stratus for $199 per month or purchasing the car for $14,996. See if you can determine the needed mathematical facts in order to make a well-

Put your new skills to work. These multi-part projects are relevant to chapter concepts as well to problems you encounter from day to day.

page 57

Explore on your own...

QUESTIONS FOR INDIVIDUAL INVESTIGATION

1. The lease is for a rental period of 48 months and requires a down payment of $1000. What is the basic minimum cost to lease the car for 4 years?
2. If the car is purchased with a down payment of $1000 and the remainder is paid for with a new car loan for 48 months with monthly payments of $370, what will it cost to purchase the car?

page 57

QUESTIONS FOR INDIVIDUAL INVESTIGATION

1. The lease is for a rental period of 48 months and requires a down payment of $1000. What is the basic minimum cost to lease the car for 4 years?
2. If the car is purchased with a down payment of $1000 and the remainder is paid for with a new car loan for 48 months with monthly payments of $370, what will it cost to purchase the car?

QUESTIONS FOR GROUP INVESTIGATION AND COOPERATIVE STUDY

Together with some members of your class, see if you can answer the following questions.

3. If the Downeys plan to keep the car at the end of the 4-year lease, they will need to purchase it from the dealer at a cost of $9352. In that case, what is the total cost of obtaining the car?
4. If the Downeys pla dealer after 4 years, charge of 15 cents p of 12,000 miles per car 64,000 miles ov the total approximat

Work cooperatively with fellow students to solve more involved problems. You will find that problems are often solved collaboratively in the workplace.

page 57

Surf the 'Net for real data that relates to the *Putting Your Skills to Work* explorations. Extend the Concepts!

INTERNET CONNECTIONS: Go to ''http://www.prenhall.com/tobey'' to be connected

Site: Microsoft Carpoint or a related site

This site includes a loan calculator which can help you determine the cost of financing a car. Use the loan calculator to answer the following questions.

6. Kimberly is buying a car for $13,000. She plans to make a 20% down payment, and she will finance the rest of the purchase price over a 3-year period. Her interest rate will be 9%. Find the amount of her monthly payment, and the total amount that she will pay for the car.
7. Jonathan wants to buy a car. He plans to make a 15% down payment and finance the rest over a 4-year period. His interest rate will be 7%. If Jonathan wants to keep his monthly car payment under $225, what is the most expensive car he can afford to buy?

page 57

These features have been included in this text to help you make connections to mathematics. Use them to explore, connect, and discover.

PREPARE YOURSELF

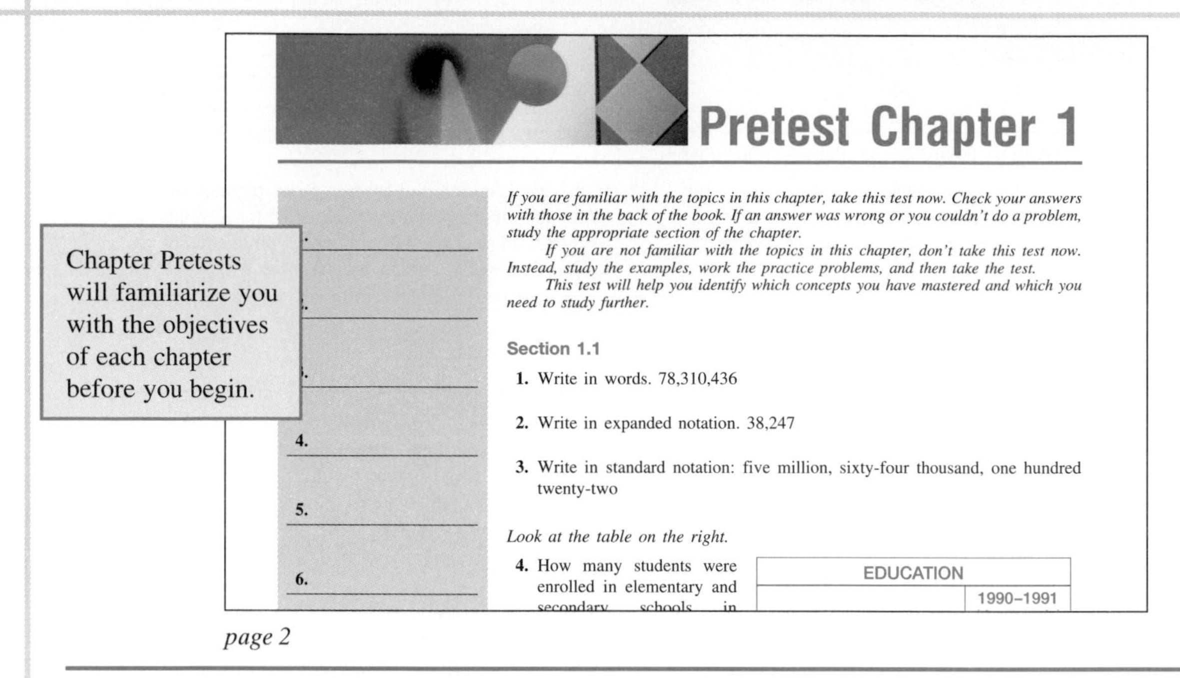

Pretest Chapter 1

If you are familiar with the topics in this chapter, take this test now. Check your answers with those in the back of the book. If an answer was wrong or you couldn't do a problem, study the appropriate section of the chapter.

If you are not familiar with the topics in this chapter, don't take this test now. Instead, study the examples, work the practice problems, and then take the test.

This test will help you identify which concepts you have mastered and which you need to study further.

Section 1.1

1. Write in words. 78,310,436

2. Write in expanded notation. 38,247

3. Write in standard notation: five million, sixty-four thousand, one hundred twenty-two

Look at the table on the right.

4. How many students were enrolled in elementary and secondary schools in

EDUCATION	
	1990–1991

4.

5.

6.

Chapter Pretests will familiarize you with the objectives of each chapter before you begin.

page 2

Have a plan—Develop a problem-solving strategy.

MATHEMATICS BLUEPRINT FOR PROBLEM SOLVING			
Gather the Facts	What Am I Asked to Do?	How Do I Proceed?	Key Points to Remember
At start of trip odometer read 28,353 miles. At end of trip odometer read 30,162 miles.	Find out how many miles Theofilos traveled.	I must subtract the two mileage readings.	Subtract the mileage at the start of the trip from the mileage at the end of the trip.

Use these features as guides to make yourself a better problem-solver.

page 75

Chapter Organizer		
Topic	Procedure	Examples
Place value of numbers, p. 4.	Each digit has a value depending on location. millions, hundred thousands, ten thousands, thousands, hundreds, tens, ones	In the number 2,896,341, w does 9 have? Ten thousan
Writing expanded notation, p. 5.	Take the number of each digit and multiply it by one, ten, hundred, thousand, . . . according to its place.	Write in expanded notation. 40,000 + 6000 + 20
Writing whole	Take the number of each group of three	Write in words. 134,718,21

page 84

Developing Your Study Skills

PROBLEMS WITH ACCURACY

Strive for accuracy. Mistakes are often made because of human error rather than lack of understanding. Such mistakes are frustrating. A simple arithmetic or sign error can lead to an incorrect answer.

These five steps will help you cut down on errors.

1. Work carefully, and take your time. Do not rush through a problem just to get it done.
2. Concentrate on the problem. Sometimes problems become mechanical, and your mind begins to wander. You become careless and make a mistake.
3. Check your problem. Be sure that you copied it correctly from the book.
4. Check your computations from step to step. Check the solution to the problem. Does it work? Does it make sense?
5. Keep practicing new skills. Remember the old saying ''Practice makes perfect.'' An increase in practice results in an increase in accuracy. Many errors are due simply to lack of practice.

There is no magic formula for eliminating all errors, but these five steps will be a tremendous help in reducing them.

page 150

CHECK YOUR UNDERSTANDING

Basic College Mathematics, Third Edition includes many different types of exercises—Apply yourself and *expand* your understanding.

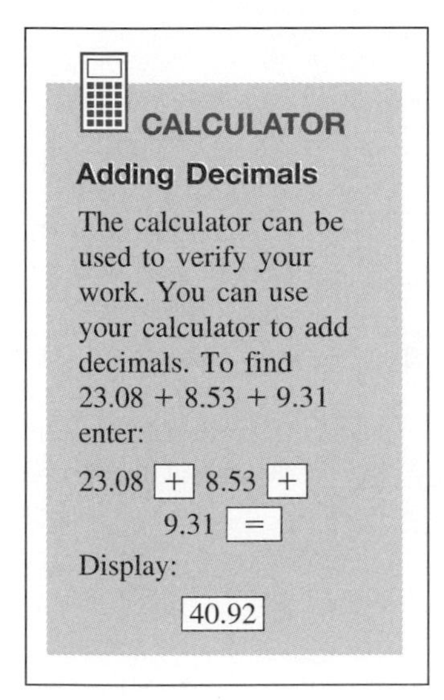

page 191

1.6 Exercises

Verbal and Writing Skills

1. Explain what the expression 5^3 means. Evaluate 5^3.
2. The ________________ tells how many times to multiply the base.
3. The ________________ is the number that is multiplied.
4. 10^5 is read as ________________________________.
5. Explain the order in which we perform mathematical operations to ensure consistency.

page 62

To Think About

75. Division of fractions is not commutative. For example, $\frac{2}{3} \div \frac{5}{7} \neq \frac{5}{7} \div \frac{2}{3}$.

In general, $\frac{a}{b} \div \frac{c}{d} \neq \frac{c}{d} \div \frac{a}{b}$ (a, b, c, d all $\neq 0$).

But sometimes there are exceptions. Can you think of any numbers a, b, c, d such that for fractions $\frac{a}{b}$ and $\frac{c}{d}$ it would be true that $\frac{a}{b} \div \frac{c}{d} = \frac{c}{d} \div \frac{a}{b}$?

76. Can you think of a way to divide $3\frac{7}{12} \div \frac{1}{4}$ *without* changing $3\frac{7}{12}$ to an

page 131

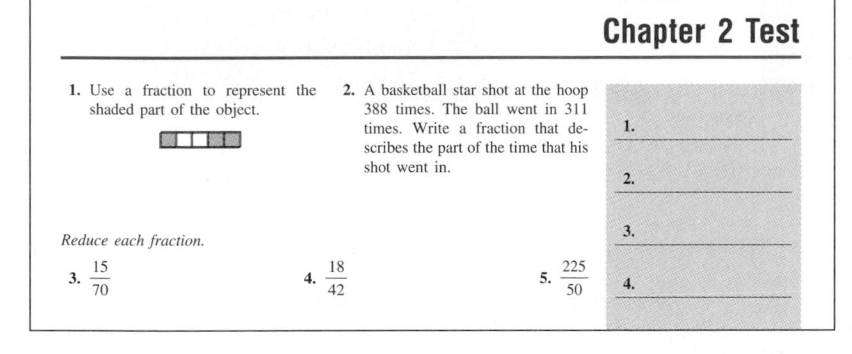

Chapter 2 Test

1. Use a fraction to represent the shaded part of the object.

2. A basketball star shot at the hoop 388 times. The ball went in 311 times. Write a fraction that describes the part of the time that his shot went in.

Reduce each fraction.

3. $\frac{15}{70}$

4. $\frac{18}{42}$

5. $\frac{225}{50}$

1. ________
2. ________
3. ________
4. ________

page 171

Cumulative Review Problems

77. Write in words. 39,576,304

78. Write in expanded form. 459,273

page 131

ENHANCE YOUR LEARNING

Basic College Mathematics, Third Edition is more than a textbook; it is an integrated package of instruction. Ask your professor about these supplements, which are a part of the **Basic College Mathematics** suite of learning materials.

Most items are keyed specifically to this text.

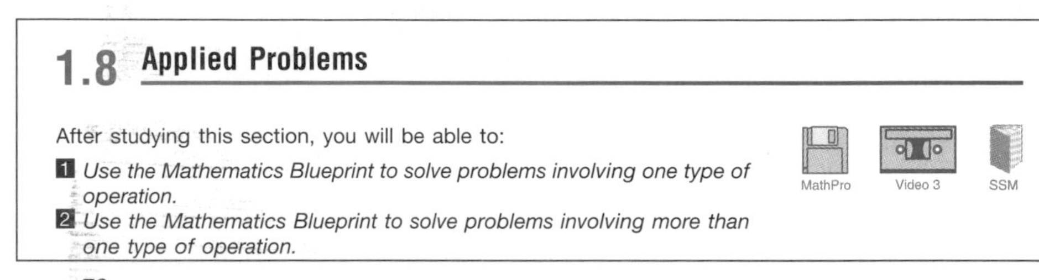

page 73

- Each section of the text begins with a reminder of the additional companion tools that have been designed to enhance your learning experience.

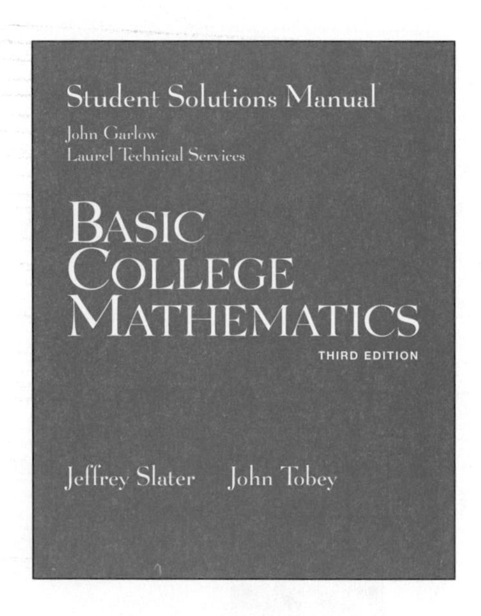

- Contains complete step-by step solutions for every odd-numbered exercise
- Contains complete step by step solutions for all Chapter Review Problems, Chapter Tests and Cumulative Tests

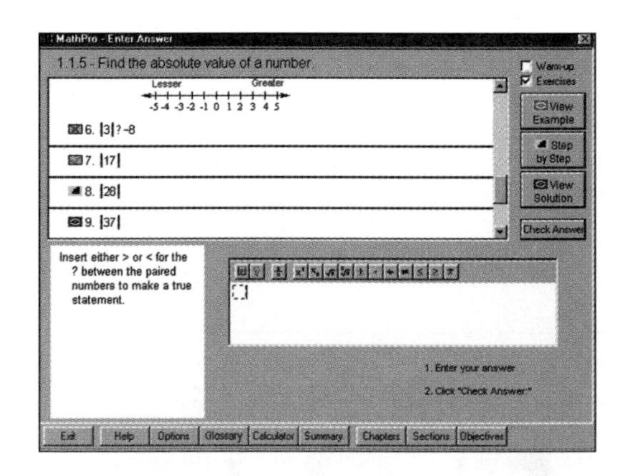

- MathPro Explorer: Interactive and Tutorial Software
- For Windows and Power Macintosh
- Includes preformatted algebra explorations
- Generates unlimited practice exercises

New York Times/Themes of the Times
Newspaper-format supplement–
ask your professor about this free supplement

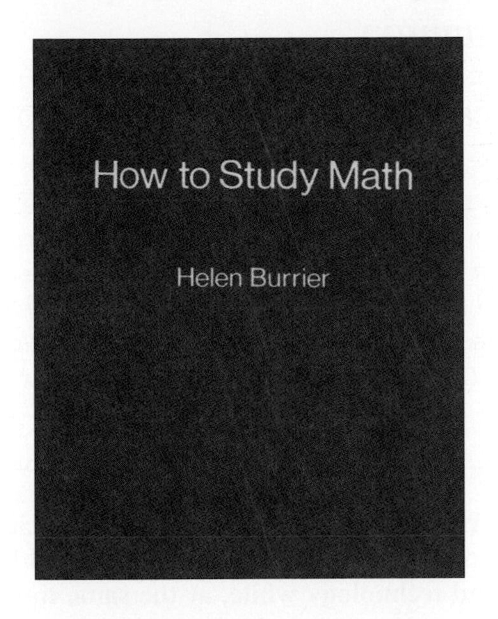

- Tips include how to prepare for class, how to study for and take tests, and how to improve your grades

- Team taught video instruction covers each section of the text

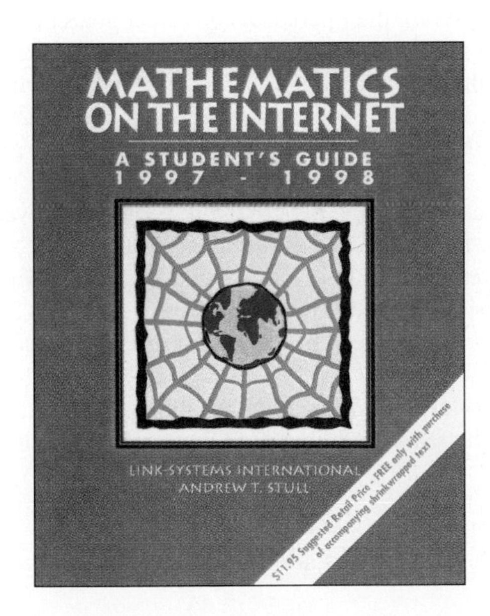

- A guide to navigation strategies through the Internet as well as practice exercises and lists of resources

- Visit the companion website for related and extended applications

www.prenhall.com/tobey

Crossroads in Mathematics

Standards for Introductory College Mathematics Before Calculus

Executive Summary

A Document Prepared by the American Mathematical Association of Two-Year Colleges (AMATYC)*

Participating in the preparation of this document were representatives from the American Mathematical Society, the Mathematical Association of America, the National Association of Developmental Education, and the National Council of Teachers of Mathematics. The preparation and dissemination of Crossroads in Mathematics *have been made possible through funding from the National Science Foundation, the Exxon Education Foundation, and Texas Instruments.*

Higher education is situated at the intersection of two major crossroads: A growing societal need exists for a well-educated citizenry and for a workforce adequately prepared in the areas of mathematics, science, engineering, and technology while, at the same time, increasing numbers of academically underprepared students are seeking entrance to postsecondary education.

Mathematics programs at two- and four-year colleges as well as at many universities serve students from diverse personal and academic backgrounds. These students begin their postsecondary education with a wide variety of educational goals and personal aspirations. Some enter college with solid mathematical backgrounds, ready to study calculus. Many others intend to study calculus but lack the preparation to do so. A third group can prepare for their personal and career goals through the study of mathematics below the level of calculus. The students in the last two groups are the ones whose needs are addressed in *Crossroads in Mathematics*. These students make up the majority of the nation's college students who are studying mathematics; they represent over 75 percent of the mathematics students at two-year colleges and 49 percent at four-year colleges and universities (*Statistical Abstract of the Undergraduate Programs in the Mathematical Sciences and Computer Science in the United States, 1990–91 CBMS Survey*).

Mathematics is a vibrant and growing discipline and is being used in more ways by more people than ever before. An alarming situation now exists, however, in the nation's postsecondary institutions. Each year greater numbers of students enter the mathematics ''pipeline'' at a point below the level of calculus, yet there is no significant gain in the numbers of students studying at higher levels. The failure of many of these students to persist in mathematics not only prevents them from pursuing their chosen careers, it also has a negative impact on our nation's economy, as fewer members of the work force are prepared for jobs in technical fields.

The purpose of *Crossroads in Mathematics* is to address the special circumstances of, establish standards for, and make recommendations about introductory college mathematics. The recommendations are based upon research evidence and the best judgment of the educators who contributed to the document.

*Inclusion of the Executive Summary of *Crossroads* does not constitute an endorsement of the text by AMATYC, or constitute AMATYCS's agreement that the text adheres to the contents of *Crossroads*.

Basic Principles

The following principles form the philosophical underpinnings of the document:

1. *All college students should grow in their knowledge of mathematics while attending college.* Students who are not prepared for college-level mathematics upon entering college will obtain the necessary knowledge by studying the Foundation.
2. *The mathematics that students study should be meaningful and relevant.* Mathematics should be introduced in the context of real, understandable problem-solving situations.
3. *Mathematics must be taught as a laboratory discipline.* Effective mathematics instruction should involve active student participation using in-depth projects.
4. *The use of technology is an essential part of an up-to-date curriculum.* Faculty and students will make effective use of appropriate technology.
5. *Students will acquire mathematics through a carefully balanced educational program that emphasizes the content and instructional strategies recommended in the standards along with the viable components of traditional instruction.* These standards emphasize problem solving, technology, intuitive understanding, and collaborative learning strategies. Skill acquisition, mathematical abstraction and rigor, and whole-class instruction, however, are still critical components of mathematics education.
6. *Increased participation in mathematics and in careers using mathematics is a critical goal in our heterogeneous society.* Mathematics education must reach out to all students.

The Standards

The standards provide goals for introductory college mathematics and guidelines for selecting content and instructional strategies for accomplishing the goals.

The *Standards for Intellectual Development* address desired modes of student thinking and represent goals for student outcomes.

Standard I-1: Problem Solving
Students will engage in substantial mathematical problem solving.

Standard I-2: Modeling
Students will learn mathematics through modeling real-world situations.

Standard I-3: Reasoning
Students will expand their mathematical reasoning skills as they develop convincing mathematical arguments.

Standard I-4: Connecting with Other Disciplines
Students will develop the view that mathematics is a growing discipline, interrelated with human culture, and understand its connections to other disciplines.

Standard I-5: Communicating
Students will acquire the ability to read, write, listen to, and speak mathematics.

Standard I-6: Using Technology
Students will use appropriate technology to enhance their mathematical thinking and understanding and to solve mathematical problems and judge the reasonableness of their results.

Standard I-7: Developing Mathematical Power
Students will engage in rich experiences that encourage independent, nontrivial exploration in mathematics, develop and reinforce tenacity and confidence in their abilities to use mathematics, and inspire them to pursue the study of mathematics and related disciplines.

The *Standards for Content* provide guidelines for the selection of content that will be taught at the introductory level.

Standard C-1: Number Sense

Students will perform arithmetic operations, as well as reason and draw conclusions from numerical information.

Standard C-2: Symbolism and Algebra

Students will translate problem situations into their symbolic representations and use those representations to solve problems.

Standard C-3: Geometry

Students will develop a spatial and measurement sense.

Standard C-4: Function

Students will demonstrate understanding of the concept of function by several means (verbally, numerically, graphically, and symbolically) and incorporate it as a central theme into their use of mathematics.

Standard C-5: Discrete Mathematics

Students will use discrete mathematical algorithms and develop combinatorial abilities in order to solve problems of finite character and enumerate sets without direct counting.

Standard C-6: Probability and Statistics

Students will analyze data and use probability and statistical models to make inferences about real-world situations.

Standard C-7: Deductive Proof

Students will appreciate the deductive nature of mathematics as an identifying characteristic of the discipline, recognize the roles of definitions, axioms, and theorems, and identify and construct valid deductive arguments.

The *Standards for Pedagogy* recommend the use of instructional strategies that provide for student activity and interaction and for student-constructed knowledge.

Standard P-1: Teaching with Technology

Mathematics faculty will model the use of appropriate technology in the teaching of mathematics so that students can benefit from the opportunities it presents as a medium of instruction.

Standard P-2: Interactive and Collaborative Learning

Mathematics faculty will foster interactive learning through student writing, reading, speaking, and collaborative activities so that students can learn to work effectively in groups and communicate about mathematics both orally and in writing.

Standard P-3: Connecting with Other Experiences

Mathematics faculty will actively involve students in meaningful mathematics problems that build upon their experiences, focus on broad mathematical themes, and build connections within branches of mathematics and between mathematics and other disciplines so that students will view mathematics as a connected whole relevant to their lives.

Standard P-4: Multiple Approaches

Mathematics faculty will model the use of multiple approaches—numerical, graphical, symbolic, and verbal—to help students learn a variety of techniques for solving problems.

Standard P-5: Experiencing Mathematics

Mathematics faculty will provide learning activities, including projects and apprenticeships, that promote independent thinking and require sustained effort and time so that students will have the confidence to access and use needed mathematics and other technical information independently, to form conjectures from an array of specific examples, and to draw conclusions from general principles.

Interpreting the Standards

The standards reflect many of the same principles found in school reform and calculus reform. They focus, however, on the needs and experiences of college students studying introductory mathematics in various instructional programs. These standards place emphasis on using technology as a tool and as an aid to instruction, developing general strategies for solving real-world problems, and actively involving students in the learning process.

In particular,

- The *Foundation* includes topics traditionally taught in ''developmental mathematics'' but also brings in additional that all students must understand and be able to use. Courses at this level should not simply be repeats of those offered in high school. Their goal is to prepare college students to study additional mathematics, thus expanding their educational and career options.
- *Technical programs* place strong emphasis on mathematics in the context of real applications. They should prepare students for the immediate needs of employment. At the same time, students should learn to appreciate the usefulness of mathematics and to use mathematics to solve problems in a variety of fields. The content and structure of the mathematics curriculum for technical students must be both rigorous and relevant. The mathematics that technical students study must broaden their options both in careers and in formal education.
- *Mathematics-intensive programs* include the study of calculus and beyond. Consequently, the mathematics that these students study must prepare them to be successful in a wide variety of calculus programs. The study of functions is the heart of precalculus education. While not departing from concerns about mathematical processes and techniques, more emphasis should be placed on developing student understanding of concepts, helping them make connections among concepts, and building their reasoning skills.
- *Liberal arts programs* are designed for bachelor degree–intending students majoring in the humanities and social sciences. These students should gain an appreciation for the roles that mathematics will play in their education, in their careers, and in their personal lives. Each institution has the responsibility of evaluating local needs and resources to determine how best to educate liberal arts majors beyond the Foundation. Options include interdisciplinary modules and introductory statistics.
- *Prospective teacher programs* should shift the emphasis from teaching isolated mathematical knowledge and skills to helping students to apply knowledge and develop in-depth understanding of the subject that goes beyond what they will be expected to teach. Courses for prospective teachers must provide an awareness of what research reveals about how children learn mathematics, models for effective pedagogy, and an understanding of the power and limitations of the use of technology in the classroom.

Implications

Although the standards focus on curriculum and pedagogy, they have wide implications for institutions and their mathematics programs.

- Professional development opportunities must be made available to all faculty members so that they may experience these reform recommendations as learners.
- Colleges must have laboratory and learning center facilities and provide adequate support personnel.
- Appropriate technology must be available for faculty and student use.
- Assessment instruments must measure the full range of what students are expected to know.
- Ongoing program evaluation must be used to make recommendations for improvement while retaining the effective aspects of the program.
- Articulation with high schools, with other colleges and universities, and with employers enables faculty at all levels to work in concert to improve mathematics education.

Implementation

Crossroads in Mathematics provides a framework for the development of improved curriculum and pedagogy. Adoption and implementation of the standards will require a systemic nationwide effort, which must be supported by postsecondary institutions, business and industry, and public and private funding agencies. Faculty, with the support of administrators, must lead the way. National professional organizations and their affiliated groups must promote reform through their conferences and by providing more extensive faculty development opportunities.

Looking to the Future

Introductory college mathematics holds the promise of opening new paths to future learning and fulfilling careers to an often neglected segment of the student population. Mathematics education at this level plays such a critical role in people's lives that its improvement is essential to our nation's vitality. *Crossroads in Mathematics* outlines a standards-based reform effort that will provide all students with a more engaging and valuable learning experience.

Crossroads in Mathematics was prepared by the Task Force and Writing Team of the Standards for Introductory College Mathematics Project with Don Cohen as Editor, Marilyn Mays as Project Director, and Karen Sharp and Dale Ewen as Project Co-Directors.

Individual copies of this document may be obtained by writing to AMATYC, State Technical Institute at Memphis, 5983 Macon Cove, Memphis, TN 38134. A limited number are available free. When the supply is exhausted, copies will be made available at a moderate charge.

Acknowledgments

This book is the product of many years of work and many contributions from faculty and students across the country. We would like to thank the many reviewers and participants in focus groups and special meetings with the authors.

Our deep appreciation to each of the following:

George J. Apostolopoulos, DeVry Institute of Technology

Katherine Barringer, Central Virginia Community College

Rita Beaver, Valencia Community College

Jamie Blair, Orange Coast College

Larry Blevins, Tyler Junior College

Joan P. Capps, Raritan Valley Community College

Brenda Callis, Rappahannock Community College

Robert Christie, Miami-Dade Community College

Mike Contino, California State University at Heyward

Judy Dechene, Fitchburg State University

Floyd L. Downs, Arizona State University

Barbara Edwards, Portland State University

Janice F. Gahan-Rech, University of Nebraska at Omaha

Colin Godfrey, University of Massachusetts, Boston

Carl Mancuso, William Paterson College

Janet McLaughlin, Montclair State College

Gloria Mills, Tarrant County Junior College

Norman Mittman, Northeastern Illinois University

Elizabeth A. Polen, County College of Morris

Ronald Ruemmler, Middlesex County College

Sally Search, Tallahassee Community College

Ara B. Sullenberger, Tarrant County Community College

Michael Trappuzanno, Arizona State University

Cora S. West, Florida Community College at Jacksonville

Jerry Wisnieski, Des Moines Community College

We have been greatly helped by a supportive group of colleagues who not only teach at North Shore Community College but who have provided a number of ideas as well as extensive help on all of our mathematics books. Also, a special word of thanks to Hank Harmeling, Tom Rourke, Wally Hersey, Bob McDonald, Judy Carter, Bob Campbell, Rick Ponticelli, Russ Sullivan, Kathy LeBlanc, Laura Connelly, Sharyn Sharaf, Donna Stefano and Nancy Tufo. Joan Peabody has done an excellent job of typing various materials for the manuscript and her help is gratefully acknowledged.

As a new edition of a book gets finalized new and fresh ideas are always helpful. We want to thank Sharyn Sharaf for contributing several new exercise problems in every section of the book. We also want to thank Louise Elton for providing several new applied problems and suggested applications. Error checking is a challenging task and few can do it well. So we especially want to thank Cindy Trimble, Sharyn Sharaf, and Rick Ponticelli for accuracy checking the content of the book at different stages of text preparation.

Each textbook is a combination of ideas, writing, and revisions from the authors and wise editorial direction and assistance from the editors. We want to thank our Prentice-Hall editor, Karin Wagner, for her administrative support and encouragement, and for her helpful insight and perspective on each phase of the revision of the textbook. Her patience, her willingness to listen, and her flexibility to adapt to changing publishing decisions has been invaluable to the production of this book. We also express our thanks to Jerome Grant and Melissa Acuña who have continued to support our writing projects and given wise direction and focus to our work.

Book writing is impossible for us without the loyal support of our families. Our deepest thanks and love to Nancy, Johnny, Melissa, Marcia, Shelley, Rusty, and Abby. Your understanding, your love and help, and your patience have been a source of great encouragement. Finally, we thank God for the strength and energy to write and the opportunity to help others through this textbook.

We have spent more than 25 years teaching mathematics. Each teaching day we find our greatest joy is helping students learn. We take a personal interest that each student has a good learning experience in taking this course. If you have some personal comments, suggestions, or ideas for future editions of this textbook, please write to us at:

Prof. John Tobey and Prof. Jeffrey Slater
Prentice Hall Publishing
Office of the College Mathematics Editor
One Lake Street
Upper Saddle River, NJ 07458

We wish you success in this course and in your future life!

John Tobey
Jeffrey Slater

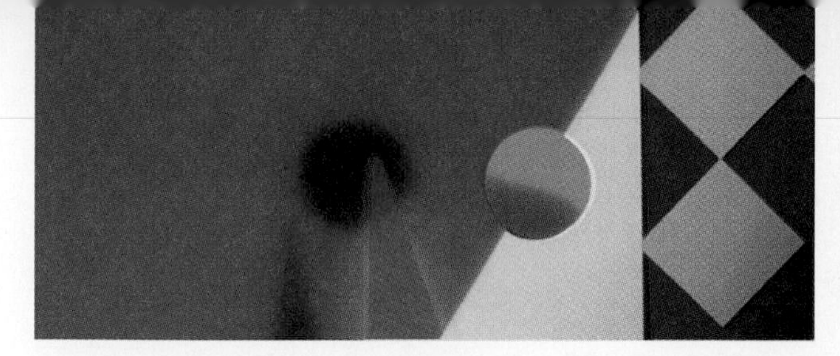

Diagnostic Pretest: Basic College Mathematics

Chapter 1

1. Add. $3846 + 527$

2. Divide. $58\overline{)1508}$

3. Subtract. $\begin{array}{r} 12,807 \\ -\ 11,679 \\ \hline \end{array}$

4. The highway department used 115 truckloads of sand. Each truck held 8 tons of sand. How many tons of sand were used?

Chapter 2

5. Add. $\frac{3}{7} + \frac{2}{5}$

6. Multiply and simplify.
$3\frac{3}{4} \times 2\frac{1}{5}$

7. Subtract. $2\frac{1}{6} - 1\frac{1}{3}$

8. Mike's car traveled 237 miles on $7\frac{9}{10}$ gallons of gas. How many miles per gallon did he achieve?

Chapter 3

9. Multiply. $\begin{array}{r} 51.06 \\ \times\ 0.307 \\ \hline \end{array}$

10. Divide. $0.026\overline{)0.0884}$

11. The copper pipe was 24.375 centimeters long. Paula had to shorten it by cutting off 1.75 centimeters. How long will the copper pipe be when it is shortened?

12. Russ bicycled 20.5 miles on Monday, 5.8 miles on Tuesday, and 14.9 miles on Wednesday. How many miles did he bicycle on those three days?

Chapter 4

Solve the following proportion problems. Round to the nearest tenth if necessary.

13. $\frac{3}{7} = \frac{n}{24}$

14. $\frac{0.5}{0.8} = \frac{220}{n}$

15. Wally's Landscape earned $600 for mowing lawns at 25 houses last week. At that rate, how much would he earn for doing 45 houses?

16. Two cities that are actually 300 miles apart appear to be 8 inches apart on the road map. How many miles apart are two cities that appear to be 6 inches apart on the map?

1. 4373

2. 26

3. 1128

4. 920 tons of sand

5. $\frac{29}{35}$

6. $8\frac{1}{4}$

7. $\frac{5}{6}$

8. 30 miles per gallon

9. 15.67542

10. 3.4

11. 22.625 centimeters

12. 41.2 miles

13. $n \approx 10.2$

14. $n = 352$

15. $1080 would be earned.

16. They are 225 miles apart.

17.	37.5%
18.	7728
19.	There are 3900 students at the college.
20.	0.3% are defective.
21.	3.75
22.	0.03
23.	3120
24.	4,900,000,000
25.	391 square meters
26.	$2747.50
27.	12 meters
28.	21,980 pounds
29.	500
30.	100
31.	1997
32.	625 cars per quarter

Chapter 5

Round all answers to the nearest tenth if necessary.

17. Change to a percent: $\frac{3}{8}$

18. 138% of 5600 is what number?

19. At Mountainview College 53% of the students are women. There are 2067 women at the college. How many students are at the college?

20. At a manufacturing plant it was discovered that 9 out of every 3000 parts made were defective. What percent of the parts are defective?

Chapter 6

Complete the following:

21. 15 qt = ____ gal

22. 3 cm = ____ meters

23. 1.56 tons = ____ lb

24. 4900 kg = ____ milligrams

Chapter 7

Round your answers to the nearest hundredth when necessary. Use $\pi \approx 3.14$ when necessary.

25. Find the area of a triangle with a base of 34 meters and an altitude of 23 meters.

26. Find the cost to install caret in a circular area with a radius of 5 yards at a cost of $35 per square yard.

27. In a right triangle the longest side is 15 meters and the shortest side is 9 meters. What is the length of the other side of the triangle?

28. How many pounds of fertilizer can be placed in a cylindrical tank that measures 4 feet tall and has a radius of 5 feet if one cubic foot of fertilizer weighs 70 pounds?

Chapter 8

The following double bar graph indicates the sale of Dodge Neons for Westover County as reported by the district sales managers. Use this graph to answer questions 29–32.

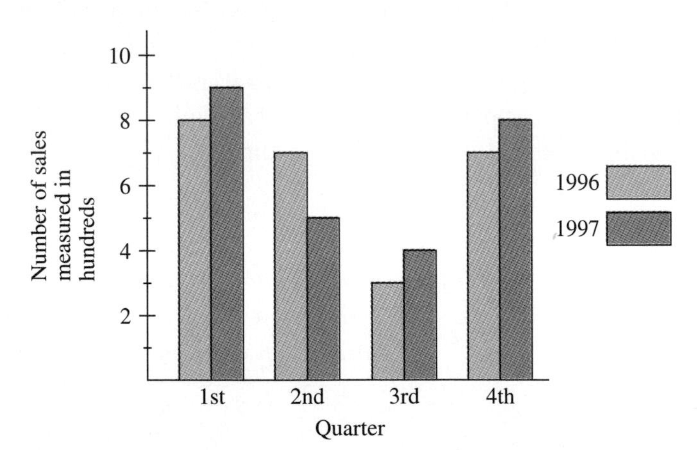

29. How many Dodge Neons were sold in the second quarter of 1997?

30. How many more Dodge Neons were sold in the fourth quarter of 1997 than were sold in the fourth quarter of 1996?

31. In which year were more Dodge Neons sold, in 1996 or in 1997?

32. What is the MEAN number of Dodge Neons sold per quarter in 1996?

Chapter 9

Perform the following operations:

33. $-5 + (-2) + (-8)$

34. $-8 - (-20)$

35. $\left(-\frac{3}{4}\right) \div \left(\frac{5}{6}\right)$

36. $(-3)(2)(-1)(-3)$

Chapter 10

Simplify.

37. $9(x + y) - 3(2x - 5y)$

Solve for x.

38. $3x - 7 = 5x - 19$

39. $2(x - 3) + 4x = -2(3x + 1)$

40. A rectangle has a perimeter of 134 meters. The length of the rectangle is 4 meters longer than double the width of the rectangle. What is the length and the width of the rectangle?

33. -15

34. 12

35. $-\frac{9}{10}$

36. -18

37. $3x + 24y$

38. $x = 6$

39. $x = \frac{1}{3}$

40. The width is 21 meters. The length is 46 meters.

CHAPTER 1

Whole Numbers

Just ten years ago hardly anyone in the United States leased a car for private use. New car leasing ads appear everywhere and growing numbers of Americans are leasing their own private car. But is this really a good decision? Could you figure out the mathematics to determine if leasing or buying would be best for you? Turn to the Putting Your Skills to Work Problem on page 57 and find out.

Pretest Chapter 1

1. Seventy-eight million, three hundred ten thousand four hundred thirty-six
2. 30,000 + 8,000 + 200 + 40 + 7
3. 5,064,122
4. 41,217,000
5. 2,237,000
6. 244
7. 50,570
8. 14,729
9. 2111
10. 76,311
11. 1,981,652
12. 108
13. 18,606
14. 6734
15. 331,420

If you are familiar with the topics in this chapter, take this test now. Check your answers with those in the back of the book. If an answer was wrong or you couldn't do a problem, study the appropriate section of the chapter.

If you are not familiar with the topics in this chapter, don't take this test now. Instead, study the examples, work the practice problems, and then take the test.

This test will help you identify which concepts you have mastered and which you need to study further.

Section 1.1

1. Write in words. 78,310,436

2. Write in expanded notation. 38,247

3. Write in standard notation: five million, sixty-four thousand, one hundred twenty-two

Look at the table on the right.

4. How many students were enrolled in elementary and secondary schools in 1990–1991?

5. How many high school graduates were there in 1991?

EDUCATION	1990–1991 (thousands)
Elementary and secondary	41,217
High school graduates	2,237

Section 1.2

Add.

6. $13 + 31 + 88 + 43 + 69$

7. $28,318 + 5,039 + 17,213$

8. $7148 + 500 + 19 + 7062$

Section 1.3

Subtract.

9. $6439 - 4328$

10. $100,450 - 24,139$

11. $45,861,413 - 43,879,761$

Section 1.4

Multiply.

12. $9 \times 6 \times 1 \times 2$

13. 2658×7

14. 91×74

15. 365×908

Section 1.5

Divide. If there is a remainder, be sure to state it as part of the answer.

16. $8\overline{)84{,}840}$ **17.** $7\overline{)51{,}633}$ **18.** $42\overline{)5838}$

Section 1.6

19. Write in exponent form. $6 \times 6 \times 6 \times 6$

20. Evaluate. 3^4

Perform the operations in their proper order.

21. $2 \times 3^3 - (4 + 1)^2$

22. $2^4 + (6^2 + 36) \div 2$

23. $6 \times 7 - 2 \times 6 - 2^3 + 17 \div 17$

Section 1.7

24. Round to the nearest thousand. 270,612

25. Round to the nearest hundred. 26,539

26. Round to the nearest million. 59,540,000

Estimate by first rounding each number so that there is only one nonzero digit and then performing the calculation.

27. $187 + 458 + 375 + 803 + 666$

28. $56{,}784 \times 459{,}202$

Section 1.8

Solve each of the following applied problems.

29. Emily planned a trip of 1483 miles from her home to Chicago. She traveled 317 miles on her first day of the trip. How many more miles must she travel?

30. Betsey purchased 84 shares of stock at a cost of $1848.00. How much was the cost per share?

31. Dr. Alfonso bought three chairs at $46 each, three lamps at $29 each, and two tables at $37 each for the waiting room of his office. How much did his purchases cost?

POPULATION FOR STATE OF ALASKA, U.S. CENSUS REPORT				
Year	1970	1980	1990	2000
Population	302,583	401,851	550,043	?

32. Using the table above, answer the following questions.
(a) By how much did the population in Alaska increase from 1980 to 1990?
(b) Some officials predict that between 1990 and 2000 the population of Alaska will increase by 134,890. If that figure is correct, what is the projected population of Alaska for the year 2000?

16. 10,605

17. 7376 R 1

18. 139

19. 6^4

20. 81

21. 29

22. 52

23. 23

24. 271,000

25. 26,500

26. 60,000,000

27. 2600

28. 30,000,000,000

29. 1166 miles

30. $22.00

31. $299

32. (a) 148,192

(b) 684,933

1.1 Understanding Whole Numbers

MathPro

Video 1

SSM

After studying this section, you will be able to:

1 *Write numbers in expanded form.*
2 *Write a word name for a number and write a number for a word name.*
3 *Read numbers in tables.*

1 *Writing Numbers in Expanded Form*

To count a number of objects or to answer the question "How many?" we use a set of numbers called **whole numbers.** These whole numbers are as follows.

0, 1, 2, 3, 4, 5, 6, 7, 8, 9, 10, 11, 12, 13, 14, 15, . . .

There is no largest whole number. The three dots . . . indicate that the set of whole numbers goes on indefinitely. Our number system is based on tens and ones and is called the *decimal system* (or the *base 10 system*). The numbers 0, 1, 2, 3, 4, 5, 6, 7, 8, 9 are called **digits.** The position, or placement, of the digits in the number tells the value of the digits. For example, in the number 521, the "5" means 5 hundreds (500). In the number 54, the "5" means 5 tens (50).

521 — 5 means 5 hundred or 500

54 — 5 means 5 tens or 50

For this reason, our number system is called a *place-value system.*

Consider the number 5643. We will use a place-value chart to illustrate the value of each digit in the number 5643.

millions			thousands			ones		
Hundred millions	Ten millions	Millions	Hundred thousands	Ten thousands	Thousands	Hundreds	Tens	Ones
					5	6	4	3

The value of the number is 5 thousands, 6 hundreds, 4 tens, 3 ones.

The place-value chart above shows the value of each place, from ones on the right to hundred millions on the left. When we write very large numbers, we place a comma after every group of three digits, moving from right to left. This makes the number easier to read. It is usually agreed that a four-digit number does not have a comma, but that numbers with five or more digits do. So 32,000 would be written with a comma but 7000 would not.

To show the value of each digit in a number, we sometimes write the number in expanded notation. For example, 56,327 in expanded notation is

$$50{,}000 + 6000 + 300 + 20 + 7$$

EXAMPLE 1 Write each number in expanded notation.

(a) 2378 **(b)** 508,271 **(c)** 980,340,654

(a) Sometimes it helps to say the number to yourself.

	two thousand		three hundred		seventy		eight
2378 =	2000	+	300	+	70	+	8

(b) When 0 is used as a placeholder, you do not include it in the expanded form.

$$508{,}271 = \underbrace{500{,}000 + 8000 + 200 + 70 + 1}_{\text{Expanded notation}}$$

(c) $980{,}340{,}654 = \underbrace{900{,}000{,}000 + 80{,}000{,}000 + 300{,}000 + 40{,}000 + 600 + 50 + 4}_{\text{Expanded notation}}$

Practice Problem 1 Write each number in expanded notation.

(a) 3182 **(b)** 520,890 **(c)** 709,680,059 ■

The number as you usually see it is called the **standard notation.** 980,340,654 is the standard notation for the number nine hundred eighty million, three hundred forty thousand, six hundred fifty-four.

EXAMPLE 2 Write each number in standard notation.

(a) 500 + 30 + 8 **(b)** 300,000 + 7000 + 40 + 7

(a) 538

(b) Be careful to keep track of the place value of each digit. You may need to use 0 as a placeholder.

$$\underbrace{300{,}000}_{\text{3 hundred thousand}} + \underbrace{7000}_{\text{7 thousand}} + 40 + 7 = 307{,}047$$

We needed to use 0 in the ten thousands place and in the hundreds place.

TEACHING TIP You can remind students that a few ancient cultures actually avoided the use of zero by never writing numbers like 40. Instead they would make the number larger by writing 41 or smaller by writing 39. Needless to say, such a culture had a hard time developing effective records for business transactions.

Practice Problem 2 Write each number in standard notation.

(a) 400 + 90 + 2 **(b)** 80,000 + 400 + 20 + 7 ■

EXAMPLE 3 Last year the population of Central City was 1,509,637. In the number 1,509,637

(a) How many ten thousands are there?

(b) How many tens are there?

(c) What is the value of the digit 5?

(d) In what place is the digit 6?

The place-value chart will help you to identify the value of each place.

(a) Look at the digit in the ten thousands place. There are 0 ten thousands.

(b) Look at the digit in the tens place. There are 3 tens.

(c) The digit 5 is in the hundred thousands place. The value of the digit is

5 hundred thousand or 500,000

(d) The digit 6 is in the hundreds place.

Practice Problem 3 The campus library has 904,759 books.

(a) What digit tells the number of hundreds?

(b) What digit tells the number of hundred thousands?

(c) What is the value of the digit 4?

(d) What is the value of the digit 9? Why does this question have two answers? ■

2 *Writing Word Names for Numbers and Numbers for Word Names*

A number has the same *value* no matter how we write it. For example, "a million dollars" means the same as "$1,000,000." In fact, any number in our number system can be written in several ways or forms:

• Standard notation	521
• Expanded notation	500 + 20 + 1
• In words	five hundred twenty-one

You may want to write a number in any of these ways. To write a check, you need to use both standard notation and words.

To write a word name for a number, think about how you might say the number.

TEACHING TIP Students may wonder why they need to write the word name for a number. Remind them that they need to do this when making out a check or withdrawal slip for a passbook savings account. Sometimes banks will delay processing a check if the incorrect word name for the number is written on it.

EXAMPLE 4 Write the word name for each number.

(a) 1695 **(b)** 200,470 **(c)** 7,003,038

Look at the place-value chart if you need help in identifying the place for each digit.

(a) Note that this number does not represent a date. It is a number. The number begins with 1 in the thousands place. The word name is

One thousand six hundred ninety-five

↑ We use a hyphen here.

(b) The number begins with 2 in the hundred thousands place.

Two hundred thousand, four hundred seventy

(c) The number begins with 7 in the millions place.

Seven million, three thousand, thirty-eight

Practice Problem 4 Write the word name for each number.

(a) 2736 **(b)** 980,306 **(c)** 12,000,021 ■

Very large numbers are used to measure quantities in some disciplines, such as distance in astronomy and the national debt in macroeconomics. We can extend the place-value chart to include these large numbers.

The national debt for the United States as of December 1996 was \$5,311,192,809,000. This number is indicated in the place-value chart below.

Trillions			Billions			Millions			Thousands			Ones		
		5	3	1	1	1	9	2	8	0	9	0	0	0

EXAMPLE 5 Write the national debt for the United States as of December 1996 in the amount of \$5,311,192,809,000 using a word name.

The national debt in December 1996 was five trillion, three hundred eleven billion, one hundred ninety-two million, eight hundred nine thousand dollars.

Practice Problem 5 The U.S. Treasury reported that the national debt in 1989 was \$2,857,406,300,000. Write the amount of this debt using a word name. ■

Occasionally you may want to write a word name as a number.

EXAMPLE 6 Write each number in standard notation.

(a) four hundred nincty-two

(b) twenty-six thousand, eight hundred fifty-four

(c) two billion, three hundred eighty-six million, five hundred forty-seven thousand, one hundred ninety

(a) 492

(b) 26,854

(c) 2,386,547,190

Practice Problem 6 Write in standard notation.

(a) eight hundred three **(b)** thirty thousand, two hundred twenty-nine ■

3 *Reading Numbers in Tables*

Sometimes numbers found in charts and tables are abbreviated. Look at the chart below from the *World Almanac.* Notice the third line tells us the numbers represent thousands. To understand what these numbers mean, think thousands. If the number 23 appears under 1740 for New Hampshire, the 23 represents 23 thousand. 23 thousand is 23,000. Note that census figures for some colonies are not available for certain years.

Estimated population of the American colonies from 1650 to 1780 (in thousands)

Source: U.S. Bureau of the Census

Colony	1650	1670	1690	1700	1720	1740	1750	1770	1780
Maine	1	★	★	★	★	★	★	31	49
New Hampshire	1	2	4	5	9	23	28	62	88
Vermont	★	★	★	★	★	★	★	10	48
Plymouth and Massachusetts	16	35	57	56	91	152	188	235	269
Rhode Island	1	2	4	6	12	25	33	58	53
Connecticut	4	13	22	26	59	90	111	184	207

EXAMPLE 7 Refer to the chart above to answer the following questions. Write each number in standard notation.

(a) What was the estimated population of Maine in 1780?

(b) What was the estimated population of Plymouth and Massachusetts in 1720?

(c) What was the estimated population of Rhode Island in 1700?

(a) To read the chart, first look for Maine on the left. Read across to the column under 1780. The number is 49. In this chart 49 means 49 thousands.

$$49 \text{ thousands} \Rightarrow 49{,}000$$

(b) Read the line of the chart for Plymouth and Massachusetts. The population of Plymouth and Massachusetts in 1720 was 91. This means 91 thousands. We will write this as 91,000.

(c) Read the line of the chart for Rhode Island. The population of Rhode Island in 1700 was 6. This means 6 thousands. We will write this as 6000.

To Think About Why do you think Plymouth and Massachusetts had the largest population for the years shown in the table?

Practice Problem 7 Refer to the chart above to answer the following questions. Write each number in standard notation.

(a) What was the estimated population of Connecticut in 1670?

(b) What was the estimated population of New Hampshire in 1780?

(c) What was the estimated population of Vermont in 1770? ■

1.1 Exercises

Write each number in expanded notation.

1. 6731
6000 + 700 + 30 + 1

2. 9519
9000 + 500 + 10 + 9

3. 108,276
100,000 + 8,000 + 200 + 70 + 6

4. 350,765
300,000 + 50,000 + 700 + 60 + 5

5. 23,761,345
20,000,000 + 3,000,000 + 700,000 + 60,000 + 1000 + 300 + 40 + 5

6. 46,198,253
40,000,000 + 6,000,000 + 100,000 + 90,000 + 8000 + 200 + 50 + 3

7. 103,260,768
100,000,000 + 3,000,000 + 200,000 + 60,000 + 700 + 60 + 8

8. 820,310,574
800,000,000 + 20,000,000 + 300,000 + 10,000 + 500 + 70 + 4

Write each number in standard notation.

9. 600 + 70 + 1
671

10. 500 + 90 + 6
596

11. 9000 + 800 + 60 + 3
9863

12. 7000 + 600 + 50 + 2
7652

13. 70,000 + 6000 + 30 + 6
76,036

14. 80,000 + 2000 + 300 + 5
82,305

15. 700,000 + 6000 + 200
706,200

16. 900,000 + 50,000 + 40 + 7
950,047

Verbal and Writing Skills

17. In the number 56,782
(a) What digit tells the number of hundreds? 7
(b) What is the value of the digit 5? 50,000

18. In the number 123,769
(a) What digit tells the number of tens? 6
(b) What is the value of the digit 2? 20,000

19. In the number 984,328
(a) What digit tells the number of ten thousands? 8
(b) What is the value of the digit? 80,000

20. In the number 6,789,345
(a) What digit tells the number of thousands? 9
(b) What is the value of the digit? 9000

Write a word name for each number.

21. 53
Fifty-three

22. 46
Forty-six

23. 8936
Eight thousand, nine hundred thirty-six

24. 4629
Four thousand, six hundred twenty-nine

25. 36,118
Thirty-six thousand, one hundred eighteen

26. 55,742
Fifty-five thousand, seven hundred forty-two

27. 105,261
One hundred five thousand, two hundred sixty-one

28. 370,258
Three hundred seventy thousand, two hundred fifty-eight

29. 23,561,248
Twenty-three million, five hundred sixty-one thousand, two hundred forty-eight

30. 19,376,584
Nineteen million, three hundred seventy-six thousand, five hundred eighty-four

31. 4,302,156,200
Four billion, three hundred two million, one hundred fifty-six thousand, two hundred

32. 7,436,210,400
Seven billion, four hundred thirty-six million, two hundred ten thousand, four hundred

Write each number in standard notation.

33. Three hundred seventy-five
375

34. Seven hundred thirty-six
736

35. Fifty-six thousand, two hundred eighty-one
56,281

36. Seventy-nine thousand, three hundred forty-six
79,346

37. One hundred million, seventy-nine thousand, eight hundred twenty-six 100,079,826

38. Four hundred fifty million, three hundred thousand, two hundred forty-nine 450,300,249

Applications

When writing a check, a person must write the word name for the dollar amount of the check.

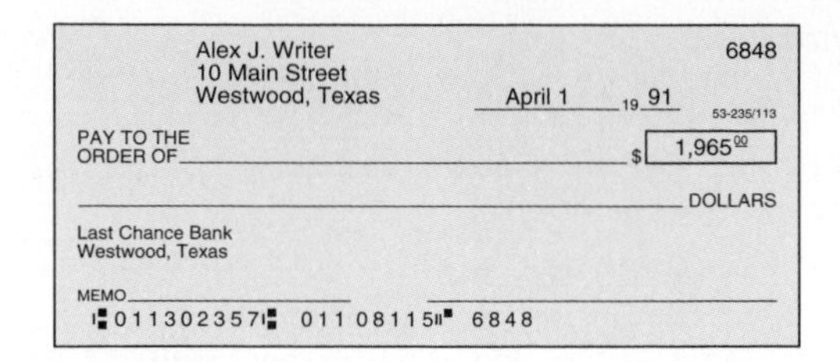
Alex J. Writer
10 Main Street
Westwood, Texas
6848
April 1 19 91
53-235/113
PAY TO THE ORDER OF
$ 1,965 00
DOLLARS
Last Chance Bank
Westwood, Texas
MEMO
0113023571 011 08115 6848

39. Alex bought new equipment for his laboratory for \$1965. What word name should he write on the check? One thousand, nine hundred sixty-five

40. Alex later bought a new personal computer for \$6383. What word name should he write on the check? Six thousand, three hundred eighty-three

In problems 41–44 use the following chart prepared with data from the World Almanac. *Notice the second line tells us that the numbers represent millions. These values are only approximate values representing numbers written to the nearest million. They are not exact census figures.*

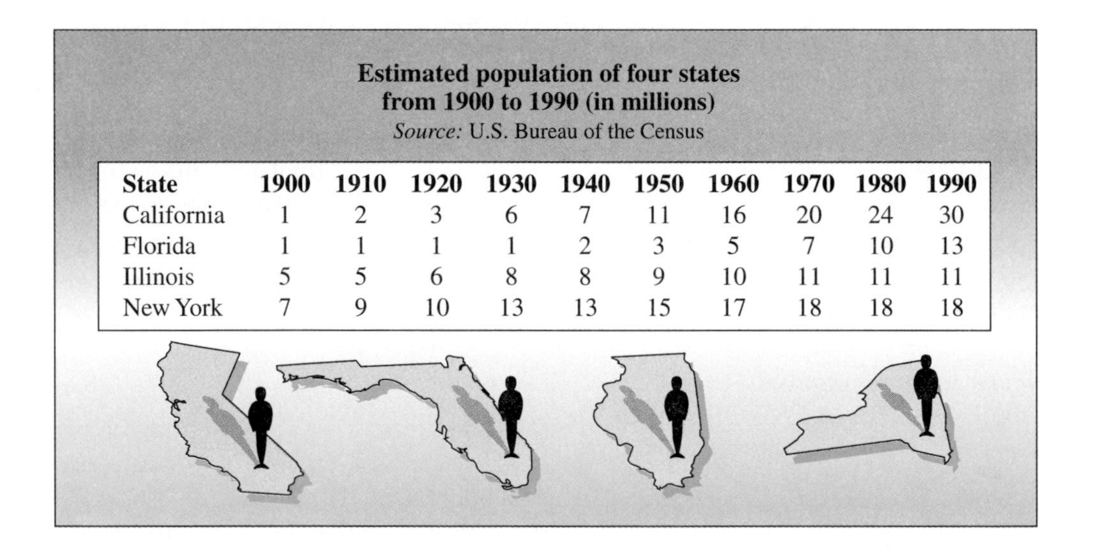
Estimated population of four states from 1900 to 1990 (in millions)
Source: U.S. Bureau of the Census

State	1900	1910	1920	1930	1940	1950	1960	1970	1980	1990
California	1	2	3	6	7	11	16	20	24	30
Florida	1	1	1	1	2	3	5	7	10	13
Illinois	5	5	6	8	8	9	10	11	11	11
New York	7	9	10	13	13	15	17	18	18	18

41. What was the estimated population of New York in 1910? 9 million or 9,000,000

42. What was the estimated population of Florida in 1970? 7 million or 7,000,000

43. What was the estimated population of California in 1960? 16 million or 16,000,000

44. What was the estimated population of Illinois in 1950? 9 million or 9,000,000

In problems 45–48 use the following chart prepared with data from the World Almanac. *Notice the third line that tells us the numbers represent billions of dollars. These values are only approximate values representing numbers written to the nearest billion. They are not exact financial figures.*

PERSONAL CONSUMPTION EXPENDITURES IN THE UNITED STATES FROM 1985 TO 1991

Source: Bureau of Economic Analysis, U.S. Department of Commerce (billions of dollars)

Category	1985	1986	1987	1988	1989	1990	1991
Transportation	360	366	380	413	437	454	438
Medical care	328	358	399	488	536	596	656
Recreation	186	201	223	247	266	281	290
Religious and welfare activities	57	63	68	86	93	102	108

45. How much did people in the United States spend in 1986 on recreation?
$201 billion or $201,000,000,000

46. How much did people in the United States spend in 1989 on medical care?
$536 billion or $536,000,000,000

47. How much did people in the United States spend in 1987 on transportation?
$380 billion or $380,000,000,000

48. How much did people in the United States spend in 1991 on religious and welfare activities? $108 billion or $108,000,000,000

To Think About

Specify the digit.

49. The speed of light is approximately 29,979,250,000 centimeters per second.
(a) What digit tells the number of ten thousands? 5
(b) What digit tells the number of ten billions? 2

50. The radius of the earth is approximately 637,814,000 centimeters.
(a) What digit tells the number of ten thousands? 1
(b) What digit tells the number of ten millions? 3

51. In 1994, more than 4,300,000 immigrants came to the United States from Mexico.
(a) Which digit tells the number of hundred thousands? 3
(b) Which digit tells the number of millions? 4

52. The world's population is expected to reach 7,900,000,000 by the year 2020, according to the U.S. Bureau of the Census.
(a) Which digit tells the number of hundred millions? 9
(b) Which digit tells the number of billions? 7

53. Write in standard notation: six hundred thirteen trillion, one billion, thirty-three million, two hundred eight thousand, three.
613,001,033,208,003

54. Write a word name for 3,682,968,009,931,960,747. (*Hint:* The digit 1 followed by 18 zeros represent the number *1 quintillion.* 1 followed by 15 zeros represents the number *1 quadrillion.*)
Three quintillion, six hundred eighty-two quadrillion, nine hundred sixty-eight trillion, nine billion, nine hundred thirty-one million, nine hundred sixty thousand, seven hundred forty-seven

1.2 Addition of Whole Numbers

MathPro

Video 1

SSM

After studying this section, you will be able to:

1. *Master basic addition facts.*
2. *Add several single-digit numbers.*
3. *Add numbers of several digits when carrying is not needed.*
4. *Add numbers of several digits when carrying is needed.*
5. *Apply addition to real-life situations.*

1 *Mastering Basic Addition Facts*

We see the addition process time and time again. Carpenters add to find the amount of lumber they need for a job. Auto mechanics add to make sure they have enough parts in the inventory. Bank tellers add to get cash totals.

What is addition? We do addition when we put sets of objects together.

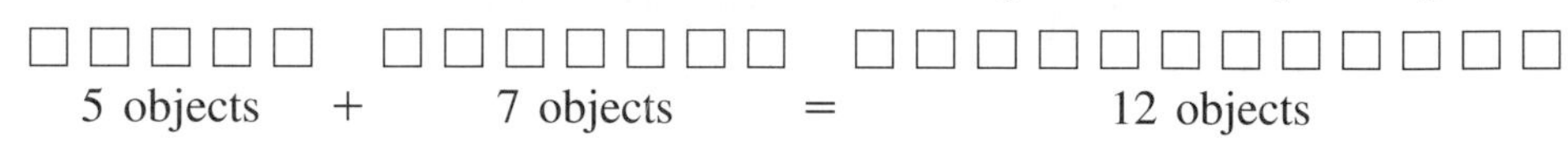

$$5 + 7 = 12$$

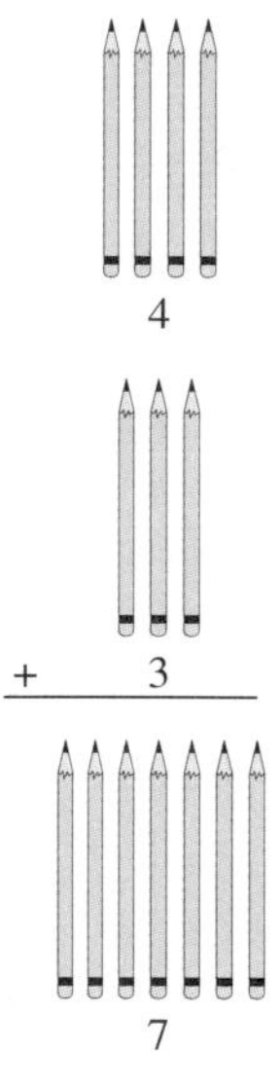

Usually when we add numbers, we put one number under the other in a column. The numbers being added are called **addends.** The result is called the **sum.**

Suppose that we have four pencils in the car and we bring three more pencils from home. How many pencils do we have with us now? We add 4 and 3 to obtain a value of 7. In this case, the numbers 4 and 3 are the addends and the answer 7 is the sum.

$$\begin{array}{rl} 4 & \text{addend} \\ +\,3 & \text{addend} \\ \hline 7 & \text{sum} \end{array}$$

Think about what we do when we add 0 to another number. We are not making a change, so whenever we add zero to another number, that number will be the sum. Since this is always true, this is called a *property.* Since the sum is identical to the number added to zero, this is called the **identity property of zero.**

EXAMPLE 1 Add.

(a) $8 + 5$ **(b)** $3 + 7$ **(c)** $9 + 0$

(a) $\begin{array}{r} 8 \\ +\,5 \\ \hline 13 \end{array}$ **(b)** $\begin{array}{r} 3 \\ +\,7 \\ \hline 10 \end{array}$ **(c)** $\begin{array}{r} 9 \\ +\,0 \\ \hline 9 \end{array}$ ← *Note:* When we add zero to any other number, that number is the sum.

Practice Problem 1 Add.

(a) $\begin{array}{r} 7 \\ +\,5 \\ \hline \end{array}$ **(b)** $\begin{array}{r} 9 \\ +\,4 \\ \hline \end{array}$ **(c)** $\begin{array}{r} 3 \\ +\,0 \\ \hline \end{array}$ ■

The table below shows the basic addition facts. You should know these facts. If any of the answers don't come to you quickly, now is the time to learn them. To check your knowledge try Exercises 1.2, problems 3 and 4.

TEACHING TIP Some students will find that there are certain number facts that they do not know. For example, some students may not remember that $7 + 8 = 15$ but rather will remember that $7 + 7 = 14$ and then add one. Stress the fact that now is the time to totally learn all the basic addition facts by mastering the content of this addition table. Some students may need to make up flash cards of addition facts in order to master them or to improve their speed in mental addition.

BASIC ADDITION FACTS

+	0	1	2	3	4	5	6	7	8	9
0	0	1	2	3	4	5	6	7	8	9
1	1	2	3	4	5	6	7	8	9	10
2	2	3	4	5	6	7	8	9	10	11
3	3	4	5	6	7	8	9	10	11	12
4	4	5	6	7	8	9	10	11	12	13
5	5	6	7	8	9	10	11	12	13	14
6	6	7	8	9	10	11	12	13	14	15
7	7	8	9	10	11	12	13	14	15	16
8	8	9	10	11	12	13	14	15	16	17
9	9	10	11	12	13	14	15	16	17	18

To use the table to find the sum $4 + 7$, read across the top of the table to the 4 column, and then read down the left to the 7 row. The box where the 4 and 7 meet is 11, which means that $4 + 7 = 11$. Now read across the top to the 7 column and down the left to the 4 row. The box where these numbers meet is also 11. We can see that the order in which we add the numbers does not change the sum. $4 + 7 = 11$, and $7 + 4 = 11$. We call this the **commutative property of addition.**

This property does not hold true for everything in our lives. When you put on your socks and then your shoes, the result is not the same as if you put on your shoes first and then your socks! Can you think of any other examples where changing the order in which you add things would change the result?

2 *Adding Several Single-Digit Numbers*

If more than two numbers are to be added, we usually add from the first number to the next number and mentally note the sum. Then we add that sum to the next number, and so on.

EXAMPLE 2 Add. $3 + 4 + 8 + 2 + 5$

We rewrite the addition problem in a column format.

Mentally, we do these steps.

$$\begin{array}{r} 3 \\ 4 \\ 8 \\ 2 \\ +\ 5 \\ \hline 22 \end{array}$$

$3 + 4 = 7$

$7 + 8 = 15$

$15 + 2 = 17$

$17 + 5 = 22$

Practice Problem 2 Add. $7 + 6 + 5 + 8 + 2$ ■

Because the order in which we add numbers doesn't matter, we can choose to add from the top down, from the bottom up, or in any other way. One shortcut is to add first any numbers that will give a sum of 10, or 20, or 30, and so on.

EXAMPLE 3 Add.

$$\begin{array}{r} 3 \\ 4 \\ 8 \\ 2 \\ +\ 6 \\ \hline \end{array}$$

We mentally group the numbers into tens.

3
4 ←
8 ← → 8 + 2 = 10 → 4 + 6 = 10
2 ←
6 ←

The sum is 10 + 10 + 3 or 23

Practice Problem 3 Add. 1 + 7 + 2 + 9 + 3 ■

3 *Adding Numbers of Several Digits When Carrying Is Not Needed*

Of course, many numbers that we need to add have more than one digit. In such cases, we must be careful to first add the digits in the ones column, then the digits in the tens column, then those in the hundreds column, and so on. Notice that we move from *right to left.*

EXAMPLE 4 Add. 4304 + 5163

$$\begin{array}{r} 4\ 3\ 0\ 4 \\ +\ 5\ 1\ 6\ 3 \\ \hline 9\ 4\ 6\ 7 \end{array}$$

- 7: sum of 4 ones + 3 ones = 7 ones
- 6: sum of 0 tens + 6 tens = 6 tens
- 4: sum of 3 hundreds + 1 hundred = 4 hundreds
- 9: sum of 4 thousands + 5 thousands = 9 thousands

Practice Problem 4 Add.

$$\begin{array}{r} 8246 \\ +\ 1702 \\ \hline \end{array}$$ ■

4 *Adding Numbers of Several Digits When Carrying Is Needed*

When you add several whole numbers, often the sum in a column is greater than 9, but we can only use *one* digit in any one place. What do we do with a two-digit sum? Look at the following example.

EXAMPLE 5 Add. 45 + 37

```
tens ones
 ┌→ 1
 │  4 5
 │ +3 7      5 ones and 7 ones = 12.
 │ ----      We rename 12 in expanded notation: 1 ten + 2 ones.
 │    2 ←─── We place the 2 ones in the ones column.
 └────────── We carry the 1 ten over to the tens column.
```

Note: Placing the 1 in the next column is often called ''carrying the one.''

```
    1
    4 5
  + 3 7      Now we can add the digits in the tens column.
  -----
    8 2
```

Thus, 45 + 37 = 82.

Practice Problem 5 Add.

```
  56
+ 36
```
■

Often you must use carrying several times by bringing the left digit into the next column to the left.

EXAMPLE 6 Add. 257 + 688 + 94

TEACHING TIP Remind students that when they carry a digit such as in Example 6, they may write down the digit they are carrying. Some students were probably criticized in elementary school for showing the carrying step. In college, students should feel free to write down the carrying step if it is needed. Of course, if students can do that part in their heads, there is no need to write down the carrying digit.

Thousands Column	Hundreds Column	Tens Column	Ones Column
	2	1	
	2	5	7
	6	8	8
+		9	4
1	0	3	9

In the ones column we add 7 + 8 + 4 = 19. Because 19 is 1 ten and 9 ones, we place 9 in the ones column and carry 1 to the top of the tens column.

In the tens column we add 1 + 5 + 8 + 9 = 23. Because 23 tens is 2 hundreds and 3 tens, we place the 3 in the tens column and carry 2 to the top of the hundreds column.

In the hundreds column we add 2 + 2 + 6 = 10 hundreds. Because 10 hundreds is 1 thousand and 0 hundreds, we place the 0 in the hundreds column and place the 1 in the thousands column.

Practice Problem 6 Add. 789 + 63 + 297 ■

We can add numbers in more than one way. To add $5 + 3 + 7$ we can first add the 5 and 3. We do this by using parentheses to show the first operation to be done. This shows us that $5 + 3$ is to be grouped together.

$$5 + 3 + 7 = (5 + 3) + 7 = 15$$
$$= 8 + 7 = 15$$

We could add the 3 and 7 first. We use parentheses to show that we group $3 + 7$ together and that we will add these two numbers first.

$$5 + 3 + 7 = 5 + (3 + 7) = 15$$
$$= 5 + 10 = 15$$

The way we group numbers to be added does not change the sum. This property is called the **associative property of addition.**

Look again at the three properties of addition we have discussed in this section.

1. **Associative Property of Addition** When we add three numbers, we can group them in any way. $(8 + 2) + 6 = 8 + (2 + 6)$; $10 + 6 = 8 + 8$; $16 = 16$
2. **Commutative Property of Addition** Two numbers can be added in either order with the same result. $5 + 12 = 12 + 5$; $17 = 17$
3. **Identity Property of Zero** When zero is added to a number, the sum is that number. $8 + 0 = 8$; $3 + 0 = 3$

Because of the commutative and associative properties of addition, we can check our addition by adding the numbers in the opposite order.

TEACHING TIP Some students lack confidence that they will be able to find their own errors. As a classroom activity, have students add 258 + 167 + 879. Then have them add 879 + 167 + 258. The sum is 1304. If you ask students how many of them made an error and detected it by adding the numbers in the opposite order and getting a different answer, there will usually be several students in the class who raise their hands.

EXAMPLE 7 **(a)** Add the numbers. $39 + 7284 + 3132$
(b) Check by reversing the order of addition.

(a)

$$\begin{array}{r} \overset{1\,1}{39} \\ 7284 \\ +\ 3132 \\ \hline 10455 \end{array}$$

Addition

(b)

$$\begin{array}{r} 3\overset{1\,1}{13}2 \\ 7284 \\ +\ \ \ 39 \\ \hline 10455 \end{array}$$

Check by reversing the order.

The sum is the same in each case.

Practice Problem 7

(a) Add.

$$\begin{array}{r} 127 \\ 9876 \\ +\ \ 342 \\ \hline \end{array}$$

(b) Check by reversing the order.

$$\begin{array}{r} 342 \\ 9876 \\ +\ \ 127 \\ \hline \end{array}$$ ■

5 *Applying Addition to Real-Life Situations*

We use addition in all kinds of situations. There are several key words in word problems that imply addition. For example, it may be stated that there are 12 math books, 9 chemistry books, and 8 biology books on a book shelf. To find the *total* number of books implies that we add the numbers $12 + 9 + 8$. Other key words are *how much, how many,* and *all.*

Sometimes a problem will have more information than you will need to answer the question. If you have too much information, to solve the problem you will need to separate out the facts that are not important. The following three steps are involved in the problem-solving process.

Step 1 Understand the problem

Step 2 Calculate and state the answer

Step 3 Check

We may not write all of these steps down, but they are the steps we use to solve all problems.

EXAMPLE 8 The bookkeeper for Smithville Trucking was examining the following data for the company checking account.

Monday: $23,416 was deposited and $17,389 was debited.

Tuesday: $44,823 was deposited and $34,089 was debited.

Wednesday: $16,213 was deposited and $20,057 was debited.

What was the total of all deposits during this period?

Step 1 *Understand the problem.*
Total implies that we will use addition. Since we don't need to know about the debits to answer this question, we use only the *deposit* amounts.

Step 2 *Calculate and state the answer.*

Monday:	$23,416 was deposited.	$\overset{1}{2}\overset{1}{3},4\overset{1}{1}6$
Tuesday:	$44,823 was deposited.	44,823
Wednesday:	$16,213 was deposited.	+ 16,213
		84,452

A total of $84,452 was deposited on those three days.

Step 3 *Check.*
You may add the numbers in reverse order to check. We leave the check up to you.

Practice Problem 8 North University has 23,413 men and 18,316 women. South University has 19,316 men and 24,789 women. East University has 20,078 men and 22,965 women. What is the total enrollment of *women* at the three schools? ■

TEACHING TIP Stress the idea of finding the perimeter of an object by adding up the lengths of all the sides. You may want to give an example of a four-sided field that has four different lengths to stress that this concept works for figures other than rectangles.

EXAMPLE 9 Mr. Ortiz has a rectangular field whose length is 400 feet and width 200 feet. What is the total number of feet of fence that would be required to fence in the field?

1. *Understand the problem.*
   To help us to get a picture of what the field looks like, we will draw a diagram.

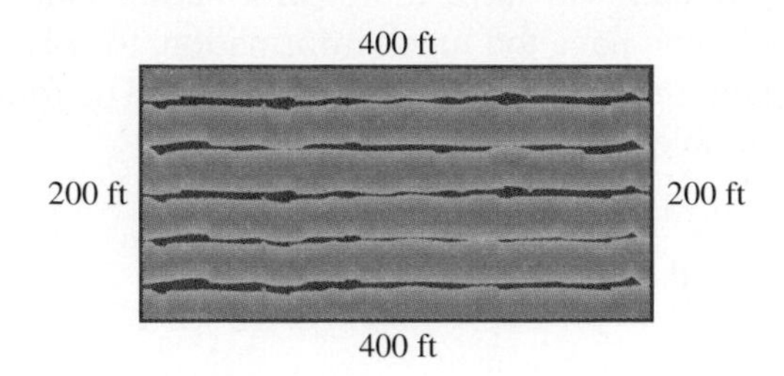

   Note that ft is the abbreviation for feet. ft means feet.

2. *Calculate and state the answer.*
   Since the fence will be along each side of the field, we add the lengths all around the field.

$$\begin{array}{r} 200 \\ 200 \\ 400 \\ +\ 400 \\ \hline 1200 \end{array}$$

   The amount of fence that would be required is 1200 feet.

3. *Check.*
   Regroup the addends and add.

$$\begin{array}{r} 200 \\ 400 \\ 200 \\ +\ 400 \\ \hline 1200 \end{array} \checkmark$$

Practice Problem 9 In Vermont, Gretchen fenced the rectangular field on which her sheep graze. The length of the field is 2000 feet and the width of the field is 1000 feet. What is the perimeter of the field? (*Hint:* The ''distance around'' an object [such as a field] is called the *perimeter.*) ■

Developing Your Study Skills

CLASS ATTENDANCE

You will want to get started in the right direction by choosing to attend class every day, beginning with the first day of class. Statistics show that class attendance and good grades go together. Classroom activities are designed to enhance learning, and therefore you must be in class to benefit from them. Each day vital information and explanations are given that can help you understand concepts. Do not be deceived into thinking that you can just find out from a friend what went on in class. There is no good substitute for firsthand experience. Give yourself a push in the right direction by developing the habit of going to class every day.

1.2 Exercises

Verbal and Writing Skills

1. Explain in your own words. Answers may vary. Samples are below.

(a) the commutative property of addition

You can change the order of the addends without changing the sum.

(b) the associative property of addition

You can group the addends in any way without changing the sum.

2. When zero is added to any number, it does not change that number. Why do you think this is called the identity property of zero? When zero is added to any number, the sum is identical to that number.

Complete the addition facts for each table. Strive for total accuracy, but work quickly. Allow a maximum of 5 minutes for each table.

3.

+	3	5	4	8	0	6	7	2	9	1
2	5	7	6	10	2	8	9	4	11	3
7	10	12	11	15	7	13	14	9	16	8
5	8	10	9	13	5	11	12	7	14	6
3	6	8	7	11	3	9	10	5	12	4
0	3	5	4	8	0	6	7	2	9	1
4	7	9	8	12	4	10	11	6	13	5
1	4	6	5	9	1	7	8	3	10	2
8	11	13	12	16	8	14	15	10	17	9
6	9	11	10	14	6	12	13	8	15	7
9	12	14	13	17	9	15	16	11	18	10

4.

+	1	6	5	3	0	9	4	7	2	8
3	4	9	8	6	3	12	7	10	5	11
9	10	15	14	12	9	18	13	16	11	17
4	5	10	9	7	4	13	8	11	6	12
0	1	6	5	3	0	9	4	7	2	8
2	3	8	7	5	2	11	6	9	4	10
7	8	13	12	10	7	16	11	14	9	15
8	9	14	13	11	8	17	12	15	10	16
1	2	7	6	4	1	10	5	8	3	9
6	7	12	11	9	6	15	10	13	8	14
5	6	11	10	8	5	14	9	12	7	13

Add.

5. $\begin{array}{r} 4 \\ 2 \\ 8 \\ +\ 9 \\ \hline 23 \end{array}$

6. $\begin{array}{r} 4 \\ 6 \\ 2 \\ +\ 7 \\ \hline 19 \end{array}$

7. $\begin{array}{r} 3 \\ 8 \\ 9 \\ +\ 6 \\ \hline 26 \end{array}$

8. $\begin{array}{r} 1 \\ 4 \\ 3 \\ +\ 8 \\ \hline 16 \end{array}$

9. $\begin{array}{r} 12 \\ 45 \\ +\ 3 \\ \hline 60 \end{array}$

10. $\begin{array}{r} 46 \\ 20 \\ +\ 2 \\ \hline 68 \end{array}$

11. $\begin{array}{r} 63 \\ 24 \\ +\ 12 \\ \hline 99 \end{array}$

12. $\begin{array}{r} 54 \\ 21 \\ +\ 23 \\ \hline 98 \end{array}$

13. $\begin{array}{r} 2847 \\ 1634 \\ +\ 98 \\ \hline 4579 \end{array}$

14. $\begin{array}{r} 4816 \\ 3015 \\ +\ 798 \\ \hline 8629 \end{array}$

15. $\begin{array}{r} 6908 \\ 2173 \\ +\ 4255 \\ \hline 13{,}336 \end{array}$

16. $\begin{array}{r} 5017 \\ 2984 \\ +\ 1328 \\ \hline 9329 \end{array}$

17. $\begin{array}{r} 8235 \\ +\ 5626 \\ \hline 13{,}861 \end{array}$

18. $\begin{array}{r} 5673 \\ +\ 3572 \\ \hline 9245 \end{array}$

19. $\begin{array}{r} 18{,}718 \\ +\ 24{,}021 \\ \hline 42{,}739 \end{array}$

20. $\begin{array}{r} 99{,}023 \\ +\ 47{,}325 \\ \hline 146{,}348 \end{array}$

Add from the top. Then check by adding in the reverse order.

21.
```
   36
   41
   25
    6
 + 13
  121
```

22.
```
   24
   39
   16
   14
 +  9
  102
```

23.
```
   106
    13
     4
    28
 + 981
  1132
```

24.
```
   463
    27
     8
    41
 + 507
  1046
```

Add.

25.
```
   126
  8142
    37
+ 9604
17,909
```

26.
```
   223
  7021
    89
+ 7634
14,967
```

27.
```
  1,362,214
  7,002,316
+ 3,214,896
 11,579,426
```

28.
```
  4,002,983
  2,134,702
+ 3,592,001
  9,729,686
```

29.
```
  837,241,000
+ 298,039,240
1,135,280,240
```

30.
```
  982,306,000
+ 583,215,320
1,565,521,320
```

31.
```
    516,208
     24,317
+ 1,763,295
  2,303,820
```

32.
```
     32,500
    763,420
+ 2,837,667
  3,633,587
```

33. 12 + 8 + 156 + 72 248

34. 85 + 3 + 407 + 26 521

35. 15,216 + 485 + 5208 20,909

36. 26,002 + 599 + 3500 30,101

Applications

Answer each question.

37. When Diana went shopping for holiday presents for her family, she spent $224 on Thursday, $387 on Friday, and $183 on Saturday. What is the total amount of money she spent on gifts? $794

38. Peter is building an extra room onto the house to accommodate the new baby. At the building supply store he made purchases of $421 on Friday, $218 on Saturday, and $79 on Sunday. How much did the new room cost to build? $718

39. Andy's annual vacation two years ago cost him $2311. Last year he spent $2502. This year his vacation cost him $3173. What is the total amount he spent on vacations for the three years? $7986

40. The taxes on Jack's house two years ago were $4658. Last year they were $5222. This year they are $6027. What is the total amount for three years? $15,907

41. Anthony has a field with the length of each side as labeled on the sketch. What is the total number of feet of fence that would be required to fence in the field? (Find the perimeter of the field.)

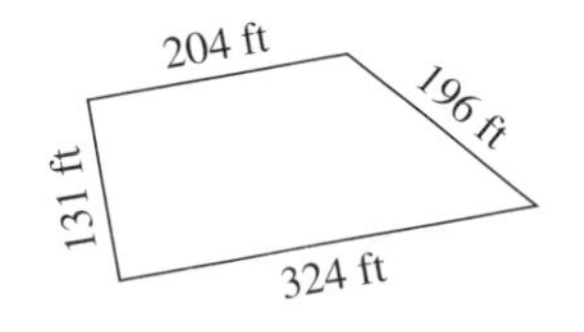


```
  204
  196
  324
+ 131
  855 feet
```

42. Jessica has a field with the length of each side as labeled on the sketch. What is the total number of feet of fence that would be required to fence in the field? (Find the perimeter of the field.)

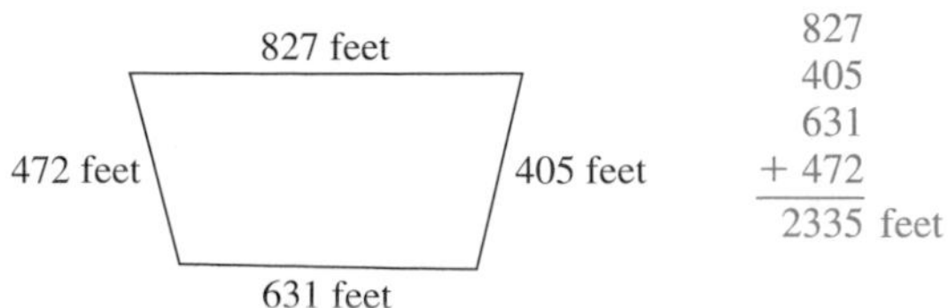


```
  827
  405
  631
+ 472
 2335 feet
```

In problems 43, 44, and 46, you will find information about the relative sizes of different bodies of water found on Earth. For each problem write the total number of square miles in each grouping.

43. The Pacific Ocean, the world's largest, has an area of 64,000,000 square miles. The Atlantic Ocean has an area of 31,800,000 square miles. The Indian Ocean has an area of 25,300,000 square miles. What is the total area for these oceans?

121,100,000 square miles

44. The Arctic Ocean has an area of 5,400,000 square miles. The Mediterranean Sea has an area of 1,100,000 square miles. The Caribbean Sea has an area of 1,000,000 square miles. What is the total area for these oceans?

7,500,000 square miles

45. 2,144,856 people in California, 307,244 people in Oregon, and 470,239 people in Washington voted for Ross Perot for president in the year 1992. What was the total vote for Perot in those three states?

2,922,339 votes

46. Lake Huron, which is bordered by the United States and Canada measures 23,010 square miles. Lake Tanganyika, which borders Tanzania and Zaire, measures 12,700 square miles. Lake Baikal in Russia measures 12,162 square miles. What is the total area for these lakes?

47,872 square miles

In problems 47–49, you will have to choose the numbers you need in order to answer each question. Then calculate the correct answer.

47. The admissions department of a competitive university is reviewing applications to see whether students are *eligible* or *ineligible* for student aid. On Monday, 415 were found eligible and 27 ineligible. On Tuesday, 364 were found eligible and 68 ineligible. On Wednesday, 159 were found eligible and 102 ineligible. On Thursday, 196 were found eligible and 61 ineligible.

(a) How many eligible students qualified for student aid during the four days?

(a) 1134 students

(b) How many students were considered in all?

(b) 1392 students

48. The quality control division of a motorcycle company classifies the final assembled bike as *passing* or *failing* final inspection. In January 14,311 vehicles passed whereas 56 failed. In February, 11,077 passed and 158 failed. In March, 12,580 passed and 97 failed.

(a) How many motorcycles passed the inspection during the three months?

(a) 37,968 motorcycles passed inspection.

(b) How many motorcycles were assembled during the three months in all?

(b) 38,279 motorcycles were assembled.

49. Answer using the information in the following Western University expense chart for the academic year 1996–1997.

Western University Yearly Expenses	In-State Student, U.S. Citizen	Out-of-State Student, U.S. Citizen	Foreign Student
Tuition	$3640	$5276	$8352
Room	1926	2437	2855
Board	1753	1840	1840

How much is the total cost for tuition, room, and board for
(a) *an out-of-state U.S. citizen?*
(b) *an in-state U.S. citizen?*
(c) *a foreign student?*

(a) Out-of-state
5276
2437
+ 1840
$9553

(b) In-state
3640
1926
+ 1753
$7319

(c) Foreign
8352
2855
+ 1840
$13,047

To Think About

Add.

50. 2,368,521,788 + 5,721,368,701 + 4,027,399,206 12,117,289,695

51. 89 + 166 + 23 + 45 + 72 + 190 + 203 + 77 + 18 + 93 + 46 + 73 + 66 1161

52. What would happen if addition were not commutative? Answers may vary. A sample is: You could not add the addends in reverse order to check the addition.

53. What would happen if addition were not associative? Answers may vary. A sample is: You could not group the addends in groups that sum to 10s to make column addition easier.

Cumulative Review Problems

Write the word name for each number.

54. 76,208,941
Seventy-six million, two hundred eight thousand, nine hundred forty-one

55. 121,000,374
One hundred twenty-one million, three hundred seventy-four

Write each number in standard notation.

56. Eight million, seven hundred twenty-four thousand, three hundred ninety-six 8,724,396

57. Nine million, fifty-one thousand, seven hundred nineteen 9,051,719

Developing Your Study Skills

CLASS PARTICIPATION

People learn mathematics through active participation, not through observation from the sidelines. If you want to do well in this course, get involved in classroom activities. Sit near the front where you can see and hear well, where your focus is on the instruction process. Ask questions, be ready to contribute toward solutions, and take part in all classroom activities. Your contributions are valuable to the class and to yourself. Class participation requires an investment of yourself in the learning process, which you will find pays huge dividends.

1.3 Subtraction of Whole Numbers

After studying this section, you will be able to:

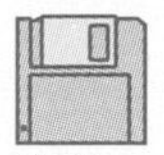

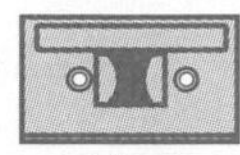

1 *Master basic subtraction facts.*
2 *Subtract whole numbers when borrowing is not necessary.*
3 *Subtract whole numbers when borrowing is necessary.*
4 *Check the answer to a subtraction problem.*
5 *Apply subtraction to real-world situations.*

1 *Mastering Basic Subtraction Facts*

Subtraction is used day after day in the business world. The owner of a bakery placed an ad for his cakes in a local newspaper to see if this might increase his profits. To learn how many cakes had been sold, at closing time he subtracted the number of cakes remaining from the number of cakes the bakery had when it opened. To figure his profits, he subtracted his costs (including the cost of the ad) from his sales. Finally, to see if the ad paid off, he subtracted the profits he usually made in that period from the profits after advertising. He needed subtraction to see whether it paid to advertise.

What is subtraction? We do subtraction when we take objects away from a group. If you have 12 objects and take away 3 of them, 9 objects remain.

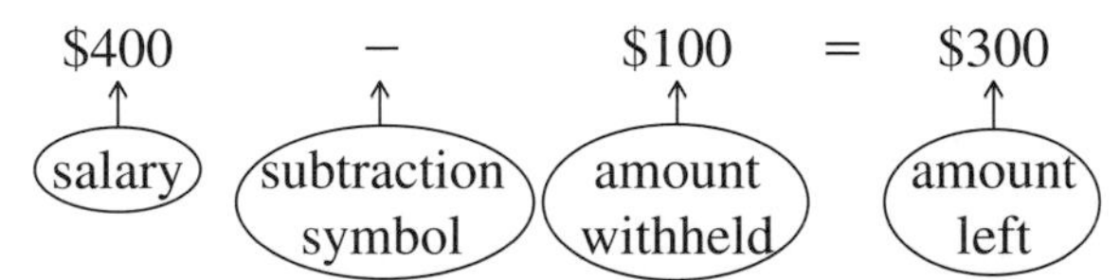

If you earn $400 per month, but have $100 taken out for taxes, how much do you have left?

$400 − $100 = $300

(salary) (subtraction symbol) (amount withheld) (amount left)

We can use addition to help with a subtraction problem.

To subtract: 200 − 196 = what number

We can think: 196 + what number = 200

Usually when we subtract numbers, we put one number under the other in a column. When we subtract one number from another, the answer is called the **difference.**

$$\begin{array}{r} 9 \\ -\,2 \\ \hline 7 \end{array} \qquad \begin{array}{r} 8 \\ -\,3 \\ \hline 5 \end{array} \qquad \begin{array}{r} 12 \\ -\,6 \\ \hline 6 \end{array} \qquad \begin{array}{r} 17 \\ -\,9 \\ \hline 8 \end{array}$$

Each of these is called the difference of the two numbers.

The other two parts of a subtraction problem have labels, although you will not often come across them. The number being subtracted is called the **subtrahend.** The number being subtracted from is called the **minuend.**

17	minuend
− 9	subtrahend
8	difference

In this case, the number 17 is called the *minuend.* The number 9 is called the *subtrahend.* The number 8 is called the *difference.*

Quick Recall of Subtraction Facts

It is helpful if you can subtract quickly. See if you can do Example 1 correctly in 15 seconds or less. Repeat again with Practice Problem 1. Strive to obtain all answers correctly in 15 seconds or less.

EXAMPLE 1 Subtract. **(a)** $8 - 2$ **(b)** $13 - 5$ **(c)** $12 - 4$ **(d)** $15 - 8$ **(e)** $16 - 0$

(a) $\begin{array}{r} 8 \\ -2 \\ \hline 6 \end{array}$ **(b)** $\begin{array}{r} 13 \\ -5 \\ \hline 8 \end{array}$ **(c)** $\begin{array}{r} 12 \\ -4 \\ \hline 8 \end{array}$ **(d)** $\begin{array}{r} 15 \\ -8 \\ \hline 7 \end{array}$ **(e)** $\begin{array}{r} 16 \\ -0 \\ \hline 16 \end{array}$

Practice Problem 1 Subtract.

(a) $\begin{array}{r} 9 \\ -6 \\ \hline \end{array}$ **(b)** $\begin{array}{r} 12 \\ -5 \\ \hline \end{array}$ **(c)** $\begin{array}{r} 17 \\ -8 \\ \hline \end{array}$ **(d)** $\begin{array}{r} 14 \\ -0 \\ \hline \end{array}$ **(e)** $\begin{array}{r} 18 \\ -9 \\ \hline \end{array}$

■

2 *Subtracting Whole Numbers When Borrowing Is Not Necessary*

When we subtract numbers with more than two digits, in order to keep track of our work, we line up the ones column, the tens column, the hundreds column, and so on. Note that we begin with the ones column, and move from right to left.

EXAMPLE 2 Subtract. $9867 - 3725$

$$\begin{array}{r} 9\,8\,6\,7 \\ -\,3\,7\,2\,5 \\ \hline 6\,1\,4\,2 \end{array}$$

- 7 ones − 5 ones = 2 ones
- 6 tens − 2 tens = 4 tens
- 8 hundreds − 7 hundreds = 1 hundred
- 9 thousands − 3 thousands = 6 thousands

Practice Problem 2 Subtract. $7695 - 3481$ ■

3 *Subtracting Whole Numbers When Borrowing Is Necessary*

In the subtraction that we have looked at so far, each digit in the upper number (the minuend) has been greater than the digit in the lower number (the subtrahend) for each place value. Many times, however, a digit in the lower number is greater than the digit in the upper number for that place value.

$$\begin{array}{r} 42 \\ -28 \\ \hline \end{array}$$

The digit in the ones place in the lower number—the 8 of 28—is greater than the number in the ones place in the upper number—the 2 of 42. To subtract, we must *rename* 42, using place values. This is called **borrowing.**

EXAMPLE 3 Subtract. 42 − 28

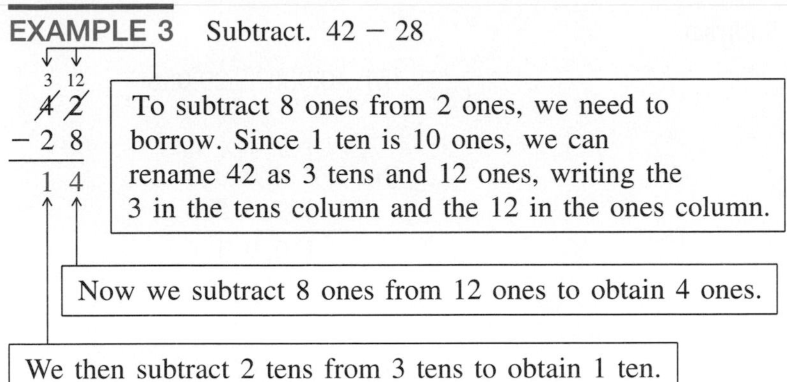

To subtract 8 ones from 2 ones, we need to borrow. Since 1 ten is 10 ones, we can rename 42 as 3 tens and 12 ones, writing the 3 in the tens column and the 12 in the ones column.
Now we subtract 8 ones from 12 ones to obtain 4 ones.
We then subtract 2 tens from 3 tens to obtain 1 ten.

Practice Problem 3 Subtract. 34 − 16 ■

EXAMPLE 4 Subtract. 864 − 548

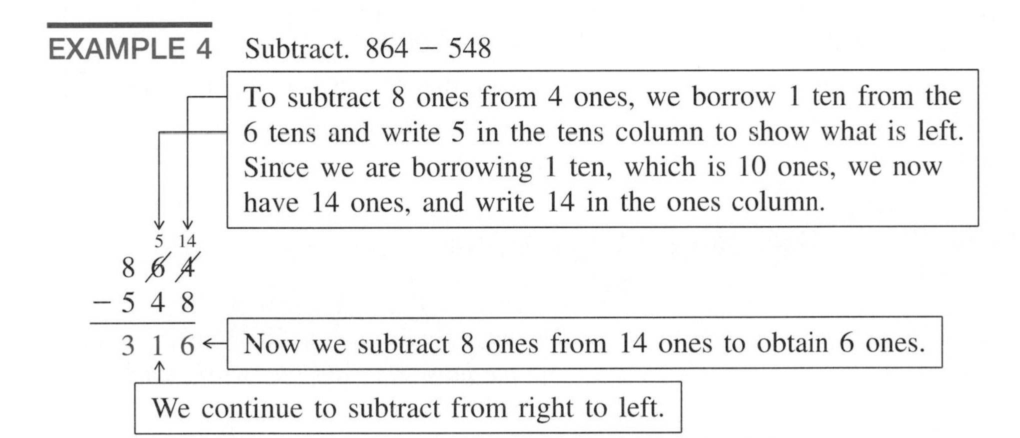

To subtract 8 ones from 4 ones, we borrow 1 ten from the 6 tens and write 5 in the tens column to show what is left. Since we are borrowing 1 ten, which is 10 ones, we now have 14 ones, and write 14 in the ones column.
Now we subtract 8 ones from 14 ones to obtain 6 ones.
We continue to subtract from right to left.

Practice Problem 4 Subtract.

$$\begin{array}{r} 693 \\ -\ 426 \\ \hline \end{array}$$ ■

EXAMPLE 5 Subtract. 8040 − 6375

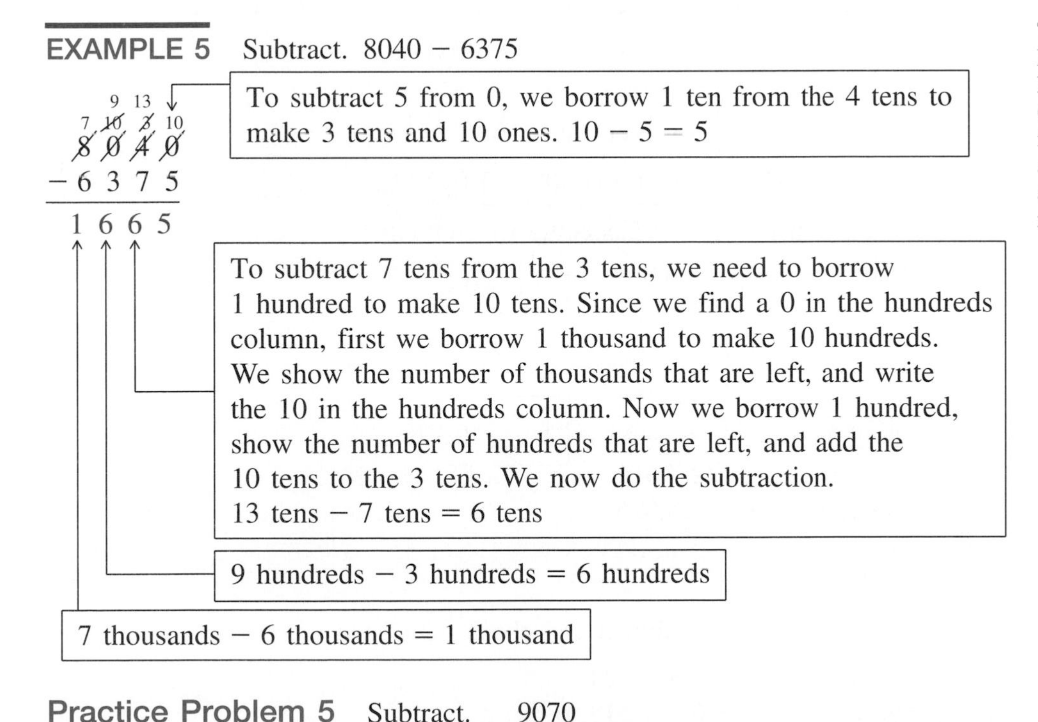

To subtract 5 from 0, we borrow 1 ten from the 4 tens to make 3 tens and 10 ones. 10 − 5 = 5
To subtract 7 tens from the 3 tens, we need to borrow 1 hundred to make 10 tens. Since we find a 0 in the hundreds column, first we borrow 1 thousand to make 10 hundreds. We show the number of thousands that are left, and write the 10 in the hundreds column. Now we borrow 1 hundred, show the number of hundreds that are left, and add the 10 tens to the 3 tens. We now do the subtraction.
13 tens − 7 tens = 6 tens
9 hundreds − 3 hundreds = 6 hundreds
7 thousands − 6 thousands = 1 thousand

TEACHING TIP After discussing Example 5 or a similar problem, have the students subtract 6087 − 5793 as a classroom activity. If they did not obtain the correct answer of 294, they probably made an error in borrowing. Show the correct steps for borrowing in this case.

Practice Problem 5 Subtract.

$$\begin{array}{r} 9070 \\ -\ 5886 \\ \hline \end{array}$$ ■

EXAMPLE 6 Subtract.

(a) 9521 − 943 **(b)** 40,000 − 29,056

(a)
```
      14 11
    8  4  1 11
    9  5  2  1
 −     9  4  3
 ------------
    8  5  7  8
```

(b)
```
   3  9  9  9 10
   4  0, 0  0  0
 − 2  9, 0  5  6
 ---------------
   1  0, 9  4  4
```

Practice Problem 6 Subtract.

(a) 8964 − 985

(b) 50,000 − 32,508 ■

4 *Checking the Answer to a Subtraction Problem*

TEACHING TIP Remind students that addition can usually be done in a lot less time than subtraction. Therefore, the checking step of adding to verify the subtraction is easier and takes less time than the original problem.

We observe that when $9 - 7 = 2$ it follows that $7 + 2 = 9$. Each subtraction problem is equivalent to a corresponding addition problem. This gives us a convenient way to check our answers to subtraction.

EXAMPLE 7 Check this subtraction problem.

$$5829 - 3647 = 2182$$

```
  5 8 2 9
− 3 6 4 7
---------
  2 1 8 2   then    3 6 4 7
                  + 2 1 8 2
                  ---------
                    5 8 2 9
```

The sum should equal 5829, *which it does. We have checked our work, and it is correct.*

Practice Problem 7 Check this subtraction problem by adding.

$$9763 - 5732 = 4031$$ ■

EXAMPLE 8 Subtract and check your answers.

(a) 156,000 − 29,326 **(b)** 1,264,308 − 1,057,612

	Subtraction	Checking by addition	
(a)	156,000 − 29,326 = 126,674	29,326 + 126,674 = 156,000	It checks.
(b)	1,264,308 − 1,057,612 = 206,696	1,057,612 + 206,696 = 1,264,308	It checks.

Practice Problem 8 Subtract and check your answers.

(a) 284,000 − 96,327 **(b)** 8,526,024 − 6,397,518 ■

Subtraction can be used to solve word problems. Some problems can be expressed (and solved) with an *equation*. An equation is a number sentence with an equal sign, such as

$$10 = 4 + x$$

TEACHING TIP Taking the time to emphasize the idea of a variable in very simple terms in such problems as $10 = 4 + x$ will make the use of variables in later chapters much easier for the students to learn.

Here we use the letter x to represent a number we do not know. When we write $10 = 4 + x$, we are stating that 10 is equal to 4 added to some other number. Since $10 - 4 = 6$, we would assume that the number is 6. If we substitute 6 for x in the equation, we have two values that are the same.

$10 = 4 + x$

$10 = 4 + 6$ *Substitute 6 for x.*

$10 = 10$ *Both sides of the equation are the same.*

We can write an equation when one of the addends is not known, then use subtraction to solve for the unknown.

EXAMPLE 9 The librarian knows that he has eight world atlases and that five of them are in full color. How many are not in full color?

We represent the number that we don't know as x and write an equation, or mathematical sentence.

$$8 = 5 + x$$

To solve an equation means to find those values that will make the equation true. We solve this equation by reasoning and by a knowledge of the relationship between addition and subtraction.

$$8 = 5 + x \text{ is equivalent to } 8 - 5 = x$$

We know that $8 - 5 = 3$. Then $x = 3$. We can check the answer by substituting 3 for x in the original equation.

$8 = 5 + x$

$8 = 5 + 3$ True ✓

We see that $x = 3$ checks, so our answer is correct. There are 3 atlases not in full color.

Practice Problem 9 Solve.

(a) $x + 12 = 17$ **(b)** $10 + x = 22$ ■

5 *Applying Subtraction to Real-Life Situations*

We use subtraction in all kinds of situations. There are several key words in word problems that imply subtraction. Words that involve comparison, such as *how much more, how much greater,* or how much a quantity *increased* or *decreased,* all imply subtraction.

EXAMPLE 10 Look at the following population table.

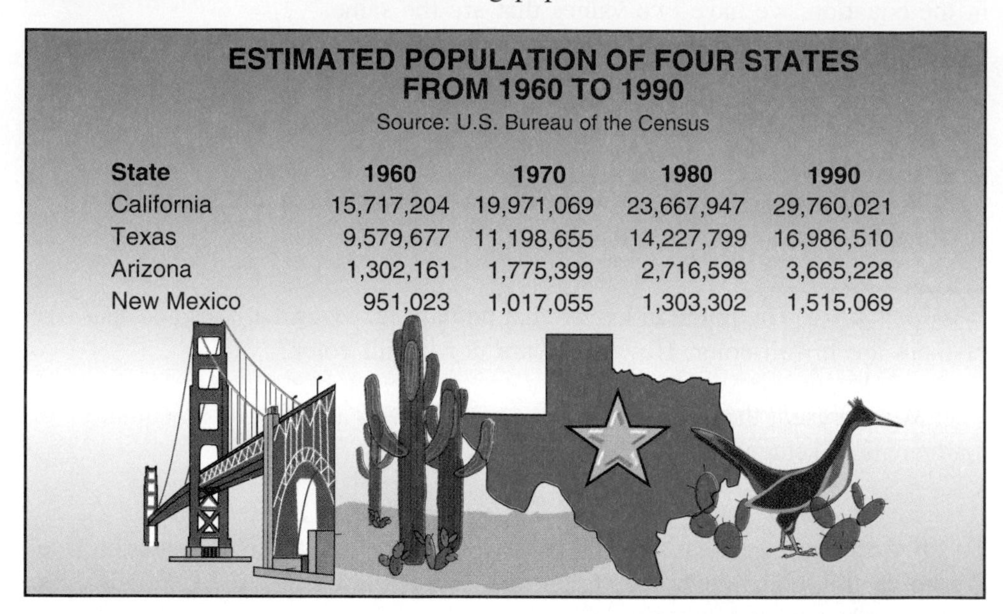

ESTIMATED POPULATION OF FOUR STATES FROM 1960 TO 1990

Source: U.S. Bureau of the Census

State	1960	1970	1980	1990
California	15,717,204	19,971,069	23,667,947	29,760,021
Texas	9,579,677	11,198,655	14,227,799	16,986,510
Arizona	1,302,161	1,775,399	2,716,598	3,665,228
New Mexico	951,023	1,017,055	1,303,302	1,515,069

(a) In 1980 how much greater was the population of Texas than that of Arizona?

(b) How much did the population of California increase from 1960 to 1990?

(c) How much greater was the population of California in 1990 than the other three states combined?

(a)

14,227,799	1980 population of Texas
− 2,716,598	1980 population of Arizona
11,511,201	difference

The population of Texas was greater by 11,511,201.

(b)

29,760,021	1990 population of California
− 15,717,204	1960 population of California
14,042,817	difference

The population of California increased by 14,042,817 in those 30 years.

(c) First we need to find the total population in 1990 of Texas, Arizona, and New Mexico.

16,986,510	1990 population of Texas
3,665,228	1990 population of Arizona
+ 1,515,069	1990 population of New Mexico
22,166,807	

We use subtraction to compare this total with the population of California.

29,760,021	1990 population of California
− 22,166,807	
7,593,214	

The population in California in 1990 is 7,593,214 more than the population in the other three states combined.

Practice Problem 10

(a) In 1980 how much greater was the population of California than the population of Texas?

(b) How much did the population of Texas increase from 1960 to 1970? ■

EXAMPLE 11 The number of real estate transfers in several towns during the years 1994 to 1996 is given in the following bar graph.

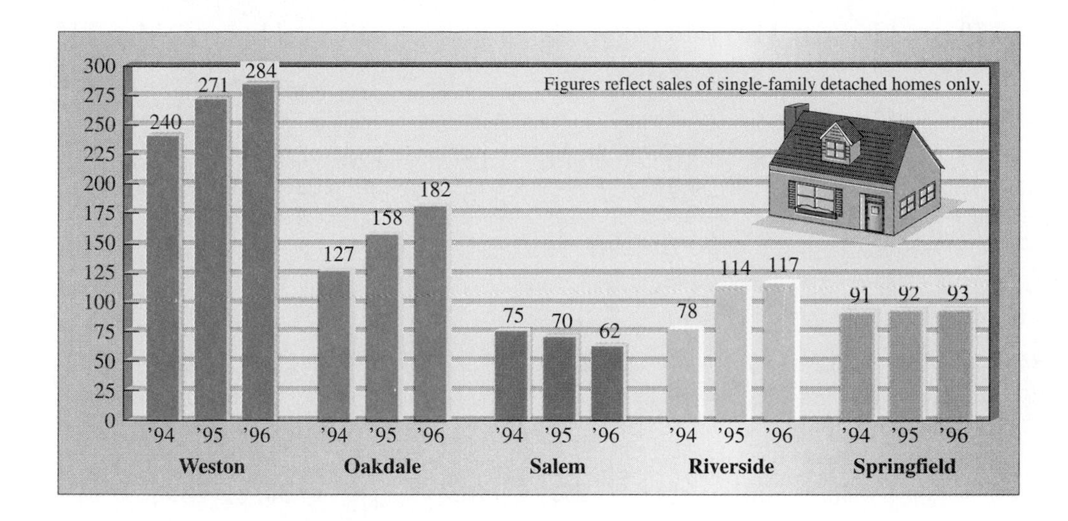

(a) What was the increase in homes sold in Weston from 1995 to 1996?

(b) What was the decrease in homes sold in Salem from 1994 to 1996?

(c) Between what two years did Oakdale have the greatest increase in sales?

(a) From the labels on the bar graph we see that 284 homes were sold in 1996 in Weston and 271 homes were sold in 1995. Thus the increase can be found by subtracting $284 - 271 = 13$. There was an increase of 13 homes sold in Weston from 1995 to 1996.

(b) In 1994, 75 homes were sold in Salem. In 1996, 62 homes were sold in Salem. The decrease in the number of homes sold is $75 - 62 = 13$. There was a decrease of 13 homes sold in Salem from 1994 to 1996.

(c) Here we will need to make two calculations in order to decide where the greatest increase occurs.

158	1995 sales	182	1996 sales
− 127	1994 sales	− 158	1995 sales
31	Sales increase from 1994 to 1995	24	Sales increase from 1995 to 1996

The greatest increase in sales in Oakdale occurred from 1994 to 1995.

Practice Problem 11 Based on the bar graph above, answer the following questions.

(a) What was the increase in homes sold in Riverside from 1994 to 1995?

(b) How many more homes were sold in Springfield in 1994 than in Riverside in 1994?

(c) Between what two years did Weston have the greatest increase in sales? ■

1.3 Exercises

Try to do problems 1–20 in 1 minute or less with no errors.

Subtract.

1. 8 − 2   6

2. 7 − 5   2

3. 13 − 5   8

4. 6 − 0   6

5. 6 − 5   1

6. 17 − 8   9

7. 15 − 6   9

8. 14 − 5   9

9. 16 − 0   16

10. 17 − 9   8

11. 18 − 9   9

12. 12 − 7   5

13. 11 − 4   7

14. 15 − 8   7

15. 13 − 7   6

16. 16 − 9   7

17. 11 − 8   3

18. 10 − 7   3

19. 15 − 6   9

20. 12 − 5   7

Subtract. Check your answers by adding.

21. 98 − 37   61   37 + 61   98

22. 84 − 32   52   32 + 52   84

23. 85 − 73   12   73 + 12   85

24. 77 − 36   41   36 + 41   77

25. 126 − 95   31   95 + 31   126

26. 193 − 72   121   72 + 121   193

27. 768 − 143   625   143 + 625   768

28. 621 − 320   301   320 + 301   621

29. 1763 − 422   1341   422 + 1341   1763

30. 1896 − 743   1153   743 + 1153   1896

31. 24,396 − 13,205   11,191   13,205 + 11,191   24,396

32. 52,980 − 31,660   21,320   31,660 + 21,320   52,980

33. 986,302 − 433,201   553,101   433,201 + 553,101   986,302

34. 807,965 − 304,214   503,751   304,214 + 503,751   807,965

Check each subtraction. If the problem has not been done correctly, find the correct answer.

35. 129 − 19 = 110   19 + 110   129 Correct

36. 186 − 45 = 141   45 + 141   186 Correct

37. 6180 − 1113 = 5067   1113 + 5067   6180 Correct

38. 5976 − 3243 = 2733   3243 + 2733   5976 Correct

39. 6030 − 5020 = 1020   6030 − 5020   1010 Incorrect

40. 7890 − 3200 = 7670   7890 − 3200   4690 Incorrect

41. 98,763 − 42,531 = 55,232   98,763 − 42,531   56,232 Incorrect

42. 47,969 − 33,846 = 14,223   47,969 − 33,846   14,123 Incorrect

Subtract. Use borrowing if necessary.

43. 93 − 47   46

44. 86 − 33   53

45. 125 − 88   37

46. 136 − 95   41

47. 451 − 376   75

48. 706 − 435   271

49. 905 − 324   581

50. 861 − 345   516

51. 10,000 − 6,704   3,296

52. 20,000 − 13,120   6,880

53. 152,000 − 117,908   34,092

54. 361,000 − 121,520   239,480

55. 42,312 − 39,998   2,314

56. 54,913 − 29,997   24,916

57. 760,108 − 536,992   223,116

58. 580,092 − 349,905   230,187

Solve.

59. $x + 14 = 19$ $x = 5$

60. $x + 35 = 50$ $x = 15$

61. $34 = x + 13$ $x = 21$

62. $25 = x + 18$ $x = 7$

63. $100 + x = 127$ $x = 27$

64. $86 + x = 120$ $x = 34$

Applications

Answer each question.

65. This year, for the college graduation ceremonies, the senior class voted for "Power of Example to Humanity." Martin Luther King received 765 votes, John F. Kennedy received 960 votes, and Mother Theresa received 778 votes. How many more votes did John F. Kennedy receive than Mother Theresa? 182 votes

66. Three of the highest elevations on Earth are Mt. Everest, in Asia, which is 8846 meters high, Mt. Aconcagua, in Argentina, which is 6960 meters high, and Mt. McKinley in the United States (Alaska), which is 6194 meters high. How much higher is Mt. Everest than Mt. Aconcagua? 1886 meters

67. In 1994, the population of Ireland was approximately 3,539,296. In the same year, the population of Portugal was approximately 10,524,210. How much less than the population of Portugal was the population of Ireland in 1994? 6,984,914

68. The principality of Monaco, located on the Mediterranean coast of France, had a budget revenue of $424,130,812 in 1994. The expenses for the year were $376,211,007. How much was left over after expenses? $47,919,805

69. James earned $475 painting a few rooms for his neighbor. He gave $142 to his assistant and paid $85 to the hardware store for materials. How much money did he actually receive after paying expenses? $248

70. The Ski Club raised $1682 at a party for their favorite charity. The band received $350 and they spent $265 on refreshments. How much money was actually raised for the charity after paying expenses? $1067

In answering problems 71–76, consider the following population table.

	1960	1970	1980	1990
Illinois	10,081,158	11,110,285	11,427,409	11,430,602
Michigan	7,823,194	8,881,826	9,262,044	9,295,297
Indiana	4,662,498	5,195,392	5,490,212	5,544,159
Minnesota	3,413,864	3,806,103	4,075,970	4,375,099

71. How much did the population of Minnesota increase from 1960 to 1990?
4,375,099 − 3,413,864 = 961,235 people

72. How much did the population of Michigan increase from 1960 to 1990?
9,295,297 − 7,823,194 = 1,472,103 people

73. In 1970, how much greater was the population of Michigan than the population of Indiana?
8,881,826 − 5,195,392 = 3,686,434 people

74. In 1980, how much greater was the population of Indiana than the population of Minnesota?
5,490,212 − 4,075,970 = 1,414,242 people

75. How much did the population of Illinois increase from 1970 to 1990?
11,430,602 − 11,110,285 = 320,317 people

76. How much did the population of Michigan increase from 1980 to 1990?
9,295,297 − 9,262,044 = 33,253 people

The number of real estate transfers in several towns during the years 1994 to 1996 is given in the following bar graph. Use the bar graph to answer questions 77–84. The figures in the bar graph reflect sales of single-family detached homes only.

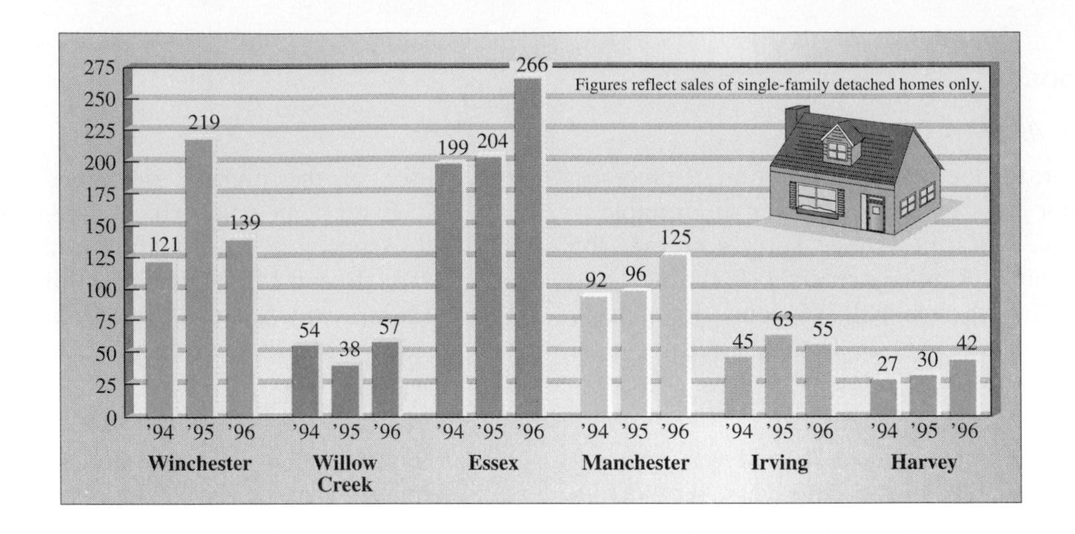

77. What was the increase in the number of homes sold in Manchester from 1995 to 1996? 29 homes

78. What was the increase in the number of homes sold in Irving from 1994 to 1995? 18 homes

79. What was the decrease in the number of homes sold in Winchester from 1995 to 1996? 80 homes

80. What was the decrease in the number of homes sold in Willow Creek from 1994 to 1995? 16 homes

81. Between what two years did the greatest change occur in the number of homes sold in Irving? Between 1994 and 1995

82. Between what two years did the greatest change occur in the number of homes sold in Winchester? Between 1994 and 1995

83. A realtor was trying to determine which two towns were closest to having the same number of sales of homes in 1994. What two towns should she select?
Willow Creek and Irving

84. A realtor was trying to determine which two towns were closest to having the same number of sales of homes in 1995. What two towns should she select?
Willow Creek and Harvey

To Think About

85. In general, subtraction is not commutative. If a and b are whole numbers, $a - b \neq b - a$. For what types of numbers would it be true that $a - b = b - a$? If a and b represent the same number. For example, if $a = 10$ and $b = 10$.

86. In general, subtraction is not associative. For example, $8 - (4 - 3) \neq (8 - 4) - 3$. In general, $a - (b - c) \neq (a - b) - c$. Can you find some numbers a, b, c for which $a - (b - c) = (a - b) - c$? (Remember, do operations inside the parentheses first.) It is true for all a and b if $c = 0$. For example, if $a = 5$, $b = 3$, and $c = 0$.

87. Subtract. 28,007,653,121,863 − 27,986,430,705,999 21,222,415,864

88. Subtract. 101,000,101,000,101 − 99,090,989,878,789 1,909,111,121,312

Cumulative Review Problems

89. Write in standard notation: eight million, four hundred sixty-six thousand, eighty-four 8,466,084

90. Write a word name for 296,308.
Two hundred ninety-six thousand, three hundred eight.

91. Add. 16 + 27 + 82 + 34 + 9 168

92. Add.

$$\begin{array}{r} 156{,}325 \\ +\ 963{,}209 \\ \hline 1{,}119{,}534 \end{array}$$

Developing Your Study Skills

HOW TO DO HOMEWORK

Set aside time each day for your homework assignments. Do not attempt to do a whole week's worth on the weekend. Two hours spent studying outside of class for each hour in class is usual for college courses. You may need more than that for mathematics.

Before beginning to solve your homework problems, read your textbook very carefully. Expect to spend much more time reading a few pages of a mathematics textbook than several pages of another text. Read for complete understanding, not just for the general idea.

As you begin your homework assignments, read the directions carefully. You need to understand what is being asked. Concentrate on each exercise or problem, taking time to solve it accurately. Rushing through your work usually causes errors. Check your answers with those given in the back of the textbook. If your answer is incorrect, check to see that you are doing the right problem. Redo the problem, watching for little errors. If it is still wrong, check with a friend. Perhaps the two of you can figure out where you are going wrong.

Work on your assignments every day and do as many problems as it takes for you to know what you are doing. Begin by doing all the problems that have been assigned. If there are more available in that section of your text, then do more. When you think you have done enough problems to understand fully the topic at hand, do a few more to be sure. This may mean that you do many more problems than the instructor assigns, but you can never practice mathematics too much. Practice improves your skills and increases your accuracy, speed, competence, and confidence.

Also, check the examples in the textbook or in your notes for a similar exercise or problem. Can this one be solved in the same way? Give it some thought. You may want to leave it for a while by taking a break or doing a different problem. But come back later and try again. If you are still unable to figure it out, ask your instructor for help during office hours or in class.

1.4 Multiplication of Whole Numbers

MathPro

Video 2

SSM

After studying this section, you will be able to:

1 *Master basic multiplication facts.*
2 *Multiply a single-digit number by a number with several digits.*
3 *Multiply a number by a number that is a power of 10.*
4 *Multiply a several-digit number by a several-digit number.*
5 *Use the properties of multiplication to perform efficient calculations.*
6 *Apply multiplication to real-life situations.*

1 *Mastering Basic Multiplication Facts*

Like subtraction, multiplication is related to addition. Suppose that the pastry chef at the Gourmet Restaurant bakes croissants on a sheet that holds four croissants across, with room for three rows. How many croissants does the sheet hold?

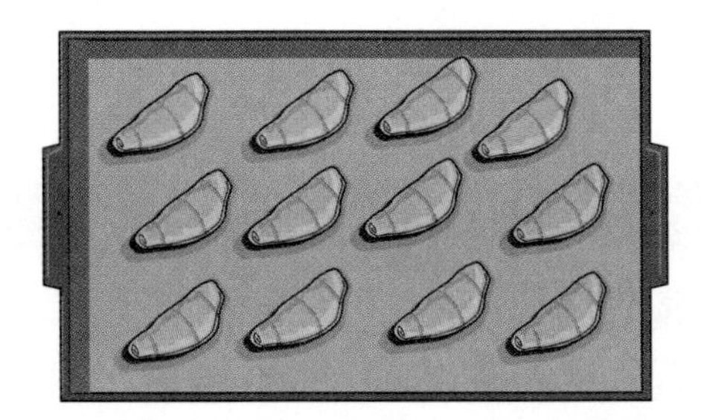

We can add 4 + 4 + 4 to get the total, or we can use a shortcut: 3 rows of 4 is the same as 3 times 4, which equals 12. This is **multiplication,** a shortcut for repeated addition.

The numbers that we multiply are called **factors.** The answer is called the **product.** For now, we will use × to show multiplication. 3 × 4, is read "3 times 4."

$$\underbrace{3}_{\text{factor}} \times \underbrace{4}_{\text{factor}} = \underbrace{12}_{\text{product}} \qquad \begin{array}{rl} 3 & \text{factor} \\ \times 4 & \text{factor} \\ \hline 12 & \text{product} \end{array}$$

Your skill in multiplication depends on how well you know the basic multiplication facts. Look at the table on page 35. You should learn these facts well enough to quickly and correctly give the products of any two factors in the table. To check your knowledge, try Exercises 1.4, problems 3 and 4.

Study the table to see if you can discover any properties of multiplication. What do you see as results when you multiply zero by any number? When you multiply any number times zero, the result is zero. That is the **multiplication property of zero.**

$$2 \times 0 = 0 \qquad 5 \times 0 = 0 \qquad 0 \times 6 = 0 \qquad 0 \times 0 = 0$$

You may recall that zero plays a special role in addition. Zero is the *identity element* for addition. When we add any number to zero, that number does not change. Is there an identity element for multiplication? Look at the table. What is the identity element for multiplication?

What other properties of addition hold for multiplication? Is multiplication commutative? Does the order in which you multiply two numbers change the results? Find the product of 3×4. Then find the product of 4×3.

$$3 \times 4 = 12$$

$$4 \times 3 = 12$$

The **commutative property of multiplication** tells us that when we multiply two numbers, changing the order of the numbers gives the same result.

TEACHING TIP Remind students that they may need to practice or relearn facts that they cannot instantly recall from the multiplication table. As with the addition table, some students will need to make and use multiplication flash cards in order to master the basic multiplication facts.

BASIC MULTIPLICATION FACTS

×	0	1	2	3	4	5	6	7	8	9
0	0	0	0	0	0	0	0	0	0	0
1	0	1	2	3	4	5	6	7	8	9
2	0	2	4	6	8	10	12	14	16	18
3	0	3	6	9	12	15	18	21	24	27
4	0	4	8	12	16	20	24	28	32	36
5	0	5	10	15	20	25	30	35	40	45
6	0	6	12	18	24	30	36	42	48	54
7	0	7	14	21	28	35	42	49	56	63
8	0	8	16	24	32	40	48	56	64	72
9	0	9	18	27	36	45	54	63	72	81

Quick Recall of Multiplication Facts

It is helpful if you can multiply quickly. See if you can do Example 1 correctly in 15 seconds or less. Repeat again with Practice Problem 1. Strive to obtain all answers correctly in 15 seconds or less.

EXAMPLE 1 Multiply.

(a) 5×7 **(b)** 8×9 **(c)** 6×8 **(d)** 9×3 **(e)** 7×8

(a) $\begin{array}{r} 7 \\ \times 5 \\ \hline 35 \end{array}$ **(b)** $\begin{array}{r} 9 \\ \times 8 \\ \hline 72 \end{array}$ **(c)** $\begin{array}{r} 8 \\ \times 6 \\ \hline 48 \end{array}$ **(d)** $\begin{array}{r} 3 \\ \times 9 \\ \hline 27 \end{array}$ **(e)** $\begin{array}{r} 8 \\ \times 7 \\ \hline 56 \end{array}$

Practice Problem 1 Multiply.

(a) $\begin{array}{r} 8 \\ \times 8 \\ \hline \end{array}$ **(b)** $\begin{array}{r} 7 \\ \times 6 \\ \hline \end{array}$ **(c)** $\begin{array}{r} 5 \\ \times 8 \\ \hline \end{array}$ **(d)** $\begin{array}{r} 9 \\ \times 7 \\ \hline \end{array}$ **(e)** $\begin{array}{r} 9 \\ \times 9 \\ \hline \end{array}$ ■

2 *Multiplying a Single-Digit Number by a Number with Several Digits*

EXAMPLE 2 Multiply. 4312 × 2

We first multiply the ones column, then the tens column, and so on, moving right to left.

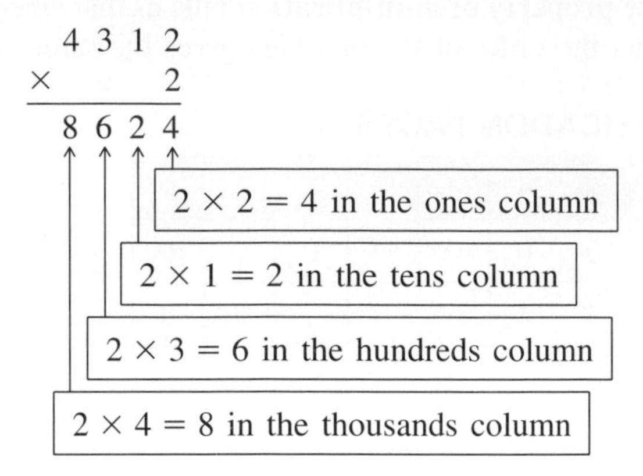

Practice Problem 2 Multiply. 3021 × 3 ■

Usually, we will have to carry one digit of the result of some of the multiplication into the next left-hand column.

EXAMPLE 3 Multiply. 36 × 7

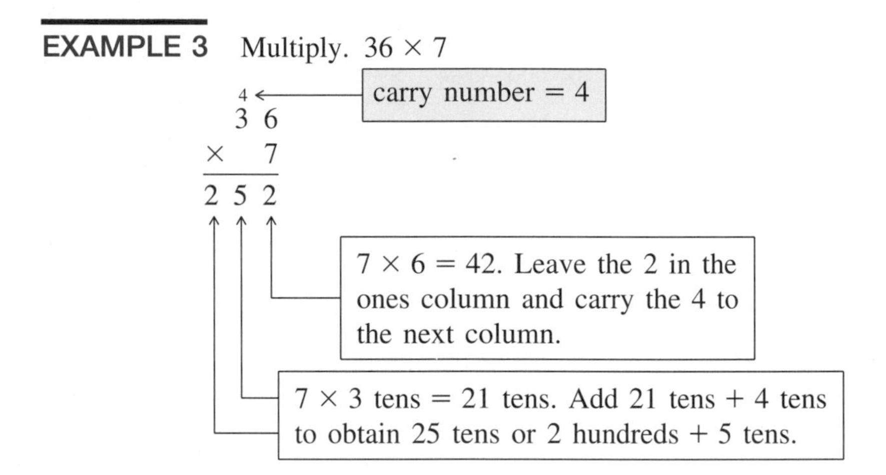

carry number = 4
4
3 6
× 7
2 5 2
7 × 6 = 42. Leave the 2 in the ones column and carry the 4 to the next column.
7 × 3 tens = 21 tens. Add 21 tens + 4 tens to obtain 25 tens or 2 hundreds + 5 tens.

Practice Problem 3 Multiply. 43 × 8 ■

EXAMPLE 4 Multipy. 359 × 9

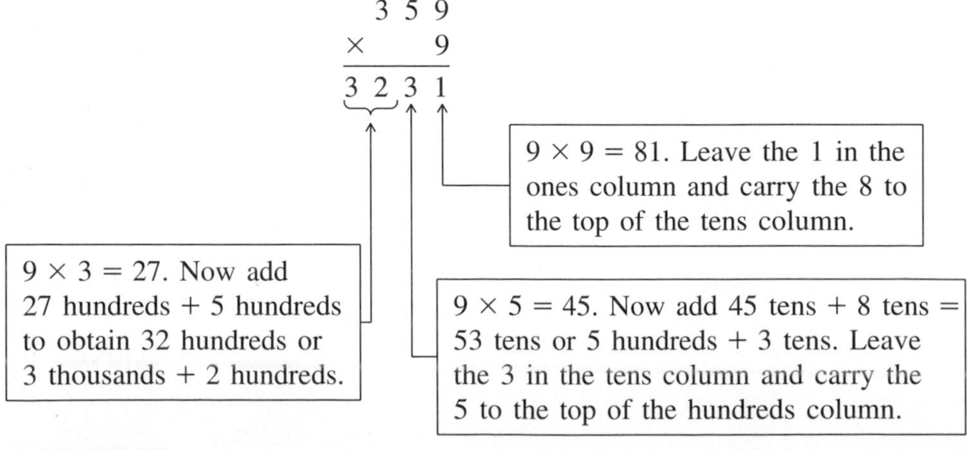

5 8
3 5 9
× 9
3 2 3 1
9 × 9 = 81. Leave the 1 in the ones column and carry the 8 to the top of the tens column.
9 × 3 = 27. Now add 27 hundreds + 5 hundreds to obtain 32 hundreds or 3 thousands + 2 hundreds.
9 × 5 = 45. Now add 45 tens + 8 tens = 53 tens or 5 hundreds + 3 tens. Leave the 3 in the tens column and carry the 5 to the top of the hundreds column.

Practice Problem 4 Multiply.

$$\begin{array}{r} 579 \\ \times \quad 7 \\ \hline \end{array}$$ ■

3 *Multiplying by a Power of 10*

Observe what happens when a number is multiplied by 10, 100, 1000, 10,000, and so on.

$56 \times 10 = 560$ (one zero) $56 \times 100 = 5600$ (two zeros)

$56 \times 1000 = 56{,}000$ (three zeros) $56 \times 10{,}000 = 560{,}000$ (four zeros)

A **power of 10** is a whole number that begins with 1 and ends in one or more zeros. The numbers 10, 100, 1000, 10,000, and so on are powers of 10.

To multiply a whole number by a power of 10:

1. Count the number of zeros in the power of 10.
2. Attach that number of zeros to the right side of the other whole number to obtain the answer.

EXAMPLE 5 Multiply 358 by each number.

(a) 10 **(b)** 100 **(c)** 1000 **(d)** 100,000

(a) $358 \times 10 = 3580$ (*one zero*)

(b) $358 \times 100 = 35{,}800$ (*two zeros*)

(c) $358 \times 1000 = 358{,}000$ (*three zeros*)

(d) $358 \times 100{,}000 = 35{,}800{,}000$ (*five zeros*)

Practice Problem 5 Multiply 1267 by each number.

(a) 10 **(b)** 1000 **(c)** 10,000 **(d)** 1,000,000 ■

How can we handle zeros in multiplication involving a number that is not 10, 100, 1000, or any other power of 10? Consider 32×400. We can rewrite 400 as 4×100, which gives us $32 \times 4 \times 100$. We can simply multiply 32×4 and then attach two zeros for the factor 100. We find that $32 \times 4 = 128$. Attaching two zeros gives us 12,800, or $32 \times 400 = 12{,}800$.

EXAMPLE 6 Multiply. **(a)** 12×3000 **(b)** 25×600 **(c)** $43 \times 40{,}000$

(a) $12 \times 3000 = 12 \times 3 \times 1000 = 36 \times 1000 = 36{,}000$

(b) $25 \times 600 = 25 \times 6 \times 100 = 150 \times 100 = 15{,}000$

(c) $43 \times 40{,}000 = 43 \times 4 \times 10{,}000 = 172 \times 10{,}000 = 1{,}720{,}000$

Practice Problem 6 Multiply. **(a)** $9 \times 60{,}000$ **(b)** 15×400 **(c)** 21×8000 ■

TEACHING TIP You may need to do several problems of the type 345×30 and 2300×50 until students grasp the shortcut of counting the number of end zeros and then adding them at the end. Show them that to multiply 2300×50, you merely multiply $23 \times 5 = 115$ and then add the three zeros to obtain the final answer $2300 \times 50 = 115{,}000$.

4 *Multiplying a Several-Digit Number by a Several-Digit Number*

EXAMPLE 7 Multiply. 234 × 21

We can consider 21 as 2 tens (20) and 1 one (1). First we multiply 234 by 1.

$$\begin{array}{r} 234 \\ \times\ \ 21 \\ \hline 234 \end{array}$$

The 234 is called a **partial product.** We next multiply 234 by 20.

$$\begin{array}{r} 234 \\ \times\ \ 21 \\ \hline 234 \\ 468 \end{array}$$

Note that 468—also called a partial product—is placed so that its ones digit appears in the tens column. This placement is necessary because we are multiplying by the digit in the tens column of the factor 21. We may add a zero to the right of the 468 to make sure that the 8 of 468 lines up in the tens column.

$$\begin{array}{r} 234 \\ \times\ \ 21 \\ \hline 234 \\ 4680 \end{array}$$

We now add the two partial products to reach the final product, which is the solution.

$$\begin{array}{rl} 234 & \\ \times\ \ 21 & \\ \hline 234 & \leftarrow \text{Multiply } 234 \times 1. \\ 4680 & \leftarrow \text{Multiply } 234 \times 20. \text{ First put a zero in the ones column to line up the digits.} \\ \hline 4914 & \leftarrow \text{Add the two partial products.} \end{array}$$

Practice Problem 7 Multiply.

$$\begin{array}{r} 323 \\ \times\ \ 32 \\ \hline \end{array}$$ ■

EXAMPLE 8 Multiply. 671 × 35

$$\begin{array}{rl} 6\ 7\ 1 & \\ \times\ \ \ 3\ 5 & \\ \hline 3\ 3\ 5\ 5 & \leftarrow \text{First multiply } 671 \times 5. \\ 2\ 0\ 1\ 3\ \textcircled{0} & \leftarrow \text{Now multiply } 671 \times 30. \\ \hline 2\ 3\ 4\ 8\ 5 & \leftarrow \text{Now add the two partial products.} \end{array}$$

Note: We could omit zero on this line and leave the ones place blank.

Practice Problem 8 Multiply.

$$\begin{array}{r} 385 \\ \times\ \ 69 \\ \hline \end{array}$$ ■

EXAMPLE 9 Multiply. 14×20

```
                          14
                        × 20
                        ----
                           0 ← Multiply 14 by 0.
                         280 ← Multiply 14 by 2 tens.
                        ----
Now add the ——————→      280   Place 28 with the 8
partial products.              in the tens column.
      |                        To line up the digits for
      ↓                        adding, we can insert a 0.
```

Notice that you will also get this result if you multiply $14 \times 2 = 28$ *and then attach a zero to multiply it by* 10: 280.

Practice Problem 9 Multiply. 34×20 ■

EXAMPLE 10 Multiply. 120×40

```
                          120
                        ×  40
                        -----
                            0 ← Multiply 120 × 0.
                         4800 ← Multiply 120 by 4 tens.
                        -----
Now add the ——————→      4800   The answer is 480 tens.
partial products.               We place the 0 of the 480
      |                         in the tens column. To
      ↓                         line up the digits for adding,
                                we can insert a 0.
```

Notice that this result is the same as $12 \times 4 = 48$ *with two zeros attached:* 4800.

Practice Problem 10 Multiply. 130×50 ■

EXAMPLE 11 Multiply. 684×763

```
      6 8 4
    × 7 6 3
    -------
    2 0 5 2 ← Multiply 684 × 3.
  4 1 0 4   ← Multiply 684 × 60. Note that we omit the final zero.
4 7 8 8     ← Multiply 684 × 700. Note we omit final two zeros.
-----------
5 2 1 8 9 2
```

Practice Problem 11 Multiply. 923×675 ■

5 *Using the Properties of Multiplication to Perform Efficient Calculations*

When we add three numbers, we use the associative property. Recall that the associative property allows us to group the three numbers in any way. Thus to add $9 + 7 + 3$, we can group the numbers as $9 + (7 + 3)$ because it is easier to find the sum. $9 + (7 + 3) = 9 + 10 = 19$. We can demonstrate that multiplication is also associative.

Is this true? $2 \times (5 \times 3) = (2 \times 5) \times 3$

$$2 \times (15) = (10) \times 3$$
$$30 = 30$$

The final product is the same in both cases.

The way we group numbers to be multiplied does not change the product. This property is called the **associative property of multiplication.**

EXAMPLE 12 Multiply. $14 \times 2 \times 5$

Since we can group any two numbers together, let's take advantage of multiplying by 10.

$$14 \times 2 \times 5 = 14 \times (2 \times 5) = 14 \times 10 = 140$$

Practice Problem 12 Multiply. $25 \times 4 \times 17$ ■

For convenience, we list the properties of multiplication that we have discussed in this section.

Property	Example
1. **Associative Property of Multiplication.** When we multiply three numbers, the multiplication can be grouped in any way.	$(7 \times 3) \times 2 = 7 \times (3 \times 2)$ $21 \times 2 = 7 \times 6$ $42 = 42$
2. **Commutative Property of Multiplication.** Two numbers can be multiplied in either order with the same result.	$9 \times 8 = 8 \times 9$ $72 = 72$
3. **Identity Property of One.** When one is multiplied by a number, the result is that number.	$7 \times 1 = 7$ $1 \times 15 = 15$
4. **Multiplication Property of Zero.** The product of any number and zero yields zero as a result.	$0 \times 14 = 0$ $2 \times 0 = 0$

Sometimes you can use several properties in one problem to make the calculation easier.

EXAMPLE 13 Multiply. $7 \times 20 \times 5 \times 6$

$$
\begin{aligned}
7 \times 20 \times 5 \times 6 &= 7 \times (20 \times 5) \times 6 && \textit{Associative property} \\
&= 7 \times 6 \times (20 \times 5) && \textit{Commutative property} \\
&= 42 \times 100 \\
&= 4200
\end{aligned}
$$

Practice Problem 13 Multiply. $8 \times 4 \times 3 \times 25$ ■

Thus far we have discussed the properties of addition and the properties of multiplication. There is one more property that links both operations.

Before we discuss that property, we will illustrate several different ways of showing multiplication. The following are all the ways to show "3 times 4."

3×4	$(3)(4)$	$3(4)$ $(3)4$	$3 \cdot 4$	$3 * 4$
with an ×	with two sets of parentheses	with a single set of parentheses	with a dot	with a star

We will use parentheses to mean multiplication when we use the **distributive property.**

SIDELIGHT Why does our method of multiplying several-digit numbers work? Why can we say that 234×21 is the same as $234 \times 1 + 234 \times 20$?

The *distributive property of multiplication over addition* allows us to distribute the multiplication and then add the results. To illustrate, 234×21 can be written as $234(20 + 1)$. By the distributive property

$$
\begin{aligned}
234(20 + 1) &= (234 \times 20) + (234 \times 1) \\
&= \quad 4680 \quad + \quad 234 \\
&= 4914
\end{aligned}
$$

This is what we actually do when we multiply.

$$
\begin{array}{r}
234 \\
\times \ \ 21 \\
\hline
234 \\
4680 \\
\hline
4914
\end{array}
$$

Distributive Property of Multiplication over Addition Multiplication can be distributed over addition without changing the result.

$$5 \times (10 + 2) = (5 \times 10) + (5 \times 2)$$

6 *Applying Multiplication to Real-Life Situations*

To use multiplication in word problems, the number of items or the value of each item must be the same. Recall that multiplication is a quick way to do repeated addition where *each addend is the same.* In the beginning of the section we showed 3 rows of 4 croissants to illustrate 3×4. The number of croissants in each row was the same, 4. Look at another example. If we had six nickels, we could use multiplication to find the total value of the coins because the value of each nickel is the same, 5¢. Since $6 \times 5 = 30$, six nickels are worth 30¢.

In the following example the word *average* is used. The word average has several different meanings. In this example, we are told that the *average annual salary* of an employee at Software Associates is $21,342. This means that we can calculate the total payroll as if each employee made $21,342 even though we know that the president probably makes more than any other employee.

EXAMPLE 14 The average annual salary of an employee at Software Associates is $21,342. There are 38 employees. What is the annual payroll?

$$\begin{array}{r} \$21{,}342 \\ \times \quad\quad 38 \\ \hline 170\,736 \\ 640\,26 \\ \hline 810{,}996 \end{array}$$

The total annual payroll is about $810,966.

Practice Problem 14 The average cost of a new car sold last year at Westover Chevrolet was $17,348. The dealership sold 378 cars. What were the total sales of cars at the dealership last year? ■

Another useful application of multiplication is area. The following example involves the area of a rectangle.

EXAMPLE 15 What is the area of a rectangular hallway that measures 4 feet wide and 8 feet long?

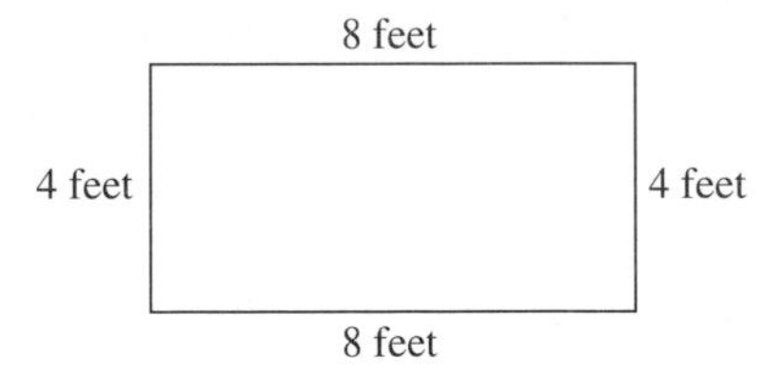

The **area** of a rectangle is the product of the length times the width. Thus for this hallway

$$\text{Area} = 8 \text{ feet} \times 4 \text{ feet} = 32 \text{ square feet}$$

The area of the hallway is 32 square feet.

Note: All measurements for area are given in square units such as square feet, square meters, square yards, etc.

Practice Problem 15 What is the area of a rectangular rug that measures 5 yards by 7 yards? ■

1.4 Exercises

Verbal and Writing Skills

1. Explain in your own words. Answers may vary. Samples are below.

(a) the commutative property of multiplication
You can change the order of the factors without changing the product.

(b) the associative property of multiplication
You can group the factors in any way without changing the product.

2. How does the distributive property of multiplication over addition help us to multiply 4×13? You can write 13 as $10 + 3$ and distribute 4 over the addition.
$4 \times (10 + 3) = (4 \times 10) + (4 \times 3)$

Complete the multiplication facts for each table. Strive for total accuracy, but work quickly. (For problems 3 and 4 allow a maximum of 5 minutes each.)

Multiply.

3.

×	6	2	3	8	0	5	7	9	1	4
5	30	10	15	40	0	25	35	45	5	20
7	42	14	21	56	0	35	49	63	7	28
1	6	2	3	8	0	5	7	9	1	4
0	0	0	0	0	0	0	0	0	0	0
6	36	12	18	48	0	30	42	54	6	24
2	12	4	6	16	0	10	14	18	2	8
3	18	6	9	24	0	15	21	27	3	12
8	48	16	24	64	0	40	56	72	8	32
4	24	8	12	32	0	20	28	36	4	16
9	54	18	27	72	0	45	63	81	9	36

4.

×	2	7	0	5	3	4	8	1	6	9
1	2	7	0	5	3	4	8	1	6	9
6	12	42	0	30	18	24	48	6	36	54
5	10	35	0	25	15	20	40	5	30	45
3	6	21	0	15	9	12	24	3	18	27
0	0	0	0	0	0	0	0	0	0	0
9	18	63	0	45	27	36	72	9	54	81
4	8	28	0	20	12	16	32	4	24	36
7	14	49	0	35	21	28	56	7	42	63
2	4	14	0	10	6	8	16	2	12	18
8	16	56	0	40	24	32	64	8	48	72

5. 32×3 = 96

6. 21×4 = 84

7. 104×2 = 208

8. 302×3 = 906

9. 6102×3 = 18,306

10. 5203×2 = 10,406

11. $101{,}204 \times 3$ = 303,612

12. $301{,}200 \times 4$ = 1,204,800

13. 14×5 = 70

14. 12×4 = 48

15. 87×6 = 522

16. 95×7 = 665

17. 326×5 = 1630

18. 732×6 = 4392

19. 1087×7 = 7609

20. 2605×9 = 23,445

21. $12{,}526 \times 8$ = 100,208

22. $48{,}761 \times 7$ = 341,327

23. $235{,}702 \times 4$ = 942,808

24. $127{,}054 \times 6$ = 762,324

Multiply by powers of 10.

25. 156 × 10 1560	**26.** 278 × 10 2780	**27.** 27,158 × 100 2,715,800	**28.** 89,361 × 100 8,936,100	**29.** 482 × 1000 482,000
30. 579 × 1000 579,000	**31.** 316,250 × 10,000 3,162,500,000	**32.** 25,346 × 10,000 253,460,000	**33.** 423 × 20 8460	**34.** 134 × 20 2680
35. 2120 × 30 63,600	**36.** 4230 × 20 84,600	**37.** 14,000 × 4000 56,000,000	**38.** 62,000 × 3000 186,000,000	

Multiply.

39. 1022 × 31 1 022 30 66 31,682	**40.** 415 × 24 1 660 8 30 9 960	**41.** 146 × 54 584 730 7884	**42.** 3021 × 12 6 042 30 21 36,252	**43.** 89 × 64 356 534 5696
44. 68 × 49 612 272 3332	**45.** 607 × 25 3 035 12 14 15,175	**46.** 817 × 84 3 268 65 36 68,628	**47.** 569 × 73 1 707 39 83 41,537	**48.** 915 × 47 6 405 36 60 43,005
49. 912 × 76 5 472 63 84 69,312	**50.** 498 × 39 4 482 14 94 19,422	**51.** 5123 × 29 46 107 102 46 148,567	**52.** 1268 × 38 10 144 38 04 48,184	**53.** 9053 × 91 9 053 814 77 823,823
54. 3078 × 72 6 156 215 46 221,616	**55.** 178 × 235 890 5 34 35 6 41,830	**56.** 493 × 416 2 958 4 93 197 2 205,088	**57.** 678 × 132 1 356 20 34 67 8 89,496	**58.** 392 × 187 2 744 31 36 39 2 73,304
59. 2076 × 105 10 380 00 00 207 6 217,980	**60.** 5092 × 302 10 184 00 00 1 527 6 1,537,784	**61.** 3561 × 403 10 683 00 00 1 424 4 1,435,083	**62.** 7023 × 2007 49 161 14 046 14,095,161	**63.** 1023 × 4005 5 115 4 092 4,097,115

64. $\begin{array}{r} 1298 \\ \times\ 304 \\ \hline \end{array}$
5 192
00 00
389 4
394,592

65. $\begin{array}{r} 260 \\ \times\ 40 \\ \hline \end{array}$
10,400

66. $\begin{array}{r} 307 \\ \times\ 300 \\ \hline \end{array}$
92,100

67. $\begin{array}{r} 307 \\ \times\ 30 \\ \hline \end{array}$
9210

68. $\begin{array}{r} 150 \\ \times\ 200 \\ \hline \end{array}$
30,000

69. $7 \cdot 2 \cdot 5$ 70

70. $8 \cdot 3 \cdot 2$ 48

71. $11 \cdot 7 \cdot 4$ 308

72. $15 \cdot 4 \cdot 4$ 240

73. $10 \cdot 7 \cdot 10$ 700

74. $20 \cdot 5 \cdot 3$ 300

75. $12 \cdot 3 \cdot 5 \cdot 2$ 360

76. $7 \cdot 6 \cdot 2 \cdot 5$ 420

77. What is x if $x = 8 \cdot 7 \cdot 6 \cdot 0$? $x = 0$

78. What is x if $x = 3 \cdot 12 \cdot 0 \cdot 5$? $x = 0$

Applications

79. Find the area of a rectangular university fitness center if the fitness center floor is 48 feet long and 67 feet wide. 3216 square feet

80. Find the area of a rectangular computer chip that is 17 millimeters long and 9 millimeters wide. 153 square millimeters

81. The Police Department is purchasing 35 bulletproof vests at $345 each. What is the total cost of purchasing these vests? $12,075

82. The Andersonville College football team is staying overnight at a local Holiday Inn. They have been quoted a rate of $43 per night per room. The team will need 27 rooms. What will be the team cost to stay overnight at this motel? $1161

83. The student commons food supply needs to purchase espresso coffee. Find the cost of purchasing 240 pounds of espresso coffee beans at $5 per pound. $1200

84. The music department of Wheaton College wishes to purchase 345 sets of headphones at the music supply store at a cost of $8 each. What will be the total amount of the purchase? $2760

85. Helen pays $266 per month for her car payment on her new Honda Civic. What is her automobile payment cost for a one-year period? $3192

86. A company rents a Ford Escort for a salesman at $276 per month for eight months. What is the cost for the car rental during this time? $8 \times 276 = \$2208$

87. Marcos has a Geo Prism that gets 34 miles per gallon during highway driving. Approximately how far can he travel if he has 18 gallons of gas in the tank? $18 \times 34 = 612$ miles

88. Cheryl has a subcompact car that gets 48 miles per gallon on highway driving. Approximately how far can she travel if she has 12 gallons of gas in the tank? 576 miles

89. Today, the Burger Basket has 12 employees who made 85 hamburgers each. How many burgers were made in total? 1020 hamburgers

90. Fun 'n' Games Amusement Park has 102 employees on staff. They are all scheduled to work 38 hours per week. How many working hours should be scheduled in one week for all employees? 3876 hours

91. The country of Haiti has an average per capita (per person) income of $800. If the approximate population of Haiti is 6,530,000, what is the approximate total yearly income of the entire country. $5,224,000,000

92. The kingdom of the Netherlands (Holland) has an approximate population of 15,500,000. The average per capita (per person) income is $17,200. What is the approximate total yearly income of the entire country? $266,600,000,000

To Think About

93. Multiply.
$4 \times 9 \times 25 \times 12 \times 5$
54,000

Find the value of x in each equation.

94. $5(x) = 40$ $x = 8$

95. $7(x) = 56$ $x = 8$

96. $72 = 8(x)$ $x = 9$

97. $63 = 9(x)$ $x = 7$

98. Would the distributive property of multiplication be true for Roman numerals such as (XII) × (IV)? Why or why not? No, it would not always be true. In our number system 62 = 60 + 2. But in roman numerals IV ≠ I + V. The digit system in roman numerals involves subtraction. Thus (XII) × (IV) ≠ (XII × I) + (XII × V).

99. We saw that multiplication is distributive over addition. Is it distributive over subtraction? Why or why not? Give examples.
Yes. $5 \times (8 - 3) = 5 \times 8 - 5 \times 3$
$a \times (b - c) = (a \times b) - (a \times c)$

Cumulative Review Problems

100. Subtract.
$$\begin{array}{r} 34{,}084 \\ -\ 27{,}328 \\ \hline 6{,}756 \end{array}$$

101. Add.
$$\begin{array}{r} 263 \\ 27 \\ 891 \\ 5 \\ +\ 63 \\ \hline 1249 \end{array}$$

102. Albert Gachet earns $156 per week. He has withheld weekly $12 for taxes, $2 for dues, and $3 for retirement. How much is left?
$139

103. Mrs. May Washington retired at $743 per month. Her cost-of-living raise increased her payment to $802 per month. What was the increase?
$59

Developing Your Study Skills

WHY IS HOMEWORK NECESSARY?

Mathematics involves mastering a set of skills that you learn by practicing, not by watching someone else do it. Your instructor may make solving a mathematics problem look very easy, but for you to learn the necessary skills, you must practice them over and over again, just as your instructor once had to. There is no other way. Learning mathematics is like learning to play a musical instrument, to type, or to play a sport. No matter how much you watch someone else do it, no matter how many books you read on "how to" do it, no matter how easy it seems to be, the key to success is practice on a regular basis.

Homework provides this practice. The amount of practice needed varies for each individual, but usually students need to do most or all of the exercises provided at the end of each section in the text. The more problems you do, the better you get. Some problems in a set are more difficult than others, and some stress different concepts. Only by working all the problems will you cover the full range of difficulty.

1.5 Division of Whole Numbers

After studying this section, you will be able to:

1. *Perform simple division using basic division facts.*
2. *Perform long division by a one-digit number.*
3. *Perform long division by a two- or three-digit number.*
4. *Apply division to real-life situations.*

1 *Mastering Basic Division Facts*

Suppose that we have eight quarters and want to divide them into two equal piles. We would discover that each pile contains four quarters.

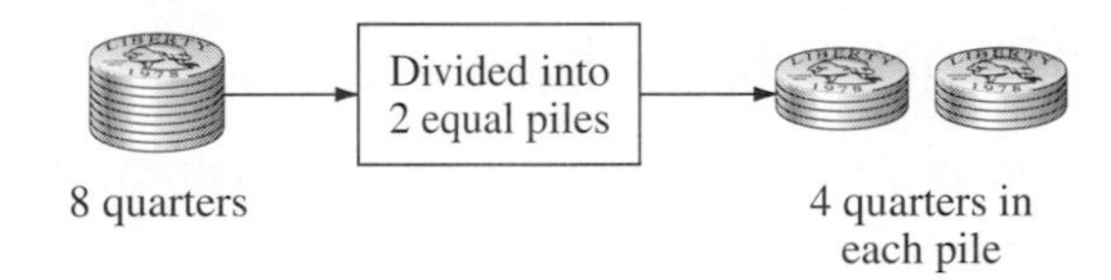

In mathematics we would express this thought by saying that

$$8 \div 2 = 4$$

We know that this answer is right because two piles of four quarters is the same dollar amount as eight quarters. In other words, we know that $8 \div 2 = 4$ because $2 \times 4 = 8$. These two mathematical sentences are called **related sentences.** The division sentence $8 \div 2 = 4$ is related to the multiplication sentence $2 \times 4 = 8$.

In fact, in mathematics we usually define division in terms of multiplication. The answer to the division problem $12 \div 3$ is that number which when multiplied by 3 yields 12. Thus

$$12 \div 3 = 4 \text{ because } 3 \times 4 = 12$$

Suppose that a surplus of \$30 in the French Club budget at the end of the year is to be equally divided among the five club members. We would want to divide the \$30 into five equal parts. We would write $30 \div 5 = 6$ because $5 \times 6 = 30$. Thus each of the five people would get \$6 in this situation.

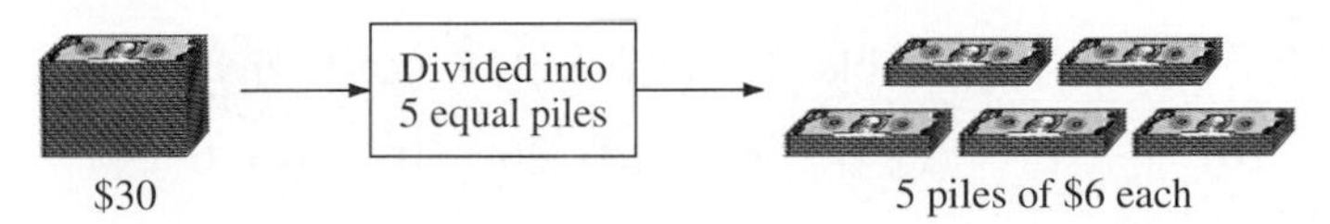

As a mathematical sentence, $30 \div 5 = 6$.

The division problem $30 \div 5 = 6$ could also be written $\frac{30}{5} = 6$ or $5\overline{)30}$ with quotient 6.

When referring to division, we sometimes use the words **divisor, dividend,** and **quotient** to identify the three parts.

$$\begin{array}{r} \text{quotient} \\ \text{divisor}\overline{)\text{dividend}} \end{array}$$

With $30 \div 5 = 6$, 30 is the dividend, 5 is the divisor, and 6 is the quotient.

$$\begin{array}{rcl} & 6 \longleftarrow & \text{quotient} \\ \text{divisor} \longrightarrow & 5\overline{)30} \longleftarrow & \text{dividend} \end{array}$$

So the quotient is the answer to a division problem. It is important that you be able to do short problems involving basic division facts quickly.

TEACHING TIP Throughout all mathematics courses from Basic Mathematics to Calculus, the words *divisor, dividend,* and *quotient* are used extensively. Stress to students that in any given division problem they need to be able to recognize which number is the divisor, which is the dividend, and which is the quotient.

EXAMPLE 1 Divide.

(a) $12 \div 4$ **(b)** $81 \div 9$ **(c)** $56 \div 8$ **(d)** $54 \div 6$

(a) $4\overline{)12}$ = 3 **(b)** $9\overline{)81}$ = 9 **(c)** $8\overline{)56}$ = 7 **(d)** $6\overline{)54}$ = 9

Practice Problem 1 Divide.

(a) $36 \div 4$ **(b)** $25 \div 5$ **(c)** $72 \div 9$ **(d)** $30 \div 6$ ■

Zero can be divided by any nonzero number, but division by zero is not possible. Why is this?

Suppose that we could divide by zero. Then $7 \div 0 =$ some number. Let us represent "some number" by the letter a.

$$\text{If } 7 \div 0 = a, \text{ then } 7 = 0 \times a$$

because every division problem has a related multiplication problem. But zero times any number is zero, $0 \times a = 0$. Thus

$$7 = 0 \times a = 0$$

That is, $7 = 0$, which we know is not true. Therefore, our assumption that $7 \div 0 = a$ is wrong. Thus we conclude that we cannot divide by zero.

It is helpful to remember the following basic concepts:

Division Problems Involving the Number 1 and the Number 0

1. Any nonzero number divided by itself is 1 ($7 \div 7 = 1$).
2. Any number divided by 1 remains unchanged ($29 \div 1 = 29$).
3. Zero may be divided by any nonzero number; the result is always zero ($0 \div 4 = 0$).
4. Zero can never be the divisor in a division problem ($3 \div 0$ cannot be done; $0 \div 0$ cannot be done).

EXAMPLE 2 Divide if possible. If not possible, state why.

(a) $8 \div 8$ **(b)** $9 \div 1$ **(c)** $0 \div 6$ **(d)** $20 \div 0$ **(e)** $0 \div 0$

(a) $\frac{8}{8} = 1$ Any number divided by itself is 1.

(b) $\frac{9}{1} = 9$ Any number divided by 1 remains unchanged.

(c) $\frac{0}{6} = 0$ Zero divided by any nonzero number is zero.

(d) $\frac{20}{0}$ cannot be done Division by zero is impossible.

(e) $\frac{0}{0}$ cannot be done Division by zero is impossible.

TEACHING TIP It is critical that all students master division with zero. They need to know that zero can be divided by any nonzero number, but that division by zero is never allowed. A simple example usually helps them to see the logic of the situation. A student club with profits of $400 and 10 members could distribute 400 divided by 10 = $40 to each member. A student club with profits of $0 and 10 members would only be able to distribute 0 divided by 10 = $0 to each member. A student club with profits of $400 and 0 members could not distribute any money to each member. 400 divided by zero cannot be done.

Practice Problem 2 Divide, if possible.

(a) $7 \div 1$ **(b)** $\frac{8}{8}$ **(c)** $\frac{0}{5}$ **(d)** $12 \div 0$ **(e)** $\frac{0}{0}$ ■

2 *Performing Division by a One-Digit Number*

Our accuracy with division is improved if we have a checking procedure. For each division fact, there is a related multiplication fact.

$$\text{If } 20 \div 4 = 5, \text{ then } 20 = 4 \times 5$$

$$\text{If } 36 \div 9 = 4, \text{ then } 36 = 9 \times 4$$

We will often use multiplication to check our answers.

When two numbers do not divide exactly, a number called the **remainder** is left over. For example, 13 cannot be divided exactly by 2. The number 1 is left over. We call this 1 the *remainder.*

$$\begin{array}{r} 6 \\ 2\overline{)13} \\ \underline{12} \\ 1 \end{array} \longleftarrow \text{remainder}$$

Thus $13 \div 2 = 6$ with a remainder of 1. We can abbreviate this answer as

$$6 \text{ R } 1$$

To check this division, we multiply $2 \times 6 = 12$ and add the remainder $12 + 1 = 13$. That is, $2 \times 6 + 1 = 13$. The result will be the dividend if the division was done correctly. The following fact shows you how to check a division that has a remainder.

TEACHING TIP Be sure to go over in detail how to check a division problem if there is a remainder. A few students will find this idea to be totally new and will need to see a few examples worked out before they understand the concept.

> (divisor × quotient) + remainder = dividend

EXAMPLE 3 Divide. $33 \div 4$. Check your answer.

$$\begin{array}{rl} 8 & \longrightarrow \text{How many times can 4 be divided into 33? } 8. \\ 4\overline{)33} & \\ \underline{32} & \longleftarrow \text{What is } 8 \times 4\text{? } 32. \\ 1 & \longleftarrow \text{What is 33 subtract 32? } 1. \end{array}$$

The answer is 8 with a remainder of 1. We abbreviate 8 R 1.

Check.

$$\begin{array}{rl} 8 & \text{Multiply. } 8 \times 4 = 32. \\ \underline{\times 4} & \\ 32 & \\ \underline{1} & \text{Add the remainder. } 32 + 1 = 33. \\ 33 & \text{Because the dividend is 33, the answer is correct.} \end{array}$$

Practice Problem 3 Divide. $45 \div 6$. Check your answer. ■

EXAMPLE 4 Divide. 158 ÷ 5. Check your answer.

```
    31
 5)158     5 divided into 15?  3.
   15  ← What is 3 × 5?  15.
   08 ← 15 subtract 15?  0. Bring down 8.
    5 ← 5 divided into 8?  1. What is 1 × 5?  5.
    3 ← 8 subtract 5?  3.
```

The answer is 31 R 3. Check.

```
   31    Multiply. 31 × 5 = 155.
 ×  5
  155
    3    Add the remainder 3.
  158    Because the dividend is 158, the answer is correct.
```

Practice Problem 4 Divide. 129 ÷ 6. Check your answer. ■

EXAMPLE 5 Divide. 3672 ÷ 7

```
    524
 7)3672     How many times can 7 be divided into 36?  5.
   35   ← What is 5 × 7?  35.
    17  ← 36 subtract 35?  1. Bring down 7.
    14  ← 7 divided into 17?  2. What is 2 × 7?  14.
     32 ← 17 subtract 14?  3. Bring down 2.
     28 ← 7 divided into 32?  4. What is 4 × 7?  28.
      4 ← 32 subtract 28?  4.
```

The answer is 524 R 4.

TEACHING TIP Some students do not find it necessary when doing Example 5 to show the steps of long division. They merely record it mentally. Remind students that if they can divide accurately by a single digit without writing out these steps, they are not obligated to show the steps of long division.

Practice Problem 5 Divide. 4237 ÷ 8 ■

3 *Performing Division by a Two- or Three-Digit Number*

When the divisor has more than one digit, an estimation technique may help. Figure how many times the first digit of the divisor goes into the first two digits of the dividend. Try this answer as the first number in the quotient.

EXAMPLE 6 Divide. 283 ÷ 41

```
First guess:    7
             41)283
                287 too large
```

How many times can the first digit of the divisor (4) be divided into the first two digits of the dividend (28)? 7. We try the answer 7 as the first number of the quotient. We multiply 7 × 41 = 287. We see that 287 is larger than 283.

Second guess:

```
     6
41)283    Because 7 is slightly too large, we try 6.
   246 <— 6 × 41?  246.
    37 <— 283 subtract 246?  37.
```

The answer is 6 R 37. (Note that the remainder must always be less than the divisor.)

Practice Problem 6 Divide. 229 ÷ 32 ■

EXAMPLE 7 Divide. 33,897 ÷ 56

First guess:

```
     60
56)33897    How many times can 33 be divided by 5?  6.
   336    <— What is 6 × 56?  336.
    29      338 subtract 336?  2. Bring down 9.
            56 cannot be divided into 29. Write 0 in quotient.
```

Second set of steps:

```
    605
56)33897    Bring down 7.
   336      How many times can 5 be divided
    297     into 29?  5.
    280 <— What is 5 × 56? 280. Subtract 297 − 280.
     17     Remainder is 17.
```

The answer is 605 R 17.

Practice Problem 7 Divide. 42,183 ÷ 33 ■

EXAMPLE 8 Divide. 5629 ÷ 134

```
     42
134)5629    How many times does 134 divide into 562?
    536     We guess by saying that 1 divides into 5 five times,
     269    but this is too large. (5 × 134 = 670!)
     268    So we try 4. What is 4 × 134?  536.
       1    Subtract 562 − 536. We obtain 26. Bring down 9.
            How many times does 134 divide into 269?
            We guess by saying that 1 divided into 2 goes 2 times.
            What is 2 × 134?  268. Subtract 269 − 268.
            The remainder is 1.
```

The answer is 42 R 1.

Practice Problem 8 Divide. 3227 ÷ 128 ■

4 *Applying Division to Real-Life Situations*

When you solve a word problem that requires division, you will be given the total number and asked to calculate the number of items in each group or to calculate the number of groups. In the beginning of this section we showed 8 quarters (the total number) and we divided them into 2 equal piles (the number of groups). Division was used to find how many quarters were in each pile (the number in each group). That is, $8 \div 2 = 4$. There were 4 quarters in each pile.

Let's look at another example. Suppose that $30 is to be divided equally among the members of a group. If each person receives $6, how many people are in the group? We use division, $30 \div 6 = 5$, to find that there are 5 people in the group.

EXAMPLE 9 City Service Realty just purchased nine identical computers for the realtors in the office. The total cost for the nine computers was $25,848. What was the cost of one computer? Check your answer.

To find the cost of one computer, we need to divide the total cost by 9. Thus we will calculate $25,848 ÷ 9.

```
   2872
9)25848
  18
  --
   78
   72
   --
    64
    63
    --
     18
     18
     --
      0
```

Therefore, the cost of one computer is $2872. In order to check our work we will need to see if 9 computers each costing $2872 will in fact result in a total of $25,848. We use multiplication to check division.

```
  2872
×    9
------
 25848 ✓
```

We did obtain 25,848. Our answer is correct.

Practice Problem 9 The Dallas police department purchased seven identical police cars at a total cost of $117,964. Find the cost of one car. Check your answer. ■

In the following example you will see the word *average* used as it applies to division. The problem states that a car traveled 1144 miles in 22 hours. The problem asks you to find the average speed in miles per hour. This means that we will treat the problem as if the speed of the car was the same during each hour of the trip. We will use division to solve.

EXAMPLE 10 A car traveled 1144 miles in 22 hours. What was the average speed in miles per hour?

When doing distance problems, it is helpful to remember that distance ÷ time = rate. We need to divide 1144 miles by 22 hours to obtain the rate or speed in miles per hour.

$$\begin{array}{r} 52 \\ 22\overline{)1144} \\ \underline{110} \\ 44 \\ \underline{44} \\ 0 \end{array}$$

The car traveled an average of 52 miles per hour.

Practice Problem 10 An airplane traveled 5138 miles in 14 hours. What was the average speed in miles per hour? ■

Developing Your Study Skills

TAKING NOTES IN CLASS

An important part of mathematics studying is taking notes. In order to take meaningful notes, you must be an active listener. Keep your mind on what the instructor is saying, and be ready with questions whenever you do not understand something.

If you have previewed the lesson material, you will be prepared to take good notes. The important concepts will seem somewhat familiar. You will have a better idea of what needs to be written down. If you frantically try to write all that the instructor says or copy all the examples done in class, you may find your notes to be nearly worthless when you look at them at home. You may find that you are unable to make sense of what you have written.

Write down *important* ideas and examples as the instructor lectures, making sure that you are listening and following the logic. Include any helpful hints or suggestions that your instructor gives you or refers to in your text. You will be amazed at how easily you will forget these if you do not write them down. Try to review your notes the *same day* sometime after class. You will find the material in your notes easier to understand if you have attended class within the last few hours.

Successful note taking requires active listening and processing. Stay alert in class. You will realize the advantages of taking your own notes over copying those of someone else.

1.5 Exercises

Verbal and Writing Skills

1. Explain in your own words what happens when you

(a) divide a nonzero number by itself

When you divide a nonzero number by itself, the result is 1.

(b) divide a number by 1

When you divide a number by 1, the result is that number.

(c) divide zero by a nonzero number

When you divide zero by a nonzero number, the result is zero.

(d) divide a number by 0

You cannot divide a number by 0.

Divide. See if you can work problems 2–30 in 3 minutes or less.

2. $\begin{array}{r} 7 \\ 5\overline{)35} \end{array}$	**3.** $\begin{array}{r} 7 \\ 6\overline{)42} \end{array}$	**4.** $\begin{array}{r} 8 \\ 4\overline{)32} \end{array}$	**5.** $\begin{array}{r} 3 \\ 8\overline{)24} \end{array}$	**6.** $\begin{array}{r} 3 \\ 9\overline{)27} \end{array}$	**7.** $\begin{array}{r} 5 \\ 8\overline{)40} \end{array}$
8. $\begin{array}{r} 8 \\ 7\overline{)56} \end{array}$	**9.** $\begin{array}{r} 4 \\ 9\overline{)36} \end{array}$	**10.** $\begin{array}{r} 3 \\ 4\overline{)12} \end{array}$	**11.** $\begin{array}{r} 3 \\ 7\overline{)21} \end{array}$	**12.** $\begin{array}{r} 9 \\ 9\overline{)81} \end{array}$	**13.** $\begin{array}{r} 7 \\ 8\overline{)56} \end{array}$
14. $\begin{array}{r} 9 \\ 6\overline{)54} \end{array}$	**15.** $\begin{array}{r} 9 \\ 7\overline{)63} \end{array}$	**16.** $\begin{array}{r} 7 \\ 4\overline{)28} \end{array}$	**17.** $\begin{array}{r} 9 \\ 8\overline{)72} \end{array}$	**18.** $\begin{array}{r} 8 \\ 8\overline{)64} \end{array}$	**19.** $\begin{array}{r} 7 \\ 9\overline{)63} \end{array}$
20. $\begin{array}{r} 8 \\ 9\overline{)72} \end{array}$	**21.** $\begin{array}{r} 4 \\ 6\overline{)24} \end{array}$	**22.** $\begin{array}{r} 8 \\ 1\overline{)8} \end{array}$	**23.** $\begin{array}{r} 7 \\ 1\overline{)7} \end{array}$	**24.** $\begin{array}{r} 0 \\ 5\overline{)0} \end{array}$	**25.** $\begin{array}{r} 0 \\ 6\overline{)0} \end{array}$
26. $\begin{array}{r} 8 \\ 6\overline{)48} \end{array}$	**27.** $54 \div 6$ 9	**28.** $42 \div 7$ 6	**29.** $6 \div 6$ 1	**30.** $5 \div 5$ 1	

Divide. Check your answer on 31–42.

31. $32 \div 6$ 5 R 2

$$\begin{array}{r} 6\overline{)32} \\ \underline{30} \\ 2 \end{array} \qquad \text{Check} \quad \begin{array}{r} 5 \\ \underline{\times\ 6} \\ 30 \\ \underline{+\ 2} \\ 32 \end{array}$$

32. $37 \div 5$ 7 R 2

$$\begin{array}{r} 5\overline{)37} \\ \underline{35} \\ 2 \end{array} \qquad \text{Check} \quad \begin{array}{r} 7 \\ \underline{\times\ 5} \\ 35 \\ \underline{+\ 2} \\ 37 \end{array}$$

33. $76 \div 8$ 9 R 4

$$\begin{array}{r} 8\overline{)76} \\ \underline{72} \\ 4 \end{array} \qquad \text{Check} \quad \begin{array}{r} 9 \\ \underline{\times\ 8} \\ 72 \\ \underline{+\ 4} \\ 76 \end{array}$$

34. $39 \div 9$ 4 R 3

$$\begin{array}{r} 9\overline{)39} \\ \underline{36} \\ 3 \end{array} \qquad \text{Check} \quad \begin{array}{r} 4 \\ \underline{\times\ 9} \\ 36 \\ \underline{+\ 3} \\ 39 \end{array}$$

35. $128 \div 5$ 25 R 3

$$\begin{array}{r} 5\overline{)128} \\ \underline{10} \\ 28 \\ \underline{25} \\ 3 \end{array} \qquad \text{Check} \quad \begin{array}{r} 25 \\ \underline{\times\ 5} \\ 125 \\ \underline{+\ 3} \\ 128 \end{array}$$

36.
$$\begin{array}{r} 17\ \text{R}\ 1 \\ 6\overline{)103} \\ \underline{6} \\ 43 \\ \underline{42} \\ 1 \end{array} \qquad \text{Check} \quad \begin{array}{r} 17 \\ \underline{\times\ 6} \\ 102 \\ \underline{+\ 1} \\ 103 \end{array}$$

37.
$$\begin{array}{r} 23\ \text{R}\ 4 \\ 7\overline{)165} \\ \underline{14} \\ 25 \\ \underline{21} \\ 4 \end{array} \qquad \text{Check} \quad \begin{array}{r} 23 \\ \underline{\times\ 7} \\ 161 \\ \underline{+\ 4} \\ 165 \end{array}$$

38.
$$\begin{array}{r} 15\ \text{R}\ 4 \\ 8\overline{)124} \\ \underline{8} \\ 44 \\ \underline{40} \\ 4 \end{array} \qquad \text{Check} \quad \begin{array}{r} 15 \\ \underline{\times\ 8} \\ 120 \\ \underline{+\ 4} \\ 124 \end{array}$$

39.
$$\begin{array}{r} 32 \\ 9\overline{)288} \\ \underline{27} \\ 18 \\ \underline{18} \\ 0 \end{array} \qquad \text{Check} \quad \begin{array}{r} 32 \\ \underline{\times\ 9} \\ 288 \end{array}$$

40.
$$\begin{array}{r} 18 \\ 7\overline{)126} \\ \underline{7} \\ 56 \\ \underline{56} \\ 0 \end{array} \qquad \text{Check} \quad \begin{array}{r} 18 \\ \underline{\times\ 7} \\ 126 \end{array}$$

41.
$$\begin{array}{r} 37 \\ 5\overline{)185} \\ \underline{15} \\ 35 \\ \underline{35} \\ 0 \end{array} \qquad \text{Check} \quad \begin{array}{r} 37 \\ \underline{\times\ 5} \\ 185 \end{array}$$

42.
$$\begin{array}{r} 28 \\ 8\overline{)224} \\ \underline{16} \\ 64 \\ \underline{64} \\ 0 \end{array} \qquad \text{Check} \quad \begin{array}{r} 28 \\ \underline{\times\ 8} \\ 224 \end{array}$$

43.
$$\begin{array}{r} 322\ \text{R}\ 1 \\ 4\overline{)1289} \\ \underline{12} \\ 8 \\ \underline{8} \\ 9 \\ \underline{8} \\ 1 \end{array}$$

44.
$$\begin{array}{r} 252\ \text{R}\ 2 \\ 3\overline{)758} \\ \underline{6} \\ 15 \\ \underline{15} \\ 8 \\ \underline{6} \\ 2 \end{array}$$

45.
$$\begin{array}{r} 127\ \text{R}\ 1 \\ 6\overline{)763} \\ \underline{6} \\ 16 \\ \underline{12} \\ 43 \\ \underline{42} \\ 1 \end{array}$$

46. $7\overline{)403}$ — 57 R 4

35
53
49
4

47. $8\overline{)6024}$ — 753

56
42
40
24
24
0

48. $9\overline{)2286}$ — 254

18
48
45
36
36
0

49. $5\overline{)6789}$ — 1357 R 4

5
17
15
28
25
39
35
4

50. $5\overline{)7324}$ — 1464 R 4

5
23
20
32
30
24
20
4

51. $7\overline{)12304}$ — 1757 R 5

7
53
49
40
35
54
49
5

52. $6\overline{)18127}$ — 3021 R 1

18
12
12
7
6
1

53. $9\overline{)22305}$ — 2478 R 3

18
43
36
70
63
75
72
3

54. $8\overline{)26037}$ — 3254 R 5

24
20
16
43
40
37
32
5

Mixed Practice

Divide.

55. $185 \div 36$

5 R 5
$36\overline{)185}$
180
5

56. $152 \div 48$

3 R 8
$48\overline{)152}$
144
8

57. $267 \div 52$

5 R 7
$52\overline{)267}$
260
7

58. $321 \div 53$

6 R 3
$53\overline{)321}$
318
3

59. $427 \div 61$

7
$61\overline{)427}$
427
0

60. $72\overline{)432}$ — 6

432
0

61. $12\overline{)1930}$ — 160 R 10

12
73
72
10
0
10

62. $13\overline{)6810}$ — 523 R 11

65
31
26
50
39
11

63. $30\overline{)1452}$ — 48 R 12

120
252
240
12

64. $40\overline{)1125}$ — 28 R 5

80
325
320
5

65. $15\overline{)9236}$ — 615 R 11

90
23
15
86
75
11

66. $16\overline{)8314}$ — 519 R 10

80
31
16
154
144
10

67. $36\overline{)7568}$ — 210 R 8

72
36
36
8
0
8

68. $32\overline{)3527}$ — 110 R 7

32
32
32
7
0
7

69. $18\overline{)3643}$ — 202 R 7

36
43
36
7

70. $19\overline{)2174}$ — 114 R 8

19
27
19
84
76
8

71. $124\overline{)500}$ — 4 R 4

496
4

72. $132\overline{)792}$ — 6

792
0

73. $322\overline{)2280}$ — 7 R 26

2254
26

74. $41\overline{)205000}$ — 5000

205
0

75. $3483 \div 129 = x$. What is the value of x?

27

76. $2214 \div 123 = x$. What is the value of x?

18

Applications

Solve each problem.

77. Western Saddle Stable uses 21,900 pounds of feed per year to feed its 30 horses. How much does each horse eat per year?
730 pounds of feed per horse.

78. A *run* in skiing is going from the top of the ski lift to the bottom. If over seven days 431,851 runs were made, what was the average number of ski runs per day? 61,693 runs per day.

79. A telethon raised \$3,677,880 over 15 hours. What was the average amount of money raised per hour? \$245,192 per hour

80. A factory that produces silver earrings made 864 pairs of hoops in 36 hours. What was the average amount of earrings produced per hour?
24 pairs of earrings per hour

81. A group of 8 old friends invested the same amount each in a beach property that sold for \$369,432. How much did each friend pay? \$46,179

82. A horse and carriage company in New York City bought 7 new carriages at exactly the same price each. The total bill was \$147,371. How much did each carriage cost? \$21,053

83. A new sorting machine sorts 26 letters per minute. The machine is fed 884 letters. How many minutes will it take to sort the letters?
$884 \div 26 = 34$ minutes

84. Chung wishes to pay off a loan of \$4104 in 24 months. How large will his monthly payments be?
$4104 \div 24 = \$171$ for each monthly payment

To Think About

85. Division is not commutative. For example, $12 \div 4 \neq 4 \div 12$. If $a \div b = b \div a$, what must be true of the numbers a and b besides the fact that $b \neq 0$ and $a \neq 0$?
a and b must represent the same number.
For example, if $a = 12$, then $b = 12$.

86. You can think of division as repeated subtraction. Show how $874 \div 138$ is related to repeated subtraction.

874	
− 138	first time
736	
− 138	second time
598	
− 138	third time
460	
− 138	fourth time
322	
− 138	fifth time
184	
− 138	sixth time
46	

$$\begin{array}{r} 6 \text{ R } 46 \\ 138\overline{)874} \\ \underline{828} \\ 46 \end{array}$$

Cumulative Review Problems

Multiply.

87. $\begin{array}{r} 128 \\ \times\ 43 \\ \hline 384 \\ 512 \\ \hline 5504 \end{array}$

88. $\begin{array}{r} 7162 \\ \times\ 145 \\ \hline 35\,810 \\ 286\,48 \\ 716\,2 \\ \hline 1{,}038{,}490 \end{array}$

Add.

89. $316{,}214 + 89{,}981$

$\begin{array}{r} 316{,}214 \\ +\ 89{,}981 \\ \hline 406{,}195 \end{array}$

Subtract.

90. $1{,}360{,}000 - 1{,}293{,}156$

$\begin{array}{r} 1{,}360{,}000 \\ -\ 1{,}293{,}156 \\ \hline 66{,}844 \end{array}$

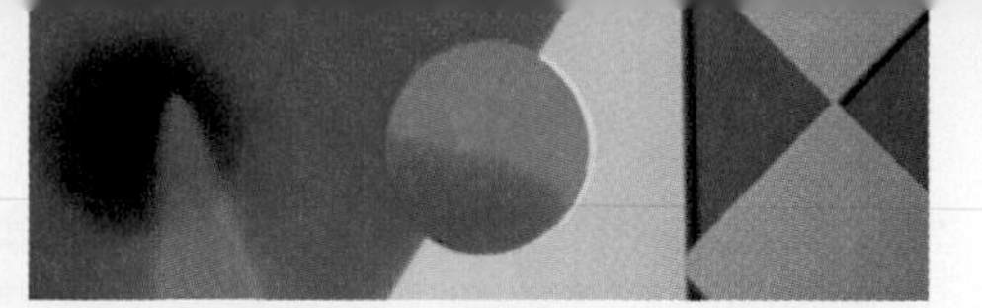

Putting Your Skills to Work

BUYING OR LEASING A CAR

Today over $\frac{1}{3}$ of all potential new car buyers decide to lease the car instead of purchasing it. What are the mathematics of buying versus leasing? How do we compare the costs?

Charlie and Sue Downey are considering leasing a new Dodge Stratus for $199 per month or purchasing the car for $14,996. See if you can determine the needed mathematical facts in order to make a well-informed decision.

PROBLEMS FOR INDIVIDUAL INVESTIGATION

1. The lease is for a rental period of 48 months and requires a down payment of $1000. What is the basic minimum cost to lease the car for 4 years? $10,552

2. If the car is purchased with a down payment of $1000 and the remainder is paid for with a new car loan for 48 months with monthly payments of $370, what will it cost to purchase the car? $18,760

PROBLEMS FOR GROUP INVESTIGATION AND COOPERATIVE STUDY

Together with some members of your class, see if you can answer the following questions.

3. If the Downeys plan to keep the car at the end of the 4-year lease, they will need to purchase it from the dealer at a cost of $9352. In that case, what is the total cost of obtaining the car? $19,904

4. If the Downeys plan to give the car back to the dealer after 4 years, they must pay a mileage surcharge of 15 cents per mile for all miles in excess of 12,000 miles per year. If they normally drive a car 64,000 miles over a 4-year period, what will be the total approximate cost for them to lease the car for 4 years? $12,952

5. If the Downeys are considering leasing and buying the car after 4 years (question 3) or purchasing the car (question 2), what would you recommend to them if they plan to drive the car about 64,000 miles during the 4-year period? It would probably be wisest to buy the car with $1000 down and 48 payments of $370, which gives a total cost of $18,760. The other option not only costs more ($19,904), it might be difficult to come up with the $9352 to pay for the car at the end of 4 years of leasing without taking out another loan, which will drive the costs up significantly.

INTERNET CONNECTIONS: Go to ``http://www.prenhall.com/tobey'' to be connected

Site: Microsoft Carpoint or a related site

This site includes a loan calculator which can help you determine the cost of financing a car. Use the loan calculator to answer the following questions.

6. Kimberly is buying a car for $13,000. She plans to make a 20% down payment, and she will finance the rest of the purchase price over a 3-year period. Her interest rate will be 9%. Find the amount of her monthly payment, and the total amount that she will pay for the car.

7. Jonathan wants to buy a car. He plans to make a 15% down payment and finance the rest over a 4-year period. His interest rate will be 7%. If Jonathan wants to keep his monthly car payment under $225, what is the most expensive car he can afford to buy?

1.6 Exponents and Order of Operations

MathPro

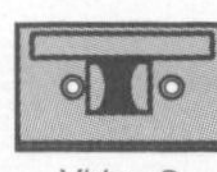
Video 2

SSM

After studying this section, you will be able to:

1 *Evaluate expressions with whole-number exponents.*
2 *Perform several arithmetic operations in the proper order.*

1 *Evaluating Whole-Number Exponents*

Sometimes a simple math idea comes ''disguised'' in technical language. For example, an **exponent** is just a ''shorthand'' number that saves writing multiplication of the same numbers.

10^3 The exponent 3 means $10 \times 10 \times 10.$

(which takes longer to write)

TEACHING TIP Students who have never used exponents before often write the exponent with the same-size digit as the base. Remind them that exponents are written with a smaller digit than the base.

The product 5×5 can be written as 5^2. The small number 2 is called the *exponent.* The exponent tells us how many factors are in the multiplication. The number 5 is called the **base.** The base is the number that is multiplied.

$$3 \times 3 \times 3 \times 3 = 3^4$$
(4 ← exponent; 3 ← base)

In 3^4 the base is 3 and the exponent is 4. (The 4 is sometimes called the *superscript.*) 3^4 is read as ''3 to the fourth power.''

EXAMPLE 1 Write each product in exponent form.

(a) $15 \times 15 \times 15$ **(b)** $7 \times 7 \times 7 \times 7 \times 7$

(a) $15 \times 15 \times 15 = 15^3$ **(b)** $7 \times 7 \times 7 \times 7 \times 7 = 7^5$

Practice Problem 1 Write each product in exponent form.

(a) $12 \times 12 \times 12 \times 12$ **(b)** $2 \times 2 \times 2 \times 2 \times 2 \times 2$ ■

EXAMPLE 2 Find the value of each expression.

(a) 3^3 **(b)** 7^2 **(c)** 2^5 **(d)** 1^8

(a) To find the value of 3^3, multiply the base 3 by itself 3 times.
$3^3 = 3 \times 3 \times 3 = 27$

(b) To find the value of 7^2, multiply the base 7 by itself 2 times.
$7^2 = 7 \times 7 = 49$

(c) $2^5 = 2 \times 2 \times 2 \times 2 \times 2 = 32$

(d) $1^8 = 1 \times 1 \times 1 \times 1 \times 1 \times 1 \times 1 \times 1 = 1$

Practice Problem 2 Find the value of each expression.

(a) 12^2 **(b)** 6^3 **(c)** 2^6 **(d)** 1^{10} ■

If a whole number does not have a visible exponent, the exponent is understood to be 1. Thus

$$3 = 3^1 \quad \text{or} \quad 10 = 10^1$$

In today's world, large numbers are often expressed as a power of 10.

$10^1 = 10 = 1$ ten $10^4 = 10{,}000 = 1$ ten thousand

$10^2 = 100 = 1$ hundred $10^5 = 100{,}000 = 1$ hundred thousand

$10^3 = 1000 = 1$ thousand $10^6 = 1{,}000{,}000 = 1$ million

What does it mean to have an exponent of zero? What is 10^0? Any whole number that is not zero can be raised to the zero power. The result is 1. Thus $10^0 = 1$, $3^0 = 1$, $5^0 = 1$, and so on. Why is this? Let's reexamine the powers of 10. As we go down one line at a time, notice the pattern that occurs.

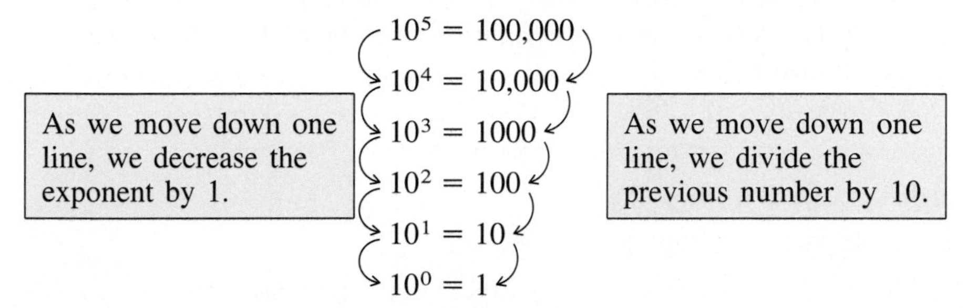

As we move down one line, we decrease the exponent by 1.
10⁵ = 100,000
10⁴ = 10,000
10³ = 1000
10² = 100
10¹ = 10
10⁰ = 1
As we move down one line, we divide the previous number by 10.

Therefore, we present the following definition.

Definition

For any whole number a other than zero, $a^0 = 1$.

If numbers having exponents are added to other numbers, it is first necessary to **evaluate,** or find the value of, the number that is raised to a power. Then we may combine the results with another number.

EXAMPLE 3 Find the value of each expression.

(a) $3^4 + 2^3$ **(b)** $5^3 + 7^0$ **(c)** $6^3 + 6$

(a) $3^4 + 2^3 = (3)(3)(3)(3) + (2)(2)(2) = 81 + 8 = 89$

(b) $5^3 + 7^0 = (5)(5)(5) + 1 = 125 + 1 = 126$

(c) $6^3 + 6 = (6)(6)(6) + 6 = 216 + 6 = 222$

Practice Problem 3 Find the value of each expression.

(a) $7^3 + 8^2$ **(b)** $9^2 + 6^0$ **(c)** $5^4 + 5$ ■

2 *Performing Several Arithmetic Operations in the Proper Order*

Sometimes the order in which we do things is not important. The order in which chefs hang up their pots and pans probably does not matter. The order in which they add and mix the elements in preparing food, however, makes all the difference in the world! If various cooks follow a recipe, though, they will get similar results. The recipe assures that the results will be consistent. It shows the **order of operations.**

In mathematics the order of operations is a list of priorities for working with the numbers in computational problems. This mathematical ''recipe'' tells how to handle certain indefinite computations. For example, how does a person solve $5 + 3 \times 2$?

A problem such as $5 + 3 \times 2$ sometimes causes students difficulty. Some people think $(5 + 3) \times 2 = 8 \times 2 = 16$. Some people think $5 + (3 \times 2) = 5 + 6 = 11$. Only one answer is right, 11. To obtain the right answer follow the steps outlined below.

TEACHING TIP Students often ask why it is necessary to learn this order of operations. Explain that scientific calculators and computers follow this order of operations in performing complex calculations. It is therefore necessary for people using computers or scientific calculators to know this or they may misinterpret how the problem can be done using one of these devices. It is equally necessary in algebra courses and any higher math courses taken in college. This given order of operations is assumed in all math classes.

Order of Operations

In the absence of grouping symbols:

Do first **1.** Simplify any expressions with exponents.
↕ **2.** Multiply or divide from left to right.
Do last **3.** Add or subtract from left to right.

EXAMPLE 4 Evaluate. $3^2 + 5 - 4 \times 2$

$3^2 + 5 - 4 \times 2 = 9 + 5 - 4 \times 2$ *Evaluate the expression with exponents.*

$= 9 + 5 - 8$ *Multiply from left to right.*

$= 14 - 8$ *Add from left to right.*

$= 6$ *Subtract.*

Practice Problem 4 Evaluate. $7 + 4^3 \times 3$ ■

EXAMPLE 5 Evaluate. $5 + 3 \times 6 - 4 + 12 \div 3$

TEACHING TIP Students often try to do all the multiplication and then all the division in a given problem. Stress the fact that both multiplication and division have equal priority. They are to be done in order as you move from left to right.

There are no numbers to raise to a power, so we first do any multiplication or division in order from *left to right.*

$5 + 3 \times 6 - 4 + 12 \div 3 = 5 + 18 - 4 + 12 \div 3$ *Multiply.*

$= 5 + 18 - 4 + 4$ *Divide.*

$= 23 - 4 + 4$ *Add.*

$= 19 + 4$ *Subtract.*

$= 23$ *Add.*

Practice Problem 5 Evaluate. $7 - 3 \times 4 \div 2 + 5$ ■

EXAMPLE 6 Evaluate. $2^3 + 3^2 - 7 \times 2$

$2^3 + 3^2 - 7 \times 2 = 8 + 9 - 7 \times 2$	*Evaluate exponent expressions* $2^3 = 8$ *and* $3^2 = 9$.
$= 8 + 9 - 14$	*Multiply.*
$= 17 - 14$	*Add.*
$= 3$	*Subtract.*

Practice Problem 6 Evaluate. $4^3 - 2 + 3^2$ ■

You can change the order in which you compute by using grouping symbols. Place the numbers you want to calculate first within parentheses. This tells you to do those calculations first.

Order of Operations

With grouping symbols:

Do first
1. Perform operations inside the parentheses.
2. Simplify any expressions with exponents.
3. Multiply or divide from left to right.
4. Add or subtract from left to right.

Do last

EXAMPLE 7 Evaluate. $2 \times (7 + 5) \div 4 + 3 - 6$

First, we combine numbers inside the parentheses by adding the 7 to the 5. Multiplication and division have equal priority, so we work from left to right doing whichever of these operations comes first.

$= 2 \times 12 \div 4 + 3 - 6$	
$= 24 \div 4 + 3 - 6$	*Multiply.*
$= 6 + 3 - 6$	*Divide.*
$= 9 - 6$	*Add.*
$= 3$	*Subtract.*

Practice Problem 7 Evaluate. $(17 + 7) \div 6 \times 2 + 7 \times 3 - 4$ ■

EXAMPLE 8 Evaluate. $4^3 + 18 \div 3 - 2^4 - 3 \times (8 - 6)$

$= 4^3 + 18 \div 3 - 2^4 - 3 \times 2$	*Work inside the parentheses.*
$= 64 + 18 \div 3 - 16 - 3 \times 2$	*Evaluate exponents.*
$= 64 + 6 - 16 - 3 \times 2$	*Divide.*
$= 64 + 6 - 16 - 6$	*Multiply.*
$= 70 - 16 - 6$	*Add.*
$= 54 - 6$	*Subtract.*
$= 48$	*Subtract.*

Practice Problem 8 Evaluate. $5^2 - 6 \div 2 + 3^4 + 7 \times (12 - 10)$ ■

1.6 Exercises

Verbal and Writing Skills

1. Explain what the expression 5^3 means. Evaluate 5^3. 5^3 means $5 \times 5 \times 5$. $5^3 = 125$.

2. The ______exponent______ tells how many times to multiply the base.

3. The ______base______ is the number that is multiplied.

4. 10^5 is read as ______10 to the 5th power______.

5. Explain the order in which we perform mathematical operations to ensure consistency.

To insure consistency we
1. perform operations inside parentheses
2. simplify any expressions with exponents
3. multiply or divide from left to right
4. add or subtract from left to right

6. Use the order of operations to solve $12 \times 5 + 3 \times 5 + 7 \times 5$. Is this the same as $5(12 + 3 + 7)$? Why or why not?

110 Yes. Because of the distributive property.

Write each number in exponent form.

7. $6 \times 6 \times 6 \times 6$
6^4

8. $2 \times 2 \times 2 \times 2 \times 2$
2^5

9. $5 \times 5 \times 5$
5^3

10. $12 \times 12 \times 12$
12^3

11. $8 \times 8 \times 8 \times 8$
8^4

12. $1 \times 1 \times 1 \times 1 \times 1 \times 1 \times 1$
1^7

13. 9
9^1

14. 27
27^1

Find the value of each expression.

15. 2^4 16
16. 3^3 27
17. 4^2 16
18. 5^2 25
19. 6^3 216
20. 10^3 1000
21. 10^4 10,000
22. 1^{20} 1

23. 1^{17} 1
24. 2^5 32
25. 2^6 64
26. 4^3 64
27. 3^4 81
28. 13^2 169
29. 15^2 225
30. 8^3 512

31. 7^3 343
32. 5^4 625
33. 4^4 256
34. 2^7 128
35. 9^0 1
36. 8^0 1
37. 25^2 625
38. 20^3 8000

39. 10^6 1,000,000
40. 8^1 8
41. 4^5 1024
42. 10^5 100,000
43. 9^1 9
44. 14^2 196
45. 7^4 2401
46. 4^5 1024

Find the value of each expression.

47. $3^4 + 7^0$
$81 + 1 = 82$

48. $4^3 + 4^0$
$64 + 1 = 65$

49. $6^3 + 3^2$
$216 + 9 = 225$

50. $7^3 + 4^2$
$343 + 16 = 359$

51. $8^3 + 8$
$512 + 8 = 520$

52. $9^2 + 9$
$81 + 9 = 90$

Work each problem, using the correct order of operations.

53. $5 \times 6 - 3 + 10$
$30 - 3 + 10$
$27 + 10 = 37$

54. $8 \times 3 - 4 \times 2$
$24 - 8$
16

55. $2 + 6 \times 12 \div 2$
$2 + 72 \div 2$
$2 + 36 = 38$

56. $6 \times 7 - 4 \div 2$
$42 - 2$
40

57. $100 \div 5^2 + 3$
$100 \div 25 + 3$
$4 + 3 = 7$

58. $24 \div 2^2 + 4$
$24 \div 4 + 4$
$6 + 4 = 10$

59. $7 \times 3^2 + 4 - 8$
$7 \times 9 + 4 - 8$
$63 + 4 - 8 = 59$

60. $2^3 \times 4 + 6 - 9$
$8 \times 4 + 6 - 9$
$32 + 6 - 9 = 29$

61. $4 \times 9^2 - 4 \times (7 - 2)$
$4 \times 81 - 4 \times 5$
$324 - 20 = 304$

62. $5 \times 2^5 - 4 \times (11 - 8)$
$5 \times 32 - 4 \times 3$
$160 - 12 = 148$

63. $(400 \div 20) \div 20$
$20 \div 20 = 1$

64. $(600 \div 30) \div 20$
$20 \div 20 = 1$

65. $950 \div (25 \div 5)$
$950 \div 5 = 190$

66. $875 \div (35 \div 7)$
$875 \div 5 = 175$

67. $(16)(4) - (16 + 4)$
$64 - 20 = 44$

68. $(8)(30) - (8 + 30)$
$240 - 38 = 202$

69. $3^2 + 4^2 \div 2^2$
$9 + 16 \div 4 = 9 + 4 = 13$

70. $7^2 + 9^2 \div 3^2$
$49 + 81 \div 9 = 49 + 9 = 58$

71. $(6)(7) - (12 - 8) \div 4$
$42 - 4 \div 4 = 42 - 1 = 41$

72. $(8)(9) - (15 - 5) \div 5$
$72 - 10 \div 5 = 72 - 2 = 70$

73. $100 - 3^2 \times 4$
$100 - 9 \times 4 = 100 - 36 = 64$

74. $130 - 4^2 \times 5$
$130 - 16 \times 5 = 130 - 80 = 50$

75. $5^2 + 2^2 + 3^3$
$25 + 4 + 27 = 56$

76. $2^3 + 3^2 + 4^3$
$8 + 9 + 64 = 81$

77. $100 \div 5 \times 3 \times 2 \div 2$
$20 \times 3 \times 2 \div 2$
$120 \div 2 = 60$

78. $36 \div 4 \times 5 \times 6 \div 6$
$9 \times 5 \times 6 \div 6$
$270 \div 6 = 45$

79. $12^2 - 6 \times 3 \times 4 \times 0$
$144 - 0$
144

80. $14^2 - 5 \times 2 \times 3 \times 0$
$196 - 0$
196

81. $7^2 \times 5 \div 5$
$49 \times 5 \div 5$
$245 \div 5 = 49$

82. $5^2 \times 4 \div 10$
$25 \times 4 \div 10$
$100 \div 10 = 10$

83. $5 + 6 \times 2 \div 4 - 1$
$5 + 3 - 1$
7

84. $7 + 5 \times 4 \div 10 - 2$
$7 + 2 - 2$
7

85. $3 + 3^2 \times 6 + 4$
$3 + 9 \times 6 + 4$
$3 + 54 + 4 = 61$

86. $5 + 4^3 \times 2 + 7$
$5 \times 64 \times 2 + 7$
$5 + 128 + 7 = 140$

87. $12 \div 2 \times 3 - 2^4$
$12 \div 2 \times 3 - 16$
$18 - 16 = 2$

88. $30 \div 10 \times 5 - 3^2$
$30 \div 10 \times 5 - 9$
$15 - 9 = 6$

89. $3^2 \times 6 \div 9 + 4 \times 3$
$9 \times 6 \div 9 + 4 \times 3$
$6 + 12 = 18$

90. $5^2 \times 3 \div 25 + 7 \times 6$
$25 \times 3 \div 25 + 7 \times 6$
$3 + 42 = 45$

91. $4(10 - 6)^3 \div (2 + 6)$
$4(4)^3 \div 8 = 4(64) \div 8 = 32$

92. $3(12 - 5)^2 \div (4 + 3)$
$3(7)^2 \div 7 = 3(49) \div 7 = 21$

93. $1200 - 2^3(3) \div 6$
$1200 - 8(3) \div 6 =$
$1200 - 4 = 1196$

94. $2150 - 3^4(2) \div 9$
$2150 - 81(2) \div 9 =$
$2150 - 18 = 2132$

95. $300 \div 25 - 10 + 2^3$
$12 - 10 + 8 = 10$

96. $400 \div 16 - 8 + 3^4$
$25 - 8 + 81 = 98$

97. $20 \div 4 \times 3 - 6 + 2 \times (1 + 4)$
$15 - 6 + 10$
19

98. $50 \div 10 \times 2 - 8 + 3 \times (17 - 11)$
$10 - 8 + 18$
20

99. $3 \times 5 + 7 \times (8 - 5) - 5 \times 4$
$15 + 21 - 20$
16

100. $(5 + 3) \times 9 + 6 \times 5 - 6 \times 6$
$72 + 30 - 36$
66

Cumulative Review Problems

101. Write in expanded notation. 156,312
100,000 + 50,000 + 6000 + 300 + 10 + 2

102. Write in standard notation. Two hundred million, seven hundred sixty-five thousand, nine hundred nine 200,765,909

103. Write in words. 261,763,002
Two hundred sixty-one million, seven hundred sixty-three thousand, two

104. Add.
$$\begin{array}{r} 1563 \\ 2736 \\ 381 \\ 25 \\ +\ 9823 \\ \hline 14{,}528 \end{array}$$

105. Subtract.
$$\begin{array}{r} 16{,}093 \\ -\ 14{,}937 \\ \hline 1156 \end{array}$$

To Think About

106. The earth rotates once every 86,164 seconds. Round this figure to the nearest ten thousand seconds. 90,000

107. The planet Saturn rotates once every 10 hours, 12 minutes. How many minutes in total is this? How many seconds?
612 minutes. It is 36,720 seconds.

1.7 Rounding and Estimation

MathPro

Video 3

SSM

After studying this section, you will be able to:

1. *Round whole numbers.*
2. *Estimate the answer to a problem involving calculation with whole numbers.*

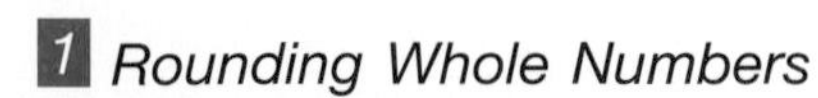

1 *Rounding Whole Numbers*

Large numbers are often expressed to the nearest hundred or to the nearest thousand, because an approximate number is ''good enough'' for certain uses.

Distances from the earth to other galaxies are measured in light-years. Although light really travels at 5,865,696,000,000 miles a year, we usually **round** this number to the nearest trillion and say it travels at 6,000,000,000,000 miles a year. To round a number, we first determine the place we are rounding to. In this case: trillion. Then we find which value is closest to the number that we are rounding. In this case, the number we want to round is closer to 6 trillion than to 5 trillion. How do we know the number is closer to 6 trillion than to 5 trillion?

To see which is the closest value, we may picture a **number line,** where whole numbers are represented by points on a line. To show how to use the number line in rounding, we will round 368 to the nearest hundred. 368 is between 300 and 400. When we round off, we pick the hundred 368 is ''closest to.'' We draw a number line to show 300 and 400. We also show the point midway between 300 and 400 to help us to determine which hundred 368 is closest to.

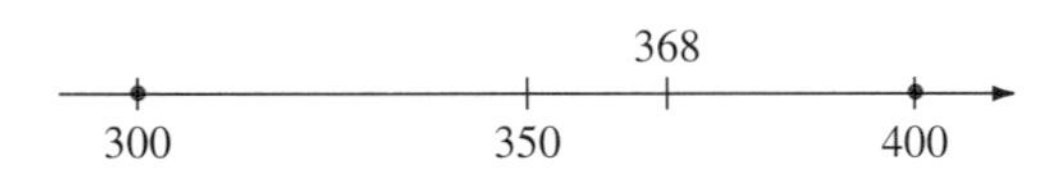

We find that the number 368 is closer to 400 than to 300, so we round 368 *up to* 400.

Let's look at another example. We will round 129 to the nearest hundred. 129 is between 100 and 200. We show this on the number line. We include the midpoint 150 as a guide.

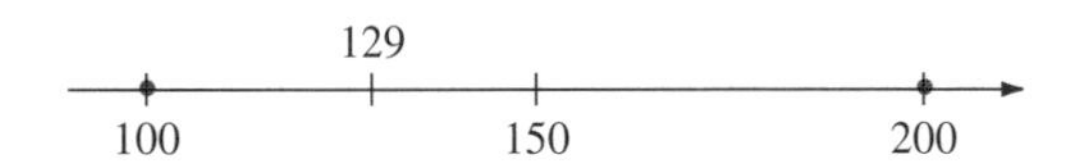

We find that the number 129 is closer to 100 than to 200, so we round 129 *down to* 100.

This leads us to this simple rule for rounding.

TEACHING TIP These rules for rounding off are used uniformly in science and mathematics classes and applications. However, a particular business or bank may employ a different rule. Inform students that if rounding off is done in a business, a different rule may be employed.

Rounding a Whole Number

1. If the first digit to the right of the round-off place is:
   (a) *less than 5,* we make no change to the digit in the round-off place. (We know it is closer to the smaller number, so we round down.)
   (b) *5 or more,* we increase the digit in the round-off place by 1. (We know it is closer to the larger number, so we round up.)
2. Then we replace the digits to the right of the round-off place by zeros.

EXAMPLE 1 Round 37,843 to the nearest thousand.

3 7,8 4 3 ↓	*According to the directions, the thousands will be the round-off place. We locate the thousands place.*
3 7(8)4 3 ↓ 3 8,0 0 0	*We see that the first digit to the right of the round-off place is 8, which is 5* or more. *We increase the thousands digit by 1, and replace all digits to the right by zero.*

We have rounded 37,843 to the nearest thousand: 38,000. This means that 37,843 is closer to 38,000 than to 37,000.

Practice Problem 1 Round 65,528 to the nearest thousand. ■

EXAMPLE 2 Round 2,445,360 to the nearest hundred thousand.

2,4 4 5,3 6 0	*Locate the hundred-thousands round-off place.*
2,4(4)5,3 6 0 ↓	*The first digit to the right of this is* less than 5, *so round down. Do not change the hundred-thousands digit.*
2,4 0 0,0 0 0	*Replace all digits to the right by zero.*

Practice Problem 2 Round 172,963 to the nearest ten thousand. ■

EXAMPLE 3 Round as indicated. **(a)** 561,328 to the nearest ten **(b)** 3,798,152 to the nearest hundred **(c)** 51,362,523 to the nearest million

(a) First locate the digit in the tens place.

561,328	*The digit to the right of the tens is greater than* 5.
561,330	*Round up.*

561,328 rounded to the nearest ten is 561,330.

(b) 3,798,152 *The digit to the right of the hundreds is* 5.

3,798,200 *Round up.*

3,798,152 rounded to the nearest hundred is 3,798,200.

(c) 51,362,523 *The digit to the right of the millions is less than* 5.

51,000,000 *Round down.*

51,362,523 rounded to the nearest million is 51,000,000.

Practice Problem 3 Round as indicated. **(a)** 53,282 to nearest ten **(b)** 164,485 to nearest thousand **(c)** 1,365,273 to nearest hundred thousand ■

TEACHING TIP As a classroom activity, ask students to round off 56,489 to the nearest thousand. They will obtain the correct answer 56,000. Then ask them what is wrong with this reasoning: Rounded to the nearest hundred, 56,489 becomes 56,500. Then rounding 56,500 to the nearest thousand gives 57,000. Why is this approach not correct? Usually students will point out the error in the reasoning. If not, draw a number line showing the location of each number.

EXAMPLE 4 Round 763,571. **(a)** To the nearest thousand **(b)** To the nearest ten thousand **(c)** To the nearest million

(a) 763,571 = 764,000 to the nearest thousand.
The digit to the right of the thousands is 5. We rounded up.

(b) 763,571 = 760,000 to the nearest ten thousand.
The digit to the right of the ten thousands is less than 5. We rounded down.

(c) 763,571 does not have any digits for millions. If it helps, you can think of this number as 0,763,571. Since the digit to the right of the millions place is 7, we round up to obtain one million or 1,000,000.

Practice Problem 4 Round 935,682 as indicated. **(a)** To the nearest thousand **(b)** To the nearest hundred thousand **(c)** To the nearest million ■

EXAMPLE 5 Astronomers use the parsec as a measurement of distance. One **parsec** is approximately 30,900,000,000,000 kilometers. Round 1 parsec to the nearest trillion kilometers.

30,900,000,000,000 km is 31,000,000,000,000 km or 31 trillion km to the nearest trillion kilometers.

Practice Problem 5 One light-year is approximately 9,460,000,000,000,000 meters. Round to the nearest hundred trillion meters. ■

2 *Estimating the Answer to a Calculation with Whole Numbers*

Often we need to quickly check the answer of a calculation to be reasonably assured that the answer is correct. If you expected your bill to be "around $40" for the groceries you had selected and the cashier's total came to $41.89, you would probably be confident that the bill is correct and pay it. If, however, the cashier rang up a bill of $367, you would not just assume that it is correct. You would know an error had been made. If the cashier's total came to $60, you might not be certain, but you would probably suspect an error and check the calculation.

In mathematics we often **estimate,** or determine the approximate value of a calculation, if we need to do a quick check. There are many ways to estimate, but in this book we will use one simple principle of estimation.

Principle of Estimation

1. Round the numbers so that there is one nonzero digit in each number.
2. Perform the calculation with the rounded numbers.

EXAMPLE 6 Estimate the sum. 163 + 237 + 846 + 922

We first determine where to round each number in our problem to leave only one nonzero digit in each. In this case, we round all numbers to the nearest hundred. Then we perform the calculation with the rounded numbers.

Actual Sum	*Estimated Sum*
163	200
237	200
846	800
+ 922	+ 900
	2100

We estimate the answer to be 2100. If we calculate using the exact numbers, we obtain a sum of 2168, so our estimate is quite close to the actual sum.

Practice Problem 6 Estimate the sum. 3456 + 9876 + 5421 + 1278 ■

When we use the principle of estimation, we will not always round each number in a problem to the same place.

EXAMPLE 7 Charles and Melinda bought their first car last week. The selling price of this compact car was $8980. The dealer preparation charge was $289 and the sales tax was $449. Estimate the total cost of the car that Charles and Melinda had to pay.

We round each number to have only one nonzero digit, and add the rounded numbers.

8980	9000
289	300
+ 449	+ 400
	9700

We estimate that the total cost is $9700. (The exact answer is $9718, so we see that our answer is quite close.)

Practice Problem 7 Jenna and Carlos purchased a new sofa for $697, plus $35 sales tax. The store also charged them $19 to deliver the sofa. Estimate their total cost. ■

Now we turn to a case where an estimate can help us discover an error.

EXAMPLE 8 Roberto added together four numbers and obtained the following result. Estimate the sum and determine if the answer seems reasonable.

$$12{,}456 + 17{,}976 + 18{,}452 + 32{,}128 = 61{,}012$$

We round each number so that there is one nonzero digit. In this case, we round them all to the nearest ten thousand.

12,456	10,000	
17,976	20,000	
18,452	20,000	
+ 32,128	+ 30,000	
	80,000	*Our estimate is* 80,000.

This is significantly different from 61,012, so we would suspect that an error has been made. In fact, Roberto did make an error. The exact sum is actually 81,012!

Practice Problem 8 Ming did the following calculation. Estimate to see if her sum appears to be correct or incorrect.

$$11{,}849 + 14{,}376 + 16{,}982 + 58{,}151 = 81{,}358$$ ■

Next we look at a subtraction example where estimation is used.

EXAMPLE 9 The profit from Techno Industries for the first quarter of the year was $642,987,000. The profit for the second quarter was $238,890,000. Estimate how much less the profit was for the second quarter than for the first quarter.

We round each number so that there is one nonzero digit. Then we subtract, using the two rounded numbers.

642,987,000	600,000,000
− 238,890,000	− 200,000,000
	400,000,000

We estimate that the profit was $400,000,000 less for the second quarter.

Practice Problem 9 The 1990 population of Florida was 12,937,926. The 1990 population of California was 29,760,021. Estimate how many more people lived in California in 1990 than in Florida. ■

We also use this principle to estimate results of multiplication and division.

EXAMPLE 10 Estimate the product. $56{,}789 \times 529$

We round each number so that there is one nonzero digit. Then we multiply the rounded numbers to obtain our estimate.

56,789	60,000
× 529	× 500
	30,000,000

We therefore estimate the product to be 30,000,000. (This is reasonably close to the exact answer of 30,041,381.)

Practice Problem 10 Estimate the product. 8945×7317 ■

EXAMPLE 11 Estimate the answer for the following division problem.

$$23\overline{)148{,}902}$$

We round each number to a number with one nonzero digit. Then we perform the division, using the two rounded numbers.

$$23\overline{)148{,}902} \qquad \begin{array}{r} 5000 \\ 20\overline{)100{,}000} \end{array}$$

Our estimate is 5000. (The exact answer is 6474. We see that our estimate is "in the ballpark" but is not very close to the exact answer. Remember, an estimate is just a rough approximation of the exact answer.)

Practice Problem 11 Estimate the answer for the following division problem.

$$39\overline{)75{,}342} \quad ■$$

Not all division estimates come out so easily. In some cases, you may need to carry out a long-division problem of several steps just to obtain the estimate.

EXAMPLE 12 Ralph drove his car a distance of 778 miles. He used 25 gallons of gas. Estimate how many miles he can travel on 1 gallon of gas.

In order to solve this problem, we need to divide 778 by 25 to obtain the number of miles Ralph gets with 1 gallon of gas. We round each number to a number with one nonzero digit and then perform the division, using the rounded numbers.

$$25\overline{)778} \qquad \begin{array}{r} 26 \\ 30\overline{)800} \\ \underline{60} \\ 200 \\ \underline{180} \\ 20 \end{array} \quad \textit{Remainder}$$

We obtain an answer of 26 with a remainder of 20. For our estimate we will use the whole number 27. Thus we estimate that the number of miles Ralph's car obtained on 1 gallon of gas was 27 miles. (This is reasonably close to the exact answer, which is just slightly more than 31 miles per gallon of gas.)

Practice Problem 12 The highway department purchased 58 identical trucks at a total cost of $1,864,584. Estimate the cost for one truck. ■

1.7 Exercises

Verbal and Writing Skills

1. Explain the rule for rounding and provide examples. Locate the rounding place. If the digit to the right of the rounding place is 5 or greater than 5, round up. If the digit to the right of the rounding place is less than 5, round down.
Note: Examples provided by students will vary. Check for accuracy.

2. What happens when you round 98 to the nearest ten? Since the digit to the right of tens is greater than 5, you round up. When you round up 9 tens, it becomes 10 tens or 100.

Round to the nearest ten.

3. 83
80

4. 45
50

5. 65
70

6. 57
60

7. 92
90

8. 138
140

9. 526
530

10. 2834
2830

11. 4235
4240

12. 8963
8960

Round to the nearest hundred.

13. 437
400

14. 735
700

15. 2781
2800

16. 1258
1300

17. 7692
7700

18. 1643
1600

Round to the nearest thousand.

19. 7621
8000

20. 3754
4000

21. 881
1000

22. 679
1000

23. 27,863
28,000

24. 94,671
95,000

Applications

Solve each problem.

25. The worst death rate from an earthquake was in Shaanxi, China, in 1556. That earthquake killed an estimated 832,400 people. Round this number to the nearest hundred thousand. 800,000

26. One light year (the distance light travels in 1 year) measures 5,878,612,843,000 miles. Round this figure to the nearest hundred million.
5,878,600,000,000 miles

27. The Hubble Space Telescope's *Guide Star Catalogue* lists 15,169,873 stars. Round this figure to the nearest million. 15,000,000 stars

28. In 1993, the countries included in the Far East and Oceanea utilized 15,817 barrels of petroleum per day. Round this value to the nearest thousand.
16,000 barrels

29. The total Native American population living on the Navajo and Trust Lands in the Arizona/New Mexico/Utah area numbered 143,405 in 1990. Round this figure to
(a) the nearest thousand 143,000
(b) the nearest hundred 143,400

30. A few years ago, there were 614,571 Catholic School pupils in the United States. Round this figure to
(a) the nearest thousand 615,000
(b) the nearest hundred 614,600

31. The total area of mainland China is 3,705,392 square miles, or, 9,596,960 square kilometers. For *both* square miles and square kilometers, round this figure to
(a) the nearest hundred thousand
3,700,000 square miles or 9,600,000 square kilometers
(b) the nearest ten thousand
3,710,000 square miles or 9,600,000 square kilometers

32. Recently, in the United States 10,957,442,856 pounds of apples were produced.
(a) Round this figure to the nearest hundred thousand. 10,957,400,000 pounds
(b) Round this figure to the nearest ten thousand.
19,957,440,000 pounds

Use the principle of estimation to find an estimate for each of the following calculations.

33. $456 + 234 + 875$

500
200
+ 900
1600

34. $341 + 178 + 987$

300
200
+ 1000
1500

35. $34 + 78 + 59 + 31$

30
80
60
+ 30
200

36. $31 + 75 + 82 + 43$

30
80
80
+ 40
230

37. $146{,}270 + 47{,}566 + 96{,}112$

100,000
50,000
+ 100,000
250,000

38. $345{,}957 + 89{,}334 + 56{,}487$

300,000
90,000
+ 60,000
450,000

39. $567{,}984 - 129{,}562$

600,000
− 100,000
500,000

40. $975{,}935 - 593{,}228$

1,000,000
− 600,000
400,000

41. $78{,}945{,}000 - 61{,}076{,}500$

80,000,000
− 60,000,000
20,000,000

42. $4{,}596{,}450 - 3{,}894{,}202$

5,000,000
− 4,000,000
1,000,000

43. $33{,}261{,}378 - 18{,}199{,}276$

30,000,000
− 20,000,000
10,000,000

44. $89{,}263{,}000 - 54{,}198{,}635$

90,000,000
− 50,000,000
40,000,000

45. 47×62

60
× 50
3000

46. 43×95

100
× 40
4000

47. 1798×7

2000
× 7
14,000

48. 3254×6

3000
× 6
18,000

49. $631{,}540 \times 312$

600,000
× 300
180,000,000

50. $374{,}193 \times 193$

400,000
× 200
80,000,000

51. $24{,}318 \div 21$

1,000
20)20,000

52. $61{,}986 \div 32$

2,000
30)60,000

53. $156{,}721 \div 42$

5,000
40)200,000

54. $581{,}361 \div 28$

20,000
30)600,000

55. $6{,}787{,}129 \div 538$

14,000
500)7,000,000
500
2000
2000

56. $7{,}863{,}127 \div 352$

20,000
400)8,000,000

Estimate the result of each of the following calculations. Some results are correct and some are incorrect. Which results appear to be correct? Which results appear to be incorrect?

57.
361 — 400
522 — 500
873 — 900
+ 164 — + 200
1320 — 2000
Incorrect

58.
476 — 500
124 — 100
516 — 500
+ 389 — + 400
1505 — 1500
Correct

59.
97,635 — 100,000
52,123 — 50,000
+ 41,986 — + 40,000
291,744 — 190,000
Incorrect

60.
26,181 — 30,000
47,998 — 50,000
+ 63,271 — + 60,000
137,450 — 140,000
Correct

61.
257,163
− 99,358
157,805
Correct
300,000
− 100,000
200,000

62.
682,351
− 96,891
585,460
Correct
700,000
− 100,000
600,000

63.
78,126,345
− 48,972,103
19,154,242
Incorrect
80,000,000
− 50,000,000
30,000,000

64.
42,765,317
− 29,318,274
23,447,043
Incorrect
40,000,000
− 30,000,000
10,000,000

65. $\begin{array}{r} 216 \\ \times\ 24 \\ \hline 6184 \end{array}$ $\begin{array}{r} 200 \\ \times\ 20 \\ \hline 4000 \end{array}$
Incorrect

66. $\begin{array}{r} 578 \\ \times\ 32 \\ \hline 10{,}496 \end{array}$ $\begin{array}{r} 600 \\ \times\ 30 \\ \hline 18{,}000 \end{array}$
Incorrect

67. $\begin{array}{r} 5896 \\ \times\ 72 \\ \hline 424{,}512 \end{array}$ $\begin{array}{r} 6000 \\ \times\ 70 \\ \hline 420{,}000 \end{array}$
Correct

68. $\begin{array}{r} 8076 \\ \times\ 89 \\ \hline 718{,}764 \end{array}$ $\begin{array}{r} 8000 \\ \times\ 90 \\ \hline 720{,}000 \end{array}$
Correct

69. $78\overline{)412{,}308}$ = 5286
$80\overline{)400{,}000}$ = 5000
Correct

70. $58\overline{)277{,}994}$ = 4793
$60\overline{)300{,}000}$ = 5000
Correct

71. $423\overline{)161{,}163}$ = 381
$400\overline{)200{,}000}$ = 500
Correct

72. $781\overline{)477{,}972}$ = 612
$800\overline{)500{,}000}$ = 625
4800
2000
1600
4000
4000
Correct

Applications

73. A huge restaurant in New York City is 40 yards wide and 110 yards long. Estimate the number of square yards in the restaurant. 4000 square yards

74. Larry and Nella are planning an outdoor wedding by the ocean. There is a beautiful meadow that is 35 feet wide and 62 feet long. Estimate the number of square feet in the meadow. 2400 square feet

75. The last four times Mrs. Watkins went food shopping for her family of six kids, two adults, one cat, one dog, and a canary, she spent \$442, \$379, \$431, and \$333. Estimate what she spent to feed her family during this time period. \$1500

76. During the first week of October, the daily number of people buying football tickets for New England Patriots home games over four days was 975, 509, 784, and 881. Estimate the total number of people who bought football tickets for that 4-day period. 3200

77. To get ready for bathing suit weather, Wendy did 85 sit-ups daily for 63 days. Estimate how many sit-ups she did during that period. 5400 sit-ups

78. The local pizzeria makes 267 pizzas on an average day. Estimate how many pizzas were made in the last 134 days. 30,000 pizzas

79. The price of a Superbowl ticket increased from \$30 in 1980 to \$175 in 1993. Estimate the increase. Determine the exact increase. Estimate = \$170, exact amount = \$145

80. U.S. air travel increased from 457,000,000 passengers in 1990 to 509,000,000 in 1994. Round each figure to the nearest ten million. Then estimate the increase. Approximately 50,000,000 passengers

81. Tina determined that she spent \$780 last year renting videos. If she rented videos each week, estimate how much she spent weekly for videos last year. About \$16 per week

82. Ted Kinney determined that he had earned \$5,125,780 as an attorney working 41 years in his law practice. Estimate his average annual salary over those 41 years. \$125,000 per year salary

To Think About

83. A space probe is sent at 23,560 miles per hour for a distance of 7,824,560,000 miles.
(a) How many *hours* will it take the space probe to travel that distance? (Estimate.) 400,000 hours
(b) How many *days* will it take the space probe to travel that distance? (Estimate.) 20,000 days

84. A space probe is sent at 28,367 miles per hour for a distance of 9,348,487,000 miles.
(a) Estimate the number of *hours* it will take the space probe to travel that distance. 300,000 hours
(b) Estimate the number of *days* it will take the space probe to travel that distance. 15,000 days

Cumulative Review Problems

Evaluate the following.

85. $26 \times 3 + 20 \div 4$
83

86. $5^2 + 3^2 - (17 - 10)$
27

87. $3 \times (16 \div 4) + 8 \times 2$
28

88. $126 + 4 - (20 \div 5)^3$
66

1.8 Applied Problems

After studying this section, you will be able to:

1. *Use the Mathematics Blueprint to solve problems involving one type of operation.*
2. *Use the Mathematics Blueprint to solve problems involving more than one type of operation.*

MathPro Video 3 SSM

1 *Solving Problems Involving One Type of Operation*

When a builder constructs a new home or office building, he often has a blueprint. This accurate drawing shows the basic structure of the building. It also shows the dimensions of the structure to be built. This blueprint serves as a useful reference throughout the construction process.

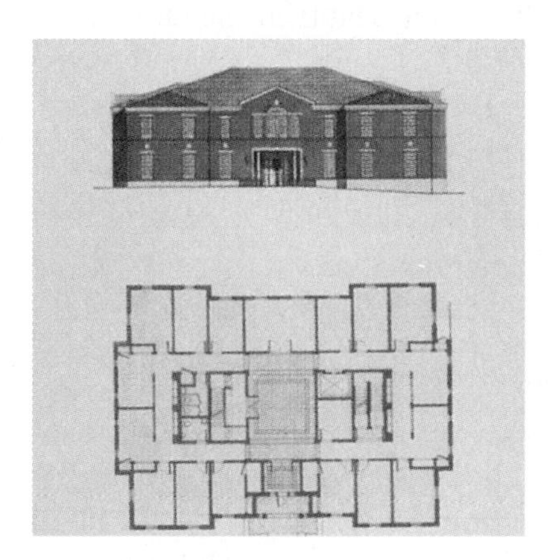

Similarly, when solving applied problems, it is helpful to have a "mathematics blueprint." This is a simple way to organize the information provided in the word problem. You record the facts you need to use and specify what you are solving for. You also record any other information that you feel will be helpful. We will use a Mathematics Blueprint for Problem Solving in the following situation.

EXAMPLE 1 Gerald made deposits of \$317, \$512, \$84, and \$161 into his checking account. He also made out checks for \$100 and \$125. What was the total of his deposits?

1. *Understand the problem.*
   First we read over the problem carefully and we fill in the Mathematics Blueprint.

MATHEMATICS BLUEPRINT FOR PROBLEM SOLVING			
Gather the Facts	What Am I Asked to Do?	How Do I Proceed?	Key Points to Remember
The **deposits** are \$317 \$512 \$84 \$161	Find the total of Gerald's four deposits.	I must add the four deposits to obtain the total.	Watch out! Don't use the **checks** of \$100 and \$125 in the calculation. We only want the total of the **deposits.**

2. *Solve and state the answer.*
We need to *add* to find the sum of the deposits.

$$\begin{array}{r} 317 \\ 512 \\ 84 \\ +\ 161 \\ \hline 1074 \end{array}$$

The total of the four deposits is $1074.

3. *Check.*
Reread the problem. Be sure you have answered the question that was asked. Did they ask for the total of the deposits? Yes. ✓

Is the calculation correct? You can use estimation to check. Here we round each of the deposits so that we have one nonzero digit.

$$\begin{array}{rr} 317 & 300 \\ 512 & 500 \\ 84 & 80 \\ +\ 161 & +\ 200 \\ \hline & 1080 \end{array}$$

Our estimate is $1080. $1074 is close to our estimated answer of $1080. Our answer is reasonable. ✓

Thus we conclude that the total of the four deposits is $1074.

Practice Problem 1 Use the Mathematics Blueprint below to solve the following problem. Diane's paycheck shows deductions of $135 for federal taxes, $28 for state taxes, $13 for FICA, and $34 for health insurance. Her gross pay (amount before deductions) is $1352. What is the total amount that is taken out of Diane's paycheck?

MATHEMATICS BLUEPRINT FOR PROBLEM SOLVING			
Gather the Facts	What Am I Asked to Do?	How Do I Proceed?	Key Points to Remember

■

EXAMPLE 2 Theofilos looked at his odometer before he began his trip from Portland, Oregon, to Kansas City, Kansas. He checked his odometer again when he arrived in Kansas City. The two readings are shown on the right. How many miles did Theofilos travel?

Portland 28353

Kansas

1. *Understand the problem.*
   Determine what information is given.
   The mileage reading before the trip began and when the trip was over.
   What do you need to find?
   The number of miles traveled.

MATHEMATICS BLUEPRINT FOR PROBLEM SOLVING			
Gather the Facts	What Am I Asked to Do?	How Do I Proceed?	Key Points to Remember
At start of trip odometer read 28,353 miles. At end of trip odometer read 30,162 miles.	Find out how many miles Theofilos traveled.	I must subtract the two mileage readings.	Subtract the mileage at the start of the trip from the mileage at the end of the trip.

2. *Solve and state the answer.*
   We need to subtract the two mileage readings to find the difference in the number of miles. This will give us the number of miles the car traveled on this trip alone.

   $30{,}162 - 28{,}353 = 1809$ The trip totaled 1809 miles.

3. *Check.*
   We estimate and compare the estimate with the answer above.

Kansas City	30,162→	→30,000	*We subtract*
Portland	28,353→	→28,000	*our rounded values.*
		2,000	

Our estimate is 2000 miles. We compare this estimate with our answer. Our answer is reasonable. ✓

Practice Problem 2 The table on the right shows the results of the 1992 presidential race in Ohio. By how many votes did the Democratic candidate beat the Republican candidate that year in Ohio?

1992 PRESIDENTIAL RACE, OHIO	
Candidate	Number of Votes
Clinton	1,964,842
Bush	1,876,495
Perot	1,024,319

MATHEMATICS BLUEPRINT FOR PROBLEM SOLVING			
Gather the Facts	What Am I Asked to Do?	How Do I Proceed?	Key Points to Remember

EXAMPLE 3 One horsepower is the power needed to lift 550 pounds a distance of 1 foot in 1 second. How many pounds can be lifted 1 foot in 1 second by 7 horsepower?

1. *Understand the problem.*
   Simplify the problem. If 1 horsepower can lift 550 pounds, how many pounds can be lifted by 7 horsepower? We draw and label a diagram.

We use the mathematics blueprint to organize the information.

MATHEMATICS BLUEPRINT FOR PROBLEM SOLVING			
Gather the Facts	What Am I Asked to Do?	How Do I Proceed?	Key Points to Remember
One horsepower will lift 550 pounds.	Find how many pounds can be lifted by 7 horsepower.	I need to multiply 550 by 7.	I do not use the information about moving one foot in one second.

2. *Solve and state the answer.*
   To solve the problem we multiply the 7 horsepower by 550 pounds for each horsepower.

$$\begin{array}{r} 550 \\ \times \quad 7 \\ \hline 3850 \end{array}$$

We find that 7 horsepower moves 3850 pounds 1 foot in 1 second. We include 1 foot in 1 second in our answer because it is part of the unit of measure.

3. *Check.*
   We estimate our answer. We round 550 to 600 pounds.

$$600 \times 7 = 4200 \text{ pounds}$$

Our estimate is 4200 pounds. Our calculations in step 2 gave us 3850. Is this reasonable? This answer is close to our estimate. Our answer is reasonable. ✓

Practice Problem 3 In a measure of liquid capacity, 1 gallon is 1024 fluid drams. How many fluid drams would be in 9 gallons?

MATHEMATICS BLUEPRINT FOR PROBLEM SOLVING			
Gather the Facts	What Am I Asked to Do?	How Do I Proceed?	Key Points to Remember

■

EXAMPLE 4 Laura can type 35 words per minute. She has to type an English theme that has 5180 words. How many minutes will it take her to type the theme? How many hours and how many minutes will it take her to type the theme?

1. *Understand the problem.*
   We draw a picture. Each "package" of 1 minute is 35 words. We want to know how many packages make up 5180 words.

35 words in 1 minute	35 words in 1 minute	35 words in 1 minute	
5180 words			

We use the mathematics blueprint to organize the information.

MATHEMATICS BLUEPRINT FOR PROBLEM SOLVING

Gather the Facts	What Am I Asked to Do?	How Do I Proceed?	Key Points to Remember
Laura can type 35 words per minute. She must type a paper with 5180 words.	Find out how many 35-word units are in 5180 words.	I need to divide 5180 by 35.	In converting minutes to hours, I will use the fact that 1 hour = 60 minutes.

2. *Solve and state the answer.*

Division
$$\begin{array}{r} 148 \\ 35\overline{)5180} \\ \underline{35} \\ 168 \\ \underline{140} \\ 280 \\ \underline{280} \end{array}$$

It will take 148 minutes.

We will change this answer to hours and minutes. Since 60 minutes = 1 hour, we divide 148 by 60. The quotient will tell us how many hours. The remainder will tell us how many minutes.

$$\begin{array}{r} 2 \text{ R } 28 \\ 60\overline{)148} \\ \underline{-120} \\ 28 \end{array}$$

Laura can type the theme in 148 minutes or 2 hours, 28 minutes.

3. *Check.*

The theme has *5180 words;* she can type 35 words per minute. 5180 words is approximately 5000 words.

5180 words →	5000 words rounded to nearest thousand.	$40\overline{)5000}$ = 125
35 words per minute →	40 words per minute rounded to nearest ten.	*We divide our estimated values.*

Our estimate is 125 minutes. This is close to our calculated answer. Our answer is reasonable. ✓

Practice Problem 4 Donna bought 45 shares of stock for $1620. How much did the stock cost her per share?

MATHEMATICS BLUEPRINT FOR PROBLEM SOLVING			
Gather the Facts	What Am I Asked to Do?	How Do I Proceed?	Key Points to Remember

■

2 *Solving Problems Involving More Than One Type of Operation*

Sometimes a chart, table, or bill of sale can be used to help us organize the data in an applied problem.

EXAMPLE 5 Cleanway Rent-A-Car bought four luxury sedans at $21,000 each, three compact sedans at $14,000 each, and seven subcompact sedans at $8000 each. What was the total cost of the purchase?

1. *Understand the problem.*

We will make an imaginary bill of sale to help us to visualize the problem.

2. *Solve and state the answer.*

We do the calculation and enter the results in the bill of sale.

CAR FLEET SALES, INC. HAMILTON, MASSACHUSETTS			
Customer: *Cleanway Rent-A-Car*			
Quantity	Type of Car	Cost per Car	Amount for This Type of Car
4	Luxury sedans	$21,000	$84,000 ($4 \times \$21{,}000 = \$84{,}000$)
3	Compact sedans	$14,000	$42,000 ($3 \times \$14{,}000 = \$42{,}000$)
7	Subcompact sedans	$8,000	$56,000 ($7 \times \$8{,}000 = \$56{,}000$)
		TOTAL	$182,000 (Sum of the three amounts)

The total cost of all 14 cars is $182,000.

3. *Check.*

You may use estimation to check. The check is left to the student.

Practice Problem 5 Anderson Dining Commons purchased 50 tables at \$200 each, 180 chairs at \$40 each, and six moving carts at \$65 each. What was the cost of the total purchase? ■

EXAMPLE 6 Dawn had a balance of \$410 in her checking account last month. She made deposits of \$46, \$18, \$150, \$379, and \$22. She made out checks for \$316, \$400, and \$89. What is her balance?

1. *Understand the problem.*
We want to *add* to get a total of all deposits and *add* to get a total of all checks.

Old balance + total of deposits − total of checks = new balance

MATHEMATICS BLUEPRINT FOR PROBLEM SOLVING

Gather the Facts	What Am I Asked to Do?	How Do I Proceed?	Key Points to Remember
Old balance \$410 New deposits \$46 \$18 \$150 \$379 \$22 New checks \$316 \$400 \$89	Find the amount of money in checking account after deposits are made and checks are withdrawn.	**(a)** I need to calculate the total of the deposits and the total of the checks. **(b)** I add the total deposits to the old balance. **(c)** Then I subtract the total of the checks from that result.	Deposits are added to a checking account. Checks are subtracted from a checking account.

2. *Solve and state the answer.*

First we find

Total sum of deposits

$$\begin{array}{r} 46 \\ 18 \\ 150 \\ 379 \\ +\ 22 \\ \hline \$615 \end{array}$$

Then the

Total sum of checks

$$\begin{array}{r} 316 \\ 400 \\ +\ 89 \\ \hline \$805 \end{array}$$

Add the deposits to the old balance and subtract the amount of the checks.

$$\begin{array}{r|r} \text{Old balance} & 410 \\ +\ \text{total deposits} & +\ 615 \\ \cline{2-2} & 1025 \\ -\ \text{total checks} & -\ 805 \\ \cline{2-2} \text{New balance} & 220 \end{array}$$

The new balance of the checking account is \$220.00.

3. *Check.*

Work backward. You can add the total checks to the new balance and then subtract the total deposits. The result should be the old balance. Try it.

$$\begin{array}{rl} 410 & \text{Old balance } \checkmark \\ -\ 615 & \\ \hline 1025 & \uparrow \\ +\ 805 & \\ \hline 220 & \text{Work backward.} \end{array}$$

Practice Problem 6 Last month Bridget had $498 in a savings account. She made two deposits: one for $607 and one for $163. The bank credited her with $36 interest. Since last month, she has made four withdrawals: $19, $158, $582, and $74. What is her balance this month?

MATHEMATICS BLUEPRINT FOR PROBLEM SOLVING			
Gather the Facts	What Am I Asked to Do?	How Do I Proceed?	Key Points to Remember

■

EXAMPLE 7 When Lorenzo began his car trip, his gas tank was full and the odometer read 76,358 miles. He ended his trip at 76,668 miles and filled the gas tank with 10 gallons of gas. How many miles per gallon did he get with his car?

1. *Understand the problem.*

MATHEMATICS BLUEPRINT FOR PROBLEM SOLVING			
Gather the Facts	What Am I Asked to Do?	How Do I Proceed?	Key Points to Remember
End of trip, 76,668 miles Start of trip, 76,358 miles Used on trip, 10 gallons of gas	Find the number of miles per gallon that the car obtained on the trip.	**(a)** I need to subtract the two odometer readings to obtain the numbers of miles traveled. **(b)** I divide the number of miles driven by the number of gallons of gas used to get the number of miles obtained per gallon of gas.	The gas tank was full at the beginning of the trip. 10 gallons fills the tank at the end of the trip.

2. *Solve and state the answer.*
First we subtract the odometer readings to obtain the miles traveled.

$$\begin{array}{r} 76{,}668 \\ -\ 76{,}358 \\ \hline 310 \end{array}$$

The trip was 310 miles.
Next we divide the miles driven by the number of gallons.

$$\begin{array}{r} 31 \\ 10\overline{)310} \\ 30 \\ \hline 10 \\ 10 \\ \hline 0 \end{array}$$

Thus Lorenzo obtained 31 miles per gallon on the trip.

3. *Check.*
We do not want to round to one nonzero digit here, because, if we do, the result will be zero when we subtract. Thus we will round to the nearest hundred for the values of mileage.

76,668 becomes 76,700

76,358 becomes 76,400

Now we subtract the estimated values.

$$\begin{array}{r} 76{,}700 \\ -\ 76{,}400 \\ \hline 300 \end{array}$$

Thus we estimate the trip to be 300 miles.
Then we divide.

$$\begin{array}{r} 30 \\ 10\overline{)300} \end{array}$$

We obtain 30 miles per gallon for our estimate. This is very close to our calculated value of 31 miles per gallon. ✓

Practice Problem 7 Deidre took a car trip with a full tank of gas. Her trip began with the odometer at 50,698 and ended at 51,118 miles. She then filled the tank with 12 gallons of gas. How many miles per gallon did her car get on the trip?

MATHEMATICS BLUEPRINT FOR PROBLEM SOLVING			
Gather the Facts	What Am I Asked to Do?	How Do I Proceed?	Key Points to Remember

■

1.8 Exercises

Applications

Use the Mathematics Blueprint for Problem Solving to help you to solve the word problems in the exercise set.

1. The Federal Nigeria game preserve has 24,111 animals, 327 full time staff, and 793 volunteers. What is the total of these three groups? 25,231

2. The number of people who flew the shuttle between New York and Boston on the 3 p.m., 4 p.m., 5 p.m., and 6 p.m. flights today was 135, 167, 89, and 152, respectively. What was the total number of passengers today? 543 passengers

3. Wei Ling had a budget of $37,650 for home renovations for Dr. Smith's residence. When all the work was done, she had actually spent $42,318. How far over budget has she gone? $4668

4. An airplane was flying at an altitude of 4760 feet. It had to clear a mountain of 4356 feet. By how many feet did it clear the mountain? 404 feet

5. There are 250 hors d'oeuvres (canapés), in a box. The chef of Dexter's Hearthside ordered 15 boxes. How many hors d'oeuvres did she order? 3750 hors d'oeuvres

6. There are 144 pencils in a gross. Mr. Jim Weston ordered 14 gross of pencils for the office. How many pencils did he order? 2016 pencils

7. A 16-ounce can of beets cost 96¢. What is the unit cost of the beets? (How much did they cost per ounce?) $96 \div 16 = 6$¢ per ounce

8. A 14-ounce can of chicken soup cost 98¢. What is the unit cost of the soup? (How much does the soup cost per ounce?) $98 \div 14 = 7$¢ per ounce

9. Mike is a member of the Dartmouth College ski team. During the ski season, Mike had his skis tuned 14 times for a total of $322. How much did it cost to tune his skis each time? $23

10. There are approximately 50,000 bison (American Buffalo) living in the United States. If Northwest Trek, the animal preserve located in Mt. Rainier National Park has 103 bison, how many bison are living elsewhere? 49,897

11. Salim can make 20 sandwiches in an hour. He has to make 180 sandwiches. How many hours will it take him to make the sandwiches? How many minutes? 9 hours. This is equivalent to 540 minutes.

12. If a new amusement park covers 43 acres and there are 44,010 square feet in 1 acre, how many square feet of land does the amusement park cover? 1,892,430 sq ft

13. Every 60 minutes, the world population increases by 100,000 persons. How many people will be born during the next 480 minutes? 800,000 people will be born

14. Roberto had $2158 in his savings account six months ago. In the last six months he made four deposits: $156, $238, $1119, and $866. The bank deposited $136 in interest over the six-month period. How much does he have in the savings account at present? $2158 + 156 + 238 + 1119 + 866 + 136 = \$4673$

15. A games arcade has recently opened in a West Chicago neighborhood. The owners were nervous about whether it would be a success. Fortunately, the gross revenues over the last four weeks were $7356, $3257, $4777, and $4992. What was the gross income this month for the arcade? $20,382

16. The two largest cities in Saudi Arabia are Riyadh, the capital, with 1,250,000 people and Jeddah, with 900,000 people. What is the difference in population between these two cities? 350,000 people

Solve each applied problem. In each case, more than one type of operation is required.

17. Harry just got promoted to assistant store manager of a Wal-Mart store. He bought two suits at \$250 each, two shirts at \$35 each, two pairs of shoes at \$75 each, and three ties at \$22 each. What was the total cost of his new work wardrobe? \$786

18. Hector bought three shirts at \$26 each, two pairs of pants at \$45 each, and three pairs of shoes at \$48 each. What was the total cost of his shopping spree? \$312

19. Wei Mai Lee had a balance in her checking account of \$61. During the last few months, she has made deposits of \$385, \$945, \$732, and \$144. She wrote checks against her account for \$223, \$29, \$98, and \$435. When all the deposits are recorded and all the checks clear, what balance will she have in her checking account? \$1482

20. Wagner had a balance in his checking account of \$13. He made deposits of \$786, \$566, \$415, and \$50. He wrote out checks for \$554, \$351, \$14, and \$87. When all the deposits are recorded and all the checks clear, what balance will he have in his checking account? \$824

21. Diana owns 85 acres of forest land in Oregon. She rents it to a timber grower for \$250 per acre per year. Her property taxes are \$57 per acre. How much profit does she make on the land each year? \$16,405

22. Todd owns 13 acres of commercially zoned land in the city of Columbus, Ohio. He rents it to a construction company for \$12,350 per acre per year. His property taxes to the city are \$7362 per acre per year. How much profit does he make on the land each year?
$13 \times 12{,}350 - 13 \times 7362 = \$64{,}844$

23. Hanna wants to determine the miles-per-gallon rating of her Chevrolet Cavalier. She filled the tank when the odometer read 14,926 miles. She then drove her car on a trip. At the end of the trip the odometer read 15,276 miles. It took 14 gallons to fill the tank. How many miles per gallon does her car deliver?

$$\begin{array}{r} 15{,}276 \\ -\ 14{,}926 \\ \hline 350 \end{array} \quad 350 \div 14 = 25 \text{ miles per gallon}$$

24. Garcia wants to determine the miles-per-gallon rating of his Ford Contour. He filled the tank when the odometer read 36,339 miles. He drove the car for a week. The odometer then read 36,781 miles and the tank required 17 gallons to be filled. How many miles per gallon did Garcia's car achieve?

$$\begin{array}{r} 36{,}781 \\ -\ 36{,}339 \\ \hline 442 \end{array} \quad 442 \div 17 = 26 \text{ miles per gallon}$$

Cumulative Review Problems

25. Evaluate. 7^3 343

26. Perform in the proper order.
$3 \times 2^3 + 15 \div 3 - 4 \times 2$
$3 \times 8 + 15 \div 3 - 4 \times 2 = 24 + 5 - 8 = 21$

27. Calculate. 126×38 4788

28. Calculate. $12\overline{)3096}$ 258

Chapter Organizer

Topic	Procedure	Examples
Place value of numbers, p. 4.	Each digit has a value depending on location. millions | hundred thousands | ten thousands | thousands | hundreds | tens | ones	In the number 2,896,341, what place value does 9 have? Ten thousands
Writing expanded notation, p. 5.	Take the number of each digit and multiply it by one, ten, hundred, thousand, . . . according to its place.	Write in expanded notation. 46,235 $40{,}000 + 6000 + 200 + 30 + 5$
Writing whole numbers in words, p. 6.	Take the number of each group of three digits and indicate if they are (millions) (thousands) (ones) xxx, xxx, xxx	Write in words. 134,718,216. One hundred thirty-four million, seven hundred eighteen thousand, two hundred sixteen
Adding whole numbers, p. 12.	Starting with right column, add each column separately. If a two-digit sum occurs, "carry" the first digit over to the next column to the left.	Add. 2 1 2 5 8 3 6 7 2 9 1 + 4 5 3 1 3 6 9
Subtracting whole numbers, p. 23.	Starting with right column, subtract each column separately. If necessary, borrow a unit from column to the left and bring it to right as a "10."	Subtract. 13 6 3 12 1 6,7 4 2 − 1 2,3 9 5 4,3 4 7
Multiplying several factors, p. 40.	Keep multiplying from left to right. Take each product and multiply by next factor to right. Continue until all factors are used once. (Since multiplication is commutative and associative, the factors can be multiplied in any order.)	Multiply. $2 \times 9 \times 7 \times 6 \times 3$ $= 18 \times 7 \times 6 \times 3$ $= 126 \times 6 \times 3$ $= 756 \times 3$ $= 2268$
Multiplying several-digit numbers, p. 38.	Multiply top factor by ones digit, then by tens digit, then by hundreds digit. Add the partial products together.	Multiply. 5 6 7 × 2 3 8 4 5 3 6 1 7 0 1 1 1 3 4 1 3 4,9 4 6
Dividing by a two- or three-digit number, p. 50.	Figure how many times the first digit of the divisor goes into the first two digits of the dividend. To try this answer, multiply it back to see if it is too large or small. Continue each step of long division until finished.	Divide. 589 238)140182 1190 2118 1904 2142 2142 0

Chapter Organizer

Topic	Procedure	Examples
Exponent form, p. 58.	To show in short form the repeated multiplication of the same number, write the number being multiplied. (This is the base.) Write in smaller print above the line the number of times it appears as a factor. (This is the exponent.) To evaluate the exponent form, write the factor the number of times shown in the exponent. Then multiply.	Write in exponent form. $10 \times 10 \times 10 \times 10 \times 10 \times 10 \times 10 \times 10$ 10^8 Evaluate. 6^3 $6 \times 6 \times 6 = 216$
Order of operations, p. 60.	**1.** Perform operations inside parentheses. **2.** First raise to a power. **3.** Then do multiplication and division in order from left to right. **4.** Then do addition and subtraction in order from left to right.	Evaluate. $2^3 + 16 \div 4^2 \times 5 - 3$ Raise to a power first. $8 + 16 \div 16 \times 5 - 3$ Then do multiplication or division from left to right. $8 + 1 \times 5 - 3$ $8 + 5 - 3$ Then do addition and subtraction. $13 - 3 = 10$
Rounding, p. 64.	**1.** If the first digit to right of round-off place is less than 5, the digit in round-off place is unchanged. **2.** If the first digit to right of round-off place is 5 or more, the digit in round-off place is increased by 1. **3.** Digits to right of round-off place are replaced by zeros.	Round to nearest hundred. 5 6,7 (4) 3 (arrow above 7) The digit 4 is less than 5. 56,700 Round to the nearest thousand. 1 2 8,5 1 7 (arrow above 8) The digit 5 is obviously 5 or greater. We increase the thousands digit by 1. 129,000
Estimating the answer to a calculation, p. 66.	**1.** Round each number so that there is one nonzero digit. **2.** Perform the calculation with the rounded numbers.	Estimate the answer. $45{,}780 \times 9453$ First we round. $50{,}000 \times 9000$ Then we multiply $50{,}000 \times 9{,}000 = 450{,}000{,}000$ We estimate the answer to be 450,000,000.

Using the Mathematics Blueprint for Problem Solving, p. 73.

In solving an applied problem, students may find it helpful to complete the following steps. You will not use all the steps all the time. Choose the steps that best fit the conditions of the problem.

1. *Understand the problem.*
  (a) Read the problem carefully.
  (b) Draw a picture if this helps you to visualize the situation. Think about what facts you are given and what you are asked to find.
  (c) Use the Mathematics Blueprint for Problem Solving to organize your work. Follow these four parts.
    1. Gather the facts. (Write down specific values given in the problem.)
    2. What am I asked to do? (Identify what you must obtain for an answer.)
    3. How do I proceed? (What calculations need to be done.)
    4. Key points to remember. (Record any facts, warnings, formulas, or concepts you think will be important as you solve the problem.)

2. *Solve and state the answer.*
  (a) Perform the necessary calculations.
  (b) State the answer, including the unit of measure.

3. *Check.*
  (a) Estimate the answer to the problem. Compare this estimate to the calculated value. Is your answer reasonable?
  (b) Repeat your calculations.
  (c) Work backward from your answer. Do you arrive at the original conditions of the problem?

Example

The Manchester highway department has just purchased two pickup trucks and three dump trucks. The cost of a pickup truck is $17,920. The cost of a dump truck is $48,670. What was the cost to purchase these five trucks?

1. *Understand the problem.*

MATHEMATICS BLUEPRINT FOR PROBLEM SOLVING			
Gather the Facts	What Am I Asked to Do?	How Do I Proceed?	Key Points to Remember
Buy 2 pickup trucks 3 dump trucks Cost Pickup: $17,920 Dump: $48,670	Find the total cost of the 5 trucks.	Find the cost of 2 pickup trucks. Find the cost of 3 dump trucks. Add to get final cost of all 5 trucks.	Multiply 2 times pickup truck cost. Multiply 3 times dump truck cost.

2. *Solve.*

Calculate cost of pickup trucks

$$\begin{array}{r} \$17{,}920 \\ \times\ 2 \\ \hline \$35{,}840 \end{array}$$

Calculate cost of dump trucks

$$\begin{array}{r} \$48{,}670 \\ \times\ 3 \\ \hline \$146{,}010 \end{array}$$

Find total cost. $35,840 + $146,010 = $181,850
The total cost of the five trucks is $181,850.

3. *Check.*

Estimate cost of pickup trucks $20{,}000 \times 2 = 40{,}000$

Estimate cost of dump trucks $50{,}000 \times 3 = 150{,}000$

Total estimate $40{,}000 + 150{,}000 = 190{,}000$

This is close to our calculated answer of $181,850. We determine that our answer is reasonable. ✓

Chapter 1 Review Problems

If you have trouble with a particular type of problem, review the examples in the section indicated for that group of problems. Answers to all exercises are located in the answer key.

1.1 *Write in words.*

1. 376
Three hundred seventy-six

2. 5082
Five thousand, eighty-two

3. 109,276
One hundred nine thousand, two hundred seventy-six

4. 423,576,055
Four hundred twenty-three million, five hundred seventy-six thousand, fifty-five

Write in expanded notation.

5. 4364
4000 + 300 + 60 + 4

6. 27,986
20,000 + 7000 + 900 + 80 + 6

7. 1,305,128
1,000,000 + 300,000 + 5000 + 100 + 20 + 8

8. 42,166,037
40,000,000 + 2,000,000 + 100,000 + 60,000 + 6000 + 30 + 7

Write in standard notation.

9. Nine hundred twenty-four 924

10. Six thousand ninety-five 6095

11. One million, three hundred twenty-eight thousand, eight hundred twenty-eight 1,328,828

12. Forty-five million, ninety-two thousand, six hundred fifty-one 45,092,651

1.2 *Add.*

13. $36 + 94$ = 130

14. $76 + 39$ = 115

15. $127 + 563$ = 690

16. $12 + 28 + 34 + 76$ = 150

17. $123 + 61 + 9 + 84 + 123$ = 400

18. $125 + 364 + 980$ = 1469

19. $937 + 405 + 256$ = 1598

20. $28{,}364 + 97{,}059$ = 125,423

21. $1356 + 2892 + 561 + 89 + 9805$ = 14,703

22. $26 + 503 + 935 + 1257 + 7861$ = 10,582

1.3 *Subtract.*

23. $36 - 19$ = 17

24. $54 - 48$ = 6

25. $126 - 99$ = 27

26. $543 - 372$ = 171

27. $1296 - 1137$ = 159

28. $9821 - 4993$ = 4828

29. $101{,}300 - 98{,}274$ = 3026

30. $201{,}010 - 137{,}864$ = 63,146

31. $1{,}986{,}312 - 1{,}761{,}555$ = 224,757

32. $7{,}216{,}003 - 5{,}985{,}312$ = 1,230,691

1.4 *Multiply.*

33. $\begin{array}{r} 12 \\ \times\ 3 \\ \hline \end{array}$ 36

34. $\begin{array}{r} 57 \\ \times\ 2 \\ \hline \end{array}$ 114

35. $\begin{array}{r} 36 \\ \times\ 0 \\ \hline \end{array}$ 0

36. $\begin{array}{r} 24 \\ \times\ 1 \\ \hline \end{array}$ 24

37. $1 \times 3 \times 6$ 18

38. $2 \times 4 \times 8$ 64

39. $5 \times 7 \times 3$ 105

40. $4 \times 6 \times 5$ 120

41. $8 \times 1 \times 9 \times 2$ 144

42. $7 \times 6 \times 0 \times 4$ 0

43. $3 \cdot 4 \cdot 2 \cdot 2 \cdot 5$ 240

44. $1 \cdot 2 \cdot 7 \cdot 3 \cdot 4$ 168

45. $26{,}121 \times 100$ 2,612,100

46. $84{,}312 \times 1000$ 84,312,000

47. $832 \times 100{,}000$ 83,200,000

48. $563 \times 1{,}000{,}000$ 563,000,000

49. $\begin{array}{r} 36 \\ \times\ 24 \\ \hline \end{array}$ 864

50. $\begin{array}{r} 58 \\ \times\ 32 \\ \hline \end{array}$ 1856

51. $\begin{array}{r} 150 \\ \times\ 27 \\ \hline \end{array}$ 4050

52. $\begin{array}{r} 360 \\ \times\ 38 \\ \hline \end{array}$ 13,680

53. $\begin{array}{r} 709 \\ \times\ 36 \\ \hline \end{array}$ 25,524

54. $\begin{array}{r} 502 \\ \times\ 48 \\ \hline \end{array}$ 24,096

55. $\begin{array}{r} 123 \\ \times\ 714 \\ \hline \end{array}$ 87,822

56. $\begin{array}{r} 431 \\ \times\ 623 \\ \hline \end{array}$ 268,513

57. $\begin{array}{r} 1782 \\ \times\ 305 \\ \hline \end{array}$ 543,510

58. $\begin{array}{r} 2057 \\ \times\ 124 \\ \hline \end{array}$ 255,068

59. $\begin{array}{r} 300 \\ \times\ 500 \\ \hline \end{array}$ 150,000

60. $\begin{array}{r} 400 \\ \times\ 600 \\ \hline \end{array}$ 240,000

61. $\begin{array}{r} 1200 \\ \times\ 6000 \\ \hline \end{array}$ 7,200,000

62. $\begin{array}{r} 2500 \\ \times\ 3000 \\ \hline \end{array}$ 7,500,000

63. $\begin{array}{r} 100{,}000 \\ \times\ 20{,}000 \\ \hline \end{array}$ 2,000,000,000

64. $\begin{array}{r} 300{,}000 \\ \times\ 40{,}000 \\ \hline \end{array}$ 12,000,000,000

1.5 *Divide, if possible.*

65. $20 \div 10$ 2

66. $40 \div 8$ 5

67. $70 \div 5$ 14

68. $36 \div 9$ 4

69. $0 \div 8$ 0

70. $12 \div 1$ 12

71. $7 \div 1$ 7

72. $0 \div 5$ 0

73. $\frac{49}{7}$ 7

74. $\frac{42}{6}$ 7

75. $\frac{5}{0}$ not possible

76. $\frac{24}{6}$ 4

77. $\frac{56}{8}$ 7

78. $\frac{48}{8}$ 6

79. $\frac{72}{9}$ 8

80. $\frac{0}{0}$ not possible

Divide. Be sure to indicate the remainder, if one exists.

81. $7\overline{)875}$ 125

82. $6\overline{)750}$ 125

83. $5\overline{)1290}$ 258

84. $4\overline{)1476}$ 369

85. $3\overline{)77{,}622}$ 25,874

86. $8\overline{)29{,}536}$ 3692

87. $6\overline{)221{,}748}$ 36,958

88. $5\overline{)184{,}605}$ 36,921

89. $8\overline{)127{,}890}$ 15,986 R 2

90. $7\overline{)250{,}485}$ 35,783 R 4

91. $67\overline{)490}$ 7 R 21

92. $72\overline{)325}$ 4 R 37

93. $21\overline{)666}$ 31 R 15

94. $22\overline{)319}$ 14 R 11

95. $68\overline{)2614}$ 38 R 30

96. $76\overline{)4142}$ 54 R 38

97. $35\overline{)9030}$ 258

98. $45\overline{)4275}$ 95

99. $132\overline{)7128}$ 54

100. $204\overline{)3876}$ 19

1.6 *Write in exponent form.*

101. 13×13 13^2

102. 24×24 24^2

103. $8 \times 8 \times 8 \times 8 \times 8$ 8^5

104. $9 \times 9 \times 9 \times 9 \times 9$ 9^5

Evaluate.

105. 2^6 64

106. 3^4 81

107. 5^3 125

108. 2^7 128

109. 7^2 49

110. 9^2 81

111. 6^3 216

112. 4^4 256

Perform each operation in proper order.

113. $6 \times 2 - 4 + 3$ 11

114. $7 + 2 \times 3 - 5$ 8

115. $2^5 + 4 - (5 + 3^2)$ 22

116. $4^3 + 20 \div (2 + 2^3)$ 66

117. $3^3 \times 4 - 6 \div 6$ 107

118. $20 \div 20 + 5^2 \times 3$ 76

119. $2^3 \times 5 \div 8 + 3 \times 4$ 17

120. $3^2 \times 6 \div 3 + 5 \times 6$ 48

121. $9 \times 2^2 + 3 \times 4 - 36 \div (4 + 5)$ 44

122. $5 \times 3 + 5 \times 4^2 - 14 \div (6 + 1)$ 93

Round to the nearest ten.

123. 5673 5670

124. 1275 1280

125. 15,305 15,310

126. 42,644 42,640

Round to the nearest thousand.

127. 12,350 12,000

128. 22,986 23,000

129. 675,800 676,000

130. 202,498 202,000

131. Round to the nearest hundred thousand. 5,668,243. 5,700,000

132. Round to the nearest ten thousand. 9,995,312. 10,000,000

Use the principle of estimation to find an estimate for each of the following calculations.

133. $589 + 622 + 933 + 864$

$$\begin{array}{r} 600 \\ 600 \\ 900 \\ +\ 900 \\ \hline 3000 \end{array}$$

134. $25{,}981 + 36{,}782 + 73{,}125$

$$\begin{array}{r} 30{,}000 \\ 40{,}000 \\ +\ 70{,}000 \\ \hline 140{,}000 \end{array}$$

135. $29{,}378 - 17{,}924$

$$\begin{array}{r} 30{,}000 \\ -\ 20{,}000 \\ \hline 10{,}000 \end{array}$$

136. $4{,}326{,}171 - 2{,}916{,}788$

$$\begin{array}{r} 4{,}000{,}000 \\ -\ 3{,}000{,}000 \\ \hline 1{,}000{,}000 \end{array}$$

137. 1763×5782

$$\begin{array}{r} 2000 \\ \times\ 6000 \\ \hline 12{,}000{,}000 \end{array}$$

138. $2{,}965{,}372 \times 893$

$$\begin{array}{r} 3{,}000{,}000 \\ \times\ 900 \\ \hline 2{,}700{,}000{,}000 \end{array}$$

139. $83{,}421 \div 24$ $\quad 20\overline{)80{,}000}$ = 4,000

140. $7{,}963{,}127 \div 378$ $\quad 400\overline{)8{,}000{,}000}$ = 20,000

Estimate the result of each of the following calculations. Some results are correct and some are incorrect. Which results appear to be correct? Which results appear to be incorrect?

141. $87 + 36 + 94 + 55 = 272$

Correct

$$\begin{array}{r} 90 \\ 40 \\ 90 \\ +\ 60 \\ \hline 280 \end{array}$$

142. $938{,}526 - 398{,}445 = 540{,}081$

Correct

$$\begin{array}{r} 900{,}000 \\ -\ 400{,}000 \\ \hline 500{,}000 \end{array}$$

143.

$$\begin{array}{r} 176{,}394 \\ \times\ 5216 \\ \hline 92{,}007{,}114 \end{array} \qquad \begin{array}{r} 200{,}000 \\ \times\ 5000 \\ \hline 1{,}000{,}000{,}000 \end{array}$$

Incorrect

144.

$$32\overline{)893{,}152} = 27{,}911 \qquad 30\overline{)900{,}000} = 30{,}000$$

Correct

1.8 *Solve each applied problem.*

145. Ward types 25 words per minute. He typed for 7 minutes at that speed. How many words did he type? 175 words

146. The soft-drink cans come six to a package. There are 34 packages in the storage room. How many soft-drink cans are there? 204 cans

147. Applepickers, Inc., bought a truck for $26,300, a car for $14,520, and a minivan for $18,650. What was the total purchase price? $59,470

148. Alfonso drove 1362 kilometers last summer, 562 km during Christmas break, and 473 km during spring break. How many total kilometers were driven? 2397 km

149. Roberta was billed $11,658 for tuition. She received a $4630 grant. How much did she have to pay after the grant was deducted? $7028

150. A plane was flying at 14,630 feet. It flew over a mountain 4329 feet high. How many feet was it from the plane to the top of the mountain? 10,301 feet

151. Vincent bought 92 shares of stock for $5888. What was the cost per share? $64 per share

152. The expedition cost a total of $32,544 for 24 paying passengers, who shared the cost equally. What was the cost per passenger? $1356

153. Marcia's checking account balance last month was $436. She made deposits of $16, $98, $125, and $318. She made out checks of $29, $128, $100, and $402. What will be her balance this month? $334

154. Melissa's savings account balance last month was $810. The bank added $24 interest. Melissa deposited $105, $36, and $177. She made withdrawals of $18, $145, $250, and $461. What will be her balance this month? $278

155. Ali began a trip on a full tank of gas with the car odometer at 56,320 miles. He ended the trip at 56,720 miles and added 16 gallons of gas. How many miles per gallon did he get on the trip? 25 miles per gallon

156. Amina began a trip on a full tank of gas with the car odometer at 24,396 miles. She ended the trip at 24,780 miles and added 16 gallons of gas. How many miles per gallon did she get on the trip? 24 miles per gallon

157. The maintenance group bought three lawn mowers at $279, four power drills at $61, and two riding tractors at $1980. What was the total purchase price for these items? $5041

158. The library bought 24 sets of shelves at $118 each, four desks at $120 each, and six chairs at $24 each. What was the total purchase price for these items? $3456

Developing Your Study Skills

EXAM TIME: GETTING ORGANIZED

Studying adequately for an exam requires careful preparation. Begin early so that you will be able to spread your review over several days. Even though you may still be learning new material at this time, you can be reviewing concepts previously learned in the chapter. Giving yourself plenty of time for review will take the pressure off. You need this time to process what you have learned and to tie concepts together.

Adequate preparation enables you to feel confident and to think clearly with less tension and anxiety.

Chapter 1 Test

Do these problems simulating test conditions.

1. Write in words. 44,007,635

2. Write in expanded notation. 26,859

3. Write in standard notation: three million, five hundred eighty-one thousand, seventy-six.

Add.

4. $126 + 83 + 5 + 294 + 192$

5. $470 + 386 + 189$

6. $135{,}484 + 2{,}376 + 81{,}004 + 100{,}113$

Subtract.

7. $7932 - 513$

8. $300{,}523 - 262{,}182$

9. $18{,}400{,}100 - 13{,}174{,}332$

Multiply.

10. $1 \times 6 \times 9 \times 7$

11. 56×39

12. 147×625

13. $18{,}491 \times 7$

1. forty-four million, seven thousand, six hundred thirty-five

2. 20,000 + 6000 + 800 + 50 + 9

3. 3,581,076

4. 700

5. 1045

6. 318,977

7. 7419

8. 38,341

9. 5,225,768

10. 378

11. 2184

12. 91,875

13. 129,437

14.	3014 R 1
15.	2189
16.	352
17.	11^3
18.	64
19.	23
20.	78
21.	79
22.	26,450
23.	6,460,000
24.	3,600,000
25.	150,000,000,000
26.	18,000
27.	$2148
28.	467 feet
29.	$127
30.	$292

Divide. If there is a remainder, be sure to state it as part of your answer.

14. $5\overline{)15{,}071}$

15. $7\overline{)15{,}323}$

16. $37\overline{)13{,}024}$

17. Write in exponent form.
$11 \times 11 \times 11$

18. Evaluate. 2^6

Perform each operation in proper order.

19. $5 + 6^2 - 2 \times (9 - 6)^2$

20. $2^3 + 4^3 + 18 \div 3$

21. $4 \times 6 + 3^3 \times 2 + 23 \div 23$

22. Round to the nearest ten. 26,453

23. Round to the nearest ten thousand. 6,462,431

24. Round to the nearest hundred thousand. 3,593,452

Estimate the answer to each of the following.

25. $4{,}867{,}010 \times 27{,}058$

26. $1423 + 4287 + 4103 + 8549$

Answer each question.

27. A cruise for 15 people cost $32,220. If each person paid the same amount, how much will it cost each individual?

28. The river is 602 feet wide at Big Bend Corner. A boy is in the shallow water, 135 feet from the shore. How far is the boy from the other side of the river?

29. At the bookstore, Hector bought three notebooks at $2 each, one textbook for $45, two lamps at $21 each, and two sweatshirts at $17 each. What was his total bill?

30. Patricia is looking at her checkbook. She had a balance last month of $31. She deposited $902 and $399. She made out checks for $885, $103, $26, $17, and $9. What will be her new balance?

CHAPTER 2

Fractions

Crop profits, farm operations, equipment maintenance—it is easy to see that working on a farm involves a lot of calculations with whole numbers. However, did you know that many decisions on farms involving thousands of dollars are made based on the careful use of fractions? Are you confident enough in your knowledge of fractions that you could base the entire profit or loss of a farming operation on your accuracy in operations with fractions? Turn to the Putting Your Skills to Work Problems on page 153.

Pretest Chapter 2

1.	$\frac{3}{8}$
2.	Answers will vary.
3.	$\frac{17}{124}$
4.	$\frac{1}{6}$
5.	$\frac{1}{3}$
6.	$\frac{1}{7}$
7.	$\frac{5}{7}$
8.	$\frac{4}{11}$
9.	$\frac{9}{7}$
10.	$\frac{55}{9}$
11.	$24\frac{1}{4}$
12.	$5\frac{4}{5}$
13.	$2\frac{2}{17}$
14.	$\frac{5}{44}$
15.	$\frac{2}{3}$
16.	$67\frac{5}{6}$

If you are familiar with the topics in this chapter, take this test now. Check your answers with those in the back of the book. If an answer was wrong or you couldn't do a problem, study the appropriate section of the chapter.

If you are not familiar with the topics in this chapter, don't take this test now. Instead, study the examples, work the practice problems, and then take the test.

This test will help you identify which concepts you have mastered and which you need to study further.

Section 2.1

1. Use a fraction to represent the shaded part of the object.

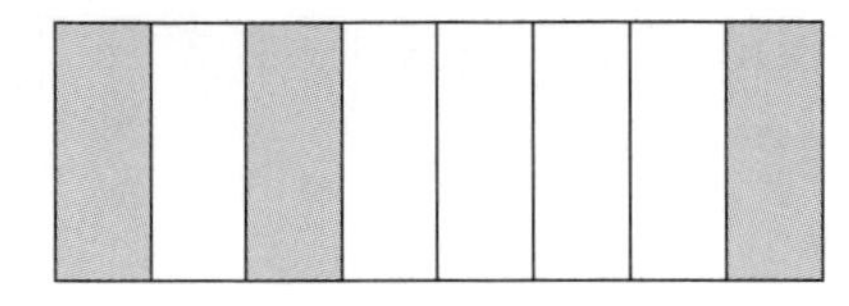

2. Draw a sketch to show $\frac{3}{6}$ of an object.

3. An inspector inspected 124 books. Of these, 17 were defective. Write a fraction that describes the part that was defective.

Section 2.2

Reduce each fraction.

4. $\frac{3}{18}$ **5.** $\frac{13}{39}$ **6.** $\frac{16}{112}$ **7.** $\frac{35}{49}$ **8.** $\frac{44}{121}$

Section 2.3

Change to an improper fraction.

9. $1\frac{2}{7}$ **10.** $6\frac{1}{9}$

Change to a mixed number.

11. $\frac{97}{4}$ **12.** $\frac{29}{5}$ **13.** $\frac{36}{17}$

Section 2.4

Multiply.

14. $\frac{5}{11} \times \frac{1}{4}$ **15.** $\frac{3}{7} \times \frac{14}{9}$ **16.** $12\frac{1}{3} \times 5\frac{1}{2}$

Section 2.5

Divide.

17. $\frac{3}{7} \div \frac{3}{7}$

18. $\frac{7}{16} \div \frac{7}{8}$

19. $6\frac{4}{7} \div 1\frac{5}{21}$

20. $8 \div \frac{12}{7}$

Section 2.6

Find the least common denominator.

21. $\frac{1}{8}, \frac{3}{4}, \frac{1}{2}$

22. $\frac{2}{9}, \frac{4}{45}$

23. $\frac{4}{11}, \frac{2}{55}$

24. $\frac{5}{24}, \frac{7}{36}$

Section 2.7

Add or subtract.

25. $\frac{7}{18} - \frac{3}{24}$

26. $\frac{5}{24} + \frac{4}{9} + \frac{1}{36}$

Section 2.8

Add or subtract.

27. $8 - 3\frac{2}{3}$

28. $1\frac{5}{6} + 2\frac{1}{7}$

29. $6\frac{4}{9} - 2\frac{3}{18}$

Section 2.9

Answer each question.

30. Miguel and Lee set out to hike $13\frac{1}{2}$ miles from Arlington to Concord. During the first 5 hours, they covered $6\frac{1}{3}$ miles going from Arlington to Bedford. How many miles are left to be covered from Bedford to Concord?

31. Robert harvested $4\frac{3}{12}$ tons of wheat. His helper harvested $3\frac{1}{18}$ tons of wheat. How much did they harvest together?

32. The students of Smith Dorm fifth floor contributed money to make a stock purchase of one share of stock each. They paid $776 in all to buy the shares of stock. The cost of buying one share was $$43\frac{1}{9}$. How many students shared in the purchase?

17. 1

18. $\frac{1}{2}$

19. $5\frac{4}{13}$

20. $4\frac{2}{3}$

21. 8

22. 45

23. 55

24. 72

25. $\frac{19}{72}$

26. $\frac{49}{72}$

27. $\frac{13}{3}$ or $4\frac{1}{3}$

28. $3\frac{41}{42}$

29. $\frac{77}{18}$ or $4\frac{5}{18}$

30. $7\frac{1}{6}$ miles

31. $7\frac{11}{36}$ tons

32. 18 students

2.1 Understanding Fractions

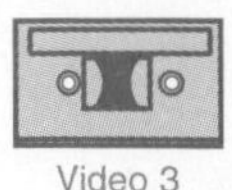

MathPro Video 3 SSM

After studying this section, you will be able to:

1. *Use a fraction to represent part of a whole.*
2. *Draw a sketch to illustrate a fraction.*
3. *Use fractions to represent the data from applied situations.*

1 *Using a Fraction to Represent Part of a Whole*

In Chapter 1 we studied whole numbers. In this chapter we will study a fractional part of a whole number. One way to represent parts of a whole is with fractions. The word *fraction* (like the word *fracture*) suggests that something is being broken. In mathematics, fractions represent the part that is ''broken off'' from a whole. The whole can be a single object (like a whole pie) or a group (the employees of a company). Here are some examples.

Single object

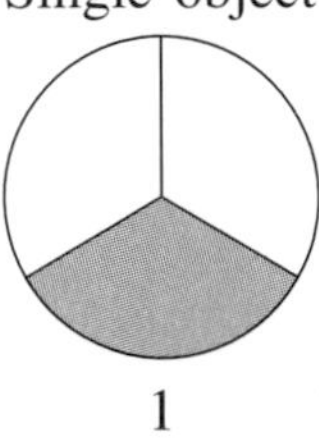

$\frac{1}{3}$

The whole is the pie on the left. The fraction $\frac{1}{3}$ represents the shaded part of the pie, 1 of 3 pieces.

A group: ACE company employs 150 men, 200 women.

$\frac{150}{350}$

The whole is the company of 350 people. The fraction $\frac{150}{350}$ represents that part of the company consisting of men. Note that the total number of employees is 150 men plus 200 women, or 350.

Recipe: Applesauce
4 apples
1/2 cup sugar
1 teaspoon cinnamon

The whole is 1 whole cup of sugar. This recipe calls for $\frac{1}{2}$ cup of sugar. Notice that in many real-life situations $\frac{1}{2}$ is written as 1/2.

When we say ''$\frac{3}{8}$ of a pizza has been eaten,'' we mean 3 of 8 equal parts of a pizza have been eaten. (See the figure on the left.) When we write the fraction $\frac{3}{8}$, the number on the top, 3, is the **numerator,** and the number on the bottom, 8, is the **denominator.**

TEACHING TIP Stress to students that they must know the names of the two parts of the fraction: the numerator and the denominator. You can tell them that an easy way to remember which is which is to recall that the *D*enominator is *D*own at the bottom of the fraction.

The numerator specifies how many parts ⟶ 3
The denominator specifies the total number of parts ⟶ 8

When we say, ''$\frac{2}{3}$ of the marbles are red,'' we mean 2 marbles out of a total of 3 are red marbles.

Part we are interested in ⟶ 2 numerator
Total number in the group ⟶ 3 denominator

EXAMPLE 1 Use a fraction to represent the part of the whole shown.

(a) 

(b) 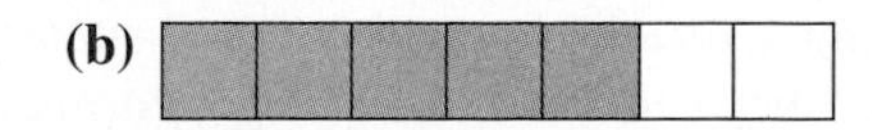

(c) 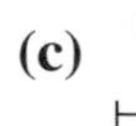

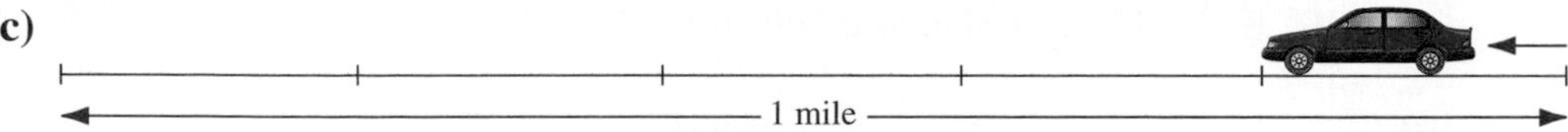

(a) Three out of four circles are red. The fraction is $\frac{3}{4}$.

(b) Five out of seven equal parts are shaded. The fraction is $\frac{5}{7}$.

(c) The mile is divided into five equal parts. The car has traveled 1 part out of 5 of the one mile distance. The fraction is $\frac{1}{5}$.

Practice Problem 1 Use a fraction to represent the shaded part of the whole.

(a) 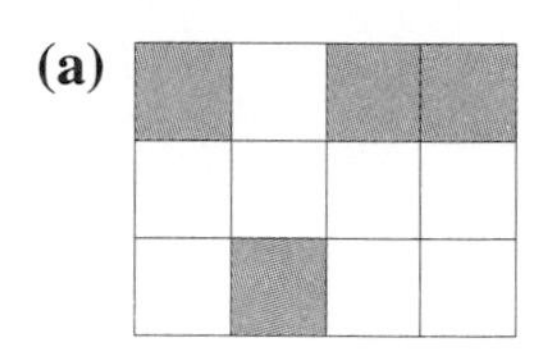

(b) 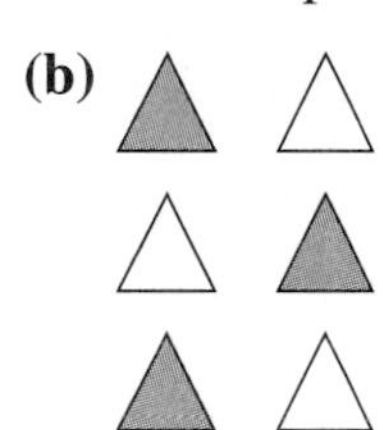

(c) 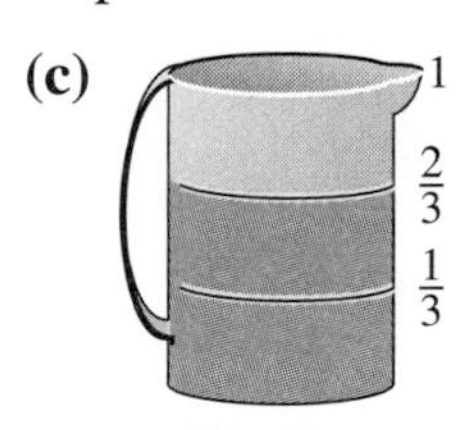

■

We can also think of a fraction as a division problem.

$$\frac{1}{3} = 1 \div 3 \quad \text{and} \quad 1 \div 3 = \frac{1}{3}$$

The division way of looking at fractions asks the questions:

What is the result of dividing one whole into three equal parts?

Thus we can say the fraction $\frac{a}{b}$ means the same as $a \div b$. However, special care must be taken with the number 0.

Suppose that we had four equal parts and we wanted to take none of them. We would want $\frac{0}{4}$ of the parts. Since $\frac{0}{4} = 0 \div 4 = 0$, we see that $\frac{0}{4} = 0$. Any fraction with a 0 numerator equals zero.

$$\frac{0}{8} = 0 \qquad \frac{0}{5} = 0 \qquad \frac{0}{13} = 0$$

What happens when zero is in the denominator? $\frac{4}{0}$ means 4 out of 0 parts. Taking 4 out of 0 does not make sense. We say $\frac{4}{0}$ is *not defined.*

$$\frac{3}{0}, \frac{7}{0}, \frac{4}{0} \quad \textit{are not defined.}$$

We cannot have a fraction with 0 in the denominator. Since $\frac{4}{0} = 4 \div 0$, we say division by zero is *undefined.* We cannot divide by 0.

2 *Drawing a Sketch to Illustrate a Fraction*

Drawing a sketch of a mathematical situation is a powerful problem-solving technique. The picture often reveals information not always apparent in the words.

EXAMPLE 2 Draw a sketch to illustrate.

(a) $\frac{7}{11}$ of an object **(b)** $\frac{2}{9}$ of a group

(a) The easiest figure to draw is a rectangular bar. We call this a *fraction bar.*

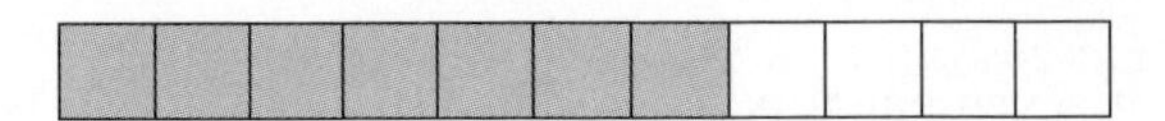

We divide the bar into 11 equal parts. We then shade in 7 parts to show $\frac{7}{11}$.

(b) We draw 9 circles of equal size to represent a group of 9.

We shade in 2 of the 9 circles to show $\frac{2}{9}$.

Practice Problem 2 Draw a sketch to illustrate.

(a) $\frac{4}{5}$ of an object **(b)** $\frac{3}{7}$ of a group ■

Recall these facts about division problems involving the number 1 and the number 0.

Division Involving the Number 1 and the Number 0

1. Any nonzero number divided by itself is 1.

$$\frac{7}{7} = 1$$

2. Any number divided by 1 remains unchanged.

$$\frac{29}{1} = 29$$

3. Zero may be divided by any nonzero number; the result is always zero.

$$\frac{0}{4} = 0$$

4. Zero can never be the denominator in a division problem.

$\frac{0}{0}$ cannot be done $\frac{3}{0}$ cannot be done

3 *Using Fractions to Represent Data from Applied Situations*

Several real-life or applied situations can be described using fractions.

EXAMPLE 3 Use a fraction to describe each of the following situations.

(a) A baseball player gets a hit 5 out of 12 times at bat.

(b) There are 156 men and 185 women taking psychology this semester. Describe the part of the class that consists of women.

(c) APEX stock went up three-eighths of a point yesterday.

(a) The baseball player got a hit $\frac{5}{12}$ of his times at bat.

(b) The total class is $156 + 185 = 341$. The fractional part that is women is 185 out of 341. Thus $\frac{185}{341}$ of the class is women.

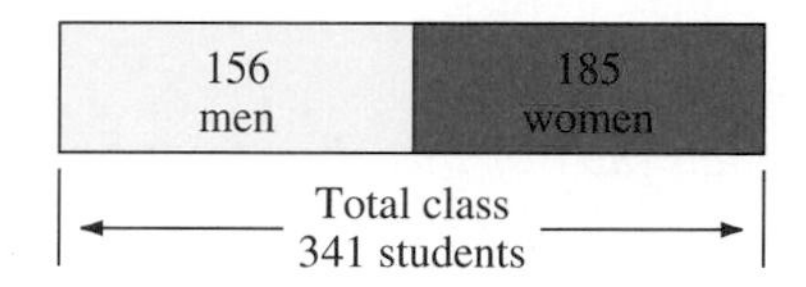

(c) The price of stock is calculated in dollars and part of a dollar. Three-eighths of a point is three-eighths of a dollar. The fraction is $\frac{3}{8}$.

Practice Problem 3 Use a fraction to describe each of the following situations.

(a) 9 out of the 17 players on the basketball team are on the Dean's List.

(b) The senior class has 382 men and 351 women. Describe the part of the class consisting of men.

(c) John needed seven-eighths of a yard of material. ■

EXAMPLE 4 Wanda made 13 calls, out of which she made five sales. Albert made 17 calls, out of which he made six sales. Write a fraction that describes for both people together the number of calls in which a sale was made compared with the total number of calls.

There are $5 + 6 = 11$ calls in which a sale was made.
There were $13 + 17 = 30$ total calls.

Thus $\dfrac{11}{30}$ of the calls resulted in a sale.

Practice Problem 4 An inspector found that one out of seven belts was defective. She also found that two out of nine shirts were defective. Write a fraction that describes what part of all the objects examined were defective. ■

2.1 Exercises

Verbal and Writing Skills

1. A fraction can be used to represent part of a whole or part of a group.
2. The numerator tells the number of parts we are interested in.
3. The denominator tells the total number of parts in the whole or in the group.
4. Describe a real-life situation that involves fractions. Answers will vary.

Name the numerator and the denominator in each fraction.

5. $\frac{3}{5}$ N: 3 D: 5
6. $\frac{9}{11}$ N: 9 D: 11
7. $\frac{2}{3}$ N: 2 D: 3
8. $\frac{3}{4}$ N: 3 D: 4
9. $\frac{1}{17}$ N: 1 D: 17
10. $\frac{1}{15}$ N: 1 D: 15

In problems 11–30, use a fraction to represent the shaded part of the object or the shaded portion of the set of objects.

11. 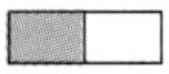 $\frac{1}{2}$
12. 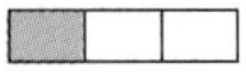 $\frac{1}{3}$
13. 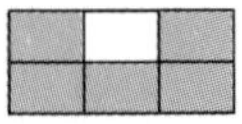 $\frac{5}{6}$
14. 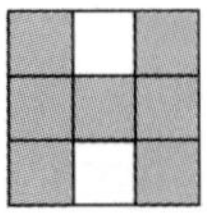 $\frac{7}{9}$
15. 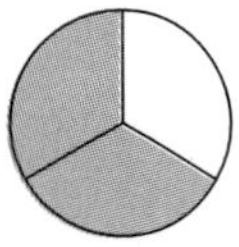 $\frac{2}{3}$
16. 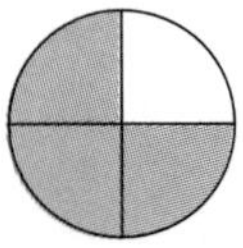 $\frac{3}{4}$
17. 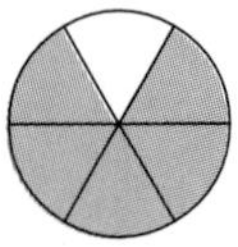 $\frac{5}{6}$
18. 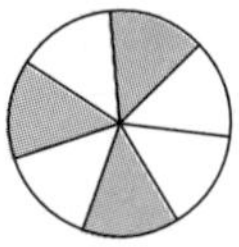 $\frac{3}{7}$
19. 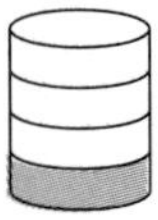 $\frac{1}{4}$
20. 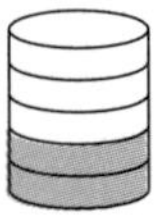 $\frac{2}{5}$
21. 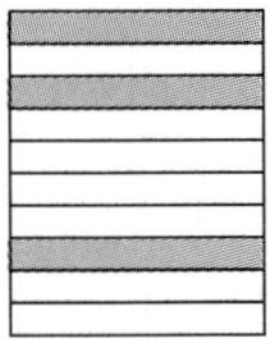 $\frac{3}{10}$
22. 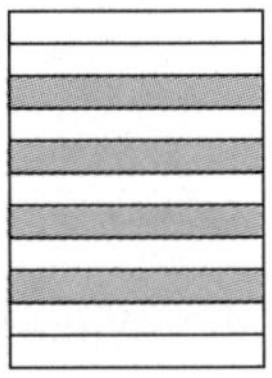 $\frac{4}{11}$
23. 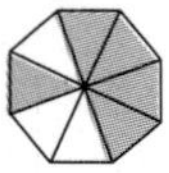 $\frac{5}{8}$
24. 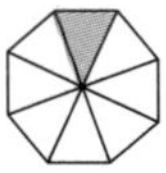 $\frac{1}{8}$
25.  $\frac{4}{7}$
26. 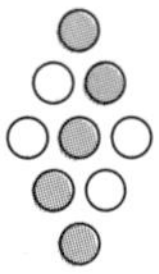 $\frac{5}{9}$
27. 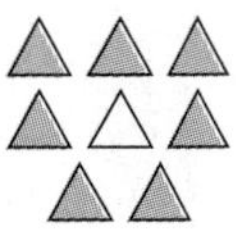 $\frac{7}{8}$
28. 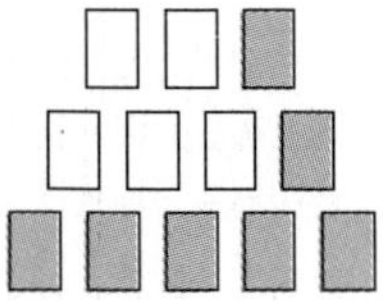 $\frac{7}{12}$
29.  5 4 3 2 1 $\frac{2}{5}$
30.  7 6 5 4 3 2 1 $\frac{5}{7}$

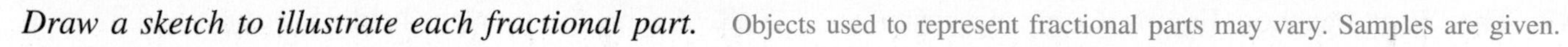

Draw a sketch to illustrate each fractional part. Objects used to represent fractional parts may vary. Samples are given.

31. $\frac{4}{5}$ of an object

32. $\frac{3}{7}$ of an object

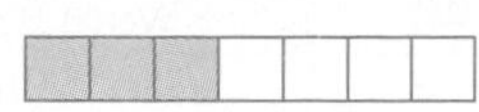

33. $\frac{11}{13}$ of an object

34. $\frac{5}{12}$ of an object

35. $\frac{7}{10}$ of an object

36. $\frac{8}{15}$ of an object

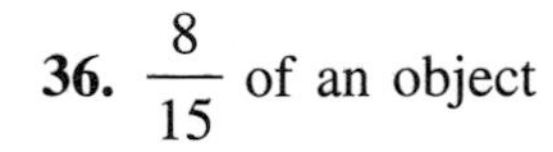

Applications

37. The Crafts Fair had 165 exhibitors, 29 of whom were toy makers. What fractional part of the exhibitors at the Crafts Fair were toy makers?

$\frac{29}{165}$

38. The total purchase amount was 83¢, of which 5¢ was sales tax. What fractional part of the total purchase price was sales tax?

$\frac{5}{83}$

39. Lance bought a 100-CD jukebox for $750. Part of it was paid for with the $209 he earned parking cars for the valet service at a local wedding reception hall. What fractional part of the jukebox was paid for by his weekend earnings?

$\frac{209}{750}$

40. Charles drove for 47 minutes to go to the Laker game. 18 minutes of his trip were spent in bumper-to-bumper traffic. What fractional part of his time was spent in traffic?

$\frac{18}{47}$

41. The Democratic National Committee fund-raising event served 122 chicken dinners and 89 roast beef dinners to its contributors. What fractional part of the guests ate roast beef?

$\frac{89}{211}$

42. The East dormitory has 111 smokers and 180 nonsmokers. What fractional part of the dormitory has nonsmoking residents?

$\frac{180}{291}$

43. In the Golden West apartment building, 17 people own cats, 14 people own dogs, 4 own birds, and 15 own fish. What fractional part of the total pet owners have cats?

$\frac{17}{50}$

44. From Roger's paycheck $100 was withheld for federal income tax, $24 was withheld for state income tax, and $63 was withheld for social security. What fractional part of the withheld money is the federal income tax?

$\frac{100}{187}$

45. The picnic table held 2 bowls of corn, 3 bowls of potato salad, 4 bowls of baked beans, and 5 bowls of ribs. What fractional part of the bowls on the table contains either ribs or beans?

$\frac{9}{14}$

46. A box of compact discs contains 5 classical CDs, 6 jazz CDs, 4 sound tracks, and 24 blues CDs. What fractional part of the total CDs is either jazz or blues?

$\frac{30}{39}$

★**47.** The West Peabody Engine Company manufactured two items last week: 101 engines and 94 lawn mowers. It was discovered that 19 engines were defective and 3 lawn mowers were defective. Of the engines that were not defective, 40 were properly constructed but 42 were not of the highest quality. Of the lawn mowers that were not defective, 50 were properly constructed but 41 were not of the highest quality.

(a) What fractional part of all items manufactured was of the highest quality?

$\frac{90}{195}$

(b) What fractional part of all items manufactured was defective?

$\frac{22}{195}$

★**48.** A Chicago tour bus held 25 women and 33 men. 12 women wore jeans. 19 men wore jeans. In the group of 25 women, a subgroup of 8 women wore sandals. In the group of 19 men, a subgroup of 10 wore sandals.

(a) What fractional part of the people on the bus wore jeans?

$\frac{31}{58}$

(b) What fractional part of the women on the bus wore sandals?

$\frac{8}{25}$

To Think About

49. Illustrate a real-world example of the fraction $\frac{0}{6}$.

The amount of money each of 6 business owners gets if the business has a profit of $0.

50. What happens when we try to illustrate a real-world example of the fraction $\frac{6}{0}$? Why?

We cannot do it. Division by zero is undefined.

Cumulative Review Problems

51. Add.

$$\begin{array}{r} 18 \\ 27 \\ 34 \\ 16 \\ 125 \\ +\ 21 \\ \hline 241 \end{array}$$

52. Subtract.

$$\begin{array}{r} 38{,}114 \\ -\ 27{,}008 \\ \hline 11{,}106 \end{array}$$

53. Multiply.

$$\begin{array}{r} 4136 \\ \times\ \ 29 \\ \hline 37224 \\ 8272\ \ \\ \hline 119{,}944 \end{array}$$

54. Divide. $12\overline{)2130}$

$$\begin{array}{r} 177 \text{ R } 6 \\ 12\overline{)2130} \\ 12\ \ \ \\ \hline 93\ \ \\ 84\ \ \\ \hline 90\ \\ 84\ \\ \hline 6\ \end{array}$$

Developing Your Study Skills

PREVIEWING NEW MATERIAL

Part of your study time each day should consist of looking ahead to those sections in your text that are to be covered the following day. You do not necessarily have to study and learn the material on your own, but if you survey the concepts, terminology, diagrams, and examples, the new ideas will seem more familiar to you when the instructor presents them. You can take note of concepts that appear confusing or difficult and be ready to listen carefully for your instructor's explanations. You can be prepared to ask the questions that will increase your understanding. Previewing new material enables you to see what is coming and prepares you to be ready to absorb it.

2.2 Simplifying Fractions

After studying this section, you will be able to:

1. *Write a number as a product of prime factors.*
2. *Reduce a fraction.*
3. *Determine if two fractions are equal.*

MathPro

Video 4

SSM

1 *Writing a Number as a Product of Prime Factors*

A **prime number** is a whole number greater than 1 that cannot be evenly divided except by itself and 1. If you examine all the whole numbers from 1 to 50, you will find 15 prime numbers.

TEACHING TIP Tell students that the ancient Greek mathematician Eratosthenes developed a method of finding prime numbers called the Sieve of Eratosthenes. This method is still used today with some slight modification by computers that generate lists of prime numbers. Suggest that they look up the topic of the Sieve of Eratosthenes in an encyclopedia if they are interested.

The First 15 Prime Numbers

2, 3, 5, 7, 11, 13, 17, 19, 23, 29, 31, 37, 41, 43, 47

A **composite number** is a whole number greater than 1 that can be divided by whole numbers other than itself. The number 12 is a composite number.

$$12 = 2 \times 6 \quad \text{and} \quad 12 = 3 \times 4$$

The number 1 is neither a prime nor a composite number. The number 0 is neither a prime nor a composite number.

Recall that factors are numbers that are multiplied together. Prime factors are prime numbers. To check to see if a number is prime or composite, simply divide the smaller primes (such as 2, 3, 5, 7, 11, . . .) into the given number. If the number can be divided exactly without a remainder by one of the smaller primes, it is a composite and not a prime.

Some students find the following rules helpful when deciding if a number can be divided by 2, 3, or 5.

Divisibility Tests

1. A number is divisible by 2 if the last digit is 0, 2, 4, 6, or 8.
2. A number is divisible by 3 if the sum of the digits is divisible by 3.
3. A number is divisible by 5 if the last digit is 0 or 5.

TEACHING TIP The text here lists only three commonly used tests of divisibility. Ask students if they know any other similar tests for divisibility by some number other than 2, 3, or 5. If they cannot think of any, show them one or more of the following:
*A number is divisible by 4 if the number formed by its last two digits is divisible by 4. For example 456,716 is divisible by 4.
*A number is divisible by 6 if it is divisible by both 2 and 3.
*A number is divisible by 8 if the number formed by its last three digits is divisible by 8. For example 26,963,984 is divisible by 8 since 984 is divisible by 8.
*A number is divisible by 9 if the sum of its digits is divisible by 9. Thus 1,428,714 is divisible by 9 since $1 + 4 + 2 + 8 + 7 + 1 + 4$ is 27, which is divisible by 9.

To illustrate:

1. 478 is divisible by 2 since it ends in 8.
2. 531 is divisible by 3 since when we add the digits of 531 $(5 + 3 + 1)$, we get 9, which is divisible by 3.
3. 985 is divisible by 5 since it ends in 5.

EXAMPLE 1 Write each whole number as the product of prime factors.

(a) 12 **(b)** 60 **(c)** 168

(a) To start, write 12 as the product of any two factors. We will write 12 as 4×3.

$$12 = 4 \times 3$$ *Now check whether the factors are prime. If not, factor these.*

$$2 \times 2 \times 3$$

$$12 = 2 \times 2 \times 3$$ *Now all factors are prime, so 12 is completely factored.*

Instead of writing $2 \times 2 \times 3$, we can write $2^2 \times 3$.

Note: To start, we could write 12 as 2×6. Begin this way and follow the steps above. Is the product of prime factors the same? Will this always be true?

(b) We follow the same steps as in (a).

$$60 = 6 \times 10$$

$$3 \times 2 \times 2 \times 5$$ *Check that all factors are prime.*

$$60 = 2 \times 2 \times 3 \times 5$$

Instead of writing $2 \times 2 \times 3 \times 5$, we can write $2^2 \times 3 \times 5$.

Note that in the final answer the prime factors are listed in order from least to greatest.

(c) $168 = 4 \times 42$

$$= 2 \times 2 \times 2 \times 21$$

$$= 2 \times 2 \times 2 \times 3 \times 7$$

$$168 = 2 \times 2 \times 2 \times 3 \times 7$$ *Check that all factors are prime.*

Instead of writing $2 \times 2 \times 2 \times 3 \times 7$, we can write $2^3 \times 3 \times 7$.

Practice Problem 1 Write each whole number as a product of primes.

(a) 18 **(b)** 72 **(c)** 400 ■

Suppose we started Example 1(c) by writing $168 = 14 \times 12$. Would we get the same answer? Would our answer be correct? Let's compare.

$$168 = 4 \times 42$$

$$= 2 \times 2 \times 2 \times 21$$

$$168 = 2 \times 2 \times 2 \times 3 \times 7$$

or $168 = 2^3 \times 3 \times 7$

$$168 = 14 \times 12$$

$$= 2 \times 7 \times 4 \times 3$$

$$= 2 \times 7 \times 2 \times 2 \times 3$$

$$168 = 2 \times 2 \times 2 \times 3 \times 7$$

or $168 = 2^3 \times 3 \times 7$

The result is the same.

The order of prime factors is not important because multiplication is commutative. No matter how we start, when we factor a composite number, we always get exactly the same prime factors.

The Fundamental Theorem of Arithmetic

Every composite number can be written in exactly one way as a product of prime numbers.

You will be able to check this theorem out for yourself. See Exercises 2.2, problems 7 through 26. Writing a number as a product of prime factors is also called **prime factorization.**

2 *Reducing Fractions*

You know that $5 + 2$ and $3 + 4$ are two ways to write the same number. We say they are *equivalent* because they are *equal* to the same *value.* They are both ways of writing the value 7.

Like whole numbers, fractions can be written in more than one way. For example, $\frac{2}{4}$ and $\frac{1}{2}$ are two ways to write the same number. The value of the fractions is the same. When we use fractions, we often need to write them in another form. If we make the numerator and denominator smaller, we *simplify* the fractions.

Compare the two fractions in the drawings on the right. In each picture the shaded part is the same size. The fractions $\frac{3}{4}$ and $\frac{6}{8}$ are called **equivalent fractions.** The fraction $\frac{3}{4}$ is in simplest form. To see how we can change $\frac{6}{8}$ to $\frac{3}{4}$, we look at a property of the number 1.

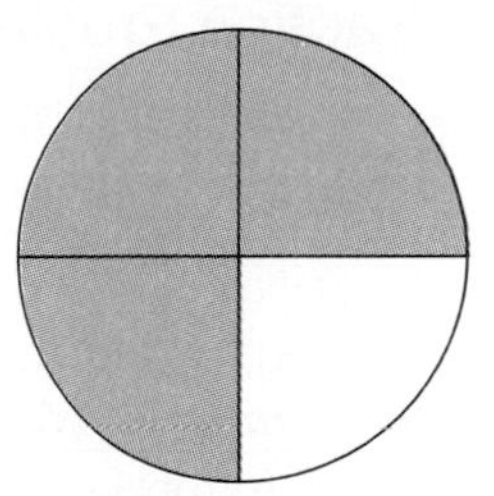

$\frac{3}{4}$ of the circle is shaded.

Any nonzero number divided by itself is 1.

$$\frac{5}{5} = \frac{17}{17} = \frac{c}{c} = 1.$$

Thus, if we multiply a fraction by $\frac{5}{5}$ or $\frac{17}{17}$ or $\frac{c}{c}$ (remember c cannot be zero), the value of the fraction is unchanged because we are multiplying by a form of 1. We can use this rule to show that $\frac{3}{4}$ and $\frac{6}{8}$ are equivalent.

$$\frac{3}{4} \times \frac{2}{2} = \frac{6}{8}$$

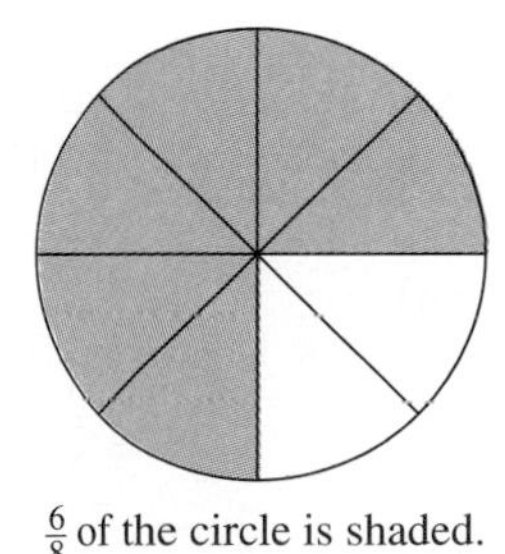

$\frac{6}{8}$ of the circle is shaded.

In general, if b and c are not zero

$$\frac{a}{b} = \frac{a \times c}{b \times c}$$

To reduce a fraction, we find a common factor in the numerator and in the denominator and divide it out. In the fraction $\frac{6}{8}$, the common factor is 2.

$$\frac{6}{8} = \frac{3 \times 2}{4 \times 2} = \frac{3}{4}$$

$$\frac{6}{8} = \frac{3}{4}$$

For all fractions (if a, b, and c are not zero), if c is a common factor,

$$\frac{a}{b} = \frac{a \div c}{b \div c}$$

A fraction is called **simplified** or **in lowest terms** if the numerator and the denominator have no common factor other than 1.

EXAMPLE 2 Simplify (write in lowest terms). **(a)** $\frac{15}{25}$ **(b)** $\frac{42}{56}$

(a) $\frac{15}{25} = \frac{15 \div 5}{25 \div 5} = \frac{3}{5}$ *The common factor is 5. Divide the numerator and the denominator by 5.*

(b) $\frac{42}{56} = \frac{42 \div 14}{56 \div 14} = \frac{3}{4}$ *The common factor is 14. Divide the numerator and the denominator by 14.*

Perhaps 14 was not the first common factor you thought of. Perhaps you did see the common factor 2. Divide out 2. Then look for another common factor, 7. Now divide out 7.

$$\frac{42}{56} = \frac{42 \div 2}{56 \div 2} = \frac{21}{28} = \frac{21 \div 7}{28 \div 7} = \frac{3}{4}$$

If we do not see large factors at first, sometimes simplifying a fraction by dividing both numerator and denominator by a smaller common factor can be done several times, until no common factors are left.

Practice Problem 2 Reduce (or simplify) by dividing out common factors.

(a) $\frac{30}{42}$ **(b)** $\frac{60}{132}$ ■

A second method to reduce or simplify fractions is called the *method of prime factors.* We factor the numerator and the denominator into prime numbers. We then divide the numerator and the denominator by any common prime factors.

EXAMPLE 3 Simplify the fractions by the method of prime factors.

(a) $\frac{35}{42}$ **(b)** $\frac{70}{110}$

(a) $\frac{35}{42} = \frac{5 \times 7}{2 \times 3 \times 7}$ *We factor 35 and 42 into prime factors. The common prime factor is 7.*

$= \frac{5 \times \overset{1}{\cancel{7}}}{2 \times 3 \times \underset{1}{\cancel{7}}}$ *Now we divide out 7.*

$= \frac{5 \times 1}{2 \times 3 \times 1} = \frac{5}{6}$ *We multiply the factors in the numerator and denominator to write the reduced or simplified form.*

Thus $\frac{35}{42} = \frac{5}{6}$, and $\frac{5}{6}$ is the simplified form.

(b) $\frac{70}{110} = \frac{2 \times 5 \times 7}{2 \times 5 \times 11} = \frac{\overset{1}{\cancel{2}} \times \overset{1}{\cancel{5}} \times 7}{\underset{1}{\cancel{2}} \times \underset{1}{\cancel{5}} \times 11} = \frac{7}{11}$

Practice Problem 3 Simplify the fractions by the method of prime factors.

(a) $\frac{120}{135}$ **(b)** $\frac{715}{880}$ ■

3 *Determining If Two Fractions Are Equal*

After we simplify, how can we check that a reduced fraction is *equivalent* to the original fraction? If two fractions are equal, their diagonal products are equal. This is called the *equality test for fractions.* If $\frac{3}{4} = \frac{6}{8}$, then

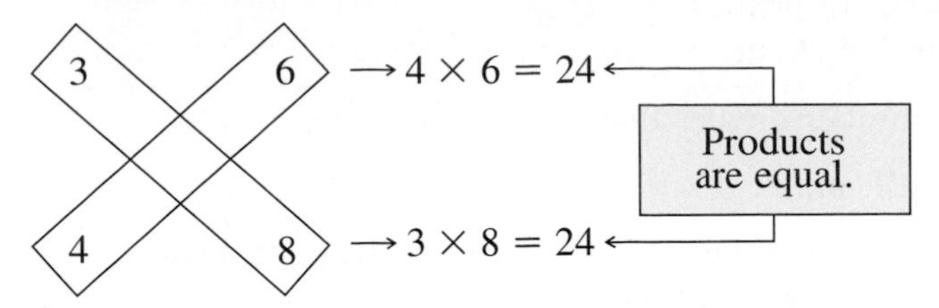

If two fractions are unequal (we use the symbol $\neq$), their diagonal products are unequal. If $\frac{5}{6} \neq \frac{6}{7}$, then

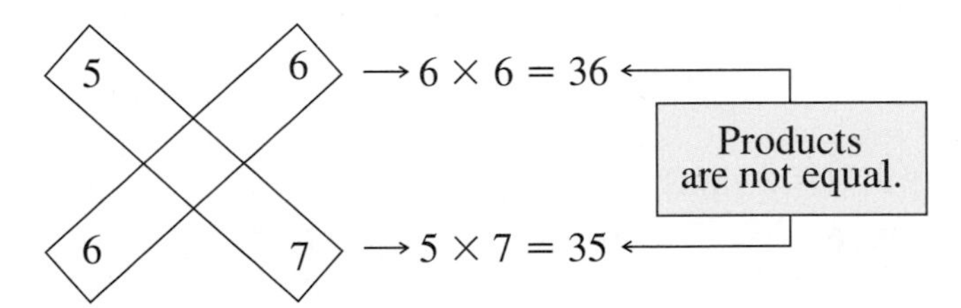

Since $36 \neq 35$, we know that $\frac{5}{6} \neq \frac{6}{7}$. The test can be described in this way.

> **Equality Test for Fractions**
>
> For any two fractions where a, b, c, and d are whole numbers and $b \neq 0$, $d \neq 0$, if $\frac{a}{b} = \frac{c}{d}$, then $a \times d = b \times c$.

EXAMPLE 4 Are these fractions equal? Use the equality test.

(a) $\frac{2}{11} \stackrel{?}{=} \frac{18}{99}$ **(b)** $\frac{3}{16} \stackrel{?}{=} \frac{12}{62}$

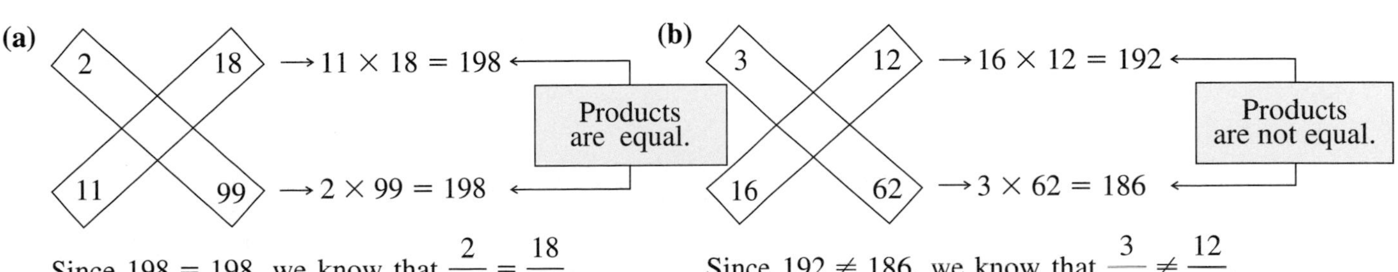

(a) Since $198 = 198$, we know that $\frac{2}{11} = \frac{18}{99}$.

(b) Since $192 \neq 186$, we know that $\frac{3}{16} \neq \frac{12}{62}$.

Practice Problem 4 Test whether the following fractions are equal.

(a) $\frac{84}{108} \stackrel{?}{=} \frac{7}{9}$ **(b)** $\frac{3}{7} \stackrel{?}{=} \frac{79}{182}$ ■

2.2 Exercises

Verbal and Writing Skills

1. Which of these whole numbers are prime?
4, 12, 11, 15, 6, 19, 1, 41, 38, 24, 5, 46   11, 19, 41, 5

2. A prime number is a whole number greater than 1 that cannot be evenly ___divided___ except by itself and 1.

3. A ___composite___ ___number___ is a whole number greater than 1 that can be divided by whole numbers other than itself and 1.

4. Every composite number can be written in exactly one way as a ___product___ of ___prime___ numbers.

5. Give an example of a composite number written as a product of primes. Answers may vary.

6. Give an example of equivalent (equal) fractions. Answers may vary.

Write each number as a product of prime factors.

7. 15 3×5
8. 9 3×3
9. 6 2×3
10. 8 2^3
11. 49 7^2
12. 25 5^2
13. 64 2^6
14. 81 3^4
15. 55 5×11
16. 42 $2 \times 3 \times 7$
17. 35 5×7
18. 36 $2^2 \times 3^2$
19. 75 3×5^2
20. 125 5^3
21. 54 2×3^3
22. 90 $2 \times 3^2 \times 5$
23. 84 $2^2 \times 3 \times 7$
24. 56 $2^3 \times 7$
25. 98 2×7^2
26. 65 5×13

Identify which of these whole numbers is prime. If a number is composite, write it as the product of prime factors.

27. 31 Prime
28. 47 Prime
29. 57 3×19
30. 51 3×17
31. 67 Prime
32. 71 Prime
33. 62 2×31
34. 91 7×13
35. 89 Prime
36. 97 Prime
37. 127 Prime
38. 137 Prime
39. 121 11×11
40. 49 7×7
41. 161 7×23
42. 169 13×13

Reduce each fraction by finding a common factor in the numerator and in the denominator and dividing by the common factor.

43. $\frac{18}{27}$ $\frac{18 \div 9}{27 \div 9} = \frac{2}{3}$

44. $\frac{16}{24}$ $\frac{16 \div 8}{24 \div 8} = \frac{2}{3}$

45. $\frac{32}{48}$ $\frac{32 \div 16}{48 \div 16} = \frac{2}{3}$

46. $\frac{30}{42}$ $\frac{30 \div 6}{42 \div 6} = \frac{5}{7}$

47. $\frac{30}{48}$ $\frac{30 \div 6}{48 \div 6} = \frac{5}{8}$

48. $\frac{48}{64}$ $\frac{48 \div 16}{64 \div 16} = \frac{3}{4}$

49. $\frac{42}{48}$ $\frac{42 \div 6}{48 \div 6} = \frac{7}{8}$

50. $\frac{35}{80}$ $\frac{35 \div 5}{80 \div 5} = \frac{7}{16}$

Reduce each fraction by the method of prime factors.

51. $\dfrac{3}{15}$ $\quad \dfrac{3 \times 1}{3 \times 5} = \dfrac{1}{5}$

52. $\dfrac{7}{21}$ $\quad \dfrac{7 \times 1}{7 \times 3} = \dfrac{1}{3}$

53. $\dfrac{39}{52}$ $\quad \dfrac{3 \times 13}{2 \times 2 \times 13} = \dfrac{3}{4}$

54. $\dfrac{63}{84}$ $\quad \dfrac{3 \times 3 \times 7}{2 \times 2 \times 3 \times 7} = \dfrac{3}{4}$

55. $\dfrac{18}{24}$ $\quad \dfrac{2 \times 3 \times 3}{2 \times 2 \times 2 \times 3} = \dfrac{3}{4}$

56. $\dfrac{65}{91}$ $\quad \dfrac{5 \times 13}{7 \times 13} = \dfrac{5}{7}$

57. $\dfrac{27}{45}$ $\quad \dfrac{3 \times 3 \times 3}{3 \times 3 \times 5} = \dfrac{3}{5}$

58. $\dfrac{28}{42}$ $\quad \dfrac{2 \times 2 \times 7}{2 \times 3 \times 7} = \dfrac{2}{3}$

Mixed Practice

Reduce each fraction by any method.

59. $\dfrac{33}{36}$ $\quad \dfrac{3 \times 11}{3 \times 12} = \dfrac{11}{12}$

60. $\dfrac{40}{96}$ $\quad \dfrac{8 \times 5}{8 \times 12} = \dfrac{5}{12}$

61. $\dfrac{65}{169}$ $\quad \dfrac{5 \times 13}{13 \times 13} = \dfrac{5}{13}$

62. $\dfrac{21}{98}$ $\quad \dfrac{3 \times 7}{14 \times 7} = \dfrac{3}{14}$

63. $\dfrac{88}{121}$ $\quad \dfrac{11 \times 8}{11 \times 11} = \dfrac{8}{11}$

64. $\dfrac{165}{180}$ $\quad \dfrac{15 \times 11}{15 \times 12} = \dfrac{11}{12}$

65. $\dfrac{150}{200}$ $\quad \dfrac{3 \times 50}{4 \times 50} = \dfrac{3}{4}$

66. $\dfrac{175}{300}$ $\quad \dfrac{7 \times 25}{12 \times 25} = \dfrac{7}{12}$

67. $\dfrac{119}{210}$ $\quad \dfrac{7 \times 17}{7 \times 30} = \dfrac{17}{30}$

68. $\dfrac{99}{189}$ $\quad \dfrac{9 \times 11}{9 \times 21} = \dfrac{11}{21}$

Are these fractions equal? Why?

69. $\dfrac{3}{17} \stackrel{?}{=} \dfrac{15}{85}$
$3 \times 85 \stackrel{?}{=} 17 \times 15$
$255 = 255$
Yes

70. $\dfrac{12}{13} \stackrel{?}{=} \dfrac{36}{39}$
$12 \times 39 \stackrel{?}{=} 13 \times 36$
$468 = 468$
Yes

71. $\dfrac{12}{40} \stackrel{?}{=} \dfrac{9}{30}$
$12 \times 30 \stackrel{?}{=} 40 \times 9$
$360 = 360$
Yes

72. $\dfrac{24}{72} \stackrel{?}{=} \dfrac{15}{45}$
$24 \times 45 \stackrel{?}{=} 72 \times 15$
$1080 = 1080$
Yes

73. $\dfrac{23}{27} \stackrel{?}{=} \dfrac{92}{107}$
$23 \times 107 \stackrel{?}{=} 27 \times 92$
$2461 \neq 2484$
No

74. $\dfrac{70}{120} \stackrel{?}{=} \dfrac{41}{73}$
$70 \times 73 \stackrel{?}{=} 120 \times 41$
$5110 \neq 4920$
No

75. $\dfrac{35}{55} \stackrel{?}{=} \dfrac{69}{110}$
$55 \times 69 \stackrel{?}{=} 35 \times 110$
$3795 \neq 3850$
No

76. $\dfrac{18}{24} \stackrel{?}{=} \dfrac{23}{28}$
$18 \times 28 \stackrel{?}{=} 24 \times 23$
$504 \neq 552$
No

77. $\dfrac{65}{70} \stackrel{?}{=} \dfrac{13}{14}$
$65 \times 14 \stackrel{?}{=} 70 \times 13$
$910 = 910$
Yes

78. $\dfrac{98}{182} \stackrel{?}{=} \dfrac{7}{13}$
$98 \times 13 \stackrel{?}{=} 182 \times 7$
$1274 = 1274$
Yes

Applications

Reduce the fractions in your answer.

79. Every few weeks, Harold and Carolyn Bossél clean out their pantry to make sure that none of the food has gone bad. This week, they found that 4 out of the 6 boxes of cereal on the shelves were stale. What fractional part of the boxes of cereal were stale?

$\frac{2}{3}$

80. Medical students frequently work long hours. James worked a 14-hour shift, spending 10 hours in the emergency room and 4 hours in surgery. What fractional part of his shift was he in the emergency room? What fractional part of his shift was he in surgery?

$\frac{5}{7}$ in the emergency room, $\frac{2}{7}$ in surgery

81. At the local movie theatre, 65 out of 95 movies lasted only one week. What fractional part of the movies were more successful (lasted more than one week?)

$\frac{6}{19}$ lasted more than one week.

82. William works for a wireless communications company that makes beepers and mobile phones. He inspected 315 beepers and found that 20 were defective. What fractional part of the beepers were not defective?

$\frac{59}{63}$ were not defective.

83. Rosie bought a used convertible for $16,000. She then paid $2000 to make it look and run like new. What fractional part of her total investment in the car went to make it look and run like new?

$\frac{1}{9}$

84. Monique's sister and her husband have been working two jobs each to put a down payment on a plot of land where they plan to build their house. The purchase price is $42,500. They have saved $5500. What fractional part of the cost of the land have they saved?

$\frac{11}{85}$

Cumulative Review Problems

85. Multiply. 386 × 425

$$\begin{array}{r} 386 \\ \times\ 425 \\ \hline 1\,930 \\ 7\,72 \\ 154\,4 \\ \hline 164{,}050 \end{array}$$

86. Divide. 15,552 ÷ 12

$$\begin{array}{r} 1296 \\ 12\overline{)15552} \\ \underline{12} \\ 35 \\ \underline{24} \\ 115 \\ \underline{108} \\ 72 \\ \underline{72} \\ 0 \end{array}$$

***Estimate** the answer to the following.*

87. 176,328 + 43,926 + 786,035 + 92,196 + 543,076

1,630,000

88. 58 × 62 × 384 × 926 × 18

25,920,000,000

2.3 Improper Fractions and Mixed Numbers

After studying this section, you will be able to:

1 *Change a mixed number to an improper fraction.*
2 *Change an improper fraction to a mixed number.*
3 *Reduce a mixed number or an improper fraction.*

MathPro

Video 4

SSM

1 *Changing a Mixed Number to an Improper Fraction*

Let's look at the value of a fraction. The value of any fraction may be

- Less than 1 or
- Equal to 1 or
- Greater than 1

We have names for different kinds of fractions. If the value of a fraction is less than 1, we call the fraction proper.

$$\frac{3}{5}, \frac{5}{7}, \frac{1}{8} \quad \textit{are called } \textbf{\textit{proper fractions.}}$$

Notice that the numerator is less than the denominator. If the numerator is less than the denominator, the fraction is a proper fraction.

If the value of a fraction is greater than 1, the quantity can be written as an improper fraction or as a mixed number.

Suppose that we have $1\frac{1}{6}$ of a pizza. We have 1 whole pizza and $\frac{1}{6}$ of a pizza. We could write this as $1\frac{1}{6}$. $1\frac{1}{6}$ is called a mixed number. A **mixed number** is the sum of a whole number greater than zero and a proper fraction. The notation $1\frac{1}{6}$ actually means $1 + \frac{1}{6}$. The plus sign is not usually shown.

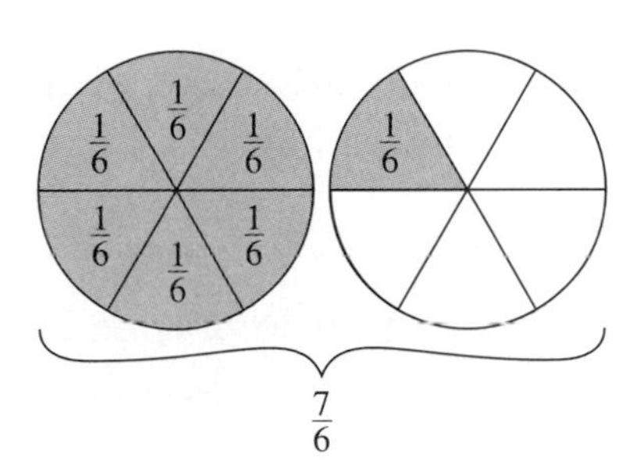

Another way of writing $1\frac{1}{6}$ pizza is to write $\frac{7}{6}$ pizza. $\frac{7}{6}$ is called an improper fraction. Notice the numerator is greater than the denominator. If the numerator is greater than or equal to the denominator, the fraction is an improper fraction.

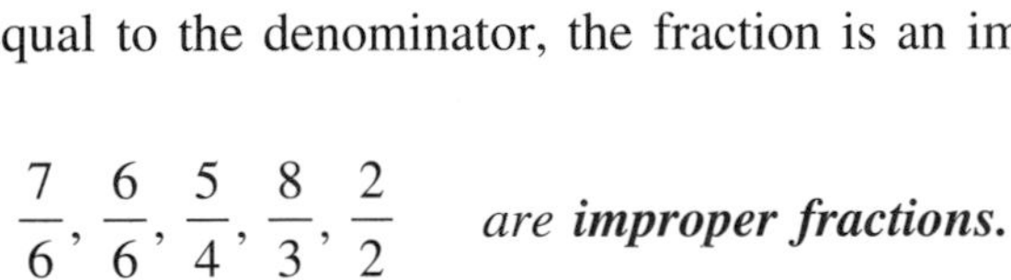

$$\frac{7}{6}, \frac{6}{6}, \frac{5}{4}, \frac{8}{3}, \frac{2}{2} \quad \textit{are } \textbf{\textit{improper fractions.}}$$

The following chart will help you to visualize these different fractions and their names.

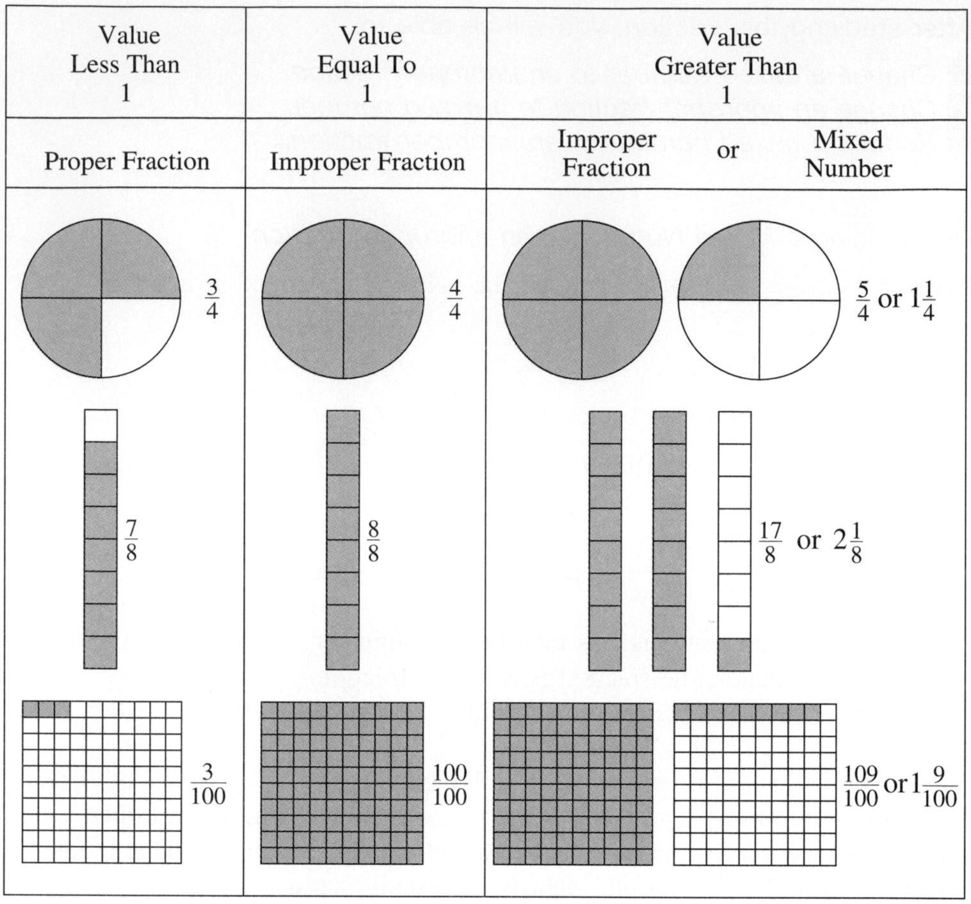

Value Less Than 1	Value Equal To 1	Value Greater Than 1
Proper Fraction	Improper Fraction	Improper Fraction or Mixed Number
$\frac{3}{4}$	$\frac{4}{4}$	$\frac{5}{4}$ or $1\frac{1}{4}$
$\frac{7}{8}$	$\frac{8}{8}$	$\frac{17}{8}$ or $2\frac{1}{8}$
$\frac{3}{100}$	$\frac{100}{100}$	$\frac{109}{100}$ or $1\frac{9}{100}$

Because improper fractions are easier to add, subtract, multiply, and divide than mixed numbers, we change mixed numbers to improper fractions.

TEACHING TIP Some students have been "drilled" in elementary school that improper fractions are "wrong" since they are not simplified. You may need to explain that there is nothing "improper" or incorrect about improper fractions. Some mathematical problems are best left with improper fractions as answers. In mathematics we will find several places where improper fractions are needed and are appropriate.

Changing a Mixed Number to an Improper Fraction

1. Multiply the whole number by the denominator of the fraction.
2. Add the numerator of the fraction to the product found in step 1.
3. Write the sum found in step 2 over the denominator of the fraction.

EXAMPLE 1 Change the mixed numbers to improper fractions.

(a) $3\frac{2}{5}$ **(b)** $5\frac{4}{9}$ **(c)** $18\frac{3}{5}$

Multiply the whole number by the denominator.

Add the numerator to the product.

(a) $3\frac{2}{5} \quad \frac{3 \times 5 + 2}{5} = \frac{15 + 2}{5} = \frac{17}{5} \longleftarrow$ Write the sum over the denominator.

(b) $5\frac{4}{9} = \frac{9 \times 5 + 4}{9} = \frac{45 + 4}{9} = \frac{49}{9}$

(c) $18\frac{3}{5} = \frac{5 \times 18 + 3}{5} = \frac{90 + 3}{5} = \frac{93}{5}$

Practice Problem 1 Change the mixed numbers to improper fractions.

(a) $4\frac{3}{7}$ **(b)** $6\frac{2}{3}$ **(c)** $19\frac{4}{7}$ ■

2 *Changing an Improper Fraction to a Mixed Number*

We often need to change an improper fraction to a mixed number.

> **Changing an Improper Fraction to a Mixed Number**
>
> 1. Divide the numerator by the denominator.
> 2. Write the quotient followed by the fraction with the remainder over the denominator.
>
> $$\text{quotient}\frac{\text{remainder}}{\text{denominator}}$$

EXAMPLE 2 Write as a mixed number.

(a) $\frac{13}{5}$ **(b)** $\frac{29}{7}$ **(c)** $\frac{105}{31}$ **(d)** $\frac{85}{17}$

(a) We divide the denominator 5 into 13.

$$\begin{array}{r} 2 \leftarrow \text{quotient} \\ 5\overline{)13} \\ \underline{10} \\ 3 \leftarrow \text{remainder} \end{array}$$

The answer is in the form $\text{quotient}\frac{\text{remainder}}{\text{denominator}}$

$$\frac{13}{5} \qquad \begin{array}{r} 2 \leftarrow \text{quotient} \\ 5\overline{)13} \\ \underline{10} \\ 3 \leftarrow \text{remainder} \end{array} \qquad 2\frac{3}{5}$$

quotient → 2; remainder → 3; denominator → 5

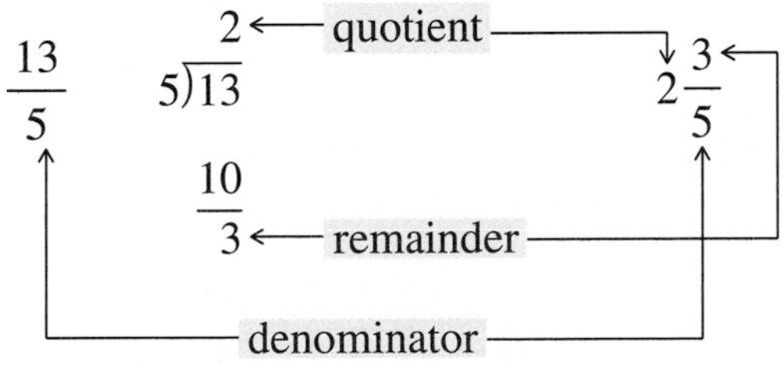

Thus $\frac{13}{5} = 2\frac{3}{5}$.

(b) $\begin{array}{r} 4 \\ 7\overline{)29} \\ \underline{28} \\ 1 \end{array}$ $\qquad \frac{29}{7} = 4\frac{1}{7}$

(c) $31\overline{)105}$ with quotient 3; 93; remainder 12 $\qquad \dfrac{105}{31} = 3\dfrac{12}{31}$

(d) $17\overline{)85}$ with quotient 5; 85; remainder 0 $\qquad$ The remainder is 0, so $\dfrac{85}{17} = 5$, a whole number.

Practice Problem 2 Write as a mixed number or a whole number.

(a) $\dfrac{17}{4}$ **(b)** $\dfrac{36}{5}$ **(c)** $\dfrac{116}{27}$ **(d)** $\dfrac{91}{13}$ ■

3 *Reducing a Mixed Number or an Improper Fraction*

Mixed numbers and improper fractions may need to be reduced if they are not in simplest form. Recall that we look for common factors in the numerator and the denominator of the fraction. Then we divide the numerator and the denominator by the common factor.

EXAMPLE 3 Reduce the improper fraction. $\dfrac{22}{8}$

$$\frac{22}{8} = \frac{\overset{1}{\cancel{2}} \times 11}{\underset{1}{\cancel{2}} \times 2 \times 2} = \frac{11}{4}$$

Practice Problem 3 Reduce the improper fraction. $\dfrac{51}{15}$ ■

EXAMPLE 4 Reduce the mixed number. $4\dfrac{21}{28}$

We do not need to reduce the whole number 4, only the fraction $\dfrac{21}{28}$.

$$\frac{21}{28} = \frac{3 \times \overset{1}{\cancel{7}}}{4 \times \underset{1}{\cancel{7}}} = \frac{3}{4}$$

Therefore, $4\dfrac{21}{28} = 4\dfrac{3}{4}$.

Practice Problem 4 Reduce the mixed number. $3\dfrac{16}{80}$ ■

If an improper fraction contains a very large numerator and denominator, it is best to change the fraction to a mixed number before reducing.

EXAMPLE 5 Reduce $\frac{945}{567}$ by first changing to a mixed number.

$$567\overline{)945} = 1 \text{ R } 378 \qquad \text{so } \frac{945}{567} = 1\frac{378}{567}$$

$$\begin{array}{r} 1 \\ 567\overline{)945} \\ \underline{567} \\ 378 \end{array}$$

To reduce the fraction we write

$$\frac{378}{567} = \frac{3 \times 3 \times 3 \times 2 \times 7}{3 \times 3 \times 3 \times 3 \times 7} = \frac{\cancel{3}^1 \times \cancel{3}^1 \times \cancel{3}^1 \times 2 \times \cancel{7}^1}{\cancel{3}_1 \times \cancel{3}_1 \times \cancel{3}_1 \times 3 \times \cancel{7}_1} = \frac{2}{3}$$

So $\frac{945}{567} = 1\frac{378}{567} = 1\frac{2}{3}$.

Problems like Example 5 can be done in several different ways. It is not necessary to follow these exact steps when reducing this fraction.

TEACHING TIP After explaining Example 5 or some similar problem, have the group do the following problem as a class activity. Have half the class reduce the fraction $\frac{663}{255}$ using this method of changing to a mixed number first. Have the remainder of the class reduce the fraction as an improper fraction. The answer is $2\frac{3}{5}$. Record which half of the class scored the greatest number of correct answers first. (As long as you divide the class into two groups of roughly equal ability, the improper fraction method will usually win.)

Practice Problem 5 Reduce $\frac{1001}{572}$ by first changing to a mixed number. ■

To Think About A student concluded that just by looking at the denominator he could tell that the fraction $\frac{1655}{97}$ cannot be reduced unless $1655 \div 97$ is a whole number. How did he come to that conclusion? Note that 97 is a prime number. The only factors of 97 are 97 and 1. Therefore, *any* fraction with 97 in the denominator can be reduced only if 97 is a factor of the numerator. Since $1655 \div 97$ is not a whole number (see the division below), it is therefore impossible to reduce $\frac{1655}{97}$.

$$\begin{array}{r} 17 \\ 97\overline{)1655} \\ \underline{97} \\ 685 \\ \underline{679} \\ 6 \end{array}$$

You may explore this idea in Exercises 2.3, problems 81 and 82.

2.3 Exercises

Change each mixed number to an improper fraction.

1. $3\frac{1}{2}$ $\frac{7}{2}$ **2.** $5\frac{1}{4}$ $\frac{21}{4}$ **3.** $4\frac{2}{3}$ $\frac{14}{3}$ **4.** $3\frac{5}{6}$ $\frac{23}{6}$ **5.** $2\frac{3}{7}$ $\frac{17}{7}$ **6.** $3\frac{3}{8}$ $\frac{27}{8}$

7. $5\frac{3}{10}$ $\frac{53}{10}$ **8.** $4\frac{7}{10}$ $\frac{47}{10}$ **9.** $4\frac{5}{6}$ $\frac{29}{6}$ **10.** $9\frac{2}{3}$ $\frac{29}{3}$ **11.** $21\frac{2}{3}$ $\frac{65}{3}$ **12.** $13\frac{1}{3}$ $\frac{40}{3}$

13. $9\frac{1}{6}$ $\frac{55}{6}$ **14.** $56\frac{1}{2}$ $\frac{113}{2}$ **15.** $28\frac{1}{6}$ $\frac{169}{6}$ **16.** $6\frac{6}{7}$ $\frac{48}{7}$ **17.** $10\frac{11}{12}$ $\frac{131}{12}$ **18.** $13\frac{5}{7}$ $\frac{96}{7}$

19. $7\frac{9}{10}$ $\frac{79}{10}$ **20.** $4\frac{1}{50}$ $\frac{201}{50}$ **21.** $8\frac{1}{25}$ $\frac{201}{25}$ **22.** $66\frac{2}{3}$ $\frac{200}{3}$ **23.** $105\frac{1}{2}$ $\frac{211}{2}$ **24.** $207\frac{2}{3}$ $\frac{623}{3}$

25. $164\frac{2}{3}$ $\frac{494}{3}$ **26.** $33\frac{1}{3}$ $\frac{100}{3}$ **27.** $7\frac{14}{15}$ $\frac{119}{15}$ **28.** $4\frac{26}{27}$ $\frac{134}{27}$ **29.** $5\frac{13}{25}$ $\frac{138}{25}$ **30.** $6\frac{18}{19}$ $\frac{132}{19}$

Change each improper fraction to a mixed number or a whole number.

31. $\frac{7}{5}$ $1\frac{2}{5}$ **32.** $\frac{10}{3}$ $3\frac{1}{3}$ **33.** $\frac{11}{4}$ $2\frac{3}{4}$ **34.** $\frac{9}{5}$ $1\frac{4}{5}$ **35.** $\frac{15}{6}$ $2\frac{1}{2}$ **36.** $\frac{23}{6}$ $3\frac{5}{6}$ **37.** $\frac{27}{8}$ $3\frac{3}{8}$

38. $\frac{48}{16}$ 3 **39.** $\frac{65}{13}$ 5 **40.** $\frac{56}{15}$ $3\frac{11}{15}$ **41.** $\frac{86}{9}$ $9\frac{5}{9}$ **42.** $\frac{47}{2}$ $23\frac{1}{2}$ **43.** $\frac{28}{13}$ $2\frac{2}{13}$ **44.** $\frac{54}{17}$ $3\frac{3}{17}$

45. $\frac{51}{16}$ $3\frac{3}{16}$ **46.** $\frac{19}{3}$ $6\frac{1}{3}$ **47.** $\frac{28}{3}$ $9\frac{1}{3}$ **48.** $\frac{100}{3}$ $33\frac{1}{3}$ **49.** $\frac{35}{2}$ $17\frac{1}{2}$ **50.** $\frac{132}{11}$ 12 **51.** $\frac{91}{7}$ 13

52. $\frac{183}{7}$ $26\frac{1}{7}$ **53.** $\frac{200}{9}$ $22\frac{2}{9}$ **54.** $\frac{156}{13}$ 12 **55.** $\frac{102}{17}$ 6 **56.** $\frac{105}{8}$ $13\frac{1}{8}$ **57.** $\frac{403}{11}$ $36\frac{7}{11}$ **58.** $\frac{212}{9}$ $23\frac{5}{9}$

Reduce each mixed number.

59. $2\frac{9}{12}$ $2\frac{3}{4}$ **60.** $2\frac{10}{15}$ $2\frac{2}{3}$ **61.** $4\frac{11}{66}$ $4\frac{1}{6}$ **62.** $3\frac{15}{90}$ $3\frac{1}{6}$ **63.** $12\frac{15}{40}$ $12\frac{3}{8}$ **64.** $15\frac{12}{36}$ $15\frac{1}{3}$

Reduce each improper fraction.

65. $\frac{24}{6}$ 4 **66.** $\frac{32}{6}$ $\frac{16}{3}$ **67.** $\frac{36}{15}$ $\frac{12}{5}$ **68.** $\frac{32}{18}$ $\frac{16}{9}$ **69.** $\frac{78}{9}$ $\frac{26}{3}$ **70.** $\frac{143}{22}$ $\frac{13}{2}$

Change to a mixed number and reduce.

71. $\frac{340}{126}$ $2\frac{88}{126} = 2\frac{44}{63}$ **72.** $\frac{386}{226}$ $1\frac{160}{226} = 1\frac{80}{113}$ **73.** $\frac{986}{424}$ $2\frac{138}{424} = 2\frac{69}{212}$ **74.** $\frac{764}{328}$ $2\frac{108}{328} = 2\frac{27}{82}$ **75.** $\frac{950}{350}$ $2\frac{250}{350} = 2\frac{5}{7}$ **76.** $\frac{1008}{1000}$ $1\frac{8}{1000} = 1\frac{1}{125}$

Applications

77. The Science Museum is hanging banners all over the building to commemorate the Apollo astronauts. The art department is using $360\frac{2}{3}$ yards of starry sky parachute fabric. Change this number to an improper fraction.
$\frac{1082}{3}$ yards

78. For the Northwestern University alumni homecoming, the students studying sculpture have made a giant replica of the school using $244\frac{3}{4}$ pounds of clay. Change this number to an improper fraction.
$\frac{979}{4}$ pounds

79. A Cape Cod cranberry bog was contaminated by waste from an abandoned army base. Damage was done to $\frac{151}{3}$ acres of land. Write this as a mixed number. $50\frac{1}{3}$ acres

80. We are helping to build housing for the homeless one weekend per month. There are plans for a porch which will use $\frac{267}{8}$ feet of lumber. Write this as a mixed number. $33\frac{3}{8}$ feet

To Think About

81. Can $\frac{5687}{101}$ be reduced? Why or why not?
No. 101 is prime and is not a factor of 5687

82. Can $\frac{9810}{157}$ be reduced? Why or why not?
No. 157 is prime and not a factor of 9810

Cumulative Review Problems

83. Add. 16,385 + 4126 + 8056
28,567

84. Subtract. 1,398,210 − 1,137,963
260,247

85. Estimate the answer.

78,964 × 229,350
16,000,000,000

86. Estimate the answer.

872,365 ÷ 286 3000

2.4 Multiplication of Fractions and Mixed Numbers

MathPro

Video 4

SSM

After studying this section, you will be able to:

1. *Multiply two fractions that are proper or improper.*
2. *Multiply a whole number by a fraction.*
3. *Multiply mixed numbers.*

1 *Multiplying Two Fractions That Are Proper or Improper*

Fudge Squares

Ingredients:

2 cups sugar	1/4 teaspoon salt
4 oz chocolate	1 teaspoon vanilla
1/2 cup butter	1 cup all-purpose flour
4 eggs	1 cup nutmeats

Suppose you want to cook an amount equal to half of what the recipe shown calls for. You would multiply the measure given for each ingredient by $\frac{1}{2}$.

$\frac{1}{2}$ of 2 cups sugar	$\frac{1}{2}$ of $\frac{1}{4}$ teaspoon salt
$\frac{1}{2}$ of 4 oz chocolate	$\frac{1}{2}$ of 1 teaspoon vanilla
$\frac{1}{2}$ of $\frac{1}{2}$ cup butter	$\frac{1}{2}$ of 1 cup all-purpose flour
$\frac{1}{2}$ of 4 eggs	$\frac{1}{2}$ of 1 cup nutmeats

We often use multiplication of fractions to describe taking a fractional part of something. To find $\frac{1}{2}$ of $\frac{3}{7}$, we multiply

$$\frac{1}{2} \times \frac{3}{7} = \frac{3}{14}$$

We begin with a bar that is $\frac{3}{7}$ shaded. To find $\frac{1}{2}$ of $\frac{3}{7}$ we divide the bar in half and take $\frac{1}{2}$ of the shaded section. $\frac{1}{2}$ of $\frac{3}{7}$ yields 3 out of 14 squares.

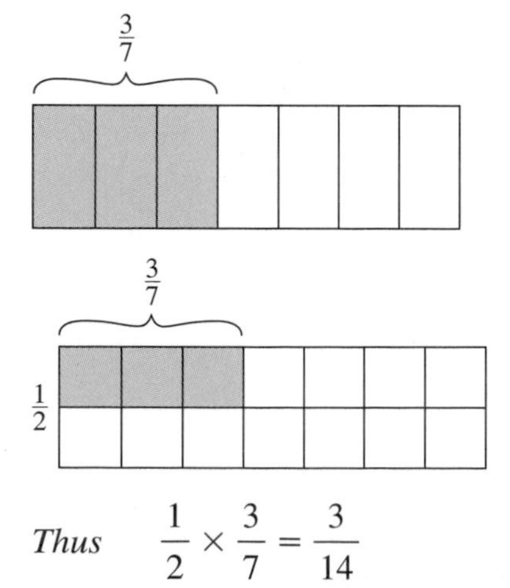

Thus $\frac{1}{2} \times \frac{3}{7} = \frac{3}{14}$

When you multiply two fractions together, you get a smaller fraction.

To multiply two fractions, we multiply the numerators and multiply the denominators.

$$\frac{2}{3} \times \frac{5}{7} = \frac{10}{21} \begin{array}{l} \longleftarrow 2 \times 5 = 10 \\ \longleftarrow 3 \times 7 = 21 \end{array}$$

In general, for all positive whole numbers a, b, c, and d,

$$\frac{a}{b} \times \frac{c}{d} = \frac{a \times c}{b \times d}$$

EXAMPLE 1 Multiply.

(a) $\frac{3}{8} \times \frac{5}{7}$ **(b)** $\frac{1}{11} \times \frac{2}{13}$

(a) $\frac{3}{8} \times \frac{5}{7} = \frac{3 \times 5}{8 \times 7} = \frac{15}{56}$ **(b)** $\frac{1}{11} \times \frac{2}{13} = \frac{1 \times 2}{11 \times 13} = \frac{2}{143}$

Practice Problem 1 Multiply.

(a) $\frac{6}{7} \times \frac{3}{13}$ **(b)** $\frac{1}{5} \times \frac{11}{12}$ ■

Some products may be reduced.

$$\frac{12}{35} \times \frac{25}{18} = \frac{300}{630} = \frac{10}{21}$$

By simplifying before multiplication, the reducing can be done more easily. In a multiplication problem, a factor in the numerator can be paired with a common factor in the denominator of the same or a different fraction. We then divide numerator and denominator by their common factor.

EXAMPLE 2 Simplify first and then multiply.

$$\frac{12}{35} \times \frac{25}{18}$$

$\frac{\overset{2}{\cancel{12}}}{35} \times \frac{25}{\underset{3}{\cancel{18}}}$ *First divide 12 by 6 and 18 by 6.*

$\frac{\overset{2}{\cancel{12}}}{\underset{7}{\cancel{35}}} \times \frac{\overset{5}{\cancel{25}}}{\underset{3}{\cancel{18}}}$ *Next divide 25 by 5 and 35 by 5.*

$\frac{2}{7} \times \frac{5}{3} = \frac{10}{21}$ *Then multiply the simplified fractions.*

The answer is $\frac{10}{21}$.

Practice Problem 2 Simplify first and then multiply.

$$\frac{55}{72} \times \frac{16}{33}$$ ■

2 *Multiplying a Whole Number by a Fraction*

When multiplying a fraction by a whole number, it is more convenient to express the whole number as a fraction with a denominator of 1. We know that $5 = \frac{5}{1}$, $7 = \frac{7}{1}$, and so on.

TEACHING TIP This skill of multiplying a whole number by a fraction is very useful in everyday life. After you have explained Examples 3 and 4, ask the class to find the answer to this problem. "There are 15,041 students at the university. Approximately $\frac{3}{13}$ of the students are over the age of 25. How many students are over age 25?" The correct answer is 3471 students.

EXAMPLE 3 Multiply.

(a) $5 \times \frac{3}{8}$ **(b)** $\frac{22}{7} \times 14$

(a) $5 \times \frac{3}{8} = \frac{5}{1} \times \frac{3}{8} = \frac{15}{8}$ or $1\frac{7}{8}$

(b) $\frac{22}{7} \times 14 = \frac{22}{\cancel{7}_1} \times \frac{\cancel{14}^2}{1} = \frac{44}{1} = 44$

Practice Problem 3 Multiply.

(a) $7 \times \frac{5}{13}$ **(b)** $\frac{13}{16} \times 8$ ■

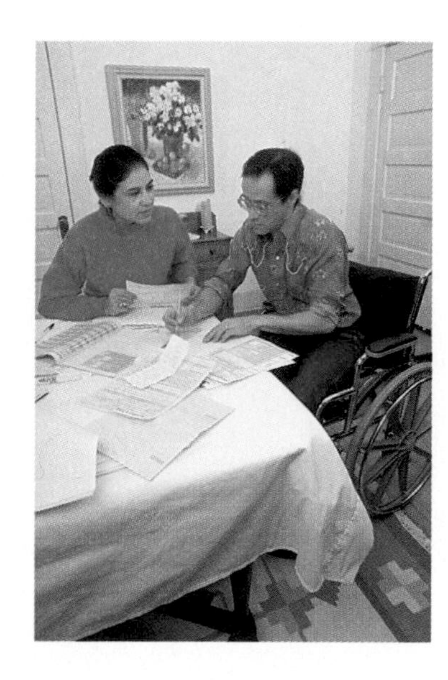

EXAMPLE 4 Mr. and Mrs. Jones found that $\frac{2}{7}$ of their income went to pay federal income taxes. Last year they earned $37,100. How much did they pay in taxes?

We need to find $\frac{2}{7}$ of \$37,100. So we must multiply $\frac{2}{7} \times 37,100$.

$$\frac{2}{\cancel{7}_1} \times \cancel{37,100}^{5300} = \frac{2}{1} \times 5300 = 10,600$$

They paid $10,600 in federal income taxes.

Practice Problem 4 Fred and Linda own 98,400 square feet of land. They found that $\frac{3}{8}$ of the land is in a wetland area and cannot be used for building. How many square feet of land are in the wetland area? ■

3 *Multiplying Mixed Numbers*

To multiply a fraction by a mixed number or to multiply two mixed numbers, first change each mixed number to an improper fraction.

EXAMPLE 5 Multiply.

(a) $\frac{5}{7} \times 3\frac{1}{4}$ **(b)** $20\frac{2}{5} \times 6\frac{2}{3}$ **(c)** $\frac{3}{4} \times 1\frac{1}{2} \times \frac{4}{7}$ **(d)** $4\frac{1}{3} \times 2\frac{1}{4}$

(a) $\frac{5}{7} \times 3\frac{1}{4} = \frac{5}{7} \times \frac{13}{4} = \frac{65}{28}$ or $2\frac{9}{28}$

(b) $20\frac{2}{5} \times 6\frac{2}{3} = \frac{\cancel{102}^{34}}{\cancel{5}_1} \times \frac{\cancel{20}^{4}}{\cancel{3}_1} = \frac{136}{1} = 136$

(c) $\dfrac{3}{4} \times 1\dfrac{1}{2} \times \dfrac{4}{7} = \dfrac{3}{\cancel{4}_1} \times \dfrac{3}{2} \times \dfrac{\cancel{4}^1}{7} = \dfrac{9}{14}$

(d) $4\dfrac{1}{3} \times 2\dfrac{1}{4} = \dfrac{13}{\cancel{3}_1} \times \dfrac{\cancel{9}^3}{4} = \dfrac{39}{4}$ or $9\dfrac{3}{4}$

Practice Problem 5 Multiply.

(a) $2\dfrac{1}{6} \times \dfrac{4}{7}$ **(b)** $10\dfrac{2}{3} \times 13\dfrac{1}{2}$ **(c)** $\dfrac{3}{5} \times 1\dfrac{1}{3} \times \dfrac{5}{8}$ **(d)** $3\dfrac{1}{5} \times 2\dfrac{1}{2}$ ■

EXAMPLE 6 Find the area in square miles of a rectangle with width $1\frac{1}{3}$ miles and length $12\frac{1}{4}$ miles.

Length = $12\frac{1}{4}$ miles

Width = $1\frac{1}{3}$ miles

We find the area of a rectangle by multiplying the width times the length.

$$1\frac{1}{3} \times 12\frac{1}{4} = \frac{\cancel{4}^1}{3} \times \frac{49}{\cancel{4}_1} = \frac{49}{3} \text{ or } 16\frac{1}{3}$$

The area is $16\frac{1}{3}$ square miles.

TEACHING TIP Remind students that the area of both a rectangle and a square can be obtained by multiplying the length times the width or, in the case of the square, by multiplying the length of one side by itself. As a class activity, have them find the area of a square that measures $14\frac{1}{5}$ miles on each side. The answer is $201\frac{16}{25}$ square miles.

Practice Problem 6 Find the area in square meters of a rectangle with width $1\frac{1}{5}$ meters and length $4\frac{5}{6}$ meters. ■

EXAMPLE 7 Find the value of x if

$$\frac{3}{7} \cdot x = \frac{15}{42}$$

The variable x represents a fraction. We know 3 times one number equals 15 and 7 times another equals 42.

Since $3 \cdot 5 = 15$ and $7 \cdot 6 = 42$ we know that $\dfrac{3}{7} \cdot \dfrac{5}{6} = \dfrac{15}{42}$

Therefore, $x = \dfrac{5}{6}$.

Practice Problem 7 Find the value of x if $\dfrac{8}{9} \cdot x = \dfrac{80}{81}$. ■

2.4 Exercises

Multiply.

1. $\frac{3}{5} \times \frac{7}{11}$ $\quad \frac{21}{55}$

2. $\frac{1}{8} \times \frac{5}{11}$ $\quad \frac{5}{88}$

3. $\frac{3}{4} \times \frac{5}{13}$ $\quad \frac{15}{52}$

4. $\frac{4}{7} \times \frac{3}{5}$ $\quad \frac{12}{35}$

5. $\frac{6}{5} \times \frac{10}{12}$ $\quad 1$

6. $\frac{7}{8} \times \frac{16}{21}$ $\quad \frac{\cancel{7}^{1}}{\cancel{8}_{1}} \times \frac{\cancel{16}^{2}}{\cancel{21}_{3}} = \frac{2}{3}$

7. $\frac{15}{7} \times \frac{8}{25}$ $\quad \frac{\cancel{15}^{3}}{7} \times \frac{8}{\cancel{25}_{5}} = \frac{24}{35}$

8. $\frac{7}{9} \times \frac{3}{21}$ $\quad \frac{\cancel{7}^{1}}{\cancel{9}_{3}} \times \frac{\cancel{3}^{1}}{\cancel{21}_{3}} = \frac{1}{9}$

9. $\frac{15}{28} \times \frac{7}{9}$ $\quad \frac{\cancel{15}^{5}}{\cancel{28}_{4}} \times \frac{\cancel{7}^{1}}{\cancel{9}_{3}} = \frac{5}{12}$

10. $\frac{5}{24} \times \frac{18}{15}$ $\quad \frac{\cancel{5}^{1}}{\cancel{24}_{4}} \times \frac{\cancel{18}^{\cancel{3}\,1}}{\cancel{15}_{\cancel{3}\,1}} = \frac{1}{4}$

11. $\frac{9}{10} \times \frac{35}{12}$ $\quad \frac{\cancel{9}^{3}}{\cancel{10}_{2}} \times \frac{\cancel{35}^{7}}{\cancel{12}_{4}} = \frac{21}{8}$ *or* $2\frac{5}{8}$

12. $\frac{12}{17} \times \frac{3}{24}$ $\quad \frac{\cancel{12}^{1}}{17} \times \frac{3}{\cancel{24}_{2}} = \frac{3}{34}$

13. $8 \times \frac{3}{7}$ $\quad \frac{8}{1} \times \frac{3}{7} = \frac{24}{7}$ *or* $3\frac{3}{7}$

14. $\frac{2}{11} \times 4$ $\quad \frac{2}{11} \times \frac{4}{1} = \frac{8}{11}$

15. $\frac{5}{16} \times 8$ $\quad \frac{5}{\cancel{16}_{2}} \times \frac{\cancel{8}^{1}}{1} = \frac{5}{2}$ *or* $2\frac{1}{2}$

16. $5 \times \frac{7}{25}$ $\quad \frac{\cancel{5}^{1}}{1} \times \frac{7}{\cancel{25}_{5}} = \frac{7}{5}$ *or* $1\frac{2}{5}$

17. $\frac{3}{7} \times \frac{2}{5} \times \frac{14}{9}$ $\quad \frac{\cancel{3}^{1}}{\cancel{7}_{1}} \times \frac{2}{5} \times \frac{\cancel{14}^{2}}{\cancel{9}_{3}} = \frac{4}{15}$

18. $\frac{9}{5} \times \frac{3}{13} \times \frac{10}{27}$ $\quad \frac{\cancel{9}^{1}}{\cancel{5}_{1}} \times \frac{\cancel{3}^{1}}{13} \times \frac{\cancel{10}^{2}}{\cancel{27}_{1}} = \frac{2}{13}$

19. $\frac{4}{5} \times \frac{1}{8} \times \frac{35}{7}$ $\quad \frac{\cancel{4}^{1}}{\cancel{5}_{1}} \times \frac{1}{\cancel{8}_{2}} \times \frac{\cancel{35}^{\cancel{7}\,1}}{\cancel{7}_{1}} = \frac{1}{2}$

20. $\frac{10}{13} \times \frac{26}{15} \times \frac{2}{3}$ $\quad \frac{\cancel{10}^{2}}{\cancel{13}_{1}} \times \frac{\cancel{26}^{2}}{\cancel{15}_{3}} \times \frac{2}{3} = \frac{8}{9}$

Multiply. Change any mixed number to an improper fraction before multiplying.

21. $3\frac{1}{5} \times \frac{7}{8}$ $\quad \frac{\cancel{16}^{2}}{5} \times \frac{7}{\cancel{8}_{1}} = \frac{14}{5}$ *or* $2\frac{4}{5}$

22. $\frac{8}{11} \times 4\frac{3}{4}$ $\quad \frac{\cancel{8}^{2}}{11} \times \frac{19}{\cancel{4}_{1}} = \frac{38}{11}$ *or* $3\frac{5}{11}$

23. $1\frac{1}{4} \times 3\frac{2}{3}$ $\quad \frac{5}{4} \times \frac{11}{3} = \frac{55}{12}$ *or* $4\frac{7}{12}$

24. $2\frac{3}{5} \times 1\frac{4}{7}$ $\quad \frac{13}{5} \times \frac{11}{7} = \frac{143}{35}$ *or* $4\frac{3}{35}$

25. $2\frac{1}{2} \times 6$ $\quad \frac{5}{2} \times \frac{6}{1} = 15$

26. $4\frac{1}{3} \times 9$ $\quad \frac{13}{3} \times \frac{9}{1} = 39$

27. $2\frac{3}{10} \times \frac{3}{5}$ $\quad \frac{23}{10} \times \frac{3}{5} = \frac{69}{50}$ *or* $1\frac{19}{50}$

28. $4\frac{3}{5} \times \frac{1}{10}$ $\quad \frac{23}{5} \times \frac{1}{10} = \frac{23}{50}$

29. $1\frac{3}{16} \times 0$ $\quad 0$

30. $3\frac{7}{8} \times 1$ $\quad 3\frac{7}{8}$

31. $4\frac{1}{5} \times 12\frac{2}{9}$ $\quad \frac{21}{5} \times \frac{110}{9} = \frac{154}{3}$ *or* $51\frac{1}{3}$

32. $5\frac{1}{4} \times 10\frac{3}{7}$ $\quad \frac{21}{4} \times \frac{73}{7} = \frac{219}{4}$ *or* $54\frac{3}{4}$

33. $8\frac{5}{6} \times \frac{2}{5}$ $\quad \frac{53}{6} \times \frac{2}{5} = \frac{53}{15}$ *or* $3\frac{8}{15}$

34. $\frac{3}{4} \times 9\frac{5}{7}$ $\quad \frac{3}{4} \times \frac{68}{7} = \frac{51}{7}$ *or* $7\frac{2}{7}$

35. $\frac{5}{5} \times 11\frac{5}{7}$ $\quad 11\frac{5}{7}$

36. $0 \times 6\frac{2}{3}$ $\quad 0$

Solve for x.

37. $\frac{2}{7} \cdot x = \frac{18}{35}$ $\quad x = \frac{9}{5}$

38. $\frac{5}{8} \cdot x = \frac{35}{88}$ $\quad x = \frac{7}{11}$

39. $\frac{11}{12} \cdot x = \frac{121}{156}$ $\quad x = \frac{11}{13}$

40. $\frac{12}{13} \cdot x = \frac{72}{91}$ $\quad x = \frac{6}{7}$

Applications

41. A spy is running from his captors in a forest that is $8\frac{3}{4}$ miles long and $4\frac{1}{3}$ miles wide. Find the area of the forest where he is hiding. (*Hint:* The area of a rectangle is the product of the length times the width.) $37\frac{11}{12}$ square miles

42. An area in the Midwest is a designated tornado danger zone. The land is $22\frac{5}{6}$ miles long and $16\frac{1}{2}$ miles wide. Find the area of the tornado danger zone. (*Hint:* The area of a rectangle is the product of the length times the width.) $376\frac{3}{4}$ square miles

43. A jeep has $11\frac{1}{6}$ gallons of gas. The jeep averages 12 miles per gallon. How far will the jeep be able to go on what is in the tank? 134 miles

44. A Lear Jet airplane has 360 gallons of fuel. The plane averages $4\frac{1}{3}$ miles per gallon. How far can the plane go? 1560 miles

45. We used $18\frac{1}{4}$ yards of wallpaper to decorate our TV room. If we were to wallpaper 8 rooms with the exact same measurements, how many yards of wallpaper would we need? 146 yards

46. A recipe from Nanette's French cookbook for a scalloped potato tart requires $90\frac{1}{2}$ grams of grated cheese. How many grams of cheese would she need if she made one tart for each of her 18 cousins? 1629 grams

47. Value Hardware Store is making a custom color paint for Susan's kitchen walls. The paint requires $12\frac{1}{4}$ ounces of pigment. How many ounces of pigment would be necessary to make $\frac{3}{4}$ of the formula? $9\frac{3}{16}$ ounces

48. The propeller on the Ipswich River Cruise Boat turns 320 revolutions per minute. How fast would it turn at $\frac{3}{4}$ of that speed?
240 revolutions per minute

49. If you wanted to buy a share of Disney stock last year, you would have paid $\$72\frac{1}{4}$. How much money would you have needed to buy 85 shares?
$\$6141\frac{1}{4}$

50. If one share of Polaroid stock costs $\$45\frac{7}{8}$, how much money will be needed to buy 64 shares?
$2936

★ **51.** Of the students eligible to apply for grants, $\frac{3}{4}$ actually sent away for applications. Of the students who received applications, only $\frac{5}{6}$ sent their completed forms in.

(a) What fractional portion of the students actually sent in grant applications?

(b) If the school has 8600 eligible students, how many people actually sent in grant applications?

(a) $\frac{5}{8}$ of the students; (b) 5375 students

★ **52.** When Sue turned 13, her wealthy grandparents gave her an allowance and put $\frac{3}{5}$ of the money into a college fund. When she turned 18, her grandparents surprised her by telling her that $\frac{3}{8}$ of *that* money was put into a long-term fund.

(a) What portion of the original allowance went into the long-term fund?

(b) What amount was put into the long-term fund if the allowance is $150 per month?

(a) $\frac{9}{40}$ of the allowance; (b) $2025

Cumulative Review Problems

53. A total of 16,399 cars used a toll bridge in January (31 days). What is the average number of cars using the bridge in one day? 529 cars

54. The Office of Investors Services has 15,456 calls made per month by the sales personnel. There are 42 sales personnel in the office. What is the average number of calls made per month by one salesperson? 368 calls

55. A computer printer can print 146 lines per minute. How many lines can it print in 12 minutes?
$146 \times 12 = 1752$ lines

56. At cruising speed a new commercial jetplane uses 12,360 gallons of fuel per hour. How many gallons will be used in 14 hours of flying time?
$12{,}360 \times 14 = 173{,}040$ gallons

2.5 Division of Fractions and Mixed Numbers

MathPro

Video 5

SSM

After studying this section, you will be able to:

1. *Divide two proper or improper fractions.*
2. *Divide a whole number and a fraction.*
3. *Divide two mixed numbers or one mixed number and a fraction.*

1 *Dividing Two Proper or Improper Fractions*

Why would you divide fractions? Consider this problem.

- A copper pipe that is $\frac{3}{4}$ of a foot long is to be cut into $\frac{1}{4}$-foot pieces. How many pieces will there be?

To find how many $\frac{1}{4}$'s are in $\frac{3}{4}$, we divide $\frac{3}{4} \div \frac{1}{4}$. We draw a sketch.

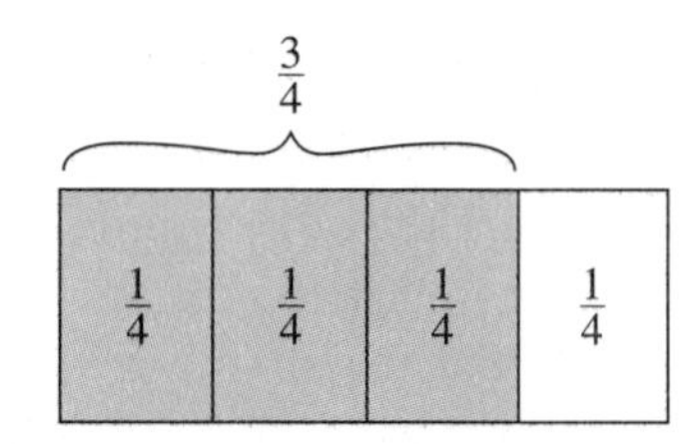

Notice there are three $\frac{1}{4}$'s in $\frac{3}{4}$.

How do we divide two fractions? We **invert** the second fraction and multiply.

$$\frac{3}{4} \div \frac{1}{4} = \frac{3}{\cancel{4}_1} \times \frac{\cancel{4}^1}{1} = \frac{3}{1} = 3$$

When we invert a fraction, we interchange the numerator and the denominator. If we invert $\frac{5}{9}$, we obtain $\frac{9}{5}$. If we invert $\frac{6}{1}$, we obtain $\frac{1}{6}$. Numbers such as $\frac{5}{9}$ and $\frac{9}{5}$ are called **reciprocals** of each other.

TEACHING TIP Tell students that in dividing two fractions it is always the second fraction that is inverted. Inform them that it is a very common mistake for students to erroneously invert the first fraction. If they have trouble remembering which one to flip, tell them to think of the following suggestion from a student who was a cook. ''You cannot flip a one-sided pancake. It has to have two sides to flip. You cannot flip fraction number one. It has to be fraction two to flip it.'' This rule seems silly, but no student who has learned it ever inverts the wrong fraction!

Rule for Division of Fractions

To divide two fractions, we invert the second fraction and multiply.

$$\frac{a}{b} \div \frac{c}{d} = \frac{a}{b} \times \frac{d}{c}$$

(when b, c, and d are not zero).

EXAMPLE 1 Divide. **(a)** $\frac{3}{11} \div \frac{2}{5}$ **(b)** $\frac{5}{8} \div \frac{25}{16}$

(a) $\frac{3}{11} \div \frac{2}{5} = \frac{3}{11} \times \frac{5}{2} = \frac{15}{22}$ **(b)** $\frac{5}{8} \div \frac{25}{16} = \frac{\cancel{5}^1}{\cancel{8}_1} \times \frac{\cancel{16}^2}{\cancel{25}_5} = \frac{2}{5}$

Practice Problem 1 Divide. **(a)** $\frac{7}{13} \div \frac{3}{4}$ **(b)** $\frac{16}{35} \div \frac{24}{25}$ ■

2 *Dividing a Whole Number and a Fraction*

When dividing with whole numbers, it is helpful to remember that for any whole number a, $a = \frac{a}{1}$.

EXAMPLE 2 Divide. **(a)** $\frac{3}{7} \div 2$ **(b)** $5 \div \frac{10}{13}$

(a) $\frac{3}{7} \div 2 = \frac{3}{7} \div \frac{2}{1} = \frac{3}{7} \times \frac{1}{2} = \frac{3}{14}$

(b) $5 \div \frac{10}{13} = \frac{5}{1} \div \frac{10}{13} = \frac{\overset{1}{\cancel{5}}}{1} \times \frac{13}{\underset{2}{\cancel{10}}} = \frac{13}{2}$ or $6\frac{1}{2}$

Practice Problem 2 Divide. **(a)** $\frac{3}{17} \div 6$ **(b)** $14 \div \frac{7}{15}$ ■

EXAMPLE 3 Divide, if possible.

(a) $\frac{23}{25} \div 1$ **(b)** $1 \div \frac{7}{5}$ **(c)** $0 \div \frac{4}{9}$ **(d)** $\frac{3}{17} \div 0$

(a) $\frac{23}{25} \div 1 = \frac{23}{25} \times \frac{1}{1} = \frac{23}{25}$ ← *Any fraction can be multiplied by 1 without changing the value of the fraction.*

(b) $1 \div \frac{7}{5} = \frac{1}{1} \times \frac{5}{7} = \frac{5}{7}$ ←

(c) $0 \div \frac{4}{9} = \frac{0}{1} \times \frac{9}{4} = \frac{0}{4} = 0$ *Zero divided by any nonzero number is zero.*

(d) $\frac{3}{17} \div 0$ *You can never divide by zero. This problem cannot be done.*

TEACHING TIP This is a good time to stress the concept of division by zero and dividing zero by a nonzero number. After covering Example 3 or one like it, have the students do the following eight problems quickly.

1. $\frac{4}{5} \div 3$ **2.** $\frac{5}{6} \div 0$
3. $\frac{8}{9} \div 1$ **4.** $\frac{12}{17} \div 4$
5. $0 \div \frac{14}{17}$ **6.** $1 \div \frac{2}{3}$
7. $\frac{14}{19} \div 0$ **8.** $0 \div \frac{5}{8}$

In order, the correct answers are $\frac{4}{15}$, undefined, $\frac{8}{9}$, $\frac{3}{17}$, 0, $\frac{3}{2}$, undefined, 0. Students find their most common mistakes are with division problems involving zero. Remind them that if they master this concept now, they will avoid a lot of problems later.

Practice Problem 3 Divide, if possible.

(a) $1 \div \frac{11}{13}$ **(b)** $\frac{14}{17} \div 1$ **(c)** $\frac{3}{11} \div 0$ **(d)** $0 \div \frac{9}{16}$ ■

SIDELIGHT Why do we divide by inverting the second fraction and multiplying? What is really going on when we do this? We are actually multiplying by 1. To see why, consider the following.

$$\frac{3}{7} \div \frac{2}{3} = \frac{\frac{3}{7}}{\frac{2}{3}}$$ *We write the division by using another fraction bar.*

$$= \frac{\frac{3}{7}}{\frac{2}{3}} \times 1$$ *Any fraction can be multiplied by 1 without changing the value of the fraction. This is the fundamental rule of fractions.*

$$= \frac{\frac{3}{7}}{\frac{2}{3}} \times \frac{\frac{3}{2}}{\frac{3}{2}}$$ *Any nonzero number divided by itself equals 1.*

$$= \frac{\frac{3}{7} \times \frac{3}{2}}{\frac{2}{3} \times \frac{3}{2}}$$ *Definition of multiplication of fractions.*

$$= \frac{\frac{3}{7} \times \frac{3}{2}}{1} = \frac{3}{7} \times \frac{3}{2}$$ *Any number can be written as a fraction with a denominator of 1 without changing its value.*

Thus

$$\frac{3}{7} \div \frac{2}{3} = \frac{3}{7} \times \frac{3}{2} = \frac{9}{14}$$

3 *Dividing Mixed Numbers*

If one or more mixed numbers are involved in the division, they should be converted to improper fractions first.

EXAMPLE 4 Divide. **(a)** $3\frac{7}{15} \div 1\frac{1}{25}$ **(b)** $\frac{3}{5} \div 2\frac{1}{7}$

(a) $3\frac{7}{15} \div 1\frac{1}{25} = \frac{52}{15} \div \frac{26}{25} = \frac{\overset{2}{\cancel{52}}}{\underset{3}{\cancel{15}}} \times \frac{\overset{5}{\cancel{25}}}{\underset{1}{\cancel{26}}} = \frac{10}{3}$ or $3\frac{1}{3}$

(b) $\frac{3}{5} \div 2\frac{1}{7} = \frac{3}{5} \div \frac{15}{7} = \frac{\overset{1}{\cancel{3}}}{5} \times \frac{7}{\underset{5}{\cancel{15}}} = \frac{7}{25}$

Practice Problem 4 Divide. **(a)** $1\frac{1}{5} \div \frac{7}{10}$ **(b)** $2\frac{1}{4} \div 1\frac{7}{8}$ ■

The division of two fractions may be indicated by a wide fraction bar.

EXAMPLE 5 Divide. **(a)** $\dfrac{10\frac{2}{9}}{2\frac{1}{3}}$ **(b)** $\dfrac{1\frac{1}{15}}{3\frac{1}{3}}$

(a) $\dfrac{10\frac{2}{9}}{2\frac{1}{3}} = 10\frac{2}{9} \div 2\frac{1}{3} = \frac{92}{9} \div \frac{7}{3} = \frac{92}{\cancelto{3}{9}} \times \frac{\cancelto{1}{3}}{7} = \frac{92}{21}$ or $4\frac{8}{21}$

(b) $\dfrac{1\frac{1}{15}}{3\frac{1}{3}} = 1\frac{1}{15} \div 3\frac{1}{3} = \frac{16}{15} \div \frac{10}{3} = \frac{\cancelto{8}{16}}{\cancelto{5}{15}} \times \frac{\cancelto{1}{3}}{\cancelto{5}{10}} = \frac{8}{25}$

Practice Problem 5 Divide. **(a)** $\dfrac{5\frac{2}{3}}{7}$ **(b)** $\dfrac{1\frac{2}{5}}{2\frac{1}{3}}$ ■

EXAMPLE 6 Find the valuc of x if $x \div \frac{8}{7} = \frac{21}{40}$.

First we will change the division problem to an equivalent multiplication problem.

$$x \div \frac{8}{7} = \frac{21}{40}$$

$$x \cdot \frac{7}{8} = \frac{21}{40}$$

x represents a fraction.

In the numerator above, we want to know what times 7 equals 21. In the denominator, we want to know what times 8 equals 40.

$$\frac{3}{5} \cdot \frac{7}{8} = \frac{21}{40}$$

Thus $x = \frac{3}{5}$.

Practice Problem 6 Find the value of x if $x \div \frac{3}{2} = \frac{22}{36}$. ■

TEACHING TIP Students have trouble identifying which number to divide into which. Ask them to solve the following problem. "A farmer has $117\frac{1}{3}$ tons of grain. He wants to store it in 12 equal piles. How many tons should be placed in each pile?" Go over carefully why we decide to divide $117\frac{1}{3}$ by 12. This will usually help build the students' confidence before they attempt to do the word problems in this section.

EXAMPLE 7 There are 117 milligrams of cholesterol in $4\frac{1}{3}$ cups of milk. How much cholesterol is contained in 1 cup of milk?

We want to divide the 117 by $4\frac{1}{3}$ to find out how much is in 1 cup.

$$117 \div 4\frac{1}{3} = 117 \div \frac{13}{3} = \frac{\overset{9}{\cancel{117}}}{1} \times \frac{3}{\underset{1}{\cancel{13}}} = \frac{27}{1} = 27$$

Thus there are 27 milligrams of cholesterol in 1 cup of milk.

Practice Problem 7 A copper pipe that is $19\frac{1}{4}$ feet long will be cut into 14 equal pieces. How long will each piece be? ■

Developing Your Study Skills

WHY IS REVIEW NECESSARY?

You master a course in mathematics by learning the concepts one step at a time. There are basic concepts like addition, subtraction, multiplication, and division of whole numbers that are considered the foundation upon which all of mathematics is built. These must be mastered first. Then the study of mathematics is built step by step upon this foundation, each step supporting the next. The process is a carefully designed procedure, and so no steps can be skipped. A student of mathematics needs to realize the importance of this building process to succeed.

Because learning new concepts depends on those previously learned, students often need to take time to review. The reviewing process will strengthen the understanding and application of concepts that are weak due to lack of mastery or passage of time. Review at the right time on the right concepts can strengthen previously learned skills and make progress possible.

Timely, periodic review of previously learned mathematical concepts is absolutely necessary in order to master new concepts. You may have forgotten a concept or grown a bit rusty in applying it. Reviewing is the answer. Make use of any review sections in your textbook, whether they are assigned or not. Look back to previous chapters whenever you have forgotten how to do something. Study the examples and practice some exercises to refresh your understanding.

Be sure that you understand and can perform the computations of each new concept. This will enable you to be able to move successfully on to the next ones.

Remember, mathematics is a step-by-step building process. Learn one concept at a time, skipping none, and reinforce and strengthen with review whenever necessary.

2.5 Exercises

Divide, if possible.

1. $\frac{2}{5} \div \frac{7}{2}$ $\quad \frac{2}{5} \times \frac{2}{7} = \frac{4}{35}$

2. $\frac{3}{4} \div \frac{5}{6}$ $\quad \frac{3}{4} \times \frac{6}{5} = \frac{9}{10}$

3. $\frac{3}{13} \div \frac{9}{26}$ $\quad \frac{3}{13} \times \frac{26}{9} = \frac{2}{3}$

4. $\frac{7}{8} \div \frac{2}{3}$ $\quad \frac{7}{8} \times \frac{3}{2} = \frac{21}{16}$ *or* $1\frac{5}{16}$

5. $\frac{5}{6} \div \frac{25}{27}$ $\quad \frac{5}{6} \times \frac{27}{25} = \frac{9}{10}$

6. $\frac{3}{14} \div \frac{6}{7}$ $\quad \frac{3}{14} \times \frac{7}{6} = \frac{1}{4}$

7. $\frac{5}{9} \div \frac{10}{27}$ $\quad \frac{5}{9} \times \frac{27}{10} = \frac{3}{2}$ *or* $1\frac{1}{2}$

8. $\frac{8}{15} \div \frac{24}{35}$ $\quad \frac{8}{15} \times \frac{35}{24} = \frac{7}{9}$

9. $\frac{4}{5} \div 1$ $\quad \frac{4}{5} \times 1 = \frac{4}{5}$

10. $1 \div \frac{3}{7}$ $\quad 1 \times \frac{7}{3} = \frac{7}{3}$ *or* $2\frac{1}{3}$

11. $2 \div \frac{7}{8}$ $\quad \frac{2}{1} \times \frac{8}{7} = \frac{16}{7}$ *or* $2\frac{2}{7}$

12. $\frac{3}{11} \div 4$ $\quad \frac{3}{11} \times \frac{1}{4} = \frac{3}{44}$

13. $\frac{2}{9} \div \frac{1}{6}$ $\quad \frac{2}{9} \times \frac{6}{1} = \frac{4}{3}$ *or* $1\frac{1}{3}$

14. $\frac{3}{4} \div \frac{2}{3}$ $\quad \frac{3}{4} \times \frac{3}{2} = \frac{9}{8}$ *or* $1\frac{1}{8}$

15. $\frac{4}{15} \div \frac{4}{15}$ $\quad \frac{4}{15} \times \frac{15}{4} = 1$

16. $\frac{2}{7} \div \frac{2}{7}$ $\quad \frac{2}{7} \times \frac{7}{2} = 1$

17. $\frac{3}{7} \div \frac{7}{3}$ $\quad \frac{3}{7} \times \frac{3}{7} = \frac{9}{49}$

18. $\frac{11}{12} \div \frac{1}{5}$ $\quad \frac{11}{12} \times \frac{5}{1} = \frac{55}{12}$ *or* $4\frac{7}{12}$

19. $\frac{4}{3} \div \frac{7}{27}$ $\quad \frac{4}{3} \times \frac{27}{7} = \frac{36}{7}$ *or* $5\frac{1}{7}$

20. $\frac{8}{9} \div \frac{5}{81}$ $\quad \frac{8}{9} \times \frac{81}{5} = \frac{72}{5}$ *or* $14\frac{2}{5}$

21. $0 \div \frac{3}{17}$ $\quad 0 \times \frac{17}{3} = 0$

22. $0 \div \frac{5}{16}$ $\quad 0 \times \frac{16}{5} = 0$

23. $\frac{18}{19} \div 0$ $\quad$ Cannot be done

24. $\frac{24}{29} \div 0$ $\quad$ Cannot be done

25. $\frac{9}{16} \div \frac{3}{4}$ $\quad \frac{9}{16} \times \frac{4}{3} = \frac{3}{4}$

26. $\frac{3}{4} \div \frac{9}{16}$ $\quad \frac{3}{4} \times \frac{16}{9} = \frac{4}{3}$ *or* $1\frac{1}{3}$

27. $\frac{3}{7} \div \frac{15}{28}$ $\quad \frac{3}{7} \times \frac{28}{15} = \frac{4}{5}$

28. $\frac{5}{6} \div \frac{15}{18}$ $\quad \frac{5}{6} \times \frac{18}{15} = 1$

29. $\frac{10}{25} \div \frac{20}{50}$ $\quad \frac{10}{25} \times \frac{50}{20} = 1$

30. $\frac{3}{25} \div \frac{24}{125}$ $\quad \frac{3}{25} \times \frac{125}{24} = \frac{5}{8}$

31. $12 \div \frac{3}{4}$ $\quad \frac{12}{1} \times \frac{4}{3} = 16$

32. $18 \div \frac{3}{7}$ $\quad \frac{18}{1} \times \frac{7}{3} = 42$

33. $\frac{7}{8} \div 4$ $\quad \frac{7}{8} \times \frac{1}{4} = \frac{7}{32}$

34. $\frac{5}{6} \div 12$ $\quad \frac{5}{6} \times \frac{1}{12} = \frac{5}{72}$

35. $6000 \div \frac{6}{5}$ $\quad \frac{6000}{1} \times \frac{5}{6} = 5000$

36. $8000 \div \frac{4}{7}$ $\quad \frac{8000}{1} \times \frac{7}{4} = 14{,}000$

37. $\dfrac{\frac{2}{5}}{3}$ $\quad \frac{2}{5} \times \frac{1}{3} = \frac{2}{15}$

38. $\dfrac{\frac{3}{7}}{4}$ $\quad \frac{3}{7} \times \frac{1}{4} = \frac{3}{28}$

39. $\dfrac{\frac{5}{8}}{\frac{25}{7}}$ $\quad \frac{5}{8} \times \frac{7}{25} = \frac{7}{40}$

40. $\dfrac{\frac{3}{16}}{\frac{5}{8}}$ $\quad \frac{3}{16} \times \frac{8}{5} = \frac{3}{10}$

41. $3\frac{1}{5} \div \frac{3}{10}$

$\frac{16}{5} \times \frac{10}{3} = \frac{32}{3}$ *or* $10\frac{2}{3}$

42. $2\frac{1}{8} \div \frac{1}{4}$

$\frac{17}{8} \times \frac{4}{1} = \frac{17}{2}$ *or* $8\frac{1}{2}$

43. $2\frac{1}{3} \div 6$

$\frac{7}{3} \times \frac{1}{6} = \frac{7}{18}$

44. $6\frac{1}{2} \div 3$

$\frac{13}{2} \times \frac{1}{3} = \frac{13}{6}$ *or* $2\frac{1}{6}$

45. $1\frac{7}{9} \div 2\frac{2}{3}$

$\frac{16}{9} \times \frac{3}{8} = \frac{2}{3}$

46. $2\frac{1}{15} \div 3\frac{1}{3}$

$\frac{31}{15} \times \frac{3}{10} = \frac{31}{50}$

47. $5 \div 1\frac{1}{4}$

$\frac{5}{1} \times \frac{4}{5} = 4$

48. $7 \div 1\frac{2}{5}$

$\frac{7}{1} \times \frac{5}{7} = 5$

49. $12\frac{1}{2} \div 5\frac{5}{6}$

$\frac{25}{2} \times \frac{6}{35} = \frac{15}{7}$ *or* $2\frac{1}{7}$

50. $14\frac{2}{3} \div 3\frac{1}{2}$

$\frac{44}{3} \times \frac{2}{7} = \frac{88}{21}$ *or* $4\frac{4}{21}$

51. $8\frac{1}{4} \div 2\frac{3}{4}$

$\frac{33}{4} \times \frac{4}{11} = 3$

52. $2\frac{3}{8} \div 5\frac{3}{7}$

$\frac{19}{8} \times \frac{7}{38} = \frac{7}{16}$

53. $3\frac{1}{2} \div 1\frac{7}{9}$

$\frac{7}{2} \times \frac{9}{16} = \frac{63}{32}$ *or* $1\frac{31}{32}$

54. $1\frac{1}{8} \div 2\frac{2}{3}$

$\frac{9}{8} \times \frac{3}{8} = \frac{27}{64}$

55. $2\frac{1}{2} \div 5$

$\frac{5}{2} \times \frac{1}{5} = \frac{1}{2}$

56. $3\frac{1}{7} \div 10$

$\frac{22}{7} \times \frac{1}{10} = \frac{11}{35}$

57. $\dfrac{4\frac{1}{2}}{\frac{3}{8}}$

$\frac{9}{2} \times \frac{8}{3} = 12$

58. $\dfrac{6\frac{1}{2}}{\frac{3}{4}}$

$\frac{13}{2} \times \frac{4}{3} = \frac{26}{3}$ *or* $8\frac{2}{3}$

59. $\dfrac{1\frac{1}{4}}{1\frac{7}{8}}$

$\frac{5}{4} \times \frac{8}{15} = \frac{2}{3}$

60. $\dfrac{2\frac{3}{5}}{1\frac{7}{10}}$

$\frac{13}{5} \times \frac{10}{17} = \frac{26}{17}$ *or* $1\frac{9}{17}$

61. $\dfrac{\frac{5}{9}}{2\frac{1}{3}}$

$\frac{5}{9} \times \frac{3}{7} = \frac{5}{21}$

62. $\dfrac{\frac{2}{10}}{3\frac{1}{5}}$

$\frac{2}{10} \times \frac{5}{16} = \frac{1}{16}$

63. $\dfrac{5\frac{1}{3}}{2\frac{1}{2}}$

$\frac{16}{3} \times \frac{2}{5} = \frac{32}{15}$ *or* $2\frac{2}{15}$

64. $\dfrac{6\frac{3}{4}}{3\frac{1}{2}}$

$\frac{27}{4} \times \frac{2}{7} = \frac{27}{14}$ *or* $1\frac{13}{14}$

Find the value of x in each of the following.

65. $x \div \frac{4}{3} = \frac{21}{20}$ $\quad x = \frac{7}{5}$

66. $x \div \frac{2}{5} = \frac{15}{16}$ $\quad x = \frac{3}{8}$

67. $x \div \frac{9}{5} = \frac{20}{63}$ $\quad x = \frac{4}{7}$

68. $x \div \frac{7}{3} = \frac{9}{28}$ $\quad x = \frac{3}{4}$

Applications

Answer each question.

69. A leather factory in Morocco tans leather. In order to make the leather soft, it has to soak in a vat of uric acid and other ingredients. The main holding storage tank holds $20\frac{1}{4}$ gallons of the tanning mixture. If the mixture is distributed evenly into 9 vats of equal size for the different colored leathers, how much will each vat hold? $2\frac{1}{4}$ gallons

70. A specially protected stretch of beach bordering the Great Barrier Reef in Australia is used for marine biology and ecological research. The beach, which is $7\frac{1}{2}$ miles long, has been broken up into 20 equal segments for comparison purposes. How long is each segment of the beach?

$\frac{3}{8}$ mile

71. Bruce drove in a snowstorm to get to his favorite mountain to do some snowboarding. He traveled 125 miles in $3\frac{1}{3}$ hours. What was his average speed (in miles per hour)?

$37\frac{1}{2}$ miles per hour

72. Roberto drove his truck to Cedarville, a distance of 200 miles, in $4\frac{1}{6}$ hours. What was his average speed (in miles per hour)?

$\frac{200}{1} \times \frac{6}{25} = 48$ miles per hour

73. The Patriotic Flag Company is making flags which need $3\frac{1}{8}$ yards of fabric. The warehouse has $87\frac{1}{2}$ yards of flag fabric available. How many flags can the factory make? 28 flags

74. The school cafeteria is making hamburgers for the annual Senior Day Festival. The cooks have decided that because hamburger shrinks on the grill, they will allow $\frac{2}{3}$ pound of meat for each student. If the kitchen has $38\frac{2}{3}$ pounds of meat, how many students will be fed? 58 students

To Think About

75. Division of fractions is not commutative. For example, $\frac{2}{3} \div \frac{5}{7} \neq \frac{5}{7} \div \frac{2}{3}$.

In general, $\frac{a}{b} \div \frac{c}{d} \neq \frac{c}{d} \div \frac{a}{b}$ $(a, b, c, d \text{ all} \neq 0)$.

But sometimes there are exceptions. Can you think of any numbers a, b, c, d such that for fractions $\frac{a}{b}$ and $\frac{c}{d}$ it would be true that $\frac{a}{b} \div \frac{c}{d} = \frac{c}{d} \div \frac{a}{b}$?

$a = 2, b = 3, c = 4, d = 6$. $\frac{2}{3} \div \frac{4}{6} = \frac{4}{6} \div \frac{2}{3}$. In general, if $\frac{a}{b} = \frac{c}{d}$, then it is true.

76. Can you think of a way to divide $3\frac{7}{12} \div \frac{1}{4}$ *without* changing $3\frac{7}{12}$ to an improper fraction? Try your method to see if it works.

$3\frac{7}{12} \div \frac{1}{4} = \left(3 + \frac{7}{12}\right) \times 4 = 12 + \frac{28}{12} = 12 + \frac{7}{3} = 14\frac{1}{3}$

Cumulative Review Problems

77. Write in words. 39,576,304

Thirty-nine million, five hundred seventy-six thousand, three hundred four.

78. Write in expanded form. 459,273

400,000 + 50,000 + 9000 + 200 + 70 + 3

79. Add. 126 + 34 + 9 + 891 + 12 + 27 1099

80. Write in standard notation. Eighty-seven million, five hundred ninety-five thousand, six hundred thirty-one 87,595,631

2.6 The Least Common Denominator and Building Up Fractions

MathPro Video 5 SSM

After studying this section, you will be able to:

1. *Find the least common denominator given two or three fractions.*
2. *Build up a fraction so that it has as a denominator a given least common denominator.*

1 *Finding the Least Common Denominator*

We need some way to determine which of two fractions is larger. Suppose that Marcia and Melissa each have some pizza left.

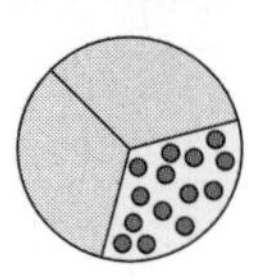

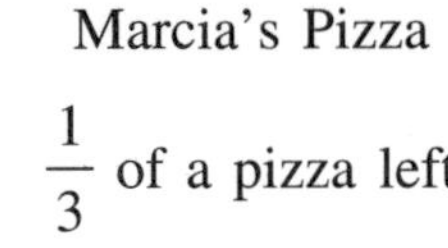

Marcia's Pizza	Melissa's Pizza
$\frac{1}{3}$ of a pizza left	$\frac{1}{4}$ of a pizza left

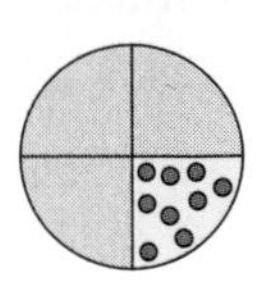

Who has more pizza left? How much more? Comparing the amounts of pizza left would be easy if each pizza had been cut into equal-sized pieces. If the original pizzas had each been cut into 12 pieces, we would be able to see that Marcia had $\frac{1}{12}$ of a pizza more than Melissa had.

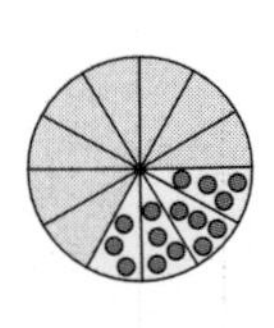

Marcia's Pizza	Melissa's Pizza
We know that $\left(\frac{4}{12} = \frac{1}{3} \text{ by reducing.}\right)$	We know that $\left(\frac{3}{12} = \frac{1}{4} \text{ by reducing.}\right)$

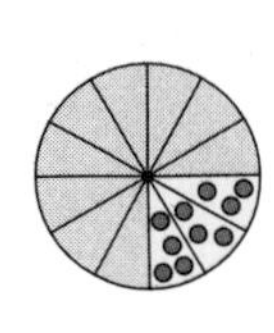

The denominator 12 is common to the fractions $\frac{4}{12}$ and $\frac{3}{12}$. We call a denominator that allows us to compare fractions the *least common denominator,* abbreviated LCD. The number 12 is the least common denominator for the fractions $\frac{1}{3}$ and $\frac{1}{4}$.

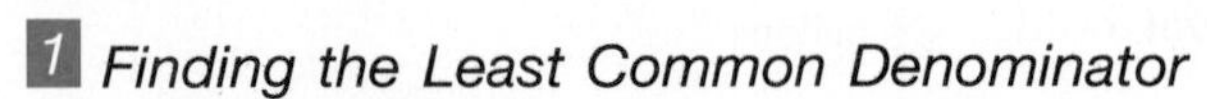

> **Definition**
>
> The least common denominator (LCD) of two or more fractions is the smallest number that can be divided evenly by each of the fractions' denominators.

Let's check. $12 \div 3 = 4$. $12 \div 4 = 3$. We have no remainder, so 12 is the LCD. Suppose we had thought the number 6 might be the LCD. $6 \div 3 = 2$ with no remainder. But we see that $6 \div 4 = 1$ with a remainder of 2, so 6 cannot be the LCD. In this case, 12 is the LCD of $\frac{1}{3}$ and $\frac{1}{4}$. The number 12 is the smallest number that can be divided evenly by 3 and by 4.

In some problems you may be able to guess the LCD quite quickly. With practice, you can often find the LCD mentally. For example, you now know that if the denominators of two fractions are 3 and 4, the LCD is 12. For the fractions $\frac{1}{2}$ and $\frac{1}{4}$, the LCD is 4; for the fractions $\frac{1}{3}$ and $\frac{1}{6}$, the LCD is 6. We can see that if the denominator of one fraction divides without remainder into the denominator of another, the LCD of the two fractions is the larger of the denominators.

EXAMPLE 1 Determine the LCD of the fractions.

(a) $\frac{7}{15}$ and $\frac{4}{5}$ **(b)** $\frac{2}{3}$ and $\frac{5}{27}$

(a) Since 5 can be divided into 15, the LCD of $\frac{7}{15}$ and $\frac{4}{5}$ is 15.

(b) Since 3 can be divided into 27, the LCD of $\frac{2}{3}$ and $\frac{5}{27}$ is 27.

Practice Problem 1 Determine the LCD of the fractions.

(a) $\frac{3}{4}$ and $\frac{11}{12}$ **(b)** $\frac{1}{7}$ and $\frac{8}{35}$ ■

In a few cases, the LCD is the product of the two denominators.

EXAMPLE 2 Find the LCD for $\frac{1}{4}$ and $\frac{3}{5}$.

We see that $4 \times 5 = 20$. Also, 20 is the *smallest* number that can be divided without remainder by 4 and by 5. So the LCD = 20.

Practice Problem 2 Find the LCD for $\frac{3}{7}$ and $\frac{5}{6}$. ■

Sometimes we need to follow steps to find the LCD.

> **Three-Step Procedure for Finding the Least Common Denominator**
>
> 1. Write each denominator as the product of prime factors.
> 2. List all the prime factors that appear in either product.
> 3. Form a product of those prime factors, using each factor the greatest number of times it appears in any one denominator.

EXAMPLE 3 Find the LCD by the three-step procedure.

(a) $\frac{5}{6}$ and $\frac{4}{15}$ **(b)** $\frac{7}{18}$ and $\frac{7}{30}$ **(c)** $\frac{10}{27}$ and $\frac{5}{18}$

(a) Step 1 Write each denominator as a product of prime factors.

$$6 = 2 \times 3 \qquad 15 = 5 \times 3$$

Step 2 The LCD will contain the factors 2, 3, and 5.

$$6 = 2 \times 3 \qquad 15 = 5 \times 3$$

$$\text{LCD} = 2 \times 3 \times 5$$
$$= 30$$

Step 3 is not needed. No factor was repeated in any one denominator.

TEACHING TIP Stress the fact that not all students will approach these problems the same way. Some students were taught in school to find the LCD, others to find the GCF, others the LCM. If a student wishes to find the LCD in a way different from the one presented in this book, that is fine as long as the student can obtain correct answers.

(b) Step 1 Write each denominator as a product of prime factors.

$$18 = 2 \times 9 = 2 \times 3 \times 3$$

$$30 = 3 \times 10 = 2 \times 3 \times 5$$

Step 2 The LCD will be a product containing 2, 3, and 5.

Step 3 The LCD will contain the factor 3 twice since it occurs twice in the denominator 18.

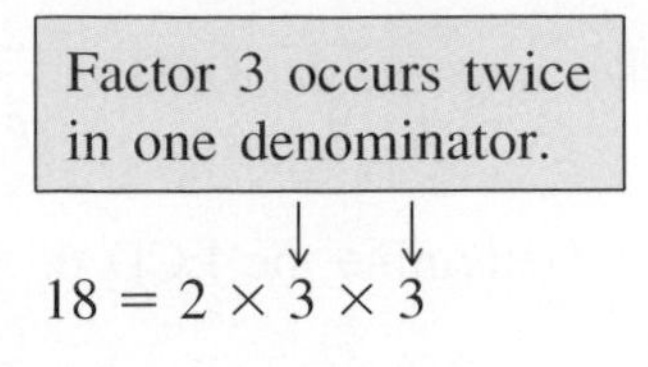
Factor 3 occurs twice
in one denominator.

$$18 = 2 \times 3 \times 3$$

$$\text{LCD} = 2 \times 3 \times 3 \times 5 = 90$$

(c) Write each denominator as a product of prime factors.

$$27 = 3 \times 3 \times 3 \qquad 18 = 3 \times 3 \times 2$$

Factor 3 occurs three times

The LCD will contain the factor 2 once but the factor 3 three times.

$$\text{LCD} = 2 \times 3 \times 3 \times 3 = 54$$

Practice Problem 3 Find the LCDs.

(a) $\dfrac{3}{14}$ and $\dfrac{1}{10}$ **(b)** $\dfrac{1}{15}$ and $\dfrac{7}{50}$ **(c)** $\dfrac{3}{16}$ and $\dfrac{5}{12}$ ■

A similar procedure can be used for three fractions.

EXAMPLE 4 Find the LCD of $\dfrac{7}{12}$, $\dfrac{1}{15}$, and $\dfrac{11}{30}$.

$$12 = 2 \times 2 \times 3$$

$$15 = 3 \times 5$$

$$30 = 2 \times 3 \times 5$$

$$\text{LCD} = 2 \times 2 \times 3 \times 5$$

$$= 60$$

TEACHING TIP Ask the students to find the LCD for the fractions $\frac{3}{7}$, $\frac{4}{21}$, $\frac{7}{24}$. The correct answer is LCD = 168.

Practice Problem 4 Find the LCD of $\dfrac{3}{49}$, $\dfrac{5}{21}$, and $\dfrac{6}{7}$. ■

2 *Building Up Fractions*

We cannot add fractions with unlike denominators. To change denominators, we must (1) find the LCD and (2) build up the addends—the fractions being added—into equivalent fractions that have the LCD as the denominator. We know now how to find the LCD. Let's look at how we build up fractions. We know, for example, that

$$\frac{1}{2} = \frac{2}{4} = \frac{50}{100} \qquad \frac{1}{4} = \frac{25}{100} \qquad \text{and} \qquad \frac{3}{4} = \frac{75}{100}$$

In these cases, we have mentally multiplied the given fraction by 1, in the form of a certain number, c, in the numerator and that same number, c, in the denominator.

$$\frac{1}{2} \times \boxed{\frac{c}{c}} = \frac{2}{4} \qquad \text{Here } c = 2, \ \frac{2}{2} = 1$$

$$\frac{1}{2} \times \boxed{\frac{c}{c}} = \frac{50}{100} \qquad \text{Here } c = 50, \ \frac{50}{50} = 1$$

This property is called the *building fraction property*.

Building Fraction Property

For whole numbers a, b, and c where $b \neq 0$, $c \neq 0$,

$$\frac{a}{b} = \frac{a}{b} \times 1 = \frac{a}{b} \times \boxed{\frac{c}{c}} = \frac{a \times c}{b \times c}$$

EXAMPLE 5 Build each fraction to an equivalent fraction with the LCD.

(a) $\frac{3}{4}$, LCD = 28 **(b)** $\frac{4}{5}$, LCD = 45 **(c)** $\frac{1}{3}$ and $\frac{4}{5}$, LCD = 15

(a) $\frac{3}{4} \times \boxed{\frac{c}{c}} = \frac{?}{28}$ *We know that $4 \times 7 = 28$, so the value c that we multiply numerator and denominator by is 7.*

$$\frac{3}{4} \times \frac{7}{7} = \frac{21}{28}$$

(b) $\frac{4}{5} \times \boxed{\frac{c}{c}} = \frac{?}{45}$ *We know that $5 \times 9 = 45$, so $c = 9$.*

$$\frac{4}{5} \times \frac{9}{9} = \frac{36}{45}$$

(c) $\frac{1}{3} = \frac{?}{15}$ *We know that $3 \times 5 = 15$, so we multiply numerator and denominator by 5.*

$$\frac{1}{3} \times \boxed{\frac{5}{5}} = \frac{5}{15}$$

$\frac{4}{5} = \frac{?}{15}$ *We know that $5 \times 3 = 15$, so we multiply numerator and denominator by 3.*

$$\frac{4}{5} \times \frac{3}{3} = \frac{12}{15}$$

Thus $\frac{1}{3} = \frac{5}{15}$ and $\frac{4}{5} = \frac{12}{15}$.

Practice Problem 5 Build each fraction to an equivalent fraction with the LCD.

(a) $\frac{3}{5}$, LCD = 40 **(b)** $\frac{7}{11}$, LCD = 44 **(c)** $\frac{2}{7}$ and $\frac{3}{4}$, LCD = 28 ■

TEACHING TIP Call students' attention to the fact that some people have learned how to find the LCD of a fraction by using least common multiples. If they are familiar with that expression, they will probably want to read the Alternative Method section. This is a good time to remind students that a good mathematician knows many ways to solve the same problem. It helps to be able to think of two different approaches to solving problems in real life as well as in math.

EXAMPLE 6

(a) Find the LCD of $\frac{1}{32}$ and $\frac{7}{48}$.

(b) Build the fractions to equivalent fractions that have the LCD as their denominators.

(a) First we find the prime factors of 32 and 48.

$$32 = 2 \times 2 \times 2 \times 2 \times 2$$

$$48 = 2 \times 2 \times 2 \times 2 \times 3$$

Thus the LCD will require a factor of 2 five times and a factor of 3 one time.

$$\text{LCD} = 2 \times 2 \times 2 \times 2 \times 2 \times 3 = 96$$

(b) $\frac{1}{32} = \frac{?}{96}$ *Since* $32 \times 3 = 96$ *we multiply by* 3.

$$\frac{1}{32} = \frac{1}{32} \times \frac{3}{3} = \frac{3}{96}$$

$\frac{7}{48} = \frac{?}{96}$ *Since* $48 \times 2 = 96$, *we multiply by* 2.

$$\frac{7}{48} = \frac{7}{48} \times \frac{2}{2} = \frac{14}{96}$$

Practice Problem 6

(a) Find the LCD of $\frac{3}{20}$ and $\frac{11}{15}$.

(b) Build the fractions to equivalent fractions that have the LCD as their denominators. ■

ALTERNATIVE METHOD Is there another method that can be used to find the least common denominator for two or more fractions? Yes. You can find the least common multiple (LCM) of the two denominators. Suppose that we want to find the least common multiple of 10 and 12. List the multiples of 10 (multiply by 1, 2, 3, 4, 5, and so on).

10, 20, 30, 40, 50, 60, 70, 80, . . .

List the multiples of 12.

12, 24, 36, 48, 60, 72, 84, 96, . . .

The first multiple that appears on both lists is the least common multiple. It is also the LCD for two fractions that have 10 and 12 for denominators.

2.6 Exercises

Find the LCD for each pair of fractions.

1. $\frac{1}{5}$ and $\frac{3}{10}$
LCD = 10

2. $\frac{3}{8}$ and $\frac{5}{16}$
LCD = 16

3. $\frac{3}{7}$ and $\frac{1}{4}$
LCD = 28

4. $\frac{5}{6}$ and $\frac{3}{5}$
LCD = 30

5. $\frac{2}{5}$ and $\frac{3}{7}$
LCD = 35

6. $\frac{1}{16}$ and $\frac{2}{3}$
LCD = 48

7. $\frac{1}{6}$ and $\frac{5}{9}$
$9 = 3 \times 3$
$6 = 2 \times 3$
LCD = 18

8. $\frac{1}{4}$ and $\frac{3}{14}$
$4 = 2 \times 2$
$14 = 2 \times 7$
LCD = 28

9. $\frac{3}{16}$ and $\frac{11}{24}$
$16 = 2 \times 2 \times 2 \times 2$
$24 = 2 \times 2 \times 2 \times 3$
LCD = 48

10. $\frac{3}{10}$ and $\frac{8}{25}$
$10 = 2 \times 5$
$25 = 5 \times 5$
LCD = 50

11. $\frac{1}{16}$ and $\frac{3}{4}$
LCD = 16

12. $\frac{2}{11}$ and $\frac{1}{44}$
LCD = 44

13. $\frac{5}{10}$ and $\frac{11}{45}$
$10 = 2 \times 5$
$45 = 3 \times 3 \times 5$
LCD = 90

14. $\frac{13}{20}$ and $\frac{17}{30}$
$20 = 2 \times 2 \times 5$
$30 = 2 \times 3 \times 5$
LCD = 60

15. $\frac{7}{12}$ and $\frac{7}{30}$
$12 = 2 \times 2 \times 3$
$30 = 2 \times 3 \times 5$
LCD = 60

16. $\frac{5}{6}$ and $\frac{7}{15}$
$6 = 2 \times 3$
$15 = 3 \times 5$
LCD = 30

17. $\frac{5}{21}$ and $\frac{8}{35}$
$21 = 3 \times 7$
$35 = 5 \times 7$
LCD = 105

18. $\frac{1}{20}$ and $\frac{7}{8}$
$20 = 2 \times 2 \times 5$
$8 = 2 \times 2 \times 2$
LCD = 40

19. $\frac{5}{18}$ and $\frac{11}{12}$
$18 = 2 \times 3 \times 3$
$12 = 2 \times 2 \times 3$
LCD = 36

20. $\frac{1}{24}$ and $\frac{7}{40}$
$40 = 2 \times 2 \times 2 \times 5$
$24 = 2 \times 2 \times 2 \times 3$
LCD = 120

Find the LCD for each set of three fractions.

21. $\frac{2}{3}, \frac{1}{2}, \frac{5}{6}$
LCD = 6

22. $\frac{1}{5}, \frac{1}{3}, \frac{7}{10}$
LCD = 30

23. $\frac{5}{24}, \frac{11}{15}, \frac{7}{30}$
$24 = 2 \times 2 \times 2 \times 3$
$15 = 3 \times 5$
$30 = 2 \times 3 \times 5$
LCD = 120

24. $\frac{5}{18}, \frac{5}{12}, \frac{13}{20}$
$12 = 2 \times 2 \times 3$
$18 = 2 \times 3 \times 3$
$20 = 2 \times 2 \times 5$
LCD = 180

25. $\frac{5}{16}, \frac{11}{18}, \frac{1}{24}$
$16 = 2 \times 2 \times 2 \times 2$
$18 = 2 \times 3 \times 3$
$24 = 2 \times 2 \times 2 \times 3$
LCD = 144

26. $\frac{5}{6}, \frac{1}{2}, \frac{5}{22}$
$6 = 2 \times 3$
$2 = 2$
$22 = 2 \times 11$
LCD = 66

27. $\frac{7}{12}, \frac{1}{21}, \frac{3}{14}$
$12 = 2 \times 2 \times 3$
$21 = 3 \times 7$
$14 = 2 \times 7$
LCD = 84

28. $\frac{1}{30}, \frac{3}{40}, \frac{7}{8}$
$30 = 2 \times 3 \times 5$
$40 = 2 \times 2 \times 2 \times 5$
$8 = 2 \times 2 \times 2$
LCD = 120

29. $\frac{7}{15}, \frac{11}{12}, \frac{7}{8}$
$15 = 3 \times 5$
$12 = 2 \times 2 \times 3$
$8 = 2 \times 2 \times 2$
LCD = 120

30. $\frac{5}{36}, \frac{2}{48}, \frac{1}{24}$
$36 = 2 \times 2 \times 3 \times 3$
$48 = 2 \times 2 \times 2 \times 2 \times 3$
$24 = 2 \times 2 \times 2 \times 3$
LCD = 144

Build each fraction to an equivalent fraction with the denominator specified. State the numerator.

31. $\frac{1}{6} = \frac{?}{18}$ 3

32. $\frac{1}{5} = \frac{?}{35}$ 7

33. $\frac{5}{6} = \frac{?}{54}$ 45

34. $\frac{4}{7} = \frac{?}{28}$ 16

35. $\frac{4}{11} = \frac{?}{55}$ 20

36. $\frac{2}{13} = \frac{?}{39}$ 6

37. $\frac{7}{24} = \frac{?}{48}$ 14

38. $\frac{3}{50} = \frac{?}{100}$ 6

39. $\frac{8}{9} = \frac{?}{108}$ 96

40. $\frac{6}{7} = \frac{?}{147}$ 126

41. $\frac{13}{42} = \frac{?}{126}$ 39

42. $\frac{14}{15} = \frac{?}{180}$ 168

The LCD of each pair of fractions is listed. Build each fraction to an equivalent fraction that has the LCD as the denominator.

43. LCD = 36, $\frac{7}{12}$ and $\frac{5}{9}$

$\frac{21}{36}$ and $\frac{20}{36}$

44. LCD = 20, $\frac{9}{10}$ and $\frac{3}{4}$

$\frac{18}{20}$ and $\frac{15}{20}$

45. LCD = 200, $\frac{3}{25}$ and $\frac{7}{40}$

$\frac{24}{200}$ and $\frac{35}{200}$

46. LCD = 72, $\frac{5}{24}$ and $\frac{7}{36}$

$\frac{15}{72}$ and $\frac{14}{72}$

47. LCD = 150, $\frac{19}{25}$ and $\frac{26}{75}$

$\frac{114}{150}$ and $\frac{52}{150}$

48. LCD = 240, $\frac{57}{120}$ and $\frac{39}{80}$

$\frac{114}{240}$ and $\frac{117}{240}$

Find the LCD. Build up the fractions to equivalent fractions having the LCD as the denominator.

49. $\frac{5}{7}$ and $\frac{7}{42}$

LCD = 42
$\frac{30}{42}$ and $\frac{7}{42}$

50. $\frac{3}{8}$ and $\frac{13}{40}$

LCD = 40
$\frac{15}{40}$ and $\frac{13}{40}$

51. $\frac{5}{12}$ and $\frac{1}{16}$

LCD = 48
$\frac{20}{48}$ and $\frac{3}{48}$

52. $\frac{7}{15}$ and $\frac{4}{25}$

LCD = 75
$\frac{35}{75}$ and $\frac{12}{75}$

53. $\frac{13}{20}$ and $\frac{11}{16}$

LCD = 80
$\frac{52}{80}$ and $\frac{55}{80}$

54. $\frac{9}{12}$ and $\frac{13}{18}$

LCD = 36
$\frac{27}{36}$ and $\frac{26}{36}$

55. $\frac{7}{12}$ and $\frac{23}{30}$

LCD = 60
$\frac{35}{60}$ and $\frac{46}{60}$

56. $\frac{3}{16}$ and $\frac{23}{24}$

LCD = 48
$\frac{9}{48}$ and $\frac{46}{48}$

57. $\frac{5}{24}, \frac{11}{36}, \frac{3}{72}$

LCD = 72
$\frac{15}{72}, \frac{22}{72}, \frac{3}{72}$

58. $\frac{1}{30}, \frac{7}{15}, \frac{1}{45}$

LCD = 90
$\frac{3}{90}, \frac{42}{90}, \frac{2}{90}$

59. $\frac{3}{56}, \frac{7}{8}, \frac{5}{7}$

LCD = 56
$\frac{3}{56}, \frac{49}{56}, \frac{40}{56}$

60. $\frac{5}{9}, \frac{1}{6}, \frac{3}{54}$

LCD = 54
$\frac{30}{54}, \frac{9}{54}, \frac{3}{54}$

61. $\frac{5}{63}, \frac{4}{21}, \frac{8}{9}$

LCD = 63
$\frac{5}{63}, \frac{12}{63}, \frac{56}{63}$

62. $\frac{3}{8}, \frac{5}{14}, \frac{13}{16}$

LCD = 112
$\frac{42}{112}, \frac{40}{112}, \frac{91}{112}$

Applications

63. Suppose that you wish to compare the lengths of the three portions of a stainless steel pin that came out of the door.

(a) What is the LCD for the three fractions?
LCD = 16

(b) Build up each fraction to an equivalent fraction that has the LCD as a denominator.

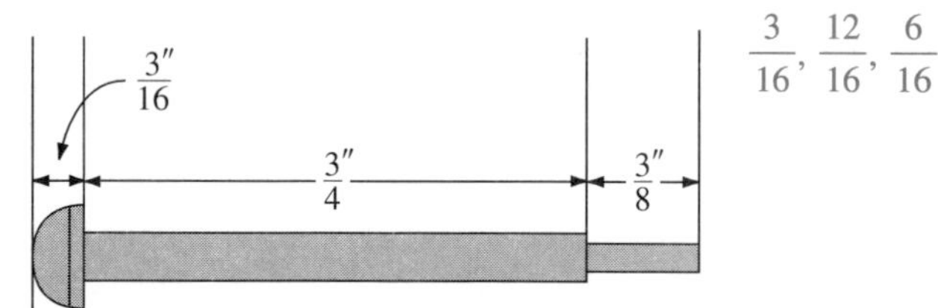

$\frac{3}{16}, \frac{12}{16}, \frac{6}{16}$

64. Suppose that you want to prepare a report on a plant that grew the following lengths during each week of a 3-week experiment.

(a) What is the LCD for the three fractions?
LCD = 96

(b) Build up each fraction to an equivalent fraction that has the LCD for a denominator.

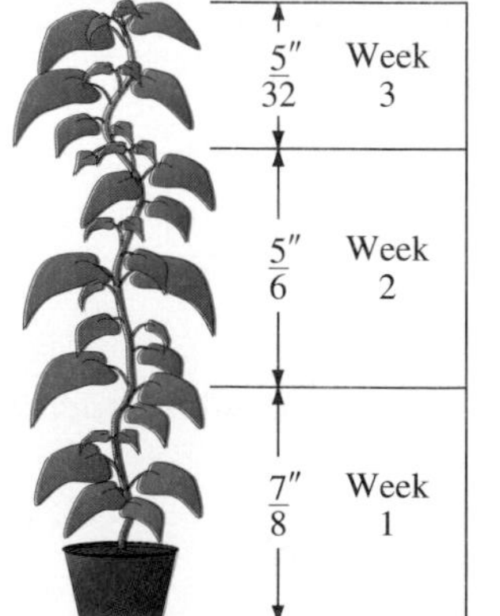

$\frac{15}{96}, \frac{80}{96}, \frac{84}{96}$

Cumulative Review Problems

65. Divide. $32\overline{)5699}$ 178 R 3

66. Divide. $182\overline{)659{,}568}$ 3624

67. Multiply. 369×27
9963

2.7 Addition and Subtraction of Fractions

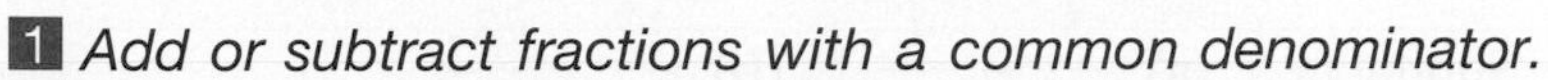

After studying this section, you will be able to:

1. *Add or subtract fractions with a common denominator.*
2. *Add or subtract fractions without a common denominator.*

1 *Adding and Subtracting Fractions with a Common Denominator*

You must have common denominators (denominators that are alike) to add or subtract fractions.

If your problem has fractions without a common denominator or if it has mixed numbers, you must use what you already know about changing the form of each fraction (how the fraction looks). Only after all the fractions have a common denominator can you add or subtract.

An important distinction: You must have common denominators to add or subtract fractions, but you need not have common denominators to multiply or divide fractions.

To add two fractions that have the same denominator, add the numerators and write the sum over the common denominator.

To illustrate we use $\frac{1}{5} + \frac{2}{5} = \frac{3}{5}$. The sketch shows that $\frac{1}{5} + \frac{2}{5} = \frac{3}{5}$.

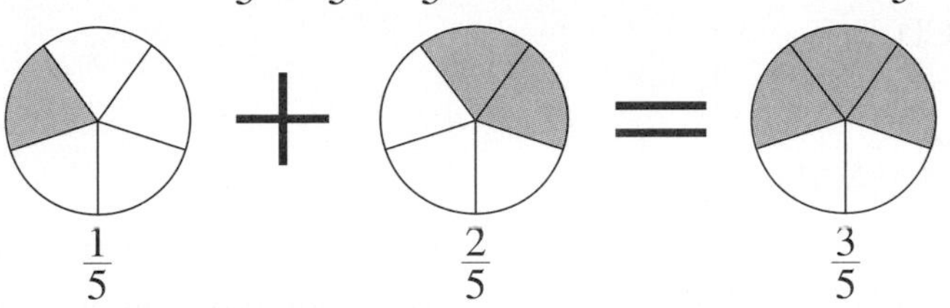

EXAMPLE 1 Add. $\frac{5}{13} + \frac{7}{13}$

$$\frac{5}{13} + \frac{7}{13} = \frac{12}{13}$$

Practice Problem 1 Add. $\frac{3}{17} + \frac{12}{17}$ ■

The answer may need to be reduced. Sometimes the answer may be written as a mixed number.

EXAMPLE 2 Add. **(a)** $\frac{4}{9} + \frac{2}{9}$ **(b)** $\frac{5}{7} + \frac{6}{7}$

(a) $\frac{4}{9} + \frac{2}{9} = \frac{6}{9} = \frac{2}{3}$ **(b)** $\frac{5}{7} + \frac{6}{7} = \frac{11}{7}$ or $1\frac{4}{7}$

Practice Problem 2 Add.

(a) $\frac{1}{12} + \frac{5}{12}$ **(b)** $\frac{13}{15} + \frac{7}{15}$ ■

A similar rule is followed for subtraction, except that the numerators are subtracted and the result placed over a common denominator. Be sure to reduce all answers when possible.

EXAMPLE 3 Subtract. **(a)** $\dfrac{5}{13} - \dfrac{4}{13}$ **(b)** $\dfrac{17}{20} - \dfrac{3}{20}$

(a) $\dfrac{5}{13} - \dfrac{4}{13} = \dfrac{1}{13}$ **(b)** $\dfrac{17}{20} - \dfrac{3}{20} = \dfrac{14}{20} = \dfrac{7}{10}$

Practice Problem 3 Subtract.

(a) $\dfrac{5}{19} - \dfrac{2}{19}$ **(b)** $\dfrac{21}{25} - \dfrac{6}{25}$ ■

2 *Adding or Subtracting Fractions without a Common Denominator*

If the two fractions do not have a common denominator, we follow the procedure in Section 2.6: Find the LCD and then build up each fraction so that its denominator is the LCD.

EXAMPLE 4 Add. $\dfrac{7}{12} + \dfrac{1}{4}$

The LCD is 12. $\frac{7}{12}$ already has the least common denominator.

$$\begin{array}{rcr} \dfrac{7}{12} & = & \boxed{\dfrac{7}{12}} \\ +\dfrac{1}{4} \times \dfrac{3}{3} & = & +\boxed{\dfrac{3}{12}} \\ \hline & & \boxed{\dfrac{10}{12}} = \dfrac{5}{6} \end{array}$$

Therefore we have found out that

$$\frac{7}{12} + \frac{1}{4} = \frac{7}{12} + \frac{3}{12} = \frac{10}{12} = \frac{5}{6}$$

It is very important to remember to reduce our final answer.

Practice Problem 4 Add. $\dfrac{2}{15} + \dfrac{2}{5}$ ■

EXAMPLE 5 Add. $\frac{7}{20} + \frac{4}{15}$

LCD = 60.

$$\frac{7}{20} \times \frac{3}{3} = \frac{21}{60} \qquad \frac{4}{15} \times \frac{4}{4} = \frac{16}{60}$$

Thus

$$\frac{7}{20} + \frac{4}{15} = \frac{21}{60} + \frac{16}{60} = \frac{37}{60}$$

Practice Problem 5 Add. $\frac{5}{12} + \frac{5}{16}$ ■

A similar procedure holds for the addition of three or more fractions.

EXAMPLE 6 Add. $\frac{3}{8} + \frac{5}{6} + \frac{1}{4}$

LCD = 24.

$$\frac{3}{8} \times \frac{3}{3} = \frac{9}{24} \qquad \frac{5}{6} \times \frac{4}{4} = \frac{20}{24} \qquad \frac{1}{4} \times \frac{6}{6} = \frac{6}{24}$$

$$\frac{3}{8} + \frac{5}{6} + \frac{1}{4} = \frac{9}{24} + \frac{20}{24} + \frac{6}{24} = \frac{35}{24} \text{ or } 1\frac{11}{24}$$

Practice Problem 6 Add. $\frac{3}{16} + \frac{1}{8} + \frac{1}{12}$ ■

EXAMPLE 7 Subtract. $\frac{17}{25} - \frac{3}{35}$

LCD = 175.

$$\frac{17}{25} \times \frac{7}{7} = \frac{119}{175} \qquad \frac{3}{35} \times \frac{5}{5} = \frac{15}{175}$$

Thus

$$\frac{17}{25} - \frac{3}{35} = \frac{119}{175} - \frac{15}{175} = \frac{104}{175}$$

Practice Problem 7 Subtract. $\frac{9}{48} - \frac{5}{32}$ ■

TEACHING TIP The ancient Egyptians, during the time of the building of the pyramids, were able to do several types of problems involving large fractions. However, the Egyptian mathematicians did not invent a way to find the LCD for any two fractions, but only certain fractional denominators. Thus they would probably be unable to solve problems like Example 7 because they could not find the LCD of 25 and 35.

EXAMPLE 8 John and Nancy have a house on $\frac{7}{8}$ acre of land. They have $\frac{1}{3}$ acre of land planted in grass. How much of the land is not planted in grass?

1. *Understand the problem.*
   Draw a picture.

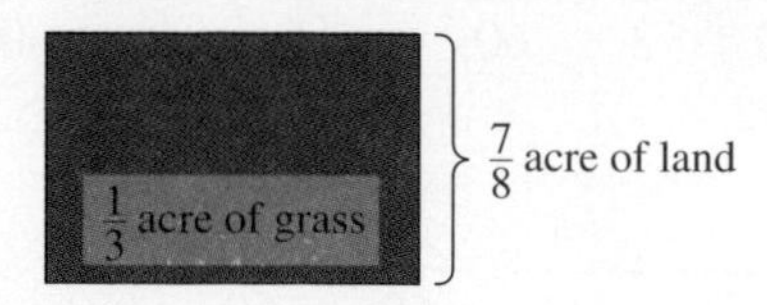

We need to subtract. $\frac{7}{8} - \frac{1}{3}$

2. *Solve and state the answer.*
   The LCD is 24.

$$\frac{7}{8} \times \frac{3}{3} = \frac{21}{24} \quad \frac{1}{3} \times \frac{8}{8} = \frac{8}{24}$$

$$\frac{7}{8} - \frac{1}{3} = \frac{21}{24} - \frac{8}{24} = \frac{13}{24}$$

We conclude that $\frac{13}{24}$ acre of the land is not planted in grass.

3. *Check.*
   The check is up to the student.

Practice Problem 8 Leon had $\frac{9}{10}$ gallon of cleaning fluid in the garage. He used $\frac{1}{4}$ gallon to clean the garage floor. How much cleaning fluid is left? ■

EXAMPLE 9 Find the value of x in the equation $x + \frac{5}{6} = \frac{9}{10}$. Reduce your answer.

The LCD for the two fractions $\frac{5}{6}$ and $\frac{9}{10}$ is 30.

$$\frac{5}{6} \times \frac{5}{5} = \frac{25}{30} \qquad \frac{9}{10} \times \frac{3}{3} = \frac{27}{30}$$

Thus we can write the equation in the equivalent form.

$$x + \frac{25}{30} = \frac{27}{30}$$

The denominators are the same. Look at the numerators. We must add 2 to 25 to get 27.

$$\frac{2}{30} + \frac{25}{30} = \frac{27}{30}$$

So if $x = \frac{2}{30}$ we reduce the fraction to obtain $x = \frac{1}{15}$.

Practice Problem 9 Find the value of x in the equation $x + \frac{3}{10} = \frac{23}{25}$. ■

2.7 Exercises

Add or subtract. Simplify all answers.

1. $\frac{5}{9}+\frac{2}{9}$

$\frac{7}{9}$

2. $\frac{5}{8}+\frac{2}{8}$

$\frac{7}{8}$

3. $\frac{11}{30}+\frac{11}{30}$

$\frac{22}{30}=\frac{11}{15}$

4. $\frac{3}{28}+\frac{7}{28}$

$\frac{10}{28}=\frac{5}{14}$

5. $\frac{21}{23}-\frac{1}{23}$

$\frac{20}{23}$

6. $\frac{5}{24}-\frac{3}{24}$

$\frac{2}{24}=\frac{1}{12}$

7. $\frac{53}{88}-\frac{19}{88}$

$\frac{34}{88}=\frac{17}{44}$

8. $\frac{103}{110}-\frac{3}{110}$

$\frac{100}{110}=\frac{10}{11}$

Add or subtract. Simplify all answers.

9. $\frac{1}{4}+\frac{1}{3}$

$\frac{3}{12}+\frac{4}{12}=\frac{7}{12}$

10. $\frac{2}{7}+\frac{1}{2}$

$\frac{4}{14}+\frac{7}{14}=\frac{11}{14}$

11. $\frac{3}{10}+\frac{3}{20}$

$\frac{6}{20}+\frac{3}{20}=\frac{9}{20}$

12. $\frac{4}{9}+\frac{1}{6}$

$\frac{8}{18}+\frac{3}{18}=\frac{11}{18}$

13. $\frac{1}{8}+\frac{3}{4}$

$\frac{1}{8}+\frac{6}{8}=\frac{7}{8}$

14. $\frac{5}{16}+\frac{1}{2}$

$\frac{5}{16}+\frac{8}{16}=\frac{13}{16}$

15. $\frac{4}{5}+\frac{7}{10}$

$\frac{8}{10}+\frac{7}{10}=\frac{3}{2}$ *or* $1\frac{1}{2}$

16. $\frac{1}{3}+\frac{4}{5}$

$\frac{5}{15}+\frac{12}{15}=\frac{17}{15}$ *or* $1\frac{2}{15}$

17. $\frac{3}{10}+\frac{7}{100}$

$\frac{30}{100}+\frac{7}{100}=\frac{37}{100}$

18. $\frac{13}{100}+\frac{7}{10}$

$\frac{13}{100}+\frac{70}{100}=\frac{83}{100}$

19. $\frac{3}{25}+\frac{1}{35}$

$\frac{21}{175}+\frac{5}{175}=\frac{26}{175}$

20. $\frac{3}{15}+\frac{1}{25}$

$\frac{15}{75}+\frac{3}{75}=\frac{18}{75}=\frac{6}{25}$

21. $\frac{7}{8}+\frac{5}{12}$

$\frac{21}{24}+\frac{10}{24}=\frac{31}{24}$ *or* $1\frac{7}{24}$

22. $\frac{5}{6}+\frac{7}{8}$

$\frac{20}{24}+\frac{21}{24}=\frac{41}{24}$ *or* $1\frac{17}{24}$

23. $\frac{3}{8}+\frac{3}{10}$

$\frac{15}{40}+\frac{12}{40}=\frac{27}{40}$

24. $\frac{12}{35}+\frac{1}{10}$

$\frac{24}{70}+\frac{7}{70}=\frac{31}{70}$

25. $\frac{5}{12}-\frac{1}{6}$

$\frac{5}{12}-\frac{2}{12}=\frac{3}{12}=\frac{1}{4}$

26. $\frac{37}{20}-\frac{2}{5}$

$\frac{37}{20}-\frac{8}{20}=\frac{29}{20}$ *or* $1\frac{9}{20}$

27. $\frac{3}{7}-\frac{1}{5}$

$\frac{15}{35}-\frac{7}{35}=\frac{8}{35}$

28. $\frac{7}{8}-\frac{5}{6}$

$\frac{21}{24}-\frac{20}{24}=\frac{1}{24}$

29. $\frac{7}{8}-\frac{5}{16}$

$\frac{14}{16}-\frac{5}{16}=\frac{9}{16}$

30. $\frac{9}{50}-\frac{2}{25}$

$\frac{9}{50}-\frac{4}{50}=\frac{5}{50}=\frac{1}{10}$

31. $\frac{5}{12}-\frac{7}{30}$

$\frac{25}{60}-\frac{14}{60}=\frac{11}{60}$

32. $\frac{9}{24}-\frac{3}{8}$

$\frac{9}{24}-\frac{9}{24}=0$

33. $\frac{11}{12} - \frac{2}{3}$

$\frac{11}{12} - \frac{8}{12} = \frac{3}{12} = \frac{1}{4}$

34. $\frac{7}{10} - \frac{2}{5}$

$\frac{7}{10} - \frac{4}{10} = \frac{3}{10}$

35. $\frac{5}{6} - \frac{1}{3}$

$\frac{5}{6} - \frac{2}{6} = \frac{3}{6} = \frac{1}{2}$

36. $\frac{7}{9} - \frac{1}{2}$

$\frac{14}{18} - \frac{9}{18} = \frac{5}{18}$

37. $\frac{5}{12} - \frac{5}{16}$

$\frac{20}{48} - \frac{15}{48} = \frac{5}{48}$

38. $\frac{2}{5} - \frac{2}{15}$

$\frac{6}{15} - \frac{2}{15} = \frac{4}{15}$

39. $\frac{10}{16} - \frac{5}{8}$

$\frac{10}{16} - \frac{10}{16} = 0$

40. $\frac{5}{6} - \frac{10}{12}$

$\frac{10}{12} - \frac{10}{12} = 0$

41. $\frac{20}{35} - \frac{4}{7}$

$\frac{20}{35} - \frac{20}{35} = 0$

42. $\frac{11}{20} - \frac{3}{8}$

$\frac{22}{40} - \frac{15}{40} = \frac{7}{40}$

43. $\frac{4}{5} + \frac{1}{20} + \frac{3}{4}$

$\frac{16}{20} + \frac{1}{20} + \frac{15}{20} = \frac{32}{20} = \frac{8}{5}$ *or* $1\frac{3}{5}$

44. $\frac{1}{3} + \frac{2}{24} + \frac{1}{6}$

$\frac{8}{24} + \frac{2}{24} + \frac{4}{24}$

$= \frac{14}{24} = \frac{7}{12}$

45. $\frac{5}{30} + \frac{3}{40} + \frac{1}{8}$

$\frac{20}{120} + \frac{9}{120} + \frac{15}{120} = \frac{44}{120} = \frac{11}{30}$

46. $\frac{1}{12} + \frac{3}{14} + \frac{4}{21}$

$\frac{7}{84} + \frac{18}{84} + \frac{16}{84} = \frac{41}{84}$

47. $\frac{1}{5} + \frac{2}{3} + \frac{11}{15}$

$\frac{3}{15} + \frac{10}{15} + \frac{11}{15} = \frac{24}{15} = \frac{8}{5} = 1\frac{3}{5}$

48. $\frac{2}{3} + \frac{1}{6} + \frac{5}{12}$

$\frac{8}{12} + \frac{2}{12} + \frac{5}{12}$

$\frac{15}{12} = \frac{5}{4}$ *or* $1\frac{1}{4}$

Find the value of x in the equation.

49. $x + \frac{1}{7} = \frac{5}{14}$

$x = \frac{3}{14}$

50. $x + \frac{1}{8} = \frac{7}{16}$

$x = \frac{5}{16}$

51. $x + \frac{2}{3} = \frac{9}{11}$

$x = \frac{5}{33}$

52. $x + \frac{3}{4} = \frac{17}{18}$

$x = \frac{7}{36}$

53. $x - \frac{1}{5} = \frac{4}{12}$

$x = \frac{8}{15}$

54. $x - \frac{2}{3} = \frac{3}{11}$

$x = \frac{31}{33}$

Applications

55. This morning at breakfast, Keith used $\frac{4}{5}$ cup of corn flakes and Danny used $\frac{2}{3}$ cup. How many cups of cereal did Keith and Danny use?
$1\frac{7}{15}$ cups

56. Jackie started running on Wednesday. She ran $\frac{1}{2}$ mile. On Friday she ran $\frac{2}{3}$ mile. How many miles has she run so far this week? $1\frac{1}{6}$ miles

57. Laurie purchased $\frac{2}{3}$ pound of bananas and $\frac{5}{6}$ pound of seedless grapes. How many pounds of fruit did she purchase? $1\frac{1}{2}$ pounds

58. Geoffrey is attaching solar energy panels to the roof of the science building. Each panel uses a bolt that goes through the panel to the frame of the roof ($4\frac{1}{3}$ inches), a washer $\frac{5}{16}$-inch thick, and a $\frac{3}{4}$-inch nut. How long a bolt does he need?
$5\frac{19}{48}$ inches

59. Consuela is a collage artist. She assembles different objects on canvases to present a message to the viewer. She is working on a collage that is attached to a dart board. She places a computer chip $\frac{1}{3}$ inch from the center bull's-eye. Then, continuing in a straight line from the bull's-eye, she places a tea bag $\frac{5}{9}$ inch from the computer chip. The last item she uses is a chewing gum wrapper placed $\frac{3}{4}$ inch from the tea bag. How far from the bull's-eye is the chewing gum wrapper?
$1\frac{23}{36}$ inches

60. An infant's father knows that straight apple juice is too strong for his daughter. Her bottle is $\frac{1}{2}$ full, and he adds $\frac{1}{3}$ of a bottle of water to dilute the apple juice.
(a) How much of the bottle is full? $\frac{5}{6}$
(b) If she drinks $\frac{2}{5}$ of the bottle, how much is left?
$\frac{13}{30}$ of a bottle is left.

61. The tomatoes in the Botany Department's organic garden needed a boost so Mary added $\frac{1}{3}$ cup of compost to every $\frac{3}{4}$ cup of soil. How many cups of the mixture did she end up with if she started with $5\frac{1}{4}$ cups of soil? $7\frac{7}{12}$ cups

62. On Saturday morning the gasoline storage tank at Dusty's station was $\frac{11}{12}$ full. During Saturday and Sunday the attendants pumped out $\frac{2}{3}$ of the tank. How much remained in the tank when they opened Monday morning?
$\frac{1}{4}$ tank

63. The manager at Fit Factory Health Club was going through his files for 1996 and discovered that only $\frac{9}{14}$ of the members actually used the club. When he checked the numbers from the previous year of 1995, he found that $\frac{5}{6}$ of the members had used the club. What fractional part of the membership represents the decrease in club usage?
$\frac{4}{21}$ of the membership

64. Travis typed $\frac{11}{12}$ of his book report on his computer. Then he printed out $\frac{3}{5}$ of his book report on his computer printer. Suddenly, there was a power outage, and he discovered that he hadn't saved his book report before the power went off. What fractional part of the book report was lost when the power failed?
$\frac{19}{60}$ of the book report was lost.

Cumulative Review Problems

65. Reduce to the lowest terms. $\frac{15}{85}$

$\frac{3}{17}$

66. Reduce to the lowest terms. $\frac{38}{57}$

$\frac{2}{3}$

67. Change to a mixed number. $\frac{123}{16}$

$7\frac{11}{16}$

68. Change to an improper fraction. $14\frac{3}{7}$

$\frac{101}{7}$

2.8 Addition and Subtraction of Mixed Numbers

MathPro

Video 6

SSM

After studying this section, you will be able to:

1. *Add two or more mixed numbers.*
2. *Subtract mixed numbers.*

1 *Adding Mixed Numbers*

When adding mixed numbers, it is best to add the fractions together and then add the whole numbers together.

EXAMPLE 1 Add. $3\frac{1}{8} + 2\frac{5}{8}$

$$\begin{array}{r} 3\frac{1}{8} \\ +\,2\frac{5}{8} \\ \hline 5\frac{6}{8} \end{array}$$

Add the whole numbers. $3 + 2 = 5$ → 5

Add the fractions. $\frac{1}{8} + \frac{5}{8} = \frac{6}{8}$ → $\frac{6}{8}$

$= 5\frac{3}{4}$ ← Reduce $\frac{6}{8} = \frac{3}{4}$

Practice Problem 1 Add. $5\frac{1}{12} + 9\frac{5}{12}$ ■

If the fraction portions of the mixed numbers do not have a common denominator, we must build up the fraction parts to obtain a common denominator before adding.

EXAMPLE 2 Add. $1\frac{2}{7} + 5\frac{1}{3}$

The LCD of $\frac{2}{7}$ and $\frac{1}{3}$ is 21.

$$\frac{2}{7} \times \frac{3}{3} = \frac{6}{21} \qquad \frac{1}{3} \times \frac{7}{7} = \frac{7}{21}$$

Thus $1\frac{2}{7} + 5\frac{1}{3} = 1\frac{6}{21} + 5\frac{7}{21}$.

$$\begin{array}{r} 1\frac{6}{21} \\ +\,5\frac{7}{21} \\ \hline 6\frac{13}{21} \end{array}$$

Add the whole numbers. $1 + 5$ → 6

Add the fractions. $\frac{6}{21} + \frac{7}{21}$ → $\frac{13}{21}$

Practice Problem 2 Add. $6\frac{1}{4} + 2\frac{2}{5}$ ■

If the sum of the fractions is an improper fraction, we convert it to a mixed number and add the whole numbers together.

EXAMPLE 3 Add. $6\frac{5}{6} + 4\frac{3}{8}$

The LCD of $\frac{5}{6}$ and $\frac{3}{8}$ is 24.

$$\begin{array}{rcl} 6\;\boxed{\frac{5}{6}\times\frac{4}{4}} & = & 6\;\frac{20}{24} \\ +\,4\;\boxed{\frac{3}{8}\times\frac{3}{3}} & = & +\,4\;\frac{9}{24} \\ \hline \end{array}$$

Add the whole numbers. → $10\frac{29}{24}$ ← *Add the fractions.*

$= 10 + \boxed{1\frac{5}{24}}$ *Since* $\frac{29}{24} = 1\frac{5}{24}$

$= 11\frac{5}{24}$ *We add the whole numbers* $10 + 1 = 11$.

Practice Problem 3 Add. $7\frac{1}{4} + 3\frac{5}{6}$ ■

2 *Subtracting Mixed Numbers*

Subtracting mixed numbers is like adding.

EXAMPLE 4 Subtract. $8\frac{5}{7} - 5\frac{5}{14}$

The LCD of $\frac{5}{7}$ and $\frac{5}{14}$ is 14.

$$\begin{array}{rcl} 8\;\boxed{\frac{5}{7}\times\frac{2}{2}} & = & 8\frac{10}{14} \\ -\,5\frac{5}{14} & = & 5\frac{5}{14} \\ \hline \end{array}$$

Subtract the whole numbers. → $3\frac{5}{14}$ ← Subtract the fractions.

Practice Problem 4 Subtract. $12\frac{5}{6} - 7\frac{5}{12}$ ■

Sometimes we must borrow before we can subtract.

TEACHING TIP Studying and understanding Example 5 or one similar to it will be most critical for the students. You may want to add a similar example on the board. We suggest working out the problem $7\frac{1}{6} - 3\frac{16}{21}$. Be sure the students see how to change this to the problem $7\frac{7}{42} - 3\frac{32}{42}$ before finally writing the expression $6\frac{49}{42} - 3\frac{32}{42} = 3\frac{17}{42}$.

EXAMPLE 5 Subtract. **(a)** $9\frac{1}{4} - 6\frac{5}{14}$ **(b)** $15 - 9\frac{3}{16}$

The following example is fairly challenging. Read through each step carefully. Be sure to have paper and pencil handy and see if you can verify each step.

(a) The LCD of $\frac{1}{4}$ and $\frac{5}{14}$ is 28.

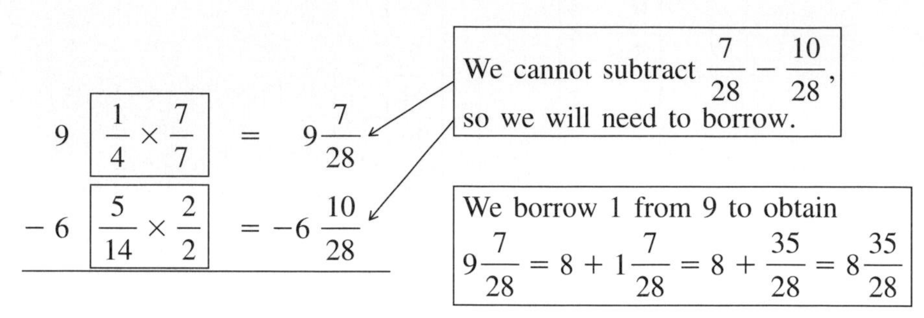

We write this as

$$
\begin{array}{rcr}
9\frac{7}{28} & = & 8\frac{35}{28} \\
-\;6\frac{10}{28} & = & -\;6\frac{10}{28} \\
\hline
\boxed{8 - 6 = 2} & \rightarrow & 2\frac{25}{28} \leftarrow \boxed{\frac{35}{28} - \frac{10}{28} = \frac{25}{28}}
\end{array}
$$

(b) The LCD = 16.

$$
\begin{array}{rcr}
15 & = & 14\frac{16}{16} \\
-\;9\frac{3}{16} & = & -\;9\frac{3}{16} \\
\hline
\boxed{14 - 9 = 5} & \rightarrow & 5\frac{13}{16} \leftarrow \boxed{\frac{16}{16} - \frac{3}{16} = \frac{13}{16}}
\end{array}
$$

We borrow 1 from 15 to obtain $15 = 14 + 1 = 14 + \frac{16}{16} = 14\frac{16}{16}$

Practice Problem 5 Subtract. **(a)** $9\frac{1}{8} - 3\frac{2}{3}$ **(b)** $18 - 6\frac{7}{18}$ ■

EXAMPLE 6 A plumber had a pipe $5\frac{3}{16}$ inches long for a fitting under the sink. He needed a pipe that was $3\frac{7}{8}$ inches long, so he cut the pipe down. How much of the pipe did he cut off?

We will need to subtract $5\frac{3}{16} - 3\frac{7}{8}$ to find the length that was cut off.

$$5\frac{3}{16} \quad = \quad 5\frac{3}{16}$$

$$-\,3\frac{7}{8} \times \frac{2}{2} = -\,3\frac{14}{16}$$

We borrow 1 from 5 to obtain $5\frac{3}{16} = 4 + 1\frac{3}{16} = 4 + \frac{19}{16}$

$$4\frac{19}{16}$$

$$-\,3\frac{14}{16}$$

$$1\frac{5}{16}$$

$4 - 3 = 1$ ⟶ $1\frac{5}{16}$ ⟵ $\frac{19}{16} - \frac{14}{16} = \frac{5}{16}$

The plumber had to cut off $1\frac{5}{16}$ inches of pipe.

Practice Problem 6 Hillary and Sam purchased $6\frac{1}{4}$ gallons of paint to paint the inside of their house. They used $4\frac{2}{3}$ gallons of paint. How much paint was left over? ■

ALTERNATIVE METHOD Can mixed numbers be added and subtracted as improper fractions? Yes. Recall Example 5(a).

$$9\frac{1}{4} - 6\frac{5}{14} = 2\frac{25}{28}$$

If we write $9\frac{1}{4} - 6\frac{5}{14}$ using improper fractions, we have $\frac{37}{4} - \frac{89}{14}$. Now we build up each of these improper fractions so that they both have the LCD for their denominators.

$$\begin{array}{rcl} \dfrac{37}{4} & \boxed{\times \dfrac{7}{7}} & = \dfrac{259}{28} \\ -\dfrac{89}{14} & \boxed{\times \dfrac{2}{2}} & = \dfrac{178}{28} \\ \hline & & \dfrac{81}{28} = 2\dfrac{25}{28} \end{array}$$

The same result is obtained as in Example 5(a). This method does not require borrowing. However, you do work with larger numbers. For more practice, see problems 47–50.

Developing Your Study Skills

PROBLEMS WITH ACCURACY

Strive for accuracy. Mistakes are often made because of human error rather than lack of understanding. Such mistakes are frustrating. A simple arithmetic or sign error can lead to an incorrect answer.

These five steps will help you cut down on errors.

1. Work carefully, and take your time. Do not rush through a problem just to get it done.
2. Concentrate on the problem. Sometimes problems become mechanical, and your mind begins to wander. You become careless and make a mistake.
3. Check your problem. Be sure that you copied it correctly from the book.
4. Check your computations from step to step. Check the solution to the problem. Does it work? Does it make sense?
5. Keep practicing new skills. Remember the old saying ''Practice makes perfect.'' An increase in practice results in an increase in accuracy. Many errors are due simply to lack of practice.

There is no magic formula for eliminating all errors, but these five steps will be a tremendous help in reducing them.

2.8 Exercises

Add or subtract. Simplify all answers. Express the answer as a mixed number.

1. $7\frac{1}{8} + 2\frac{5}{8}$ $\quad 9\frac{6}{8} = 9\frac{3}{4}$

2. $6\frac{3}{10} + 4\frac{1}{10}$ $\quad 10\frac{4}{10} = 10\frac{2}{5}$

3. $15\frac{3}{14} - 11\frac{1}{14}$ $\quad 4\frac{2}{14} = 4\frac{1}{7}$

4. $8\frac{3}{4} - 3\frac{1}{4}$ $\quad 5\frac{2}{4} = 5\frac{1}{2}$

5. $12\frac{1}{3} + 5\frac{1}{6}$ $\quad 12\frac{2}{6} + 5\frac{1}{6} = 17\frac{3}{6} = 17\frac{1}{2}$

6. $20\frac{1}{4} + 3\frac{1}{8}$ $\quad 20\frac{2}{8} + 3\frac{1}{8} = 23\frac{3}{8}$

7. $5\frac{4}{5} + 10\frac{3}{10}$ $\quad 5\frac{8}{10} + 10\frac{3}{10} = 15\frac{11}{10} = 16\frac{1}{10}$

8. $6\frac{3}{8} + 4\frac{1}{16}$ $\quad 6\frac{6}{16} + 4\frac{1}{16} = 10\frac{7}{16}$

9. $1 - \frac{3}{7}$ $\quad \frac{7}{7} - \frac{3}{7} = \frac{4}{7}$

10. $1 - \frac{9}{11}$ $\quad \frac{11}{11} - \frac{9}{11} = \frac{2}{11}$

11. $1\frac{5}{6} + \frac{7}{8}$ $\quad 1\frac{20}{24} + \frac{21}{24} = 1\frac{41}{24} = 2\frac{17}{24}$

12. $1\frac{2}{3} + \frac{5}{18}$ $\quad 1\frac{12}{18} + \frac{5}{18} = 1\frac{17}{18}$

13. $7\frac{1}{2} + 8\frac{3}{4}$ $\quad 7\frac{2}{4} + 8\frac{3}{4} = 15\frac{5}{4} = 16\frac{1}{4}$

14. $9\frac{1}{6} + 2\frac{2}{3}$ $\quad 9\frac{1}{6} + 2\frac{4}{6} = 11\frac{5}{6}$

15. $8\frac{2}{3} - 5\frac{1}{6}$ $\quad 8\frac{4}{6} - 5\frac{1}{6} = 3\frac{3}{6} = 3\frac{1}{2}$

16. $7\frac{4}{5} - 2\frac{1}{10}$ $\quad 7\frac{8}{10} - 2\frac{1}{10} = 5\frac{7}{10}$

17. $18\frac{1}{6} - 10\frac{3}{4}$ $\quad 17\frac{28}{24} - 10\frac{18}{24} = 7\frac{10}{24} = 7\frac{5}{12}$

18. $12\frac{4}{9} - 7\frac{5}{6}$ $\quad 11\frac{26}{18} - 7\frac{15}{18} = 4\frac{11}{18}$

19. $30 - 15\frac{3}{7}$ $\quad 29\frac{7}{7} - 15\frac{3}{7} = 14\frac{4}{7}$

20. $25 - 14\frac{2}{11}$ $\quad 24\frac{11}{11} - 14\frac{2}{11} = 10\frac{9}{11}$

Add or subtract. Express the answer as a mixed number.

21. $15\frac{4}{15} + 26\frac{8}{15}$ $\quad 41\frac{4}{5}$

22. $22\frac{1}{8} + 14\frac{3}{8}$ $\quad 36\frac{1}{2}$

23. $4\frac{1}{3} + 2\frac{1}{4}$ $\quad 4\frac{4}{12} + 2\frac{3}{12} = 6\frac{7}{12}$

24. $6\frac{1}{8} + 7\frac{3}{4}$ $\quad 6\frac{1}{8} + 7\frac{6}{8} = 13\frac{7}{8}$

25. $2\frac{1}{15} + 14\frac{3}{5}$ $\quad 2\frac{1}{15} + 14\frac{9}{15} = 16\frac{2}{3}$

26. $8\frac{1}{7} + 3\frac{11}{14}$ $\quad 8\frac{2}{14} + 3\frac{11}{14} = 11\frac{13}{14}$

27. $47\frac{3}{10} + 26\frac{5}{8}$ $\quad 47\frac{12}{40} + 26\frac{25}{40} = 73\frac{37}{40}$

28. $34\frac{1}{20} + 45\frac{8}{15}$ $\quad 34\frac{3}{60} + 45\frac{32}{60} = 79\frac{35}{60} = 79\frac{7}{12}$

29. $19\frac{5}{6} - 14\frac{1}{3}$ $\quad 19\frac{5}{6} - 14\frac{2}{6} = 5\frac{3}{6} = 5\frac{1}{2}$

30. $27\frac{11}{12} - 21\frac{1}{3}$ $\quad 27\frac{11}{12} - 21\frac{4}{12} = 6\frac{7}{12}$

31. $3\frac{5}{6} - 3\frac{10}{12}$ $\quad 3\frac{10}{12} - 3\frac{10}{12} = 0$

32. $7\frac{1}{14} - 4\frac{3}{7}$ $\quad 6\frac{15}{14} - 4\frac{6}{14} = 2\frac{9}{14}$

33. $12\frac{3}{20} - 7\frac{7}{15}$ $\quad 11\frac{69}{60} - 7\frac{28}{60} = 4\frac{41}{60}$

34. $8\frac{5}{12} - 5\frac{9}{10}$ $\quad 7\frac{85}{60} - 5\frac{54}{60} = 2\frac{31}{60}$

35. $12 - 3\frac{7}{15}$ $\quad 11\frac{15}{15} - 3\frac{7}{15} = 8\frac{8}{15}$

36. $19 - 6\frac{3}{7}$ $\quad 18\frac{7}{7} - 6\frac{3}{7} = 12\frac{4}{7}$

37. $\begin{array}{r} 120 \\ -\ 17\frac{3}{8} \\ \hline \end{array}$ $\quad \begin{array}{r} 119\frac{8}{8} \\ -\ 17\frac{3}{8} \\ \hline 102\frac{5}{8} \end{array}$

38. $\begin{array}{r} 98 \\ -\ 89\frac{15}{17} \\ \hline \end{array}$ $\quad \begin{array}{r} 97\frac{17}{17} \\ -\ 89\frac{15}{17} \\ \hline 8\frac{2}{17} \end{array}$

39. $\begin{array}{r} 3\frac{1}{8} \\ 2\frac{1}{3} \\ +7\frac{3}{4} \\ \hline \end{array}$ $\quad \begin{array}{r} 3\frac{3}{24} \\ 2\frac{8}{24} \\ +\ 7\frac{18}{24} \\ \hline 12\frac{29}{24} = 13\frac{5}{24} \end{array}$

40. $\begin{array}{r} 4\frac{2}{3} \\ 3\frac{1}{5} \\ +6\frac{3}{4} \\ \hline \end{array}$ $\quad \begin{array}{r} 4\frac{40}{60} \\ 3\frac{12}{60} \\ +6\frac{45}{60} \\ \hline 13\frac{97}{60} = 14\frac{37}{60} \end{array}$

Applications

Solve each problem.

41. Lee Hong rode his mountain bike through part of the Sangre de Cristo Mountains in New Mexico. On Wednesday he rode $20\frac{3}{4}$ miles. On Thursday he rode $22\frac{3}{8}$ miles. What was his total biking distance during those two days? $43\frac{1}{8}$ miles

42. Kathryn traveled on a flight that took $7\frac{5}{6}$ hours. She connected with another flight that took $5\frac{4}{5}$ hours. What was her total time in the air?
$13\frac{19}{30}$ hours

43. Renaldo bought $6\frac{3}{8}$ pounds of cheese for a party. The guests ate $4\frac{1}{3}$ pounds of the cheese. How much cheese did Renaldo have left over?
$2\frac{1}{24}$ pounds

44. Shanna purchased stock at $\$21\frac{3}{8}$ per share. When her son was ready for college, she sold the stock at $\$93\frac{5}{8}$ per share. How much did she make per share for her son's tuition? $\$72\frac{1}{4}$ per share

45. Jack's dog is too heavy. According to the veterinarian, the dog needs to lose 16 pounds over the next three months. During the first month he lost $5\frac{1}{8}$ pounds. The second month he lost $4\frac{2}{3}$ pounds.
(a) How much weight did Jack's dog lose during the first two months?
(b) How much weight does his dog need to lose in the third month to reach the goal advised by the veterinarian?
(a) $9\frac{19}{24}$ pounds; (b) $6\frac{5}{24}$ pounds

46. A young man has been under a doctor's care to lose weight. His doctor wanted him to lose 46 pounds in the first three months. He lost $17\frac{5}{8}$ pounds the first month and $13\frac{1}{2}$ pounds the second month.
(a) How much did he lose during the first two months?
(b) How much would he need to lose in the third month to reach the goal?
(a) $17\frac{5}{8} + 13\frac{1}{2} = 31\frac{1}{8}$ pounds
(b) $46 - 31\frac{1}{8} = 14\frac{7}{8}$ pounds

To Think About

Use improper fractions as discussed in the text to perform the following calculations.

47. $\frac{379}{8} + \frac{89}{5}$
$\frac{1895}{40} + \frac{712}{40} = \frac{2607}{40}$ *or* $65\frac{7}{40}$

48. $\frac{151}{6} - \frac{130}{7}$
$\frac{1057}{42} - \frac{780}{42} = \frac{277}{42}$ *or* $6\frac{25}{42}$

49. $\frac{200}{3} - \frac{153}{7}$
$\frac{1400}{21} - \frac{459}{21} = \frac{941}{21}$ *or* $44\frac{17}{21}$

50. $\frac{400}{9} + \frac{126}{6}$
$\frac{800}{18} + \frac{378}{18} = \frac{589}{9}$ *or* $65\frac{4}{9}$

Cumulative Review Problems

Multiply.

51. $\begin{array}{r} 12{,}367 \\ \times\quad 9 \\ \hline 111{,}303 \end{array}$

52. $\begin{array}{r} 304 \\ \times\ 128 \\ \hline 2\ 432 \\ 6\ 08 \\ 30\ 4 \\ \hline 38{,}912 \end{array}$

53. $\begin{array}{r} 6737 \\ \times\quad 76 \\ \hline 40\ 422 \\ 471\ 59 \\ \hline 512{,}012 \end{array}$

54. $\begin{array}{r} 4050 \\ \times\ 2106 \\ \hline 24\ 300 \\ 405\ 00 \\ 8\ 100 \\ \hline 8{,}529{,}300 \end{array}$

Putting Your Skills to Work

FARMING ON THE HALVES

An interesting use of fractions is found among the rural farm areas of the Maritime Provinces (New Brunswick, Nova Scotia, and Prince Edward Island) of Canada. They have a custom of ''farming on the halves'' and farming on other fractional values as well.

PROBLEMS FOR INDIVIDUAL INVESTIGATION AND ANALYSIS

1. Everett Hatfield raises a blueberry crop in Nova Scotia on 40 acres of land. He has arranged with the local farmers' cooperative to ''farm on the halves.'' He will share one half of the expenses of raising the crop and one half of the profits with the local cooperative. It cost \$1300 to burn a portion of the fields each year with a portable propane burner to maximize the yield of the blueberry crop. If his fields bring in 23 tons of blueberries, the price of blueberries is $\$2\frac{1}{2}$ per pound, and the workers are paid $\$1\frac{3}{4}$ per pound to pick the blueberries, how much profit will he take in? \$16,600

2. Dick Wightman owns 100 acres of potato fields in Prince Edward Island. He will ask the farming cooperative to take three fifths of the expenses of raising the crop, and they will therefore also take three fifths of the profits from raising the crop. It cost $\$1\frac{1}{2}$ per acre to till the soil and plant and fertilize the potatoes. It cost $\$3\frac{1}{4}$ per acre to water and weed the potato fields. Dick sold $26\frac{1}{4}$ tons of potatoes from his fields. The cost of the laborers who harvested the potatoes was $\$\frac{1}{2}$ per pound. The price that he sold his potatoes for this year was $\$\frac{3}{4}$ per pound. How much profit will Dick take in?
\$5060

PROBLEMS FOR GROUP INVESTIGATION AND COOPERATIVE PROBLEM SOLVING

Together with some other members of your class determine the following.

3. Loring Kerr needs to raise \$24,750 from farming wheat. He hopes to produce 300 bushels of wheat per acre on his 60-acre wheat fields. The expected price of wheat this year is $\$2\frac{1}{2}$ per bushel. The cost to harvest the wheat is \$200 per acre. He does the rest of the labor himself. The local cooperative has offered to pay for $\frac{3}{8}$ of the costs and take $\frac{3}{8}$ of the profits. Can he afford to take their offer? If not what fractional part of the costs and the profits should he have as a goal?
No; He needs them to take $\frac{1}{4}$ of the costs and $\frac{1}{4}$ of the profits.

4. Betty and Walter Ames want to raise corn on their farm this year. They have reached an agreement with the local farming cooperative to pay for $\frac{2}{3}$ of the expenses and also to receive $\frac{2}{3}$ of the profits. The hope to produce 210 bushels of corn per acre from their farm. The expected price of corn this year is $\$1\frac{1}{2}$ per bushel. The cost to harvest corn is \$150 per acre. Betty and Walter do the rest of the labor themselves. They want to raise \$19,250 from the corn crop this year. They own 1000 acres of corn fields but only want to plant the minimum number of acres to achieve their financial goal. How many acres of corn should they plant?
350 acres

INTERNET CONNECTIONS: Go to ''http://www.prenhall.com/tobey'' to be connected

Site: Farm Workers, by Type, 1910–1995 (United States Department of Agriculture) or a related site

This site shows a graph of the number of family workers and hired workers on American farms. Use the graph to answer the following questions.

5. Estimate the number of family workers in 1950.

6. Estimate the number of hired workers in 1925.

7. Estimate the total number of farm workers in 1985.

8. Tell what patterns you see in the graph.

2.9 Applied Problems Involving Fractions

MathPro

Video 6

SSM

After studying this section, you will be able to:

1 *Solve applied problems with fractions.*

1 *Solving Applied Problems with Fractions*

All problem solving requires the same kind of thinking. In this section we will combine problem-solving skills with our new computational skills with fractions. Sometimes the difficulty is in figuring out what must be done. Sometimes it is in doing the computation. Remember that *estimating* is important in problem solving. We may use the following steps.

1. *Understand the problem.*
 (a) Read the problem carefully.
 (b) Draw a picture if this helps you.
 (c) Fill in the mathematics blueprint.

2. *Solve.*
 (a) Perform the calculations.
 (b) State the answer, including the unit of measure.

3. *Check.*
 (a) Estimate the answer.
 (b) Compare the exact answer with the estimate to see if your answer is reasonable.

TEACHING TIP The stock market provides many interesting examples of calculations with fractions. However, some students are unfamiliar with the stock market and how prices there are quoted with fractions. It is good to go over Example 1 and ask students if they have ever purchased stock. Can they think of other types of problems where you have to calculate with fractions when dealing with the stock market? Usually one or two students will give some good suggestions. Be sure students who are unfamiliar with stocks feel free to ask questions.

EXAMPLE 1 Maria bought Digital Equipment Corporation stock at $\$26\frac{1}{8}$. The stock's value went up $\$2\frac{5}{8}$ the first week. The second week it went up $\$3\frac{3}{4}$. What was its value after these two increases?

1. *Understand the problem.*
We draw a picture to help us.

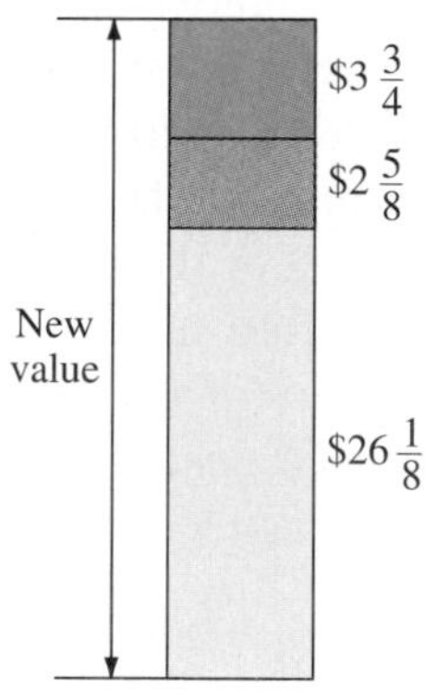

Then we fill in the Mathematics Blueprint.

MATHEMATICS BLUEPRINT FOR PROBLEM SOLVING

Gather the Facts	What Am I Asked to Do?	How Do I Proceed?	Key Points to Remember
Purchase Price: $\$26\frac{1}{8}$ Week One: up $\$2\frac{5}{8}$ Week Two: up $\$3\frac{3}{4}$	Find the value of the stock after the two increases.	Add the two increases to the purchase price.	When adding mixed numbers, add the whole numbers first and then add the fractions.

2. *Solve and state the answer.*

Add the three amounts. $26\frac{1}{8} + 2\frac{5}{8} + 3\frac{3}{4}$

$$\text{LCD} = 8 \qquad \begin{aligned} 26\frac{1}{8} &= 26\frac{1}{8} \\ 2\frac{5}{8} &= 2\frac{5}{8} \\ + \ 3\boxed{\frac{3}{4} \times \frac{2}{2}} &= + \ 3\frac{6}{8} \\ \hline & \ \ 31\frac{12}{8} = 32\frac{4}{8} = 32\frac{1}{2} \end{aligned}$$

The value of the stock is now $\$32\frac{1}{2}$.

3. *Check.*

Estimate the sum by rounding each fraction to one nonzero digit.

$$\begin{aligned} &\textit{Thus} && 26\frac{1}{8} + 2\frac{5}{8} + 3\frac{3}{4} \\ &\textit{becomes} && 30 + 3 + 4 = 37 \end{aligned}$$

This is close to our answer, $32\frac{1}{2}$. Our answer seems reasonable.

One of the most important uses of estimation in mathematics is in the calculation of problems involving fractions. People find it easier to detect significant errors when working with whole numbers. However, the extra steps involved in the calculation with fractions and mixed numbers often distract our attention from an error that we should have detected.

Thus it is particularly critical to take the time to check your answer by estimating the results of the calculation by using whole numbers. Be sure to ask yourself, is this answer reasonable? Does this answer seem realistic? Only by estimating our results by using whole numbers will we be able to answer that question. It is this estimating skill that you will find more useful in your own life as a consumer and as a citizen.

Practice Problem 1 Nicole purchased the following amounts of gas for her car in the last three fill-ups: $18\frac{7}{10}$ gallons, $15\frac{2}{5}$ gallons, and $14\frac{1}{2}$ gallons. How many gallons did she buy altogether? ■

The word *diameter* has two common meanings. First, it means a line segment that passes through the center of and intersects a circle twice. Second, it means the *length* of this segment.

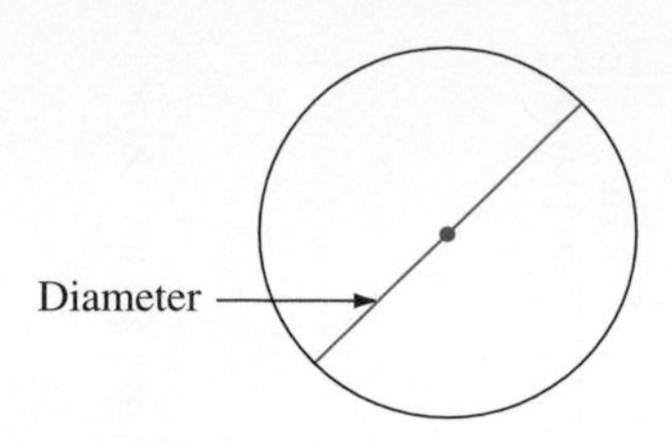

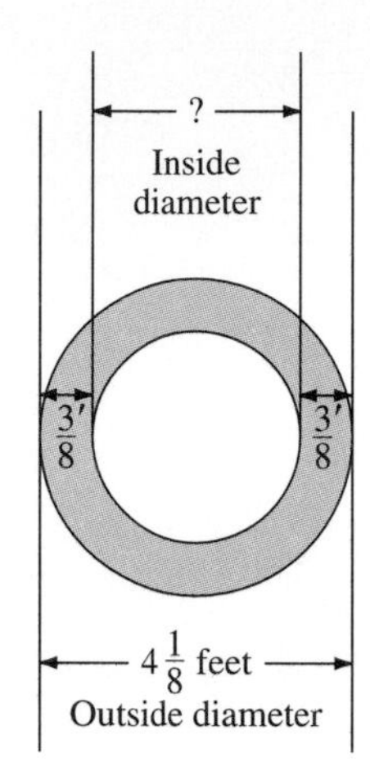

EXAMPLE 2 What is the inside diameter (distance across) of a cement storm drain pipe that has an outside diameter of $4\frac{1}{8}$ feet and is $\frac{3}{8}$ feet thick?

1. *Understand the problem.*

Read the problem carefully. Draw a picture. The picture is in the margin on the left. Now fill in the Mathematics Blueprint.

MATHEMATICS BLUEPRINT FOR PROBLEM SOLVING			
Gather the Facts	What Am I Asked to Do?	How do I Proceed?	Key Points to Remember
Outside diameter is $4\frac{1}{8}$ feet. Thickness is 3/8 foot on both ends of the diameter.	Find the *inside* diameter of the pipe.	Add the two measures of thickness. Then subtract this total from the outside diameter.	Since the LCD = 8, all fractions must have this denominator.

2. *Solve and state the answer.*

Add the two thickness measurements together.

Adding $\frac{3}{8} + \frac{3}{8} = \frac{6}{8}$ gives the total thickness of the pipe, $\frac{6}{8}$ foot. We will not reduce $\frac{6}{8}$ since the LCD is 8.

We subtract the total of the two thickness measurements from the outside diameter.

$$\begin{array}{rcr} 4\frac{1}{8} & = & 3\frac{9}{8} \\ -\ \frac{6}{8} & = & -\ \frac{6}{8} \\ \hline & & 3\frac{3}{8} \end{array}$$

We borrow 1 from 4 to get $3 + 1\frac{1}{8}$ or $3\frac{9}{8}$.

The inside diameter is $3\frac{3}{8}$ feet.

3. *Check.*

We will work backward to check. We will use the exact values.

If we have done our work correctly, $\frac{3}{8}$ foot + $3\frac{3}{8}$ feet + $\frac{3}{8}$ foot should add up to the outside diameter, $4\frac{1}{8}$ feet.

$$\frac{3}{8} + 3\frac{3}{8} + \frac{3}{8} \stackrel{?}{=} 4\frac{1}{8}$$

$$3\frac{9}{8} \stackrel{?}{=} 4\frac{1}{8}$$

$$4\frac{1}{8} = 4\frac{1}{8} \checkmark$$

Our answer of $3\frac{3}{8}$ feet is correct.

Practice Problem 2 A poster is $12\frac{1}{4}$ inches long. We want a $1\frac{3}{8}$-inch border on the top and a 2-inch border on the bottom. What is the length of the inside portion of the poster? ■

EXAMPLE 3 On Tuesday Michael earned $\$8\frac{1}{4}$ per hour working for 8 hours. He also earned overtime pay, which is $1\frac{1}{2}$ times his regular rate of $\$8\frac{1}{4}$, for 4 hours on Tuesday. How much pay did he earn altogether on Tuesday?

1. *Understand the problem.*

We draw a picture of the parts of Michael's pay on Tuesday.

Michael's earnings on Tuesday are the sum of two parts:

Pay at regular pay rate	+	Pay at overtime pay rate	=	Total pay for the day

Now fill in the Mathematics Blueprint.

MATHEMATICS BLUEPRINT FOR PROBLEM SOLVING

Gather the Facts	What Am I Asked to Do?	How Do I Proceed?	Key Points to Remember
He works 8 hours at $\$8\frac{1}{4}$ per hour. He works 4 hours at the overtime rate, $1\frac{1}{2}$ times the regular rate.	Find his total pay for Tuesday.	Find out how much he is paid for regular time. Find out how much he is paid for overtime. Then add the two.	The overtime rate is $1\frac{1}{2}$ multiplied by the regular rate.

2. *Solve and state the answer.*
Find his overtime pay rate.

$$1\frac{1}{2} \times 8\frac{1}{4} = \frac{3}{2} \times \frac{33}{4} = \frac{\$99}{8} \text{ per hour}$$

We leave our answer as an improper fraction because we will need to multiply it by another fraction.
How much was he paid for regular time? For overtime?

For 8 regular hours, he earned $8 \times 8\frac{1}{4} = \overset{2}{\cancel{8}} \times \frac{33}{\underset{1}{\cancel{4}}} = \$66.$

For 4 overtime hours, he earned $\overset{1}{\cancel{4}} \times \frac{99}{\underset{2}{\cancel{8}}} = \frac{99}{2} = \$49\frac{1}{2}.$

Now we add to find the total pay.

$66	Pay at regular pay rate
$\$49\frac{1}{2}$	Pay at overtime rate

Michael earned $\$115\frac{1}{2}$ working on Tuesday.

3. *Check.*
We estimate his regular pay rate at $8 per hour.
We estimate his overtime pay rate at $1\frac{1}{2} \times 8 = \frac{3}{2} \times 8 = 12$ or $12 per hour.

$$8 \text{ hours} \times \$8 \text{ per hour} = \$64 \text{ regular pay}$$
$$4 \text{ hours} \times \$12 \text{ per hour} = \$48 \text{ overtime pay}$$

Estimated sum. $\$64 + \$48 \approx 60 + 50 = 110$
$110 is close to our calculated value, $\$115\frac{1}{2}$, so our answer is reasonable. ✓

Practice Problem 3 A tent manufacturer uses $8\frac{1}{4}$ yards of waterproof duck cloth to make a regular tent. She uses $1\frac{1}{2}$ times that amount to make a large tent. How many yards of cloth will she need to make six regular tents and 16 large tents? ■

EXAMPLE 4 Alicia is buying some 8-foot boards for shelving. She wishes to make two bookcases, each with three shelves. Each shelf will be $3\frac{1}{4}$ feet long.

(a) How many boards does she need to buy?

(b) How many linear feet of shelving are actually needed to build the bookcases?

(c) How many linear feet of shelving are left over?

1. *Understand the problem.*

Draw a sketch of a bookcase. Each bookcase will have 3 shelves. Alicia is making 2 such bookcases. (Alicia's boards are for the shelves, not the sides.)

Now fill in the Mathematics Blueprint.

MATHEMATICS BLUEPRINT FOR PROBLEM SOLVING

Gather the Facts	What Am I Asked to Do?	How Do I Proceed?	Key Points to Remember
She needs three shelves for each bookcase. Each shelf is $3\frac{1}{4}$ feet long. She will make two bookcases. Shelves are cut from 8-foot boards.	Find out how many boards to buy. Find out how many feet of board is needed for shelves and how many feet will be left over.	First find out how many $3\frac{1}{4}$-foot shelves she can get from one board. Then see how many boards she needs to make all six shelves.	Each time she cuts up an 8-foot board, she will get some shelves and some leftover wood.

2. *Solve and state the answer.*

We want to know how many $3\frac{1}{4}$-foot boards are in an 8-foot board.

$3\frac{1}{4}$	$3\frac{1}{4}$	

← 8 feet →

By drawing a rough sketch, we would probably guess the answer is 2. To find exactly how many $3\frac{1}{4}$-foot long pieces are in 8 feet, we will use division.

$$8 \div 3\frac{1}{4} = \frac{8}{1} \div \frac{13}{4} = \frac{8}{1} \times \frac{4}{13} = \frac{32}{13} = 2\frac{6}{13} \text{ boards}$$

She will get 2 shelves from each board, and some wood will be left over.

(a) How many boards does Alicia need to build two bookcases?

For 2 bookcases, she needs 6 shelves. This will provide 2 shelves out of each board. $6 \div 2 = 3$. She will need three of the 8-foot boards.

(b) How many linear feet of shelving are actually needed to build the bookcases?

She needs 6 shelves at $3\frac{1}{4}$ feet $= 6 \times 3\frac{1}{4} = \overset{3}{\cancel{6}} \times \frac{13}{\underset{2}{\cancel{4}}} = \frac{39}{2} = 19\frac{1}{2}$ linear feet. A total of $19\frac{1}{2}$ linear feet of shelving is needed.

(c) How many linear feet of shelving are left over?

Each time she uses one board she has $8 - 3\frac{1}{4} - 3\frac{1}{4} = 8 - 6\frac{1}{2} = 1\frac{1}{2}$ feet left over. Each of the three boards has $1\frac{1}{2}$ feet left over.

$3 \times 1\frac{1}{2} = 3 \times \frac{3}{2} = \frac{9}{2} = 4\frac{1}{2}$ linear feet. A total of $4\frac{1}{2}$ linear feet of shelving is left over.

TEACHING TIP Many building and construction problems are similar to Example 4. After you have covered Example 4, ask the students if they can think of similar types of problems that they have encountered in their daily lives. Often you will get some interesting problems. If they have nothing to share, give them the following problem. "A man had three rooms with wall-to-wall carpeting. One room had $12\frac{1}{2}$ square yards, a second room had $13\frac{1}{4}$ square yards, and the third room had $15\frac{7}{8}$ square yards. The carpet-cleaning company charged $\$3\frac{1}{2}$ per square yard or any 3 rooms cleaned for \$159. The man could not determine which was a better buy. Can you?" The answer is that it is cheaper to get the $\$3\frac{1}{2}$ per square yard rate because then the cleaning job will only come to $\$145\frac{11}{16}$.

3. *Check.*

Work backward. See if you can check that with three 8-foot boards you

(a) can make the 6 shelves for the two bookcases

(b) will use exactly $19\frac{1}{2}$ linear feet to make the shelves

(c) will have exactly $4\frac{1}{2}$ linear feet left over

The check is left to you.

Practice Problem 4 Michael is purchasing 12-foot boards for shelving. He wishes to make two bookcases, each with four shelves. Each shelf will be $2\frac{3}{4}$ feet long.

(a) How many boards does he need to buy?

(b) How many linear feet of shelving are actually needed to build the bookcases?

(c) How many linear feet of shelving are left over? ■

Another useful method for solving applied problems is called "Do a similar, simpler problem." When a problem seems difficult to understand because of the fractions, change the problem to an easier but similar problem. Then decide how to solve the simpler problem and use the same steps to solve the original problem. For example:

> How many gallons of water can a tank hold if its volume is $58\frac{2}{3}$ cubic feet? (1 cubic foot holds about $7\frac{1}{2}$ gallons.)

A similar, easier problem would be: "If 1 cubic foot holds about 8 gallons and a tank holds 60 cubic feet, how many gallons of water does the tank hold?"

The easier problem can be read more quickly and seems to make more sense. Probably we will see how to solve the easier problem right away: "I can find the number of gallons by multiplying 8×60." Therefore we can solve the first problem by multiplying $7\frac{1}{2} \times 58\frac{2}{3}$ to obtain the number of gallons of water. See the next example.

EXAMPLE 5 A fishing boat traveled $69\frac{3}{8}$ nautical miles in $3\frac{3}{4}$ hours. How many knots (nautical miles per hour) did the fishing boat average?

1. *Understand the problem.*

   Let us think of a simpler problem. If a boat traveled 70 nautical miles in 4 hours, how many knots did it average? We would divide distance by time.

   $$70 \div 4 = \text{average speed}$$

   Likewise in our original problem we need to divide distance by time.

   $$69\frac{3}{8} \div 3\frac{3}{4} = \text{average speed}$$

   Now fill in the Mathematics Blueprint.

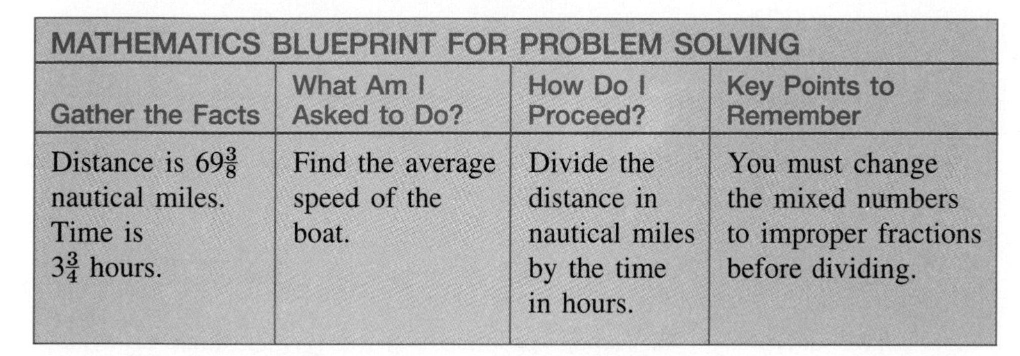

MATHEMATICS BLUEPRINT FOR PROBLEM SOLVING			
Gather the Facts	**What Am I Asked to Do?**	**How Do I Proceed?**	**Key Points to Remember**
Distance is $69\frac{3}{8}$ nautical miles. Time is $3\frac{3}{4}$ hours.	Find the average speed of the boat.	Divide the distance in nautical miles by the time in hours.	You must change the mixed numbers to improper fractions before dividing.

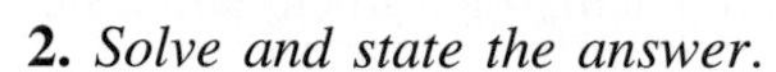

2. *Solve and state the answer.*

   Divide distance by time to get speed in knots.

   $$69\frac{3}{8} \div 3\frac{3}{4} = \frac{555}{8} \div \frac{15}{4} = \frac{\overset{37}{\cancel{555}}}{\underset{2}{\cancel{8}}} \cdot \frac{\overset{1}{\cancel{4}}}{\underset{1}{\cancel{15}}}$$

   $$= \frac{37}{2} \cdot \frac{1}{1} = \frac{37}{2} = 18\frac{1}{2} \text{ knots}$$

   The speed of the boat was $18\frac{1}{2}$ knots.

3. *Check.*

   We estimate $69\frac{3}{8} \div 3\frac{3}{4}$.

   $$\text{Use } 70 \div 4 = 17\frac{1}{2} \text{ knots}$$

   Our estimate is close to the calculated value.
   Our answer is reasonable. ✓

Practice Problem 5 Alfonso traveled $199\frac{3}{4}$ miles in his car and used $8\frac{1}{2}$ gallons of gas. How many miles per gallon did he get? ■

2.9 Exercises

You may use the Mathematics Blueprint for problem solving to help you solve the following problems.

Applications

1. A truck hauled $5\frac{1}{4}$ tons of gravel on Monday, $4\frac{7}{8}$ tons on Tuesday, and $6\frac{1}{6}$ tons on Wednesday. How many tons were hauled over the three days?

$5\frac{1}{4} + 4\frac{7}{8} + 6\frac{1}{6} = 16\frac{7}{24}$ tons

2. A deli used $8\frac{1}{4}$ pounds of roast beef on Monday, $7\frac{1}{2}$ pounds on Tuesday, and $6\frac{7}{10}$ pounds on Wednesday. How many pounds were used over the three days?

$8\frac{1}{4} + 7\frac{1}{2} + 6\frac{7}{10} = 21\frac{29}{20} = 22\frac{9}{20}$ pounds

3. The coastal area of North Carolina experienced devastating erosion caused by fierce storms. The National Guard assisted by hauling sand to try and rebuild the dunes. During three days in September 1996, the National Guard hauled $7\frac{5}{6}$ tons of sand on the first day, $8\frac{1}{8}$ tons on the second day, and $9\frac{1}{2}$ tons on the third. How many tons of sand were hauled?

$25\frac{11}{24}$ tons

4. Nancy wonders if a tree in her front yard will grow to be 8 feet tall. When the tree was planted, it was $2\frac{1}{6}$ feet above the ground in height. It grew $1\frac{1}{4}$ feet the first year after being planted. How many more feet does it need to grow to reach a height of 8 feet?

$4\frac{7}{12}$ feet

5. A bolt extends through $\frac{3}{4}$-inch-thick plywood, two washers that are each $\frac{1}{16}$ inch thick, and a nut that is $\frac{3}{16}$ inch thick. The bolt must be $\frac{1}{2}$ inch longer than the sum of the thicknesses of plywood, washers, and nut. What is the minimum length of the bolt?

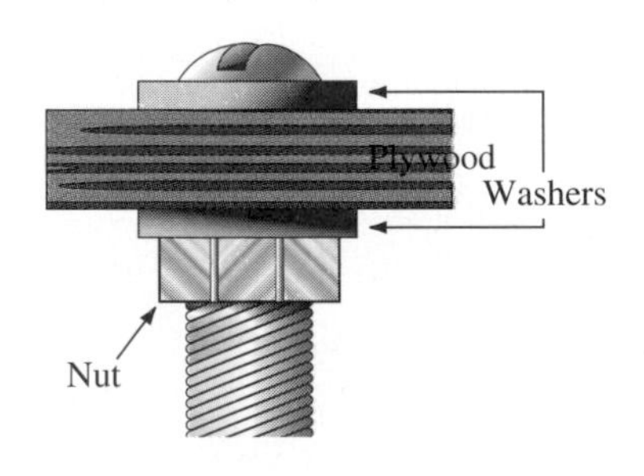

$\frac{1}{16} + \frac{3}{4} + \frac{1}{16} + \frac{3}{16} + \frac{1}{2} = \frac{25}{16}$

$= 1\frac{9}{16}$ inches

6. Repeat problem 5 with $\frac{11}{16}$-inch-thick plywood, two washers that are each $\frac{1}{8}$ inch thick, and a nut that is $\frac{1}{4}$ inch thick.

$\frac{1}{8} + \frac{11}{16} + \frac{1}{8} + \frac{1}{4} + \frac{1}{2} = \frac{27}{16} = 1\frac{11}{16}$ inches

7. A cable wire is $65\frac{1}{6}$ inches long. It is to be cut into pieces that are $3\frac{5}{6}$ inch long to be packaged and sold separately in electronics stores. How many pieces will be made? 17 pieces

8. Hank is practicing to run in the Boston Marathon. His goal this week is to run 55 miles. Today he ran $15\frac{1}{4}$ miles. Yesterday he ran $14\frac{3}{8}$ miles. How many more miles must he run this week to meet his goal?

$25\frac{3}{8}$ miles more

9. For a party of the British Literature Club using all "English foods," Nancy bought a $10\frac{2}{3}$-pound wheel of Stilton cheese, to go with the pears and the apples, at $\$8\frac{3}{4}$ per pound. How much did the wheel of Stilton cheese cost?

$\$93\frac{1}{3}$

10. The following exaggerated story was repeated by certain basketball fans: Michael Jordan wasn't always 7 feet tall. When he was a little boy, at the height of $3\frac{1}{2}$ feet, no one knew that he would grow so quickly. The following year he grew $1\frac{5}{6}$ feet, and the year after that he grew $1\frac{1}{5}$ feet. How many more feet did he grow to reach 7 feet tall?

$\frac{7}{15}$ of a foot

11. How many gallons can a tank hold that has a volume of $36\frac{3}{4}$ cubic feet? (Assume that 1 cubic foot holds about $7\frac{1}{2}$ gallons.)

$36\frac{3}{4} \times 7\frac{1}{2} = \frac{147}{4} \times \frac{15}{2} = 275\frac{5}{8}$ gallons

12. A tank can hold a volume of $7\frac{1}{4}$ cubic feet. If it is filled with water, how much does the water weigh? (Assume that 1 cubic foot of water weighs $62\frac{1}{2}$ pounds.)

$7\frac{1}{4} \times 62\frac{1}{2} = \frac{29}{4} \times \frac{125}{2} = 453\frac{1}{8}$ pounds

13. The night of the *Titanic* cruise ship disaster, the captain decided to run his ship at $22\frac{1}{2}$ knots, (nautical miles per hour). The *Titanic* traveled at that speed for $4\frac{3}{4}$ hours before it met its tragic demise. How far did the *Titanic* travel at this excessive speed before the disaster?

$106\frac{7}{8}$ nautical miles

14. Barbara earns $450 per week. She has $\frac{1}{5}$ of her income withheld for federal taxes, $\frac{1}{15}$ of her income withheld for state taxes, and $\frac{1}{25}$ of her income withheld for medical coverage. How much per week is left for Barbara after those three deductions? $312 per week is left.

15. Noriko earns $660 per week. She has $\frac{1}{5}$ of her income withheld for federal taxes, $\frac{1}{15}$ of her income withheld for state taxes, and $\frac{1}{20}$ of her income withheld for medical coverage. How much per week is left for Noriko after these three deductions? $451 per week is left.

16. When it was new, Ginny Sue's car got $28\frac{1}{6}$ miles per gallon. It now gets $1\frac{5}{6}$ miles per gallon less. How far can she drive now if the car has $10\frac{3}{4}$ gallons in the tank?

$283\frac{1}{12}$ miles

17. The crafts store bought $36\frac{3}{4}$ ounces of potpourri (a type of air freshener) for $2. The shop divides the potpourri into $\frac{7}{8}$-ounce bags to put into drawers and closets, and sells each bag for $$3\frac{1}{2}$.

(a) How many $\frac{7}{8}$-ounce bags of potpourri can the shop make? 42

(b) If all of the bags are sold, how much money will the store take in? $147

(c) How much profit will it make? $145

18. A candle company purchased $48\frac{1}{8}$ pounds of wax to make specialty candles. It takes $\frac{5}{8}$ pound of wax to make one candle. The owners of the business plan to sell the candles for $12 each. The specialty wax cost them $2 per pound.

(a) How many candles can they make?

(b) How much does it cost to make one candle?

(c) How much profit will they make if they sell all of the candles?

(a) 77; (b) $$1\frac{1}{4}$; (c) $$827\frac{3}{4}$

19. Cecilia bought a loaf of sourdough bread that was made by a local gourmet bakery. The label said that the bread, plus its fancy box, weighed $18\frac{1}{2}$ ounces in total. Of this, $1\frac{1}{4}$ ounces turned out to be the weight of the ribbon. The box weighed $3\frac{1}{8}$ ounces.

(a) How many ounces of bread did she actually buy?

(b) The box stated its net weight as $14\frac{3}{4}$ ounces. (This means that she should have found $14\frac{3}{4}$ ounces of gourmet sourdough bread in the box.) How much in error was this measurement?

(a) $14\frac{1}{8}$ ounces of bread; (b) $\frac{5}{8}$ of an ounce

20. Tony was making brownies one night. The recipe calls for $3\frac{3}{4}$ cups of chocolate chips. He had started to put in the chips (he put in $1\frac{1}{8}$ cups) when his little brother came into the kitchen and threw in $1\frac{3}{4}$ cups of chocolate chips.

(a) How many more cups of chocolate chips does Tony need to add according to the recipe?

(b) He has $\frac{3}{4}$ cup of chocolate chips left in the bowl. Does he have enough to add the amount needed in (a)?

(a) $\frac{7}{8}$ of a cup of chocolate chips; (b) No

21. The largest Coast Guard boat stationed at San Diego can travel $160\frac{1}{8}$ nautical miles in $5\frac{1}{4}$ hours.

(a) At how many knots is the boat traveling?

$$\frac{1281}{8} \div \frac{21}{4} = \frac{1281}{8} \cdot \frac{4}{21} = \frac{61}{2} = 30\frac{1}{2} \text{ knots}$$

(b) At this speed, how long would it take the Coast Guard boat to travel $213\frac{1}{2}$ nautical miles?

$$\frac{427}{2} \div \frac{61}{2} = \frac{427}{2} \cdot \frac{2}{61} = 7 \text{ hours}$$

22. Russ and Norma's Mariah water ski boat can travel $72\frac{7}{8}$ nautical miles in $2\frac{3}{4}$ hours.

(a) At how many knots is the boat traveling?

$$\frac{583}{8} \div \frac{11}{4} = \frac{\overset{53}{\cancel{583}}}{\underset{2}{\cancel{8}}} \times \frac{\overset{1}{\cancel{4}}}{\underset{1}{\cancel{11}}} = \frac{53}{2} = 26\frac{1}{2} \text{ knots}$$

(b) At this speed, how long would it take their water ski boat to travel $92\frac{3}{4}$ nautical miles?

$$92\frac{3}{4} \div 26\frac{1}{2} = \frac{371}{4} \div \frac{53}{2} = \frac{\overset{7}{\cancel{371}}}{\underset{2}{\cancel{4}}} \times \frac{\overset{1}{\cancel{2}}}{\underset{1}{\cancel{53}}} = \frac{7}{2} = 3\frac{1}{2} \text{ hours}$$

★ **23.** A Kansas wheat farmer has a storage bin with a capacity of $6856\frac{1}{4}$ cubic feet.

(a) If a bushel of wheat is $1\frac{1}{4}$ cubic feet, how many bushels can it hold?

(b) If a farmer wants to make a new storage bin $1\frac{3}{4}$ times larger, how many cubic feet will it hold?

(c) How many bushels will the new bin hold?

(a) $6856\frac{1}{4} \div 1\frac{1}{4} = \frac{27,425}{4} \times \frac{4}{5} = 5485$ bushels

(b) $6856\frac{1}{4} \times 1\frac{3}{4} = \frac{27,425}{4} \times \frac{7}{4} = 11,998\frac{7}{16}$ cubic feet

(c) $11,998\frac{7}{16} \div 1\frac{1}{4} = \frac{191,975}{16} \times \frac{4}{5} = 9598\frac{3}{4}$ bushels

★ **24.** A Greyhound bus made a long-distance run in upstate New York, stopping several times. The entire trip took $24\frac{1}{6}$ hours. The bus stopped a total of $2\frac{1}{3}$ hours for restroom and meal stops. The bus also stopped $1\frac{1}{2}$ hours for repairs. The bus traveled 854 miles. What was its average speed in miles per hour when it was moving? (Do not count stop times.)

$2\frac{1}{3} + 1\frac{1}{2} = 3\frac{5}{6}$ $\quad 24\frac{1}{6} - 3\frac{5}{6} = 23\frac{7}{6} - 3\frac{5}{6} = 20\frac{1}{3}$ hours

$854 \div 20\frac{1}{3} = 854 \div \frac{61}{3} = 854 \times \frac{3}{61} = 42$ miles per hour

Cumulative Review Problems

25. Add. 16,846 + 19,321 + 8,078

44,245

26. Subtract. 193,705 − 165,891

27,814

27. Multiply. 1683 × 27

11 781
33 66
45,441

28. Divide. 19)5285

278 R 3

38
148
133
155
152
3

Developing Your Study Skills

WHY STUDY MATHEMATICS?

Students often question the value of mathematics. They see little real use for it in their everyday lives. However, mathematics is often the key that opens the door to a better paying job.

In our present-day technological world, many people use mathematics daily. Many vocational and professional areas—such as the fields of business, statistics, economics, psychology, finance, computer science, chemistry, physics, engineering, electronics, nuclear energy, banking, quality control, and teaching—require a certain level of expertise in mathematics. Those who want to work in these fields must be able to function at a given mathematical level. Those who cannot will not be able to enter this job area.

So, whatever your field, be sure to realize the importance of mastering the basics of this course. It is very likely to help you advance to the career of your choice.

Chapter Organizer

Topic	Procedure	Examples
Concept of a fractional part, p. 96.	The numerator is the number of parts selected. The denominator is the number of total parts.	What part of this sketch is shaded? $\frac{7}{10}$
Prime factorization, p. 103.	Prime factorization is the writing of a number as the product of prime numbers.	Write the prime factorization of 36. $36 = 4 \times 9$; $4 = 2 \times 2$, $9 = 3 \times 3$; $= 2 \times 2 \times 3 \times 3$
Reducing fractions, p. 105.	1. Factor numerator and denominator into prime factors. 2. Divide out factors common to numerator and denominator.	Reduce. $\frac{54}{90}$ $\frac{54}{90} = \frac{\cancel{3}^1 \times \cancel{3}^1 \times 3 \times \cancel{2}^1}{\cancel{3}_1 \times \cancel{3}_1 \times \cancel{2}_1 \times 5} = \frac{3}{5}$
Changing mixed numbers to improper fractions, p. 111.	1. Multiply whole number by denominator. 2. Add product to numerator. 3. Place sum over denominator.	Write as an improper fraction. $7\frac{3}{4} = \frac{4 \times 7 + 3}{4} = \frac{28 + 3}{4} = \frac{31}{4}$
Changing an improper fraction to a mixed number, p. 113.	1. Divide denominator into numerator. 2. The quotient is the whole number. 3. The fraction is the remainder over the divisor.	Change to a mixed number. $\frac{32}{5}$ $5\overline{)32}$ = 6 remainder 2 ($32 - 30 = 2$); $= 6\frac{2}{5}$
Multiplying fractions, p. 118.	1. Divide out common factors from the numerators and denominators whenever possible. 2. Multiply numerators. 3. Multiply denominators.	Multiply. $\frac{3}{7} \times \frac{5}{13} = \frac{15}{91}$ Multiply. $\frac{\cancel{5}^1}{\cancel{8}_1} \times \frac{\cancel{16}^2}{\cancel{15}_3} = \frac{2}{3}$
Multiplying mixed and/or whole numbers, p. 120.	1. Change any whole numbers to a fraction with a denominator of 1. 2. Change any mixed numbers to improper fractions. 3. Use multiplication rule for fractions.	Multiply. $7 \times 3\frac{1}{4}$ $\frac{7}{1} \times \frac{13}{4} = \frac{91}{4}$ $= 22\frac{3}{4}$
Dividing fractions, p. 124.	To divide two fractions, we invert the second fraction and multiply.	Divide. $\frac{3}{7} \div \frac{2}{9} = \frac{3}{7} \times \frac{9}{2} = \frac{27}{14}$ *or* $1\frac{13}{14}$
Dividing mixed numbers and/or whole numbers, p. 126.	1. Change any whole numbers to a fraction with a denominator of 1. 2. Change any mixed numbers to improper fractions. 3. Use rule for division of fractions.	Divide. $8\frac{1}{3} \div 5\frac{5}{9} = \frac{25}{3} \div \frac{50}{9}$ $= \frac{\cancel{25}^1}{\cancel{3}_1} \times \frac{\cancel{9}^3}{\cancel{50}_2} = \frac{3}{2}$ *or* $1\frac{1}{2}$

Chapter Organizer

Topic	Procedure	Examples
Finding the least common denominator, p. 132.	1. Write each denominator as the product of prime factors. 2. List all the prime factors that appear in both products. 3. Form a product of those factors, using each factor the greatest number of times it appears in any denomination.	Find LCD of $\frac{1}{10}$, $\frac{3}{8}$, and $\frac{7}{25}$. $10 = 2 \times 5$ $8 = 2 \times 2 \times 2$ $25 = 5 \times 5$ $\text{LCD} = 2 \times 2 \times 2 \times 5 \times 5$ $= 200$
Building fractions, p. 134.	1. Find how many times the original denominator can be divided into the new denominator. 2. Multiply that value by numerator and denominator of original fraction.	Build $\frac{5}{7}$ to an equivalent fraction with a denominator of 42. First we find $7\overline{)42}$ = 6. Then we multiply by 6 for numerator and denominator. $\frac{5}{7} \times \frac{6}{6} = \frac{30}{42}$
Adding or subtracting fractions with a common denominator, p. 139.	1. Add or subtract the numerators. 2. Keep the common denominator.	Add. $\frac{3}{13} + \frac{5}{13} = \frac{8}{13}$ Subtract. $\frac{15}{17} - \frac{12}{17} = \frac{3}{17}$
Adding or subtracting fractions without a common denominator, p. 140.	1. Find the LCD of the fractions. 2. Build up each fraction, if needed, to obtain the LCD in the denominator. 3. Follow the steps for adding and subtracting fractions with the same denominator.	Add. $\frac{1}{4} + \frac{3}{7} + \frac{5}{8}$ LCD = 56 $\quad \frac{1 \times 14}{4 \times 14} + \frac{3 \times 8}{7 \times 8} + \frac{5 \times 7}{8 \times 7}$ $= \frac{14}{56} + \frac{24}{56} + \frac{35}{56} = \frac{73}{56} = 1\frac{17}{56}$
Adding mixed numbers, p. 146.	1. Change fractional parts to equivalent fractions with LCD as a denominator, if needed. 2. Add whole numbers and fractions separately. 3. If improper fractions occur, change to mixed numbers and simplify.	Add. $6\frac{3}{4} + 2\frac{5}{8}$ $6\,\boxed{\frac{3}{4} \times \frac{2}{2}} = 6\frac{6}{8}$ $+2\frac{5}{8} \qquad = 2\frac{5}{8}$ $8\frac{11}{8} = 9\frac{3}{8}$
Subtracting mixed numbers, p. 147.	1. Change fractional parts to equivalent fractions with LCD as a denominator, if needed. 2. If necessary, borrow from whole number to subtract fractions. 3. Subtract whole numbers and fractions separately.	Subtract. $8\frac{1}{5} - 4\frac{2}{3}$ $8\,\boxed{\frac{1}{5} \times \frac{3}{3}} = 8\frac{3}{15}$ $-4\,\boxed{\frac{2}{3} \times \frac{5}{5}} = -4\frac{10}{15}$ $= 7\frac{18}{15}$ $-4\frac{10}{15}$ $3\frac{8}{15}$

Using the Mathematics Blueprint for Problem Solving, p. 154.

In solving an applied problem with fractions, students may find it helpful to complete the following steps. You will not use the steps all of the time. Choose the steps that best fit the conditions of the problem.

1. *Understand the problem.*
  (a) Read the problem carefully.
  (b) Draw a picture if this helps you to visualize the situation. Think about what facts you are given and what you are asked to find.
  (c) It may help to write a similar, simpler problem to get started and to determine what operation to use.
  (d) Use the Mathematics Blueprint for Problem Solving to organize your work. Follow these four parts.
    1. Gather the facts. (Write down specific values given in the problem.)
    2. What am I asked to do? (Identify what you must obtain for an answer.)
    3. How do I proceed? (What calculations need to be done.)
    4. Key points to remember. (Record any facts, warnings, formulas, or concepts you think will be important as you solve the problem.)

2. *Solve and state the answer.*
  (a) Perform the necessary calculations.
  (b) State the answer, including the unit of measure.

3. *Check.*
  (a) Estimate the answer to the problem. Compare this estimate to the calculated value. Is your answer reasonable?
  (b) Repeat your calculations.
  (c) Work backward from your answer. Do you arrive at the original conditions of the problem?

EXAMPLE

A wire is $95\frac{1}{3}$ feet long. It is cut up into smaller, equal-sized pieces, each $4\frac{1}{3}$ feet long. How many pieces will there be?

1. *Understand the problem.*
Let us draw a picture of the situation.

$95\frac{1}{3}$ ft wire

$4\frac{1}{3}$ $4\frac{1}{3}$?

How will we find the number of pieces?
Now we will use a simpler problem to clarify the idea. A wire 100 feet long is cut up into smaller pieces each 4 feet long. How many pieces will there be? We readily see that we would divide 100 by 4. Thus in our original problem we should divide $95\frac{1}{3}$ feet by $4\frac{1}{3}$ feet. This will tell us the number of pieces. Now we fill in the Mathematics Blueprint.

MATHEMATICS BLUEPRINT FOR PROBLEM SOLVING

Gather the Facts	What Am I Asked to Do?	How Do I Proceed?	Key Points to Remember
Wire is $95\frac{1}{3}$ feet. It is cut into equal pieces $4\frac{1}{3}$ feet long.	Determine how many pieces of wire there will be.	Divide $95\frac{1}{3}$ by $4\frac{1}{3}$.	Change mixed numbers to improper fractions before carrying out the division.

2. *Solve and state the answer.*
We need to divide $95\frac{1}{3} \div 4\frac{1}{3}$

$$\frac{286}{3} \div \frac{13}{3} = \frac{\overset{22}{\cancel{286}}}{\underset{1}{\cancel{3}}} \times \frac{\overset{1}{\cancel{3}}}{\underset{1}{\cancel{13}}} = \frac{22}{1} = 22$$

There will be 22 pieces of wire.

3. *Check.*
Estimate. Round $95\frac{1}{3}$ to nearest ten = 100.

Round $4\frac{1}{3}$ to nearest integer = 4.

$100 \div 4 = 25$
This is close to our estimate. Our answer is reasonable. ✓

Chapter 2 Review Problems

2.1 *In each problem, use a fraction to represent the shaded part of the object.*

1. 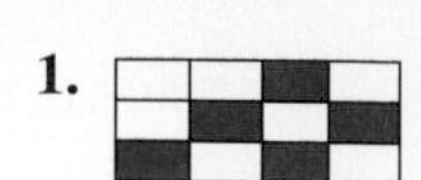 $\frac{5}{12}$

2. $\frac{3}{8}$

Draw a sketch to illustrate the fraction.

3. $\frac{4}{7}$ of an object Answers will vary.

4. $\frac{7}{9}$ of an object Answers will vary.

5. An inspector looked at 31 parts and found 6 of them to be defective. What fractional part of these items was defective? $\frac{6}{31}$

6. The dean asked 100 of the freshman if they would be staying in the dorm over the holidays. A total of 87 said they would not. What fractional part of the freshmen said they would not? $\frac{87}{100}$

2.2 *Express each number as a product of prime factors.*

7. 42 $2 \times 3 \times 7$

8. 54 2×3^3

9. 168 $2^3 \times 3 \times 7$

Identify which of the following numbers is prime. If it is composite, express it as the product of prime factors.

10. 59 Prime

11. 78 $2 \times 3 \times 13$

12. 167 Prime

Reduce the fraction.

13. $\frac{13}{52}$ $\frac{1}{4}$

14. $\frac{12}{42}$ $\frac{2}{7}$

15. $\frac{21}{36}$ $\frac{7}{12}$

16. $\frac{26}{34}$ $\frac{13}{17}$

17. $\frac{168}{192}$ $\frac{7}{8}$

18. $\frac{51}{105}$ $\frac{17}{35}$

2.3 *Change each mixed number to an improper fraction.*

19. $4\frac{3}{8}$ $\frac{35}{8}$

20. $2\frac{19}{23}$ $\frac{65}{23}$

Change each improper fraction to a mixed number.

21. $\frac{19}{7}$ $2\frac{5}{7}$

22. $\frac{63}{13}$ $4\frac{11}{13}$

23. Reduce and leave your answer as a mixed number.
$3\frac{15}{55}$ $3\frac{3}{11}$

24. Reduce and leave your answer as an improper fraction.
$\frac{234}{16}$ $\frac{117}{8}$

25. Change to a mixed number and then reduce.
$\frac{385}{240}$ $1\frac{29}{48}$

2.4 *Multiply each fraction.*

26. $\frac{4}{7} \times \frac{5}{11}$ $\frac{20}{77}$

27. $\frac{7}{9} \times \frac{21}{35}$ $\frac{7}{15}$

28. $12 \times \frac{3}{7} \times 0$ 0

29. $\frac{3}{5} \times \frac{2}{7} \times \frac{10}{27}$ $\frac{4}{63}$

30. $12 \times 8\frac{1}{5}$ $\frac{492}{5}$ *or* $98\frac{2}{5}$

31. $5\frac{3}{8} \times 3\frac{4}{5}$ $\frac{817}{40}$ *or* $20\frac{17}{40}$

32. $5\frac{1}{4} \times 4\frac{6}{7}$ $\frac{51}{2}$ *or* $25\frac{1}{2}$

33. $35 \times \frac{7}{10}$ $24\frac{1}{2}$

34. One share of stock costs $\$37\frac{5}{8}$. How much money will 18 shares cost?

$\$667\frac{1}{4}$

35. Find the area of a rectangle that is $18\frac{1}{5}$ inches wide and $26\frac{3}{4}$ inches long.

$486\frac{17}{20}$ square inches

2.5 *Divide, if possible.*

36. $\frac{3}{7} \div \frac{2}{5}$ $\quad \frac{15}{14}$ *or* $1\frac{1}{14}$

37. $\frac{9}{17} \div \frac{18}{5}$ $\quad \frac{5}{34}$

38. $1200 \div \frac{5}{8}$ $\quad 1920$

39. $\dfrac{2\frac{1}{4}}{3\frac{1}{3}}$ $\quad \frac{27}{40}$

40. $\dfrac{20}{4\frac{4}{5}}$ $\quad \frac{25}{6}$ *or* $4\frac{1}{6}$

41. $2\frac{1}{8} \div 20\frac{1}{2}$ $\quad \frac{17}{164}$

42. $0 \div 3\frac{7}{5}$ $\quad 0$

43. $4\frac{2}{11} \div 3$ $\quad \frac{46}{33}$ *or* $1\frac{13}{33}$

44. There are 420 calories in $2\frac{1}{4}$ cans of grape soda. How many calories are contained in 1 can of soda?

$186\frac{2}{3}$ calories

45. A dress requires $3\frac{1}{8}$ yards of fabric. Amy has $21\frac{7}{8}$ yards available. How many dresses can she make?

7 dresses

2.6 *Find the LCD for each set of fractions.*

46. $\frac{7}{14}$ and $\frac{3}{49}$ $\quad 98$

47. $\frac{7}{40}$ and $\frac{11}{30}$ $\quad 120$

48. $\frac{5}{18}, \frac{1}{6}, \frac{7}{45}$ $\quad 90$

Build each fraction to an equivalent fraction with the specified denominator.

49. $\frac{3}{7} = \frac{?}{56}$ $\quad \frac{24}{56}$

50. $\frac{11}{24} = \frac{?}{72}$ $\quad \frac{33}{72}$

51. $\frac{9}{43} = \frac{?}{172}$ $\quad \frac{36}{172}$

52. $\frac{17}{18} = \frac{?}{198}$ $\quad \frac{187}{198}$

2.7 *Add or subtract.*

53. $\frac{3}{7} - \frac{5}{14}$ $\quad \frac{1}{14}$

54. $\frac{1}{2} + \frac{1}{3} + \frac{1}{4}$ $\quad \frac{13}{12}$ *or* $1\frac{1}{12}$

55. $\frac{4}{7} + \frac{7}{9}$ $\quad \frac{85}{63}$ *or* $1\frac{22}{63}$

56. $\frac{7}{8} - \frac{3}{5}$ $\quad \frac{11}{40}$

57. $\frac{7}{30} + \frac{2}{21}$ $\quad \frac{23}{70}$

58. $\frac{5}{18} + \frac{5}{12}$ $\quad \frac{25}{36}$

59. $\frac{14}{15} - \frac{3}{25}$ $\quad \frac{61}{75}$

60. $\frac{15}{16} - \frac{13}{24}$ $\quad \frac{19}{48}$

2.8 *Add or subtract.*

61. $1 - \frac{17}{23}$ $\quad \frac{6}{23}$

62. $6 - \frac{5}{9}$ $\quad 5\frac{4}{9}$

63. $3 + 5\frac{2}{3}$ $\quad 8\frac{2}{3}$

64. $8 + 12\frac{5}{7}$ $\quad 20\frac{5}{7}$

65. $3\frac{1}{4} + 1\frac{5}{8}$ $\quad \frac{39}{8}$ *or* $4\frac{7}{8}$

66. $7\frac{3}{16} - 2\frac{5}{6}$ $\quad \frac{209}{48}$ *or* $4\frac{17}{48}$

67. $120 - 16\frac{2}{3}$ $\quad \frac{310}{3}$ *or* $103\frac{1}{3}$

68. $22\frac{2}{3} + 48\frac{3}{4}$ $\quad \frac{857}{12}$ *or* $71\frac{5}{12}$

69. Bob jogged $1\frac{7}{8}$ miles on Monday, $2\frac{3}{4}$ miles on Tuesday, and $4\frac{1}{10}$ miles on Wednesday. How many miles did he jog on these three days?

$8\frac{29}{40}$ miles

70. On a recent day, Coca-Cola opened at $\$36\frac{5}{8}$ a share and closed at $\$38\frac{1}{4}$. How much did a share of Coca-Cola gain that day?

$\$1\frac{5}{8}$

71. A recipe calls for $3\frac{1}{3}$ cups of sugar and $4\frac{1}{4}$ cups of flour. How much sugar and how much flour would be needed for $\frac{1}{2}$ of that recipe?

$1\frac{2}{3}$ cups sugar, $2\frac{1}{8}$ cups flour

72. Rafael traveled in a car that gets $24\frac{1}{4}$ miles per gallon. He had $8\frac{1}{2}$ gallons of gas in the gas tank. Approximately how far could he drive?

$206\frac{1}{8}$ miles

73. How many lengths (pieces) of pipe $3\frac{1}{5}$ inches long can be cut from a pipe 48 inches long? 15 lengths

74. A car radiator holds $15\frac{3}{4}$ liters. If it contains $6\frac{1}{8}$ liters of antifreeze and the rest is water, how much is water?

$9\frac{5}{8}$ liters

75. Delbert types 366 words in $12\frac{1}{5}$ minutes. How many words per minute did he type?
30 words per minute

76. Alicia earns $\$4\frac{1}{2}$ dollars per hour for regular pay and $1\frac{1}{2}$ times that rate of pay for overtime. On Monday she worked 8 hours at regular pay and 3 hours at overtime. How much did she earn on Monday?

$\$56\frac{1}{4}$

77. George purchased stock at $\$88\frac{3}{8}$ a share. He sold it in four months at $\$79\frac{5}{8}$ a share. How much did the value of the stock decrease during that time period?

$\$8\frac{3}{4}$

78. A 3-inch bolt passes through $1\frac{1}{2}$ inches of pine board, a $\frac{1}{16}$-inch washer, and a $\frac{1}{8}$-inch nut. How many inches extend beyond the board, washer, and nut if the head of the bolt is $\frac{1}{4}$ inch long.

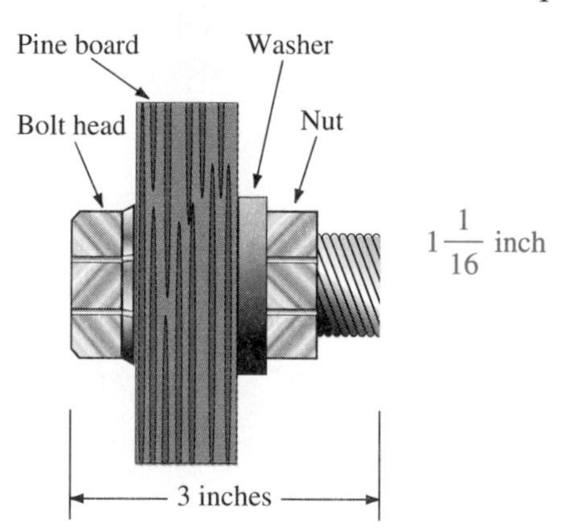

$1\frac{1}{16}$ inch

79. Francine has a take-home pay of \$880 per month. She gives $\frac{1}{10}$ of it to her church, spends $\frac{1}{2}$ of it for rent and food, and spends $\frac{1}{8}$ of it on electricity, heat, and telephone. How many dollars per month does she have left for other things? \$242

80. Manuel's new car used $18\frac{2}{5}$ gallons of gas on a 460-mile trip.

(a) How many miles can his car travel on 1 gallon of gas? 25 miles per gallon

(b) How much did his trip cost him in gasoline expense if the average cost of gasoline was $\$1\frac{1}{5}$ per gallon?

$\$22\frac{2}{25}$

Chapter 2 Test

1. Use a fraction to represent the shaded part of the object.

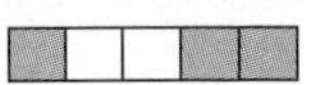

2. A basketball star shot at the hoop 388 times. The ball went in 311 times. Write a fraction that describes the part of the time that his shot went in.

Reduce each fraction.

3. $\dfrac{15}{70}$

4. $\dfrac{18}{42}$

5. $\dfrac{225}{50}$

6. Change to an improper fraction. $6\dfrac{4}{5}$

7. Change to a mixed number. $\dfrac{114}{14}$

Multiply.

8. $42 \times \dfrac{2}{7}$

9. $\dfrac{7}{9} \times \dfrac{2}{5}$

10. $4\dfrac{1}{3} \times 7\dfrac{1}{5}$

Divide.

11. $\dfrac{7}{8} \div \dfrac{5}{11}$

12. $5\dfrac{1}{7} \div 3$

13. $7\dfrac{1}{5} \div 1\dfrac{1}{25}$

14. $\dfrac{12}{31} \div \dfrac{8}{13}$

Find the least common denominator of each set of fractions.

15. $\dfrac{5}{24}$ and $\dfrac{7}{18}$

16. $\dfrac{3}{16}$ and $\dfrac{1}{24}$

17. $\dfrac{1}{4}, \dfrac{3}{8}, \dfrac{5}{6}$

18. Build the fraction to an equivalent fraction with the specified denominator. $\dfrac{5}{12} = \dfrac{?}{72}$

1.	$\frac{3}{5}$
2.	$\frac{311}{388}$
3.	$\frac{3}{14}$
4.	$\frac{3}{7}$
5.	$\frac{9}{2}$
6.	$\frac{34}{5}$
7.	$8\frac{1}{7}$
8.	12
9.	$\frac{14}{45}$
10.	$31\frac{1}{5}$
11.	$\frac{77}{40}$ or $1\frac{37}{40}$
12.	$\frac{12}{7}$ or $1\frac{5}{7}$
13.	$6\frac{12}{13}$
14.	$\frac{39}{62}$
15.	72
16.	48
17.	24
18.	$\frac{30}{72}$

19.	$\frac{13}{36}$
20.	$\frac{11}{20}$
21.	$\frac{25}{28}$
22.	$14\frac{6}{35}$
23.	$4\frac{13}{14}$
24.	$2\frac{4}{9}$
25.	$21\frac{1}{4}$
26.	42 square yards
27.	8 packages
28.	$\frac{7}{10}$ mile
29.	$14\frac{1}{24}$ miles
30. (a)	40 oranges
(b)	\$9.60 or 960¢

Add or subtract.

19. $\frac{7}{9} - \frac{5}{12}$

20. $\frac{2}{15} + \frac{5}{12}$

21. $\frac{1}{4} + \frac{3}{7} + \frac{3}{14}$

22. $8\frac{3}{5} + 5\frac{4}{7}$

23. $18\frac{6}{7} - 13\frac{13}{14}$

24. $6 - 3\frac{5}{9}$

25. $30 - 8\frac{3}{4}$

Answer each question.

26. A hallway measures $8\frac{1}{6}$ yards by $5\frac{1}{7}$ yards. How many square yards is the area of the hallway?

27. A butcher has $18\frac{2}{3}$ pounds of steak that he wishes to place into packages that average $2\frac{1}{3}$ pounds each. How many packages can he make?

28. From central parking it is $\frac{9}{10}$ of a mile to the science building. Bob started at central parking and walked $\frac{1}{5}$ of a mile toward the science building. He stopped for coffee. When he finishes, how far does he have to walk?

29. Robin jogged $4\frac{1}{8}$ miles on Monday, $3\frac{1}{6}$ miles on Tuesday, and $6\frac{3}{4}$ miles on Wednesday. How far did she jog on those three days?

30. Mr. and Mrs. Samuel visited Florida and purchased 120 oranges. They gave $\frac{1}{4}$ to relatives, ate $\frac{1}{12}$ of them in the hotel, and gave $\frac{1}{3}$ of them to friends. They shipped the rest home to Illinois.

(a) How many oranges did they ship?

(b) If it costs 24¢ for each orange to be shipped to Illinois, what was the total of the shipping bill?

Cumulative Test for Chapters 1–2

One-half of this test is based on Chapter 1 material. The remainder is based on material covered in Chapter 2.

1. Write in words. 84,361,208

2. Add.

$$\begin{array}{r} 128 \\ 452 \\ 178 \\ 34 \\ +\ 77 \\ \hline \end{array}$$

3. Add.

$$\begin{array}{r} 156{,}200 \\ 364{,}700 \\ +\ 198{,}320 \\ \hline \end{array}$$

4. Subtract.

$$\begin{array}{r} 5718 \\ -\ 3643 \\ \hline \end{array}$$

5. Subtract.

$$\begin{array}{r} 1{,}000{,}361 \\ -\ \ 983{,}145 \\ \hline \end{array}$$

6. Multiply.

$$\begin{array}{r} 126 \\ \times\ 38 \\ \hline \end{array}$$

7. Multiply.

$$\begin{array}{r} 16{,}908 \\ \times\ \ 12 \\ \hline \end{array}$$

8. Divide. $7\overline{)30{,}150}$

9. Divide. $18\overline{)6642}$

10. Evaluate. 7^2

11. Round to the nearest thousand. 6,037,452

12. Perform the operations in their proper order.

$$4 \times 3^2 + 12 \div 6$$

13. Roone bought three shirts at \$26 and two pairs of pants at \$48. What was his total bill?

14. Leslie had a balance of \$64 in her checking account. She deposited \$1160. She made checks out for \$516, \$199, and \$203. What will be her new balance?

15. Eighty-four students enrolled in psychology. Fifty-five were women. Write a fraction that describes the part of the class that was made up of women.

1. Eighty-four million, three hundred sixty-one thousand, two hundred eight

2. 869

3. 719,220

4. 2075

5. 17,216

6. 4788

7. 202,896

8. 4307 R 1

9. 369

10. 49

11. 6,037,000

12. 38

13. \$174

14. \$306

15. $\frac{55}{84}$

16. $\frac{7}{13}$

17. $\frac{75}{4}$

18. $14\frac{2}{7}$

19. $\frac{527}{48}$ *or* $10\frac{47}{48}$

20. $\frac{12}{35}$

21. 39

22. $\frac{31}{54}$

23. $\frac{71}{8}$ *or* $8\frac{7}{8}$

24. $\frac{113}{15}$ *or* $7\frac{8}{15}$

25. $\frac{13}{28}$

26. $23\frac{1}{8}$ tons

27. $24\frac{3}{5}$ miles per gallon

28. $8\frac{1}{8}$ cups of sugar, $5\frac{5}{6}$ cups of flour

29. 60,000,000 miles

16. Reduce. $\dfrac{28}{52}$

17. Write as an improper fraction. $18\dfrac{3}{4}$

18. Write as a mixed number. $\dfrac{100}{7}$

19. Multiply. $3\dfrac{7}{8} \times 2\dfrac{5}{6}$

20. Divide. $\dfrac{44}{49} \div 2\dfrac{13}{21}$

21. Find the least common denominator of $\dfrac{6}{13}$ and $\dfrac{5}{39}$.

Add or subtract.

22. $\dfrac{7}{18} + \dfrac{5}{27}$

23. $2\dfrac{1}{8} + 6\dfrac{3}{4}$

24. $12\dfrac{1}{5} - 4\dfrac{2}{3}$

25. $\dfrac{11}{14} - \dfrac{9}{28}$

26. A truck hauled $9\frac{1}{2}$ tons of gravel on Monday. On Tuesday, it hauled $6\frac{3}{8}$ tons, and on Wednesday, $7\frac{1}{4}$ tons. How many tons were hauled on the three days?

27. Melinda traveled $221\frac{2}{5}$ miles on 9 gallons of gas. How many miles per gallon did her car get?

28. A recipe requires $3\frac{1}{4}$ cups of sugar and $2\frac{1}{3}$ cups of flour. Marcia wants to make $2\frac{1}{2}$ times that recipe. How many cups of each will she need?

29. A space probe travels at 28,356 miles per hour for 2142 hours. *Estimate* how many miles it travels.

CHAPTER 3

Decimals

There is a major problem that affects almost half of the college students of the United States during their first year of school. This problem costs them extra money and sometimes has serious financial consequences. You probably cannot guess what it is! Believe it or not—almost half of the first-year college students in the United States cannot balance their checkbook!

If you think you are above problems like that, please turn to the Putting Your Skills To Work on page 198.

Pretest Chapter 3

1. forty-seven and eight hundred thirteen thousandths
2. 0.0567
3. $2\frac{11}{100}$
4. $\frac{21}{40}$
5. 1.59, 1.6, 1.601, 1.61
6. 10.5
7. 1.053
8. 19.45
9. 27.191
10. 10.59
11. 8.892

If you are familiar with the topics in this chapter, take this test now. Check your answers with those in the back of the book. If an answer was wrong or you couldn't do a problem, study the appropriate section of the chapter.

If you are not familiar with the topics in this chapter, don't take this test now. Instead, study the examples, work the practice problems, and then take the test.

This test will help you identify which concepts you have mastered and which you need to study further.

Section 3.1

1. Write a word name for the decimal. 47.813

2. Express this fraction as a decimal. $\frac{567}{10{,}000}$

Write as a fraction or a mixed number. Reduce whenever possible.

3. 2.11 **4.** 0.525

Section 3.2

5. Place the set of numbers in the proper order from smallest to largest. 1.6, 1.59, 1.61, 1.601

6. Round to the nearest tenth. 10.472935

7. Round to the nearest thousandth. 1.053458

Section 3.3

Add.

8.
$$\begin{array}{r} 5.12 \\ 4.7 \\ 8.03 \\ +\ 1.6 \\ \hline \end{array}$$

9. 24.613 + 0.273 + 2.305

Subtract.

10.
$$\begin{array}{r} 42.16 \\ -\ 31.57 \\ \hline \end{array}$$

11. 13 − 4.108

Section 3.4

Multiply.

12. $\begin{array}{r} 11.67 \\ \times\ 0.03 \\ \hline \end{array}$ **13.** 4.7805×1000

14. 0.0005129×10^4

Section 3.5

Divide.

15. $0.09\overline{)0.03186}$

16. $0.3328 \div 2.6$

Section 3.6

17. Write as a decimal. $\frac{7}{16}$

18. Write as repeating decimal. $\frac{5}{22}$

19. Perform the operations in the correct order. $(0.2)^2 + 8.7 \times 0.3 - 1.68$

Section 3.7

20. When Marcia began her trip, the odometer on her car read 57,124.8. When she ended the trip, it read 57,312.8. She used 10.5 gallons of gas. How many miles per gallon did her car get? Round your answer to the nearest tenth.

21. A rectangular room is 10.5 yards long and 3.6 yards wide. How much will it cost to install carpeting for the entire room at $12.95 per square yard?

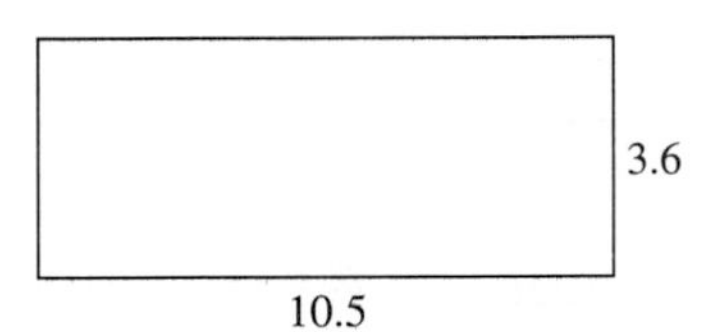

22. Tony worked 35 hours last week. He was paid $298.55. How much was he paid per hour?

12. 0.3501

13. 4780.5

14. 5.129

15. 0.354

16. 0.128

17. 0.4375

18. $0.2\overline{27}$ or 0.22727 . . .

19. 0.97

20. 17.9 miles per gallon

21. $489.51

22. $8.53

3.1 Decimal Notation

MathPro

Video 6

SSM

After studying this section, you will be able to:

1. *Write a word name for a decimal fraction.*
2. *Change from fractional notation to decimal notation.*
3. *Change from decimal notation to fractional notation.*

1 *Writing a Word Name for a Decimal Fraction*

In Chapter 2 we discussed *fractions*—the set of numbers such as $\frac{1}{2}$, $\frac{2}{3}$, $\frac{1}{10}$, $\frac{6}{7}$, $\frac{18}{100}$, and so on. Now we will take a closer look at **decimal fractions**—that is, fractions with 10, 100, 1000, and so on, in the denominator, such as $\frac{1}{10}$, $\frac{18}{100}$, and $\frac{43}{1000}$.

Why, of all fractions, do we take special notice of these? Our hands have *ten* digits. Our U.S. money system is based on the dollar, which has 100 equal parts, or cents. And the international system of measurement called the *metric system* is based on 10 and powers of 10.

As with other numbers, these decimal fractions can be written in different ways (forms). For example, the shaded part of the whole in the drawing can be written

In words (one-tenth)
In fraction form ($\frac{1}{10}$)
In decimal form (0.1)

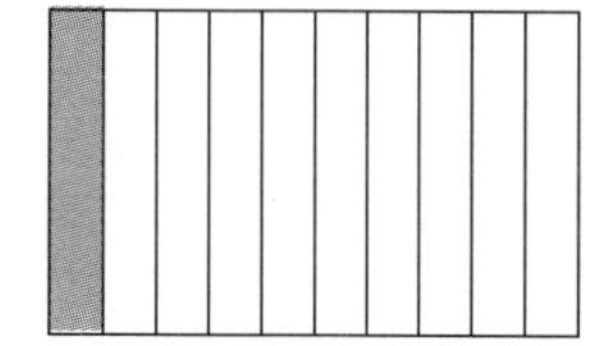

All mean the same quantity, namely 1 out of 10 equal parts of the whole. We'll see that when we use decimal notation, computations can be easily done based on the old rules for whole numbers and a few new rules about where to place the decimal point. In a world where calculators and computers are commonplace, many of the fractions we encounter are decimal fractions. A decimal fraction is a fraction whose denominator is a power of 10.

$\frac{7}{10}$ is a decimal fraction. $\frac{89}{10^2} = \frac{89}{100}$ is a decimal fraction.

Decimal fractions can be written with numerals in two ways: fractional form or decimal form.

FRACTIONAL FORM		DECIMAL FORM
$\frac{3}{10}$	=	0.3
$\frac{59}{100}$	=	0.59
$\frac{171}{1000}$	=	0.171

TEACHING TIP Students often wonder if they should write .45 rather than 0.45 because they think the 0 is unnecessary. Remind them that the extra zero in front of a decimal point reduces the chances for error. Also, most calculators display this extra 0.

The zero in front of the decimal point is not actually required. We place it there simply to make sure that we don't miss seeing the decimal point. When a number is written in decimal form, the first digit to the right of the decimal point represents tenths, the next digit hundredths, the next digit thousandths, and so on. 0.9 means nine tenths and is equivalent to $\frac{9}{10}$. 0.51 means fifty-one hundredths and is equivalent to $\frac{51}{100}$. Some decimals are larger than 1. For example, 1.683 means one and six hundred eighty-three thousandths. It is equivalent to $1\frac{683}{1000}$. Note that the word *and* is used to indicate the decimal point. A place-value chart is helpful.

Decimal Place Values

Hundreds	Tens	Ones	Decimal Point	Tenths	Hundredths	Thousandths	Ten Thousandths
1	5	6	.	2	8	7	4
100	10	1	"and"	$\frac{1}{10}$	$\frac{1}{100}$	$\frac{1}{1,000}$	$\frac{1}{10,000}$

So, for 156.2874, we can write in words one hundred fifty-six and two thousand eight hundred seventy-four ten thousandths. We say ten thousandths because it is the name of the last decimal place on the right.

EXAMPLE 1 Write a word name for each decimal.

(a) 0.79 **(b)** 0.5308 **(c)** 1.6 **(d)** 23.765

(a) 0.79 = Seventy-nine hundredths

(b) 0.5308 = Five thousand three hundred eight ten thousandths

(c) 1.6 = One and six tenths

(d) 23.765 = Twenty-three and seven hundred sixty-five thousandths

Practice Problem 1 Write a word name for each decimal.

(a) 0.073 **(b)** 4.68 **(c)** 0.0017 **(d)** 561.78 ■

Sometimes, decimals are used where we would not expect them. For example, we commonly say that there are 365 days in a year with 366 days in every fourth year (or leap year). However, this is not quite correct. In fact from time to time further adjustments need to be made to the calendar to adjust for these inconsistencies. Astronomers know that a more accurate measure of a year is called a **tropical year** (measured from one equinox to the next). Rounded to the nearest hundred thousandth, 1 tropical year = 365.24122 days. This is read three hundred sixty-five and twenty-four thousand, one hundred twenty-two hundred thousandths. This approximate value is a more accurate measurement of the amount of time it takes the earth to complete one orbit around the sun.

TEACHING TIP Students often think that writing a word name for 4.36 is "Four point three six" instead of the correct answer, which is four and thirty-six hundredths. Explain to them the correct answer, and emphasize that if they gain the ability to write word names for decimals correctly, it will help them to make out checks correctly.

Decimal notation is commonly used with money. When writing a check, we often write the amount that is less than 1 dollar, such as 23¢, as $\frac{23}{100}$ dollar.

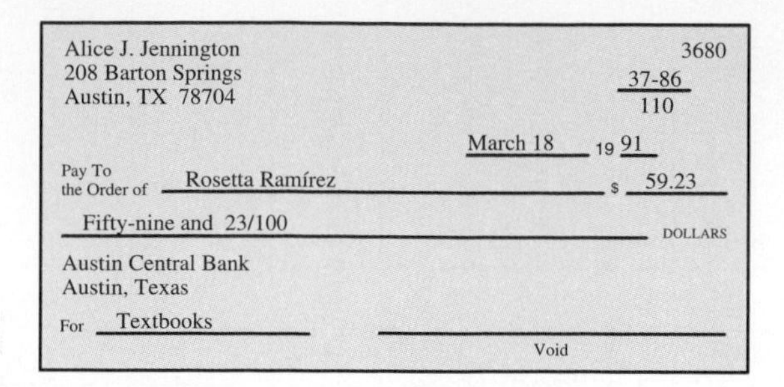

EXAMPLE 2 Write a word name for the amount on a check made out for $672.89.

$$\text{Six hundred seventy-two and } \frac{89}{100} \text{ dollars}$$

Practice Problem 2 Write a word name for the amount of a check made out for $7863.04. ■

2 *Changing from Fractional Notation to Decimal Notation*

It is helpful to be able to write decimals in both decimal notation and fractional notation. First we illustrate changing a fraction with a denominator of 10, 100, or 1000 into decimal form.

EXAMPLE 3 Write each fraction as a decimal.

(a) $\frac{8}{10}$ **(b)** $\frac{74}{100}$ **(c)** $1\frac{3}{10}$ **(d)** $2\frac{56}{1000}$

(a) $\frac{8}{10} = 0.8$ **(b)** $\frac{74}{100} = 0.74$

(c) $1\frac{3}{10} = 1.3$ **(d)** $2\frac{56}{1000} = 2.056$

Practice Problem 3 Write each fraction as a decimal.

(a) $\frac{9}{10}$ **(b)** $\frac{136}{1000}$ **(c)** $2\frac{56}{100}$ **(d)** $34\frac{86}{1000}$ ■

3 *Changing from Decimal Notation to Fractional Notation*

EXAMPLE 4 Write in fractional notation.

(a) 0.51 **(b)** 18.1 **(c)** 0.7611 **(d)** 1.363

(a) $0.51 = \frac{51}{100}$ **(b)** $18.1 = 18\frac{1}{10}$

(c) $0.7611 = \frac{7611}{10{,}000}$ **(d)** $1.363 = 1\frac{363}{1000}$

Practice Problem 4 Write in fractional notation.

(a) 0.37 **(b)** 182.3 **(c)** 0.7131 **(d)** 42.019 ■

When we convert from decimal form to fractional form, we reduce whenever possible.

EXAMPLE 5 Write in fractional form. Reduce whenever possible.

(a) 2.6 **(b)** 0.38 **(c)** 0.525 **(d)** 361.007

(a) $2.6 = 2\frac{6}{10} = 2\frac{3}{5}$ **(b)** $0.38 = \frac{38}{100} = \frac{19}{50}$

(c) $0.525 = \frac{525}{1000} = \frac{105}{200} = \frac{21}{40}$

(d) $361.007 = 361\frac{7}{1000}$ (cannot be reduced)

Practice Problem 5 Write in fraction form. Reduce whenever possible.

(a) 8.5 **(b)** 0.58 **(c)** 36.25 **(d)** 106.013 ■

TEACHING TIP You may want to do for the class a few simple problems like 5 parts per thousand, or 15 parts per ten thousand, before doing the type of numbers contained in Example 6.

EXAMPLE 6 A chemist found that the concentration of lead in a water sample was 5 parts per million. What fraction would represent the concentration of lead?

Five parts per million means 5 parts out of 1,000,000. As a fraction, this is $\frac{5}{1{,}000{,}000}$. We can reduce this by dividing numerator and denominator by five. Thus

$$\frac{5}{1{,}000{,}000} = \frac{1}{200{,}000}$$

The concentration of lead in the water sample is $\frac{1}{200{,}000}$.

Practice Problem 6 A chemist found that the concentration of PCBs in a water sample was 2 parts per billion. What fraction would represent the concentration of PCBs? ■

Developing Your Study Skills

STEPS TOWARD SUCCESS IN MATHEMATICS

Mathematics is a building process, mastered one step at a time. The foundation of this process is built on a few basic requirements. Those who are successful in mathematics realize the absolute necessity for building a study of mathematics on the firm foundation of these six minimum requirements.

1. Attend class every day.
2. Read the textbook.
3. Take notes in class.
4. Do assigned homework every day.
5. Get help immediately when needed.
6. Review regularly.

3.1 Exercises

Verbal and Writing Skills

1. Describe a decimal fraction and provide examples. A decimal fraction is a fraction whose denominator is a power of 10. *Note:* Check student's examples.

2. What word is used to describe the decimal point when writing the word name for a decimal that is greater than one? The word that describes the decimal point is *and*.

3. What is the name of the last decimal place on the right for the decimal 132.45678? Hundred thousandths

4. When writing \$82.75 on a check, we write 75¢ as $\frac{75}{100}$.

Write a word name for each decimal.

5. 0.57
Fifty-seven hundredths

6. 0.78
Seventy-eight hundredths

7. 3.8
Three and eight tenths

8. 5.2
Five and two tenths

9. 0.124
One hundred twenty-four thousandths

10. 0.367
Three hundred sixty-seven thousandths

11. 28.0007
Twenty-eight and seven ten thousandths

12. 54.0003
Fifty-four and three ten thousandths

Write a word name as you would on a check.

13. \$36.18
Thirty-six and $\frac{18}{100}$ dollars

14. \$25.99
Twenty-five and $\frac{99}{100}$ dollars

15. \$1236.08
One thousand two hundred thirty-six and $\frac{8}{100}$ dollars

16. \$7652.02
Seven thousand, six hundred fifty-two and $\frac{2}{100}$ dollars

17. \$10,000.76
Ten thousand and $\frac{76}{100}$ dollars

18. \$20,000.67
Twenty thousand and $\frac{67}{100}$ dollars

Write each of the following in decimal notation.

19. seven tenths
0.7

20. twelve hundredths
0.12

21. thirty-nine thousandths
0.039

22. seventy-five thousandths
0.075

23. sixty-five ten thousandths
0.0065

24. twenty-nine ten thousandths
0.0029

25. two hundred eighty-six millionths
0.000286

26. seven hundred sixteen millionths
0.000716

Write each fraction as a decimal.

27. $\frac{3}{10}$ 0.3 **28.** $\frac{7}{10}$ 0.7 **29.** $\frac{76}{100}$ 0.76 **30.** $\frac{84}{100}$ 0.84 **31.** $\frac{771}{1000}$ 0.771 **32.** $\frac{652}{1000}$ 0.652

33. $\frac{53}{1000}$ 0.053 **34.** $\frac{42}{1000}$ 0.042 **35.** $\frac{26}{10{,}000}$ 0.0026 **36.** $\frac{36}{10{,}000}$ 0.0036 **37.** $8\frac{3}{10}$ 8.3 **38.** $3\frac{1}{10}$ 3.1

39. $84\frac{13}{100}$ 84.13 **40.** $52\frac{77}{100}$ 52.77 **41.** $1\frac{19}{1000}$ 1.019 **42.** $2\frac{23}{1000}$ 2.023 **43.** $126\frac{571}{10{,}000}$ 126.0571 **44.** $198\frac{333}{10{,}000}$ 198.0333

Write in fractional notation. Reduce whenever possible.

45. 0.02 $\frac{2}{100} = \frac{1}{50}$

46. 0.05 $\frac{5}{100} = \frac{1}{20}$

47. 3.6 $3\frac{6}{10} = 3\frac{3}{5}$

48. 7.4 $7\frac{4}{10} = 7\frac{2}{5}$

49. 0.121 $\frac{121}{1000}$

50. 0.143 $\frac{143}{1000}$

51. 12.625 $12\frac{625}{1000} = 12\frac{5}{8}$

52. 29.875 $29\frac{875}{1000} = 29\frac{7}{8}$

53. 7.0015 $7\frac{15}{10{,}000} = 7\frac{3}{2000}$

54. 4.0016 $4\frac{16}{10{,}000} = 4\frac{1}{625}$

55. 235.1254 $235\frac{1254}{10{,}000} = 235\frac{627}{5000}$

56. 581.2406 $581\frac{2406}{10{,}000} = 581\frac{1203}{5000}$

57. 0.0187 $\frac{187}{10{,}000}$

58. 0.0209 $\frac{209}{10{,}000}$

59. 8.0108 $8\frac{108}{10{,}000} = 8\frac{27}{2500}$

60. 7.0605 $7\frac{605}{10{,}000} = 7\frac{121}{2000}$

61. 289.376 $289\frac{376}{1000} = 289\frac{47}{125}$

62. 423.814 $423\frac{814}{1000} = 423\frac{407}{500}$

63. 0.9889 $\frac{9889}{10{,}000}$

64. 0.3211 $\frac{3211}{10{,}000}$

Applications

65. Every year turtles lay eggs on the islands of South Carolina. Unfortunately, due to illegal polluting, a lot of the eggs are contaminated. If the turtle eggs contain more than 2 parts per one hundred million of chemical pollutants they will not hatch and the population will continue to head toward extinction. Write the above as a fraction in the lowest terms.

$\frac{2}{100{,}000{,}000} = \frac{1}{50{,}000{,}000}$

66. American Bald Eagles have been fighting extinction due to environmental hazards such as DDT, PCBs, and dioxin. The problem is with the food chain. Fish or rodents eat or drink contaminated food and/or water. Then, the eagles ingest the poison, which in turn affects the durability of the eagles' eggs. If it takes only 4 parts per million of certain chemicals to ruin an eagle egg, write this number as a fraction in lowest terms. In 1994 the Bald Eagle was removed from the endangered species list.

$\frac{4}{1{,}000{,}000} = \frac{1}{250{,}000}$

Cumulative Review Problems

67. Add. 156 + 84 + 39 + 463 + 76 = 818

68. Subtract. 12,843 − 11,905 = 938

69. Round to the nearest *hundred.* 56,758 56,800

70. Round to the nearest *thousand.* 8,069,482 8,069,000

3.2 Compare, Order, and Round Decimals

MathPro

Video 7

SSM

After studying this section, you will be able to:

1 *Compare decimals.*
2 *Place decimals in order from smallest to largest.*
3 *Round decimals to a specified decimal place.*

1 *Comparing Decimals*

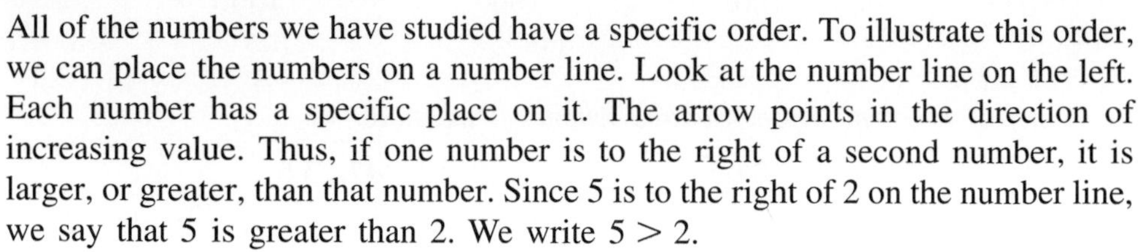

0 1 2 3 4 5 6 7 8 9
Number Line

All of the numbers we have studied have a specific order. To illustrate this order, we can place the numbers on a number line. Look at the number line on the left. Each number has a specific place on it. The arrow points in the direction of increasing value. Thus, if one number is to the right of a second number, it is larger, or greater, than that number. Since 5 is to the right of 2 on the number line, we say that 5 is greater than 2. We write $5 > 2$.

Since 4 is to the left of 6 on the number line, we say that 4 is less than 6. We write $4 < 6$.

The symbols "$>$" and "$<$" are called **inequality symbols.**

$a < b$ is read "a is less than b."

$a > b$ is read "a is greater than b."

TEACHING TIP This is a good time to stress the use of the number line. Remind the students that they can make any scale they want for a number line. It is just as easy to make a number line with units labeled 1.34, 1.35, 1.36, 1.37, 1.38, etc., as it is to make a number line with units labeled 1, 2, 3, 4, 5, 6, etc. This will help them whenever they compare decimals.

We can assign exactly one point on the number line to each decimal number. When two decimal numbers are placed on a number line, the one farther to the right is the larger. Thus we can say that $3.4 > 2.7$ and $4.3 > 4.0$. We can also say that $0.5 < 1.0$ and $1.8 < 2.2$. Why?

0.5 1.8 2.2 2.7 3.4 4.3
0 1.0 2.0 3.0 4.0 5.0

To compare or order decimals without using a number line, we compare each digit.

> **Comparing Two Numbers in Decimal Notation**
>
> **1.** Start at the left and compare corresponding digits. If the digits are the same, move one place to the right.
>
> **2.** When two digits are different, the larger number is the one with the larger digit.

EXAMPLE 1 Write an inequality statement with 0.167 and 0.166.

The number 1 in the tenths place is the same.

0.1 6 7 0.1 6 6

The number 6 in the hundredths place is the same.

0.1 6 7 0.1 6 6

The numbers in the thousandths place differ.

0. 1 6 7 0.1. 6 6

Since $7 > 6$, we know that $0.167 > 0.166$.

Practice Problem 1 Write an inequality statement with 5.74 and 5.75. ■

Whenever necessary, extra zeros can be written to the right of the last digit—that is, to the right of the decimal point—without changing the value of the decimal. Thus

$$0.56 = 0.56000 \qquad \text{and} \qquad 0.7768 = 0.77680$$

The zero to the left of the decimal point is optional. Thus 0.56 = .56. Both notations are used. You are encouraged to place a zero to the left of the decimal point so that you don't miss the decimal point when you work with decimals.

EXAMPLE 2 Fill in the blank with one of the symbols <, =, or >.

0.77 ____ 0.777

If we add a zero to the first decimal

$$0.77\underline{0} \qquad\qquad 0.77\underline{7}$$

we see that the tenths and hundredths digits are equal. But the thousandths digits differ. Since $0 < 7$, we have $0.770 < 0.777$.

TEACHING TIP You may want to mention that in a mathematics class it does not make any difference whether you write 0.56 or 0.5600 since both numbers have exactly the same value. You might point out that in physics, chemistry, and engineering classes, however, the extra zeros to the right of the 56 are often used to indicate the level of accuracy of measurement.

Practice Problem 2 Fill in the blank with one of the symbols <, =, or >. ■

0.894 ____ 0.89

2 *Placing Decimals in Order*

You can place three or more decimals in order. If you are asked to order the decimals from smallest to largest, look for the smallest decimal and place it first.

EXAMPLE 3 Place the following five decimal numbers in order from smallest to largest.

1.834, 1.83, 1.381, 1.38, 1.8

First we add zeros to make the comparison easier.

1.834, 1.830, 1.381, 1.380, 1.800

Now we rearrange with smallest first.

1.380, 1.381, 1.800, 1.830, 1.834

Practice Problem 3 Place the following five decimal numbers in order from smallest to largest.

2.45, 2.543, 2.46, 2.54, 2.5 ■

TEACHING TIP You may want to ask the students if they are familiar with other rules for rounding off numbers. Particularly if you have students from other countries or cultures, you may see some unusual round-off rules. Remind them that this rule for rounding decimals is used uniformly in mathematics courses in the United States and that they will need to be able to use it.

3 *Rounding Decimals*

Sometimes in calculations involving money we see numbers like \$386.432 and \$29.5986. To make these useful, we usually round them to the nearest cent. \$386.432 is rounded to \$386.43. \$29.5986 is rounded to \$29.60. A general rule for rounding decimals follows.

> **Rounding Decimals**
>
> **1.** Find the decimal place (units, tenths, hundredths, and so on) for which rounding off is required.
>
> **2.** If the first digit to the right of this place is less than 5, drop it and all digits to the right of it.
>
> **3.** If the first digit to the right of this place is 5 or greater, increase the digit by one. Drop all digits to the right of this place.

EXAMPLE 4 Round 156.37 to the nearest tenth.

156.3 7
↑ We find the tenths place.

Note that 7, the next place to the right, is greater than 5. We round up to 156.4 and drop the digits to the right. The answer is 156.4.

Practice Problem 4 Round 723.88 to the nearest tenth. ■

EXAMPLE 5 Round to the nearest thousandth.

(a) 0.06358 **(b)** 128.37448

(a) 0.06 3 58
↑ We locate the thousandths place.

Note that the digit to the right of the thousandths place is 5. We round up to 0.064 and drop all the digits to the right.

(b) 128.37 4 48
↑ We locate the thousandths place.

Note that the digit to the right of thousandths place is less than 5. We round to 128.374 and drop all the digits to the right.

Practice Problem 5 Round to the nearest thousandth.

(a) 12.92647 **(b)** 0.007892 ■

Remember that rounding up to the next digit in a position may result in several digits being changed.

EXAMPLE 6 Round to the nearest hundredth. Fred and Linda used 203.9964 kilowatt hours of electricity in their house in May.

We locate the hundredths place ↓

203.9964

Since the digit to the right of hundredths is greater than 5, we round up. This affects the next two positions. Do you see why? The result is 204.00 kilowatt hours.

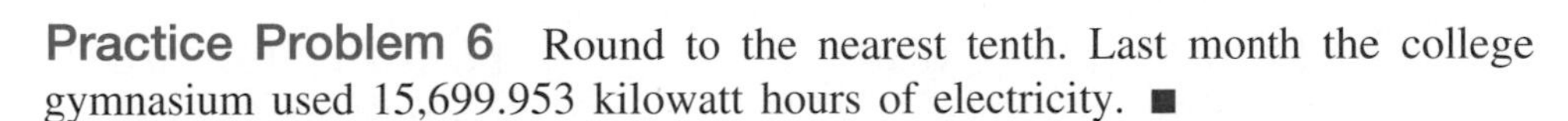

Practice Problem 6 Round to the nearest tenth. Last month the college gymnasium used 15,699.953 kilowatt hours of electricity. ■

Sometimes we round a decimal to the nearest whole number. For example, when writing figures on income tax forms, a taxpayer may round all figures to the nearest dollar.

EXAMPLE 7 To complete her income tax return, Marge needs to round these figures to the nearest whole dollar.

Medical bills $779.86 Taxes $563.49

Retirement contributions $674.38 Contributions to charity $534.77

Round the amounts.

	ORIGINAL FIGURE	ROUNDED TO NEAREST DOLLAR
Medical bills	779.86	780
Taxes	563.49	563
Retirement	674.38	674
Charity	534.77	535

Practice Problem 7 Round to the nearest whole dollar the following figures.

Medical bills $375.50 Taxes $981.39

Retirement contributions $980.49 Contributions to charity $817.65 ■

WARNING Why is it so important to consider only *one* digit to the right of the desired round-off position? What is wrong with rounding in steps? Suppose that Mark rounds 1.349 to the nearest tenth in steps. First he rounds 1.349 to 1.35 (nearest hundredth). Then he rounds 1.35 to 1.4 (nearest tenth). What is wrong with this reasoning?

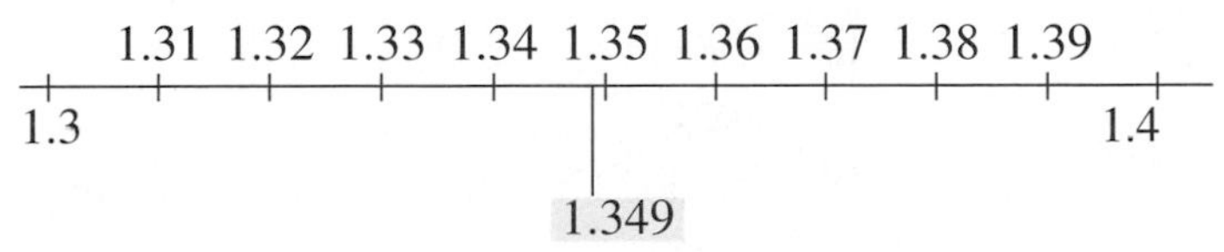

To round 1.349 to the nearest tenth, we ask if 1.349 is closer to 1.3 or to 1.4. It is closer to 1.3. Mark got 1.4, so he is not correct. He "rounded in steps" by first moving to 1.35, thus increasing the error and moving in the wrong direction. To control rounding errors, we consider *only* the first digit to the right of the decimal place to which we are rounding.

3.2 Exercises

Fill in the blank with one of the symbols <, =, or >.

1. 1.3 > 1.29 **2.** 2.6 > 2.58 **3.** 0.68 < 0.681 **4.** 72.54 < 72.56

5. 18.92 < 18.93 **6.** 0.460 = 0.46 **7.** 0.0006 > 0.0005 **8.** 0.0037 > 0.003

9. 1.0024 < 1.003 **10.** 1.003 < 1.004 **11.** 126.34 > 125.35 **12.** 406.78 < 407.75

13. 16.0572 < 16.0574 **14.** 18.00039 > 18.00038 **15.** $\frac{8}{10}$ > 0.08 **16.** $\frac{5}{100}$ > 0.005

Arrange each set of decimals from smallest to largest.

17. 12.6, 12.8, 12.65
12.6, 12.65, 12.8

18. 18.32, 18.038, 18.04
18.038, 18.04, 18.32

19. 0.0071, 0.05, 0.007
0.007, 0.0071, 0.05

20. 0.0025, 0.0052, 0.002
0.002, 0.0025, 0.0052

21. 1.8, 1.1, 1.81, 1.79
1.1, 1.79, 1.8, 1.81

22. 2.7, 2.5, 2.53, 2.48
2.48, 2.5, 2.53, 2.7

23. 26.034, 26.003, 26.04, 26.033
26.003, 26.033, 26.034, 26.04

24. 33.082, 33.02, 33.088, 33.079
33.02, 33.079, 33.082, 33.088

25. 18.006, 18.060, 18.066, 18.606, 18.065
18.006, 18.060, 18.065, 18.066, 18.606

26. 15.020, 15.002, 15.001, 15.018, 15.0019
15.001, 15.0019, 15.002, 15.018, 15.020

Round each decimal to the nearest tenth.

27. 5.67
5.7

28. 8.35
8.4

29. 29.49
29.5

30. 38.48
38.5

31. 578.064
578.1

32. 311.091
311.1

33. 2176.83
2176.8

34. 4082.74
4082.7

Round each decimal to the nearest hundredth.

35. 26.032
26.03

36. 47.071
47.07

37. 5.76582
5.77

38. 2.98613
2.99

39. 156.1197
156.12

40. 283.2168
283.22

41. 2786.706
2786.71

42. 4609.285
4609.29

Round to the nearest given place.

43. 1.06132 thousandths
1.061

44. 8.10263 thousandths
8.103

45. 0.047357 ten thousandths
0.0474

46. 0.063148 ten thousandths
0.0631

47. 5.00761238 hundred thousandths
5.00761

48. 4.01062378 hundred thousandths
4.01062

49. 0.00753682 millionths
0.007537

50. 0.00964983 millionths
0.009650

51. 129.08939 nearest whole number
129

52. 208.4372 nearest whole number
208

Round each amount to the nearest dollar.

53. \$2536.85
\$2537

54. \$5319.62
\$5320

55. \$10,098.47
\$10,098

56. \$20,159.48
\$20,159

Round each amount to the nearest cent.

57. \$56.9832
\$56.98

58. \$28.7619
\$28.76

59. \$5783.716
\$5783.72

60. \$3928.649
\$3928.65

Applications

61. If an average is taken of all the men and women on the planet, the average man consumes 0.095 kilogram of protein per day, while the average woman consumes 0.066 kilogram of protein per day. Round these values to the nearest hundredth. 0.10 kilogram, 0.07 kilogram

62. In the 1960s, the average person in the United States ate 10.62 pounds of chocolate per year. In the 1990s, the average person eats 21.78 pounds of chocolate per year. Round these values to the nearest tenth. 10.6 pounds, 21.8 pounds

To Think About

63. Arrange in order from smallest to largest.

$$0.61,\ 0.062,\ \frac{6}{10},\ 0.006,\ 0.0059,\ \frac{6}{100},\ 0.0601,\ 0.0519,\ 0.0612$$

$0.0059, 0.006, 0.0519, \frac{6}{100}, 0.0601, 0.0612, 0.062, \frac{6}{10}, 0.61$

64. Arrange in order from smallest to largest.

$$1.05,\ 1.512,\ \frac{15}{10},\ 1.0513,\ 0.049,\ \frac{151}{100},\ 0.0515,\ 0.052,\ 1.051$$

$0.049, 0.0515, 0.052, 1.05, 1.051, 1.0513, \frac{15}{10}, \frac{151}{100}, 1.512$

65. A person wants to round 86.23498 to the nearest hundredth. He first rounds 86.23498 to 86.2350. He then rounds to 86.235. Finally, he rounds to 86.24. What is wrong with his reasoning? You should consider only one digit to the right of the decimal place that you wish to round to. 86.23498 is closer to 86.23 than to 86.24.

66. Fred is checking the calculations on his monthly bank statement. An interest charge of \$16.3724 was rounded to \$16.38. An interest charge of \$43.7214 was rounded to \$43.73. What rule does the bank use for rounding off to the nearest cent? The bank rounds up for any fractional part of a cent.

Cumulative Review Problems

67. Add. $3\frac{1}{4} + 2\frac{1}{2} + 6\frac{3}{8}$

$12\frac{1}{8}$

68. Subtract. $27\frac{1}{5} - 16\frac{3}{4}$

$10\frac{9}{20}$

69. Mary drove her Dodge Caravan on a trip. At the start of the trip the odometer (which measures distance) read 46,381. At the end of the trip it read 47,073. How many miles long was the trip? 692 miles

70. The cost of four pieces of equipment for the college computer center was \$1736, \$2714, \$892, and \$4316. What was their total cost? \$9658

3.3 Addition and Subtraction of Decimals

MathPro

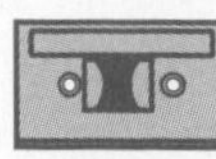
Video 7

SSM

After studying this section, you will be able to:

1 *Add two or more decimals.*
2 *Subtract two decimals.*

1 *Adding Decimals*

The Apple Crate Restaurant	
2 apple pie specials	$4.95
2 cups coffee	$1.85
	$6.80
tax	.34
total	$7.14
Please pay cashier	

We often add decimals when we check the addition of our bill at a restaurant or at a store. We can relate addition of decimals to addition of fractions. For example,

$$\frac{3}{10}+\frac{6}{10}=\frac{9}{10} \quad \text{and} \quad 1\frac{1}{10}+2\frac{8}{10}=3\frac{9}{10}$$

These same problems can be written more efficiently as decimals.

$$\begin{array}{r} 0.3 \\ +\,0.6 \\ \hline 0.9 \end{array} \qquad \begin{array}{r} 1.1 \\ +\,2.8 \\ \hline 3.9 \end{array}$$

The steps to follow when adding decimals are listed below.

> **Adding Decimals**
>
> 1. Write the numbers to be added vertically and line up the decimal points. Extra zeros may be placed to right of the decimal points if needed.
> 2. Add all the digits with the same place value, starting with the right column and moving to the left.
> 3. Place the decimal point of the sum in line with the decimal points of the numbers added.

TEACHING TIP This is a good time to assign a classroom activity. Ask your class to add the following: 7.21 + 13.346 + 0.0008 + 9.6. The correct answer is 30.1568. Usually you will find that almost all of the students who made an error in adding these numbers did not insert the extra zeros to the right. Students will be quite convinced of the merit of this suggestion to add in the extra zeros to the right when they do Example 1(c) or any similar example.

EXAMPLE 1 Add.

(a) 2.8 + 5.6 + 3.2 **(b)** 158.26 + 200.07 + 315.98

(c) 5.3 + 26.182 + 0.0007 + 6.24

(a)
$$\begin{array}{r} \scriptstyle 1 \\ 2.8 \\ 5.6 \\ +\,3.2 \\ \hline 11.6 \end{array}$$

(b)
$$\begin{array}{r} \scriptstyle 1\,1\,2 \\ 158.26 \\ 200.07 \\ +\,315.98 \\ \hline 674.31 \end{array}$$

(c)
$$\begin{array}{r} \scriptstyle 11 \\ 5.3000 \\ 26.1820 \\ 0.0007 \\ +\;\;6.2400 \\ \hline 37.7227 \end{array}$$

Extra zeros have been added to make the problem easier.

Practice Problem 1 Add.

(a)
$$\begin{array}{r} 9.8 \\ 3.6 \\ +\,5.4 \\ \hline \end{array}$$

(b)
$$\begin{array}{r} 300.72 \\ 163.75 \\ +\,291.08 \\ \hline \end{array}$$

(c) 8.9 + 37.056 + 0.0023 + 9.45 ■

SIDELIGHT When we add decimals like 3.1 + 2.16 + 4.007, we may write in zeros as below.

$$\begin{array}{r} 3.100 \\ 2.160 \\ +\ 4.007 \\ \hline 9.267 \end{array}$$

What are we really doing here? What is the advantage of adding these extra zeros?

"Decimals" means "decimal fractions." If we look at the number as fractions, we see that we are actually using the property of multiplying a fraction by 1 in order to obtain common denominators. Look at the problem this way:

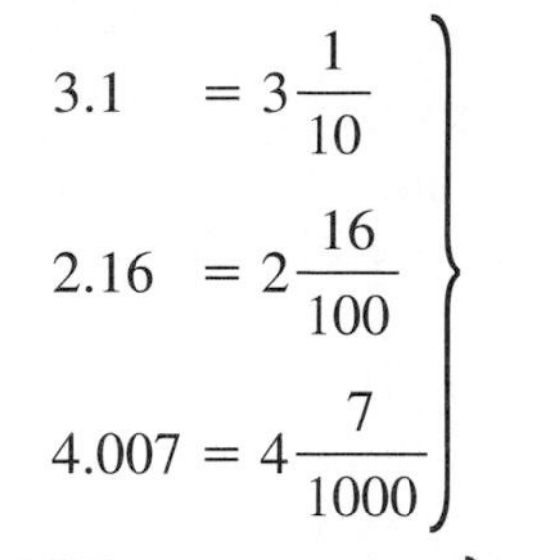

$$\left.\begin{aligned} 3.1 &= 3\frac{1}{10} \\ 2.16 &= 2\frac{16}{100} \\ 4.007 &= 4\frac{7}{1000} \end{aligned}\right\}$$

The least common denominator is 1000. *To obtain the common denominator for the first two fractions, we multiply.*

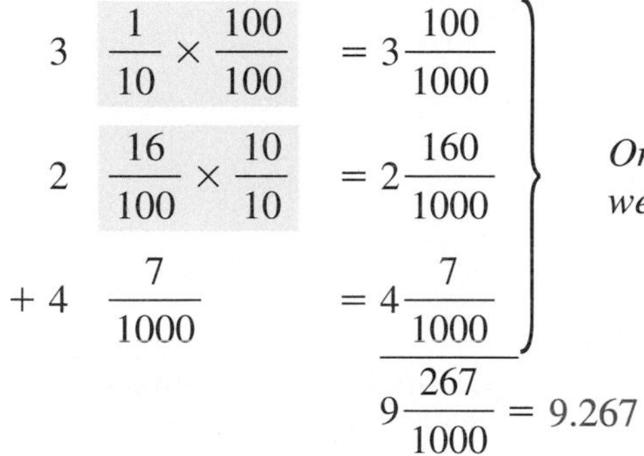

$$\left.\begin{aligned} 3\ \frac{1}{10} \times \frac{100}{100} &= 3\frac{100}{1000} \\ 2\ \frac{16}{100} \times \frac{10}{10} &= 2\frac{160}{1000} \\ +\,4\ \frac{7}{1000} \qquad &= 4\frac{7}{1000} \end{aligned}\right\}$$

$$9\frac{267}{1000} = 9.267$$

Once we obtain a common denominator, we can add the three fractions.

This is the answer we arrived at above using the decimal form for each number. Thus writing in zeros in a decimal fraction is really an easy way to transform fractions to equivalent fractions with a common denominator. Working with decimal fractions is easier than working with other fractions.

CALCULATOR

Adding Decimals

The calculator can be used to verify your work. You can use your calculator to add decimals. To find 23.08 + 8.53 + 9.31 enter:

23.08 [+] 8.53 [+] 9.31 [=]

Display:

[40.92]

The final digit of an odometer measures tenths of a mile. The odometer reading shown in the odometer on the right is 38,516.2 miles.

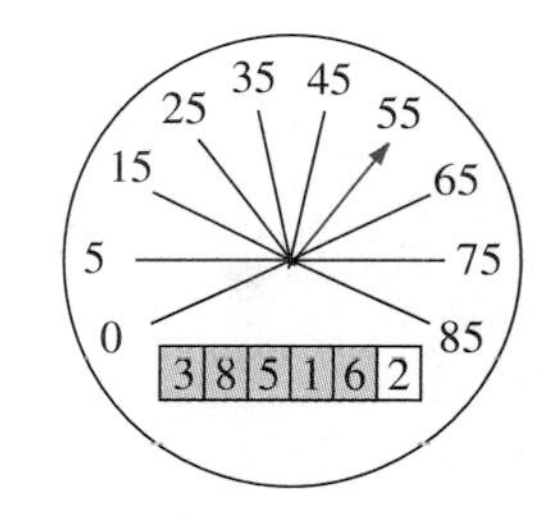

EXAMPLE 2 Barbara checked her odometer before the summer began. It read 49,645.8 miles. She traveled 3852.6 miles that summer in her car. What was the odometer reading at the end of the summer?

$$\begin{array}{r} {}^{1\,1\quad 1} \\ 49{,}645.8 \\ +\ \ 3{,}852.6 \\ \hline 53{,}498.4 \end{array}$$

The odometer read 53,498.4 miles.

Practice Problem 2 A car odometer read 93,521.8 miles before a trip of 1634.8 miles. What was the final odometer reading? ■

EXAMPLE 3 During his first semester at Tarrant County Community College, Kelvey deposited checks into his checking account in the amounts of \$98.64, \$157.32, \$204.81, \$36.07, and \$229.89. What was the sum of his five checks?

$$\begin{array}{r} {}^{2\,3\,2\;2} \\ \$\ 98.64 \\ 157.32 \\ 204.81 \\ 36.07 \\ +\ 229.89 \\ \hline \$726.73 \end{array}$$

Practice Problem 3 Beforing buying textbooks for the fall semester, Will deposited the following checks into his account: \$80.95, \$133.91, \$256.47, \$53.08, and \$381.32. What was the sum of his five checks? ■

2 *Subtracting Decimals*

It is important to see the relationship between the decimal form of a mixed number and the fractional form of a mixed number. This relationship helps us to understand why calculations with decimals are done the way they are. When we subtract mixed numbers with common denominators, sometimes we must borrow from the whole number.

$$\begin{array}{rcr} 5\frac{4}{10} & = & 4\frac{14}{10} \\ -\ 2\frac{7}{10} & = & -\ 2\frac{7}{10} \\ \hline & & 2\frac{7}{10} \end{array}$$

We could write the same problem in decimal form:

$$\begin{array}{r} \overset{4\ 14}{\cancel{5}.\cancel{4}} \\ -\ 2.7 \\ \hline 2.7 \end{array}$$

Subtraction of decimals is thus similar to subtraction of fractions (we get the same result), but it's usually easier to subtract with decimals than to subtract with fractions.

Subtracting Decimals

1. Write the decimals to be subtracted vertically and line up the decimal points. Additional zeros may be placed to the right if not all numbers have the same number of decimal places.
2. Subtract all digits with the same place value, starting with the right column and moving to the left. Borrow when necessary.
3. Place the decimal point of the difference in line with the decimal point of the two numbers being subtracted.

EXAMPLE 4 Subtract.

(a) $\begin{array}{r} 84.8 \\ -\ 27.3 \\ \hline \end{array}$ **(b)** $\begin{array}{r} 1076.320 \\ -\ \ 983.518 \\ \hline \end{array}$

(a)
$$\begin{array}{rrrr} & 7 & 14 & \\ & \cancel{8} & \cancel{4} & .8 \\ - & 2 & 7 & .3 \\ \hline & 5 & 7 & .5 \end{array}$$

(b)
$$\begin{array}{rrrrrrrr} & & 9 & & & & & \\ & & \cancel{10} & 17 & 5 & 13 & 1 & 10 \\ & \cancel{1} & \cancel{0} & \cancel{7} & \cancel{6}. & \cancel{3} & \cancel{2} & \cancel{0} \\ - & & 9 & 8 & 3. & 5 & 1 & 8 \\ \hline & & & 9 & 2. & 8 & 0 & 2 \end{array}$$

Practice Problem 4 Subtract.

(a) $\begin{array}{r} 38.8 \\ -\ 26.9 \\ \hline \end{array}$ **(b)** $\begin{array}{r} 2034.908 \\ -\ 1986.325 \\ \hline \end{array}$ ■

TEACHING TIP Some students get very nervous about showing the borrowing steps as in Example 4. You will reduce their anxiety if you remind them that writing out the borrowing steps is always allowed and is very helpful for some students. They should never hesitate to add these steps if it will help them. If, however, they find that they can accurately subtract without writing the borrowing steps, then it is not necessary to write down the borrowing notation.

When the two numbers being subtracted do not have the same number of decimal places, write in zeros as needed.

EXAMPLE 5 Subtract. **(a)** $12 - 8.362$ **(b)** $156.381 - 99.82$

(a)
$$\begin{array}{rrrrrr} & & & 9 & 9 & \\ & & 11 & \cancel{10} & \cancel{10} & 10 \\ & \cancel{1} & \cancel{2}. & \cancel{0} & \cancel{0} & \cancel{0} \\ - & & 8 & .3 & 6 & 2 \\ \hline & & 3 & .6 & 3 & 8 \end{array}$$

(b)
$$\begin{array}{rrrrrrr} & & 14 & 15 & & & \\ & & \cancel{4} & \cancel{5} & \cancel{13} & & \\ & \cancel{1} & \cancel{5} & \cancel{6}. & \cancel{3} & 8 & 1 \\ - & & 9 & 9 & .8 & 2 & 0 \\ \hline & & 5 & 6 & .5 & 6 & 1 \end{array}$$

Practice Problem 5 Subtract. **(a)** $19 - 12.579$ **(b)** $283.076 - 96.38$ ■

TEACHING TIP Students are usually surprised to realize how easy it is to get a common denominator when adding fractions or subtracting fractions that are written in decimal form. Remind them that the word *decimal* means *decimal fraction.* They usually remember that common denominators are needed for adding normal fractions but not decimal fractions. This concept is very interesting to students.

EXAMPLE 6 On Tuesday Don Ling filled the gas tank in his car. The odometer read 56,098.5. He drove for four days. The next time he filled the tank, the odometer read 56,420.2. How many miles had he driven?

$$\begin{array}{r} \overset{3}{\cancel{4}}\overset{11}{\cancel{1}}\ \ \overset{9}{\cancel{2}}\ \overset{12}{\cancel{0}\cancel{2}} \\ 5\ 6{,}4\ 2\ 0.2 \\ -\ 5\ 6{,}0\ 9\ 8.5 \\ \hline 3\ 2\ 1.7 \end{array}$$

He had driven 321.7 miles.

Practice Problem 6 Abdul had his car oil changed when his car odometer read 82,370.9 miles. When he changed oil again, the odometer read 87,160.1 miles. How many miles did he drive between oil changes? ■

EXAMPLE 7 Find the value of x if $x + 3.9 = 14.6$.

Recall that the letter x is a variable. It represents a number that is added to 3.9 to obtain 14.6. We can find the number x if we subtract $14.6 - 3.9$.

$$\begin{array}{r} \overset{3}{\cancel{4}}\ \overset{16}{\cancel{6}} \\ 14.6 \\ -\ \ 3.9 \\ \hline 10.7 \end{array}$$

Thus $x = 10.7$.

Check

Is this true? If we replace x by 10.7, do we get a true statement?

$$x + 3.9 = 14.6$$

$$10.7 + 3.9 \stackrel{?}{=} 14.6$$

$$14.6 = 14.6 \quad ✓$$

Practice Problem 7 Find the value of x if $x + 10.8 = 15.3$. ■

Developing Your Study Skills

MAKING A FRIEND IN THE CLASS

Attempt to make a friend in your class. You may find that you enjoy sitting together and drawing support and encouragement from each other. Exchange phone numbers so you can call each other whenever you get stuck in your study. Set up convenient times to study together on a regular basis, to do homework, and to review for exams.

You must not depend on a friend or fellow student to tutor you, do your work for you, or in any way be responsible for your learning. However, you will learn from each other as you seek to master the course. Studying with a friend and comparing notes, methods, and solutions can be very helpful. And it can make learning mathematics a lot more fun!

3.3 Exercises

Add.

1. $44.6 + 28.2$ — 72.8

2. $18.6 + 23.2$ — 41.8

3. $718.98 + 496.57$ — 1215.55

4. $813.47 + 629.86$ — 1443.33

5. $2.107 + 4.918$ — 7.025

6. $5.306 + 8.822$ — 14.128

7. $79.061 + 57.783$ — 136.844

8. $26.905 + 87.453$ — 114.358

9. $6.5 + 12.6 + 304.8$ — 323.9

10. $18.2 + 7.6 + 199.8$ — 225.6

11. $5.6 + 9.23 + 8.17$ — 23.00

12. $2.65 + 3.2 + 7.76$ — 13.61

13. $4.9637 + 28.12 + 3.645$ — 36.7287

14. $7.0276 + 3.451 + 16.98$ — 27.4586

15. $12. + 3.62 + 51.8$ — 67.42

16. $13. + 4.52 + 63.7$ — 81.22

17. $753.61 + 28.75 + 162.3 + 100.5 + 67.05$ 1112.21

18. $432.51 + 16.08 + 892.1 + 301.2 + 84.07$ 1725.96

Applications

Calculate the perimeter of each triangle.

19. 6.1 m, 5.62 m, 8.14 m

6.1
5.62
+ 8.14
19.86 m

20.

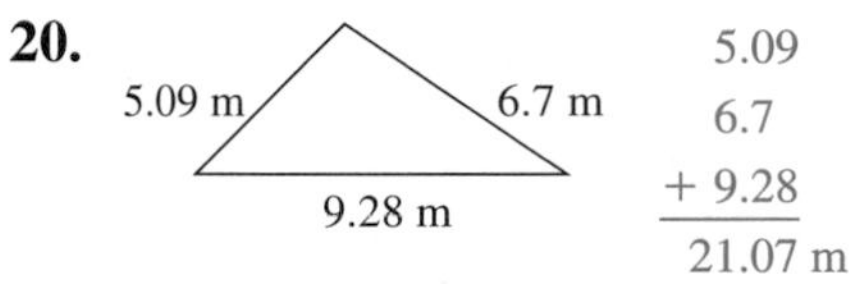

5.09
6.7
+ 9.28
21.07 m

21. Fred and Barney are driving to their high school reunion in New Jersey from their home in Boulder, Colorado. They make purchases of 7.8, 12.1, 9.7, 10.4, 8.8, and 11.3 gallons of fuel. How many gallons of fuel did they buy? 60.1 gallons

22. Greg, Peter, and Bobby drove from Los Angeles up the Pacific Coast Highway to visit their sister Marcia, who goes to college in Santa Cruz. On the way they purchased 7.7, 8.9, 9.5, and 11.8 gallons of gas. How many total gallons did they purchase? 37.9 gallons

23. Mick and Keith have arrived in Miami and are going to the beach. They buy sun-block for \$4.99, beverages for \$12.50, sandwiches for \$11.85, towels for \$28.50, bottled water for \$3.29, and two novels for \$16.99. After they got what they needed, what was Mick and Keith's bill for their day at the beach? \$78.12

24. Brent purchased items at a grocery store just off-campus. They cost \$7.18, \$5.29, \$0.61, \$1.15, and \$1.49. What was the total of Brent's grocery bill? \$15.72

A portion of a bank checking account deposit slip is shown below. Add the numbers to determine the total deposit in each case. The line drawn between the dollars and the cents column serves as the decimal point.

25. FOOTHILLS BANK, AUSTIN, TEXAS — DEPOSIT SLIP

	Dollars	Cents
Cash	18	42
Checks	706	15
	21	03
	45	00
	621	37
Total	\$1411	97

26. SUNSHINE BANK, PHOENIX, ARIZONA — DEPOSIT SLIP

	Dollars	Cents
Cash	52	89
Checks	105	37
	76	04
	25	00
	167	82
Total	\$427	12

Subtract.

27. $\begin{array}{r} 6.8 \\ -\ 2.9 \\ \hline \end{array}$ 3.9

28. $\begin{array}{r} 3.6 \\ -\ 2.8 \\ \hline \end{array}$ 0.8

29. $\begin{array}{r} 123.51 \\ -\ 96.34 \\ \hline \end{array}$ 27.17

30. $\begin{array}{r} 161.78 \\ -\ 89.29 \\ \hline \end{array}$ 72.49

31. $\begin{array}{r} 76.8 \\ -\ 12.62 \\ \hline \end{array}$ 64.18

32. $\begin{array}{r} 82.5 \\ -\ 43.93 \\ \hline \end{array}$ 38.57

33. $\begin{array}{r} 586.513 \\ -\ 78.2 \\ \hline \end{array}$ 508.313

34. $\begin{array}{r} 243.967 \\ -\ 84.2 \\ \hline \end{array}$ 159.767

35. $\begin{array}{r} 1.00782 \\ -\ 0.98631 \\ \hline \end{array}$ 0.02151

36. $\begin{array}{r} 7.00278 \\ -\ 6.34125 \\ \hline \end{array}$ 0.66153

37. $\begin{array}{r} 24.0079 \\ -\ 19.3614 \\ \hline \end{array}$ 4.6465

38. $\begin{array}{r} 52.0708 \\ -\ 41.9312 \\ \hline \end{array}$ 10.1396

39. $\begin{array}{r} 8. \\ -\ 1.263 \\ \hline \end{array}$ 6.737

40. $\begin{array}{r} 12. \\ -\ 7.981 \\ \hline \end{array}$ 4.019

41. $\begin{array}{r} 7362.14 \\ -\ 6173.07 \\ \hline \end{array}$ 1189.07

42. $\begin{array}{r} 4986.71 \\ -\ 3615.93 \\ \hline \end{array}$ 1370.78

43. $\begin{array}{r} 1.5 \\ -\ 0.0365 \\ \hline \end{array}$ 1.4635

44. $\begin{array}{r} 2.8 \\ -\ 0.07763 \\ \hline \end{array}$ 2.72237

Applications

Solve each problem.

45. A toilet seat should not cost more than \$29.95. Someone purchased one, without looking at the bill, for \$63.45. How much did that person overpay? \$33.50

46. Jack and Rita are building a garage. The builder is charging them \$11,597.00. If they pay cash in advance, the builder will reduce the cost by \$1593.25. How much will the garage cost if they pay in advance? \$10,003.75

47. During the three months of August, September, and October, baby Kathryn gained 7.675 kilograms. During the three months of November, December, and January, she gained another 9.986 kilograms. How much weight has the baby gained in the last six months? 17.661 kilograms

48. The average length of an Emerald Tree boa constrictor snake is 1.8 meters. A native found one that measured 3.264 meters long. How much longer was this snake than the average Emerald Tree boa constrictor? 1.464 meters longer

49. A child's beginner telescope is priced at \$79.49. The price of a certain professional telescope is \$37,026.65. How much more does the professional telescope cost? \$36,947.16 more

50. Tamika drove on a summer trip. When she began, the odometer read 26,052.3 miles. At the end of the trip, the odometer read 28,715.1 miles. How long was the trip?

$\begin{array}{r} 28{,}715.1 \\ -\ 26{,}052.3 \\ \hline 2{,}662.8 \end{array}$ miles

51. Malcolm took a taxi from John F. Kennedy Airport in New York to his hotel in the city. His fare was \$47.70 and he tipped the driver \$7.00. How much change should Malcolm get back if he gives the driver a \$100 bill? \$45.30 charge

52. Michael bought \$76.49 worth of supplies at the hardware store. He gave the clerk \$80.00. How much change should he receive?

$\begin{array}{r} 80.00 \\ -\ 76.49 \\ \hline \$3.51 \end{array}$

53. The outside radius of a pipe is 9.39 centimeters. The inside radius is 7.93 centimeters. What is the thickness of the pipe?

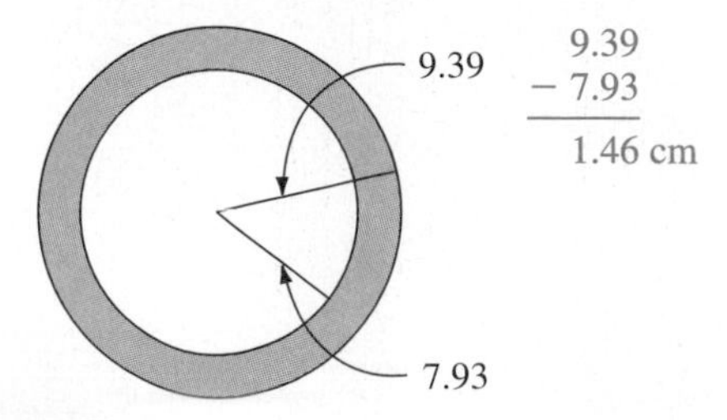

$\begin{array}{r} 9.39 \\ -\ 7.93 \\ \hline 1.46 \end{array}$ cm

54. An insulated wire measures 12.62 centimeters. The last 0.98 centimeter of the wire is exposed. How long is the part of the wire that is not exposed?

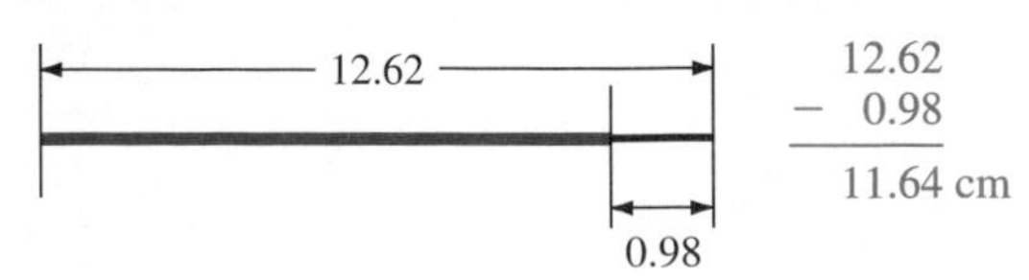

$\begin{array}{r} 12.62 \\ -\ 0.98 \\ \hline 11.64 \end{array}$ cm

55. The state budget called for \$12,563,784.56 in revenue. Unfortunately, the state's actual revenue for the year was only \$11,962,375.49. What was the revenue shortage for the state?

$$\begin{array}{r} 12{,}563{,}784.56 \\ -\ 11{,}962{,}375.49 \\ \hline \end{array}$$

\$601,409.07 revenue shortage

★ **56.** Everyone is becoming aware of the rapid loss of the Earth's rainforests. Between 1981 and 1990, tropical South America lost a substantial amount of natural resources due to deforestation and development. In 1981, there were 797,100,000 hectares of rainforest. In 1990 there were 729,300,000 hectares. If 1 hectare is approximately equal to 2.47 acres, how much rainforest, in acres, was destroyed?

Approximately 167,466,000 acres were destroyed.

The federal water safety standard requires that drinking water contain no more than 0.015 milligrams of lead per liter of water.

57. Fred and Donna use water provided by the city for the drinking water in their home. A sample of their tap water contained 0.023 milligrams of lead per liter of water. What is the difference between their sample and the federal safety standard? Is it safe for them to drink the water? 0.008 milligrams, no

58. Carlos and Maria had the well that supplies their home analyzed for safety. A sample of well water contained 0.0089 milligrams of lead per liter of water. What is the difference between their sample and the federal safety standard? Is it safe for them to drink the water? 0.0061 milligrams, yes

The table below shows the income of the United States by industry. Use this table for problems 59 to 62. Write each answer as a decimal and as a whole number. The table values are recorded in billions of dollars.

TYPE OF INDUSTRY	INCOME IN 1980	INCOME IN 1990
Agriculture, forestry, fisheries	61.4	97.1
Mining	43.8	38.1
Construction	126.6	234.4
Communication	48.1	96.4

59. How many more dollars were earned in mining in 1980 than in 1990?

\$5.7 billion; 5,700,000,000 dollars

60. How many more dollars were earned in construction in 1990 than in 1980?

\$107.8 billion; 107,800,000,000 dollars

61. In 1980, how many more dollars were earned in agriculture, forestry, and fisheries than in communication? \$13.3 billion; 13,300,000,000 dollars

62. In 1990, how many more dollars were earned in communication than in mining?

\$58.3 billion; 58,300,000,000 dollars

Find the value of x.

63. $x + 7.1 = 15.5$ $x = 8.4$

64. $x + 4.8 = 23.1$ $x = 18.3$

65. $156.9 + x = 200.6$ $x = 43.7$

66. $210.3 + x = 301.2$ $x = 90.9$

67. $4.162 = x + 2.053$ $x = 2.109$

68. $7.076 = x + 5.602$ $x = 1.474$

Cumulative Review Problems

Multiply.

69. $\begin{array}{r} 2536 \\ \times\ \ 8 \\ \hline 20{,}288 \end{array}$

70. $\begin{array}{r} 467 \\ \times\ 39 \\ \hline 18{,}213 \end{array}$

71. $\dfrac{22}{7} \times \dfrac{49}{50}$ $\quad \dfrac{77}{25}$ or $3\dfrac{2}{25}$

72. $2\dfrac{1}{3} \times 3\dfrac{3}{4}$ $\quad \dfrac{35}{4}$ or $8\dfrac{3}{4}$

Putting Your Skills to Work

PERSONAL CHECKING ACCOUNT

If you have a personal checking account, you will keep a record of your banking on a check register. When the bank sends you a copy of their records (showing the amount in your account), you should compare the bank's total to your register and be sure the two are the same. This is called *balancing* the checkbook.

Regina's checking account balance was $23.69. She made deposits of $116.84 on March 3 and $89.46 on March 15. Then she wrote check 156 to Joe's Market for $16.80 on March 8 and check 157 to Marshall Field for $129.95 on March 18. She recorded her transactions as follows:

Check Number	Date	Transaction	Amount of Withdrawal (−)		Amount of Deposit (+)		New Balance	
							23	69
	3/3	Deposit			116	84	140	53
156	3/8	Joe's Market	16	80			123	73
	3/15	Deposit			89	46	213	19
157	3/18	Marshall Field	129	95			83	24

PROBLEMS FOR INDIVIDUAL INVESTIGATION

1. What was Regina's balance on March 18? $83.24
2. Regina's balance on March 18 was $83.24. Use the check register below to record the following transactions Regina made. Determine the new balance after each transaction.
   (a) A monthly service fee of $3.50 is charged to her account on March 20.
   (b) A deposit of $301.64 is made on March 22.
   (c) Check 158 is made out to Sears for $159.95 on March 30.
   (d) A deposit of $450.36 is made on March 31.
   (e) Check 159 is made out to Gulf Oil for $46.52 on April 2.
   (f) Check 160 is made out to Western Bell for $39.64 on April 3.
   (g) Check 161 is made out to IRS for $561.18 on April 14.
   (h) A bank charge for $6.95 is billed to Regina for printing new checks on April 18.

	Check Number	Date	Transaction	Amount of Withdrawal (−)		Amount of Deposit (+)		New Balance	
								83	24
(a)		3/20	Service fee	3	50			79	74
(b)		3/22	Deposit			301	64	381	38
(c)	158	3/30	Sears	159	95			221	43
(d)		3/31	Deposit			450	36	671	79
(e)	159	4/2	Gulf Oil	46	52			625	27
(f)	160	4/3	Western Bell	39	64			585	63
(g)	161	4/14	IRS	561	18			24	45
(h)		4/18	New Checks	6	95			17	50

3. On June 30, Regina's balance was $382.54. On the next Monday, she made an ATM withdrawal for $120.00 using her bank card. On the next Tuesday, she deposited $98.96. On the next Wednesday, she used her bank card at an ATM to withdraw $70.00. Her bank charges her $0.65 for each ATM withdrawal. What was her final balance? $290.20

QUESTIONS FOR COOPERATIVE GROUP INVESTIGATION

Together with some other members of your class, investigate the following.

4. Regina usually uses about 15 checks per month and makes about 5 withdrawals per month from the ATM machine near her apartment. The bank offers two checking plans. The Economy Account costs $2.00 per month, plus $0.35 for every check she writes and $0.65 for each ATM withdrawal. The Advanced Service Account costs $10.00 per month, but there is no charge for ATM withdrawals and no service charge for checks unless she uses more than 30 checks per month. In that case, the cost is $0.20 for each additional check. Which type of checking account would be better for Regina? Why? Advanced Service. Economy would cost her $10.50 and Advanced Service would cost her $10.00 per month.

5. During the month of August, Regina found that she could not balance her checking account. Her records showed she had $56.78 in the account, but the bank statement listed $53.53 in the account. Give four possible reasons that could account for this discrepancy. What should Regina do to find out which reason is the cause in her case? Regina (1) hasn't included some bank charges in her account, (2) hasn't recorded a check or a deposit, or (3) has recorded a check or deposit in the wrong amount. Another possibility is that she has made an arithmetic error.

She should post all charges the bank makes in her own records. Then she should compare the amount of each check and deposit on the bank's statement with the amounts listed in her records. She should also check her arithmetic.

INTERNET CONNECTIONS: Go to ''http://www.prenhall.com/tobey'' to be connected

Site: Balancing Your Checkbook (College Board Online) or a related site

This site gives information about balancing a checkbook. It includes a worksheet that you can use to balance a checkbook. Print a copy of this worksheet and use it to answer the question below.

6. Sara just received a bank statement with an ending balance of $643.94. The statement does not include a recent deposit of $275.00, nor does it include two checks she made out for $28.35 and $64.89. What should be the balance in her checkbook?

3.4 Multiplication of Decimals

MathPro

Video 7

SSM

After studying this section, you will be able to:

1. *Multiply a decimal by a decimal or a whole number.*
2. *Multiply a decimal by a power of 10.*

1 *Multiplying a Decimal by a Decimal or a Whole Number*

We learned previously that the product of two fractions is the product of the numerators over the product of the denominators. For example,

$$\frac{3}{10} \times \frac{7}{100} = \frac{21}{1000}$$

In decimal form this product would be written

$$\underbrace{0.3}_{\textit{one decimal place}} \times \underbrace{0.07}_{\textit{two decimal places}} = \underbrace{0.021}_{\textit{three decimal places}}$$

> **Multiplication of Decimals**
>
> 1. Multiply the numbers just as you would multiply whole numbers.
> 2. Find the sum of the decimal places in the two factors.
> 3. Place the decimal point in the product so that the product has the same number of decimal places as the sum in step 2. You may need to write zeros to the left of the number found in step 1.

Now use these steps to do the multiplication problem above.

EXAMPLE 1 Multiply. 0.07×0.3

0.07	2 decimal places
$\times$ 0.3	1 decimal place
0.021	3 decimal places in product $(2 + 1 = 3)$

Practice Problem 1 Multiply. 0.09×0.6 ■

When performing the calculation, it is usually easier to place the factor with the fewest number of nonzero digits underneath the other factor.

EXAMPLE 2 Multiply.

(a) 0.8×2.6 **(b)** 0.3×1.672 **(c)** 0.38×0.26 **(d)** 12.64×0.572

(a)

2.6	1 decimal place
$\times$ 0.8	1 decimal place
2.08	2 decimal places $(1 + 1 = 2)$

(b)

1.672	3 decimal places
$\times$ 0.3	1 decimal place
0.5016	4 decimal places $(3 + 1 = 4)$

(c)

0.38	2 decimal places
× 0.26	2 decimal places
228	
76	
0.0988	4 decimal places (2 + 2 = 4)

Note that we need to insert a zero before the 988.

(d)

12.64	2 decimal places
× 0.572	3 decimal places
2528	
8848	
6 320	
7.23008	5 decimal places (2 + 3 = 5)

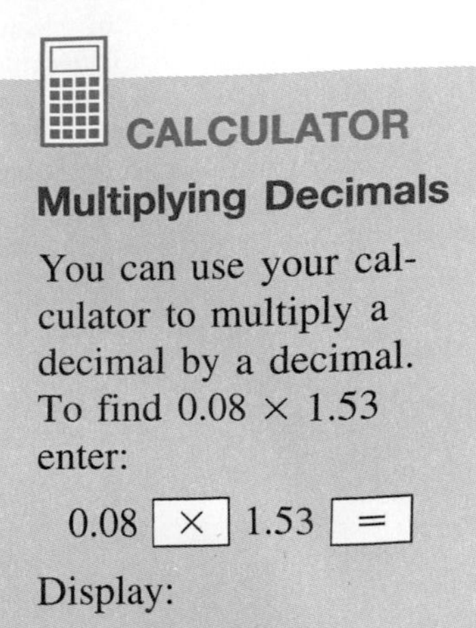

CALCULATOR

Multiplying Decimals

You can use your calculator to multiply a decimal by a decimal. To find 0.08 × 1.53 enter:

0.08 [×] 1.53 [=]

Display:

[0.1224]

Practice Problem 2 Multiply.

(a) 0.9 × 5.3 **(b)** 2.831 × 0.7 **(c)** 0.47 × 0.28 **(d)** 0.436 × 18.39 ■

When multiplying decimal fractions by a whole number, you need to remember that a whole number has no decimal places.

EXAMPLE 3 Multiply. 5.261 × 45

5.261	3 decimal places
× 45	0 decimal places
26 305	
210 44	
236.745	3 decimal places (3 + 0 = 3)

Practice Problem 3 Multiply. 0.4264 × 38 ■

EXAMPLE 4 Uncle Roger's rectangular front lawn measures 50.6 yards wide and 71.4 yards long. What is the area of the lawn in square yards?

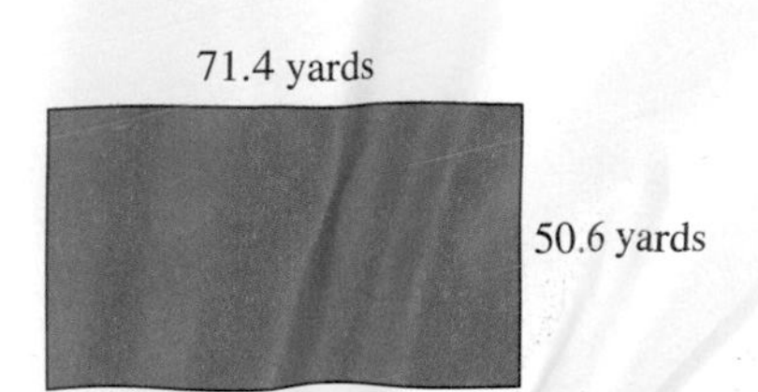

Since the lawn is rectangular in shape, we will use the fact that to find the area of a rectangle we multiply the length by the width.

71.4	(1 decimal place)
× 50.6	(1 decimal place)
42 84	
3570 0	
3612.84	(2 decimal places)

The area of the lawn is 3612.84 square yards.

Practice Problem 4 A rectangular computer chip measures 1.26 millimeters wide and 2.3 millimeters long. What is the area of the chip in square millimeters? ■

TEACHING TIP Multiplication of a decimal by a power of 10 is a simple idea. However, it will be totally new to some students. Be sure to do an example like: Multiply $123.8765 \times 100{,}000 = 12{,}387{,}650$ "in your head." Ask students what is the rule that allowed them to do that problem "in your head"? You will usually get several enthusiastic suggestions from the class. Students like the idea of this type of "shortcut."

2 Multiplying a Decimal by a Power of 10

Observe the following pattern.

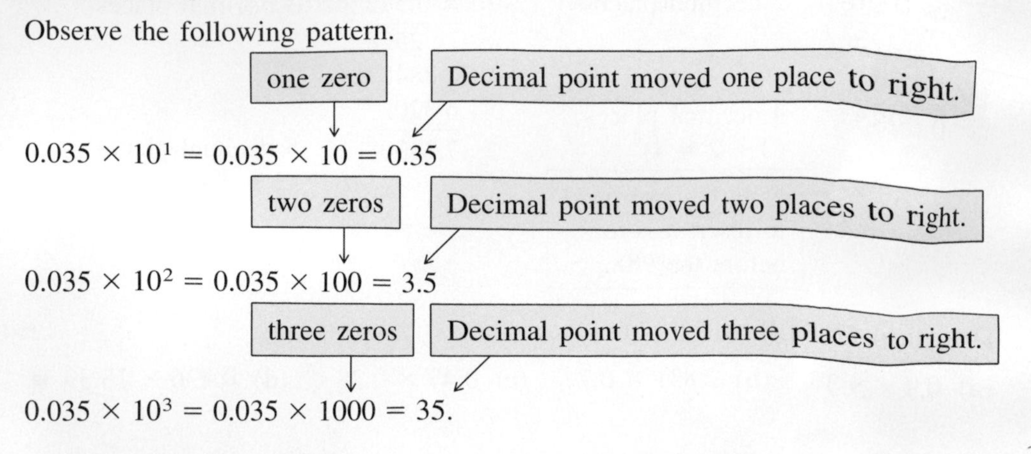

Multiplication of a Decimal by a Power of 10

To multiply a decimal by a power of 10, move the decimal point to the right the same number of places as the number of zeros in the power of 10.

EXAMPLE 5 Multiply. **(a)** 2.671×10 **(b)** 37.85×100

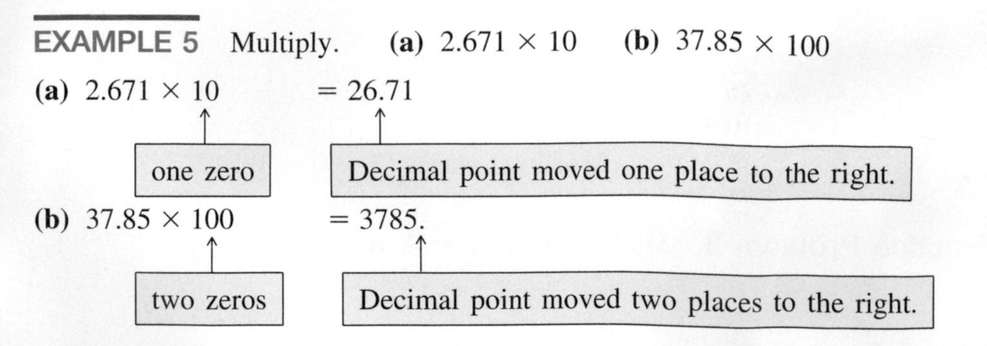

Practice Problem 5 Multiply. **(a)** 0.0561×10 **(b)** 1462.37×100 ■

Sometimes it is necessary to add extra zeros before placing the decimal point in the answer.

EXAMPLE 6 Multiply. **(a)** 4.8×1000 **(b)** $0.076 \times 10{,}000$

(a) $4.8 \times \underline{1000} = 4800.$

(three zeros)

Decimal point moved three places to the right. Two extra zeros were needed.

(b) $0.076 \times \underline{10{,}000} = 760.$

(four zeros)

Decimal point moved four places to the right. One extra zero was needed.

Practice Problem 6 Multiply. **(a)** 0.26×1000 **(b)** $5862.89 \times 10{,}000$ ■

If the number that is a power of 10 is in exponent form, move the decimal point to the right the same number of places as the number that is the exponent.

EXAMPLE 7 Multiply. **(a)** 3.68×10^3 **(b)** 0.00027×10^4

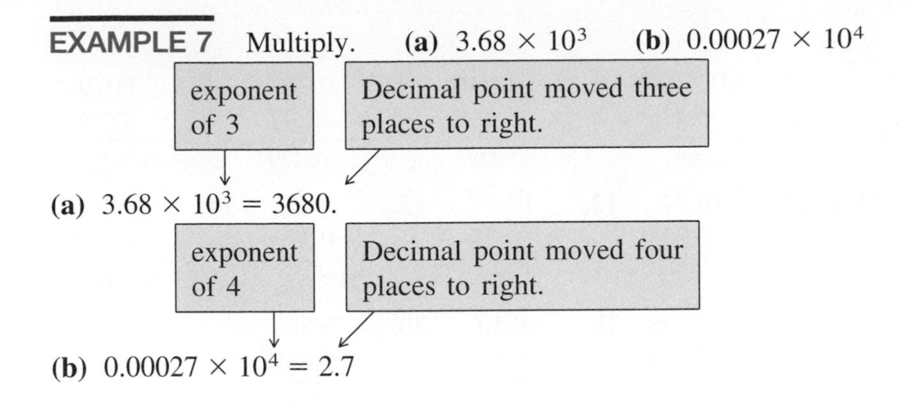

Practice Problem 7 Multiply. **(a)** 7.684×10^4 **(b)** 0.00073×10^2 ■

SIDELIGHT Can you devise a quick rule to use when multiplying a decimal fraction by $\frac{1}{10}$, $\frac{1}{100}$, $\frac{1}{1000}$, and so on? How is it like the rules developed in this section? Consider a few examples.

Original Problem	Change Fraction to Decimal	Decimal Multiplication	Observation
$86 \times \frac{1}{10}$	86×0.1	$86 \times 0.1 = 8.6$	Decimal point moved one place to left
$86 \times \frac{1}{100}$	86×0.01	$86 \times 0.01 = 0.86$	Decimal point moved two places to left
$86 \times \frac{1}{1000}$	86×0.001	$86 \times 0.001 = 0.086$	Decimal point moved three places to left

Can you think of a way to describe a rule that you could use in solving this type of problem without going through all the foregoing steps?

You use multiplying by a power of 10 when you change a larger unit to a smaller unit in the metric system of measure.

EXAMPLE 8 Change 2.96 kilometers to meters.

Since we are going from a larger unit to a smaller unit, we multiply. There are 1000 meters in 1 kilometer. Multiply 2.96 by 1000.

$$2.96 \times 1000 = 2960$$

2.96 kilometers is equal to 2960 meters.

Practice Problem 8 Change 156.2 kilometers to meters. ■

TEACHING TIP You can expand on the discussion in Example 8 with the following example: "To convert from kilometers to centimeters, multiply by 100,000. How many centimeters are there in 3457.39 kilometers?" The answer is 345,739,000 centimeters. Ask the students to imagine how difficult the problem would have been had it been formulated in terms of changing miles to inches. This will help them to appreciate the advantages of the metric system.

3.4 Exercises

Multiply.

1. 0.6×0.2 — 0.12

2. 0.9×0.3 — 0.27

3. 0.12×0.5 — 0.06

4. 0.17×0.4 — 0.068

5. 0.0036×0.8 — 0.00288

6. 0.067×0.07 — 0.00469

7. 0.079×0.09 — 0.00711

8. 0.0034×0.5 — 0.0017

9. 0.025×0.081 — 0.002025

10. 0.071×0.031 — 0.002201

11. 10.97×0.06 — 0.6582

12. 18.07×0.05 — 0.9035

13. 7986×0.32 — 2555.52

14. 5167×0.19 — 981.73

15. 1.892×0.007 — 0.013244

16. 2.163×0.008 — 0.017304

17. 0.7613×1009 — 768.1517

18. 0.6178×5004 — 3091.4712

19. 9630×0.51 — 4911.3

20. 7980×0.46 — 3670.8

21. 126×3.5
441

22. 709×4.6
3261.4

23. 5030×8.62
43,358.6

24. 8703×5.06
44,037.18

25. 0.07×0.0034
0.000238

26. 0.026×0.05
0.0013

27. 0.001×6523.7
6.5237

28. 0.01×826.75
8.2675

Applications

29. Melissa earns $5.85 per hour for a 40-hour week. How much does she make in one week?

5.85×40 = $234.00

30. Kenny is making payments on his Ford Escort of $155.40 per month for the next 60 months. How much will he have spent in car payments after he sends in his final payment? $9324

31. Mei Lee works for a company that manufactures electric and electronic equipment and earns $9.55 per hour for a 40-hour week. How much does she earn in one week? (The average wage, for 1987, for U.S. electric/electronic equipment manufacturers was $9.90 per hour for an average of 40.9 hours, for a total of $404.91 per week.)

9.55×40 = $382.00

32. Elva works for a company that manufactures textile products. She earns $7.20 per hour for a 40-hour week. How much does she earn in 1 week? (The average wage, for 1987, for U.S. textile mill industries was $7.18 per hour for an average of 41.9 hours, for a total of $300.84 weekly earnings.)

7.20×40 = $288.00

33. The dimensions of a living room are 20.3 feet and 11.6 feet. What is the area of the living room in square feet?

20.3×11.6 = 235.48 square feet

34. The dimensions of a sheet of paper are 8.2 inches and 10.7 inches. What is the area of the paper in square inches?

10.7×8.2 = 87.74 square inches

35. Dwight is paying off a student loan at Westmont College with payments of $36.90 per month for the next 18 months. How much will he pay off during the next 18 months?

36.90×18 = $664.20

36. Marcia is making car payments to Highfield Center Chevrolet of $230.50 per month for 16 more months. How much will she pay for car payments in the next 16 months?

230.50×16 = $3688.00

37. Maria bought 1.4 pounds of bananas at \$0.55 per pound. How much did she pay for the bananas?

$$\begin{array}{r} 0.55 \\ \times\ 1.4 \\ \hline \$0.77 \end{array}$$

38. Angela purchased 2.6 pounds of tomatoes at \$0.65 per pound. How much did she pay for the tomatoes?

$$\begin{array}{r} 0.65 \\ \times\ 2.6 \\ \hline \$1.69 \end{array}$$

39. Steve's car gets approximately 26.4 miles per gallon. His gas tank holds 19.5 gallons. Approximately how many miles can he travel on a full tank of gas?

$$\begin{array}{r} 26.4 \\ \times\ 19.5 \\ \hline 514.8 \end{array}$$ miles

40. Jim's 4×4 truck gets approximately 18.6 miles per gallon. His gas tank holds 20.5 gallons. Approximately how many miles can he travel on a full tank of gas?

$$\begin{array}{r} 18.6 \\ \times\ 20.5 \\ \hline 381.3 \end{array}$$ miles

Multiply.

41. 2.86×10
28.6

42. 1.98×10
19.8

43. 0.701×100
70.1

44. 0.236×100
23.6

45. 128.65×1000
128,650

46. 204.37×1000
204,370

47. $5.60982 \times 10{,}000$
56,098.2

48. $1.27986 \times 10{,}000$
12,798.6

49. $280{,}560.2 \times 10^2$
28,056,020

50. 7163.241×10^2
716,324.1

51. 816.32×10^3
816,320

52. 763.49×10^4
7,634,900

53. 0.6718×10^3
671.8

54. 0.7153×10^4
7153

55. 0.00081376×10^5
81.376

56. 0.0007163×10^5
71.63

Applications

57. To convert from meters to centimeters, multiply by 100. How many centimeters are in 5.932 meters? $5.932 \times 100 = 593.2$ centimeters

58. A store purchased 100 items at \$19.64 each. How much did the order cost? $19.64 \times 100 = \$1964$

59. To convert from kilometers to meters, multiply by 1000. How many meters are in 2.98 kilometers?
$2.98 \times 1000 = 2980$ meters

60. To convert from kilograms to grams, multiply by 1000. How many grams are in 9.64 kilograms?
$9.64 \times 1000 = 9640$ grams

61. The city school system pays approximately \$3640.50 per student for schooling. What would the city spend for 10,000 students?
$3640.50 \times 10{,}000 = \$36{,}405{,}000.00$

62. An automobile manufacturer makes a profit of \$2984.60 per car. What would be the profit for 10,000 cars?
$2984.60 \times 10{,}000 = \$29{,}846{,}000.00$

★ **63.** The college is purchasing new carpeting for the learning center. What is the price of a carpet that is 19.6 yards wide and 254.2 yards long if the cost is \$12.50 per square yard?

$$\begin{array}{r} 254.2 \\ \times\ 19.6 \\ \hline 4982.32 \end{array}$$ square yards

$$\begin{array}{r} 4982.32 \\ \times\ 12.5 \\ \hline \$62{,}279.00 \end{array}$$

★ **64.** A jewelry store purchased long lengths of gold chain, which will be cut and made into necklaces and bracelets. The store purchased 3220 grams of gold chain at \$3.50 per gram.

(a) How much did the jewelry store spend? \$11,270

(b) If they sell a 28-gram gold necklace for \$17.75 per gram, how much profit will they make on the necklace? \$399 profit

Cumulative Review Problems

Divide. Be sure to include any remainder as part of your answer.

65. $12\overline{)1176}$ 98

66. $14\overline{)1204}$ 86

67. $37\overline{)4629}$ 125 R 4

68. $29\overline{)3745}$ 129 R 4

3.5 Division of Decimals

MathPro

Video 8

SSM

After studying this section, you will be able to:

1 *Divide a decimal by a whole number.*
2 *Divide a decimal by a decimal.*

TEACHING TIP This is a good time to remind students that they will be very glad they learned these three words: DIVISOR, DIVIDEND, QUOTIENT. They are referred to frequently in mathematics. You may want to put the chart on the board as a reminder:

$$\text{DIVISOR}\overline{)\text{DIVIDEND}}^{\;\text{QUOTIENT}}$$

1 *Dividing a Decimal by a Whole Number*

When you divide a decimal by a whole number, place the decimal point for the quotient directly above the decimal point in the dividend. Then divide as if the numbers were whole numbers.

To divide 26.8 ÷ 4, we place the decimal point of our answer (the quotient) directly *above* the decimal point in the dividend.

$$4\overline{)26.8}$$

The decimal points are directly above each other.

Then we divide as if we were dividing whole numbers.

```
    6.7
 4)26.8
   24
    28
    28
     0
```

The quotient is 6.7.

The quotient to a problem may have all digits to the right of the decimal point. In some cases you will have to put a zero in the quotient as a "place holder." Thus, if we divide 0.268 ÷ 4,

We have

```
   0.067
 4)0.268
     24
      28
      28
       0
```

Note that we must have a zero after the decimal point in 0.067.

EXAMPLE 1 Divide. **(a)** $9\overline{)0.3204}$ **(b)** $14\overline{)36.12}$

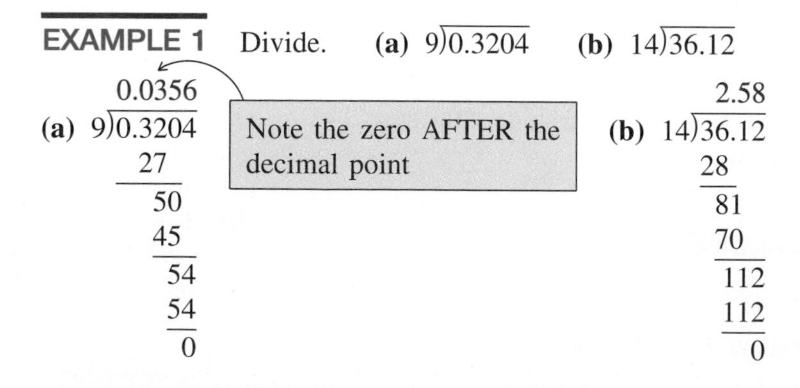

0.0356
(a) 9)0.3204
27
50
45
54
54
0
Note the zero AFTER the decimal point
2.58
(b) 14)36.12
28
81
70
112
112
0

Practice Problem 1 Divide. **(a)** $7\overline{)1.806}$ **(b)** $16\overline{)0.0928}$ ■

Some division problems do not yield a remainder of zero. In such cases, we may be asked to round off the answer to a specified place. To round off, we carry out the division until our answer contains a digit that is one place to the right of that to which we intend to round. Then we round our answer to the specified place. For example, to round to the nearest thousandth, we carry out the division to the ten-thousandths place. In some division problems, you will need to write in zeros after the dividend so that this division can be carried out.

TEACHING TIP It is helpful for most students to see a variety of division problems with decimals where the divisor is an integer and to master the ability to perform each division before going onward. Be sure to discuss carefully Examples 1 and 2 or similar problems before moving on to the case where the divisor is a decimal fraction.

EXAMPLE 2 Divide and round the quotient to the nearest thousandth.

$$12.67 \div 39$$

We will carry out our division to the ten-thousandths place. Then we round our answer to the nearest thousandth.

```
     0.3248
39)12.6700  ← Two extra zeros are written here to
   11 7        carry out the division to the required
      97       place.
      78
      190
      156
       340
       312
        28   Note that the remainder is not zero.
```

Now we round 0.3248 to 0.325. The answer is rounded to the nearest thousandth.

Practice Problem 2 Divide and round the quotient to the nearest hundredth. 23.82 ÷ 46 ■

EXAMPLE 3 Maria paid \$5.92 for 16 pounds of tomatoes. How much did she pay per pound?

The cost of one pound of tomatoes equals the total cost, \$5.92, divided by 16 pounds. Thus we will divide.

```
    0.37
16)5.92
   4 8
   112
   112
```

Maria paid \$0.37 per pound for the tomatoes.

Practice Problem 3 Won Lin will pay off his auto loan for \$3538.75 over 19 months. If the monthly payments are equal, how much will he pay each month? ■

2 *Dividing a Decimal by a Decimal*

When the divisor is not a whole number, we can convert the division problem to an equivalent problem that has a whole number as a divisor. Think about the reasons why this procedure will work. We will ask you about it after you study Examples 4 and 5.

> **Dividing by a Decimal**
>
> 1. Make the divisor a whole number by moving the decimal point to the right. Mark that position with a caret (‸). Count the number of places the decimal point moved.
> 2. Move the decimal point in the dividend to the right the same number of places. Mark that position with a caret.
> 3. Place the decimal point of your answer directly above the caret marking the decimal point of the dividend.
> 4. Divide as with whole numbers.

EXAMPLE 4 **(a)** Divide. $0.08\overline{)1.632}$ **(b)** Divide. $1.352 \div 0.026$

(a) $0.08.\overline{)1.63.2}$ *Move each decimal point two places to the right.*

Place the decimal point of the answer directly above the caret.

$0.08_\wedge\overline{)1.63_\wedge 2}$ *Mark the new position by a caret (‸).*

$$\begin{array}{r} 20.4 \\ 0.08_\wedge\overline{)1.63_\wedge 2} \\ \underline{16} \\ 3\ 2 \\ \underline{3\ 2} \\ 0 \end{array}$$

Perform the division.

The answer is 20.4.

(b)
$$\begin{array}{r} 52. \\ 0.026_\wedge\overline{)1.352_\wedge} \\ \underline{1\ 30} \\ 52 \\ \underline{52} \\ 0 \end{array}$$

Move the decimal point three places to the right and mark the new position by a caret (‸).

The answer is 52.

Practice Problem 4 Divide. **(a)** $0.09\overline{)0.1008}$ **(b)** $1.702 \div 0.037$ ■

To Think About Why do we move the decimal point so many places to the right in the divisor and the dividend? What rule allows us to do this? How do we know the answer will be valid? We are actually using the property that multiplication of a fraction by 1 leaves the fraction unchanged. This is called the *multiplication identity.* Let us examine Example 4(b) again. We will write $1.352 \div 0.026$ as a fraction.

$\dfrac{1.352}{0.026} \times 1$ — *Multiplication of a fraction by 1 does not change the value of the fraction.*

$= \dfrac{1.352}{0.026} \times \dfrac{1000}{1000}$ — *We know that* $\dfrac{1000}{1000} = 1.$

$= \dfrac{1352}{26}$ — *Multiplication by 1000 can be done by moving the decimal point three places to the right.*

$= 52$ — *Divide the whole numbers*

Thus in Example 4(b) when we moved the decimal point three places to the right in the divisor and the dividend, we were actually creating an equivalent fraction where numerator and denominator were multiplied by 1000.

EXAMPLE 5 Divide. **(a)** $1.7\overline{)0.0323}$ **(b)** $0.0032\overline{)7.68}$

(a)

```
        0.019
1.7^)0.0^323
        17
        153
        153
          0
```

Move the decimal point in the divisor and dividend one place to the right and mark that position with a caret.

(b)

```
          2400.
0.0032^)7.6800^
         6 4
         1 28
         1 28
          000
```

Note that two extra zeros are needed in the dividend as we move the decimal point four places to the right.

TEACHING TIP Sometimes students find that they need to work out some additional examples like Example 5 until they are sure of themselves. If you see a need for this, have them do $0.03154 \div 0.019$ (the answer is 1.66) and $1.2999 \div 0.0007$ (the answer is 1857).

Practice Problem 5 Divide. **(a)** $1.8\overline{)0.0414}$ **(b)** $0.0036\overline{)8.316}$ ■

CALCULATOR

Dividing decimals

You can use your calculator to divide a decimal by a decimal. To find $21.83\overline{)54.53}$ rounded to the nearest hundredth enter:

54.53 [÷] 21.38 [=]

Display:

[2.5505145]

This is an approximation. Some calculators will round to eight digits. The answer rounded to the nearest hundredth is 2.55.

EXAMPLE 6 **(a)** Find $2.9\overline{)431.2}$ rounded to the nearest tenth.
(b) Find $2.17\overline{)0.08}$ rounded to the nearest thousandth.

$$\begin{array}{r} 148.68 \\ 2.9_\wedge\overline{)431.2_\wedge00} \\ 29 \\ \hline 141 \\ 116 \\ \hline 25\ 2 \\ 23\ 2 \\ \hline 2\ 00 \\ 1\ 74 \\ \hline 260 \\ 232 \\ \hline 28 \end{array}$$

a) *Calculate to the hundredths place and round the answer to the nearest tenth.*

The answer rounded to the nearest tenth is 148.7.

$$\begin{array}{r} .0368 \\ 2.17_\wedge\overline{)0.08_\wedge0000} \\ 6\ 51 \\ \hline 1\ 490 \\ 1\ 302 \\ \hline 1880 \\ 1736 \\ \hline 144 \end{array}$$

(b) *Calculate to the ten-thousandths place and then round the answer. Rounding* 0.0368 *to the nearest thousandth, we obtain* 0.037.

Practice Problem 6 **(a)** Find $3.8\overline{)521.6}$ rounded to the nearest tenth.
(b) Find $8.05\overline{)0.17}$ rounded to the nearest thousandth. ■

EXAMPLE 7 John drove his 1987 Cavalier 420.5 miles to Chicago. He used 14.5 gallons of gas on the trip. How many miles per gallon did his car get on the trip?

To find miles per gallon we need to divide the number of miles, 420.5, by the number of gallons, 14.5.

$$\begin{array}{r} 29. \\ 14.5_\wedge\overline{)420.5_\wedge} \\ 290 \\ \hline 1305 \\ 1305 \\ \hline \end{array}$$

John's car achieved 29 miles per gallon on the trip to Chicago.

Practice Problem 7 Sarah rented a large truck to move to Boston. She drove 454.4 miles yesterday. She used 28.5 gallons of gas on the trip. How many miles per gallon did the rental truck get? ■

EXAMPLE 8 Find the value of n if $0.8 \times n = 2.68$.

Here 0.8 is multiplied by some number n to obtain 2.68. What is this number n? If we divide 2.68 by 0.8, we will find the value of n.

$$\begin{array}{r} 3.35 \\ 0.8_{\wedge})\overline{2.6_{\wedge}80} \\ \underline{2\,4} \\ 2\,8 \\ \underline{2\,4} \\ 40 \\ \underline{40} \end{array}$$

Thus the value of n is 3.35.

Check Is this true? Are we sure the value of $n = 3.35$?

We substitute the value of $n = 3.35$ into the equation to see if it makes the statement true.

$$\begin{aligned} 0.8 \times n &= 2.68 \\ 0.8 \times 3.35 &\stackrel{?}{=} 2.68 \\ 2.68 &= 2.68 \ \checkmark \end{aligned}$$ Yes. It is true.

Practice Problem 8 Find the value of n if $0.12 \times n = 0.696$. ■

EXAMPLE 9 The level of sulfur dioxide emissions in the air has slowly been decreasing over the last 15 years, as can be seen in the accompanying bar graph. Find the average amount of sulfur dioxide emissions in the air over these four specific years.

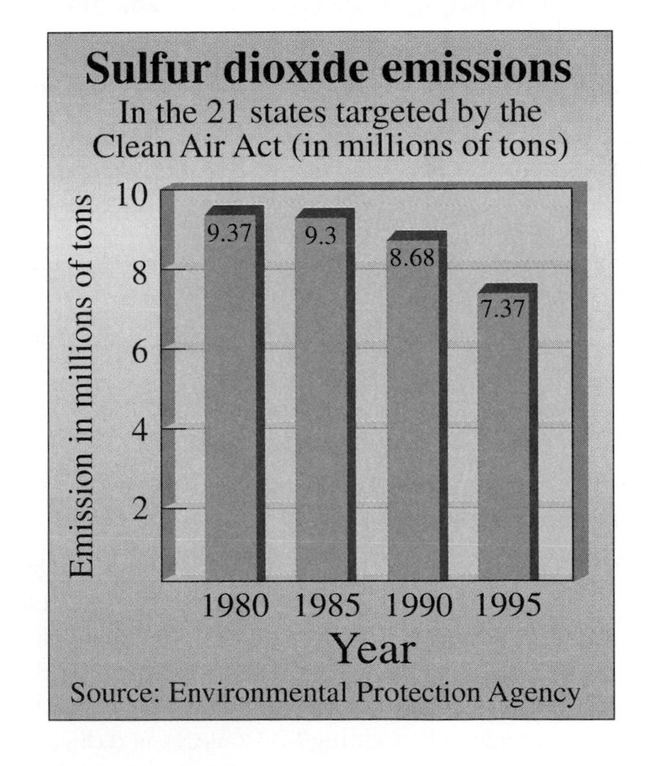

First we take the sum of the four years.

$$\begin{array}{r} 9.37 \\ 9.3 \\ 8.68 \\ +\ 7.37 \\ \hline 34.72 \end{array}$$

Then we divide by four to obtain the average.

$$\begin{array}{r} 8.68 \\ 4)\overline{34.72} \\ \underline{32} \\ 2\,7 \\ \underline{2\,4} \\ 32 \\ \underline{32} \end{array}$$

Thus the yearly average is 8.68 in millions of tons of sulfur dioxide in these 21 states.

Practice Problem 9 Use the accompanying bar graph to find the average level of sulfur dioxide for the three years: 1985, 1990, and 1995. By how much does the 3-year average differ from the 4-year average? ■

3.5 Exercises

Divide until there is a remainder of zero.

1. $6\overline{)12.6}$ 2.1
2. $8\overline{)17.28}$ 2.16
3. $4\overline{)0.1476}$ 0.0369
4. $5\overline{)0.0129}$ 0.00258
5. $7\overline{)128.17}$ 18.31
6. $9\overline{)238.86}$ 26.54
7. $64\overline{)3.616}$ 0.0565
8. $56\overline{)1.624}$ 0.029
9. $21\overline{)0.0609}$ 0.0029
10. $23\overline{)0.0805}$ 0.0035
11. $0.8\overline{)9.76}$ 12.2
12. $0.6\overline{)8.16}$ 13.6
13. $0.5\overline{)32.15}$ 64.3
14. $0.4\overline{)47.28}$ 118.2
15. $0.09\overline{)0.7209}$ 8.01
16. $0.8113 \div 0.07$ 11.59
17. $75.6 \div 3.6$ 21
18. $68.4 \div 3.8$ 18
19. $82.8 \div 0.36$ 230
20. $40.30 \div 0.31$ 130

Divide and round your answer to the nearest tenth.

21. $7\overline{)36.92}$ 5.3
22. $9\overline{)47.31}$ 5.3
23. $1.8\overline{)4.16}$ 2.3
24. $1.9\overline{)2.36}$ 1.2
25. $0.95\overline{)32.067}$ 33.8
26. $0.85\overline{)41.901}$ 49.3

Divide and round your answer to the nearest hundredth.

27. $4\overline{)263.82}$ 65.96
28. $5\overline{)471.03}$ 94.21
29. $1.7\overline{)20.8}$ 12.24
30. $1.6\overline{)36.5}$ 22.81
31. $0.27\overline{)6.729}$ 24.92
32. $0.41\overline{)8.378}$ 20.43

Divide and round your answer to the nearest thousandth.

33. $8\overline{)248.162}$ 31.020
34. $6\overline{)409.387}$ 68.231
35. $0.69\overline{)8.45}$ 12.246
36. $0.87\overline{)79.40}$ 91.264

Divide and round your answer to the nearest whole number.

37. $0.075\overline{)3.729}$ 50
38. $0.065\overline{)4.398}$ 68
39. $0.55\overline{)7.00}$ 13
40. $0.39\overline{)5.00}$ 13

Applications

41. The college cafeteria staff is portioning out butterscotch pudding for the next meal. The cook has 235.6 ounces of pudding and wants to end up with 38 equal portions. How many ounces will each portion have if the cook uses up the entire batch of butterscotch pudding? 6.2 ounces in each portion

42. The Miller family wants to use the latest technology to access the Internet from their home television system. The equipment needed to upgrade their existing equipment will cost $992.76. If the Millers make 12 equal monthly payments, how much will they pay per month? $82.73 per month

43. Steve knew that he would sacrifice some gas mileage when he sold his car and bought a 4×4 Ford Explorer. His new truck can go 312 miles on 15.6 gallons of gas. How many miles per gallon does his Explorer achieve? 20 miles per gallon

44. Wally owns a Plymouth Breeze that travels 360 miles on 13.2 gallons of gas. How many miles per gallon does it achieve? (Round your answer to the nearest tenth).
Approximately 27.3 miles per gallon

45. The local concert promoter paid a total of \$3388.50 for 9 bands to play in the outdoor festival. Each band received an equal amount. How much did he pay per band? \$376.50

46. Peter paid \$21.78 for 11 pounds of papayas. How much did he pay per pound? (Round your answer to the nearest cent.) \$1.98 per pound

47. Demitri had a contractor build an outdoor deck for his back porch. He now has \$1131.75 to pay off, and he agreed to pay \$125.75 per month. How many more payments on the outdoor deck must he make? 9 payments

48. Juanita has to pay off \$4533.13 in car payments that amount to \$146.23 per month. How many more car payments must she make?

$146.23_{\wedge}\overline{)4533.13_{\wedge}}$ = 31; 31 payments

49. Pascal works as a supervisor in a billiard supply business. He was inspecting a box of blue cue chalk, which players rub on the end of their playing sticks (cues) to keep from slipping when they hit the ball. Each small square of chalk weighs 7.9 grams. The contents of the entire box of chalk weigh 300.2 grams. How many blocks of chalk are in the box? If the box is labeled CONTENTS: 40 CUE CHALKS, how great an error was made packing the box? There are 38 blocks of chalk in the box. The error was made by placing 2 fewer blocks of chalk in the box than what was listed on the label.

50. Yoshi is working as an inspector for a company that makes snowboards. A Mach 1 snowboard weighs 3.8 kilograms. How many of these snowboards are contained in a box in which the contents weigh 87.40 kilograms? If the box is labeled CONTENTS: 24 SNOWBOARDS, how great an error was made in packing the box? The box contains 23 snowboards. The error was in putting 1 less snowbaord in the box than was required.

In each equation find the value of n.

51. $0.7 \times n = 0.0861$

0.123

52. $0.6 \times n = 3.948$

6.58

53. $1.6 \times n = 110.4$

69

54. $1.4 \times n = 821.8$

587

55. $n \times 0.063 = 2.835$

45

56. $n \times 0.098 = 4.312$

44

57. $n \times 0.008 = 6.48$

810

58. $n \times 0.009 = 46.8$

5200

To Think About

Multiply the numerator and denominator of each fraction below by 10,000. Then divide the numerator by the denominator. Is the result the same as if we divided the original numerator by the original denominator? Why? Yes. Multiplying and dividing the numerator and denominator by 10,000 is the same as multiplying by $\frac{10{,}000}{10{,}000}$, which is 1.

59. $\frac{3.8702}{0.0523}$

$\times \frac{10{,}000}{10{,}000} = \frac{38{,}702}{523} = 74$

60. $\frac{2.9356}{0.0716}$

$\times \frac{10{,}000}{10{,}000} = \frac{29{,}356}{716} = 41$

Cumulative Review Problems

61. Add. $\frac{3}{8} + 1\frac{2}{5}$

$1\frac{31}{40}$

62. Subtract. $2\frac{13}{16} - 1\frac{7}{8}$

$\frac{15}{16}$

63. Multiply. $3\frac{1}{2} \times 2\frac{1}{6}$

$\frac{91}{12}$ or $7\frac{7}{12}$

64. Divide. $4\frac{1}{3} \div 2\frac{3}{5}$

$\frac{5}{3}$ or $1\frac{2}{3}$

3.6 Converting Fractions to Decimals and Order of Operations

After studying this section, you will be able to:

1 *Convert any fraction to decimal form.*

2 *Use the correct order of operations when several decimals are combined with various operations.*

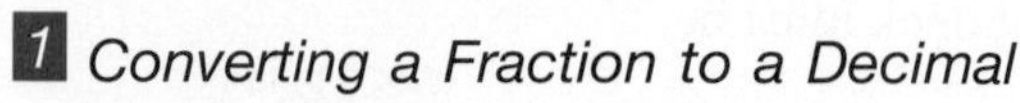

1 *Converting a Fraction to a Decimal*

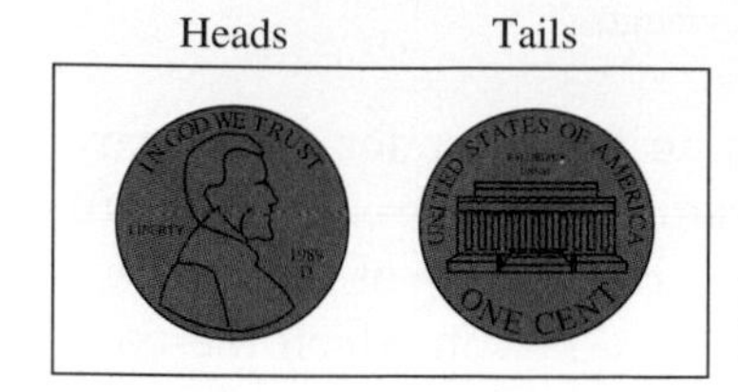

For a coin, heads and tails are two sides of the same object, two ways of identifying a single quantity. For numbers, fractions and decimals are two equivalent forms for the same quantity.

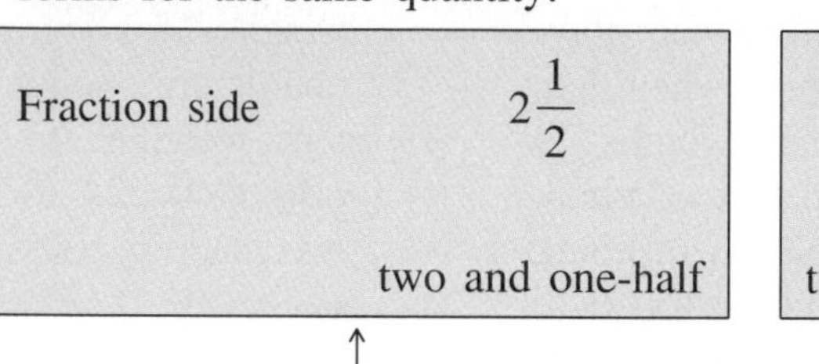

Every decimal can be expressed as an equivalent fraction. For example,

Decimal form $\Rightarrow$ fraction form

$$0.75 = \frac{75}{100} \quad \text{or} \quad \frac{3}{4}$$

$$0.5 = \frac{5}{10} \quad \text{or} \quad \frac{1}{2}$$

$$2.5 = 2\frac{5}{10} = 2\frac{1}{2} \quad \text{or} \quad \frac{5}{2}$$

And every fraction can be expressed as an equivalent decimal, as we will learn in this section. For example,

Fraction form $\Rightarrow$ decimal form

$$\frac{1}{5} = 0.20 \quad \text{or} \quad 0.2$$

$$\frac{3}{8} = 0.375$$

$$\frac{5}{11} = 0.4545\ldots \quad \text{(the ``45'' keeps repeating)}$$

Some of these decimal equivalents are so common that people find it helpful to memorize them. You would be wise to memorize the following equivalents.

$$\frac{1}{2} = 0.5 \qquad \frac{1}{4} = 0.25 \qquad \frac{1}{5} = 0.2 \qquad \frac{1}{10} = 0.1$$

We previously studied how to transfer some fractions with a denominator of 10, 100, 1000, and so on, to decimal form. For example, $\frac{3}{10} = 0.3$ and $\frac{7}{100} = 0.07$. We need to develop a procedure to write other fractions, such as $\frac{3}{8}$ and $\frac{5}{16}$, in decimal form.

TEACHING TIP Sometimes a student will not see the difference in the three results discussed in Converting a Fraction to an Equivalent Decimal. You may want to give an example of each one right next to the rule.

a. $\frac{5}{8} = 0.625$. The remainder becomes zero.

b. $\frac{1}{3} = 0.333\ldots$. The remainder repeats itself.

c. Round to the nearest thousandth $\frac{13}{19} = 0.684$. The desired number of decimal places is achieved.

Converting a Fraction to an Equivalent Decimal

Divide the denominator into the numerator until

a. the remainder becomes zero, or

b. the remainder repeats itself, or

c. the desired number of decimal places is achieved.

EXAMPLE 1 Write as an equivalent decimal. **(a)** $\frac{3}{8}$ **(b)** $\frac{31}{40}$

Divide the denominator into the numerator until the remainder becomes zero.

(a)
$$\begin{array}{r} 0.375 \\ 8\overline{)3.000} \\ \underline{2\,4} \\ 60 \\ \underline{56} \\ 40 \\ \underline{40} \\ 0 \end{array}$$

Therefore, $\frac{3}{8} = 0.375$

(b)
$$\begin{array}{r} 0.775 \\ 40\overline{)31.000} \\ \underline{280} \\ 300 \\ \underline{280} \\ 200 \\ \underline{200} \\ 0 \end{array}$$

Therefore, $\frac{31}{40} = 0.775$

Practice Problem 1 Write as an equivalent decimal.

(a) $\frac{5}{16}$ **(b)** $\frac{11}{80}$ ■

Decimals such as 0.375 and 0.775 are called **terminating decimals.** When converting $\frac{3}{8}$ to 0.375 or $\frac{31}{40}$ to 0.775, the division operation eventually yields a remainder of zero. Other fractions yield a repeating pattern. For example, $\frac{1}{3} = 0.3333\ldots$ and $\frac{2}{3} = 0.6666\ldots$ have a pattern of repeating digits. Decimals that have a digit or a group of digits that repeats are called **repeating decimals.** We often indicate the repeating pattern with a bar over the repeating group of digits:

$$0.3333\ldots = 0.\overline{3} \qquad 0.74\ 74\ 74\ldots = 0.\overline{74}$$

$$0.218\ 218\ 218\ldots = 0.\overline{218} \qquad 0.8942\ 8942\ldots = 0.\overline{8942}$$

If when converting fractions to decimal form the remainder repeats itself, we know that we have a repeating decimal.

TEACHING TIP Help the students to see the difference between the decimal 0.4787878 . . . and the decimal 0.78787878 . . . and the fact in general that a decimal can be a repeating decimal if all of its digits repeat like 1.23123123123 . . . or if only some of its digits repeat like 1.23232323. . . . Make sure the students put the bar over only the repeating digit or digits.

EXAMPLE 2 Write as an equivalent decimal.

(a) $\frac{5}{11}$ **(b)** $\frac{13}{22}$ **(c)** $\frac{5}{37}$

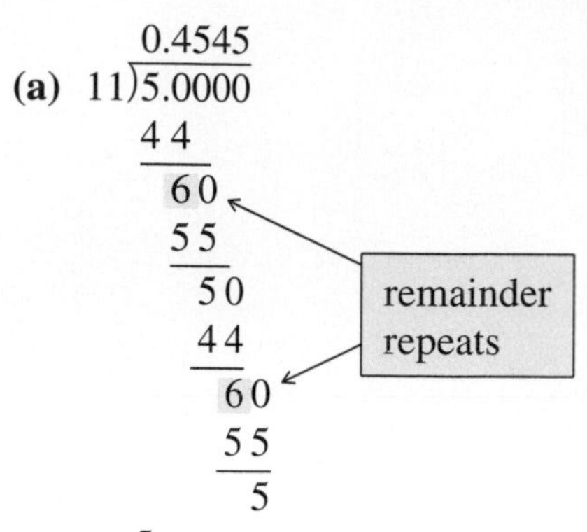

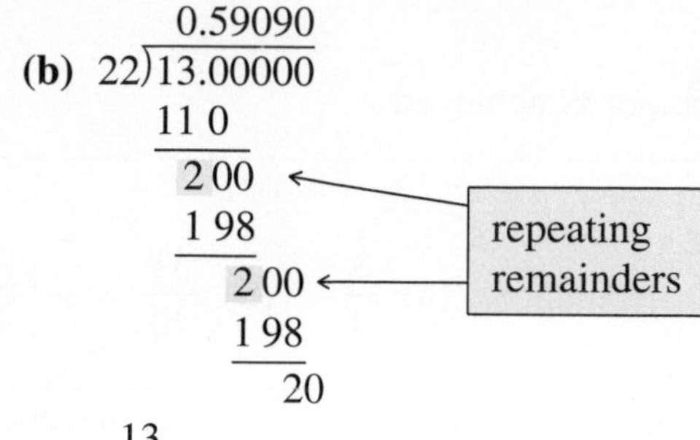

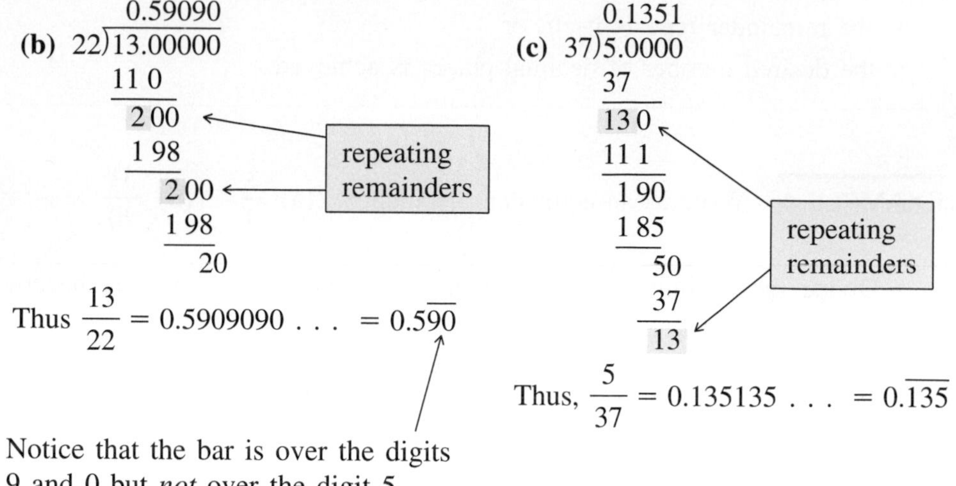

Thus $\frac{5}{11} = 0.4545\ldots = 0.\overline{45}$

Thus $\frac{13}{22} = 0.5909090\ldots = 0.5\overline{90}$

Notice that the bar is over the digits 9 and 0 but *not* over the digit 5.

Thus, $\frac{5}{37} = 0.135135\ldots = 0.\overline{135}$

> **CALCULATOR**
>
> **Fraction to Decimal**
>
> You can use a calculator to change $\frac{5}{8}$ to a decimal.
>
> Enter
>
> 5 [÷] 8 [=]
>
> The display should read
>
> [0.625]
>
> *Try*
>
> **(a)** $\frac{17}{25}$ **(b)** $\frac{2}{9}$
>
> **(c)** $\frac{13}{10}$ **(d)** $\frac{15}{19}$
>
> *Note:* 0.7894737 is an approximation for $\frac{15}{19}$. Some calculators only round to 8 places.

Practice Problem 2 Write as an equivalent decimal.

(a) $\frac{7}{11}$ **(b)** $\frac{8}{15}$ **(c)** $\frac{13}{44}$ ■

EXAMPLE 3 Write as an equivalent decimal. **(a)** $3\frac{7}{15}$ **(b)** $\frac{20}{11}$

(a) $3\frac{7}{15}$ means $3 + \frac{7}{15}$

$$
\begin{array}{r}
0.466 \\
15\overline{)7.000} \\
\underline{6\,0} \\
1\,00 \\
\underline{90} \\
100 \\
\underline{90} \\
10
\end{array}
$$

Thus $\frac{7}{15} = 0.4\overline{6}$ and $3\frac{7}{15} = 3.4\overline{6}$

(b)

$$
\begin{array}{r}
1.818 \\
11\overline{)20.000} \\
\underline{11} \\
9\,0 \\
\underline{8\,8} \\
20 \\
\underline{11} \\
90 \\
\underline{88} \\
2
\end{array}
$$

Thus $\frac{20}{11} = 1.818181\ldots = 1.\overline{81}$

Practice Problem 3 Write as an equivalent decimal.

(a) $2\frac{11}{18}$ **(b)** $\frac{28}{27}$ ■

In some cases, the pattern of repeating is quite long. For example,

$$\frac{1}{7} = 0.142857142857\ldots = 0.\overline{142857}$$

Such problems are often rounded to a certain value.

EXAMPLE 4 Express $\frac{5}{7}$ as a decimal rounded to the nearest thousandth.

$$\begin{array}{r} 0.7142 \\ 7\overline{)5.0000} \\ \underline{4\,9} \\ 10 \\ \underline{7} \\ 30 \\ \underline{28} \\ 20 \\ \underline{14} \\ 6 \end{array}$$

Rounding to the nearest thousandth, we round 0.7142 to 0.714. (In repeating form, $\frac{5}{7} = 0.714285714285\ldots = 0.\overline{714285}$.)

Practice Problem 4 Express $\frac{19}{24}$ as a decimal rounded to the nearest thousandth. ■

2 *Order of Operations with Decimals*

The rules for order of operations that we discussed in Section 1.6 apply to operations with decimals.

TEACHING TIP If you sense that many of the class members are rusty on the rules for order of operations covered in Section 1.6, you may want to do a couple of simple examples with whole numbers first. You may want to try:

$25 \times 2 - 15 + 3 \times 4$

(The answer is 47.)

$2^3 + 20 \div 10 \times 3 - 4$

(The answer is 10.)

Order of Operations

Do first	**1.** Perform operations inside parentheses.
	2. Simplify any expressions with exponents.
	3. Multiply or divide from left to right.
Do last	**4.** Add or subtract from left to right.

EXAMPLE 5 Evaluate. $0.5 + 0.7 \times 0.21 - 0.63$

$= 0.5 + 0.147 - 0.63$ We multiplied. $\begin{array}{r} 0.21 \\ \times\ \ 0.7 \\ \hline 0.147 \end{array}$

$= 0.647 - 0.63$ We added. $\begin{array}{r} 0.5 \\ +\ 0.147 \\ \hline 0.647 \end{array}$

$= 0.017$ We subtracted. $\begin{array}{r} 0.647 \\ -\ 0.630 \\ \hline 0.017 \end{array}$

Practice Problem 5 Evaluate. $0.9 - 0.3 + 0.8 \times 0.3$ ■

Sometimes exponents are used with decimals. In such cases, we merely evaluate using repeated multiplication.

$$(0.2)^2 = 0.2 \times 0.2 = 0.04$$
$$(0.2)^3 = 0.2 \times 0.2 \times 0.2 = 0.008$$
$$(0.2)^4 = 0.2 \times 0.2 \times 0.2 \times 0.2 = 0.0016$$

EXAMPLE 6 Evaluate. $(0.3)^3 + 0.6 \times 0.2 + 0.013$

First we need to evaluate $(0.3)^3 = 0.3 \times 0.3 \times 0.3 = 0.027$. Thus

$(0.3)^3 + 0.6 \times 0.2 + 0.013$
$= 0.027 + 0.6 \times 0.2 + 0.013$
$= 0.027 + 0.12 + 0.013$ ⟵ When addends have a different number of decimal places, writing the problem in column form makes adding easier.

$$\begin{array}{r} 0.027 \\ 0.120 \\ +\ 0.013 \\ \hline 0.160 \end{array}$$

$= 0.16$

Practice Problem 6 Evaluate. $0.3 \times 0.5 + (0.4)^3 - 0.036$ ■

In the next example all four steps of the rules for order of operations will be used.

EXAMPLE 7 Evaluate. $(8 - 0.12) \div 2^3 + 5.68 \times 0.1$

$= 7.88 \div 2^3 + 5.68 \times 0.1$ *First do subtraction inside the parentheses.*

$= 7.88 \div 8 + 5.68 \times 0.1$ *Simplify the expression with exponents.*

$= 0.985 + 0.568$ *From left to right do division and multiplication.*

$= 1.553$ *Add the final two numbers.*

Practice Problem 7 Evaluate. $6.56 \div (2 - 0.36) + (8.5 - 8.3)^2$ ■

Developing Your Study Skills

KEEP TRYING

We live in a highly technical world, and you cannot afford to give up on the study of mathematics. Dropping mathematics may prevent you from entering certain career fields that you may find interesting. You may not have to take math courses as high-level as calculus, but such courses as intermediate algebra, finite math, college algebra, and trigonometry may be necessary. Learning mathematics can open new doors for you.

Learning mathematics is a process that takes time and effort. You will find that regular study and daily practice are necessary to strengthen your skills and to help you grow academically. This process will lead you toward success in mathematics. Then, as you become more successful, your confidence in your ability to do mathematics will grow.

3.6 Exercises

Verbal and Writing Skills

1. 0.75 and $\frac{3}{4}$ are different forms for the ___same___ ___quantity___.

2. To convert a fraction to an equivalent decimal, divide the ___denominator___ into the numerator.

3. Why is $0.\overline{8942}$ called a repeating decimal? The digits 8942 repeat.

4. The order of operations for decimals is the same as the order of operations for whole numbers. Write the steps for the order of operations.

1. Perform operations inside parentheses.
2. Simplify any expressions with exponents.
3. Multiply or divide from left to right.
4. Add or subtract from left to right.

Write as an equivalent decimal. If a repeating decimal is obtained, use notation such as $0.\overline{7}$, $0.\overline{16}$, *or* $0.\overline{245}$.

5. $\frac{1}{4}$ 0.25 **6.** $\frac{3}{4}$ 0.75 **7.** $\frac{7}{8}$ 0.875 **8.** $\frac{5}{8}$ 0.625 **9.** $\frac{7}{16}$ 0.4375 **10.** $\frac{3}{16}$ 0.1875 **11.** $\frac{7}{20}$ 0.35 **12.** $\frac{3}{40}$ 0.075

13. $\frac{31}{50}$ 0.62 **14.** $\frac{23}{25}$ 0.92 **15.** $\frac{9}{4}$ 2.25 **16.** $\frac{14}{5}$ 2.8 **17.** $2\frac{1}{8}$ 2.125 **18.** $3\frac{13}{16}$ 3.8125 **19.** $1\frac{7}{16}$ 1.4375 **20.** $1\frac{1}{40}$ 1.025

21. $\frac{2}{3}$ $0.\overline{6}$ **22.** $\frac{1}{3}$ $0.\overline{3}$ **23.** $\frac{5}{9}$ $0.\overline{5}$ **24.** $\frac{7}{9}$ $0.\overline{7}$ **25.** $\frac{5}{6}$ $0.8\overline{3}$ **26.** $\frac{13}{18}$ $0.7\overline{2}$ **27.** $\frac{10}{11}$ $0.\overline{90}$ **28.** $\frac{4}{15}$ $0.2\overline{6}$

29. $\frac{14}{15}$ $0.9\overline{3}$ **30.** $\frac{3}{18}$ $0.1\overline{6}$ **31.** $\frac{5}{33}$ $0.\overline{15}$ **32.** $\frac{7}{22}$ $0.3\overline{18}$ **33.** $\frac{28}{27}$ $1.\overline{037}$ **34.** $\frac{41}{36}$ $1.13\overline{8}$ **35.** $2\frac{5}{18}$ $2.2\overline{7}$ **36.** $6\frac{1}{6}$ $6.1\overline{6}$

Write as an equivalent decimal or a decimal approximation. Round your answer to the nearest thousandth if needed.

37. $\frac{4}{7}$ 0.571 **38.** $\frac{13}{27}$ 0.481 **39.** $\frac{19}{21}$ 0.905 **40.** $\frac{20}{21}$ 0.952 **41.** $\frac{7}{48}$ 0.146 **42.** $\frac{5}{48}$ 0.104 **43.** $\frac{35}{27}$ 1.296 **44.** $\frac{37}{23}$ 1.609

45. $\frac{20}{81}$ 0.247 **46.** $\frac{7}{82}$ 0.085 **47.** $\frac{17}{18}$ 0.944 **48.** $\frac{5}{13}$ 0.385 **49.** $\frac{22}{7}$ 3.143 **50.** $\frac{17}{14}$ 1.214 **51.** $\frac{9}{19}$ 0.474 **52.** $\frac{11}{17}$ 0.647

Applications

53. A computer disc is $\frac{5}{24}$ inch thick. Write the thickness as a decimal.
$0.208\overline{3}$ inch thick

54. In the winter in Maine a car engine will run safely if the radiator solution is $\frac{5}{8}$ antifreeze. Write this value as a decimal. 0.625 antifreeze

55. An ice climber had the local mountaineering shop install boot heaters to the back of his hiking boots. The installer drilled a hole $\frac{3}{8}$ inch in diameter. The hole for the boot heater should have been 0.5 inch in diameter. Is the hole too large or too small? By how much? It is too small by 0.125 inch.

56. A master carpenter is re-creating a room for the set of a movie being filmed. He is using a burled maple veneer $\frac{7}{16}$ inch thick. The designer specified maple veneer 0.45 inch thick. Is the veneer he is using too thick or too thin? By how much?
It is too thin by 0.0125 inch.

57. A machinist in a factory is using sheet metal $\frac{17}{32}$ inch thick. The plans call for sheet metal 0.53 inch thick. Is the metal he is using too thick or too thin? By how much? too thick, 0.00125 inch

58. An auto body shop technician drilled a hole $\frac{7}{16}$ inch in diameter. The hole should have been 0.44 inch in diameter. Is the hole too large or too small? By how much? too small, 0.0025 inch

Evaluate.

59. $(0.2)^3 + 5.9 \times 1.3 - 2.6$
$0.008 + 7.67 - 2.6 = 5.078$

60. $(0.3)^2 + 0.6 \times (0.2 + 5.8)$
$0.09 + 0.6 \times 6 = 0.09 + 3.6 = 3.69$

61. $12.2 \times 9.4 - 2.68 + 1.6 \div 0.8$
$114.68 - 2.68 + 2 = 114$

62. $2.3 \times 12.6 - 1.98 + 12.6 \div 1.4$
$28.98 - 1.98 + 9 = 36$

63. $12 \div 0.03 - 15 \times (0.6 + 0.7)^2$
$400 - 25.35 = 374.65$

64. $61.95 \div 1.05 + 3.6 \times (2.1 + 0.31)$
$59 + 8.676 = 67.676$

65. $116.32 + (0.12)^2 + 18.06 \times 2.2$
$116.32 + 0.0144 + 39.732 = 156.0664$

66. $105.08 + (0.21)^2 - 0.05 \times 123.4$
$105.08 + 0.0441 - 6.17 = 98.9541$

67. $(1.1)^3 + 2.6 \div 0.13 + 0.083$
$1.331 + 20 + 0.083 = 21.414$

68. $(1.1)^3 + 8.6 \div 2.15 - 0.086$
$1.331 + 4 - 0.086 = 5.245$

69. $(56.3 - 56.1)^2 \div 0.016$
2.5

70. $(23.8 - 23.5)^2 \div 0.36$
0.25

71. $(0.5)^3 + (3 - 2.6) \times 0.5$
0.325

72. $(0.6)^3 + (7 - 6.3) \times 0.07$
0.265

73. $(0.76 + 4.24) \div 0.25 + 8.6$
28.6

74. $(2.4)^2 + 3.6 \div (1.2 - 0.7)$
12.96

Evaluate.

★ 75. $(1.6)^3 + (2.4)^2 + 18.666 \div 3.05 + 4.86$
$4.096 + 5.76 + 6.12 + 4.86 = 20.836$

★ 76. $5.9 \times 3.6 \times 2.4 - 0.1 \times 0.2 \times 0.3 \times 0.4$
$50.976 - 0.0024 = 50.9736$

Write each fraction as a decimal. Round your answer to six decimal places.

★ 77. $\dfrac{5236}{8921}$
0.586930

★ 78. $\dfrac{17{,}359}{19{,}826}$
0.875567

To Think About

★ 79. Subtract. $0.\overline{16} - 0.00\overline{16}$
(a) What do you obtain?
$0.1616\overline{16} - 0.0016\overline{16} = 0.16$
(b) Now subtract $0.\overline{16} - 0.01\overline{6}$. What do you obtain?
$0.1616\overline{16} - 0.0166\overline{66} = 0.1449\overline{49}$
(c) What is different about these results?
The repeating patterns line up differently.

★ 80. Subtract. $1.\overline{89} - 0.0\overline{189}$
(a) What do you obtain?
$1.8989\overline{89} - .018989 = 1.88$
(b) Now subtract $1.\overline{89} - 0.1\overline{89}$. What do you obtain?
$1.898989\overline{89} - 0.189999\overline{99} = 1.708989\overline{89}$
(c) What is different about these results?
The repeating patterns line up differently.

Cumulative Review Problems

81. What is the area of a rectangle that measures 12 feet by 26 feet?

12 feet

26 feet

$12 \times 26 = 312$ square feet

82. In the 1996 presidential election, 1,070,990 voters in Virginia voted for President Clinton and 1,071,859 voters in Wisconsin voted for President Clinton. What was the difference in the number of votes for President Clinton between these two states? 869 votes

83. If a person makes deposits of \$56, \$81, \$42, and \$198, what is the total amount deposited in the account?
$56 + 81 + 42 + 198 = \$377$ was deposited.

84. If Central Texas Catering equally divides \$3320 in tips from this weekend among its 40 employees, how much will each person receive?
\$83 per employee

3.7 Applied Problems Using Decimals

After studying this section, you will be able to:

1 *Solve applied problems using operations with decimals.*

MathPro Video 9 SSM

1 *Solving Applied Problems Using Operations with Decimals*

We use the basic plan of solving applied problems that we discussed in Section 1.8 and Section 2.9. Let us review how we analyze applied problem situations.

1. *Understand the problem.*
2. *Solve and state the answer.*
3. *Check.*

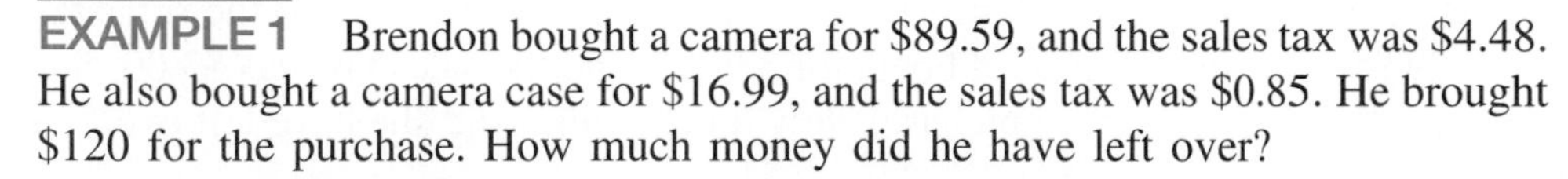

EXAMPLE 1 Brendon bought a camera for $89.59, and the sales tax was $4.48. He also bought a camera case for $16.99, and the sales tax was $0.85. He brought $120 for the purchase. How much money did he have left over?

1. *Understand the problem.*
   Fill in the Mathematics Blueprint for Problem Solving.

MATHEMATICS BLUEPRINT FOR PROBLEM SOLVING

Gather the Facts	What Am I Asked to Do?	How Do I Proceed?	Key Points to Remember
Camera cost: $89.59 Tax on camera: $4.48 Camera case cost: $16.99 Tax on case: $0.85 Brendon has $120 to spend.	Find out how much money Brendon has left after making these purchases.	Add the cost of each item and the tax on each item to obtain the total purchase price. Subtract that total from $120.	Be sure to line up the decimal points when adding and subtracting the decimal values.

2. *Solve and state the answer.*
   Add the four amounts.

$$\begin{array}{r} \$89.59 \\ 4.48 \\ 16.99 \\ +\quad 0.85 \\ \hline \$111.91 \end{array}$$

Subtract all expenses from $120.

$$\begin{array}{r} \$120.00 \\ -\quad 111.91 \\ \hline \$\quad 8.09 \end{array}$$

Brendon has $8.09 left after purchasing the camera and the case and paying the tax on each.

3. *Check.*

Estimate the answer to see if it is reasonable. First, we take each of the four amounts and round them to one significant digit. Then we add the four numbers together.

$$\$90 + \$4 + \$20 + \$1 = \$115$$

Next, we subtract our estimated sum from \$120. Since the calculation is so easy to do, we do not estimate the subtraction. Rather we do the calculation directly.

$$\begin{array}{r} \$120 \\ -\ 115 \\ \hline \$5 \end{array}$$

Thus we estimate Brendon will have about \$5 left after the purchase. Our answer of \$8.09 is fairly close to the estimate, so our answer is reasonable. ✓

Practice Problem 1 Renée purchased a coat for \$198.95, and the sales tax was \$9.95. She also purchased boots for \$59.95, and the sales tax was \$3.00. She brought \$280 to make the purchase. How much money did she have left over? ■

TEACHING TIP Although overtime pay is a very simple concept, some students are not familiar with it. Some students even find out after taking this course that, by law, they should have been getting overtime pay in their previous job but never did. You may want to give a couple of simple examples of how overtime pay works.

In the United States for almost all jobs where you are paid an hourly wage, if you work more than 40 hours in one week, you should be paid overtime, at a rate of 1.5 times the normal hourly rate, for the extra hours worked in that week. The next problem deals with overtime wages.

EXAMPLE 2 A laborer is paid \$7.38 per hour for a 40-hour week and 1.5 times that wage for any hours worked beyond the standard 40. If he works 47 hours in a week, what will he earn?

1. *Understand the problem.*

MATHEMATICS BLUEPRINT FOR PROBLEM SOLVING

Gather the Facts	What Am I Asked to Do?	How Do I Proceed?	Key Points to Remember
He works 47 hours. He gets paid \$7.38 per hour for 40 hours. He gets paid 1.5 × \$7.38 per hour for 7 hours.	Find the earnings of the laborer if he works 47 hours in one week.	Add the earnings of 40 hours at \$7.38 per hour to the earnings of 7 hours at overtime pay.	Multiply 1.5 × \$7.38 to find the pay he earns for overtime.

2. *Solve and state the answer.*

We want to compute his regular pay and his overtime pay and add the results.

Regular pay + Overtime pay = Total pay

Regular pay Calculate his pay for 40 hours of work.

$$\begin{array}{r} 7.38 \\ \times\ \ 40 \\ \hline 295.20 \end{array}$$

He earns \$295.20 at \$7.38 per hour.

Overtime pay Calculate his overtime pay rate. This is 7.38×1.5.

$$\begin{array}{r} 7.38 \\ \times\ 1.5 \\ \hline 3\,690 \\ 7\,38 \\ \hline 11.070 \end{array}$$

He earns $11.07 per hour in overtime.

Calculate how much he earned doing 7 hours of overtime work.

$$\begin{array}{r} 11.07 \\ \times\quad 7 \\ \hline 77.49 \end{array}$$

For 7 hours overtime he earns $77.49.

Total pay Add the two amounts.

$$\begin{array}{rl} \$295.20 & \text{Regular 40-hour week earnings} \\ +\ \ 77.49 & \text{Overtime earnings} \\ \hline \$372.69 & \text{Total earnings} \end{array}$$

The total earnings of the laborer for a 47-hour workweek were $372.69.

3. *Check.*

Estimate his regular pay.

$$40 \times \$7 = \$280$$

Estimate his overtime rate of pay, and then his overtime pay.

$$1.5 \times \$7 = \$10.50$$

$$7 \times \$11 = \$77$$

Then add.

$$\begin{array}{r} \$280 \\ +\ \ 77 \\ \hline \$357 \end{array}$$

Our estimate of $357 is close to our answer of $372.69. Our answer is reasonable. ✓

Practice Problem 2 Melinda works for the phone company as a line repair technician. She earns $9.36 per hour. She worked 51 hours last week. If she gets time and a half for all hours worked above 40 hours per week, how much did she earn last week? ■

EXAMPLE 3 A chemist is testing 36.85 liters of cleaning fluid. She wishes to pour it into several smaller containers that each hold 0.67 liters of fluid. **(a)** How many containers will she need? **(b)** If each liter of this fluid costs $3.50, how much does the cleaning fluid in one container cost? (Round your answer to the nearest cent.)

MATHEMATICS BLUEPRINT FOR PROBLEM SOLVING

Gather the Facts	What Am I Asked to Do?	How Do I Proceed?	Key Points to Remember
The total amount of cleaning fluid is 36.85 liters. Each small container holds 0.67 liters. Each liter of fluid costs $3.50.	(a) Find out how many containers the chemist needs. (b) Find the cost of cleaning fluid in each small container.	(a) Divide the total, 36.85, by the amount in each small container, 0.67, to find the number of containers. (b) Multiply the cost of one liter, $3.50, by the amount of liters in one container, 0.67.	If you are not clear as to what to do at any stage of the problem, then do a similar, simpler problem.

(a) How many containers will the chemist need?

She has 36.85 liters of cleaning fluid and she wants to put it into several equal-sized containers each holding 0.67 liter. Suppose we are not sure what to do. Let's do a similar, simpler problem. If we had 40 liters of cleaning fluid and we wanted to put it into little containers each holding 0.5 liter, what would we do? Since two little containers would only hold 1 liter, we would need 80 containers. We know that $40 \div 0.5 = 80$. So we see that, in general, we divide the total number of liters by the amount in the small container. Thus $36.85 \div 0.67$ will give us the number of containers in this case.

$$\begin{array}{r} 55. \\ 0.67_\wedge\overline{)36.85_\wedge} \\ 33\ 5 \\ \hline 3\ 35 \\ 3\ 35 \\ \hline \end{array}$$

The chemist will need 55 containers to hold this amount of cleaning fluid.

(b) How much does the cleaning fluid in each container cost? Each container will hold only 0.67 liter. If one liter costs $3.50, then to find the cost of one container we multiply $0.67 \times \$3.50$.

$$\begin{array}{r} 3.50 \\ \times\ 0.67 \\ \hline 2450 \\ 2100 \\ \hline 2.3450 \end{array}$$

We round our answer to the nearest cent. Thus each container would cost $2.35.

Check.

(a) Is it really true that 55 containers each holding 0.67 liter will hold a total of 36.85 liters? To check we multiply.

$$\begin{array}{r} 55 \\ \times\ 0.67 \\ \hline 385 \\ 330 \\ \hline 36.85 \end{array} \checkmark$$

(b) One liter of cleaning fluid costs \$3.50. We would expect the cost of 0.67 liter to be less than \$3.50. \$2.35 is less than \$3.50. ✓

We use estimation to check more closely.

$$\begin{array}{rcr} \$3.50 & \longrightarrow & \$4.00 \\ \times\ 0.67 & \longrightarrow & \times\ \ 0.7 \\ \hline & & \$2.800 \end{array}$$

\$2.80 is fairly close to \$2.35. Our answer is reasonable. ✓

Practice Problem 3 A butcher divides 17.4 pounds of prime steak into small equal-sized packages. Each package contains 1.45 pounds of prime steak. **(a)** How many packages of steak will he have? **(b)** Prime steak sells for \$4.60 per pound. How much will each package of prime steak cost? ■

Developing Your Study Skills

APPLICATIONS OR WORD PROBLEMS

Applications or word problems are the very life of mathematics! They are the reason for doing mathematics, because they teach you how to put into use the mathematical skills you have developed. Learning mathematics without ever doing word problems is similar to learning all the skills of a sport without ever playing a game or learning all the notes on an instrument without ever playing a song.

The key to success is practice. Make yourself do as many problems as you can. You may not be able to do them all correctly at first, but keep trying. Do not give up whenever you reach a difficult one. If you cannot solve it, just try another one. Then come back and try it again later. Ask for help from your teacher or the tutoring lab. Ask other classmates how they solved the problem.

A misconception among students when they begin studying word problems is that each problem is different. At first the problems may seem this way, but as you practice more and more, you will begin to see the similarities, the different "types." You will see patterns in solving problems, which will enable you to solve problems of a given type more easily.

3.7 Exercises

Applications

1. A single-engine plane has a passenger weight limit enforced by the Federal Aviation Association. A pilot asked his passengers to step on the scale to check their respective weights. The four passengers weighed 167.2 pounds, 136.7 pounds, 99.3 pounds, and 218 pounds. What was the total weight of the passengers? 621.2 pounds

2. Angela buys costume jewelry and watches at a wholesale price, and resells them in her retail shop. During the last four weeks, she spent \$412.65, \$78.90, \$331.47, and \$146.03. How much did she spend on merchandise? \$969.05

3. The local baseball team took in \$11,022.45 in ticket sales yesterday. Today the team took in \$19,891.33. How much more money did the box office receive today than yesterday? \$8868.88

4. Renee LeBlanc owns a mail-order company. Last year the company made a profit of \$233,598.72. This year the company made a profit of \$314,255.33. How much more profit did the company make this year? \$80,656.61

5. A parking lot in front of the Orlando Outback Steak House measures 103.4 meters long by 76.3 meters wide. What is the area in square meters of this parking lot? 7889.42 square meters

6. A mouse pad measures 22.4 centimeters by 26.3 centimeters. What is the area of the mouse pad in square centimeters? 589.12 square centimeters

7. Cynthia is making holiday cookies, which she will give as gifts to her friends and co-workers. She has made 34.5 pounds of cookies, which need to be divided up into equal packages containing 0.75 pound of cookies. How many packages can she make? 46 packages

8. Hans is making gourmet chocolate in Switzerland. He has 11.52 liters of liquid white chocolate that will be poured into molds that hold 0.12 liter. How many individual molds can Hans make with his 11.52 liters of liquid white chocolate? 96 molds

9. Donna went grocery shopping with her three children. She spent \$71.32 on groceries. On the way out she realized that she had forgotten to buy milk. She went back in and paid \$2.16 for milk. Just as she got back to the car, she realized that there was only one egg in the house, so she went back in and found out that eggs cost \$1.19. Donna started out with \$75.00 in her wallet. Did she have enough to buy the eggs? How much was left over? Yes. After she bought the eggs she had \$0.33.

10. David is planning to propose to his girlfriend today. He bought perfume for her for \$68.49 and paid \$3.94 in tax. Next, he went to the florist and bought a huge bouquet of assorted Holland flowers for \$72.39 and paid \$4.16 in tax. If he started out with \$150.00 in his wallet, how much money did he have left over? \$1.02

11. One year in Mount Waialeale, Hawaii, considered the ''rainiest place in the world,'' the yearly rainfall totaled 11.68 meters. The next year, the yearly rainfall on this mountain totaled 10.42 meters. The third year it was 12.67 meters. On average, how much rain fell on Mount Waialeale, Hawaii, per year. 11.59 meters of rainfall per year

12. Emma and Jennie took a trip in their Chrysler Concorde from Saskatoon, Saskatchewan, to calgary, Alberta, in Canada to check out the glacier lakes. When they left, their odometer read 54,089. When they returned home, the odometer read 55,401. They used 65.6 gallons of gas. How many miles per gallon did they get on the trip? (Round your answer to the nearest tenth.) 20 miles per gallon

13. The York Steak House uses 3.5 boxes of customized paper towels per day for its customers. If the master box holds 42 regular boxes, how many days will the master box last? 12 days

14. Hanu's outdoor gas grill uses 2.4 pounds of propane per hour. If the tank holds 20.64 pounds of propane, how many hours can the barbecue run on a full tank? 8.6 hours

15. The local Police Athletic League raised enough money to renovate the local youth hall and turn it into a coffeehouse/activity center so that there is a safe place to hang out. The room that holds the ping-pong table needs 43.9 square yards of new carpeting. The entryway needs 11.3 square yards and the stage/seating area needs 63.4 yards. The carpeting will cost $10.65 per square yard. What will be the total bill for carpeting these three areas of the coffeehouse? $1263.09

16. The Williams family is installing insulation in their house. They are installing 98.6 square yards of insulation in the attic, 24.3 square yards of insulation in the basement, and 36.1 square yards of insulation in the family room. The insulation costs $6.40 per square yard. How much will all the insulation cost? $1017.60

17. An electrician is paid $14.30 per hour for a 40-hour week. She is paid time and a half for overtime (1.5 times the hourly wage) for every hour more than 40 hours worked in the same week. If she works 48 hours in one week, what will she earn for that week? $743.60

18. Shirley works at the local pizza franchise and is paid $6.20 per hour for a 40-hour week. She is paid time and a half for overtime, (1.5 times the hourly wage), for every hour more than 40 hours worked in the same week. If she works 52 hours in one week, what will she earn for that week? $359.60

19. Barbara had $420.13 in her savings account last month. Since then she has made deposits of $116.32 and $318.57. The bank credited her with interest of $1.86 for the month. She wrote three checks, for $16.50, $36.89, and $376.94. What is her new balance? (Assume that no bank fees have been charged.) $426.55

20. Harvey had $113.08 in his checking account last month. He made two deposits: $612.83 and $86.54. The bank deducted a fee this month of $4.60. He made out three checks: $100.00, $216.34, and $398.92. What will his new balance be? $92.59

21. Charlie borrowed $11,500 to purchase a new car. His loan requires him to pay $288.65 each month over the next 60 months (5 years). How much will he pay over the 5 years? How much more will he pay back than the amount of the loan? $17,319. He will pay back $5819 more than the loan.

22. Hector and Junita borrowed $80,000 to buy their new home. They make monthly payments to the bank of $450.25 to repay the loan. They will be making these monthly payments for the next 30 years. How much money will they pay to the bank in the next 30 years? How much more will they pay back than they borrowed? $162,090; $82,090

23. A famous athletic shoe store at the Heartland Shopping Mall buys basketball shoes at a cost of $7920 for 12 dozen pairs. They sell the sneakers for $68 per pair. How much profit will they make if they sell all 144 pairs? $1872

24. Concert T-shirts are usually sold for $24.00 at the show. If the tour promoter buys 5000 shirts for $56,250 and sells every shirt, how much profit will the promoter realize? Profit of $63,750

25. The EPA standard for safe drinking water is a maximum of 1.3 milligrams of copper per liter of water. A study was conducted on a sample of 7 liters of water drawn from Jeff Slater's house. The analysis revealed 8.06 milligrams of copper in the sample. Is the water safe or not? By how much?
yes, by 0.149 milligram per liter

26. The EPA standard for safe drinking water is a maximum of 0.015 milligram of lead per liter of water. A study was conducted on 6 liters of water from West Towers Dormitory. The analysis revealed 0.0795 milligram of lead in the sample. Is the water safe or not? By how much?
yes, by .00175 milligram per liter

27. A jet fuel tank containing 17316.8 gallons is being emptied at the rate of 126.4 gallons per minute. How many minutes will it take to empty the tank?
137 minutes

28. If General Motors sells 6340 new Chevrolet Camaros next month and makes a profit of $2318.20 on each car, what profit will GM make on Camaros next month? $14,697,388

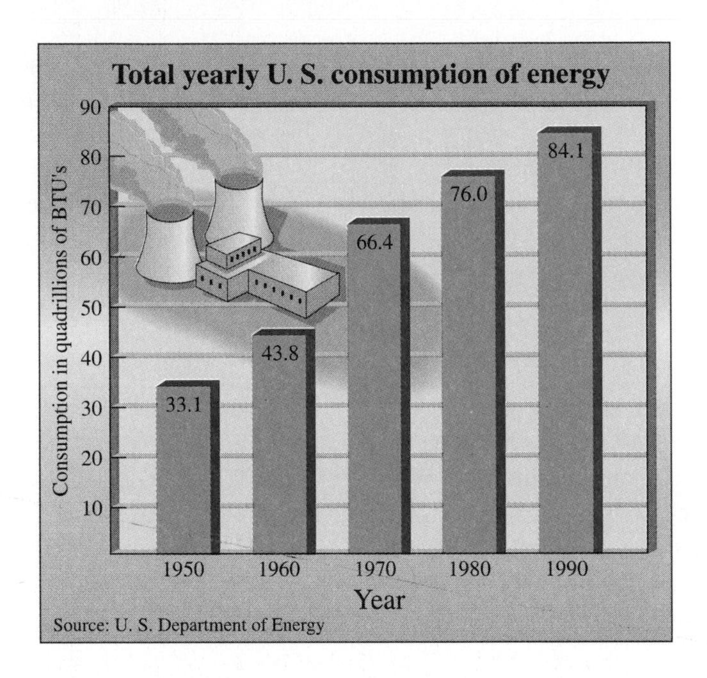

Use the bar graph above to answer questions 29–32.

29. How many more Btus were consumed in the United States during the year 1990 than in the year 1970? 17.7 quadrillion Btus

30. What was the greatest increase in consumption of energy in a 10-year period? When did it occur?
It increased by 22.6 quadrillion Btus from 1960 to 1970.

31. What was the average consumption of energy per year in the United States for the 20-year period including the years 1950, 1960, and 1970? Write your answer in quadrillion Btus and then write your answer in Btus. (Remember a quadrillion is 1000 trillions.) Approximately 47.8 quadrillion Btus. This is the same as 47,800,000,000,000,000 Btus (14 zeros in the answer).

32. What was the average consumption of energy per year in the United States for the 20-year period including the years 1970, 1980, and 1990? Write your answer in quadrillion Btus and then write your answer in Btus. (Remember a quadrillion is 1000 trillions.) Approximately 75.5 quadrillion Btus. This is the same as 75,500,000,000,000,000 Btus (14 zeros in the answer).

Cumulative Review Problems

Add.

33. $\frac{1}{5} + \frac{3}{7} + \frac{1}{2}$
$\frac{79}{70}$ or $1\frac{9}{70}$

34. $\frac{10}{17} + \frac{1}{34} - \frac{5}{17}$ $\frac{11}{34}$

Multiply.

35. $\frac{5}{12} \times \frac{36}{27}$ $\frac{5}{9}$

36. $2\frac{2}{3} \times 8\frac{1}{4}$
22

Chapter Organizer

Topic	Procedure	Examples
Word names for decimals, p. 178.	Hundreds: 3 Tens: 4 Ones: 1 Decimal Point: . Tenths: 6 Hundredths: 7 Thousandths: 8 Ten Thousandths: 3	The word name for 341.6783 is three hundred forty-one and six thousand seven hundred eighty-three ten thousandths.
Writing a decimal as a fraction, p. 180.	**1.** Read the decimal in words. **2.** Write it in fraction form. **3.** Reduce if possible.	Write 0.36 as a fraction. **1.** 0.36 is read thirty-six hundredths **2.** Write the fractional form. $\frac{36}{100}$ **3.** Reduce. $\frac{36}{100} = \frac{9}{25}$
Determining which of two decimals is larger, p. 184.	**1.** Start at the left and compare corresponding digits. Write in extra zeros if needed. **2.** When two digits are different, the larger number is the one with the larger digit.	Which is larger? 0.138 or 0.13 0.138 ? 0.130 $8 > 0$ So $0.138 > 0.130$.
Rounding decimals, p. 186.	**1.** Locate the place (units, tenths, hundredths, etc.) for which the round-off is required. **2.** If the first digit to the right of this place is less than 5, drop it and all the digits to the right of it. **3.** If the first digit to the right of this place is 5 or greater, increase the digit by one. Drop all digits to the right.	Round to the nearest hundredth. 0.8652 0.87 Round to the nearest thousandth. 0.21648 0.216
Adding and subtracting decimals, pp. 190.	**1.** Write the numbers vertically and line up the decimal points. Extra zeros may be written to the right of the decimal points if needed. **2.** Add or subtract all the digits with the same place value, starting with the right column, moving to the left. Use carrying or borrowing as needed. **3.** Place the decimal point of the sum in line with the decimal points of all the numbers added.	Add. 36.3 + 8.007 + 5.26 $\begin{array}{r} \overset{1}{3}6.300 \\ 8.007 \\ +\ 5.260 \\ \hline 49.567 \end{array}$ Subtract. 82.5 − 36.843 $\begin{array}{r} \overset{7}{\not 8}\ \overset{11}{\not 2}.\overset{14}{\not 5}\ \overset{9}{\not 0}\ \overset{10}{\not 0} \\ -\ 3\ 6.8\ 4\ 3 \\ \hline 4\ 5.6\ 5\ 7 \end{array}$
Multiplying decimals, p. 200.	**1.** Multiply the numbers just as you would multiply whole numbers. **2.** Find the sum of the decimal places in the two factors. **3.** Place the decimal point in the product so that the product has the same number of decimal places as the sum in step 2. You may need to insert zeros to the left of the number found in step 1.	Multiply. $\begin{array}{r} 0.2 \\ \times\ 0.6 \\ \hline 0.12 \end{array}$ $\begin{array}{r} 0.3174 \\ \times\ 0.8 \\ \hline 0.25392 \end{array}$ $\begin{array}{r} 0.0064 \\ \times\ 0.21 \\ \hline 64 \\ 128\ \\ \hline 0.001344 \end{array}$ $\begin{array}{r} 1364 \\ \times\ 0.7 \\ \hline 954.8 \end{array}$
Multiplying a decimal by a power of 10, p. 202.	Move the decimal point to the right the same number of places as there are zeros in the power of 10. (Sometimes it is necessary to write extra zeros before placing the decimal point in the answer.)	Multiply. $5.623 \times 10 = 56.23$ $0.597 \times 10^4 = 5970$ $0.0082 \times 1000 = 8.2$ $0.075 \times 10^6 = 75{,}000$ $28.93 \times 10^2 = 2893$

Chapter Organizer

Topic	Procedure	Examples
Dividing by a decimal, p. 208.	**1.** Make the divisor a whole number by moving the decimal point to the right. Mark that position with a caret (‸). **2.** Move the decimal point in the dividend to the right the same number of places. Mark that position with a caret. **3.** Place the decimal point of your answer directly above the caret in the dividend. **4.** Divide as with whole numbers.	Divide. **(a)** $0.06\overline{)0.162}$ **(b)** $0.003\overline{)85.8}$ **(a)** 2.7 0.06‸)0.16‸2 12 42 42 0 **(b)** 28 600. 0.003‸)85.800‸ 6 25 24 18 18 0
Converting a fraction to a decimal, p. 214.	Divide the denominator into the numerator until **1.** the remainder is zero, or **2.** the decimal repeats itself, or **3.** the desired number of decimal places is achieved.	Find the decimal equivalent for **(a)** $\frac{13}{22}$ **(b)** $\frac{5}{7}$, rounded to the nearest ten thousandth **(a)** 0.5909 22)13.0000 110 200 198 200 198 2 $\frac{13}{22} = 0.5\overline{90}$ or 0.5909090 . . . **(b)** 0.71428 7)5.00000 49 10 7 30 28 20 14 60 56 4 0.71428 rounded to the nearest ten thousandth is 0.7143.
Order of operations with decimal numbers, p. 217.	Same as order of operations of whole numbers. **1.** Perform operation inside parentheses. **2.** Simplify any expressions with exponents. **3.** Multiply or divide from left to right. **4.** Add or subtract from left to right.	Evaluate. $(0.4)^3 + 1.26 \div 0.12 - 0.12 \times (1.3 - 1.1)$ $(0.4)^3 + 1.26 \div 0.12 - 0.12 \times 0.2$ $0.064 + 1.26 \div 0.12 - 0.12 \times 0.2$ $0.064 + 10.5 - 0.024$ $10.564 - 0.024$ 10.54

Using the Mathematical Blueprint for Problem Solving, p. 221.

In solving an applied problem with decimals, students may find it helpful to complete the following steps. You will not use all the steps all of the time. Choose the steps that best fit the conditions of the problem.

1. *Understand the problem.*
- **(a)** Read the problem carefully.
- **(b)** Draw a picture if this helps you to visualize the situation. Think about what facts you are given and what you are asked to find.
- **(c)** It may help to write a similar, simpler problem to get started and to determine what operation to use.
- **(d)** Use the Mathematics Blueprint for Problem Solving to organize your work. Follow these four parts.
  - **1.** Gather the facts. (Write down specific values given in the problem.)
  - **2.** What am I asked to do? (Identify what you must obtain for an answer.)
  - **3.** How do I proceed? (What calculations need to be done?)
  - **4.** Key points to remember. (Record any facts, warnings, formulas, or concepts you think will be important as you solve the problem.)

2. *Solve and state the answer.*
- **(a)** Perform the necessary calculations.
- **(b)** State the answer, including the unit of measure.

3. *Check.*
- **(a)** Estimate the answer to the problem. Compare this estimate to the calculated value. Is your answer reasonable?
- **(b)** Repeat your calculations.
- **(c)** Work backward from your answer. Do you arrive at the original conditions of the problem?

Example

Fred has a rectangular living room that is 3.5 yards wide and 6.8 yards long. He has a hallway that is 1.8 yards wide and 3.5 yards long. He wants to carpet each area using carpeting that costs $12.50 per square yard. What will the carpeting cost him?

1. *Understand the problem.*
It is helpful to draw a sketch.

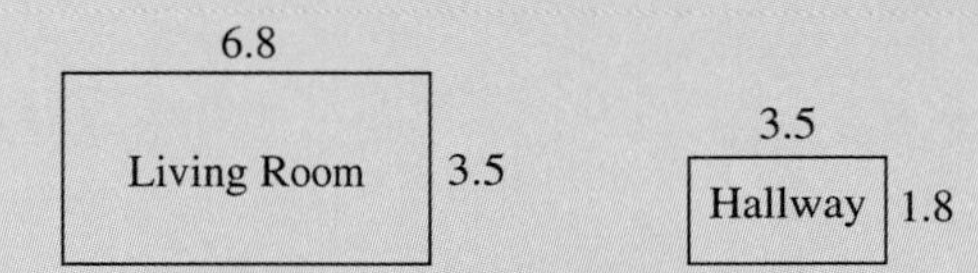

MATHEMATICS BLUEPRINT FOR PROBLEM SOLVING			
Gather the Facts	What Am I Asked to Do?	How Do I Proceed?	Key Points to Remember
Living room: 6.8 yards by 3.5 yards Hallway: 3.5 yards by 1.8 yards Cost of carpet: $12.50 per square yard	Find out what the carpeting will cost Fred.	Find the area of each room. Add the two areas. Multiply the total area by $12.50.	Multiply the length by the width to get the area of the room. Remember, area is measured in square yards.

To find the area of each room, we multiply the dimensions for each room.

Living room $6.8 \times 3.5 = 23.80$ square yards　　*Hallway* $3.5 \times 1.8 = 6.30$ square yards

Add the two areas.

$$\begin{array}{r} 23.80 \\ +\ 6.30 \\ \hline 30.10 \end{array}$$ square yards

Multiply the total area by the cost per square yard.

$$30.1 \times 12.50 = \$376.25$$

Estimate to check. You may be able to do some of this mentally.

$7 \times 4 = 28$ square yards　　$4 \times 2 = 8$ square yards

$$\begin{array}{r} 28 \\ +\ 8 \\ \hline 36 \end{array}$$

$36 \times 10 = \$360$　　$360 is close to $376.25. ✓

Chapter 3 Review Problems

3.1 *Write a word name for each decimal.*

1. 13.672
Thirteen and six hundred seventy-two thousandths

2. 0.00084
Eighty-four hundred thousandths

Write each fraction as a decimal.

3. $\frac{7}{10}$
0.7

4. $\frac{81}{100}$
0.81

5. $1\frac{523}{1000}$
1.523

6. $\frac{79}{10,000}$
0.0079

Write each decimal as a fraction or a mixed number.

7. 0.17

$\frac{17}{100}$

8. 0.365

$\frac{73}{200}$

9. 26.88

$26\frac{22}{25}$

10. 1.00025

$1\frac{1}{4000}$

3.2 *Fill in the blank with <, =, or >.*

11. $\frac{13}{100}$ ___>___ 0.103

12. 0.716 ___>___ 0.706

Arrange the set of numbers from smallest to largest.

13. 0.981, 0.918, 0.98, 0.901

0.901, 0.918, 0.98, 0.981

14. 5.62, 5.2, 5.6, 5.26, 5.59

5.2, 5.26, 5.59, 5.6, 5.62

15. Round to the nearest tenth. 0.613

0.6

16. Round to the nearest hundredth. 19.2076

19.21

17. Round to the nearest thousandth. 1.09952

1.100

18. Round to the nearest dollar. $156.48

$156.00

3.3 *Add.*

19. $\begin{array}{r} 1.8 \\ 2.603 \\ 0.52 \\ +\ 1.716 \\ \hline 6.639 \end{array}$

20. $\begin{array}{r} 9.6 \\ 11.5 \\ 21.8 \\ +\ 34.7 \\ \hline 77.6 \end{array}$

21. Fred needs 2.9 gallons of paint for the living room, 1.6 gallons for the kitchen, and 4 gallons for the bedrooms.

(a) Exactly how many gallons and decimal parts of a gallon does he need? 8.5 gallons

(b) How many whole gallons should he buy if he allows for a little extra? 9 gallons

Subtract.

22. $\begin{array}{r} 5.19 \\ -\ 1.296 \\ \hline 3.894 \end{array}$

23. $\begin{array}{r} 199.703 \\ -\ 108.964 \\ \hline 90.739 \end{array}$

24. Serina drove to Cincinnati. The car odometer read 86,804.7 miles when she started and 87,041.6 miles when she got there. How long was the trip?

236.9 miles

3.4 *Multiply.*

25. $\begin{array}{r} 0.76 \\ \times\ 0.03 \\ \hline 0.0228 \end{array}$

26. $\begin{array}{r} 1.2874 \\ \times\ 0.7 \\ \hline 0.90118 \end{array}$

27. $\begin{array}{r} 0.026 \\ \times\ 0.014 \\ \hline 0.000364 \end{array}$

28. $\begin{array}{r} 126.83 \\ \times\ 7 \\ \hline 887.81 \end{array}$

29. $\begin{array}{r} 54 \\ \times\ 1.6 \\ \hline 86.4 \end{array}$

30. $\begin{array}{r} 7053 \\ \times\ 0.34 \\ \hline 2398.02 \end{array}$

31. 7.86×10

78.6

32. 156.371×100

15,637.1

33. 0.000613×10^3

0.613

34. 1.2354×10^5

123,540

35. Hillary purchased 3.6 pounds of bananas at $0.25 per pound. How much did the purchase cost?

$0.90

3.5 *Divide until there is a remainder of zero.*

36. $0.4\overline{)9630}$ 24,075

37. $0.07\overline{)0.0001806}$ 0.00258

38. $5.2\overline{)191.36}$ 36.8

39. $8\overline{)1863.2}$ 232.9

40. Divide and round your answer to the nearest tenth.

$1.3\overline{)746.75}$ 574.4

41. Divide and round your answer to the nearest thousandth.

$0.06\overline{)0.003539}$ 0.059

42. My stereo cost $283.95 and it lasted for five years. I had it repaired once during the five years for $49.82. If I spread the total cost over five years, how much did it cost me per year for the stereo? (Round to the nearest cent.) $66.75

3.6 *Write each fraction as an equivalent decimal.*

43. $\frac{5}{18}$
$0.2\overline{7}$

44. $\frac{7}{40}$
0.175

45. $1\frac{5}{6}$
$1.8\overline{3}$

46. $\frac{15}{16}$
0.9375

Write each fraction as a decimal rounded to the nearest thousandth.

47. $\frac{11}{14}$
0.786

48. $2\frac{5}{17}$
2.294

Evaluate by doing the operations in proper order.

49. $1.6 \times 2.3 + 0.4 - 0.6 \times 0.8$
3.6

50. $2.5 \div 0.5 \times 3.2 + 1.8 - 3.7$
14.1

51. $0.03 + (1.2)^2 - 5.3 \times 0.06$
1.152

52. $(1.02)^3 + 5.76 \div 1.2 \times 0.05$
1.301208

53. $6.63 + 8.24 \div (5.76 - 5.68) - 22.5$
87.13

54. $(6 - 5.25)^2 \div 0.25 + 3.75$
6

Mixed Practice

Calculate each of the following.

55. 0.38×0.56
0.2128

56. $1.38 - 0.83$
0.55

57. $0.204 \div 0.34$
0.6

58. $8.03 + 0.562 + 4.53$
13.122

59. $2398.26 - 1959.07$
439.19

60. 67.036×0.006
0.402216

61. $0.061 + 0.0023 + 0.777$
0.8403

62. $110.72 \div 1.6$
69.2

63. $8 \div 0.4 + 0.1 \times (0.2)^2$
20.004

64. $(3.8 - 2.8)^3 \div (0.5 + 0.3)$
1.25

Applications

3.7 *Solve each problem.*

65. An office manager paid $18.50 for a desk pad set. She paid $29.95 for a lamp. She also paid $2.42 in sales tax for the two items. She took $60 from petty cash to make the purchase at a local supply store. How much will she have left after the purchase? $9.13

66. Terri had a balance of $62.36 in her checking account. She made deposits of $108.19 and $73.56. She made out checks for $102.14, $37.56, and $19.95. What will her new balance be? $84.46

67. Jill earns $6.88 per hour in her new job. She earns 1.5 times that amount for every hour over 40 hours she works per week. She worked 49 hours last week. How much did she earn? $368.08

68. Dan drove to the mountains. His odometer read 26,005.8 miles at the start, and 26,325.8 miles at the end of the trip. He used 12.9 gallons of gas on the trip. How many miles per gallon did his car get? (Round your answer to the nearest tenth.) 24.8 miles

69. Robert is considering buying a car and making installment payments of $189.60 for 48 months. The cash price of the car is $6930.50. How much extra does he pay if he uses the installment plan instead of buying the car with one payment? $2170.30

70. Mr. Zeno has a choice of working as an assistant manager at ABC Company at $315.00 per week or receiving an hourly salary of $8.26 per hour at a different company. He learned from several previous assistant managers that they usually worked 38 hours per week. At which company will he probably earn more money? He will earn more at ABC Company.

71. The EPA standard for safe drinking water is a maximum of 0.002 milligram of mercury in one liter of water. The town wells at Winchester were tested. The test was done on 12 liters of water. The entire 12-liter sample contained 0.03 milligram of mercury. Is the water safe or not? By how much? no; by 0.0005 mg per liter

72. A chemist wishes to take a sample of 0.322 liter of acid and place it in small test tubes each containing 0.046 liter. How many test tubes will be needed? 7 test tubes

Developing Your Study Skills

EXAM TIME: HOW TO REVIEW

Reviewing adequately for an exam enables you to bring together the concepts you have learned over several sections. For your review, you will need to:

1. Reread your textbook. Make a list of any terms, rules, or formulas you need to know for the exam. Be sure you understand them all.
2. Reread your notes. Go over returned homework and quizzes. Redo the problems you missed.
3. Practice some of each type of problem covered in the chapter(s) you are to be tested on.
4. Use the end-of-chapter materials provided in your textbook. Read carefully through the Chapter Organizer. Do the Review Problems. Take the Chapter Test. When you are finished, check your answers. Redo any problems you missed.
5. Get help if any concepts give you difficulty.

Chapter 3 Test

1. Write a word name for the decimal. 0.157

2. Write as a decimal. $\dfrac{3{,}977}{10{,}000}$

Write in fractional notation. Reduce whenever possible.

3. 7.15

4. 0.261

5. Arrange from smallest to largest.
   2.19, 2.91, 2.9, 2.907

6. Round to the nearest hundredth.
   78.6562

7. Round to the nearest ten thousandth.
   0.0341752

Add.

8. $\begin{array}{r} 42.718 \\ 0.191 \\ +\ 1.08 \\ \hline \end{array}$

9. $17 + 2.1 + 16.8 + 0.04 + 1.59$

Subtract.

10. $\begin{array}{r} 1.0075 \\ -\ 0.9096 \\ \hline \end{array}$

11. $72.3 - 1.145$

Multiply.

12. $\begin{array}{r} 8.31 \\ \times\ 0.07 \\ \hline \end{array}$

13. 2.189×100

1. One hundred fifty-seven thousandths
2. 0.3977
3. $7\frac{3}{20}$
4. $\frac{261}{1000}$
5. 2.19, 2.9, 2.907, 2.91
6. 78.66
7. 0.0342
8. 43.989
9. 37.53
10. 0.0979
11. 71.155
12. 0.5817
13. 218.9

Divide.

14. $0.004\overline{)0.1028}$

15. $0.69\overline{)32.43}$

Write as a decimal.

16. $\frac{11}{9}$

17. $\frac{9}{16}$

Perform the operations in the proper order to evaluate.

18. $(0.3)^3 + 1.02 \div 0.5 - 0.58$

19. $19.36 \div (0.24 + 0.26) \times (0.4)^2$

20. A beef roast weighing 7.8 pounds costs $4.25 per pound. How much will the roast cost?

21. Frank traveled from the city to the shore. His odometer read 42,620.5 miles at the start and 42,780.5 at the end of the trip. He used 8.5 gallons of gas. How many miles per gallon did his car achieve? Round to the nearest tenth.

22. The rainfall for March in Central City was 8.01 centimeters, for April, 5.03 centimeters, and for May, 8.53 centimeters. The normal rainfall for these three months is 25 centimeters. How much less rain fell during these 3 months than usual; that is, how does this year's figure compare with the figure for normal rainfall?

14. 25.7

15. 47

16. $1.\overline{2}$

17. 0.5625

18. 1.487

19. 6.1952

20. $33.15

21. 18.8 miles per gallon

22. 3.43 centimeters less

Cumulative Test for Chapters 1–3

Approximately one-half of this test is based on Chapter 3 material. The remainder is based on material covered in Chapters 1 and 2.

1. Write in words. 38,056,954

2. Add.
$$\begin{array}{r} 156{,}028 \\ 301{,}579 \\ +\ 21{,}980 \\ \hline \end{array}$$

3. Subtract.
$$\begin{array}{r} 1{,}091{,}000 \\ -\ 1{,}036{,}520 \\ \hline \end{array}$$

4. Multiply.
$$\begin{array}{r} 589 \\ \times\ 67 \\ \hline \end{array}$$

5. Divide. $17\overline{)4386}$

6. Evaluate. $20 \div 4 + 2^5 - 7 \times 3$

7. Reduce. $\frac{33}{88}$

8. Add. $4\frac{1}{3} + 3\frac{1}{6}$

9. Subtract. $\frac{23}{35} - \frac{2}{5}$

10. Multiply. $\frac{7}{10} \times \frac{5}{3}$

11. Divide. $52 \div 3\frac{1}{4}$

12. Divide. $1\frac{3}{8} \div \frac{5}{12}$

13. Estimate. 58,216 × 438,207

14. Write as a decimal. $\frac{571}{1000}$

15. Arrange from smallest to largest.
2.1, 20.1, 2.01, 2.12, 2.11

16. Round to the nearest thousandth.
26.07984

Answers

1. Thirty-eight million, fifty-six thousand, nine hundred fifty-four
2. 479,587
3. 54,480
4. 39,463
5. 258
6. 16
7. $\frac{3}{8}$
8. $7\frac{1}{2}$
9. $\frac{9}{35}$
10. $\frac{7}{6}$ or $1\frac{1}{6}$
11. 16
12. $\frac{33}{10}$ or $3\frac{3}{10}$
13. 24,000,000,000
14. 0.571
15. 2.01, 2.1, 2.11, 2.12, 20.1
16. 26.080

17. 19.54

18. 8.639

19. 1.136

20. 36,512.3

21. 1.058

22. 0.8125

23. 13.597

24. 456 miles

25. $195.57

17.
$$\begin{array}{r} 1.9 \\ 2.36 \\ 15.2 \\ +\ 0.08 \\ \hline \end{array}$$

18.
$$\begin{array}{r} 28.007 \\ -\ 19.368 \\ \hline \end{array}$$

19.
$$\begin{array}{r} 56.8 \\ \times\ 0.02 \\ \hline \end{array}$$

20. 365.123×100

21. $0.06\overline{)0.06348}$

22. Write as a decimal. $\dfrac{13}{16}$

23. Perform the operations in the proper order.

$$1.44 \div 0.12 + (0.3)^3 + 1.57$$

24. A car gets 28.5 miles per gallon. Its gas tank holds 16.0 gallons of gas. Approximately how many miles can the car travel on one tank of gas?

25. Sue's savings account balance is $199.36. This month she earned interest of $1.03. She deposited $166.35 and $93.50. She withdrew money three times, in the amounts for $90.00, $37.49, and $137.18. What will her balance be at the start of the next month?

CHAPTER 4

Ratio and Proportion

Have you ever lived in or visited a city that was so crowded that you could hardly move? Other cities and towns have many people, but somehow in them you never feel crowded.

Mathematics helps us to be more precise about this "crowded" feeling in a city. If you were to compare two cities, could you make some mathematical calculations to determine which of the two was more densely populated? Turn to the Putting Your Skills to Work on page 270.

Pretest Chapter 4

1.	$\frac{13}{18}$
2.	$\frac{1}{5}$
3.	$\frac{9}{2}$
4.	$\frac{11}{12}$
5. (a)	$\frac{7}{24}$
(b)	$\frac{11}{120}$
6.	$\frac{\$41}{12 \text{ cabinets}}$
7.	$\frac{31 \text{ gallons}}{42 \text{ square feet}}$
8.	$\frac{30.5 \text{ miles}}{1 \text{ hour}}$ or 30.5 miles/hour
9.	$\frac{\$37}{1 \text{ radio}}$ or \$37/radio
10.	$\frac{13}{40} = \frac{39}{120}$
11.	$\frac{42}{78} = \frac{21}{39}$
12.	true
13.	false

If you are familiar with the topics in this chapter, take this test now. Check your answers with those in the back of the book. If an answer was wrong or you couldn't do a problem, study the appropriate section of the chapter.

If you are not familiar with the topics in this chapter, don't take this test now. Instead, study the examples, work the practice problems, and then take the test.

This test will help you identify which concepts you have mastered and which you need to study further.

Section 4.1

Write each ratio *in simplest form.*

1. 13 to 18 **2.** 44 to 220

3. \$72 to \$16 **4.** 121 kilograms to 132 kilograms

5. Sam's take-home pay is \$240 per week. \$70 per week is withheld for federal taxes and \$22 per week is withheld for state taxes.
(a) Find the ratio of federal withholding to take-home pay.
(b) Find the ratio of state withholding to take-home pay.

Write each rate *in simplest form.*

6. \$164 for 48 cabinets

7. 620 gallons of water for each 840 square feet of lawn

Write as a unit rate. *Round to the nearest tenth if necessary.*

8. 122 miles are traveled in 4 hours. What is the rate in miles per hour?

9. 16 radios are purchased for \$592. What is the cost per radio?

Section 4.2

Write a proportion for each of the following.

10. 13 is to 40 as 39 is to 120

11. 42 is to 78 as 21 is to 39

Determine if each proportion is true or false.

12. $\frac{14}{31} = \frac{42}{93}$ **13.** $\frac{17}{33} = \frac{19}{45}$

Section 4.3

Solve for n in each equation.

14. $9 \times n = 153$ **15.** $234 = 13 \times n$

Solve for n in each proportion.

16. $\frac{36}{16} = \frac{9}{n}$ **17.** $\frac{3}{144} = \frac{n}{336}$

18. $\frac{n}{900} = \frac{15}{22.5}$

Section 4.4

Solve each problem using a proportion.

19. A recipe for six portions calls for 1.5 cups of flour. How many cups of flour are needed to make 14 portions?

20. Maria's car can travel 81 miles on 2 gallons of gas. How many can she travel on 9 gallons of gas?

21. Two cities are 5 inches apart on a map, but the actual distance between them is 365 miles. What is the actual distance between two other cities that are 2 inches apart on the map?

22. A shipment of 121 light bulbs had 6 defective bulbs. How many defective bulbs would we expect, at the same rate, in a shipment of 1089 light bulbs?

14. $n = 17$

15. $n = 18$

16. $n = 4$

17. $n = 7$

18. $n = 600$

19. 3.5 cups

20. 364.5 miles

21. 146 miles

22. 54 defective bulbs

4.1 Ratios and Rates

MathPro

Video 9

SSM

After studying this section, you will be able to:

1 *Use a ratio to compare two quantities with the same unit.*
2 *Use a rate to compare two quantities with different units.*

1 *Using a Ratio to Compare Two Quantities with the Same Unit*

Assume that you earn 13 dollars an hour and your friend earns 10 dollars per hour. The *ratio* 13:10 compares what you and your friend make. This ratio means that for every 13 dollars you earn your friend earns 10. The *rate* you are paid is 13 dollars per hour, which compares 13 dollars to 1 hour. In this section we see how to use both rates and ratios to solve many everyday problems.

Suppose that we want to compare an object weighing 20 pounds to an object weighing 23 pounds. The ratio of their weight would be 20 to 23. A **ratio** is the comparison of two quantities that have the *same units.*

TEACHING TIP You will need to remind students to reduce fractions to lowest terms when expressing ratios. They often forget this step or reduce only partially.

We can express the ratio three ways.

We can write "the ratio of 20 to 23."
We can write "20:23" using a colon.
We can write "$\frac{20}{23}$" using a fraction.

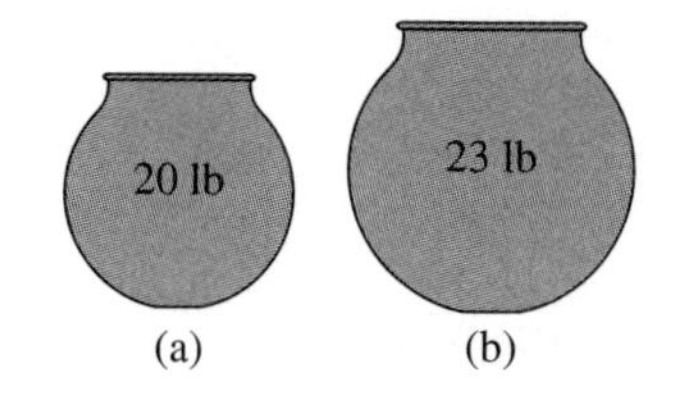

(a) (b)

All are valid ways to *compare* 20 to 23. Each is read as "20 to 23."

> We always want to write a ratio in simplest form. A ratio is in *simplest form* when the two numbers do not have a common factor.

TEACHING TIP Some students will ask why the fractional form of the ratio is emphasized in this text. They may wonder why the colon 2:3 ratio is not used more often. Explain that the fractional form is the most useful form to use when solving ratio or proportion problems.

EXAMPLE 1 Write each ratio in simplest form. Express your answer as a fraction.

(a) The ratio of 15 hours to 20 hours
(b) The ratio of 36 hours to 30 hours
(c) 125:150

(a) $\frac{15}{20} = \frac{3}{4}$ **(b)** $\frac{36}{30} = \frac{6}{5}$ **(c)** $\frac{125}{150} = \frac{5}{6}$

Notice that in each case the two numbers *do* have a common factor. When we form the fraction—that is, the ratio—we take the extra step of *reducing* the fraction.

Practice Problem 1 Write each ratio in simplest form. Express your answer as a fraction.

(a) The ratio of 36 feet to 40 feet
(b) The ratio of 18 feet to 15 feet
(c) 220:270 ■

EXAMPLE 2 Martin earns \$350 weekly. However, he takes home only \$250 per week in his paycheck.

\$350.00	gross pay (what Martin earns)	
45.00	withheld for federal tax	(what is taken out of Martin's earnings)
20.00	withheld for state tax	
35.00	withheld for retirement	
\$250.00	take-home pay (what Martin has left)	

(a) What is the ratio of the amount withheld for federal tax to gross pay?

(b) What is the ratio of the amount withheld for state tax to the amount withheld for federal tax?

(a) The ratio of the amount withheld for federal tax compared to gross pay is

$$\frac{45}{350} = \frac{9}{70}$$

(b) The ratio of the amount withheld for state tax compared to the amount withheld for federal tax is

$$\frac{20}{45} = \frac{4}{9}$$

Practice Problem 2 Professor Fowler has 90 students in his course on introductory psychology. The class has 18 freshman students, 25 sophomore students, 36 junior students, and 11 senior students.

(a) What is the ratio of freshman students to junior students?

(b) What is the ratio of sophomore students to the entire number of students in the class? ■

To Think About Perhaps you have heard statements like "a certain jet plane travels at Mach 2.2." What does that mean? A Mach number is a ratio that compares the velocity (speed) of an object to the velocity of sound. Sound travels at about 330 meters per second. This ratio is written in decimal form.

What is the Mach number of a jet traveling at 690 meters per second?

$$\text{Mach number of jet} = \frac{690 \text{ meters per second}}{330 \text{ meters per second}}$$

$$= 2.09090909\ldots \text{ or } 2.\overline{09}$$

Rounded to the nearest tenth, the Mach number of the jet is 2.1.

Problems 71 and 72 in Exercises 4.1 deal with Mach numbers.

2 *Using a Rate to Compare Two Quantities with Different Units*

A **rate** is a comparison of two quantities with *different units.* Usually, to avoid misunderstanding, we express a rate as a fraction with the units included.

EXAMPLE 3 John's car traveled 371 miles on 14 gallons of gasoline. What is the rate of miles to gallons?

The rate is

$$\frac{371 \text{ miles}}{14 \text{ gallons}} = \frac{53 \text{ miles}}{2 \text{ gallons}}$$

Practice Problem 3 A farmer is charged a $44 storage fee for every 900 tons of grain he stores. What is the rate of the storage fee in dollars to tons of grain? ■

Often we want to know the rate for a single unit, which is the unit rate. A **unit rate** is a rate in which the denominator is the number 1.

EXAMPLE 4 A car traveled 301 miles in 7 hours. Find the unit rate.

$\frac{301}{7}$ can be simplified. We divide $301 \div 7 = 43$.

Thus

$$\frac{301 \text{ miles}}{7 \text{ hours}} = \frac{43 \text{ miles}}{1 \text{ hour}}$$

The denominator is 1. We write our answer as 43 miles/hour. The fraction line is read as the word "per," so our answer here is read "43 miles per hour." "Per" means "for every," so a rate of 43 miles per hour means 43 miles traveled for every hour traveled.

Practice Problem 4 A car traveled 212 miles in 4 hours. Find the unit rate. ■

EXAMPLE 5 A grocer purchased 200 pounds of apples for $68. He sold the 200 pounds of apples for $86. How much profit did he make per pound of apples?

$$\begin{array}{rl} \$86 & \text{selling price} \\ -\ 68 & \text{cost} \\ \hline \$18 & \text{profit} \end{array}$$

The rate that compares profit to pounds of apples sold is $\frac{18 \text{ dollars}}{200 \text{ pounds}}$. We will divide $18 \div 200$.

$$\begin{array}{r} 0.09 \\ 200\overline{)18.00} \\ 18\ 00 \\ \hline 0 \end{array}$$

The unit rate of profit is $0.09 per pound.

Practice Problem 5 A retailer purchased 120 nickel-cadmium batteries for flashlights for $129.60. She sold them for $170.40. What was her profit per battery? ■

EXAMPLE 6 Hamburger at a local butcher is packaged in large and extra-large packages. A large package costs $7.86 for 6 pounds and an extra-large package is $10.08 for 8 pounds.

(a) What is the unit rate in dollars per pound for each size package?

(b) How much per pound does a consumer save by buying the extra-large package?

(a) $\frac{7.86 \text{ dollars}}{6 \text{ pounds}} = \$1.31/\text{pound}$ for the large package

$\frac{10.08 \text{ dollars}}{8 \text{ pounds}} = \$1.26/\text{pound}$ for the extra-large package

(b)
$$\begin{array}{r} \$1.31 \\ -\ 1.26 \\ \hline \$0.05 \end{array}$$
A person saves $0.05/pound by buying the extra-large package.

TEACHING TIP Ask students if they are familiar with unit pricing in a supermarket. Explain how unit pricing helps them to decide which package is the best buy. (The largest size is not always the best buy.)

Practice Problem 6 A 12-ounce package of Fred's favorite cereal costs $2.04. A 20-ounce package of the same cereal costs $2.80.

(a) What is the cost per ounce of each size of cereal?

(b) How much per ounce would Fred save by buying the larger size? ■

4.1 Exercises

Verbal and Writing Skills

1. A ___ratio___ is a comparison of two quantities that have the same unit.
2. A rate is a comparison of two quantities that have ___different___ units.
3. The ratio 5:8 is read ___5 to 8___.
4. Marion compares the number of loaves of bread she bakes to the number of pounds of flour she needs to make the bread. Is this a ratio or a rate? Why?
   This is a rate because it compares different units: loaves of bread to pounds of flour.

Write each ratio in simplest form. Express your answer as a fraction.

5. 18:24 $\frac{3}{4}$
6. 15:12 $\frac{5}{4}$
7. 21:18 $\frac{7}{6}$
8. 42:51 $\frac{14}{17}$
9. 55:121 $\frac{5}{11}$
10. 78:104 $\frac{3}{4}$
11. 150:225 $\frac{2}{3}$
12. 360:480 $\frac{3}{4}$
13. 60 to 64 $\frac{15}{16}$
14. 33 to 57 $\frac{11}{19}$
15. 28 to 42 $\frac{2}{3}$
16. 21 to 98 $\frac{3}{14}$
17. 32 to 20 $\frac{8}{5}$
18. 90 to 54 $\frac{5}{3}$
19. 8 ounces to 12 ounces $\frac{2}{3}$
20. 24 minutes to 52 minutes $\frac{6}{13}$
21. 16 feet to 24 feet $\frac{2}{3}$
22. 20 miles to 45 miles $\frac{4}{9}$
23. 50 years to 85 years $\frac{10}{17}$
24. 39 kilograms to 26 kilograms $\frac{3}{2}$
25. $86 to $120 $\frac{43}{60}$
26. $54 to $63 $\frac{6}{7}$
27. 153 inches to 17 inches $\frac{9}{1}$
28. 143 feet to 13 feet $\frac{11}{1}$
29. 91 tons to 133 tons $\frac{13}{19}$
30. 99 pounds to 143 pounds $\frac{9}{13}$

Use the following table to answer problems 31–34.

ROBIN'S WEEKLY PAYCHECK

Total (Gross) Pay	Federal Withholding	State Withholding	Retirement	Insurance	Take-Home Pay
$285	$35	$20	$28	$16	$165

31. What is the ratio of take-home pay to total (gross) pay?
    $\frac{165}{285} = \frac{11}{19}$
32. What is the ratio of retirement to insurance?
    $\frac{28}{16} = \frac{7}{4}$
33. What is the ratio of federal withholding to take-home pay?
    $\frac{35}{165} = \frac{7}{33}$
34. What is the ratio of retirement to total (gross) pay?
    $\frac{28}{285}$

An automobile insurance company prepared the following analysis for its clients. Use this table for problems 35–38.

Analysis of number of years that four-door sedans are driven

Sedans that lasted 2 years or less	Sedans that lasted 4 years or less but more than 2 years	Sedans that lasted 6 years or less but more than 4 years	Sedans that lasted more than 6 years	Total number of sedans
205	255	450	315	1225

35. What is the ratio of sedans that lasted two years or less to the total number of sedans?

$\frac{205}{1225} = \frac{41}{245}$

36. What is the ratio of sedans that lasted more than six years to the total number of sedans?

$\frac{315}{1225} = \frac{9}{35}$

37. What is the ratio of the number of sedans that lasted six years or less but more than four years to the number of sedans that lasted two years or less?

$\frac{450}{205} = \frac{90}{41}$

38. What is the ratio of the number of sedans that lasted more than six years to the number of sedans that lasted four years or less but more than two years?

$\frac{315}{255} = \frac{21}{17}$

39. A builder constructs a new home for $128,000. His labor costs were $72,000. What is the ratio of labor cost to total cost?

$\frac{72{,}000}{128{,}000} = \frac{9}{16}$

40. A company produces a new-model television set for a total cost of $405.00. The research and development cost for each set is $90. What is the ratio of research and development cost to the total cost?

$\frac{90}{405} = \frac{2}{9}$

Write as a rate in simplest form.

41. $60 for 18 plants

$\frac{\$10}{3 \text{ plants}}$

42. $135 for 20 cabinets

$\frac{\$27}{4 \text{ cabinets}}$

43. $170 for 12 bushes

$\frac{\$85}{6 \text{ bushes}}$

44. 98 pounds for 22 people

$\frac{49 \text{ pounds}}{11 \text{ people}}$

45. 310 gallons of water for every 625 square feet of lawn

$\frac{62 \text{ gallons}}{125 \text{ sq ft}}$

46. 460 gallons of water for every 825 square feet of lawn

$\frac{92 \text{ gallons}}{165 \text{ sq ft}}$

47. 6150 revolutions for every 15 miles

$\frac{410 \text{ revolutions}}{1 \text{ mile}}$ *or* 410 rev/mile

48. 9540 revolutions for every 18 miles

$\frac{530 \text{ revolutions}}{1 \text{ mile}}$ *or* 530 rev/mile

49. Snowplow 18 miles of road every 8 hours

$\frac{9 \text{ miles}}{4 \text{ hours}}$

50. 48 pies for 340 people

$\frac{12 \text{ pies}}{85 \text{ people}}$

Write as a unit rate. Round to the nearest tenth when necessary.

51. Earn \$520 in 40 hours

$\frac{\$520}{40 \text{ hours}} = \$13/\text{hr}$

52. Earn \$304 in 38 hours

$\frac{\$304}{38 \text{ hours}} = \$8/\text{hr}$

53. Travel 192 miles on 12 gallons of gas

$\frac{192 \text{ miles}}{12 \text{ gallons}} = 16 \text{ mi/gal}$

54. Travel 322 miles on 14 gallons of gas

$\frac{322 \text{ miles}}{14 \text{ gallons}} = 23 \text{ mi/gal}$

55. Pump 2480 gallons in 16 hours

155 gal/hr

56. Pump 2880 gallons in 24 hours

120 gal/hr

57. 2760 words on 12 pages

$\frac{2760 \text{ words}}{12 \text{ pages}} = 230 \text{ words/page}$

58. 1935 words on 9 pages

$\frac{3440 \text{ words}}{16 \text{ pages}} = 215 \text{ words/page}$

59. Travel 3619 kilometers in 7 hours

$\frac{3619 \text{ kilometers}}{7 \text{ hours}} = 517 \text{ km/hr}$

60. Travel 2992 kilometers in 11 hours

$\frac{2992 \text{ kilometers}}{11 \text{ hours}} = 272 \text{ km/hr}$

Applications

61. \$3870 was spent for 129 shares of Mattel stock. Find the cost per share. \$30/share

62. \$6150 was spent for 150 shares of Polaroid stock. Find the cost per share. \$41/share

63. A jeweler purchased 245 watches for \$6370. She sold them for \$9800. How much profit did she make per watch? \$14 profit per watch

64. A retailer purchased 24 compact disc players for \$3380. She sold them for \$4030. How much profit did she make per compact disc player? Round your answers to the nearest dollar.

Profit = 4030 − 3380 = \$650 $\frac{\$650}{24 \text{ players}} = \$27/\text{player}$

65. A 16-ounce box of dry pasta costs \$1.28. A 24-ounce box of the same pasta costs \$1.68.

(a) What is the cost of each box of pasta per ounce? \$0.08/oz small box, \$0.07/oz large box

(b) How much does the educated consumer save by buying the larger box?

1¢ per ounce or \$0.01 per ounce

(c) How much does the consumer save by buying 2 large boxes instead of 3 small boxes?

The consumer saves \$0.48

66. A 16-ounce can of beef stew costs \$2.88. A 26-ounce can of the same beef stew costs \$4.16.

(a) What is the cost of each can of stew per ounce?

$\frac{\$2.88}{16 \text{ ounces}} = \$0.18/\text{oz}$ $\frac{\$4.16}{26 \text{ ounces}} = \$0.16/\text{oz}$

(b) How much does the consumer save per ounce by buying the larger can?

2¢ per ounce *or* \$0.02/oz

67. Dr. Robert Tobey completed a count of two herds of moose in the central regions of Alaska. He recorded 3978 moose on the North Slope and 5520 moose on the South Slope. There are 306 acres on the North Slope and 460 acres on the South Slope.

(a) How many moose per acre were found on the North Slope? 13 moose

(b) How many moose per acre were found on the South Slope? 12 moose

(c) In which region are the moose more closely crowded together? North Slope

68. In Melbourne, Australia, 27,900 people live in the suburb of St. Kilda and 38,700 live in the suburb of Caulfield. The area of St. Kilda is 6500 acres. The area of Caulfield is 9200 acres. Round your answers to the nearest tenth.

(a) How many people per acre live in St. Kilda? 4.3 people per acre

(b) How many people per acre live in Caulfield? 4.2 people per acre

(c) Which suburb is more crowded? St. Kilda

★**69.** Mr. Jenson bought 525 shares of Zenith stock for \$12,876.50. How much did he pay per share? Round your answer to the nearest cent. \$24.53

★**70.** A Kentucky horse breeder bought 59 horses for \$4,350,500. How much did she pay per horse? Round to the nearest cent. \$73,737.29

To Think About

For Problems 71 and 72, recall that the speed of sound is about 330 meters per second. Round your answers to the nearest tenth.

71. A jet plane was originally designed to fly at 750 meters per second. It was modified to fly at 810 meters per second. By how much was its Mach number increased?

$$\frac{750 \text{ meters per second}}{330 \text{ meters per second}} = \text{Mach } 2.3$$

$$\frac{810 \text{ meters per second}}{330 \text{ meters per second}} = \text{Mach } 2.5$$

Increased by Mach 0.2

72. A rocket was first flown at 1960 meters per second. It proved unstable and unreliable at that speed. It is now flown at a maximum of 1920 meters per second. By how much was its Mach number decreased?

$$\frac{1960 \text{ meters per second}}{330 \text{ meters per second}} = \text{Mach } 5.9$$

$$\frac{1920 \text{ meters per second}}{330 \text{ meters per second}} = \text{Mach } 5.8$$

Decreased by Mach 0.1

Cumulative Review Problems

Calculate.

73. $2\frac{1}{4} + \frac{3}{8}$

$2\frac{5}{8}$

74. $\frac{5}{7} \div \frac{3}{21}$

5

75. $\frac{8}{23} \times \frac{5}{16}$

$\frac{5}{46}$

76. $3\frac{1}{16} - 2\frac{1}{24}$

$1\frac{1}{48}$

77. A room 12 yards × 5.2 yards had a carpet installed. The bill was \$764.40. What was the cost of the installed carpet per yard?

$12 \times 5.2 = 62.4$ sq yd $\quad \frac{\$764.40}{62.4 \text{ sq yd}} = \$12.25/\text{sq yd}$

78. An electronics superstore bought 1050 computer games for \$23 each. How much did the store pay in all for these games? The store sold the games for \$39 each. How much profit did the store make?

The store paid \$24,150 for the games. They made a profit of \$16,800.

4.2 The Concept of Proportions

MathPro

Video 9

SSM

After studying this section, you will be able to:

1. *Write a proportion.*
2. *Determine if a statement is a proportion.*

1 *Writing a Proportion*

A **proportion** states that two ratios or two rates are equal. For example, $\frac{5}{8} = \frac{15}{24}$ is a proportion and $\frac{7 \text{ feet}}{8 \text{ dollars}} = \frac{35 \text{ feet}}{40 \text{ dollars}}$ is also a proportion. A proportion can be read two ways. The proportion $\frac{5}{8} = \frac{15}{24}$ can be read "five eighths equals fifteen twenty-fourths," or it can be read "five *is to* eight *as* fifteen *is to* twenty-four."

EXAMPLE 1 Write the proportion 5 is to 7 as 15 is to 21.

$$\frac{5}{7} = \frac{15}{21}$$

Practice Problem 1 Write the proportion 6 is to 8 as 9 is to 12. ■

TEACHING TIP Doing simple proportions like Example 2 will help students find the applications in Section 4.4 less difficult.

EXAMPLE 2 Write a proportion to express the following: If 4 rolls of wallpaper measure 300 feet, then 8 rolls of wallpaper will measure 600 feet.

When you write a proportion, order is important. Be sure that the units for each ratio are in the same position in the fraction.

$$\frac{4 \text{ rolls}}{300 \text{ feet}} = \frac{8 \text{ rolls}}{600 \text{ feet}}$$

Practice Problem 2 Write a proportion to express the following: If it takes 2 hours to drive 72 miles, then it will take 3 hours to drive 108 miles. ■

2 *Determining If a Statement Is a Proportion*

By definition, a proportion states that two ratios are equal. $\frac{2}{7} = \frac{4}{14}$ is a proportion because $\frac{2}{7}$ and $\frac{4}{14}$ are equivalent fractions. You might say that $\frac{2}{7} = \frac{4}{14}$ is a *true* proportion. It is easy enough to see that $\frac{2}{7} = \frac{4}{14}$ is true. However, is $\frac{4}{14} = \frac{6}{21}$ true? Is $\frac{4}{14} = \frac{6}{21}$ a proportion? To determine whether or not a statement is a proportion, we use the equality test for fractions.

Equality Test for Fractions

For any two fractions where $b \neq 0$, $d \neq 0$

$$\text{If } \frac{a}{b} = \frac{c}{d}, \text{ then } a \times d = b \times c$$

Thus, to see if $\frac{4}{14} = \frac{6}{21}$, we can multiply

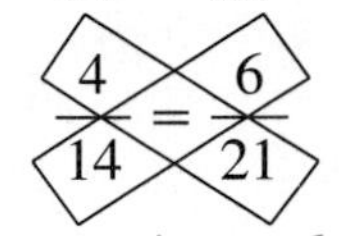

$14 \times 6 = 84$
$4 \times 21 = 84$ — Products are equal.

$\frac{4}{14} = \frac{6}{21}$ is true. $\frac{4}{14} = \frac{6}{21}$ is a proportion.

EXAMPLE 3 Determine which of the following is a proportion.

(a) $\frac{14}{18} \stackrel{?}{=} \frac{35}{45}$ **(b)** $\frac{16}{21} \stackrel{?}{=} \frac{174}{231}$

(a) $\frac{14}{18} \stackrel{?}{=} \frac{35}{45}$

$18 \times 35 = 630$

$14 \times 45 = 630$

products equal. Thus $\frac{14}{18} = \frac{35}{45}$. This is a proportion.

(b) $\frac{16}{21} \stackrel{?}{=} \frac{174}{231}$

$21 \times 174 = 3654$

$16 \times 231 = 3696$

products not equal. Thus $\frac{16}{21} \neq \frac{174}{231}$. This is *not* a proportion.

TEACHING TIP Remind students that they must complete the whole product on each side to verify that a proportion is true. However, to determine if a proportion is false, they can sometimes just check the final digits of the product. If these do not match, the proportion is false. Show them this new example.

$$\frac{19}{23} = \frac{26}{29}$$

Explain that since the final digit of the cross products is not the same, the entire products are different and the proportion is false. 551 is not equal to 598 simply because the final digit 1 is not equal to the final digit 8.

Practice Problem 3 Determine which of the following is a proportion.

(a) $\frac{10}{18} \stackrel{?}{=} \frac{25}{45}$ **(b)** $\frac{42}{100} \stackrel{?}{=} \frac{22}{55}$ ■

Proportions may involve fractions or decimals.

EXAMPLE 4 Determine which of the following is a proportion.

(a) $\frac{5.5}{7} \stackrel{?}{=} \frac{33}{42}$ **(b)** $\frac{5}{8\frac{3}{4}} \stackrel{?}{=} \frac{40}{72}$

(a) $\frac{5.5}{7} \stackrel{?}{=} \frac{33}{42}$

$7 \times 33 = 231$

$5.5 \times 42 = 231$

products equal. Thus $\frac{5.5}{7} = \frac{33}{42}$. This is a proportion.

(b) $\dfrac{5}{8\frac{3}{4}} \stackrel{?}{=} \dfrac{40}{72}$. To multiply $8\frac{3}{4} \times 40$ we use

$$\frac{35}{\not{4}_1} \times \overset{10}{\not{40}} = 35 \times 10 = 350$$

Thus

$$\frac{5}{8\frac{3}{4}} \times \frac{40}{72} \qquad 8\frac{3}{4} \times 40 = 350 \quad \text{products not equal} \quad 5 \times 72 = 360$$

Thus $\dfrac{5}{8\frac{3}{4}} \neq \dfrac{40}{72}$. This is *not* a proportion.

Practice Problem 4

Determine which of the following is a proportion.

(a) $\dfrac{2.4}{3} \stackrel{?}{=} \dfrac{12}{15}$ **(b)** $\dfrac{2\frac{1}{3}}{6} \stackrel{?}{=} \dfrac{14}{38}$ ■

EXAMPLE 5

(a) Is the rate $\dfrac{\$86}{13 \text{ tons}}$ equal to the rate $\dfrac{\$79}{12 \text{ tons}}$?

(b) Is the rate $\dfrac{3 \text{ American dollars}}{2 \text{ British pounds}}$ equal to the rate $\dfrac{27 \text{ American dollars}}{18 \text{ British pounds}}$?

(a) We want to know whether $\dfrac{86}{13} \stackrel{?}{=} \dfrac{79}{12}$.

$$\frac{86}{13} \times \frac{79}{12} \qquad 13 \times 79 = 1027 \quad \text{products not equal} \quad 86 \times 12 = 1032$$

Thus the two rates are not equal. This is not a proportion.

(b) We want to know whether $\dfrac{3}{2} \stackrel{?}{=} \dfrac{27}{18}$.

$$\frac{3}{2} \times \frac{27}{18} \qquad 2 \times 27 = 54 \quad \text{products equal} \quad 3 \times 18 = 54$$

Thus the two rates are equal. This is a proportion.

Practice Problem 5

(a) Is the rate $\dfrac{1260 \text{ words}}{7 \text{ pages}}$ equal to the rate $\dfrac{3530 \text{ words}}{20 \text{ pages}}$?

(b) Is the rate $\dfrac{2 \text{ American dollars}}{11 \text{ French francs}}$ equal to the rate $\dfrac{16 \text{ American dollars}}{88 \text{ French francs}}$? ■

4.2 Exercises

Verbal and Writing Skills

1. A proportion states that two ratios or rates are ___equal___.

2. Explain in your own words how we use the equality test for fractions to determine if a statement is a proportion. Give an example. Answers may vary. Check explanations and examples for accuracy.

Write a proportion for each.

3. 48 is to 32 as 3 is to 2.
$\frac{48}{32} = \frac{3}{2}$

4. 40 is to 36 as 10 is to 9.
$\frac{40}{36} = \frac{10}{9}$

5. 8 is to 3 as 32 is to 12.
$\frac{8}{3} = \frac{32}{12}$

6. 32 is to 48 as 18 is to 27.
$\frac{32}{48} = \frac{18}{27}$

7. 20 is to 36 as 5 is to 9.
$\frac{20}{36} = \frac{5}{9}$

8. 42 is to 48 as 14 is to 16.
$\frac{42}{48} = \frac{14}{16}$

9. 27 is to 15 as 9 is to 5.
$\frac{27}{15} = \frac{9}{5}$

10. 42 is to 28 as 6 is to 4.
$\frac{42}{28} = \frac{6}{4}$

11. 22 is to 30 as 11 is to 15.
$\frac{22}{30} = \frac{11}{15}$

12. 75 is to 100 as 9 is to 12.
$\frac{75}{100} = \frac{9}{12}$

13. 45 is to 135 as 9 is to 27.
$\frac{45}{135} = \frac{9}{27}$

14. 44 is to 121 as 8 is to 22.
$\frac{44}{121} = \frac{8}{22}$

15. 5.5 is to 10 as 11 is to 20.
$\frac{5.5}{10} = \frac{11}{20}$

16. 6.5 is to 14 as 13 is to 28.
$\frac{6.5}{14} = \frac{13}{28}$

Applications

Write a proportion for each.

17. If 12 pounds of flour cost \$4, then 33 pounds will cost \$11.
$\frac{12 \text{ pounds}}{\$4} = \frac{33 \text{ pounds}}{\$11}$

18. If we travel 110 miles on Interstate I-90 in 2 hours, then we will travel 165 miles in 3 hours.
$\frac{110 \text{ miles}}{2 \text{ hours}} = \frac{165 \text{ miles}}{3 \text{ hours}}$

19. If Ty hit 10 home runs in 45 baseball games, then he should hit 36 home runs in 162 games.
$\frac{10 \text{ runs}}{45 \text{ games}} = \frac{36 \text{ runs}}{162 \text{ games}}$

20. If Amelia drove 450 miles in her Toyota Camry on 25 gallons of gas, then she should drive 1080 miles on 60 gallons of gas.
$\frac{450 \text{ miles}}{25 \text{ gallons}} = \frac{1080 \text{ miles}}{60 \text{ gallons}}$

21. If 20 pounds of pistachio nuts cost \$75, then 30 pounds will cost \$112.50.
$\frac{20 \text{ pounds}}{\$75} = \frac{30 \text{ pounds}}{\$112.50}$

22. If 3 credit hours at El Paso Community College costs \$525, then 7 credit hours should cost \$1225.
$\frac{3 \text{ hours}}{\$525} = \frac{7 \text{ hours}}{\$1225}$

23. Last year the Little League Baseball finals held 2200 people on 100 benches. This year we will have 2750 on 125 benches.

$$\frac{2200 \text{ people}}{100 \text{ benches}} = \frac{2750 \text{ people}}{125 \text{ benches}}$$

24. There are 3 teaching assistants for every 40 children in the elementary school. If we have 280 students, then we will have 21 teaching assistants.

$$\frac{3 \text{ teaching assistants}}{40 \text{ children}} = \frac{21 \text{ teaching assistants}}{280 \text{ children}}$$

25. If 16 pounds of fertilizer covers 1520 square feet of lawn, then 19 pounds of fertilizer should cover 1805 square feet of lawn.

$$\frac{16 \text{ pounds}}{1520 \text{ square feet}} = \frac{19 \text{ pounds}}{1805 \text{ square feet}}$$

26. When New City had 4800 people, it had 3 restaurants. Now New City has 11,200 people, so it should have 7 restaurants.

$$\frac{4800 \text{ people}}{3 \text{ restaurants}} = \frac{11{,}200 \text{ people}}{7 \text{ restaurants}}$$

Determine which of the following is a proportion.

27. $\frac{21}{35} \stackrel{?}{=} \frac{15}{25}$

$21 \times 25 \stackrel{?}{=} 35 \times 15$
$525 = 525$ True

28. $\frac{12}{42} \stackrel{?}{=} \frac{10}{35}$

$12 \times 35 \stackrel{?}{=} 42 \times 10$
$420 = 420$ True

29. $\frac{18}{12} \stackrel{?}{=} \frac{42}{28}$

$18 \times 28 \stackrel{?}{=} 12 \times 42$
$504 = 504$ True

30. $\frac{14}{10} \stackrel{?}{=} \frac{56}{40}$

$14 \times 40 \stackrel{?}{=} 10 \times 56$
$560 = 560$ True

31. $\frac{48}{56} \stackrel{?}{=} \frac{40}{45}$

$48 \times 45 \stackrel{?}{=} 56 \times 40$
$2160 \neq 2240$ False

32. $\frac{27}{45} \stackrel{?}{=} \frac{22}{40}$

$27 \times 40 \stackrel{?}{=} 45 \times 22$
$1080 \neq 990$ False

33. $\frac{9}{12} \stackrel{?}{=} \frac{15}{20}$

$9 \times 20 = 12 \times 15$
$180 = 180$ True

34. $\frac{8}{9} \stackrel{?}{=} \frac{48}{54}$

$8 \times 54 \stackrel{?}{=} 9 \times 48$
$432 = 432$ True

35. $\frac{99}{100} \stackrel{?}{=} \frac{49}{50}$

$99 \times 50 = 100 \times 49$
$4950 \neq 4900$ False

36. $\frac{17}{75} \stackrel{?}{=} \frac{22}{100}$

$17 \times 100 \stackrel{?}{=} 75 \times 22$
$1700 \neq 1650$ False

37. $\frac{315}{2100} \stackrel{?}{=} \frac{15}{100}$

$315 \times 100 = 2100 \times 15$
$31{,}500 = 31{,}500$ True

38. $\frac{102}{120} \stackrel{?}{=} \frac{85}{100}$

$102 \times 100 \stackrel{?}{=} 120 \times 85$
$10{,}200 = 10{,}200$ True

39. $\frac{6}{14} \stackrel{?}{=} \frac{4.5}{10.5}$

$6 \times 10.5 \stackrel{?}{=} 14 \times 4.5$
$63 = 63$ True

40. $\frac{7}{9} \stackrel{?}{=} \frac{10.5}{13.5}$

$7 \times 13.5 = 9 \times 10.5$
$94.5 = 94.5$ True

41. $\frac{11}{12} \stackrel{?}{=} \frac{9.5}{10}$

$11 \times 10 \stackrel{?}{=} 12 \times 9.5$
$110 \neq 114$ False

42. $\frac{3}{17} \stackrel{?}{=} \frac{4.5}{24.5}$

$3 \times 24.5 \stackrel{?}{=} 17 \times 4.5$
$73.5 \neq 76.5$ False

43. $\frac{2}{4\frac{1}{3}} \stackrel{?}{=} \frac{6}{13}$

$2 \times 13 \stackrel{?}{=} 6 \times 4\frac{1}{3}$
$26 = 26$ True

44. $\frac{2\frac{1}{3}}{3} \stackrel{?}{=} \frac{7}{15}$

$2\frac{1}{3} \times 15 \stackrel{?}{=} 3 \times 7$
$35 \neq 21$ False

45. $\frac{8}{19} \stackrel{?}{=} \frac{2}{4\frac{3}{4}}$

$8 \times 4\frac{3}{4} \stackrel{?}{=} 19 \times 2$
$38 = 38$ True

46. $\frac{9}{22} \stackrel{?}{=} \frac{3}{7\frac{1}{3}}$

$9 \times 7\frac{1}{3} \stackrel{?}{=} 22 \times 3$
$66 = 66$ True

47. $\frac{55 \text{ feet}}{4 \text{ rolls}} \stackrel{?}{=} \frac{220 \text{ feet}}{16 \text{ rolls}}$

$55 \times 16 = 4 \times 220$
$880 = 880$ True

48. $\frac{68 \text{ feet}}{3 \text{ rolls}} \stackrel{?}{=} \frac{204 \text{ feet}}{9 \text{ rolls}}$

$68 \times 9 \stackrel{?}{=} 3 \times 204$
$612 = 612$ True

49. $\frac{286 \text{ gallons}}{12 \text{ acres}} \stackrel{?}{=} \frac{429 \text{ gallons}}{18 \text{ acres}}$

$286 \times 18 \stackrel{?}{=} 12 \times 429$
$5148 = 5148$ True

50. $\frac{166 \text{ gallons}}{14 \text{ acres}} \stackrel{?}{=} \frac{249 \text{ gallons}}{21 \text{ acres}}$

$166 \times 21 \stackrel{?}{=} 14 \times 249$
$3486 = 3486$ True

51. $\frac{48 \text{ points}}{56 \text{ games}} \stackrel{?}{=} \frac{40 \text{ points}}{45 \text{ games}}$

$48 \times 45 \stackrel{?}{=} 56 \times 40$
$2160 \neq 2240$ False

52. $\frac{27 \text{ points}}{45 \text{ games}} \stackrel{?}{=} \frac{22 \text{ points}}{40 \text{ games}}$

$27 \times 40 \stackrel{?}{=} 45 \times 22$
$1080 \neq 990$ False

Applications

53. At the Amy Grant concert on Friday there were 9600 female fans and 8200 male fans. The concert on Saturday had 12,480 female fans and 10,660 male fans. Is the ratio of female fans to male fans the same for each night of the concert? Yes

54. In the beginner scuba diving course, there are 96 male students and 54 female students. In the intermediate scuba diving course, there are 144 male students and 81 female students. Is the ratio of male students to female students the same in each scuba class? Yes

55. A machine folds 730 boxes in 6 hours. Another machine folds 1090 boxes in 9 hours. Do they fold boxes at the same rate?

$\frac{730 \text{ boxes}}{6 \text{ hours}} \stackrel{?}{=} \frac{1090}{9 \text{ hours}}$

$6570 \neq 6540$ No

56. A car traveled 550 miles in 15 hours. A bus traveled 230 miles in 6 hours. Did they travel at the same rate?

$\frac{550 \text{ miles}}{15 \text{ hours}} \stackrel{?}{=} \frac{230 \text{ miles}}{6 \text{ hours}}$

$3300 \neq 3450$ No

To Think About

57. Determine if $\frac{63}{161} = \frac{171}{437}$.

(a) By reducing each side to lowest terms.

$\frac{63}{161} = \frac{9}{23}$ $\frac{171}{437} = \frac{9}{23}$ True

(b) By using the equality test for fractions.

$63 \times 437 \stackrel{?}{=} 161 \times 171$

$27{,}531 = 27{,}531$ True

(c) Which method was faster? Why?

For most students it is faster to multiply than to reduce fractions.

58. Determine if $\frac{169}{221} = \frac{247}{323}$.

(a) By reducing each side to lowest terms.

$\frac{169}{221} = \frac{13}{17}$ $\frac{247}{323} = \frac{13}{17}$ True

(b) By using the equality test for fractions.

$169 \times 323 \stackrel{?}{=} 221 \times 247$

$54{,}587 = 54{,}587$ True

(c) Which method was faster? Why?

For most students it is faster to multiply than to reduce fractions.

Cumulative Review Problems

Perform each calculation.

59. $9.6 + 7.8 + 2.56 + 3.004 + 0.1765$ 23.1405

60. 2.83×5.002 14.15566

61. $29{,}366.215 - 28{,}963.807$ 402.408

62. $7.03\overline{)181.374}$ 25.8

4.3 Solving Proportions

MathPro

Video 10

SSM

After studying this section, you will be able to:

1 *Understand the idea of a variable and be able to solve for n in an equation of the form a × n = b.*
2 *Find the missing number in a proportion.*

1 Solving for a Variable in an Equation of the Form a × n = b

Consider this expression: "3 times a number yields 15. What is the number?" We could write this as

$$3 \times \boxed{?} = 15$$

and guess that the number $\boxed{?} = 5$. There is a better way of solving this problem, a way that eliminates the guesswork. We will begin by using a **variable.** That is, we will use a letter to represent a number we do not yet know. We briefly used variables in Chapters 1–3. Now we use them more extensively.

TEACHING TIP Some instructors prefer to introduce the notation $3n$ to indicate a product instead of $3 \times n$. Either is acceptable, but students who have never had algebra before may have trouble with the notation.

Let the letter n represent the unknown number. We write

$$3 \times n = 15$$

This is called an **equation.** An equation has an equals sign. We want to find the number n in this equation without guessing. We will not change the value of n in the equation if we divide both sides of the equation by 3. Thus if

$$3 \times n = 15$$

we can say

$$\frac{3 \times n}{3} = \frac{15}{3}$$

which is

$$\frac{3}{3} \times n = 5$$

or

$$1 \times n = 5$$

Since 1 × any number is the same number, we know that $n = 5$. Any equation of the form $a \times n = b$ can be solved in this way. We divide both sides of an equation of the form $a \times n = b$ by the number that is multiplied by n. (This method will not work for $3 + n = 15$, since here the 3 is added to n and not multiplied by n.)

EXAMPLE 1 Solve each equation to find the value of n.

(a) $16 \times n = 80$ **(b)** $24 \times n = 240$

(a) $16 \times n = 80$

$$\frac{16 \times n}{16} = \frac{80}{16} \quad \text{Divide each side by 16.}$$

$$n = 5 \quad \text{Because } 16 \div 16 = 1 \text{ and } 80 \div 16 = 5.$$

(b) $24 \times n = 240$

$$\frac{24 \times n}{24} = \frac{240}{24} \quad \text{Divide each side by 24.}$$

$$n = 10 \quad \text{Because } 24 \div 24 = 1 \text{ and } 240 \div 24 = 10.$$

Practice Problem 1 Solve each equation to find the value of n.

(a) $5 \times n = 45$ **(b)** $7 \times n = 84$ ■

The same procedure is followed if the variable n is on the right side of the equation.

EXAMPLE 2 Solve each equation to find the value of n.

(a) $66 = 11 \times n$ **(b)** $143 = 13 \times n$

(a) $66 = 11 \times n$

$$\frac{66}{11} = \frac{11 \times n}{11}$$ Divide each side by 11.

$$6 = n$$

(b) $143 = 13 \times n$

$$\frac{143}{13} = \frac{13 \times n}{13}$$ Divide each side by 13.

$$11 = n$$

TEACHING TIP Remind students that $n = 6$ and $6 = n$ mean exactly the same thing. This is called the *symmetric property of equality.*

Practice Problem 2 Solve each equation to find the value of n.

(a) $108 = 9 \times n$ **(b)** $210 = 14 \times n$ ■

The numbers in the equations are not always whole numbers, and the answer to an equation is not always a whole number.

EXAMPLE 3 Solve each equation to find the value of n.

(a) $16 \times n = 56$ **(b)** $18.2 = 2.6 \times n$

(a) $16 \times n = 56$

$$\frac{16 \times n}{16} = \frac{56}{16}$$ Divide each side by 16.

$$n = 3.5$$

$$\begin{array}{r} 3.5 \\ 16\overline{)56.0} \\ \underline{48} \\ 8\,0 \\ \underline{8\,0} \\ 0 \end{array}$$

(b) $18.2 = 2.6 \times n$

$$\frac{18.2}{2.6} = \frac{2.6 \times n}{2.6}$$ Divide each side by 2.6.

$$7 = n$$

$$\begin{array}{r} 7. \\ 2.6_\wedge\overline{)18.2_\wedge} \\ \underline{18\,2} \\ 0 \end{array}$$

Practice Problem 3 Solve each equation to find the value of n.

(a) $15 \times n = 63$ **(b)** $39.2 = 5.6 \times n$ ■

2 *Finding the Missing Number in a Proportion*

Sometimes one of the pieces of a proportion is unknown. We can use an equation such as $a \times n = b$ and solve for n to find the unknown quantity. Suppose that we want to know the value of n in the proportion

$$\frac{5}{12} = \frac{n}{144}$$

Since this is a proportion, we know that $5 \times 144 = 12 \times n$. Simplifying, we have

$$720 = 12 \times n$$

Next we divide both sides by 12.

$$\frac{720}{12} = \frac{12 \times n}{12}$$

$$60 = n$$

We check to see if this is correct. Do we have a true proportion?

$$\frac{5}{12} \stackrel{?}{=} \frac{60}{144}$$

$$\frac{5}{12} \times \frac{60}{144} \qquad \begin{array}{l} 12 \times 60 = 720 \\ 5 \times 144 = 720 \end{array} \quad \text{products equal}$$

Thus $\frac{5}{12} = \frac{60}{144}$ is true. We have checked our answer.

To Solve for a Missing Number in a Proportion

1. Find the two diagonal products (also called "cross-multiplying" or "forming a cross product").
2. Divide each side of the equation by the number multiplied by n.
3. Simplify the result.
4. Check your answer.

EXAMPLE 4 Find the value of n in $\frac{25}{4} = \frac{n}{12}$.

$25 \times 12 = 4 \times n$ Find the two diagonal products.

$300 = 4 \times n$

$\frac{300}{4} = \frac{4 \times n}{4}$ *Divide each side by* 4.

$75 = n$

Check whether the proportion is true.

$$\frac{25}{4} \stackrel{?}{=} \frac{75}{12}$$

$$25 \times 12 \stackrel{?}{=} 4 \times 75$$

$$300 = 300 \checkmark$$

The proportion is true. The answer $n = 75$ is correct.

Practice Problem 4 Find the value of n in $\frac{24}{n} = \frac{3}{7}$. ■

Some answers will be exact values. In other cases we will obtain answers that are rounded to a certain decimal place. If a value is rounded we sometimes use the $\approx$ symbol which indicates "is approximately equal to."

The answer to the next problem is not a whole number.

EXAMPLE 5 Find the value of n in $\frac{125}{2} = \frac{150}{n}$.

$125 \times n = 2 \times 150$ Find the two diagonal products.

$125 \times n = 300$

$\frac{125 \times n}{125} = \frac{300}{125}$ Divide each side by 125.

$n = 2.4$

$$\begin{array}{r} 2.4 \\ 125\overline{)300.0} \\ \underline{250} \\ 50\,0 \\ \underline{50\,0} \\ 0 \end{array}$$

Check. $\frac{125}{2} \stackrel{?}{=} \frac{150}{2.4}$

$125 \times 2.4 \stackrel{?}{=} 2 \times 150$

$300 = 300$ ✓

Practice Problem 5 Find the value of n in $\frac{176}{4} = \frac{286}{n}$. ■

TEACHING TIP Stress the importance of labeling the answer with the proper unit in applied problems. Students tend to avoid this, so remind them of the need to establish good habits now.

In applied situations it is helpful to write the units of measure in the proportion. Remember, order is important. The same units should be in the same position in each fraction.

EXAMPLE 6 If 5 grams of a nonicing additive are placed in 8 liters of diesel fuel, how many grams n should be added to 12 liters of diesel fuel?

We need to find the value of n in $\frac{n \text{ grams}}{12 \text{ liters}} = \frac{5 \text{ grams}}{8 \text{ liters}}$.

$$8 \times n = 12 \times 5$$

$$8 \times n = 60$$

$$\frac{8 \times n}{8} = \frac{60}{8}$$

$$n = 7.5$$

The answer is 7.5 grams. 7.5 grams of the additive should be added to 12 liters of the diesel fuel.

Check. $\frac{7.5 \text{ grams}}{12 \text{ liters}} = \frac{5 \text{ grams}}{8 \text{ liters}}$

$7.5 \times 8 \stackrel{?}{=} 12 \times 5$

$60 = 60$ ✓

Practice Problem 6 Find the value of n in $\frac{80 \text{ dollars}}{5 \text{ tons}} = \frac{n \text{ dollars}}{6 \text{ tons}}$. ■

Some proportions contain decimals.

EXAMPLE 7 Find the value of n in $\dfrac{141.75 \text{ miles}}{4.5 \text{ hours}} = \dfrac{63 \text{ miles}}{n \text{ hours}}$.

$$141.75 \times n = 63 \times 4.5$$

$$141.75 \times n = 283.5$$

$$\frac{141.75 \times n}{141.75} = \frac{283.5}{141.75}$$

$$n = 2$$

The answer is $n = 2$. The check is up to you.

Practice Problem 7 Find the value of n in $\dfrac{264.25 \text{ meters}}{3.5 \text{ seconds}} = \dfrac{n \text{ meters}}{2 \text{ seconds}}$. ■

To Think About Suppose that the proportion contains fractions or mixed numbers. Could you still follow all the steps? For example, find n when

$$\frac{n}{3\frac{1}{4}} = \frac{5\frac{1}{6}}{2\frac{1}{3}} \qquad \text{We would have} \qquad 2\frac{1}{3} \times n = 5\frac{1}{6} \times 3\frac{1}{4}$$

This can be written as

$$\frac{7}{3} \times n = \frac{31}{6} \times \frac{13}{4}$$

$$\boxed{\frac{7}{3} \times n = \frac{403}{24}} \qquad \text{equation (1)}$$

Now we divide each side of equation (1) by $\frac{7}{3}$. Why?

$$\frac{\frac{7}{3} \times n}{\frac{7}{3}} = \frac{\frac{403}{24}}{\frac{7}{3}}$$

Be careful here. The right-hand side means $\frac{403}{24} \div \frac{7}{3}$, which we evaluate by *inverting* the second fraction and multiplying.

$$\frac{403}{\underset{8}{\cancel{24}}} \times \frac{\overset{1}{\cancel{3}}}{7} = \frac{403}{56}$$

Thus $n = \frac{403}{56}$ or $7\frac{11}{56}$. Think about all the steps to solving this problem. Can you follow them? There is another way to do the problem. We could multiply each side of equation (1) by $\frac{3}{7}$. Try it. Why does this work? Now try problems 65–68 in Exercises 4.3.

4.3 Exercises

Solve each equation to find the value of n.

1. $5 \times n = 40$

$\frac{5 \times n}{5} = \frac{40}{5}$ $n = 8$

2. $7 \times n = 56$

$\frac{7 \times n}{7} = \frac{56}{7}$ $n = 8$

3. $76 = 2 \times n$

$\frac{76}{2} = \frac{2 \times n}{2}$ $n = 38$

4. $84 = 3 \times n$

$\frac{84}{3} = \frac{3 \times n}{3}$ $n = 28$

5. $6 \times n = 96$

$\frac{6 \times n}{6} = \frac{96}{6}$ $n = 16$

6. $8 \times n = 136$

$\frac{8 \times n}{8} = \frac{136}{8}$ $n = 17$

7. $117 = 9 \times n$

$\frac{117}{9} = \frac{9 \times n}{9}$ $13 = n$

8. $119 = 7 \times n$

$\frac{119}{7} = \frac{7 \times n}{7}$ $n = 17$

9. $7 \times n = 182$

$\frac{7 \times n}{7} = \frac{182}{7}$ $n = 26$

10. $6 \times n = 144$

$\frac{6 \times n}{6} = \frac{144}{6}$ $n = 24$

11. $3 \times n = 16.8$

$\frac{3 \times n}{3} = \frac{16.8}{3}$ $n = 5.6$

12. $2 \times n = 19.6$

$\frac{2 \times n}{2} = \frac{19.6}{2}$ $n = 9.8$

13. $2.6 \times n = 13$

$\frac{2.6 \times n}{2.6} = \frac{13}{2.6}$ $n = 5$

14. $3.2 \times n = 48$

$\frac{3.2 \times n}{3.2} = \frac{48}{3.2}$ $n = 15$

15. $40.6 = 5.8 \times n$

$\frac{40.6}{5.8} = \frac{5.8 \times n}{5.8}$ $n = 7$

16. $50.4 = 6.3 \times n$

$\frac{50.4}{6.3} = \frac{6.3 \times n}{6.3}$ $n = 8$

17. $260 \times n = 1430$

$\frac{260 \times n}{260} = \frac{1430}{260}$ $n = 5.5$

18. $180 \times n = 936$

$\frac{180 \times n}{180} = \frac{936}{180}$ $n = 5.2$

Find the value of n in each proportion. Check your answer.

19. $\frac{n}{20} = \frac{3}{4}$

$4 \times n = 60$
$n = 15$

20. $\frac{n}{28} = \frac{3}{7}$

$n \times 7 = 84$
$n = 12$

21. $\frac{6}{n} = \frac{3}{8}$

$3 \times n = 48$
$n = 16$

22. $\frac{4}{n} = \frac{2}{7}$

$28 = n \times 2$
$14 = n$

23. $\frac{12}{40} = \frac{n}{25}$

$300 = 40 \times n$
$7.5 = n$

24. $\frac{13}{30} = \frac{n}{15}$

$195 = 30 \times n$
$6.5 = n$

25. $\frac{25}{100} = \frac{8}{n}$

$25 \times n = 800$
$n = 32$

26. $\frac{18}{150} = \frac{6}{n}$

$18 \times n = 900$
$n = 50$

27. $\frac{n}{32} = \frac{3}{4}$

$4 \times n = 96$
$n = 24$

28. $\frac{n}{8} = \frac{225}{3}$

$3 \times n = 1800$
$n = 600$

29. $\frac{n}{9} = \frac{49}{63}$

$n \times 63 = 441$
$n = 7$

30. $\frac{n}{42} = \frac{5}{6}$

$n \times 6 = 210$
$n = 35$

31. $\frac{18}{n} = \frac{3}{11}$

$198 = n \times 3$
$66 = n$

32. $\frac{16}{n} = \frac{2}{7}$

$112 = n \times 2$
$n = 56$

Find the value of n in each proportion. Round your answer to the nearest tenth when necessary.

33. $\frac{n}{5} = \frac{7}{4}$

$4 \times n = 35$
$n \approx 8.8$

34. $\frac{n}{9} = \frac{5}{4}$

$n \times 4 = 45$
$n \approx 11.3$

35. $\frac{12}{n} = \frac{3}{5}$

$3 \times n = 60$
$n = 20$

36. $\frac{21}{n} = \frac{2}{3}$

$63 = n \times 2$
$n = 31.5$

37. $\frac{35}{n} = \frac{7}{5}$

$7 \times n = 175$
$n = 25$

38. $\frac{35}{n} = \frac{4}{3}$

$4 \times n = 105$
$n \approx 26.3$

39. $\frac{9}{26} = \frac{n}{52}$

$26 \times n = 468$
$n = 18$

40. $\frac{12}{16} = \frac{n}{40}$

$16 \times n = 480$
$n = 30$

41. $\frac{12}{8} = \frac{21}{n}$

$12 \times n = 168$
$n = 14$

42. $\frac{15}{12} = \frac{10}{n}$

$15 \times n = 120$
$n = 8$

43. $\frac{n}{18} = \frac{3.5}{1}$

$n \times 1 = 63$
$n = 63$

44. $\frac{n}{36} = \frac{4.5}{1}$

$n \times 1 = 162$
$n = 162$

45. $\frac{2.5}{n} = \frac{0.5}{10}$

$0.5 \times n = 25$
$n = 50$

46. $\frac{1.5}{n} = \frac{0.3}{8}$

$0.3 \times n = 12$
$n = 40$

47. $\frac{3}{4} = \frac{n}{3.8}$

$4 \times n = 11.4$
$n \approx 2.9$

48. $\frac{7}{8} = \frac{n}{4.2}$

$8 \times n = 29.4$
$n \approx 3.7$

49. $\frac{12.5}{16} = \frac{n}{12}$

$16 \times n = 150$
$n \approx 9.4$

50. $\frac{13.8}{15} = \frac{n}{6}$

$15 \times n = 82.8$
$n \approx 5.5$

51. $\frac{16}{100} = \frac{5}{n}$

$16 \times n = 500$
$n \approx 31.3$

52. $\frac{22\frac{2}{9}}{100} = \frac{2}{n}$

$\frac{200}{9} \times n = 200$
$n = 9$

Applications

Find the value of n. Round to the nearest hundredth when necessary.

53. $\dfrac{n \text{ grams}}{10 \text{ liters}} = \dfrac{7 \text{ grams}}{25 \text{ liters}}$

$25 \times n = 70 \quad n = 2.8$

54. $\dfrac{n \text{ grams}}{14 \text{ liters}} = \dfrac{6 \text{ grams}}{22 \text{ liters}}$

$22 \times n = 84 \quad n \approx 3.82$

55. $\dfrac{76 \text{ dollars}}{5 \text{ tons}} = \dfrac{n \text{ dollars}}{8 \text{ tons}}$

$5 \times n = 608 \quad n = 121.60$

56. $\dfrac{84 \text{ dollars}}{10 \text{ tons}} = \dfrac{n \text{ dollars}}{7 \text{ tons}}$

$10 \times n = 588 \quad n = 58.80$

57. $\dfrac{50 \text{ gallons}}{12 \text{ acres}} = \dfrac{36 \text{ gallons}}{n \text{ acres}}$

$50 \times n = 432 \quad n = 8.64$

58. $\dfrac{80 \text{ gallons}}{26 \text{ acres}} = \dfrac{42 \text{ gallons}}{n \text{ acres}}$

$80 \times n = 1092 \quad n = 13.65$

59. $\dfrac{10 \text{ miles}}{16.1 \text{ kilometers}} = \dfrac{n \text{ miles}}{7 \text{ kilometers}}$

$70 = n \times 16.1 \quad n = 4.35$

60. $\dfrac{3 \text{ kilometers}}{1.86 \text{ miles}} = \dfrac{n \text{ kilometers}}{4 \text{ miles}}$

$12 = n \times 1.86 \quad n = 6.45$

61. $\dfrac{5 \text{ pounds}}{2.27 \text{ kilograms}} = \dfrac{3 \text{ pounds}}{n \text{ kilograms}}$

$6.81 = 5 \times n \quad n = 1.36$

62. $\dfrac{6 \text{ ounces}}{170.1 \text{ grams}} = \dfrac{5 \text{ ounces}}{n \text{ grams}}$

$850.5 = 6 \times n \quad n = 141.75$

63. $\dfrac{6.3 \text{ acres}}{2 \text{ people}} = \dfrac{n \text{ acres}}{5 \text{ people}}$

$31.5 = 2 \times n \quad n = 15.75$

64. $\dfrac{136 \text{ miles}}{9 \text{ gallons}} = \dfrac{n \text{ miles}}{10.5 \text{ gallons}}$

$1428 = 9 \times n \quad n = 158.67$

To Think About

Study the "To Think About" example in the text. Then solve for n in each of the following problems. Express n as a mixed number.

65. $\dfrac{n}{2\frac{1}{3}} = \dfrac{4\frac{5}{6}}{3\frac{1}{9}}$

$n \times 3\frac{1}{9} = 11\frac{5}{18} \quad n = 3\frac{5}{8}$

66. $\dfrac{n}{7\frac{1}{4}} = \dfrac{2\frac{1}{5}}{4\frac{1}{8}}$

$n \times 4\frac{1}{8} = 15\frac{19}{20} \quad n = 3\frac{13}{15}$

67. $\dfrac{8\frac{1}{6}}{n} = \dfrac{5\frac{1}{2}}{7\frac{1}{3}}$

$n \times 5\frac{1}{2} = 59\frac{8}{9} \quad n = 10\frac{8}{9}$

68. $\dfrac{9\frac{3}{4}}{n} = \dfrac{8\frac{1}{2}}{4\frac{1}{3}}$

$n \times 8\frac{1}{2} = 42\frac{1}{4} \quad n = 4\frac{33}{34}$

Cumulative Review Problems

Evaluate by doing each operation in the proper order.

69. $4^3 + 20 \div 5 + 6 \times 3 - 5 \times 2$

$64 + 4 + 18 - 10 = 76$

70. $(1.6)^2 - 0.12 \times 3.5 + 36.8 \div 2.5$

$2.56 - 0.42 + 14.72 = 16.86$

71. Write a word name for the decimal 0.563

five hundred sixty-three thousandths

72. Write in decimal notation thirty-four ten thousandths

0.0034

4.4 Applications of Proportions

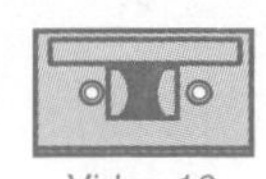

After studying this section, you will be able to:

1 *Solve applied problems using proportions.*

1 *Solving Applied Problems Using Proportions*

Let us examine a variety of applied problems that can be solved by proportions.

EXAMPLE 1 A recent citywide survey discovered that 37 pairs of eyeglasses in a sample of 120 pairs of eyeglasses were defective. If this rate remains the same each year, how many of the 36,000 pairs of eyeglasses made in the city each year are defective?

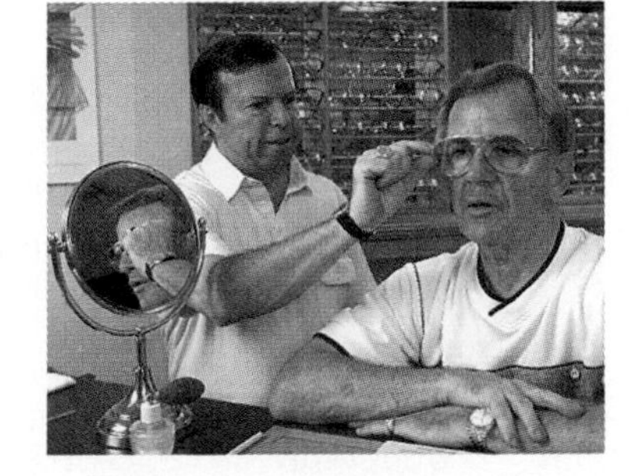

We will use the letter n to represent the number of defective eyeglasses in the city.

We compare the sample to the total number in the city

$$\frac{37 \text{ defective pairs}}{120 \text{ total pairs of eyeglasses}} = \frac{n \text{ defective pairs}}{36{,}000 \text{ total pairs of eyeglasses}}$$

$37 \times 36{,}000 = 120 \times n$ *Form the cross product.*

$1{,}332{,}000 = 120 \times n$ *Simplify.*

$\frac{1{,}332{,}000}{120} = \frac{120 \times n}{120}$ *Divide each side by* 120.

$11{,}100 = n$

Thus, if the rate of defective eyeglasses holds steady, there are about 11,100 defective pairs of eyeglasses made in the city each year.

Practice Problem 1 An automobile assembly line produced 243 engines yesterday, of which 27 were defective. If the same rate is true each day, how many of the 4131 engines produced this month would be defective? ■

TEACHING TIP This is a good time to stress the helpfulness of making an estimate in solving a proportion problem. Usually, a mistake will be obvious when compared with an estimate.

EXAMPLE 2 A survey conducted on a college campus showed that approximately 3 out of every 11 students were left-handed. If there are 2310 students on campus, how many are left-handed?

Let n represent the number of left-handed students.

We compare the measured rate to the campus-wide situation

$$\frac{3 \text{ left-handed students in survey}}{11 \text{ students in survey}} = \frac{n \text{ left-handed students on campus}}{2310 \text{ total students on campus}}$$

We assume that the measured rate may be applied to the entire campus.

$$3 \times 2310 = 11 \times n \quad \textit{Form the cross product.}$$

$$6930 = 11 \times n \quad \textit{Simplify.}$$

$$\frac{6930}{11} = \frac{11 \times n}{11} \quad \textit{Divide each side by 11.}$$

$$630 = n$$

Thus there are approximately 630 left-handed students on campus.

Practice Problem 2 A survey conducted on campus showed that 5 out of every 19 students had quit smoking during the last five years. Approximately how many of the 7410 students now on campus quit smoking in the last five years? ■

Looking back at Example 2, perhaps it occurred to you that the fractions in the proportion could be set up in an alternative way. You can set up this problem in several different ways as long as the units are in correctly corresponding positions. It would be correct to set up the problem in the form

$$\frac{\text{left-handed students in survey}}{\text{left-handed students on campus}} = \frac{\text{total students in survey}}{\text{total students on campus}}$$

or

$$\frac{\text{total students on campus}}{\text{total students in survey}} = \frac{\text{left-handed students on campus}}{\text{left-handed students in survey}}$$

But we **cannot** set up the problem this way.

$$\frac{\text{left-handed students}}{\text{total students}} \neq \frac{\text{total students}}{\text{left-handed students}}$$

This is *not* correct. Do you see why? See Exercises 4.4, problems 35–38.

TEACHING TIP Sometimes students ask why we do not find that Ted's car gets 35 miles per gallon first in Example 3. Respond by showing them how they would obtain the same answer. Point out that in real life ratios may not reduce so nicely and that the general approach shown in Example 3 is more useful for solving everyday problems.

EXAMPLE 3 Ted's car can go 245 miles on 7 gallons of gas. Ted wants to take a trip of 455 miles. Approximately how many gallons of gas will this take?

Let $n =$ the unknown number of gallons.

$$\frac{245 \text{ miles}}{7 \text{ gallons}} = \frac{455 \text{ miles}}{n \text{ gallons}}$$

$$245 \times n = 7 \times 455 \quad \textit{Form the cross product.}$$

$$245 \times n = 3185 \quad \textit{Simplify.}$$

$$\frac{245 \times n}{245} = \frac{3185}{245}$$

$$n = 13$$

Ted will need approximately 13 gallons of gas for the trip.

Practice Problem 3 Cindy's car travels 234 miles on 9 gallons of gas. How many gallons of gas will Cindy need to take a 312-mile trip? ■

EXAMPLE 4 In a certain gear, Alice's 10-speed bicycle has a gear ratio of 3 revolutions of the pedal for every 2 revolutions of the bicycle wheel. If her bicycle wheel is turning at 65 revolutions per minute, how many times must she pedal per minute?

Let n = the number of revolutions of the pedal.

$$\frac{3 \text{ revolutions of the pedal}}{2 \text{ revolutions of the wheel}} = \frac{n \text{ revolutions of the pedal}}{65 \text{ revolutions of the wheel}}$$

$$3 \times 65 = 2 \times n \qquad \textit{Cross-multiply.}$$

$$195 = 2 \times n \qquad \textit{Simplify.}$$

$$\frac{195}{2} = \frac{2 \times n}{2} \qquad \textit{Divide both sides by 2.}$$

$$97.5 = n$$

Alice will pedal at the rate of 97.5 revolutions in 1 minute.

Practice Problem 4 Alicia must pedal at 80 revolutions per minute to ride her bicycle at 16 miles per hour. If she pedals at 90 revolutions per minute, how fast will she be riding? ■

TEACHING TIP Gear ratio problems often confuse students. Ask them why a bicycle might have a gear ratio of 3 to 2 as well as 4 to 5. Ask them which ratio would help them to ride at a higher rate of speed. Usually, this type of topic will generate a good class discussion.

EXAMPLE 5 Tim operates a bicycle rental center during the summer months on the island of Martha's Vineyard. He discovered that when the ferryboats brought 8500 passengers a day to the island his center rented 340 bicycles a day. Next summer the ferryboats plan to bring 10,300 passengers a day to the island. How many bicycles a day should Tim plan to rent?

Two important cautions are necessary before we solve the proportion. We need to be sure that the bicycle rentals are directly related to the number of people on the ferryboat. (Presumably, people who fly to the island or who take small pleasure boats to the island also rent bicycles.) Next we need to be sure that the people who represent the increase in passengers per day would be as likely to rent bicycles as the present number of passengers do. For example, if the new visitors to the island are all senior citizens, they are not as likely to rent bicycles as younger people would be. If we assume those two conditions are satisfied, then we can solve the problem as follows.

$$\frac{8500 \text{ passengers per day now}}{340 \text{ bike rentals per day now}} = \frac{10{,}300 \text{ passengers per day later}}{n \text{ bike rentals per day later}}$$

$$8500 \times n = 340 \times 10{,}300$$

$$8500 \times n = 3{,}502{,}000$$

$$\frac{8500 \times n}{8500} = \frac{3{,}502{,}000}{8500}$$

$$n = 412$$

If the two conditions are satisfied, we would predict 412 bicycle rentals.

Practice Problem 5 For every 4050 people who walk into Tom's Souvenir Shop, 729 make a purchase. Assuming the same conditions, if 5500 people walk into Tom's Souvenir Shop, how many people may be expected to make a purchase? ■

Biologists and others who observe or protect wildlife sometimes use the capture-mark-recapture method to determine how many animals are in a certain region. In this approach some animals are caught and tagged in a way that does not harm them. They are then released into the wild, where they mix with their kind. It is assumed (usually correctly) that the tagged animals will mix throughout the entire population in that region, so that when they are recaptured in a future sample, the biologists can use them to make reasonable estimates about the total population. We will employ the capture-mark-recapture method in the next example.

EXAMPLE 6 A biologist catches 42 fish in a lake and tags them. She then quickly returns them to the lake. In a few days she catches a new sample of 50 fish. Of those 50 fish, 7 have her tag. Approximately how many fish are in the lake?

$$\frac{42 \text{ fish tagged in 1st sample}}{n \text{ fish in lake}} = \frac{7 \text{ fish tagged in 2nd sample}}{50 \text{ fish caught in 2nd sample}}$$

$$42 \times 50 = 7 \times n$$

$$2100 = 7 \times n$$

$$\frac{2100}{7} = \frac{7 \times n}{7}$$

$$300 = n$$

Assuming that no tagged fish died and that the tagged fish mixed throughout the population of fish in the lake, we estimate the number of fish in the lake as 300.

Practice Problem 6 A park ranger in Alaska captures and tags 50 bears. He then releases them to range through the forest. Sometime later he captures 50 bears. Of the 50, 4 have tags from the previous capture. Estimate the number of bears in the forest. ■

Developing Your Study Skills

GETTING HELP

Getting the right kind of help at the right time can be a key ingredient in being successful in mathematics. When you have gone to class on a regular basis, taken careful notes, methodically read your textbook, and diligently done your homework—in other words, when you have made every effort possible to learn the mathematics—you may still find that you are having difficulty. If this is the case, then you need to seek help. Make an appointment with your instructor to find out what help is available to you. The instructor, tutoring services, a mathematics lab, videotapes, and computer software may be among resources you can draw on.

4.4 Exercises

Applications

1. The policy at the Collonaide Hotel is to have 19 desserts for every 16 people if a buffet is being served. If the Saturday buffet has 320 people, how many desserts must be available? 380 desserts

2. An automobile dealership has found that for every 140 cars sold, 23 will be brought back to the dealer for major repairs. If the dealership sells 980 cars this year, approximately how many cars will be brought back for major repairs? 161 cars

3. A coffee bar is constantly making ''shots'' of espresso coffee. For every 90 shots of coffee that are purchased, 5 are thrown away because they get too cold while the milk for cappuccino is being steamed. If the coffee bar sells 162 shots of coffee per day, how many shots will be poured down the sink? 9 shots of coffee

4. Cynthia's car uses 0.5 quarts of oil for every 850 miles she drives. She drove 12,325 miles last year. At that rate, how many quarts of oil did she use? 7.25 quarts of oil

5. A cosmetics company uses the ratio of 35 drops of clear polish to 4 drops of pigment when it makes red nail polish. How many drops of pigment should it take if the company has a vat containing 8015 drops of clear polish? 916 drops of pigment

6. When Donna lived in Paris, France, the exchange rate was 8 French francs for every U.S. dollar. If she exchanged 355 U.S. dollars, how many French francs should she receive? 2840 French francs

7. Jason went to Canada when the exchange rate was \$7.50 U.S. for every \$10.50 Canadian. When he exchanges \$240.00 U.S., how many Canadian dollars should he receive? \$336 Canadian

8. When Douglas flew to London, the exchange rate was 7 British pounds for every 15 U.S. dollars. Douglas exchanged 255 U.S. dollars. How many British pounds did he receive? 119 British pounds

9. At the reclining stationary bike at a health club, Angela pedals at 80 revolutions per minute (rpm) to travel at an indicated speed of 18 miles per hour. If she were to pedal at 75 rpm, how fast would be her indicated speed? Round your answer to the nearest tenth. 16.9 miles per hour

10. Rod uses a computerized rowing machine at the gymnasium where he works out. He completes 17 strokes per minute and the machine indicates a speed of 5 miles per hour. If he were to increase his stroke to 34 strokes per minute, how fast would the machine indicate is his rowing speed? 10 miles per hour

In problems 11 and 12 the shadow of one object is in the same ratio to the height of that object as a second shadow is to the height of a second object at the same time of day.

11. A pro football offensive tackle who stands 6.5 feet tall casts a 5-foot shadow. At the same time, the football stadium, which he is standing next to, casts a shadow of 152 feet. How tall is the stadium? Round your answer to the nearest tenth. 197.6 feet

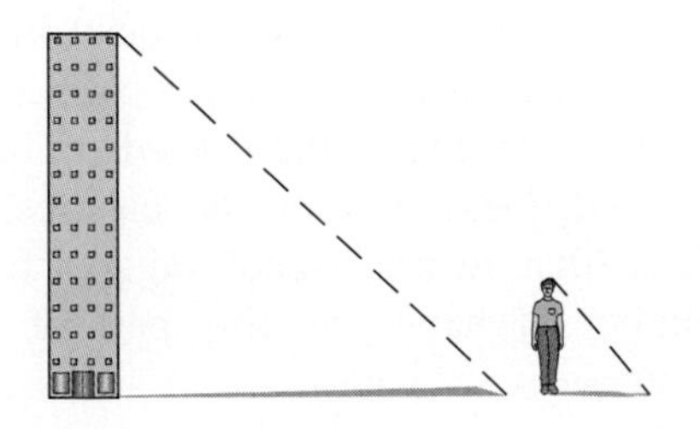

12. A souvenir life-sized cardboard cut-out of Queen Elizabeth (for the tourists) sits outside of a London office building. It is 5 feet tall, and casts a shadow that is 3 feet long. The office building casts a shadow that is 165 feet long. How tall is the building? 275 feet

13. Jacqueline can write 4 pages of a test book (blue book) in 12 minutes. How long would it take her to write an essay that takes up 11 pages of a blue book? 33 minutes

14. Marcia can type the final copy of a 4-page psychology paper in 27 minutes. How long will it take her to type a 15-page paper? Round your answer to the nearest minute. 101 minutes

15. On Dr. Jennings' map of Antarctica, the scale states that 4 inches represent 250 miles of actual distance. Two Antarctic mountains are 5.7 inches apart on the map. What is the approximate distance in miles between the two mountains? Round your answer to the nearest mile. 356 miles

16. On a tour guide map of Madagascar, the scale states that 3 inches represent 125 miles. Two beaches are 5.2 inches apart on the map. What is the approximate distance in miles between the two beaches? Round your answer to the nearest mile. 217 miles

17. A 100-watt stereo system needs copper speaker wire that is 30 millimeters thick to handle the output of sound clearly. How thick would the speaker wire need to be if you had a 140-watt stereo and you wanted the same ratio of watts to millimeters? 42 millimeters

18. An electrical engineer specializing in rock concerts uses a 21-foot wire with 0.6 ohm of resistance. (The electrical resistance in a wire is measured in ohms.) How much resistance would be in the same type of wire 230 feet long? Round your answer to the nearest tenth. 6.6 ohms

19. Jenny's blueberry pancake recipe uses 5 cups of blueberries for every 12 people. How many cups of blueberries does she need for the same recipe prepared for 28 people?

$11\frac{2}{3}$ cups or approx. 11.7 cups

20. In his lasagna recipe, Giovanni uses 2 cups of marinara sauce for every seven people. How much sauce will he need to make lasagna for 62 people?

Round your answer to the nearest tenth. 17.7 cups

21. A Mercedes-Benz salesperson sold 7 cars after having spoken to 24 customers. To maintain the same ratio of sales, how many Mercedes-Benz vehicles will she have to sell if she meets with 50 potential customers? Round to the nearest whole number. 15 cars

22. The Lincoln Lancers basketball team has won nine out of every eleven games played over the last few years. The team has played 143 times. How many times have they won? 117 times

23. A tennis player makes 14 unforced errors during 110 played points. At that rate, how many errors would he make in several matches where he and his opponent played 275 points? 35 errors

24. A baseball pitcher gave up 52 earned runs in 260 innings of pitching. At that rate, how many runs would he give up in a 9-inning game? (This decimal is called the pitcher's *earned run average*.) 1.8

25. The Chicago-O'Hare Taxi Drivers Association has discovered that for every 7 plane flights arriving at O'Hare Airport, 79 people need a taxi ride from the airport. If tomorrow 434 plane flights are scheduled to arrive at O'Hare, how many people would you estimate will need a taxi ride from the airport? 4898 people

26. The local super drugstore has determined that for every 30 customers who come into the store and buy shampoo, 24 of them will also buy conditioner/cream rinse. If the average number of customers who buy shampoo is 830 per week, how many people per week will buy shampoo and conditioner? Round to the nearest whole number. 661

27. In Kenya, a worker at the game preserve captures 26 giraffes, tags them, and then releases them back into the preserve. The next month, he captures 18 giraffes and finds that 6 of them have already been tagged. Estimate the number of giraffes on the preserve. 78 giraffes

28. A ornithologist is studying hawks in the Adirondack Mountains. She catches 24 hawks over a period of one month, tags them, and frees them back into the wild. The next month, she catches 20 hawks and finds that 12 are already tagged. Estimate the number of hawks in this part of the mountains. 40 hawks

29. Bill and Shirley Grant are raising tomatoes to sell at the local co-op. The farm has a yield of 425 pounds of tomatoes for every 3 acres. The farm has 14 acres of good tomatoes. The crop this year should bring $1.80 per pound for each pound of tomatoes. How much will the Grants get from the sale of the tomato crop? $3570

30. A paint manufacturer suggests 2 gallons of flat latex paint for every 750 square feet of wall. A painter is going to paint 7875 square feet of wall in a Dallas office building with paint that costs $8.50 per gallon. How much will the painter spend for the paint? $178.50

31. An 8-foot oak table weighs 86 pounds. The same type of table also comes in a 12-foot model. A delivery truck loads seven of the 12-foot models into a truck. How much does the cargo weigh? 903 pounds

32. A 40-foot roll of fence wire weighs 37 pounds. The same type of fence wire also comes in 200-foot rolls. Frank put six of the 200-foot rolls in his pickup truck. How much did his cargo weigh? 1110 pounds

33. It has been found that 22 out of every 100 people in the United States mailed a survey will respond. If 15,500 are mailed a survey, estimate how many will respond. 3410 people will respond

34. A study shows that with a certain breed of domesticated birds, 46 out of 80 will be able to mimick their masters by ''talking.'' If there are 2200 such birds as pets in the United States, estimate how many of them will be able to talk. 1265 birds

To Think About

The speed 60 miles per hour is equivalent to the speed 88 feet per second. A car is traveling at 49 miles per hour. Several students attempt to solve the problem of how many feet per second the car is traveling.

35. Student A uses the proportion $\frac{88}{n} = \frac{60}{49}$.

Is student A right? Why?

Yes. The units show that the parts of the proportion correspond correctly.

$$\frac{88 \text{ feet per second}}{n \text{ feet per second}} = \frac{60 \text{ miles per hour}}{49 \text{ miles per hour}}$$

36. Student B uses the proportion $\frac{60}{88} = \frac{49}{n}$.

Is student B right? Why?

Yes. The units show the parts of the proportion correspond correctly.

$$\frac{60 \text{ miles per hour}}{88 \text{ feet per second}} = \frac{49 \text{ miles per hour}}{n \text{ feet per second}}$$

37. Student C uses the proportion $\frac{n}{88} = \frac{60}{49}$.

Is student C right? Why?

No. Look at the denominators: 49 miles per hour does not correspond to 88 feet per second; 60 miles per hour does. This should be

$$\frac{n \text{ feet per second}}{88 \text{ feet per second}} = \frac{49 \text{ miles per hour}}{60 \text{ miles per hour}}$$

38. Student D uses the proportion $\frac{60}{88} = \frac{n}{49}$.

Is student D right? Why?

No. 49 miles per hour should be in the numerator on the right. Then the parts of the proportion will correspond.

$$\frac{60 \text{ miles per hour}}{88 \text{ feet per second}} = \frac{49 \text{ miles per hour}}{n \text{ feet per second}}$$

Cumulative Review Problems

39. Round to the nearest hundred. 56,179
56,200

40. Round to the nearest ten thousand. 196,379,910
196,380,000

41. Round to the nearest tenth. 56.148
56.1

42. Round to the nearest hundredth. 1.96341
1.96

43. Round to the nearest ten thousandth. 0.07615382
0.0762

44. Round to the nearest thousandth. 598.32149
598.321

Putting Your Skills to Work

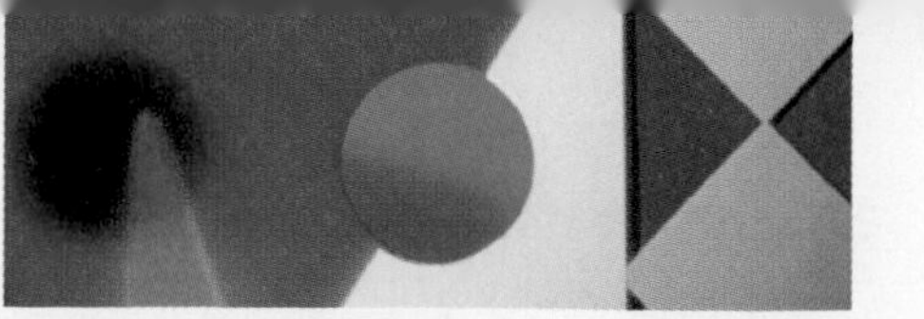

HOW CROWDED IS THIS CITY? A MATHEMATICAL APPROACH

People can have different impressions as to how crowded a city is. Hardly anyone can travel to New York City and not have the impression that it is extremely crowded. But what about other cities? How do you measure how ''crowded'' a city really is?

Mathematically, a good way to approach the problem is to calculate the number of people per square mile. The following chart lists the population of area in square miles of several major cities in the United States. Source: Bureau of the Census

City	1990 population	Area in square miles
Anaheim, California	266,406	41
Anchorage, Alaska	226,338	1732
Austin, Texas	465,648	116
Boston, Massachusetts	574,283	46
Chicago, Illinois	2,783,726	228
Cincinnati, Ohio	364,114	78
Detroit, Michigan	1,027,974	136
New Orleans, Louisiana	496,938	199
New York City, New York	7,322,564	301
Phoenix, Arizona	983,403	324

PROBLEMS FOR INDIVIDUAL INVESTIGATION

Round all answers to the nearest whole number.

1. Find the number of people per square mile for Cincinnati and Austin. According to your calculations which city is more densely populated? There are 4668 people per square mile in Cincinnati and 4014 people per square mile in Austin. Cincinnati is more densely populated.

2. Find the number of people per square mile for Chicago and Boston. According to your calculations which city is more densely populated? There are 12,209 people per square mile in Chicago and 12,484 people per square mile in Boston. Boston is more densely populated.

PROBLEMS FOR GROUP INVESTIGATION AND ANALYSIS

Together with some members of your class investigate the following.

3. The 1980 census figures for Anaheim are 219,494 and for Detroit 1,203,368. Assuming that the ratio of change from 1980 to 1990 will be the same for 1990 to 2000, what will be the estimated number of people per square mile for Anaheim and for Detroit in the year 2000? In the year 2000 there will be approximately 6457 people per square mile in Detroit and 7886 people per square mile in Anaheim.

4. The 1980 census figures for New Orleans are 557,927 and for Phoenix 739,704. Assuming that the ratio of change from 1980 to 1990 will continue to be the same for the period 1990 to 2000, what will be the estimated number of people per square mile in each of these cities in the year 2000? In the year 2000 there will be approximately 2224 people per square mile in New Orleans and 4035 people per square mile in Phoenix.

INTERNET CONNECTIONS: Go to ''`http://www.prenhall.com/tobey`'' to be connected

Site: U. S. Census Bureau County Population Census Counts or a related site

5. Find the number of people in your county in 1980 and in 1990. Then find the area of your county.

6. Calculate the number of people per square mile for your county in 1980 and 1990. Assuming that the ratio of change from 1990 to 2000 will be the same as it was for 1980 to 1990, estimate the number of people per square mile in your county in 2000.

Chapter Organizer

Topic	Procedure	Examples
Forming a ratio, p. 242.	A *ratio* is the comparison of two quantities that have the same units. A ratio is usually expressed as a fraction. The fractions should be in reduced form.	**1.** Find the ratio of 20 books to 35 books. $\frac{20}{35} = \frac{4}{7}$ **2.** Find the ratio in simplest form of 88:99. $\frac{88}{99} = \frac{8}{9}$ **3.** Bob earns \$250 each week, but \$15 is withheld for medical insurance. Find the ratio of medical insurance to total pay. $\frac{\$15}{\$250} = \frac{3}{50}$
Forming a rate, p. 244.	A *rate* is a comparison of two quantities that have different units. A rate is usually expressed as a fraction in reduced form.	A college has 2520 students with 154 faculty. What is the rate of students to faculty? $\frac{2520 \text{ students}}{154 \text{ faculty}} = \frac{180 \text{ students}}{11 \text{ faculty}}$
Forming a unit rate, p. 244.	A *unit rate* is a rate with a denominator of 1. Divide the denominator into the numerator to obtain the unit rate.	A car traveled 416 miles in 8 hours. Find the unit rate. $\frac{416 \text{ miles}}{8 \text{ hours}} = 52 \text{ miles/hour}$ Bob spread 50 pounds of fertilizer over 1870 square feet of land. Find the unit rate of square feet per pound. $\frac{1870 \text{ square feet}}{50 \text{ pounds}} = 37.4 \text{ square feet/pound}$
Writing proportions, p. 250.	A *proportion* is a statement that two rates or two ratios are equal. A proportion statement *a* is to *b* as *c* is to *d* can be written $\frac{a}{b} = \frac{c}{d}$	Write a proportion for 17 is to 34 as 13 is to 26. $\frac{17}{34} = \frac{13}{26}$
Determining if proportions are true or false, p. 250.	For any two fractions where $b \neq 0$, $d \neq 0$. $\frac{a}{b} = \frac{c}{d}$ if and only if $a \times d = b \times c$	**1.** Is the proportion $\frac{7}{56} = \frac{3}{24}$ true or false? $7 \times 24 \stackrel{?}{=} 56 \times 3$ $168 = 168$ ✓ The proportion is *true.* **2.** Is the proportion $\frac{64 \text{ gallons}}{5 \text{ acres}} = \frac{89 \text{ gallons}}{7 \text{ acres}}$ true or false? $64 \times 7 \stackrel{?}{=} 5 \times 89$ $448 \neq 445$ The proportion is *false.*

Chapter Organizer

Topic	Procedure	Examples
Solving a proportion, p. 257.	To solve a proportion where the value n is not known: **1.** Cross-multiply. **2.** Divide both sides of the equation by the number multiplied by n.	**1.** Solve for n. $\frac{17}{n} = \frac{51}{9}$ $17 \times 9 = 51 \times n$ Cross-multiply. $153 = 51 \times n$ Simplify. $\frac{153}{51} = \frac{51 \times n}{51}$ Divide by 51. $3 = n$
Solving applied problems, p. 263.	**1.** Write a proportion with n representing the unknown value. **2.** Solve the proportion.	Bob purchased eight notebooks for \$19. How much would 14 notebooks cost? $\frac{8 \text{ notebooks}}{\$19} = \frac{14 \text{ notebooks}}{n}$ $8 \times n = 19 \times 14$ $8 \times n = 266$ $\frac{8 \times n}{8} = \frac{266}{8}$ $n = 33.25$ The 14 notebooks would cost \$33.25.

Chapter 4 Review Problems

4.1 *Write each ratio in simplest form. Express your answer as a fraction.*

1. 88:40 $\frac{11}{5}$

2. 65:39 $\frac{5}{3}$

3. 28:35 $\frac{4}{5}$

4. 250:475 $\frac{10}{19}$

5. 50 to 124 $\frac{25}{62}$

6. 27 to 81 $\frac{1}{3}$

7. 156 to 441 $\frac{52}{147}$

8. 280 to 651 $\frac{40}{93}$

9. 26 tons to 65 tons $\frac{2}{5}$

10. 34 tons to 170 tons $\frac{1}{5}$

11. 150 kilograms to 200 kilograms $\frac{3}{4}$

12. 115 grams to 130 grams $\frac{23}{26}$

Bob earns \$215 per week and has \$35 per week withheld for federal taxes and \$20 per week withheld for state taxes.

13. Write the ratio of federal taxes withheld to earned income.
$\frac{7}{43}$

14. Write the ratio of state taxes withheld to earned income.
$\frac{4}{43}$

Write as a rate in simplest form.

15. 10 gallons of water for every 18 people

$\frac{5 \text{ gallons}}{9 \text{ people}}$

16. 44 revolutions every 121 minutes

$\frac{4 \text{ revolutions}}{11 \text{ minutes}}$

17. 188 vibrations every 16 seconds

$\frac{47 \text{ vibrations}}{4 \text{ seconds}}$

18. 12 cups of flour for every 38 people

$\frac{6 \text{ cups}}{19 \text{ people}}$

Write as a unit rate. Round to the nearest tenth when necessary.

19. \$2125 was paid for 125 shares of stock. Find the cost per share. \$17.00/share

20. \$2244 was paid for 132 chairs. Find the cost per chair. \$17.00/chair

21. \$742.50 was spent for 55 square yards of carpet. Find the cost per square yard.
\$13.50/square yard

22. The Baseball Boosters Club spent \$768.80 for 62 tickets to the ball game. Find the cost per ticket.
\$12.40/ticket

23. A 4-ounce jar of instant coffee costs \$2.96. A 9-ounce jar of the same brand of instant coffee costs \$5.22.

(a) What is the cost per ounce of the 4-ounce jar? \$0.74

(b) What is the cost per ounce of the 9-ounce jar? \$0.58

(c) How much per ounce do you save by buying the larger jar? \$0.16

24. A 12.5-ounce can of white tuna costs \$2.75. A 7.0-ounce can of the same brand of white tuna costs \$1.75.

(a) What is the cost per ounce of the large can?
\$0.22

(b) What is the cost per ounce of the small can?
\$0.25

(c) How much per ounce do you save by buying the larger can? \$0.03

4.2 *Write as a proportion.*

25. 12 is to 48 as 7 is to 28

$\frac{12}{48} = \frac{7}{28}$

26. 10 is to 21 as 30 is to 63

$\frac{10}{21} = \frac{30}{63}$

27. 7.5 is to 45 as 22.5 is to 135

$\frac{7.5}{45} = \frac{22.5}{135}$

28. 8.6 is to 43 as 17.2 is to 86

$\frac{8.6}{43} = \frac{17.2}{86}$

29. 136 is to 17 as 408 is to 51

$\frac{136}{17} = \frac{408}{51}$

30. 117 is to 61 as 351 is to 183

$\frac{117}{61} = \frac{351}{183}$

31. If 15 pounds cost \$4.50, then 27 pounds will cost \$8.10.

$\frac{4.50 \text{ dollars}}{15 \text{ pounds}} = \frac{8.10 \text{ dollars}}{27 \text{ pounds}}$

32. If 3 buses can transport 138 passengers, then 5 buses can transport 230 passengers.

$\frac{138 \text{ passengers}}{3 \text{ buses}} = \frac{230 \text{ passengers}}{5 \text{ buses}}$

Determine if each proportion is true or false.

33. $\frac{16}{48} = \frac{2}{12}$ False

34. $\frac{20}{25} = \frac{8}{10}$ True

35. $\frac{24}{20} = \frac{18}{15}$ True

36. $\frac{84}{48} = \frac{14}{8}$ True

37. $\frac{37}{33} = \frac{22}{19}$ False

38. $\frac{15}{18} = \frac{18}{22}$ False

39. $\frac{84 \text{ miles}}{7 \text{ gallons}} = \frac{108 \text{ miles}}{9 \text{ gallons}}$ True

40. $\frac{156 \text{ revolutions}}{6 \text{ minutes}} = \frac{181 \text{ revolutions}}{7 \text{ minutes}}$ False

41. $\frac{1.6 \text{ pounds}}{32 \text{ feet}} = \frac{4.8 \text{ pounds}}{96 \text{ feet}}$ True

42. $\frac{3.9 \text{ pounds}}{45 \text{ feet}} = \frac{7.9 \text{ pounds}}{90 \text{ feet}}$ False

4.3 *Solve for n in each equation.*

43. $7 \times n = 161$ $n = 23$

44. $8 \times n = 256$ $n = 32$

45. $558 = 18 \times n$ $n = 31$

46. $663 = 39 \times n$ $n = 17$

Solve each proportion. Round to the nearest tenth when necessary.

47. $\frac{3}{11} = \frac{9}{n}$ $n = 33$

48. $\frac{2}{7} = \frac{12}{n}$ $n = 42$

49. $\frac{n}{28} = \frac{6}{24}$ $n = 7$

50. $\frac{n}{32} = \frac{15}{20}$ $n = 24$

51. $\frac{3}{7} = \frac{n}{9}$ $n \approx 3.9$

52. $\frac{4}{9} = \frac{n}{8}$ $n \approx 3.6$

53. $\frac{54}{72} = \frac{n}{4}$ $n = 3$

54. $\frac{45}{135} = \frac{n}{3}$ $n = 1$

55. $\frac{6}{n} = \frac{2}{29}$ $n = 87$

56. $\frac{8}{n} = \frac{2}{81}$ $n = 324$

57. $\frac{25}{7} = \frac{60}{n}$ $n = 16.8$

58. $\frac{60}{9} = \frac{31}{n}$ $n \approx 4.7$

59. $\frac{35 \text{ miles}}{28 \text{ gallons}} = \frac{15 \text{ miles}}{n \text{ gallons}}$ $n = 12$

60. $\frac{8 \text{ defective parts}}{100 \text{ perfect parts}} = \frac{44 \text{ defective parts}}{n \text{ perfect parts}}$ $n = 550$

61. $\frac{7 \text{ tons}}{5.5 \text{ horsepower}} = \frac{16 \text{ tons}}{n \text{ horsepower}}$ $n \approx 12.6$

62. $\frac{27 \text{ feet}}{4 \text{ quarts}} = \frac{30.5 \text{ feet}}{n \text{ quarts}}$ $n \approx 4.5$

4.4 *Solve each problem by using a proportion. Round your answer to the nearest hundredth when necessary.*

63. The school volunteers used 3 gallons of paint to paint two rooms. How many gallons would they need to paint 10 rooms of the same size? 15 gallons

64. Fred paid \$77 for three chairs. How much would he pay for nine chairs? \$231.00

65. Chaneyville Hospital has 84 patients for seven registered nurses. To keep this ratio, how many registered nurses would be needed for 108 patients? 9 nurses

66. South Park College has 1680 students and 100 faculty. To keep this ratio, how many students should the college have if it increases the faculty to 130? 2184 students

67. When Marguerite traveled, the rate of French francs to American dollars was 24 francs to 5 dollars. How many francs did Marguerite receive for 420 dollars? 2016 francs

68. A catering service recommends providing 7.5 pounds of cold cuts for every 18 people at the reception. The men's glee club is having a reception for 120 people. How many pounds of cold cuts should they order? 50 pounds

69. Two cities located 225 miles apart appear 3 inches apart on a map. If two other cities appear 8 inches apart on the map, how many miles apart are the cities? 600 miles

70. If Melissa pedals her bicycle at 84 revolutions per minute, she travels at 14 miles per hour. How fast does she go if she pedals at 96 revolutions per minute? 16 miles per hour

71. Michael conducted a science experiment and found that sound travels in air at 34,720 feet in 31 seconds. How many feet would sound travel in 60 seconds? 67,200 feet

72. After college, Roberto began a chain of Pizza Villas in his home state. He eventually became a millionaire. He told his friends that for every \$3 he invested he earned \$985 in 10 years. How much did he invest to earn \$1 million? \$3045.69

73. In the setting sun, a 6-foot man casts a shadow 16 feet long. At the same time a building casts a shadow of 320 feet. How tall is the building? 120 feet

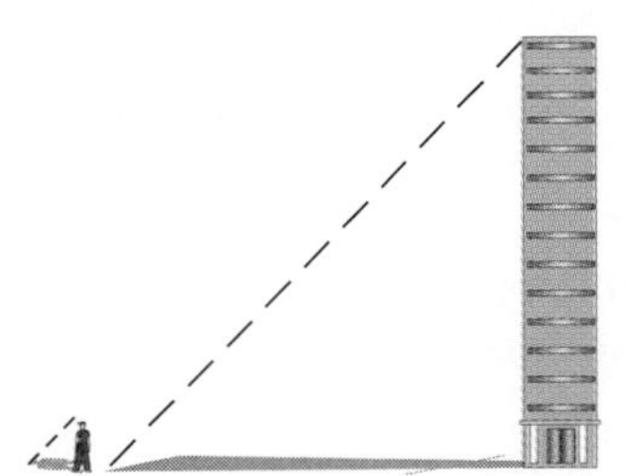

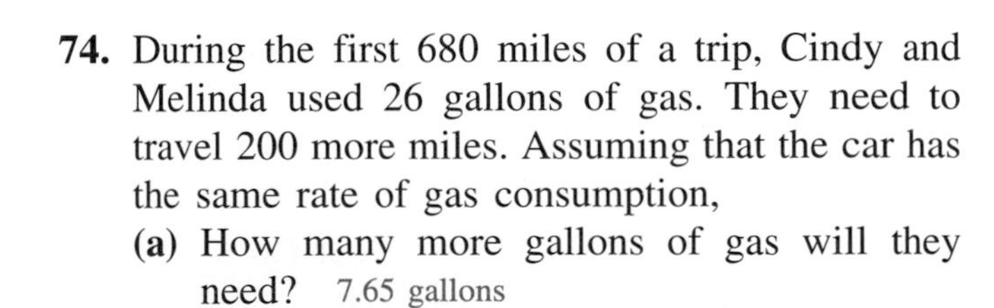

74. During the first 680 miles of a trip, Cindy and Melinda used 26 gallons of gas. They need to travel 200 more miles. Assuming that the car has the same rate of gas consumption,

(a) How many more gallons of gas will they need? 7.65 gallons

(b) If gas costs \$1.15 per gallon, what will fuel cost them for the last 200 miles? \$8.80

Putting Your Skills to Work

BUDGET PROBLEMS

Whether you are a full-time college student or the president of a small company, the issue of balancing the budget often comes up. Often, when faced with price increases, people are required to make decisions about future spending practices based on a careful analysis of how the money has been spent in the past. After John spent his first year at Wheaton College he analyzed his first-year expenses and created a spending plan for the second year.

Budget Categories	Actual Budget of First-Year College Expenses	Proposed Budget of Second-Year College Expenses
Tuition	12,350	13,500
Room	2,400	2,700
Board	2,700	3,000
Telephone	250	?
Clothing	650	?
Transportation	500	?
Entertainment	800	?
Churches and Charities	350	380
Total Expenses	20,000	22,000

PROBLEMS FOR INDIVIDUAL INVESTIGATION

Use this budget to answer the following questions.

1. Is the ratio of Tuition costs to Total Expenses higher for the second year than for the first year? No, the ratio is slightly less.

2. After he made up the proposed budget for the second year, John realized that his Total Expenses could only increase in the same ratio as the ratio of increase of the Tuition. On that basis, what should be his proposed budget for Total Expenses for the second year? (Round your answer to the nearest whole dollar.) $21,862

PROBLEMS FOR GROUP INVESTIGATION AND COOPERATIVE GROUP ACTIVITY

3. John allowed that his Telephone expenses would double for the second year, but his Clothing and Transportation expenses would be the same for the second year. Using your answer for question **2,** determine what his new entertainment budget will be for the second year. $632

4. Of the four categories Tuition, Room, Board, and Churches and Charities, which category had the greatest ratio of increase from the first year to the second year? Which had the least ratio of increase?
Room expenses had the greatest ratio.
Churches and Charities had the least ratio.

INTERNET CONNECTIONS: Go to ''http://www.prenhall.com/tobey'' to be connected
Site: The Government vs. a Personal Budget or a related site
This Australian site compares a personal budget to the budget of a government.
Read the information about the waitress, Maxine, and answer the following questions.

5. What is the ratio of Maxine's entertainment costs to clothing costs? What is the ratio of Maxine's rent to her income?

6. What is the ratio of Maxine's total expenses to her income? How much money can Maxine save each week? How long will it take her to save $3000?

Chapter 4 Test

Write as a ratio *in simplest form.*

1. 18:52

2. 70 to 185

Write as a rate *in simplest form. Express your answer as a fraction.*

3. 784 miles per 24 gallons

4. 2100 square feet per 45 pounds

Write as a unit rate. *Round to the nearest hundredth when necessary.*

5. 19 tons in five days

6. $57.96 for 7 hours

7. 5400 feet per 22 telephone poles

8. $9373 for 110 shares of stock

Write as a proportion.

9. 17 is to 29 as 51 is to 87.

10. 12 is to 19 as 18 is to 28.5.

11. 490 miles is to 21 gallons as 280 miles is to 12 gallons.

12. 5 tablespoons of flour is to 18 people as 15 tablespoons of flour is to 54 people.

Determine if each proportion is true or false.

13. $\frac{50}{24} = \frac{34}{16}$

14. $\frac{18.4}{20} = \frac{46}{50}$

15. $\frac{\text{32 smokers}}{\text{46 nonsmokers}} = \frac{\text{160 smokers}}{\text{230 nonsmokers}}$

16. $\frac{\$0.74}{\text{16 ounces}} = \frac{\$1.84}{\text{40 ounces}}$

Answers

1. $\frac{9}{26}$
2. $\frac{14}{37}$
3. $\frac{\text{98 miles}}{\text{3 gallons}}$
4. $\frac{\text{140 sq. ft.}}{\text{3 pounds}}$
5. 3.8 tons/day
6. $8.28/hour
7. 245.45 feet/pole
8. $85.21/share
9. $\frac{17}{29} = \frac{51}{87}$
10. $\frac{12}{19} = \frac{18}{28.5}$
11. $\frac{\text{490 miles}}{\text{21 gallons}} = \frac{\text{280 miles}}{\text{12 gallons}}$
12. $\frac{\text{5 tablespoons}}{\text{18 people}} = \frac{\text{15 tablespoons}}{\text{54 people}}$
13. False
14. True
15. True
16. False

Solve each proportion. Round to the nearest tenth when necessary.

17. $\frac{n}{20} = \frac{4}{5}$

18. $\frac{9}{2} = \frac{63}{n}$

19. $\frac{33}{n} = \frac{11}{4}$

20. $\frac{4.2}{11} = \frac{n}{77}$

21. $\frac{45 \text{ women}}{15 \text{ men}} = \frac{n \text{ women}}{40 \text{ men}}$

22. $\frac{3.5 \text{ ounces}}{4.2 \text{ grams}} = \frac{7 \text{ ounces}}{n \text{ grams}}$

23. $\frac{n \text{ inches of snow}}{14 \text{ inches of rain}} = \frac{12 \text{ inches of snow}}{1.4 \text{ inches of rain}}$

24. $\frac{28 \text{ pounds of bananas}}{\$n} = \frac{3 \text{ pounds of bananas}}{\$0.55}$

Solve each problem by using a proportion. Round your answer to the nearest hundredth when necessary.

25. Bob's recipe for pancakes calls for 3 eggs and will serve 11 people. If he wants to feed 22 people, how many eggs will he need?

26. A steel cable 42 feet long weighs 170 pounds. How much will 20 feet of this cable weigh?

27. If 9 inches on a map represents 57 miles what distance does 3 inches represent?

28. John and Nancy found it would cost $240 per year to fertilize their front lawn of 4000 square feet. How much would it cost to fertilize 6000 square feet?

29. If Jenny's car uses 1.5 quarts of oil every 3000 miles, how many quarts will it use in 8000 miles?

30. Stephen traveled 570 kilometers in 9 hours. At this rate, how far could he go in 11 hours?

Answers:

17. $n = 16$

18. $n = 14$

19. $n = 12$

20. $n = 29.4$

21. $n = 120$

22. $n = 8.4$

23. $n = 120$

24. $n = 5.13$

25. 6 eggs

26. 80.95 pounds

27. 19 miles

28. $360

29. 4 quarts

30. 696.67 kilometers

Cumulative Test for Chapters 1–4

Approximately one-half of this test is based on Chapter 4 material. The remainder is based on material covered in Chapters 1–3.

1. Write in words. 26,597,089

2. Add.
$$\begin{array}{r} 86 \\ 124 \\ 38 \\ 107 \\ +\ 56 \\ \hline \end{array}$$

3. Multiply.
$$\begin{array}{r} 208 \\ \times\ 67 \\ \hline \end{array}$$

4. Divide. $23\overline{)1564}$

5. Add. $\frac{1}{5} + \frac{1}{8} + \frac{3}{4}$

6. Subtract. $2\frac{1}{5} - 1\frac{3}{7}$

7. Multiply. $4\frac{1}{2} \times 3\frac{1}{4}$

8. Divide. $\dfrac{2\frac{1}{8}}{\frac{3}{4}}$

9. Round to the nearest hundredth. 163.578314

10. Subtract.
$$\begin{array}{r} 12.1 \\ -\ 3.8416 \\ \hline \end{array}$$

11. Multiply.
$$\begin{array}{r} 0.8163 \\ \times\ 0.22 \\ \hline \end{array}$$

12. Divide. $0.06\overline{)0.84}$

13. Write a *ratio* in simplest form to compare 81 miles to 27 miles.

14. Write as a *rate* in simplest form. $1.68 for 12 bananas

15. Write as a *unit rate* in simplest form. 12 yen for 3 pesos

1. Twenty-six million, five hundred ninety-seven thousand, eighty-nine

2. 411

3. 13,936

4. 68

5. $\frac{43}{40}$ *or* $1\frac{3}{40}$

6. $\frac{27}{35}$

7. $\frac{117}{8}$ *or* $14\frac{5}{8}$

8. $\frac{17}{6}$ *or* $2\frac{5}{6}$

9. 163.58

10. 8.2584

11. 0.179586

12. 14

13. $\frac{3}{1}$

14. $\frac{\$0.14}{1 \text{ banana}}$

15. $\frac{4 \text{ yen}}{1 \text{ peso}}$

Determine if each proportion is true or false.

16. $\frac{12}{17} = \frac{30}{42.5}$

17. $\frac{5}{7} = \frac{15}{21}$

Solve each proportion. Round to the nearest tenth when necessary.

18. $\frac{9}{2.1} = \frac{n}{0.7}$

19. $\frac{50}{20} = \frac{5}{n}$

20. $\frac{n}{56} = \frac{16}{7}$

21. $\frac{7}{n} = \frac{28}{36}$

22. $\frac{n}{11} = \frac{5}{16}$

23. $\frac{3\frac{1}{3}}{7} = \frac{10}{n}$

Solve each problem by using a proportion. Round your answers to the nearest hundredth when necessary.

24. Two cities that are located 300 miles apart appear 4 inches apart on a map. If two other cities are 625 miles apart, how far apart will they appear on the same map?

25. Last week Bob earned $84. He had $7 withheld for federal income tax. He earned $9000 last year. Assuming the same rate, how much did he have withheld for federal income tax last year?

26. Mom's lasagna recipe feeds 14 people and calls for 3.5 pounds of sausage. If she wants to feed 20 people, how much sausage does she need?

16. True

17. True

18. $n = 3$

19. $n = 2$

20. $n = 128$

21. $n = 9$

22. $n \approx 3.4$

23. $n = 21$

24. 8.33 inches

25. $750.00

26. 5 pounds

CHAPTER 5
Percent

Do you have any idea how many soft drinks are consumed per person in the United States? Is it more than the amount of coffee consumed per person or less? What are the future trends? Some of the answers may surprise you. Turn to the Putting Your Skills to Work on page 324.

Pretest Chapter 5

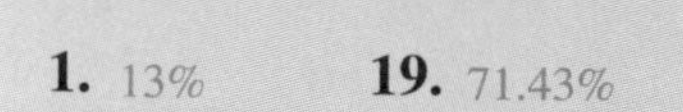

1. 13%
2. 21%
3. 18.5%
4. 37.2%
5. 134%
6. 894%
7. 0.2%
8. 0.4%
9. 17%
10. 27%
11. 13.4%
12. 19.8%
13. $6\frac{1}{2}\%$
14. $1\frac{3}{8}\%$
15. 60%
16. 2.5%
17. 115%
18. 106.25%
19. 71.43%
20. 28.57%
21. 95.65%
22. 68.42%
23. 440%
24. 275%
25. 0.33%
26. 0.25%
27. $\frac{11}{50}$
28. $\frac{19}{50}$
29. $\frac{53}{100}$
30. $\frac{41}{100}$
31. $\frac{3}{2}$ or $1\frac{1}{2}$
32. $\frac{8}{5}$ or $1\frac{3}{5}$
33. $\frac{19}{300}$
34. $\frac{7}{150}$
35. $\frac{41}{80}$
36. $\frac{7}{16}$

If you are familiar with the topics in this chapter, take this test now. Check your answers with those in the back of the book. If an answer was wrong or you couldn't do a problem, study the appropriate section of the chapter.

If you are not familiar with the topics in this chapter, don't take this test now. Instead, study the examples, work the practice problems, and then take the test.

This test will help you identify which concepts you have mastered and which you need to study further.

Section 5.1

Write as a percent.

1. 0.13 **2.** 0.21 **3.** 0.185 **4.** 0.372

5. 1.34 **6.** 8.94 **7.** 0.002 **8.** 0.004

9. $\frac{17}{100}$ **10.** $\frac{27}{100}$ **11.** $\frac{13.4}{100}$ **12.** $\frac{19.8}{100}$

13. $\frac{6\frac{1}{2}}{100}$ **14.** $\frac{1\frac{3}{8}}{100}$

Section 5.2

Change each fraction to a percent. Round to the nearest hundredth of a percent when necessary.

15. $\frac{6}{10}$ **16.** $\frac{1}{40}$ **17.** $\frac{23}{20}$ **18.** $\frac{17}{16}$

19. $\frac{5}{7}$ **20.** $\frac{2}{7}$ **21.** $\frac{22}{23}$ **22.** $\frac{13}{19}$

23. $4\frac{2}{5}$ **24.** $2\frac{3}{4}$ **25.** $\frac{1}{300}$ **26.** $\frac{1}{400}$

Write each percent as a fraction in simplified form.

27. 22% **28.** 38% **29.** 53% **30.** 41%

31. 150% **32.** 160% **33.** $6\frac{1}{3}\%$ **34.** $4\frac{2}{3}\%$

35. $51\frac{1}{4}\%$ **36.** $43\frac{3}{4}\%$

Section 5.3

Solve each percent problem. Round to the nearest hundredth when necessary.

37. Find 24% of 230.

38. What is 52% of 43?

39. 68 is what percent of 72?

40. What percent of 22 is 18?

41. 8% of what number is 240?

42. 296 is 16% of what number?

Section 5.4

Answer each question. Round to the nearest hundredth when necessary.

43. The home team won 24 of 38 basketball games. What percent of the basketball games did the home team win?

44. A salesperson earns a commission rate of 24%. How much commission would be paid if the salesperson sold $22,500 worth of goods?

45. Robert paid sales tax of $0.72 on his dinner. The sales tax rate is 5%. What was the cost of his dinner without the tax?

46. A television set that sold originally for $480 was sold for $336. What was the percent decrease?

37. 55.2

38. 22.36

39. 94.44%

40. 81.82%

41. 3000

42. 1850

43. They won 63.16% of the games.

44. $5400 commission would be paid.

45. The dinner without the tax was $14.40.

46. The percent decrease was 30%.

5.1 Understanding Percent

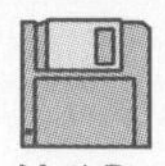

After studying this section, you will be able to:

1 *Express a fraction whose denominator is 100 as a percent.*
2 *Write a percent as a decimal.*
3 *Write a decimal as a percent.*

1 *Expressing a Fraction Whose Denominator Is 100 as a Percent*

"My raise came through. I got a 6% increase!"

"The leading economic indicators show inflation rising at a 4.3% rate."

"Babe Ruth and Hank Aaron each hit quite a few home runs. But I wonder who has the higher percentage of home runs per at bat?"

We use percents often in our everyday lives. In business, in sports, in shopping, and in many areas of life, percentages play an important role. In this section we introduce the idea of percent, which means "*per centum*" or "per hundred." We then show how to use percentages.

In previous chapters, when we described parts of a whole, we used fractions or decimals. Using a percent is another way to describe a part of a whole. Percents can be described as ratios whose denominators are 100. The word *percent* means per 100. This sketch has 100 rectangles.

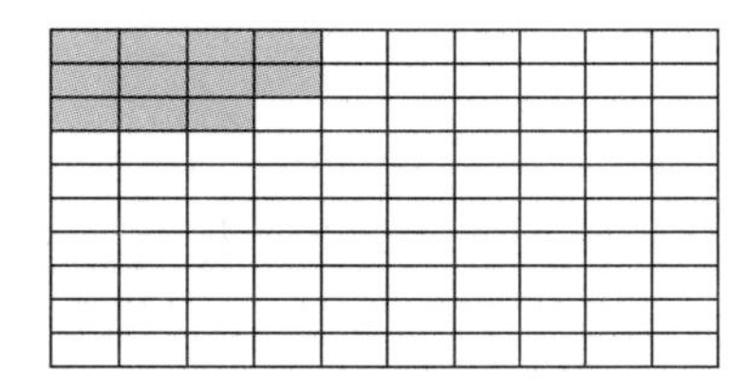

Of 100 rectangles, 11 are shaded. We can say that 11 percent of the whole is shaded. We use the symbol % for percent. It means "parts per 100." When we write 11 percent as 11%, we understand that it means 11 parts per one hundred, or, as a fraction, $\frac{11}{100}$.

EXAMPLE 1 Write each as a percent.

(a) 17 out of 100 students were left-handed.

(b) 31 out of 100 library books returned were overdue.

(c) 9 out of 100 male students were taller than 6 feet.

(d) 86 out of 100 college students had at least one parent who attended college.

(a) $\frac{17}{100} = 17\%$ **(b)** $\frac{31}{100} = 31\%$

(c) $\frac{9}{100} = 9\%$ **(d)** $\frac{86}{100} = 86\%$

Practice Problem 1 Write each as a percent.

(a) 51 out of 100 students in the class were women.

(b) 68 out of 100 cars in the parking lot have front-wheel drive.

(c) 7 out of 100 students in the dorm quit smoking.

(d) 26 out of 100 students did not vote in class elections. ■

Some percents are larger than 100. Consider the following situations.

EXAMPLE 2

(a) Write $\dfrac{386}{100}$ as a percent.

(b) Last year's enrollment in psychology classes was 100 students. This year's enrollment is 136. Write this year's enrollment as a percent of last year's.

(c) Twenty years ago four car tires for a full-sized car cost $100. Now the average price for four car tires for a full-sized car is $270. Write the present cost as a percent of the cost 20 years ago.

(a) $\dfrac{386}{100} = 386\%$

(b) The ratio is $\dfrac{136 \text{ students this year}}{100 \text{ students last year}}$. $\quad \dfrac{136}{100} = 136\%$

This year's enrollment is 136% of last year's enrollment.

(c) The ratio is $\dfrac{\$270 \text{ for four tires now}}{\$100 \text{ for four tires then}}$. $\quad \dfrac{270}{100} = 270\%$

The present cost of four car tires for a full-sized car is 270% of the cost 20 years ago.

Practice Problem 2

(a) Write $\dfrac{238}{100}$ as a percent.

(b) Write $\dfrac{505}{100}$ as a percent.

(c) Last year 100 students tried out for varsity baseball. This year 121 students tried out. Write this year's number as a percent of last year's number. ■

Some percents are smaller than 1%.

$\dfrac{0.7}{100}$ can be written as 0.7%.

$\dfrac{0.3}{100}$ can be written as 0.3%.

$\dfrac{0.04}{100}$ can be written as 0.04%.

TEACHING TIP Students usually have trouble writing $\frac{245}{100}$ as a percent. It is a good idea to put four simple examples on the board, ordered from smallest to largest, and show them in both fractional and decimal form. Try a display on the board like this:

FRACTIONAL FORM	PERCENT NOTATION FORM
$\dfrac{0.6}{100}$	0.6%
$\dfrac{7}{100}$	7%
$\dfrac{34}{100}$	34%
$\dfrac{245}{100}$	245%

Comparing one form to the other usually clarifies the idea for the student.

EXAMPLE 3 Write each as a percent.

(a) $\dfrac{0.9}{100}$ **(b)** $\dfrac{0.2}{100}$ **(c)** $\dfrac{0.07}{100}$

(a) $\dfrac{0.9}{100} = 0.9\%$ **(b)** $\dfrac{0.2}{100} = 0.2\%$ **(c)** $\dfrac{0.07}{100} = 0.07\%$

Practice Problem 3 Write each as a percent.

(a) $\dfrac{0.5}{100}$ **(b)** $\dfrac{0.06}{100}$ **(c)** $\dfrac{0.003}{100}$ ■

2 *Writing a Percent as a Decimal*

Suppose we have a percent such as 59%. What would be the equivalent in decimal form? Using our definition of percent, $59\% = \frac{59}{100}$. This fraction could be written in decimal form as 0.59. In a similar way, we could write 21% as $\frac{21}{100} = 0.21$. This pattern allows us to quickly change the form of a number from a percent to a fraction whose denominator is 100 to a decimal.

EXAMPLE 4 Write each percent as a decimal.

(a) 38% **(b)** 6% **(c)** 70%

(a) $38\% = \frac{38}{100} = 0.38$ **(b)** $6\% = \frac{6}{100} = 0.06$

(c) $70\% = \frac{70}{100} = 0.70$ or 0.7

The final zero is not really necessary.

Practice Problem 4 Write each percent as a decimal.

(a) 47% **(b)** 2% **(c)** 90% ■

The results of Example 4 suggest the following rule.

Changing a Percent to a Decimal

1. Move the decimal point two places to the left.
2. Drop the % symbol.

EXAMPLE 5 Write each percent as a decimal.

(a) 26.9% **(b)** 7.2% **(c)** 0.13% **(d)** 158%

In each case, we move the decimal point two places to the left and drop the percent symbol.

(a) 26.9% = 0.269

(b) 7.2% = 0.072 *Note we need to add an extra zero to the left of the seven.*

(c) 0.13% = 0.0013 *Here we added two zeros to the left of the* 1.

(d) 158% = 1.58

Practice Problem 5 Write each percent as a decimal.

(a) 80.6% **(b)** 2.5% **(c)** 0.29% **(d)** 231% ■

3 *Writing a Decimal as a Percent*

In Example 4(a) we changed 38% to $\frac{38}{100}$ to 0.38. We can reverse the process. We can start with $0.38 = \frac{38}{100} = 38\%$. Study all the parts of Examples 4 and 5. You will see that the steps are reversible. Thus 0.38 = 38%, 0.06 = 6%, 0.70 = 70%, 0.269 = 26.9%, 0.072 = 7.2%, 0.0013 = 0.13%, and 1.58 = 158%. In each part we see that the decimal point is moved two places to the right. Then the percent symbol is written after the number.

Changing a Decimal to a Percent

1. Move the decimal point two places to the right.
2. Then write the % symbol at the end of the number.

EXAMPLE 6 Write each decimal as a percent.

(a) 0.47 **(b)** 0.08 **(c)** 6.31 **(d)** 0.055 **(e)** 0.001

In each part we move the decimal point two places to the right and write the percent symbol at the end of the number.

(a) 0.47 = 47% **(b)** 0.08 = 8% **(c)** 6.31 = 631%
(d) 0.055 = 5.5% **(e)** 0.001 = 0.1%

Practice Problem 6 Write each decimal as a percent.

(a) 0.78 **(b)** 0.02 **(c)** 5.07 **(d)** 0.029 **(e)** 0.006 ■

TEACHING TIP It's a good idea to stress the reversibility of the operation. Write a problem like this on the board and ask the class to complete it.

DECIMAL FORM	PERCENT FORM
0.58	—
--	62%
0.03	—
0.006	—
—	0.78%
—	321%

CALCULATOR

Percent to Decimal

You can use a calculator to change 52% to a decimal.
Enter

52 [%]

The display should read

[0.52]

Try

(a) 46% **(b)** 137%
(c) 9.3% **(d)** 6%

Note: The calculator divides by 100 when the percent key is pressed. If you do not have a [%] key then you can use the keystrokes [÷] 100 [=].

To Think About What is really happening when we change a decimal to a percent? Suppose that we wanted to change 0.59 to a percent.

$0.59 = \frac{59}{100}$ *Definition of a decimal.*

$= 59 \times \frac{1}{100}$ *Definition of multiplying fractions.*

= 59 percent *Because "per 100" means percent.*

= 59% *Writing the symbol for percent.*

Can you see why each step is valid? Since we know the reason behind each step, we know we can always move the decimal point two places to the right and write the percent symbol. See Exercises 5.1, problems 77 and 78.

5.1 Exercises

Verbal and Writing Skills

1. In this section we introduced percent, which means "per centum" or "per ___hundred___."

2. The number 1 written as a percent is ___100%___.

3. To change a percent to a decimal, move the decimal point ___two___ places to the ___left___. ___Drop___ the % symbol.

4. To change a decimal to a percent, move the decimal point ___two___ places to the ___right___. ___Write___ the % symbol at the end of the number.

Write each fraction as a percent.

5. $\frac{45}{100}$ 45%

6. $\frac{48}{100}$ 48%

7. $\frac{7}{100}$ 7%

8. $\frac{9}{100}$ 9%

9. $\frac{80}{100}$ 80%

10. $\frac{60}{100}$ 60%

11. $\frac{245}{100}$ 245%

12. $\frac{110}{100}$ 110%

13. $\frac{5.3}{100}$ 5.3%

14. $\frac{8.3}{100}$ 8.3%

15. $\frac{0.6}{100}$ 0.6%

16. $\frac{0.4}{100}$ 0.4%

Applications

Write a percent to express each of the following.

17. 29 out of 100 power boats had a radar navigation system. 29%

18. 13 out of 100 students in the class were seniors. 13%

19. 78 out of 100 new car owners were satisfied with their purchase. 78%

20. 64 out of 100 college sophomores are confident of their selection of a major. 64%

21. 32 out of 100 people who ate out chose to eat at a fast-food restaurant. 32%

22. 8 out of 100 people overdraw their checking account at least twice per year. 8%

Write each percent as a decimal.

23. 51% 0.51 **24.** 42% 0.42 **25.** 7% 0.07 **26.** 6% 0.06 **27.** 20% 0.2 **28.** 40% 0.4 **29.** 43.6% 0.436 **30.** 81.5% 0.815

31. 0.3% 0.003 **32.** 0.9% 0.009 **33.** 0.72% 0.0072 **34.** 0.61% 0.0061 **35.** 126% 1.26 **36.** 175% 1.75 **37.** 366% 3.66 **38.** 398% 3.98

Write each decimal as a percent.

39. 0.74 74% **40.** 0.66 66% **41.** 0.50 50% **42.** 0.40 40% **43.** 0.08 8% **44.** 0.03 3% **45.** 0.563 56.3% **46.** 0.408 40.8%

47. 0.002 0.2% **48.** 0.009 0.9% **49.** 0.0057 0.57% **50.** 0.0026 0.26% **51.** 1.35 135% **52.** 1.86 186% **53.** 2.72 272% **54.** 3.04 304%

55. Bob paid 0.27 of his income for federal income tax. 27%

56. Sally spends 0.31 of her income for housing. 31%

Mixed Practice

Write as a percent.

57. $\frac{36}{100}$ 36%

58. 0.25 25%

59. $\frac{143}{100}$ 143%

60. 1.48 148%

61. 0.3 30%

62. $\frac{40}{100}$ 40%

63. $\frac{0.5}{100}$ 0.5%

64. 0.005 0.5%

Write as a decimal.

65. 62% 0.62

66. $\frac{12}{100}$ 0.12

67. 128% 1.28

68. $\frac{210}{100}$ 2.1

69. 0.5% 0.005

70. $\frac{0.8}{100}$ 0.008

71. $\frac{80}{100}$ 0.8

72. 70% 0.7

Applications

Write as a percent.

73. For every 100 people who voted in the 1996 presidential election, 49 voted for Bill Clinton. 49%

74. For every 100 people who voted in the 1996 presidential election, 41 voted for Robert Dole. 41%

75. In Alaska, 0.03413 of the state is covered by water. 3.413%

76. In Florida, 0.07689 of the state is covered by water. 7.689%

To Think About

77. Suppose that we want to change 36% to 0.36 by moving the decimal point two places to the left and dropping the % symbol. Explain the steps to show what is really involved in changing 36% to 0.36. Why does the rule work? 36% = 36 percent = 36 "per one hundred" = $36 \times \frac{1}{100} = \frac{36}{100} = 0.36$. The rule is using the fact that 36% means 36 per one hundred.

78. Suppose that we want to change 10.65 to 1065%. Give a complete explanation of the steps. 10.65 = $1065 \times \frac{1}{100}$ = 1065 "per one hundred" = 1065 percent = 1065%. We change 10.65 to $1065 \times \frac{1}{100}$ and use the idea that percent means "per one hundred."

Take the value in each case and write it (a) as a decimal, (b) as a fraction with a denominator of 100, and (c) as a reduced fraction.

79. 55,562%

(a) 555.62 **(b)** $\frac{55{,}562}{100}$ **(c)** $\frac{27{,}781}{50}$

80. 60,724%

(a) 607.24 **(b)** $\frac{60{,}724}{100}$ **(c)** $\frac{15{,}181}{25}$

Cumulative Review Problems

Write as a fraction in simplest form.

81. 0.56 $\frac{14}{25}$

82. 0.72 $\frac{18}{25}$

Write as a decimal.

83. $\frac{11}{16}$ 0.6875

84. $\frac{13}{16}$ 0.8125

5.2 Changing Between Percents, Decimals, and Fractions

MathPro

Video 11

SSM

After studying this section, you will be able to:

1 *Change a percent to a fraction.*
2 *Change a fraction to a percent.*
3 *Change a percent, a decimal, or a fraction to equivalent forms.*

1 *Changing a Percent to a Fraction*

By using the definition of percent, we can write any percent as a fraction whose denominator is 100. Thus when we change a percent to a fraction, we remove the percent symbol and write the number over 100. If possible, we simplify the fraction.

EXAMPLE 1 Write as a fraction in simplest form.

(a) 37% **(b)** 75% **(c)** 2%

(a) $37\% = \frac{37}{100}$ **(b)** $75\% = \frac{75}{100} = \frac{3}{4}$ **(c)** $2\% = \frac{2}{100} = \frac{1}{50}$

Practice Problem 1 Write as a fraction in simplest form.

(a) 71% **(b)** 25% **(c)** 8% ■

In some cases, it may be helpful to write the percent as a decimal before you write it as a fraction in simplest form.

EXAMPLE 2 Write as a fraction in simplest form.

(a) 43.5% **(b)** 5.8% **(c)** 36.75%

(a) $43.5\% = 0.435$ *Change the percent to a decimal.*

$= \frac{435}{1000}$ *Change the decimal to a fraction.*

$= \frac{87}{200}$ *Reduce the fraction.*

(b) $5.8\% = 0.058 = \frac{58}{1000} = \frac{29}{500}$

(c) $36.75\% = 0.3675 = \frac{3675}{10{,}000} = \frac{147}{400}$

TEACHING TIP Stress the fact that some percents have a simple fraction notation, while others have a rather large and unwieldy one.

Practice Problem 2 Write each percent as a fraction in simplest form.

(a) 8.4% **(b)** 55.25% **(c)** 28.5% ■

If the percent is greater than 100%, the simplified fraction is usually changed to a mixed number.

TEACHING TIP Some students will forget to reduce after they change the percent to a mixed fraction. Remind them that if they write $375\% = 3\frac{75}{100}$ the answer is not completely finished (and thus not correct) until it is reduced to $3\frac{3}{4}$.

EXAMPLE 3 Write each percent as a mixed number.

(a) 225% **(b)** 138%

(a) $225\% = 2.25 = 2\frac{25}{100} = 2\frac{1}{4}$ **(b)** $138\% = 1.38 = 1\frac{38}{100} = 1\frac{19}{50}$

Practice Problem 3 Write each percent as a mixed number.

(a) 170% **(b)** 288% ■

Sometimes a percent is not a whole number, such as 9% or 10%. Instead, it contains a fraction, such as $9\frac{1}{12}\%$ or $9\frac{3}{8}\%$. Extra steps will be needed to write such a percent as a simplified fraction.

EXAMPLE 4 Convert $9\frac{3}{8}\%$ to a fraction in simplest form.

$9\frac{3}{8}\% = \frac{9\frac{3}{8}}{100}$ *Change the percent to a fraction.*

$= 9\frac{3}{8} \div \frac{100}{1}$ *Write the division horizontally.* $\frac{9\frac{3}{8}}{100}$ *means* $9\frac{3}{8}$ *divided by* 100.

$= \frac{75}{8} \div \frac{100}{1}$ *Write* $9\frac{3}{8}$ *as a fraction.*

$= \frac{75}{8} \times \frac{1}{100}$ *Use the definition of division of fractions.*

$= \frac{\overset{3}{\cancel{75}}}{8} \times \frac{1}{\underset{4}{\cancel{100}}}$ *Simplify.*

$= \frac{3}{32}$ *Reduce the fraction.*

Practice Problem 4 Convert $7\frac{5}{8}\%$ to a fraction in simplest form. ■

EXAMPLE 5 In the fiscal 1994 budget of the United States, approximately $20\frac{3}{8}\%$ of the budget was designated for defense. Write this percent as a fraction.

$$20\frac{3}{8}\% = \frac{20\frac{3}{8}}{100} = 20\frac{3}{8} \div 100 = \frac{163}{8} \times \frac{1}{100} = \frac{163}{800}$$

Thus we could say $\frac{163}{800}$ of the fiscal 1994 budget was designated for defense. That is, for every \$800 in the budget, \$163 was spent for defense.

Practice Problem 5 In the fiscal 1994 budget of the United States, approximately $21\frac{1}{4}\%$ was designated for Health and Human Services. Write this percent as a fraction. ■

Certain percents occur very often, especially in money matters. Here are some common equivalents that you may already know. If not, be sure to memorize them.

$$25\% = \frac{1}{4} \qquad 33\frac{1}{3}\% = \frac{1}{3} \qquad 10\% = \frac{1}{10}$$

$$50\% = \frac{1}{2} \qquad 66\frac{2}{3}\% = \frac{2}{3}$$

$$75\% = \frac{3}{4}$$

2 *Changing a Fraction to a Percent*

A convenient way to change a fraction to a percent is to write the fraction in decimal form first and then convert the decimal to a percent.

EXAMPLE 6 Write $\frac{3}{8}$ as a percent.

We see that $\frac{3}{8} = 0.375$ by dividing out $3 \div 8$.

```
   0.375
8)3.000
   24
    60
    56
     40
     40
      0
```

Thus $\frac{3}{8} = 0.375 = 37.5\%$.

Practice Problem 6 Write $\frac{5}{8}$ as a percent. ■

EXAMPLE 7 Write each fraction as a percent.

(a) $\frac{4}{5}$ **(b)** $\frac{7}{40}$ **(c)** $\frac{39}{50}$

(a) $\frac{4}{5} = 0.8 = 80\%$ **(b)** $\frac{7}{40} = 0.175 = 17.5\%$

(c) $\frac{39}{50} = 0.78 = 78\%$

Practice Problem 7 Write each fraction as a percent.

(a) $\frac{3}{5}$ **(b)** $\frac{21}{25}$ **(c)** $\frac{7}{16}$ ■

Changing some fractions to decimal form results in an infinitely repeating decimal. In such cases, we usually round to the nearest hundredth of a percent.

TEACHING TIP Remind students that if we want to write an exact value, we must say $\frac{2}{3} = 0.666666\ldots$ or use the bar notation over the 6. They must realize that $\frac{2}{3} = 0.67$ is a value rounded to the nearest hundredth. In contrast, $\frac{2}{3} = 0.66$ is a truncated value and is not an accurate approximation.

EXAMPLE 8 Write each fraction as a percent. Round to the nearest hundredth of a percent.

(a) $\frac{1}{6}$ **(b)** $\frac{15}{33}$

(a) $\frac{1}{6} = 0.16666\ldots$ by dividing out $1 \div 6$.

$$\begin{array}{r} 0.166 \\ 6\overline{)1.000} \\ \underline{6} \\ 40 \\ \underline{36} \\ 40 \end{array}$$

The repeating quotient can be written as $0.1\overline{6}$. If we round the decimal to the nearest ten thousandth, we have $\frac{1}{6} = 0.1667$. If we change this to a percent, we have

$$\frac{1}{6} = 16.67\%$$

This is correct to the nearest hundredth of a percent.

(b) By dividing out $15 \div 33$, we see that $\frac{15}{33} = 0.45454545\ldots$ This can be written as $0.\overline{45}$. If we round to the nearest ten-thousandth, we have

$$\frac{15}{33} = 0.4545 = 45.45\%$$

This rounded value is correct to the nearest hundredth of a percent.

Practice Problem 8 Write each fraction as a percent. Round to the nearest hundredth of a percent.

(a) $\frac{7}{9}$ **(b)** $\frac{19}{30}$ ■

In some cases, a percent is written with a fraction.

EXAMPLE 9 Express $\frac{11}{12}$ as a percent containing a fraction.

We will terminate the division after two steps, and write the remainder in fraction form.

$$\begin{array}{r} 0.91 \\ 12\overline{)11.00} \\ \underline{108} \\ 20 \\ \underline{12} \\ 8 \end{array}$$

This division tells us that we can write

$$\frac{11}{12} \quad \text{as} \quad 0.91\frac{8}{12} \quad \text{or} \quad 0.91\frac{2}{3}$$

We now have a decimal with a fraction. When we express this decimal as a percent, we move the decimal point two places to the right. We do not write the decimal in front of the fraction.

$$0.91\frac{2}{3} = 91\frac{2}{3}\%$$

Practice Problem 9 Express $\frac{7}{12}$ as a percent containing a fraction. ■

3 *Changing a Percent, a Decimal, or a Fraction to Equivalent Forms*

We have seen so far that a fraction, a decimal, and a percent are three different forms (notations) for the same number. We can illustrate this in a chart.

EXAMPLE 10 Complete the following table of equivalent notations. Round decimals to the nearest ten thousandth. Round percents to the nearest hundredth of a percent.

Fraction	Decimal	Percent
$\frac{11}{16}$		
	0.265	
		$17\frac{1}{5}\%$

Begin with the first row. The number is written as a fraction. We will change the fraction to a decimal and then to a percent.

The Fraction is changed to a Decimal is changed to a Percent

$$\frac{11}{16} \longrightarrow 16\overline{)11.0000}^{\,0.6875} \longrightarrow 68.75\%$$

In the second row the number is written as a decimal. This can easily be written as a percent.

$$0.265 \longrightarrow 26.5\%$$

To write the number as a fraction, write 0.265 as a fraction and simplify.

$$0.265$$
$$\downarrow$$
$$\frac{53}{200} \longleftarrow \frac{265}{1000}$$

In the third row the number is written as a percent. Proceed from right to left—that is, write the number as a decimal and then as a fraction.

$$\frac{17\frac{1}{5}}{100} \leftarrow 17\frac{1}{5}\%$$
$$\downarrow$$
$$\frac{86}{5} \times \frac{1}{100}$$
$$\downarrow$$
$$0.172 \longleftarrow \frac{86}{500} \qquad \textit{Divide.}\ 500\overline{)86.000}^{\,0.172}$$

and

$$\frac{43}{250} \longleftarrow \frac{86}{500}$$

Thus the completed table is as follows.

Fraction	Decimal	Percent
$\frac{11}{16}$	0.6875	68.75%
$\frac{53}{200}$	0.265	26.5%
$\frac{43}{250}$	0.172	$17\frac{1}{5}\%$

Practice Problem 10 Complete the following table of equivalent notations. Round decimals to the nearest ten thousandth. Round percents to the nearest hundredth of a percent.

Fraction	Decimal	Percent
$\frac{23}{99}$	0.2323	23.23%
$\frac{129}{250}$	0.516	51.6%
$\frac{97}{250}$	0.388	$38\frac{4}{5}\%$

■

ALTERNATIVE METHOD Another way to convert a fraction to a percent is to use a proportion. To change $\frac{7}{8}$ to a percent, write the proportion

$$\frac{7}{8} = \frac{n}{100}$$

$$7 \times 100 = 8 \times n \quad \textit{Cross-multiply.}$$

$$700 = 8 \times n \quad \textit{Simplify.}$$

$$\frac{700}{8} = \frac{8 \times n}{8} \quad \textit{Divide each side by 8.}$$

$$87.5 = n \quad \textit{Simplify.}$$

Thus $\frac{7}{8} = 87.5\%$. You will use this approach in Exercises 5.2, problems 93 and 94.

CALCULATOR

Fraction to Decimal

You can use a calculator to change $\frac{3}{5}$ to a decimal.

Enter

3 [÷] 5 [=]

The display should read

[0.6]

Try

(a) $\frac{17}{25}$ **(b)** $\frac{2}{9}$

(c) $\frac{13}{10}$ **(d)** $\frac{15}{19}$

Note: 0.7894737 is an approximation for $\frac{15}{19}$. Some calculators only round to 8 places.

Developing Your Study Skills

READING THE TEXTBOOK

Homework time each day should begin with the careful reading of the section(s) assigned in your textbook. Much time and effort have gone into the selection of a particular text, and your instructor has chosen a book that will help you to become successful in this mathematics class. Expensive textbooks can be a wise investment if you take advantage of them by reading them.

Reading a mathematics textbook is unlike reading many other types of books that you may use in your literature, history, psychology, or sociology courses. Mathematics texts are technical books that provide you with exercises to practice on. Reading a mathematics text requires slow and careful reading of each word, which takes time and effort.

Begin reading your textbook with a paper and pencil in hand. As you come across a new definition, or concept, underline it in the text and/or write it down in your notebook. Whenever you encounter an unfamiliar term, look it up and make a note of it. When you come to an example, work through it step by step. Be sure to read each word and to follow directions carefully.

Notice the helpful hints the author provides to guide you to correct solutions and prevent you from making errors. Take advantage of these pieces of expert advice.

Be sure that you understand what you are reading. Make a note of any of those things that you do not understand and ask your instructor about them. Do not hurry through the material. Learning mathematics takes time.

5.2 Exercises

Write each percent as a fraction or as a mixed number.

1. 88% $\quad \frac{88}{100} = \frac{22}{25}$
2. 52% $\quad \frac{52}{100} = \frac{13}{25}$
3. 7% $\quad \frac{7}{100}$
4. 9% $\quad \frac{9}{100}$
5. 33% $\quad \frac{33}{100}$
6. 47% $\quad \frac{47}{100}$
7. 55% $\quad \frac{55}{100} = \frac{11}{20}$
8. 35% $\quad \frac{35}{100} = \frac{7}{20}$

9. 75% $\quad \frac{75}{100} = \frac{3}{4}$
10. 25% $\quad \frac{25}{100} = \frac{1}{4}$
11. 20% $\quad \frac{20}{100} = \frac{1}{5}$
12. 40% $\quad \frac{40}{100} = \frac{2}{5}$
13. 14.5% $\quad \frac{14.5}{100} = \frac{29}{200}$
14. 18.5% $\quad \frac{18.5}{100} = \frac{37}{200}$
15. 17.6% $\quad \frac{17.6}{100} = \frac{22}{125}$
16. 78.4% $\quad \frac{78.4}{100} = \frac{98}{125}$

17. 64.8% $\quad \frac{64.8}{100} = \frac{81}{125}$
18. 12.2% $\quad \frac{12.2}{100} = \frac{61}{500}$
19. 71.25% $\quad \frac{71.25}{100} = \frac{57}{80}$
20. 38.75% $\quad \frac{38.75}{100} = \frac{31}{80}$
21. 176% $\quad \frac{176}{100} = 1\frac{19}{25}$
22. 228% $\quad \frac{228}{100} = 2\frac{7}{25}$
23. 340% $\quad \frac{340}{100} = 3\frac{2}{5}$
24. 420% $\quad \frac{420}{100} = 4\frac{1}{5}$

25. 1200% $\quad \frac{1200}{100} = 12$
26. 3600% $\quad \frac{3600}{100} = 36$
27. $2\frac{1}{6}\%$ $\quad \frac{\frac{13}{6}}{100} = \frac{13}{600}$
28. $7\frac{5}{6}\%$ $\quad \frac{\frac{47}{6}}{100} = \frac{47}{600}$
29. $12\frac{3}{8}\%$ $\quad \frac{\frac{99}{8}}{100} = \frac{99}{800}$
30. $15\frac{5}{8}\%$ $\quad \frac{\frac{125}{8}}{100} = \frac{5}{32}$
31. $8\frac{4}{5}\%$ $\quad \frac{\frac{44}{5}}{100} = \frac{11}{125}$
32. $9\frac{3}{5}\%$ $\quad \frac{\frac{48}{5}}{100} = \frac{12}{125}$

Applications

33. In the 1996 presidential election, Ross Perot received $2\frac{2}{13}\%$ of the vote in the District of Columbia. Write this percent as a fraction.

$\frac{\frac{28}{13}}{100} = \frac{7}{325}$

34. During an inspection $17\frac{3}{4}\%$ of the labels printed had the wrong colors. What *fraction* of the labels was in the wrong color?

$\frac{\frac{71}{4}}{100} = \frac{71}{400}$

35. In the 1994 budget of the United States, approximately $4\frac{1}{11}\%$ was designated for the Department of Agriculture. Write this percent as a fraction.

$\frac{\frac{45}{11}}{100} = \frac{9}{220}$

36. In the 1994 budget of the United States, approximately $2\frac{6}{11}\%$ was designated for the Department of Transportation. Write this percent as a fraction.

$\frac{\frac{28}{11}}{100} = \frac{14}{550} = \frac{7}{275}$

Write each fraction or mixed number as a percent. Round to the nearest hundredth of a percent when necessary.

37. $\frac{3}{4}$ $\quad$ 75%
38. $\frac{1}{4}$ $\quad$ 25%
39. $\frac{1}{3}$ $\quad$ 33.33%
40. $\frac{2}{3}$ $\quad$ 66.67%
41. $\frac{5}{16}$ $\quad$ 31.25%
42. $\frac{9}{16}$ $\quad$ 56.25%
43. $\frac{7}{25}$ $\quad$ 28%
44. $\frac{9}{25}$ $\quad$ 36%

45. $\frac{11}{40}$ $\quad$ 27.5%
46. $\frac{13}{40}$ $\quad$ 32.5%
47. $\frac{5}{12}$ $\quad$ 41.67%
48. $\frac{8}{12}$ $\quad$ 66.67%
49. $\frac{18}{5}$ $\quad$ 360%
50. $\frac{7}{4}$ $\quad$ 175%
51. $2\frac{5}{6}$ $\quad$ 283.33%
52. $3\frac{1}{6}$ $\quad$ 316.67%

53. $4\frac{1}{8}$ $\quad$ 412.5%
54. $2\frac{5}{8}$ $\quad$ 262.5%
55. $\frac{3}{7}$ $\quad$ 42.86%
56. $\frac{2}{7}$ $\quad$ 28.57%
57. $\frac{15}{16}$ $\quad$ 93.75%
58. $\frac{11}{16}$ $\quad$ 68.75%
59. $\frac{26}{50}$ $\quad$ 52%
60. $\frac{43}{50}$ $\quad$ 86%

61. $\frac{47}{137}$ $\quad$ 34.31%
62. $\frac{53}{129}$ $\quad$ 41.09%
63. $\frac{316}{907}$ $\quad$ 34.84%
64. $\frac{427}{891}$ $\quad$ 47.92%

Applications

Round to the nearest hundredth of a percent.

65. The brain represents approximately $\frac{1}{40}$ of an average person's weight. Express this fraction as a percent. 2.5%

66. Carbon represents approximately $\frac{9}{50}$ of an average person's weight. Express this fraction as a percent. 18%

67. The U.S. mail service estimates that people do not send as many personal letters through the mail as they did 20 years ago. In fact in 1996 only 37/3000 of all mail delivered consisted of personal letters. Express this fraction as a percent. 1.23%

68. During waking hours, a person blinks 9/2000 of the time. Express this fraction as a percent. 0.45%

Express each fraction as a percent containing a fraction. (See Example 9.)

69. $\frac{5}{6}$ $83\frac{1}{3}\%$ **70.** $\frac{1}{6}$ $16\frac{2}{3}\%$ **71.** $\frac{11}{16}$ $68\frac{3}{4}\%$ **72.** $\frac{15}{16}$ $93\frac{3}{4}\%$ **73.** $\frac{3}{8}$ $37\frac{1}{2}\%$ **74.** $\frac{5}{8}$ $62\frac{1}{2}\%$ **75.** $\frac{3}{40}$ $7\frac{1}{2}\%$ **76.** $\frac{11}{90}$ $12\frac{2}{9}\%$

Mixed Practice

Complete the following table of equivalents. Round decimals to the nearest ten thousandth. Round percents to the nearest hundredth of a percent.

	Fraction	Decimal	Percent
77.	$\frac{5}{12}$	0.4167	41.67%
79.	$\frac{3}{50}$	0.06	6%
81.	$\frac{2}{5}$	0.4	40%
83.	$\frac{69}{200}$	0.345	34.5%
85.	$\frac{3}{200}$	0.015	1.5%
87.	$\frac{5}{9}$	0.5556	55.56%
89.	$\frac{1}{32}$	0.0313	$3\frac{1}{8}\%$

	Fraction	Decimal	Percent
78.	$\frac{7}{12}$	0.5833	58.33%
80.	$\frac{2}{25}$	0.08	8%
82.	$\frac{3}{5}$	0.6	60%
84.	$\frac{5}{8}$	0.625	62.5%
86.	$\frac{9}{200}$	0.045	4.5%
88.	$\frac{7}{9}$	0.7778	77.78%
90.	$\frac{21}{800}$	0.0263	$2\frac{5}{8}\%$

91. Write $28\frac{15}{16}\%$ as a fraction.

$\frac{463}{16} \times \frac{1}{100} = \frac{463}{1600}$

★ **92.** Write $18\frac{7}{12}\%$ as a fraction.

$\frac{223}{12} \times \frac{1}{100} = \frac{223}{1200}$

Change each fraction to a percent by using a proportion.

★ **93.** $\frac{123}{800}$ $\frac{123}{800} = \frac{n}{100}$ $n = 15.375$ 15.375%

★ **94.** $\frac{417}{600}$ $\frac{417}{600} = \frac{n}{100}$ $n = 69.5$ 69.5%

Cumulative Review Problems

Solve for n in each problem.

95. $\frac{15}{n} = \frac{8}{3}$ $n = 5.625$

96. $\frac{32}{24} = \frac{n}{3}$ $n = 4$

97. $\frac{n}{11} = \frac{32}{4}$ $n = 88$

98. $\frac{1}{3} = \frac{32}{n}$ $n = 96$

5.3A Solving Percent Problems Using an Equation

MathPro Video 11 SSM

After studying this section, you will be able to:

1. *Translate a percent problem into an equation.*
2. *Solve a percent problem by solving an equation.*

1 *Translating a Percent Problem into an Equation*

In word problems, like the ones in this section, we can translate from words to mathematical symbols and back again. After we have the mathematical symbols arranged in an *equation,* we solve the equation. When we find the values that make the equation true, we have also found the answer to our word problem.

To solve a percent problem, express it as an equation with an unknown quantity. We use the letter n to represent the number we do not know. The following table is helpful when translating from a percent problem to an equation.

Word	Mathematical Symbol
of	Any multiplication symbol: $\times$ or ( ) or $\cdot$
is	$=$
what	Any letter; for example, n
find	$n =$

In Examples 1–6 we show how to translate words into an equation. Please do **not** solve the problem. Translate into an equation only.

EXAMPLE 1 Translate into an equation.

What is 5% of 19.00?

↓ ↓ ↓ ↓ ↓

$$n = 5\% \times 19.00$$

Practice Problem 1 Translate into an equation. What is 26% of 35? ■

EXAMPLE 2 Translate into an equation.

142% of 35 is what?

↓ ↓ ↓ ↓ ↓

$$142\% \times 35 = n$$

Practice Problem 2 Translate into an equation. 155% of 20 is what? ■

TEACHING TIP The practice of translation for several examples is an excellent way for students to develop confidence with percent problems. Encourage them not to skip the step of practicing translation.

EXAMPLE 3 Translate into an equation.

Find 0.6% of 400.

Notice here that the words *what is* are missing. The word *find* is equivalent to *what is.*

Find 0.6% of 400.

↓ ↓ ↓ ↓

$n = 0.6\% \times 400$

Practice Problem 3 Translate into an equation. Find 0.08% of 350. ■

The unknown quantity, *n*, does not always stand alone in an equation.

EXAMPLE 4 Translate into an equation.

(a) 35% of what is 60? **(b)** 7.2 is 120% of what?

(a) 35% of what is 60?

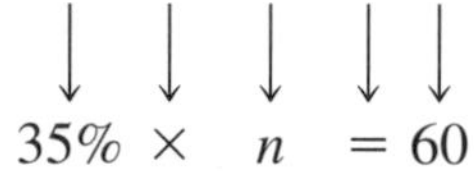

$35\% \times n = 60$

(b) 7.2 is 120% of what?

↓ ↓ ↓ ↓ ↓

$7.2 = 120\% \times n$

Practice Problem 4 Translate into an equation.

(a) 58% of what is 400? **(b)** 9.1 is 135% of what? ■

EXAMPLE 5 Translate into an equation.

What percent of 50 is 10?

$n \times 50 = 10$

We see here that the words *what percent* are represented by the letter *n*.

Practice Problem 5 Translate into an equation. What percent of 250 is 36? ■

EXAMPLE 6 Translate into an equation.

(a) 30 is what percent of 16? **(b)** What percent of 3000 is 2.6?

(a) 30 is what percent of 16?

$30 = n \times 16$

(b) What percent of 3000 is 2.6?

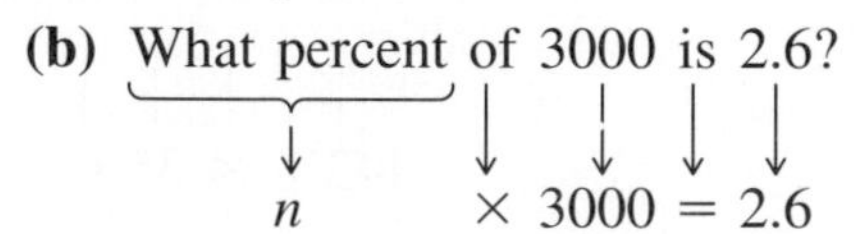

$n \times 3000 = 2.6$

Practice Problem 6 Translate into an equation.

(a) 50 is what percent of 20? **(b)** What percent of 2000 is 4.5? ■

2 *Solve a Percent Problem by Solving an Equation*

The percent problems we have translated are of three types. Consider the equation $60 = 20\% \times 300$. This problem has the form

$$\text{amount} = \text{percent} \times \text{base}$$

Any of these quantities—amount, percent, or base—may be unknown.

1. When *we do not know the amount,* we have an equation like
$$n = 20\% \times 300$$
2. When *we do not know the base,* we have an equation like
$$60 = 20\% \times n$$
3. When *we do not know the percent,* we have an equation like
$$60 = n \times 300$$

We will study each type separately. It is not necessary to memorize the three types, but it is helpful to look carefully at the examples we give of each. In each example, do the computation in a way that is easiest for you. This may be using a pencil and paper, using a calculator, or, in some cases, doing the problem mentally.

Solving Percent Problems When We Do Not Know the Amount

EXAMPLE 7 What is 45% of 590?

↓ ↓ ↓ ↓ ↓

$n = 45\% \times 590$ *Translate into an equation.*

$n = (0.45)(590)$ *Change the percent to decimal form and multiply.*

$n = 265.5$ *Multiply* 0.45×590.

TEACHING TIP Some students may insist that they do not need to use an equation to solve problems where the amount is not known. Remind them that it is good to do a few problems using an equation even though they do not need it, just so they will be better able to correctly identify which of the three types of percent problems they have encountered.

Practice Problem 7 What is 82% of 350? ■

EXAMPLE 8

(a) 3.8% of 600 is what? **(b)** Find 160% of 500.

(a) 3.8% of 600 is what?

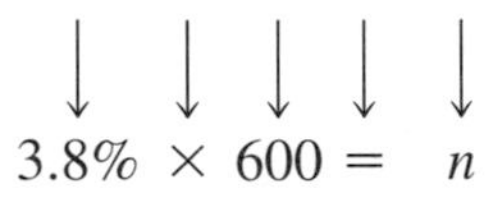

$3.8\% \times 600 = n$ *Translate into an equation.*

$(0.038)(600) = n$ *Change the percent to decimal form and multiply.*

$22.8 = n$ *Multiply* 0.038×600.

(b) Find 160% of 500. *When you translate, remember that the word* find *is equivalent to* what is.

↓ ↓ ↓ ↓

$n = 160\% \times 500$

$n = (1.60)(500)$ *Change the percent to decimal form and multiply.*

$n = 800$ *Multiply* 1.6×500.

Practice Problem 8

(a) 5.2% of 800 is what? **(b)** Find 230% of 400. ■

CALCULATOR

Percent of a Number

You can use a calculator to find 12% of 48. Enter

12 [%] [×] 48 [=]

The display should read

[5.76]

If your calculator *does not* have a percent key, use the keystrokes

0.12 [×] 48 [=]

Try

What is 54% of 450?

EXAMPLE 9 When Rick bought a new Dodge Neon, he had to pay a sales tax of 5% on the cost of the car, which was \$12,000. What was the sales tax?

This problem is asking

What is 5% of \$12,000?

$$n = 5\% \times \$12{,}000$$
$$n = 0.05 \times 12{,}000$$
$$n = \$600$$

The sales tax was \$600.

Practice Problem 9 When Oprah bought an airplane ticket, she had to pay a tax of 8% on the cost of the ticket, which was \$350. What was the tax? ■

Solving Percent Problems When We Do Not Know the Base

If a number is multiplied by the letter n, this can be indicated by a multiplication sign, parentheses, a dot, or by placing the number in front of the letter. Thus $3 \times n = 3(n) = 3 \cdot n = 3n$. In this section we use equations like $3n = 9$ and $0.5n = 20$. To solve these equations we use the procedures developed in Chapter 4. We divide each side by the number multiplied by n.

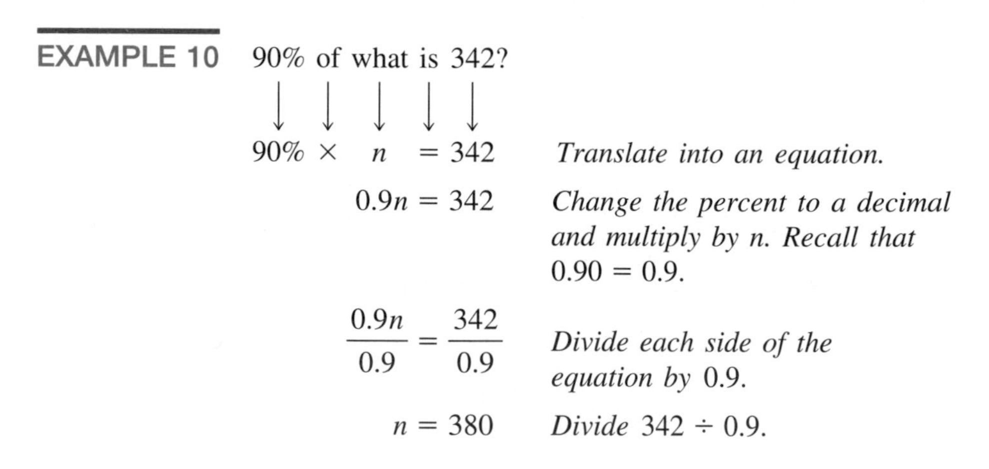

EXAMPLE 10 90% of what is 342?

$90\% \times n = 342$ *Translate into an equation.*

$0.9n = 342$ *Change the percent to a decimal and multiply by n. Recall that* $0.90 = 0.9$.

$\dfrac{0.9n}{0.9} = \dfrac{342}{0.9}$ *Divide each side of the equation by* 0.9.

$n = 380$ *Divide* $342 \div 0.9$.

Practice Problem 10 45% of what is 162? ■

EXAMPLE 11 12 is 0.6% of what?

$12 = 0.6\% \times n$ *Translate into an equation.*

$12 = 0.006n$ *Change* 0.6% *to a decimal and multiply by n.*

$\dfrac{12}{0.006} = \dfrac{0.006n}{0.006}$ *Divide each side of the equation by* 0.006.

$2000 = n$ *Divide* $12 \div 0.006$.

Practice Problem 11 32 is 0.4% of what? ■

EXAMPLE 12 Dave and Elsie went out to dinner. They gave the waiter a tip that was 15% of the total bill. The waiter received \$6. What was the total bill (not including the tip)?

This problem is asking

$$\begin{array}{ccccc} 15\% & \text{of} & \text{what} & \text{is} & \$6? \\ \downarrow & \downarrow & \downarrow & \downarrow & \downarrow \\ 15\% & \times & n & = & 6 \end{array}$$

$$0.15n = 6$$

$$\frac{0.15n}{0.15} = \frac{6}{0.15}$$

$$n = 40$$

The total bill for the meal (not including the tip) was \$40.

Practice Problem 12 The coach of the university baseball team said that 30% of the players on his team are left-handed. Six people on the team are left-handed. How many people are on the team? ■

Solving Percent Problems When We Do Not Know the Percent

In solving these problems, we notice that there is no % symbol in the problem. The percent is what we are trying to find. Therefore, our answer for this type of problem will always have a percent symbol.

EXAMPLE 13 What percent of 5000 is 3.8?

$$\begin{array}{ccccc} \underbrace{\text{What percent}} & \text{of} & 5000 & \text{is} & 3.8? \\ \downarrow & \downarrow & \downarrow & \downarrow & \downarrow \\ n & \times & 5000 & = & 3.8 \end{array}$$ *Translate into an equation.*

$$5000n = 3.8$$ *Multiplication is commutative.* $n \times 5000 = 5000 \times n$.

$$\frac{5000n}{5000} = \frac{3.8}{5000}$$ *Divide each side by 5000.*

$$n = 0.00076$$ *Divide* $3.8 \div 5000$.

$$n = 0.076\%$$ *Express the decimal as a percent.*

Practice Problem 13 What percent of 9000 is 4.5? ■

TEACHING TIP In problems like Example 14, some students will want to reduce 90 divided by 20 before they divide, and others will divide directly. You might mention that either approach is OK as long as the student can do the problem fairly quickly.

CALCULATOR

Finding the Percent

You can use a calculator to find a missing percent. What percent of 95 is 19?

1. Enter as a fraction.

is [÷] of

19 [÷] 95

2. Change to a percent.

19 [÷] 95 [×] 100 [=]

The display should read

[20]

This means 20%.

Try

What percent of 625 is 250?

EXAMPLE 14 90 is what percent of 20?

$$90 = n \times 20$$ *Translate into an equation.*

$$90 = 20n$$ *Multiplication is commutative.* $n \times 20 = 20 \times n$.

$$\frac{90}{20} = \frac{20n}{20}$$ *Divide each side by 20.*

$$4.5 = n$$ *Divide* $90 \div 20$.

$$450\% = n$$ *Express the decimal as a percent.*

Practice Problem 14 198 is what percent of 33?

EXAMPLE 15 In a recent basketball game for the Chicago Bulls, Scotty Pippin made 10 of his 24 shots. What percent of his shots did he make? (Round your answer to the nearest tenth of a percent.)

This is equivalent to

10 is what percent of 24?

$$10 = n \times 24$$

$$10 = 24n$$

$$\frac{10}{24} = \frac{24n}{24}$$

$$0.41666\ldots = n$$

To the nearest tenth of a percent we have

$$41.7\% = n$$

He made 41.7% of his shots at this game.

Practice Problem 15 In a basketball game for the Houston Rockets, Charles Barkley made 5 of his 16 shots. What percent of his shots did he make? (Round your answer to the nearest tenth of a percent.) ■

5.3A Exercises

Translate into a mathematical equation. Use the letter n *for the unknown quantity. Do* **not** *solve.*

1. What is 38% of 500?
$n = 38\% \times 500$

2. What is 46% of 700?
$n = 46\% \times 700$

3. 50% of what is 7?
$50\% \times n = 7$

4. 75% of what is 43?
$75\% \times n = 43$

5. 17 is what percent of 85?
$17 = n \times 85$

6. 24 is what percent of 144? $24 = n \times 144$

7. Find 128% of 4000.
$n = 128\% \times 4000$

8. Find 320% of 350.
$n = 320\% \times 350$

9. What percent of 400 is 15? $n \times 400 = 15$

10. What percent of 600 is 424? $n \times 600 = 424$

11. 136 is 145% of what?
$136 = 145\% \times n$

12. 200 is 160% of what?
$200 = 160\% \times n$

Solve each percent problem for which we do not know the amount.

13. What is 60% of 250?
$n = 60\% \times 250$ $n = 150$

14. What is 80% of 175?
$n = 80\% \times 175$ $n = 140$

15. Find 152% of 600.
$n = 152\% \times 600$ $n = 912$

16. Find 136% of 500.
$n = 136\% \times 500$ $n = 680$

Applications

17. Clarice rented a new Ford Escort for one week. Her bill before the 6% tax on rental cars was \$106.50. What was the amount of her tax?
$6\% \times \$106.50 = \$6.39$

18. The Center City check-cashing service charges a fee that is 7% of the amount of the check they cash. How much is the fee for cashing a check for \$240? $7\% \times \$240 = \$16.80$

Solve each percent problem for which we do not know the base.

19. 26% of what is 312?
$26\% \times n = 312$ $n = 1200$

20. 32% of what is 448?
$32\% \times n = 448$
$n = 1400$

21. 52 is 4% of what?
$52 = 4\% \times n$ $n = 1300$

22. 36 is 6% of what?
$36 = 6\% \times n$ $n = 600$

Applications

23. In Australia, all general sales (except for food) have a hidden tax of 22% built into the final price. Walter is planning to purchase a camera while in Australia. He wants to know the ‘‘before tax price’’ that the dealer is charging before he adds on the ‘‘hidden tax’’ of \$33. Can you determine the amount of the ‘‘before tax price?’’
$22\% \times n = 33$ $n = \$150$

24. Jocelyn is teaching her golden retriever puppy how to fetch. Jack, her puppy, fetched the ball 72% of the time today. Jack fetched the ball 54 times today. How many times did Jocelyn actually throw the ball for Jack?
$72\% \times n = 54$ $n = 75$ throws

Solve each percent problem for which we do not know the percent.

25. What percent of 30 is 18?
$n \times 30 = 18$ $n = 60\%$

26. What percent of 40 is 26?
$n \times 40 = 26$ $n = 65\%$

27. 56 is what percent of 200?
$56 = n \times 200$ $28\% = n$

28. 45 is what percent of 300?
$45 = n \times 300$ $15\% = n$

Applications

29. Simon inspected 280 waterbed mattresses. He found that 14 were defective. What percent of the waterbed mattresses were defective?
$14 = n \times 280$ $n = 5\%$

30. John paid \$12 tax when he bought a mountain bike for \$300. What percent tax did he pay?
$\$12 = n \times \$300$ $n = 4\%$

Mixed Practice

Solve each percent problem. All the types are represented.

31. 18% of 280 is what?
$18\% \times 280 = n \quad 50.4 = n$

32. 12% of 260 is what?
$12\% \times 260 = n \quad 31.2 = n$

33. 150% of what is 102?
$150\% \times n = 102 \quad n = 68$

34. 150% of what is 138?
$150\% \times n = 138 \quad n = 92$

35. 84 is what percent of 700?
$84 = n \times 700 \quad 12\% = n$

36. 72 is what percent of 900?
$72 = n \times 900 \quad 8\% = n$

37. Find 0.4% of 820.
$n = 0.4\% \times 820 \quad n = 3.28$

38. Find 0.3% of 540.
$n = 0.3\% \times 540 \quad n = 1.62$

39. What percent of 35 is 22.4?
$n \times 35 = 22.4 \quad n = 64\%$

40. What percent of 45 is 16.2?
$n \times 45 = 16.2 \quad n = 36\%$

41. 89 is 20% of what?
$89 = 20\% \times n \quad 445 = n$

42. 42 is 40% of what?
$42 = 40\% \times n \quad n = 105$

43. 42 is what percent of 120?
$42 = n \times 120 \quad 35\% = n$

44. 26 is what percent of 160?
$26 = n \times 160 \quad n = 16.25\%$

45. What is 16.5% of 240?
$n = 16.5\% \times 240 \quad n = 39.6$

46. What is 18.5% of 360?
$n = 18.5\% \times 360 \quad n = 66.6$

47. Find 0.06% of 2400.
$n = 0.06\% \times 2400 \quad n = 1.44$

48. Find 0.03% of 3200.
$n = 0.03\% \times 3200 \quad n = 0.96$

49. What is 38.5% of 2345?
$n = 38.5\% \times 2345 \quad n = 902.825$

50. What is 0.00345% of 567,000?
$n = 0.00345\% \times 567{,}000 \quad n = 19.5615$

51. 3458 is what percent of 5832?
$3458 = n \times 5832 \quad n = 59.29\%$ (rounded)

52. 16.8% of what is 5994.24?
$16.8\% \times n = 5994.24 \quad n = 35{,}680$ (rounded)

Applications

53. The special effects office at Magic Movie Studio fabricated 80 minimodels of office buildings for the disaster sequence of a new film. The director of photography discovered that on camera, only 68 of them really looked authentic. What percent of the models were acceptable? 85%

54. An Olympic equestrian rider practiced jumping over a water hazard. In 400 attempts, she and her horse touched the water 15 times. What percent of her jump attempts were not perfect? 3.75%

55. At Brookstone State College 44% of the freshman class is employed part-time. There are 1260 freshman this year. How many of them are employed part-time? Round your answer to the nearest whole number. 554 students

56. A survey discovered that 28% of all registered nurses working at Woodview General Hospital are vegetarians. If there are 430 registered nurses working at Woodview, how many of them don't eat meat? Round your answer to the nearest whole number. 120 nurses

57. The swim team at Stonybrook College has gone on to the state championships 24 times over the years. If that translates to 60% of the time in which the team has qualified for the finals how many years has the swim team qualified for the finals? 40 years

58. North Shore Community College found that 60% of its graduates go on for further education. Last year 570 of the graduates went on for further education. How many students graduated from the college last year?
$570 = 60\% \times n \quad n = 950$ students

59. Find 12% of 30% of \$1600.
57.6

60. Find 90% of 15% of 2700.
364.5

Cumulative Review Problems

Multiply.

61. 1.36×1.8
2.448

62. 2.04×7.3
14.892

Divide.

63. $0.06\overline{)170.04}$
2834

64. $0.08\overline{)233.36}$
2917

5.3B Solving Percent Problems Using a Proportion

After studying this section, you will be able to:

1 *Identify the parts of the percent proportion.*

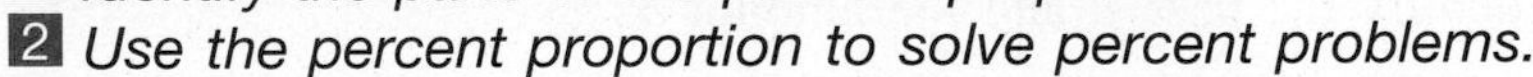

2 *Use the percent proportion to solve percent problems.*

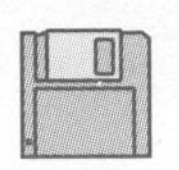

MathPro Video 12 SSM

TEACHING TIP Some students hate solving percent problems by the proportion method; other students find it very helpful.

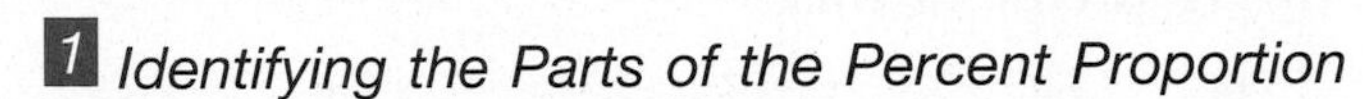

1 *Identifying the Parts of the Percent Proportion*

In Section 5.3A we showed you how to use an equation to solve a percent problem. Some students find it easier to use proportions to solve percent problems. We will show you how to use proportions in this section. The two methods work equally well. Using percent proportions allows you to see another of the many uses of proportions that we studied in Chapter 4.

Consider the following relationship.

$$\frac{19}{25} = 76\%$$

This can be written as

$$\frac{19}{25} = \frac{76}{100}$$

In general, we can write this relationship using the percent proportion

$$\frac{\text{amount}}{\text{base}} = \frac{\text{percent number}}{100}$$

To use this equation effectively, we need to find the amount, base, and percent number in a word problem. The easiest of these three parts to find is the percent number. We use the letter p (a variable) to represent the percent number.

EXAMPLE 1 Identify the percent number p in each of the following.

(a) Find 16% of 370. **(b)** 28% of what is 25?

(c) What percent of 18 is 4.5?

(a) Find 16% of 370.
The value of p is 16.

(b) 28% of what number is 25?
The value of p is 28.

(c) $\underbrace{\text{What percent}}_{p}$ of 18 is 4.5?

We let p represent the unknown percent number.

Practice Problem 1 Identify the percent number p.

(a) Find 83% of 460. **(b)** 18% of what number is 90?

(c) What percent of 64 is 8? ■

We use the letter b to represent the base number. The base is the entire quantity or the total involved. The number that is the base usually appears after the word *of*. The amount, which we represent by the letter a, is the part being compared to the whole.

EXAMPLE 2 Identify the base b and the amount a.

(a) 20% of 320 is 64. **(b)** 12 is 60% of what?

(a) 20% of 320 is 64

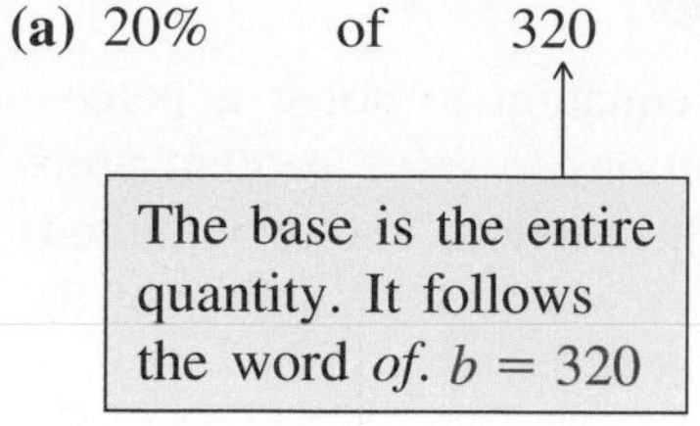

The base is the entire quantity. It follows the word *of*. $b = 320$

The amount is the part compared to the whole. Here $a = 64$.

(b) 12 is 60% of what?

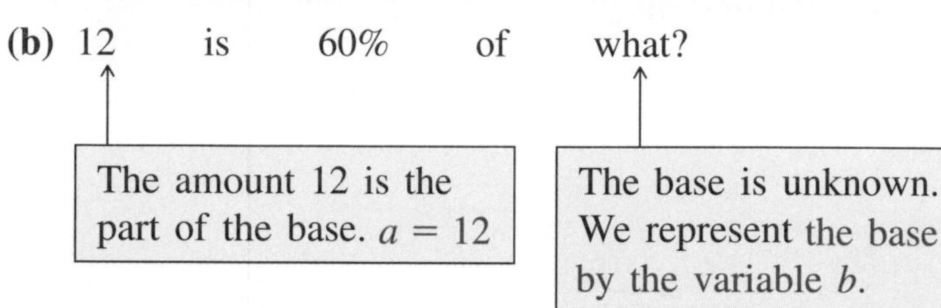

The amount 12 is the part of the base. $a = 12$
The base is unknown. We represent the base by the variable b.

Practice Problem 2 Identify the base b and the amount a.

(a) 30% of 52 is 15.6. **(b)** 170 is 85% of what? ■

When identifying p, b, and a in a problem, it is easiest to identify p and b first. The remaining quantity or variable is a.

TEACHING TIP The key to getting students to use the proportion method successfully is seeing that they can correctly identify the three parts p, b, and a. Give this aspect special emphasis.

EXAMPLE 3 Find p, b, and a.

(a) What is 52% of 300? **(b)** What percent of 30 is 18?

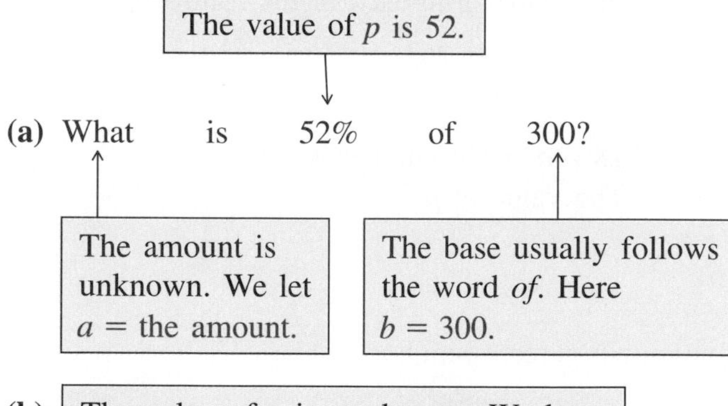

The value of p is 52.
(a) What is 52% of 300?
The amount is unknown. We let a = the amount.
The base usually follows the word *of*. Here $b = 300$.

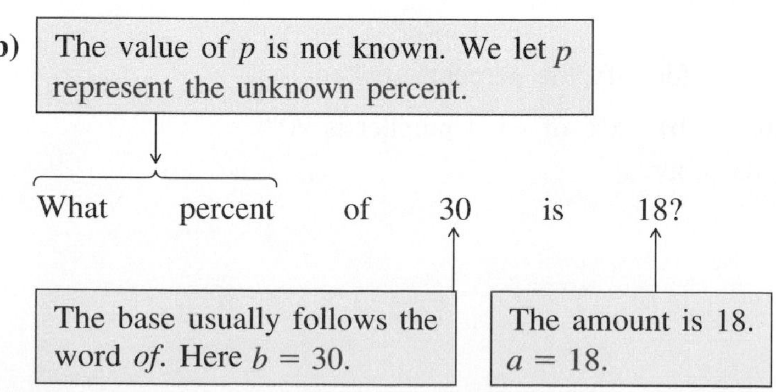

Practice Problem 3 Find p, b, and a.

(a) What is 18% of 240? **(b)** What percent of 64 is 4? ■

2 Using the Percent Proportion to Solve Percent Problems

When we solve the percent proportion, we will have enough information to state the numerical value for two of the three variables a, b, p in the equation

$$\frac{a}{b} = \frac{p}{100}$$

We first identify those two values, and then substitute those values into the equation. Then we will use the skills that we acquired for solving proportions in Chapter 4 to find the value we do not know.

EXAMPLE 4 What is 16% of 380?

The percent $p = 16$. The base $b = 380$. The amount is unknown. We use the variable a. Thus

$$\frac{a}{b} = \frac{p}{100} \quad \text{becomes} \quad \frac{a}{380} = \frac{16}{100}$$

If we reduce the fraction on the right-hand side, we have

$$\frac{a}{380} = \frac{4}{25}$$

$25a = (4)(380)$ *Cross-multiply.*

$25a = 1520$ *Simplify.*

$\frac{25a}{25} = \frac{1520}{25}$ *Divide each side by 25.*

$a = 60.8$ *Divide* $1520 \div 25$.

Thus 16% of 380 is 60.8.

Practice Problem 4 What is 32% of 550? ■

EXAMPLE 5 Find 260% of 40.

The percent $p = 260$. The number that is the base usually appears after the word *of*. The base $b = 40$. The amount is unknown. We use the variable a. Thus

$$\frac{a}{b} = \frac{p}{100} \quad \text{becomes} \quad \frac{a}{40} = \frac{260}{100}$$

If we reduce the fraction on the right-hand side, we have

$$\frac{a}{40} = \frac{13}{5}$$

$5a = (40)(13)$ *Cross-multiply.*

$5a = 520$ *Simplify.*

$\frac{5a}{5} = \frac{520}{5}$ *Divide each side of the equation by 5.*

$a = 104$

Thus 260% of 40 is 104.

Practice Problem 5 Find 340% of 70. ■

EXAMPLE 6 85% of what is 221?

The percent $p = 85$. The base is unknown. We use the variable b. The amount a is 221. Thus

$$\frac{a}{b} = \frac{p}{100} \quad \text{becomes} \quad \frac{221}{b} = \frac{85}{100}$$

If we reduce the fraction on the right-hand side, we have

$$\frac{221}{b} = \frac{17}{20}$$

$$(221)(20) = 17b \quad \textit{Cross-multiply.}$$

$$4420 = 17b \quad \textit{Simplify.}$$

$$\frac{4420}{17} = \frac{17b}{17} \quad \textit{Divide each side by 17.}$$

$$260 = b. \quad \textit{Divide } 4420 \div 17.$$

Thus 85% of 260 is 221.

Practice Problem 6 68% of what is 476? ■

EXAMPLE 7 53 is 0.2% of what?

The percent $p = 0.2$. The base is unknown. We use the variable b. The amount $a = 53$. Thus

$$\frac{a}{b} = \frac{p}{100} \quad \text{becomes} \quad \frac{53}{b} = \frac{0.2}{100}$$

When we cross-multiply, we obtain

$$(53)(100) = 0.2b$$

$$5300 = 0.2b$$

$$\frac{5300}{0.2} = \frac{0.2b}{0.2}$$

$$26{,}500 = b$$

Thus 53 is 0.2% of 26,500.

Practice Problem 7 216 is 0.3% of what? ■

EXAMPLE 8 What percent of 4000 is 160?

The percent is unknown. We use the variable p. The base $b = 4000$. The amount $a = 160$. Thus

$$\frac{a}{b} = \frac{p}{100} \quad \text{becomes} \quad \frac{160}{4000} = \frac{p}{100}$$

If we reduce the fraction on the left-hand side, we have

$$\frac{1}{25} = \frac{p}{100}$$

$$100 = 25p \qquad \textit{Cross-multiply.}$$

$$\frac{100}{25} = \frac{25p}{25} \qquad \textit{Divide each side by 25.}$$

$$4 = p \qquad \textit{Divide } 100 \div 25.$$

Thus 4% of 4000 is 160.

Practice Problem 8 What percent of 3500 is 105? ■

EXAMPLE 9 19 is what percent of 95?

The percent is unknown. We use the variable p. The base $b = 95$. The amount $a = 19$. Thus

$$\frac{a}{b} = \frac{p}{100} \quad \text{becomes} \quad \frac{19}{95} = \frac{p}{100}$$

Cross-multiplying, we have

$$(19)(100) = 95p$$

$$1900 = 95p$$

$$\frac{1900}{95} = \frac{95p}{95} \qquad \textit{Divide each side of the equation by 95.}$$

$$20 = p$$

Thus 19 is 20% of 95.

Practice Problem 9 42 is what percent of 140? ■

5.3B Exercises

Identify p, b, and a. Do not solve for the unknown.

	p	b	a
1. 75% of 660 is 495.	75	660	495
2. 65% of 820 is 532.	65	820	532
3. What is 42% of 400?	42	400	a
4. What is 56% of 600?	56	600	a
5. 49% of what is 2450?	49	b	2450
6. 38% of what is 2280?	38	b	2280
7. 30 is what percent of 50?	p	50	30
8. 50 is what percent of 250?	p	250	50
9. What percent of 25 is 10?	p	25	10
10. What percent of 24 is 6?	p	24	6
11. 400 is 160% of what?	160	b	400
12. 900 is 225% of what?	225	b	900

Solve each problem by using the percent proportion

$$\frac{a}{b} = \frac{p}{100}$$

In problems 13–18 the amount a is not known.

13. 24% of 300 is what?

$\frac{a}{300} = \frac{24}{100}$ $a = 72$

14. 54% of 500 is what?

$\frac{a}{500} = \frac{54}{100}$ $a = 270$

15. Find 250% of 30.

$\frac{a}{30} = \frac{250}{100}$ $a = 75$

16. Find 320% of 60.

$\frac{a}{60} = \frac{320}{100}$ $a = 192$

17. 0.7% of 8000 is what?

$\frac{a}{8000} = \frac{0.7}{100}$ $a = 56$

18. 0.8% of 9000 is what?

$\frac{a}{9000} = \frac{0.8}{100}$ $a = 72$

In problems 19–24 the base b is not known.

19. 45 is 60% of what?

$\frac{45}{b} = \frac{60}{100}$ $b = 75$

20. 96 is 80% of what?

$\frac{96}{b} = \frac{80}{100}$ $b = 120$

21. 150% of what is 90?

$\frac{90}{b} = \frac{150}{100}$ $b = 60$

22. 125% of what is 75?

$\frac{75}{b} = \frac{125}{100}$ $b = 60$

23. 3000 is 0.5% of what?

$\frac{3000}{b} = \frac{0.5}{100}$ $b = 600{,}000$

24. 6000 is 0.4% of what?

$\frac{6000}{b} = \frac{0.4}{100}$ $b = 1{,}500{,}000$

In problems 25–30 the percent p is not known.

25. 70 is what percent of 280?

$\frac{70}{280} = \frac{p}{100}$ $p = 25$

26. 90 is what percent of 450?

$\frac{90}{450} = \frac{p}{100}$ $p = 20$

27. What percent of 140 is 3.5?

$\frac{3.5}{140} = \frac{p}{100}$ $p = 2.5$

28. What percent of 170 is 3.4?

$\frac{3.4}{170} = \frac{p}{100}$ $p = 2$

29. What percent of \$5000 is \$90?

$\frac{90}{5000} = \frac{p}{100}$ $p = 1.8$

30. What percent of \$4000 is \$64?

$\frac{64}{4000} = \frac{p}{100}$ $p = 1.6$

Mixed Practice

31. 26% of 350 is what?

$\frac{a}{350} = \frac{26}{100}$ $a = 91$

32. 56% of 650 is what?

$\frac{a}{650} = \frac{56}{100}$ $a = 364$

33. 180% of what is 540?

$\frac{540}{b} = \frac{180}{100}$ $b = 300$

34. 160% of what is 320?

$\frac{320}{b} = \frac{160}{100}$ $b = 200$

35. 82 is what percent of 500?

$\frac{82}{500} = \frac{p}{100}$ $p = 16.4$ 16.4%

36. 75 is what percent of 600?

$\frac{75}{600} = \frac{p}{100}$ $p = 12.5$ 12.5%

37. Find 0.7% of 520.

$\frac{a}{520} = \frac{0.7}{100}$ $a = 3.64$

38. Find 0.4% of 650.

$\frac{a}{650} = \frac{0.4}{100}$ $a = 2.6$

39. What percent of 66 is 16.5?

$\frac{16.5}{66} = \frac{p}{100}$ $p = 25$ 25%

40. What percent of 49 is 34.3?

$\frac{34.3}{49} = \frac{p}{100}$ $p = 70$ 70%

41. 68 is 40% of what?

$\frac{68}{b} = \frac{40}{100}$ $b = 170$

42. 52 is 40% of what?

$\frac{40}{100} = \frac{52}{b}$ $b = 130$

43. 94.6 is what percent of 220?

$\frac{94.6}{220} = \frac{p}{100}$ $p = 43$ 43%

44. 85.8 is what percent of 260?

$\frac{85.8}{260} = \frac{p}{100}$ $p = 33$ 33%

45. What is 12.5% of 380?

$\frac{a}{380} = \frac{12.5}{100}$ $a = 47.5$

46. What is 20.5% of 320?

$\frac{a}{320} = \frac{20.5}{100}$ $a = 65.6$

47. Find 0.05% of 5600.

$\frac{a}{5600} = \frac{0.05}{100}$ $a = 2.8$

48. Find 0.04% of 8700.

$\frac{a}{8700} = \frac{0.04}{100}$ $a = 3.48$

Solve each percent problem. Round to the nearest hundredth.

49. What percent of 4550 is 720?

15.82%

50. What is 16.25% of 65,250?

10,603.13

51. 4.25% of 256.75 is what number?

10.91

52. 2760 is 5.5% of what number?

50,181.82

★ **53.** What is $19\frac{1}{4}$% of 798?

$\frac{a}{798} = \frac{19.25}{100}$ $a = 153.62$

★ **54.** $140\frac{1}{2}$% of what number is 10,397?

$\frac{10{,}397}{b} = \frac{140.5}{100}$ $b = 7400$

★ **55.** Find 18% of 20% of \$3300. (*Hint:* First find 20% of \$3300.)

\$118.80

★ **56.** Find 42% of 16% of \$5500. (*Hint:* First find 16% of \$5500.)

\$369.60

Cumulative Review Problems

Simplify.

57. $\frac{4}{5} + \frac{8}{9}$ $1\frac{31}{45}$

58. $\frac{7}{13} - \frac{1}{2}$ $\frac{1}{26}$

59. $\left(2\frac{4}{5}\right)\left(1\frac{1}{2}\right)$ $4\frac{1}{5}$

60. $1\frac{2}{5} \div \frac{3}{4}$ $1\frac{13}{15}$

5.4 Solving Applied Percent Problems

MathPro

Video 12

SSM

After studying this section, you will be able to:

1. *Solve general applied percent problems.*
2. *Solve commission problems.*
3. *Solve problems when percents are added.*
4. *Solve percent of increase or decrease problems.*
5. *Solve discount problems.*
6. *Solve simple interest problems.*

1 *Solving General Applied Percent Problems*

In Sections 5.3A and 5.3B we learned the three types of applied percent problems. Some problems ask you to find a percent of a number. Some problems give you an amount and a percent and ask you to find the base (or whole). Other problems give an amount and a base and ask you to find the percent. We will now see how the three types of percent problems occur in real life.

EXAMPLE 1 Of all the computers manufactured last month, an inspector found 18 that were defective. This is 2.5% of all the computers manufactured last month. How many computers were manufactured last month?

Method A Translating to an equation

The problem is equivalent to: 2.5% of the number of computers is 18. Let n = the number of computers.

$$\underbrace{2.5\%}_{2.5\%}\ \underbrace{\text{of}}_{\times}\ \underbrace{\text{the number of computers}}_{n}\ \underbrace{\text{is}}_{=}\ \underbrace{18}_{18}$$

$$0.025n = 18$$

$$\frac{0.025n}{0.025} = \frac{18}{0.025}$$

$$n = 720$$

$$0.025_{\wedge}\overline{)18.000_{\wedge}} = 720.$$
175
50
50
00

720 computers were manufactured last month.

Method B Use the Percent Proportion $\frac{a}{b} = \frac{p}{100}$.

The percent $p = 2.5$. The base is unknown. We will use the variable b. The amount $a = 18$. Thus

$$\frac{a}{b} = \frac{p}{100} \quad \text{becomes} \quad \frac{18}{b} = \frac{2.5}{100}$$

Using cross-multiplication, we have

$$(18)(100) = 2.5b$$

$$1800 = 2.5b$$

$$\frac{1800}{2.5} = \frac{2.5b}{2.5}$$

$$720 = b$$

$$2.5_{\wedge}\overline{)1800.0_{\wedge}} = 720.$$
175
50
50
00

720 computers were manufactured last month.

By either Method A or Method B, we obtain the same number of computers, 720.

Substitute 720 into the original problem to check.

2.5% of 720 computers are defective.

$$(0.025)(720) = 18 \quad \checkmark$$

Practice Problem 1 4800 people, or 12% of all passengers holding tickets for American Airlines flights in one month, did not show up for their flight. How many people held tickets that month? ■

EXAMPLE 2 How much sales tax will you pay on a color television priced at $299.00 if the sales tax is 5%?

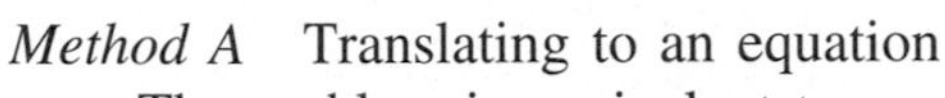

Method A Translating to an equation

The problem is equivalent to:

$$\begin{array}{llll} \text{What is} & 5\% & \text{of} & \$299? \\ \downarrow \quad \downarrow & \downarrow & \downarrow & \downarrow \\ n \quad = & 5\% & \times & 299 \end{array}$$

$$n = (0.05)(299)$$

$$n = 14.95$$

$$\begin{array}{r} 299 \\ \times\ 0.05 \\ \hline 14.95 \end{array}$$

The tax is $14.95.

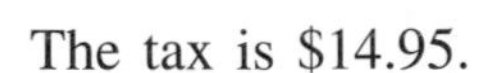

Method B Using the percent proportion $\frac{a}{b} = \frac{p}{100}$

The percent $p = 5$. The base $b = 299$. The amount is unknown. We use the variable a. Thus

$$\frac{a}{b} = \frac{p}{100} \quad \text{becomes} \quad \frac{a}{299} = \frac{5}{100}$$

If we reduce the fraction on the right-hand side, we have

$$\frac{a}{299} = \frac{1}{20}$$

We then cross-multiply to obtain

$$20a = 299$$

$$\frac{20a}{20} = \frac{299}{20}$$

$$a = 14.95$$

$$\begin{array}{r} 14.95 \\ 20\overline{)299.00} \\ 20 \\ \hline 99 \\ 80 \\ \hline 190 \\ 180 \\ \hline 100 \\ 100 \\ \hline \end{array}$$

The tax is $14.95.

Thus, by either method, the amount of the sales tax is $14.95.

Let's see if our answer is reasonable. Is 5% of $299 really $14.95? If we round each figure to one significant digit, we have 5% of 300 = 15. Since 15 is quite close to our value of $14.95, our answer seems reasonable.

Practice Problem 2 A salesperson rented a hotel room for $62.30 per night. The tax in her state is 8%. What tax does she pay for one night at the hotel? ■

EXAMPLE 3 A failing student attended class 39 times out of the 45 times the class met last semester. What percent of the classes did he attend? Round your answer to the nearest tenth of a percent.

Method A Translating to an equation

This problem is equivalent to:

$$\begin{array}{ccccc} 39 & \text{is} & \underbrace{\text{what percent}} & \text{of} & 45? \\ \downarrow & \downarrow & \downarrow & \downarrow & \downarrow \\ 39 & = & n & \times & 45 \end{array}$$

$$39 = 45n$$

$$\frac{39}{45} = \frac{45n}{45}$$

$$0.8666\ldots = n$$

To the nearest tenth of a percent we have $n = 86.7\%$.

Method B Using the percent proportion $\frac{a}{b} = \frac{p}{100}$

The percent is unknown. We use the variable p. The base b is 45. The amount a is 39. Thus

$$\frac{a}{b} = \frac{p}{100} \quad \text{becomes} \quad \frac{39}{45} = \frac{p}{100}$$

When we cross-multiply, we get

$$(39)(100) = 45p$$

$$3900 = 45p$$

$$\frac{3900}{45} = \frac{45p}{45}$$

$$86.666\ldots = p$$

To the nearest tenth, the answer is 86.7%.

By using either method, we discover that the failing student attended approximately 86.7% of the classes.

Verify by estimating that the answer is reasonable.

Practice Problem 3 Of the 130 flights at Orange County Airport yesterday only 105 of them were on time. What percent of the flights were on time? (Round your answer to the nearest tenth of a percent.) ■

Now you have some experience using the three types of percent problems in real-life applications. You can use either Method A or Method B to solve applied percent problems. In the following sections we will present more percent applications. We will not list all the steps of Method A or Method B. Most students will find after a careful study of Examples 1–3 that they do not need to write out all the steps of Method A or Method B when solving applied percent problems.

2 *Solving Commission Problems*

If you work as a salesperson, your earnings may be in part or in total a certain percentage of the sales you make. The amount of money you get that is a percentage of the value of your sales is called your **commission.** It is calculated by multiplying the percentage (called the **commission rate**) by the value of the sales.

TEACHING TIP Some students are familiar with commission sales and others are not. A few simple examples are always helpful. Ask your students if any of them have ever earned a commission on sales in one of their jobs.

Commission = commission rate × value of sales

EXAMPLE 4 A salesperson has a commission rate of 17%. She sells $32,500 worth of goods in a department store in two months. What is her commission?

$$\text{Commission} = \text{commission rate} \times \text{value of sales}$$

$$\begin{aligned}\text{Commission} &= 17\% \times \$32{,}500\\ &= 0.17 \times 32{,}500\\ &= 5525\end{aligned}$$

Her commission is $5525.00.

Does this answer seem reasonable? Check by estimating.

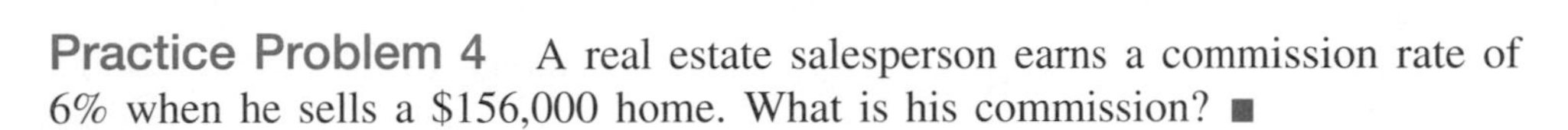

Practice Problem 4 A real estate salesperson earns a commission rate of 6% when he sells a $156,000 home. What is his commission? ■

3 *Solving Problems When Percents Are Added*

Percents can be added if the base (whole) is the same. For example, 50% of your salary added to 20% of your salary = 70% of your salary. 100% of your cost added to 15% of your cost = 115% of your cost. Problems like this are often called *markup problems.* If we add 15% of the cost of an item to the original cost, the markup is 15%. We will add percents in some applied situations.

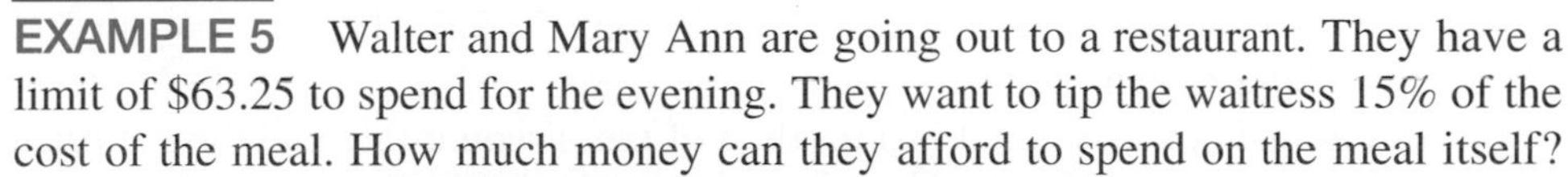

EXAMPLE 5 Walter and Mary Ann are going out to a restaurant. They have a limit of $63.25 to spend for the evening. They want to tip the waitress 15% of the cost of the meal. How much money can they afford to spend on the meal itself?

Let n = the cost of the meal. 15% of the cost = the amount of the tip. We want to add the percents of the meal.

Cost of meal n	+	tip of 15% of the cost	=	$63.25
100% of n	+	15% of n	=	$63.25

Note that 100% of n added to 15% of n is 115% of n.

$$115\% \text{ of } n = \$63.25$$

$$1.15 \times n = 63.25$$

$$\frac{1.15 \times n}{1.15} = \frac{63.25}{1.15} \qquad \textit{Divide both sides by } 1.15.$$

$$n = 55$$

$$\begin{array}{r} 55.\\ 1.15_{\wedge}\overline{)63.25_{\wedge}}\\ \underline{575}\\ 575\\ \underline{575}\\ 0 \end{array}$$

They can spend up to $55.00 on the meal itself.

Does this answer seem reasonable?

Practice Problem 5 Sue and Sam have $46.00 to spend at a restaurant, including a 15% tip. How much can they spend on the meal itself? ■

4 Solving Percent of Increase or Decrease Problems

We sometimes need to find the percent by which a number increases or decreases. If a car costs \$7000 and the price decreases \$1750, we say that the percent of decrease is $\frac{1750}{7000} = 0.25 = 25\%$.

$$\text{Percent of decrease} = \frac{\text{amount of decrease}}{\text{original amount}}$$

Similarly, if a population of 12,000 people increases by 1920 people, we say that the percent of increase is $\frac{1920}{12{,}000} = 0.16 = 16\%$.

$$\text{Percent of increase} = \frac{\text{amount of increase}}{\text{original amount}}$$

Note that for these types of problems the base is always the *original amount.*

TEACHING TIP If students have trouble with percent of increase and decrease problems, remind them that the denominator is always the original amount. Usually, students make a mistake in this type of problem because they forgot that fact.

EXAMPLE 6 The population of Center City increased from 50,000 to 59,500. What was the percent of increase?

Amount of increase

$$\begin{array}{r} 59{,}500 \\ -\,50{,}000 \\ \hline 9{,}500 \end{array}$$

$$\text{Percent of increase} = \frac{\text{amount of increase}}{\text{original amount}} = \frac{9500}{50{,}000}$$

$$= 0.19 = 19\%$$

The percent of increase is 19%.

Practice Problem 6 A new car is sold for \$15,000. A year later its price had decreased to \$10,500. What is the percent of decrease? ■

5 Solving Discount Problems

Frequently, we see signs urging us to buy during a sale when the list price is discounted by a certain percent.

The amount of a discount is a product of the discount rate and the list price.

$$\text{Discount} = \text{discount rate} \times \text{list price}$$

EXAMPLE 7 Jeff purchased a compact disc player on sale at 35% discount. The list price was \$430.00.

(a) What was the amount of the discount?

(b) How much did Jeff pay for the CD player?

(a) Discount = discount rate × list price

$= 35\% \times 430$

$= 0.35 \times 430$

$= 150.5$

The discount is \$150.50.

(b) We subtract the discount from the list price to get the selling price.

\$430.00	*list price*
− \$150.50	*discount*
\$279.50	*selling price*

Jeff paid \$279.50 for the CD player.

Practice Problem 7 Betty bought a car that lists for \$13,600 at a 7% discount.

(a) What was the discount? **(b)** What did she pay for the car? ■

6 *Solving Simple Interest Problems*

Interest is money paid for the use of money. If you deposit money in a bank, the bank uses that money and pays you interest. If you borrow money, you pay the bank interest for the use of that money. The *principal* is the amount deposited or borrowed. Interest is usually expressed as a percent rate of the principal. The interest rate is assumed to be per year, unless otherwise stated. The formula used in business to compute simple interest is

$$\text{Interest} = \text{principal} \times \text{rate} \times \text{time}$$
$$I = P \times R \times T$$

If the interest rate is *per year,* the time T *must* be in *years.*

EXAMPLE 8 Find the interest on a loan of \$7500 borrowed at 13% for one year.

$$I = P \times R \times T$$

P = principal = \$7500 $\quad R$ = rate = 13% $\quad T$ = time = 1 year

$$I = 7500 \times 13\% \times 1 = 7500 \times 0.13 = 975$$

The interest is \$975.00.

Practice Problem 8 Find the interest on a loan of \$5600 borrowed at 12% for one year. ■

Our formula is based on a yearly interest rate. Time periods of more than one year or a fractional part of a year are sometimes needed.

EXAMPLE 9 Find the interest on a loan of \$2500 that is borrowed at 9% for

(a) 3 years **(b)** 3 months

(a)

$$I = P \times R \times T$$

P = \$2500 $\quad R$ = 9% $\quad T$ = 3 years

$$I = 2500 \times 0.09 \times 3 = 225 \times 3 = 675$$

The interest for 3 years is \$675.

(b) Now 3 months = $\frac{1}{4}$ year. The period must be in years to use the formula.

Since $T = \frac{1}{4}$ year, we have

$$I = 2500 \times 0.09 \times \frac{1}{4}$$
$$= 225 \times \frac{1}{4}$$
$$= \frac{225}{4} = 56.25$$

The interest for $\frac{1}{4}$ year is \$56.25.

Practice Problem 9 Find the interest on a loan of \$1800 that is borrowed at 11% for **(a)** 4 years **(b)** 6 months ■

CALCULATOR

Interest

You can use a calculator to find simple interest. Find the interest on \$450 invested at 6.5% for 15 months. Notice the time is in months. Since the interest formula is in years, $I = P \times R \times T$, you need to change 15 months to years by dividing 15 by 12. Enter

15 [÷] 12 [=]

Display

[1.25]

Leave this on the display and multiply as follows:

1.25 [×] 450 [×] 6.5 [%] [=]

The display should read

[36.5625]

which would round to \$36.56.

Try

(a) \$9516 invested at 12% for 30 months

(b) \$593 borrowed at 8% for 5 months

5.4 Exercises

Applications

Problems 1–22 present the three types of percent problems. They are similar to Examples 1–3. Take the time to master problems 1–22 before going on to the next ones. Round your answer to the nearest hundredth when necessary.

1. No graphite was found in 4500 pencils shipped to Sureway School Supplies. This was 2.5% of the total number of pencils received by Sureway. How many pencils in total were in the order? 180,000 pencils

2. A high-jumper on the track and field team hit the bar 58 times last week. This means that he did not succeed in 29% of his jump attempts. How many total attempts did he have last week? 200 attempts

3. How much tax will Colleen pay on a surround-sound system that costs $1885.00 if the tax is 5%. $94.52

4. In a high school in London, 714 out of 1020 students speak two languages fluently. What percent of students speak two languages? 70%

5. A mechanic runs a car through a computer diagnostic check for its 60,000-mile check-up. 76 out of 95 car parts did not need any adjustment. What percentage did the mechanic need to adjust? 20%

6. 40 out of 65 people in a room can ''curl'' their tongues. What percent of these people can curl successfully? Approximately 61.54%

7. Scott's phone bill averages $48.60 per month. This is 115% of what his average monthly bill was last year. What was his average monthly phone bill last year? Approximately $42.26

8. Renata now earns $9.50 per hour. This is 125% of what she earned last year. What did she earn per hour last year? $7.60 per hour

9. An island in the South Pacific imposes a 7% room tax. If a room in a luxury hotel costs $92 per night, what is the tax? $6.44

10. The caterer estimates that 85% of the people at a banquet will request coffee. If the banquet has 340 guests, how many people are expected to request coffee? 289 people

11. The Incredible Chocolate Chip Company has discovered that 36 out of 400 chocolate chip cookies do not contain enough chocolate chips. What percent of the chocolate chip cookies do not have enough chips? 9%

12. Every day this year, Sam ordered either cappuccino or espresso from the coffee bar downstairs. He had 78 espressos and 287 cappuccinos. What percent of the coffees were espressos? 21.37%

13. In a survey taken by the Department of Agriculture of 450 people, 74% admitted that they hated brussels sprouts. How many people in this survey did not care for this vegetable? 333 people

14. At the backstage entrace to the Hollywood Bowl Concert Arena, 85% of the people who came to the door after a recent concert requested autographs. If 510 people asked for autographs, how many people came backstage? 600 people

15. Dean bought a new mountain bike. The sales tax in his state is 4%, he paid $9.60 in tax. What was the price of the mountain bike before the tax? $240

16. Maha received a raise of 8% of her monthly salary. The amount of her raise was $48.16 per month. What was her monthly salary before the raise? $602

17. Peter and Judy together earn $5060 per month. Their mortgage payment is $1265 per month. What percentage of their household income goes toward paying the mortgage? 25%

18. Jackie puts aside $65.50 per week for her monthly car payment. She earns $327.50 per week. What percent of her income is set aside for car payments. 20%

19. The Children's Wish Charity raised 75% of its funds from sporting promotions. Last year the charity received $7,200,000 from its sporting promotions. What was the charity's total income last year? $9,600,000

20. Shannon paid $8400 in federal and state income taxes as a lab technician, which amounted to 28% of her annual income. What was her income last year? $30,000

21. In a bin of 21,000 M&Ms, 6.5% are the color blue. How many M&Ms are blue?
1365 M&Ms are blue

22. Out of 26,400 daily telephone calls made in Meridian, 85% of the calls were made between relatives. How many phone calls were made by people calling relatives? 22,440 calls

Problems 23–46 include commissions, percents that are added, percent of increase or decrease, discounts, and simple interest. Solve each problem. Round each answer to the nearest hundredth when necessary.

23. Laserlight Corporation offers its outside salespeople 3% commission on any industrial laser applications sold in North America. If Trevor sold $2,500,000 worth of lasers last year, how much commission did he make? $75,000

24. Davisville Dodge pays its sales personnel a commission of 25% of the dealer profit on each car. The dealer made a profit of $18,500 on the cars Linda sold last month. What was her commission last month? $4625

25. In Flagstaff, Arizona, 0.9% of all babies walk before they reach the age of 11 months. If 24,000 babies were born in Flagstaff in the last twenty years, how many of them will walk before they reach the age of 11 months? 216 babies

26. At a Boston Celtics basketball game scheduled for 8:30 P.M., 0.8% of the spectators were children under 12 years old. If 28,000 showed up for the game, how many children under the age of 12 were in the stands? 224 children

27. Beachfront property is very expensive. A real estate agent in South Carolina earned $26,000 in commissions. The property she sold was worth $650,000. What was her commission rate? 4%

28. Adam sold encyclopedias last summer to raise tuition money. He sold $83,500 worth of encyclopedias and was paid $5010 in commissions. What commission rate did he earn? 6%

29. Manuel is taking Cynthia out to dinner. He has $66 to spend. If he wants to tip the server 20%, how much can be afford to spend on their meal? $55

30. Belinda asked Martin out to dinner. She has $47.50 to spend. She wants to tip the waitress 15% of the cost of their meal. How much money can she afford to spend on their meal itself? $41.30

31. The Democratic National Committee has a budget of $33,000,000 to spend on the inauguration of the new president. 15% of the costs will be paid to personnel, 12% of the costs will go toward food, and 10% will go to decorations. How much money will go for personnel, food, and decorations? How much will be left over to cover security, facility rental, and all other expenses?
$12,210,000 for personnel, food, and decorations
$20,790,000 for security, facility rental and all other expenses

32. A major research facility has developed an experimental drug to treat Alzheimer's disease. Twenty percent of the research costs was paid to the staff. Sixteen percent of the research costs was paid to rent the building where the research was conducted. The company has spent $6,000,000 in research on this new drug. How much was paid to cover the cost of staff and rental of the building? $2,160,000

33. Bryce bought a new VCR for $236.00. The tax in Bryce's city is 6%.
(a) What is the sales tax? $14.16
(b) What is the final price of the VCR? $250.16

34. Finn bought a used Honda Accord for $10,500. The tax in his state is 8%.
(a) What is the sales tax? $840
(b) What is the final price of the Honda Accord? $11,340

35. The daycare enrollment in the town of Harvey increased from 5550 to 7600 children per day. What was the percent of increase? 36.94%

36. The cost of a certain Chevrolet Camaro increased from $16,000 last year to $17,280 this year. What was the percent of increase? 8%

37. A stereo originally priced at $985 was sold for $394. What was the percent of decrease? 60%

38. Tay weighed 285 pounds two years ago. After careful supervision at a weight loss center, he reduced his weight to 171 pounds. What was the percent of decrease in his weight? 40%

39. Susie bought her first Harley-Davidson motorcycle. The list price was $15,990, but the dealer gave her a discount of 8%.
(a) What was the discount? $1279.20
(b) How much did she pay for the motorcycle? $14,710,80

40. Jane bought a leather sofa that had been used as the floor model. The list price of the sofa was $1250, but the dealer gave her a discount of 35%.
(a) What was the discount? $437.50
(b) How much did she pay for the sofa? $812.50

41. A Compaq laptop computer listed for $2125. This week, Lisa heard that the manufacturer is offering a rebate of 12%.
(a) What is the rebate? $255
(b) How much will Lisa pay for the computer? $1870

42. Mark and Julie wanted to buy a prefabricated log cabin to put on their property in the Colorado Rockies. The price of the kit is listed at $24,000. At the after-holiday cabin sale, a discount of 14% was offered.
(a) What was the discount? $3360
(b) How much did they pay for the carbin? $20,640

43. Manuel took out an education loan of $2500 at an interest rate of 8% for one year. He then paid back the loan.
(a) How much interest did he pay? $200
(b) How much did it cost to pay off the loan totally? $2700

44. Dina received a car loan so that she could purchase a used Ford Taurus. Her loan was for $4500 at an interest rate of 7 percent for one year. She then paid the loan back in full, with interest.
(a) How much interest did she pay? $315
(b) How much did it cost to pay off the loan totally? $4815

45. Adam deposited $3700 in his savings account for one year. His savings account earns 3.2% interest annually. He did not add any more money within the year, and at the end of that time, he withdrew all funds.
(a) How much interest did he earn? $118.40
(b) How much money did he withdraw from the bank? $3818.40

46. Nikki had $1258 outstanding on her MasterCard, which charges 2% monthly interest. At the end of this month Nikki paid off the loan.
(a) How much interest did Nikki pay for one month? $25.16
(b) How much did it cost to pay off the loan totally? $1283.16

In problems 47–52 round your answer to the nearest cent.

47. How much sales tax would you pay to purchase a new Mazda 626 that cost $18,456.82 if the sales tax rate is 4.6%? $849.01

48. The Hartling family purchased a new living room set. The list price was $1249.95. However, they got a discount of 29%. How much did they pay for the new living room set? $887.46

49. Paolo's credit card statement has a balance of $698.44. The statement said that the *monthly* rate is 1.4575%. How much interest was he charged for the month? How much would it take to pay off the total balance with interest? $10.18; $708.62

50. Juan traveled 22,437 miles in his car last year. He is a salesperson, and 68% of his mileage is for business purposes. How many miles did he travel on business last year? If he can deduct 31¢ per mile for his income tax return how much can he deduct for business mileage?
Approximately 15257.2 miles
$4729.72 may be deducted.

51. The new Nascar concept store bought two video games so that the customers can pretend that they are really driving Nascar racing cars. The games listed for $9346.80 each. The video game dealer gave the store a 12% discount on the list price. The store had to pay a 4% sales tax on the purchase price of the video games. How much did it cost the store to buy the two games? Round the answer to the nearest cent. $17,108.38

52. Pterodactyl Computers sold 100 CD-ROMs for $280.00. FedEx charged $68.00 for overnight rush delivery to the wholesaler. The wholesaler increased the price by 30% over his total cost and sold them to a retail computer store. The retailer increased the price by 16%. One customer bought all 100 CD-ROMs. The clerk received a 2% tip from the customer for delivering the CD-ROMs to his office. Including the tip, how much did the customer pay? Round your answer to the nearest cent. $535.28

Cumulative Review Problems

53. Round to the nearest thousand. 1,698,481
1,698,000

54. Round to the nearest hundred. 2,452,399
2,452,400

55. Round to the nearest hundredth. 1.63474
1.63

56. Round to the nearest thousandth. 0.793468
0.793

57. Round to the nearest ten thousandth. 0.055613
0.0556

58. Round to the nearest ten thousandth. 0.079152
0.0792

Putting Your Skills to Work

PERCENTAGES OF BEVERAGES CONSUMED IN THE UNITED STATES

Did you know that in 1970 in the United States coffee was consumed at the rate of 33.4 gallons per person? In contrast, during that same year only 1.3 gallons of skim milk were consumed for every person in the United States. The following table shows some interesting trends in the drinking of nonalcoholic beverages in this country.

U.S. consumption of selected beverages per person measured in gallons per year

Type	1970	1975	1980	1985	1990	1995
Whole milk	25.5	21.1	17.0	14.3	10.5	9.2
Lowfat milk	4.4	7.1	9.2	10.9	12.6	12.7
Skim milk	1.3	1.3	1.3	1.5	2.6	3.3
Tea	6.8	7.5	7.3	7.1	6.8	7.1
Coffee	33.4	31.4	26.7	27.4	27.0	26.2
Bottled water	NA	NA	2.4	4.5	8.0	9.4
Soft drinks	24.3	28.2	35.1	35.7	43.7	46.9
Selected juices	NA	6.6	7.2	7.7	6.9	8.8

Source: U. S. Department of Agriculture   NA = Not Available

Use the above chart to answer the following questions. Round to the nearest tenth.

PROBLEMS FOR INDIVIDUAL STUDY AND INVESTIGATION

1. What has been the percent of increase in the drinking of skim milk from 1970 to 1995? To what do you attribute this increase? An increase of 153.8%; health benefits of reducing amount of fat consumed in milk

2. What has been the percent of decrease in the drinking of coffee from 1970 to 1995? To what do you attribute this decrease? Decrease of 21.6%; health benefits in reducing amount of caffeine consumed

PROBLEMS FOR GROUP INVESTIGATION AND ANALYSIS

Together with some other members of your class see if you can determine the following.

3. The consumption of whole milk during the period 1970 to 1995 decreased while the consumption of lowfat milk and skim milk increased. During this 25-year period did the consumption of all types of milk actually increase or decrease? By what percent? Overall use of milk decreased by 19.2%.

4. From 1975 to 1995 the approximate population of the United States grew from 214,800,000 to 259,709,000. What was the percent of increase in the population? What was the percent of decrease in the use of tea during this same time? What can we conclude based on these calculations? The population increased by 20.9% while the amount of tea consumed per person decreased by 5.3%. So the actual amount of tea consumed in the United States increased during this period even though the rate of consumption per person decreased.

INTERNET CONNECTIONS: Go to ''http://www.prenhall.com/tobey'' to be connected

Site: Figures and Facts: Eating (Weight Watchers) or a related site

This site gives information about the eating and drinking habits of Americans. Answer the following questions.

5. Use the table of Major Frozen Foods Markets Annual Per Capita Spending to tell how much more a typical French person spends on frozen food than a typical Swiss person. Give your answer as a percent.

6. Use the table of Per Capita Snack Consumption to find the average price (in dollars per pound) paid for potato chips, tortilla chips, and pretzels, respectively.

Chapter Organizer

Topic	Procedure	Examples
Converting a decimal to a percent, p. 287.	**1.** Move the decimal point two places to the right. **2.** Add the percent sign.	0.19 = 19% 0.516 = 51.6% 0.04 = 4% 1.53 = 153% 0.006 = 0.6%
Converting a fraction with a denominator of 100 to a percent, p. 284.	**1.** Use the numerator only. **2.** Add the percent sign.	$\frac{29}{100} = 29\%$ $\frac{5.6}{100} = 5.6\%$ $\frac{3}{100} = 3\%$ $\frac{7\frac{1}{3}}{100} = 7\frac{1}{3}\%$ $\frac{231}{100} = 231\%$
Changing a fraction (whose denominator is not 100) to a percent, p. 292.	**1.** Divide the numerator by the denominator and obtain a decimal. **2.** Change the decimal to a percent.	$\frac{13}{50} = 0.26 = 26\%$ $\frac{1}{20} = 0.05 = 5\%$ $\frac{3}{800} = 0.00375 = 0.375\%$ $\frac{312}{200} = 1.56 = 156\%$
Changing a mixed number to a percent, p. 290.	**1.** Change the mixed number to decimal form. **2.** Transform the decimal to an equivalent percent.	$3\frac{1}{4} = 3.25 = 325\%$ $6\frac{4}{5} = 6.8 = 680\%$ $7\frac{3}{8} = 7.375 = 737.5\%$
Changing a percent to a decimal, p. 286.	**1.** Move the decimal point two places to the left. **2.** Drop the percent sign.	49% = 0.49 2% = 0.02 0.5% = 0.005 196% = 1.96 1.36% = 0.0136
Changing a percent to a fraction, p. 290.	**1.** If the percent contains a decimal, first remove the % sign by moving the decimal point two places to the left. Then write the decimal fraction, and reduce the fraction if possible. **2.** If the percent does not contain a decimal, change the percent to a fraction with a denominator of 100. Reduce the fraction if possible. **3.** If the numerator contains a fraction, change the numerator to an improper fraction. Next simplify by the "invert and multiply" rule. Then reduce the fraction if possible.	$5.8\% = 0.058$ $= \frac{58}{1000} = \frac{29}{500}$ $2.72\% = 0.0272$ $= \frac{272}{10{,}000} = \frac{17}{625}$ $25\% = \frac{25}{100} = \frac{1}{4}$ $38\% = \frac{38}{100} = \frac{19}{50}$ $130\% = \frac{130}{100} = \frac{13}{10}$ $7\frac{1}{8}\% = \frac{7\frac{1}{8}}{100}$ $= 7\frac{1}{8} \div \frac{100}{1}$ $= \frac{57}{8} \times \frac{1}{100} = \frac{57}{800}$

Chapter Organizer

Topic	Procedure	Examples
Solving percent problems by translating to equations, p. 299.	**1.** Translate by replacing "of" by x "is" by = "what" by n "find" by $n =$ **2.** Solve the resulting equation.	**(a)** What is 3% of 56? $n = 3\% \times 56$ $n = (0.03)(56)$ $= 1.68$ **(b)** 16% of what is 208? $16\% \times n = 208$ $0.16n = 208$ $\frac{0.16n}{0.16} = \frac{208}{0.16}$ $n = 1300$ **(c)** What percent of 70 is 30? $n \times 70 = 30$ $70n = 30$ $\frac{70n}{70} = \frac{30}{70}$ $n = 0.4285714\ldots$ n is approximately 42.86%.
Solving percent problems by using proportions, p. 307.	**1.** Identify the parts of the percent proportion. a = the amount b = the base (the whole; it usually appears after the word "of") p = the percent number **2.** Write the percent proportion $\frac{a}{b} = \frac{p}{100}$ using the values obtained in step 1 and solve.	**(a)** What is 28% of 420? The percent p = 28. The base b = 420. The amount a is unknown. We use the variable a. $\frac{a}{b} = \frac{p}{100}$ becomes $\frac{a}{420} = \frac{28}{100}$ If we reduce the fraction on the right-hand side, we have $\frac{a}{420} = \frac{7}{25}$ $25a = (7)(420)$ $25a = 2940$ $\frac{25a}{25} = \frac{2940}{25}$ $a = 117.6$ **(b)** 64% of what is 320? The percent p = 64. The base is unknown. We use the variable b. The amount a = 320. $\frac{a}{b} = \frac{p}{100}$ $\frac{320}{b} = \frac{64}{100}$ If we reduce the fraction on the right-hand side, we have $\frac{320}{b} = \frac{16}{25}$ $(320)(25) = 16b$ $8000 = 16b$ $\frac{8000}{16} = \frac{16b}{16}$ $500 = b$
Solving commission problems, p. 317.	Commission = commission rate × value of sales	A housewares salesperson gets a 16% commission on sales he makes. How much commission does he earn if he sells \$12,000 in housewares? Commission = (0.16)(12,000) = \$1920

Chapter Organizer

Topic	Procedure	Examples
Percent of increase or decrease problems, p. 318.	Percent of increase or decrease = amount of increase or decrease ÷ base	A car that costs \$16,500 now cost only \$15,000 last year. What is the percent of increase? $\begin{array}{r} 16{,}500 \\ -\ 15{,}000 \\ \hline 1{,}500 \end{array}$ increase $\dfrac{1500}{15{,}000} = 0.10$ Rate of increase = 10%
Solving discount problems, p. 318.	Discount = discount rate × list price	Carla purchased a color TV set that lists for \$350 at an 18% discount. **(a)** How much was the discount? **(b)** How much did she pay for the TV set? **(a)** Discount = (0.18)(350) = \$63 **(b)** 350 − 63 = 287 She paid \$287 for the color TV set.
Solving simple interest problems, p. 319.	Interest = principal × rate × time $I = P \times R \times T$	Hector borrowed \$3000 for 4 years at a simple interest rate of 12%. How much interest did he owe after 4 years? $I = P \times R \times T$ $I = (3000)(0.12)(4)$ $= (360)(4)$ $= 1440$ Hector owed \$1440 in interest.

Chapter 5 Review Problems

5.1 *Write as a percent. Round to the nearest hundredth of a percent when necessary.*

1. 0.87 — 87%
2. 0.59 — 59%
3. 0.276 — 27.6%
4. 0.329 — 32.9%
5. 0.0713 — 7.13%
6. 0.0608 — 6.08%

7. 2.52 — 252%
8. 4.37 — 437%
9. 1.036 — 103.6%
10. 1.052 — 105.2%
11. 0.006 — 0.6%
12. 0.002 — 0.2%

13. 0.0029 — 0.29%
14. 0.00534 — 0.53%
15. $\frac{72}{100}$ — 72%
16. $\frac{61}{100}$ — 61%
17. $\frac{19.5}{100}$ — 19.5%
18. $\frac{21.6}{100}$ — 21.6%

19. $\frac{0.24}{100}$ — 0.24%
20. $\frac{0.98}{100}$ — 0.98%
21. $\frac{4\frac{1}{12}}{100}$ — $4\frac{1}{12}\%$
22. $\frac{3\frac{5}{12}}{100}$ — $3\frac{5}{12}\%$
23. $\frac{317}{100}$ — 317%
24. $\frac{225}{100}$ — 225%

5.2 *Change each fraction to a percent. Round to the nearest hundredth of a percent when necessary.*

25. $\frac{16}{25}$ — 64%
26. $\frac{22}{25}$ — 88%
27. $\frac{18}{20}$ — 90%
28. $\frac{14}{20}$ — 70%
29. $\frac{5}{11}$ — 45.45%
30. $\frac{4}{9}$ — 44.44%
31. $2\frac{1}{4}$ — 225%

32. $3\frac{3}{4}$ — 375%
33. $4\frac{3}{7}$ — 442.86%
34. $5\frac{5}{9}$ — 555.56%
35. $\frac{152}{80}$ — 190%
36. $\frac{165}{90}$ — 183.33%
37. $\frac{3}{800}$ — 0.38%
38. $\frac{5}{800}$ — 0.63%

Change each percent to decimal form.

39. 0.2% — 0.002
40. 0.7% — 0.007
41. 21.9% — 0.219
42. 43.1% — 0.431
43. 166% — 1.66
44. 139% — 1.39
45. $32\frac{1}{8}\%$ — 0.32125
46. $26\frac{3}{8}\%$ — 0.26375

Change each percent to fractional form.

47. 82% — $\frac{41}{50}$
48. 36% — $\frac{9}{25}$
49. 185% — $\frac{37}{20}$
50. 225% — $\frac{9}{4}$
51. 16.4% — $\frac{41}{250}$

52. 30.5% — $\frac{61}{200}$
53. $31\frac{1}{4}\%$ — $\frac{5}{16}$
54. $43\frac{3}{4}\%$ — $\frac{7}{16}$
55. 0.05% — $\frac{1}{2000}$
56. 0.06% — $\frac{3}{5000}$

Complete the following chart.

	Fraction	Decimal	Percent
57.	$\frac{3}{5}$	0.6	60%
58.	$\frac{7}{8}$	0.875	87.5%
59.	$\frac{3}{8}$	0.375	37.5%
60.	$\frac{9}{16}$	0.5625	56.25%
61.	$\frac{1}{125}$	0.008	0.8%
62.	$\frac{9}{20}$	0.45	45%

5.3 *Solve each percent problem. Round to the nearest hundredth when necessary.*

63. What is 96% of 300?
288

64. What is 85% of 400?
340

65. 3 is 20% of what number?
15

66. 90 is 40% of what number?
225

67. 50 is what percent of 125?
40%

68. 70 is what percent of 175?
40%

69. Find 125% of 46.
57.5

70. Find 118% of 60.
70.8

71. 92% of what number is 147.2?
160

72. 68% of what number is 95.2?
140

73. What percent of 70 is 14?
20%

74. What percent of 60 is 28?
46.67%

75. 0.5% of 2600 is what number?
13

76. 0.8% of 3500 is what number?
28

77. What percent of 28 is 130?
464.29%

78. What percent of 37 is 120?
324.32%

5.4 *Solve each applied percent problem. Round your answer to the nearest hundredth when necessary.*

79. Professor Wonson found that 34% of his class is left-handed. He has 150 students in his class. How many are left-handed? 51 students

80. A Vermont truck dealer found that 64% of all the trucks he sold were four-wheel drive. If he sold 150 trucks, how many had four-wheel drive? 96 trucks

81. Six computers in a shipment of 144 computers were found to be defective. What percent were defective? 4.17%

82. Rafael obtained 16 incorrect answers on a test containing 102 questions. What percent of his answers were incorrect? 15.69%

83. Today Yvonne's car has 61% of the value that it had 2 years ago. Today it is worth $6832. What was it worth 2 years ago? $11,200

84. A charity organization spent 12% of its budget for administrative expenses. It spent $9624 on administrative expenses. What was the total budget? $80,200

85. In a test of car tires, three of 20 tires failed the test. What percent failed the test? 15%

86. Moorehouse Industries received 600 applications and hired 45 of the applicants. What percent of the applicants obtained a job? 7.5%

87. Roberta earns a commission at the rate of 7.5%. Last month she sold $16,000 worth of goods. How much commission did she make last month? $1200

88. Gary purchased a car for $18,600. The sales tax in his state is 3%. What did he pay in sales tax? $558

89. The average temperature in Winchester during the last 10 years has been 45.8°F. In the previous 10 years the average temperature was 44°F. What percent of increase is this? 4.09%

90. Sally invested $6000 in mutual funds earning 11% simple interest. How much interest will she earn in

(a) Six months? $330

(b) Two years? $1320

91. Irene purchased a dining room set at a 20% discount. The list price was $1595.

(a) What was the discount? $319

(b) What did she pay for the set? $1276

92. Joan and Michael budget 38% of their income for housing. They spend $684 per month for housing. What is their monthly income? $1800

Chapter 5 Test

Write as a percent. Round to the nearest hundredth of a percent when necessary.

1. 0.42

2. 0.01

3. 0.006

4. 0.139

5. 2.18

6. $\frac{71}{100}$

7. $\frac{2.7}{100}$

8. $\frac{3\frac{1}{7}}{100}$

Change each fraction to a percent. Round to the nearest hundredth of a percent when necessary.

9. $\frac{19}{40}$

10. $\frac{180}{450}$

11. $\frac{105}{75}$

12. $1\frac{3}{4}$

Write each decimal as a percent.

13. 0.1713

14. 3.024

Write each percent as a fraction in simplified form.

15. 152%

16. $7\frac{3}{4}\%$

1. 42%

2. 1%

3. 0.6%

4. 13.9%

5. 218%

6. 71%

7. 2.7%

8. $3\frac{1}{7}\%$

9. 47.5%

10. 40%

11. 140%

12. 175%

13. 17.13%

14. 302.4%

15. $1\frac{13}{25}$

16. $\frac{31}{400}$

17.	26.69
18.	130
19.	55.56%
20.	200
21.	5000
22.	46%
23.	699.6
24.	2.29%
25.	$6092 commission
26. (a)	$150.81 discount
(b)	They paid $306.19.
27.	89.29% were not defective.
28.	8.93% is the percent of decrease.
29.	12,000 registered voters
30. (a)	$240 in interest in 6 months
(b)	$960 in interest in 2 years

Solve each percent problem. Round to the nearest hundredth if necessary.

17. What is 17% of 157?

18. 33.8 is 26% of what number?

19. What percent of 72 is 40?

20. Find 0.8% of 25,000.

21. 16% of what number is 800?

22. 92 is what percent of 200?

23. 132% of 530 is what number?

24. What percent is 8 of 350?

Solve each applied percent problem. Round to the nearest hundredth if necessary.

25. A real estate agent sells a house for $152,300. She gets a commission of 4% on the sale. What is her commission?

26. Julia and Charles bought a new dishwasher at a 33% discount. The list price was $457.
(a) What was the discount?
(b) How much did they pay for the dishwasher?

27. An inspector found that 75 out of 84 parts were not defective. What percent of the parts were not defective?

28. At Cedars University the number of hours a three-credit college course meets in a semester is 51 hours. Twenty years ago a three-credit course met for 56 hours. What is the percent of decrease?

29. A total of 5160 people voted in the city election. This was 43% of the registered voters. How many registered voters are in the city?

30. Wanda borrowed $3000 at a 16% simple rate of interest.
(a) How much interest did she pay in six months?
(b) How much interest did she pay in two years?

Cumulative Test for Chapters 1–5

Approximately one-half of this test is based on Chapter 5 material. The remainder is based on material covered in Chapters 1–4.

Do each problem. Simplify your answer.

1. Add. $\begin{array}{r} 38 \\ 196 \\ +\ 2007 \\ \hline \end{array}$

2. Subtract. $\begin{array}{r} 23{,}007 \\ -\ 14{,}563 \\ \hline \end{array}$

3. Multiply. $\begin{array}{r} 126 \\ \times\ 42 \\ \hline \end{array}$

4. Divide. $36\overline{)3204}$

5. Add. $2\frac{1}{4} + 3\frac{1}{3}$

6. Subtract. $\frac{11}{12} - \frac{5}{6}$

7. Multiply. $3\frac{17}{36} \times \frac{21}{25}$

8. Divide. $\frac{5}{12} \div 1\frac{3}{4}$

9. Round to the nearest tenth.
5731.652

10. Add. $\begin{array}{r} 5.6 \\ 3.21 \\ 18.3 \\ +\ 7.008 \\ \hline \end{array}$

11. Multiply. $\begin{array}{r} 5.62 \\ \times\ 0.3 \\ \hline \end{array}$

12. Divide. $1.4\overline{)0.5152}$

13. Write as a rate in simplest form.
78 pounds to 130 square feet

14. Is this proportion true or false?

$$\frac{20}{25} = \frac{300}{375}$$

15. Solve the proportion. $\frac{8}{2.5} = \frac{n}{7.5}$

16. A college has a ratio of 3 faculty for every 19 students. The student body presently has 4263 students. How many faculty are there? Round to the nearest whole number.

1. 2241

2. 8444

3. 5292

4. 89

5. $\frac{67}{12}$ *or* $5\frac{7}{12}$

6. $\frac{1}{12}$

7. $\frac{35}{12}$ *or* $2\frac{11}{12}$

8. $\frac{5}{21}$

9. 5731.7

10. 34.118

11. 1.686

12. 0.368

13. $\frac{3 \text{ pounds}}{5 \text{ square feet}}$

14. True

15. $n = 24$

16. 673 faculty

17.	2.3%
18.	46.8%
19.	198%
20.	3.75%
21.	2.43
22.	0.0675
23.	17.76%
24.	114.58
25.	300
26.	718.2
27.	$8370
28.	3200 students
29.	11.31% increase
30.	$352

In Problems 17–30 round to the nearest hundredth when necessary.

Write as a percent.

17. 0.023

18. $\frac{46.8}{100}$

19. 1.98

20. $\frac{3}{80}$

Write as a decimal.

21. 243%

22. $6\frac{3}{4}\%$

23. What percent of 214 is 38?

24. Find 1.7% of 6740.

25. 219 is 73% of what number?

26. 114% of 630 is what number?

27. Alice bought a new car. She got a 7% discount. The car listed for $9000. How much did she pay for the car?

28. A total of 896 freshmen were admitted to King Frederich College. Freshmen make up 28% of the student body. How big is the student body?

29. The air pollution level in Centerville is 8.86. Ten years ago it was 7.96. What is the percent of increase of the air pollution level?

30. Fred borrowed $1600 for two years. He was charged simple interest at a rate of 11%. How much interest did he pay?

CHAPTER 6
Measurement

You might hear a person mentioning that their new computer has a 2.1-GB hard drive and has 16K of RAM. The Pentium chip that powers the computer runs at 150 MHz. Just what do all those letters mean? What is being measured? Can you solve a problem with these kind of units? Turn to the Putting Your Skills to Work on page 364.

Pretest Chapter 6

1. 204	**17.** 0.0563
2. 56	**18.** 4800
3. 3520	**19.** 0.568
4. 6400	**20.** 8900
5. 1320	**21.** 4.73
6. 24	**22.** 1.28
7. 5320	**23.** 59.52
8. 4680	**24.** 1826.78
9. 98.6	**25.** 39.69
10. 0.027	**26.** 103.4
11. 529.6	**27.** 55.8 feet
12. 0.123	**28.** **(a)** 95°F
13. 2376 m	**(b)** No
14. 94.262 m	**29.** 12.2 miles farther
15. 3820	**30.** 22.5 gal/hr
16. 3.162	

If you are familiar with the topics in this chapter, take this test now. Check your answers with those in the back of the book. If an answer was wrong or you couldn't do a problem, study the appropriate section of the chapter.

If you are not familiar with the topics in this chapter, don't take this test now. Instead, study the examples, work the practice problems, and then take the test.

This test will help you identify which concepts you have mastered and which you need to study further.

Section 6.1

Convert. When necessary, express your answer as a decimal rounded to the nearest hundredth.

1. 17 ft = ____ in. **2.** 7 gal = ____ pt **3.** 2 mi = ____ yd

4. 3.2 tons = ____ lb **5.** 22 min = ____ sec **6.** 6 gal = ____ qt

Section 6.2

Perform each conversion.

7. 5.32 km = ____ m **8.** 46.8 m = ____ cm **9.** 986 mm = ____ cm

10. 27 mm = ____ m **11.** 5296 mm = ____ cm **12.** 123 m = ____ km

Convert to meters and add.

13. 1.2 km + 192 m + 984 m **14.** 3862 cm + 9342 mm + 46.3 m

Section 6.3

Perform each conversion.

15. 3.82 L = ____ mL **16.** 3162 g = ____ kg **17.** 56.3 kg = ____ t

18. 4.8 kL = ____ L **19.** 568 mg = ____ g **20.** 8.9 L = ____ cm^3

Section 6.4

Perform each conversion. Round to the nearest hundredth when necessary.

21. 12 cm = ____ in. **22.** 4.2 ft = ____ m **23.** 96 km = ____ mi

24. 482 gal = ____ L **25.** 1.4 oz = ____ g **26.** 47 kg = ____ lb

Section 6.5

Solve each problem. Round your answer to the nearest hundredth when necessary.

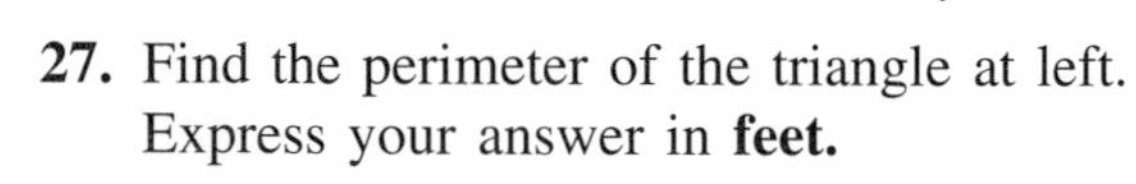

27. Find the perimeter of the triangle at left. Express your answer in **feet.**

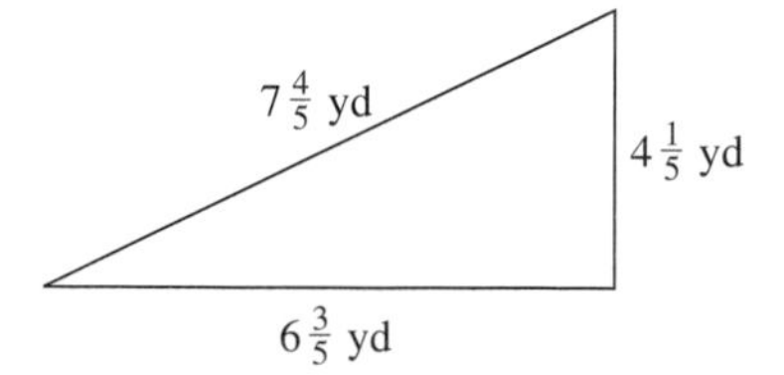

28. The radio reported the temperature today at 35°C. The record high temperature for this day is 98°F.
(a) What was the Fahrenheit temperature today?
(b) Did the temperature today set a new record?

29. Juanita traveled in Mexico for 2 hours at 95 kilometers per hour. She had to travel a distance of 130 miles. How far does she still need to travel? (Express your answer in **miles**.)

30. A pump is running at 1.5 quarts per minute. What is this rate in gallons per hour?

6.1 American Units

MathPro

Video 12
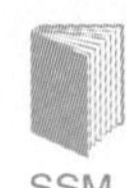
SSM

After studying this section, you will be able to:

1. *Identify the basic unit equivalencies in the American system.*
2. *Convert from one unit of measure to another.*

1 *Identifying the Basic Unit Equivalencies in the American System*

We often ask questions about measurements. How far is it to work? How much does this bottle hold? What is the weight of this box? How long will it be until exams? To answer these questions we need to agree on a unit of measure for each type of measurement.

At present there are two main systems of measurement, each with its own set of units: the metric system and the American system. Nearly all countries in the world use the metric system. In the United States, however, except in science, most measurements are made in American units.

The United States is using the metric system more and more frequently, however, and may eventually convert to metric units as the standard. But for now we need to be familiar with both systems. We cover American units in this section.

One of the most familiar measuring devices is a ruler that measures lengths as great as 1 foot. It is divided into 12 inches; that is,

$$1 \text{ foot} = 12 \text{ inches}$$

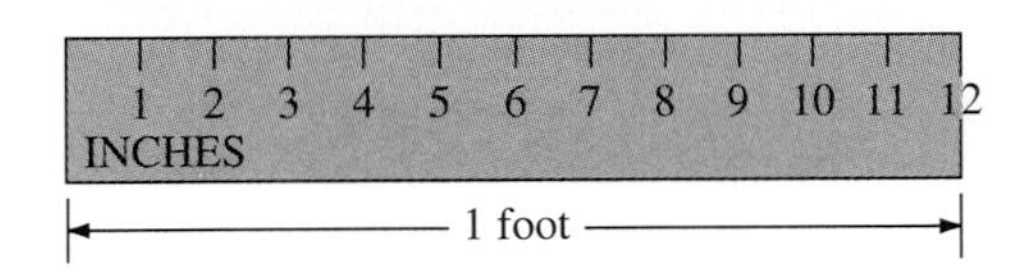

There are several other important relationships you need to know. Your instructor may require you to memorize the following facts.

Length
12 inches = 1 foot
3 feet = 1 yard
5280 feet = 1 mile
1760 yards = 1 mile

Time
60 seconds = 1 minute
60 minutes = 1 hour
24 hours = 1 day
7 days = 1 week

Note that time is measured in the same units in both the metric and the American system.

Weight
16 ounces = 1 pound
2000 pounds = 1 ton

Volume
2 cups = 1 pint
2 pints = 1 quart
4 quarts = 1 gallon

We can choose to measure an object—say, a bridge—using a small unit (an inch), a larger unit (a foot), or a still larger unit (a mile). We may say that the bridge spans 7920 inches, 660 feet, or an eighth of a mile. Although we probably would not choose to express our measurement in inches because this is not a convenient measurement to work with for an object as long as a bridge, the bridge length is the same whatever unit of measurement we use.

Notice that the smaller the measuring unit, the larger the number of those units in the final measurement. The inch is the smallest unit in our example, and the inch measurement has the greatest number of units (7920). The mile is the largest unit, and it has the smallest number of units (an eighth equals 0.125). Whatever measuring system you use, and whatever you measure (length, volume, and so on), the smaller the unit you use, the greater the number of those units.

After studying the values in the length, time, weight, and volume tables, see if you can quickly do Example 1.

TEACHING TIP An effective approach is to ask students to open their books and fill in the answers to Practice Problem 1. If they make several mistakes, or just cannot remember several facts, ask them to memorize the chart values showing the length, time, weight, and volume equivalencies. It wastes a lot of time looking these up when doing problems, so students should know them by heart.

EXAMPLE 1 Answer rapidly the following questions.

(a) How many inches in a foot? **(b)** How many yards in a mile?

(c) How many seconds in a minute? **(d)** How many hours in a day?

(e) How many pounds in ton? **(f)** How many cups in a pint?

(a) 12 **(b)** 1760 **(c)** 60 **(d)** 24 **(e)** 2000 **(f)** 2

Practice Problem 1 Answer rapidly the following questions.

(a) How many feet in a yard? **(b)** How many feet in a mile?

(c) How many minutes in an hour? **(d)** How many days in a week?

(e) How many ounces in a pound? **(f)** How many pints in a quart?

(g) How many quarts in a gallon? ■

2 *Converting One Measurement to Another*

To convert or change one measurement to another, we simply multiply by 1 since multiplying by 1 does not change the value of a quantity. For example, to convert 180 inches to feet, we look for a name for 1 that has inches and feet.

$$1 = \frac{1 \text{ foot}}{12 \text{ inches}}$$

$$180 \text{ inches} \times \frac{1 \text{ foot}}{12 \text{ inches}} = \frac{180 \text{ feet}}{12} = 15 \text{ feet}$$

Notice that when we multiplied, the inches divided out. We are left with the unit feet.

What name for 1 should we choose if we want to change from feet to inches? Convert 4 feet to inches.

TEACHING TIP The chief advantage of the "multiply by a fractional name for 1 and remove the common unit on the top of the first fraction and the denominator of the second fraction" method is that the students using it do not multiply by the wrong number. This method is very useful in science courses.

$$1 = \frac{12 \text{ inches}}{1 \text{ foot}}$$

$$4 \text{ feet} \times \frac{12 \text{ inches}}{1 \text{ foot}} = \frac{48 \text{ inches}}{1} = 48 \text{ inches}$$

Notice that in this example feet divided out and we were left with the desired unit, inches.

When choosing a fractional name for 1, the unit we want to change to should be in the *numerator*. The unit we start with should be in the *denominator*. This unit will cancel out.

EXAMPLE 2 Convert each measurement.

(a) 8800 yards to miles **(b)** 26 feet to inches

(a) $8800 \text{ yards} \times \dfrac{1 \text{ mile}}{1760 \text{ yards}} = \dfrac{8800}{1760} \text{ miles} = 5 \text{ miles}$

(b) $26 \text{ feet} \times \dfrac{12 \text{ inches}}{1 \text{ foot}} = 26 \times 12 \text{ inches} = 312 \text{ inches}$

Practice Problem 2 Convert each measurement.

(a) 15,840 feet to miles **(b)** 17 feet to inches ■

Some conversions involve fractions or decimals. Whether you want to measure the area of a living room or the dimensions of a piece of property, it is helpful to be able to make conversions like these.

EXAMPLE 3 Convert each measurement.

(a) 26.48 miles to yards **(b)** $3\frac{2}{3}$ feet to yards

(a) $26.48 \text{ miles} \times \dfrac{1760 \text{ yards}}{1 \text{ mile}} = 46{,}604.8 \text{ yards}$

(b) $3\frac{2}{3} \text{ feet} \times \dfrac{1 \text{ yard}}{3 \text{ feet}} = \dfrac{11}{3} \times \dfrac{1}{3} \text{ yard} = \dfrac{11}{9} \text{ yards} = 1\frac{2}{9} \text{ yards}$

Practice Problem 3 Convert each measurement.

(a) 18.93 miles to feet **(b)** $16\frac{1}{2}$ inches to yards ■

EXAMPLE 4 Lynda's new car weighs 2.43 tons. How many pounds is that?

$$2.43 \text{ tons} \times \frac{2000 \text{ pounds}}{1 \text{ ton}} = 4860 \text{ pounds}$$

Practice Problem 4 A package weighs 760.5 pounds. How many ounces does it weigh? ■

EXAMPLE 5 The chemistry lab has 34 quarts of weak hydrochloric acid. How many gallons of this acid are in the lab? (Express your answer as a decimal.)

$$34 \text{ quarts} \times \frac{1 \text{ gallon}}{4 \text{ quarts}} = \frac{34}{4} \text{ gallons} = 8.5 \text{ gallons}$$

Practice Problem 5 19 pints of milk is the same as how many quarts? (Express your answer as a decimal.) ■

EXAMPLE 6 A computer printout shows that a particular job took 144 seconds. How many minutes is that? (Express your answer as a decimal.)

$$144 \text{ seconds} \times \frac{1 \text{ minute}}{60 \text{ seconds}} = \frac{144}{60} \text{ minutes} = 2.4 \text{ minutes}$$

Practice Problem 6 Joe's time card read "Hours worked today: 7.2." How many minutes are in 7.2 hours? ■

EXAMPLE 7 A window is 4 feet 5 inches wide. How many inches is that?

4 feet 5 inches means 4 feet and 5 inches. Change the 4 feet to inches and add the 5 inches.

$$4 \text{ feet} \times \frac{12 \text{ inches}}{1 \text{ foot}} = 48 \text{ inches}$$

$$48 \text{ inches} + 5 \text{ inches} = 53 \text{ inches}$$

The window is 53 inches wide.

Practice Problem 7 A bag of potatoes weighs 7 pounds 12 ounces. How many ounces is that? ■

TEACHING TIP It is important to carefully go over the type of problem that requires students to multiply by two unit fractions. You may also do an example of the type "The supermarket charges $2.40 per pound for lean ground beef. If you bought 33 ounces, how much would you pay?" (The answer is $4.95.)

EXAMPLE 8 The all-night Charlotte garage charges $1.50 per hour for parking both day and night. A businessman left his car there for $2\frac{1}{4}$ days. How much was he charged?

1. *Understand the problem.*
   Here it might help to look at a simpler problem. If the businessman had left his car for 2 hours, we would multiply.

   $$\text{The rule means "per"} \rightarrow \frac{1.50 \text{ dollars}}{1 \text{ hour}} \times 2 \text{ hours} = 3.00 \text{ dollars or } \$3.00$$

   Thus, if the businessman leaves his car for 2 hours, he would be charged $3. We see that we need to multiply by the number of hours the car was in the garage to solve the problem.

   Since the original problem gave the time in days, not hours, we will need to change the days to hours. Remember, the garage charges $1.50 per *hour*.

2. *Solve and state the answer.*
   Now that we know that the way to solve the problem is to multiply hours, we will begin. To make our calculations easier we will write $2\frac{1}{4}$ as 2.25. Change days to hours.

   $$2.25 \text{ days} \times \frac{24 \text{ hours}}{1 \text{ day}} \times \frac{1.50 \text{ dollars}}{1 \text{ hour}} = 81 \text{ dollars or } \$81$$

   The businessman was charged $81.

3. *Check.*
Is our answer in the desired unit? Yes. The answer is in dollars and we would expect it to be in dollars. ✓

You may want to redo the calculation or use a calculator to check. The check is up to you.

Practice Problem 8 A businesswoman parked her car at a garage for $1\frac{3}{4}$ days. The garage charges $1.50 per hour. How much did she pay to park the car? ■

ALTERNATIVE METHOD USING PROPORTIONS How did people first come up with the idea of "multiplying by a fractional name for 1"? What mathematical principles are involved here? Actually, this is the same as solving a proportion. Consider Example 5, where we changed 34 quarts to 8.5 gallons by multiplying.

$$34 \cancel{\text{ quarts}} \times \frac{1 \text{ gallon}}{4 \cancel{\text{ quarts}}} = \frac{34}{4} \text{ gallons} = 8.5 \text{ gallons}$$

What we were actually doing is setting up the proportion

1 gallon is to 4 quarts as n gallons is to 34 quarts

and solving.

$$\frac{1 \text{ gallon}}{4 \text{ quarts}} = \frac{n \text{ gallons}}{34 \text{ quarts}}$$

Cross-multiply.

$$1 \text{ gallon} \times 34 \text{ quarts} = 4 \text{ quarts} \times n \text{ gallons}$$

Divide both sides of the equation by 4 quarts.

$$\frac{1 \text{ gallon} \times 34 \cancel{\text{ quarts}}}{4 \cancel{\text{ quarts}}} = \frac{\cancel{4} \cancel{\text{ quarts}} \times n \text{ gallons}}{\cancel{4} \cancel{\text{ quarts}}}$$

$$1 \text{ gallon} \times \frac{34}{4} = n \text{ gallons}$$

$$8.5 \text{ gallons} = n \text{ gallons}$$

Thus the number of gallons is 8.5. Using proportions takes a little longer, so multiplying by a fractional name for 1 is the more popular method. For more about using proportions in converting units, see Exercises 6.1, problems 67 and 68.

6.1 Exercises

From memory, write the equivalent values.

1. 1 foot = 12 inches

2. 1 yard = 3 feet

3. 1760 yards = 1 mile

4. 5280 feet = 1 mile

5. 1 ton = 2000 pounds

6. 1 pound = 16 ounces

7. 4 quarts = 1 gallon

8. 2 cups = 1 pint

9. 1 quart = 2 pints

10. 1 day = 24 hours

11. 60 seconds = 1 minute

12. 60 minutes = 1 hour

Convert. When necessary, express your answer as a decimal.

13. 15 feet = 5 yards

14. 21 feet = 7 yards

15. 84 inches = 7 feet

16. 180 inches = 15 feet

17. 10,560 feet = 2 miles

18. 5280 yards = 3 miles

19. 7 miles = 12,320 yards

20. 6 miles = 31,680 feet

21. 12 feet = 144 inches

22. 16 feet = 192 inches

23. 41 yards = 123 feet

24. 36 yards = 108 feet

25. 75 inches = 6.25 feet

26. 87 inches = 7.25 feet

27. 192 ounces = 12 pounds

28. 144 ounces = 9 pounds

29. 13 tons = 26,000 pounds

30. 17 tons = 34,000 pounds

31. 2.25 pounds = 36 ounces

32. 4.25 pounds = 68 ounces

33. 7 gallons = 28 quarts

34. 5 gallons = 20 quarts

35. 18 pints = 9 quarts

36. 24 pints = 12 quarts

37. 31 pints = 62 cups

38. 27 pints = = 54 cups

39. 8 gallons = 64 pints

40. 6 gallons = 48 pints

41. 12 weeks = 84 days

42. 7 weeks = 49 days

43. 660 minutes = 11 hours

44. 780 minutes = 13 hours

45. 11 days = 264 hours

46. 7 days = 168 hours

47. 70 minutes = 4200 seconds

48. 50 minutes = 3000 seconds

49. 18 hours = 64,800 seconds

50. 15 hours = 54,000 seconds

51. 180 seconds = 3 minutes

52. 3462.56 feet = 41,550.72 inches

53. 6755 yards = 20,265 feet

54. 65.62 pounds = 1049.92 ounces

Applications

55. A stockbroker left his car in an all-night garage for $2\frac{1}{2}$ days. The garage charges $2.25 per hour. How much did he pay for parking?

$$2.5\ \text{days} \times \frac{24\ \text{hours}}{1\ \text{day}} \times \frac{\$2.25}{1\ \text{hour}} = \$135$$

56. When Barry went overseas on business, he left his car in the long-term parking lot for $3\frac{1}{2}$ days. Long-term parking costs $2.15 per hour. How much did Barry pay when he came back from his trip?

$$3.5\ \text{days} \times \frac{24\ \text{hours}}{1\ \text{day}} \times \frac{\$2.15}{1\ \text{hour}} = \$180.60$$

57. Judy is making a wild mushroom sauce for pasta tonight with a large group of friends. She bought 26 ounces of wild mushrooms at $6.00 per pound. How much were the mushrooms?

$$26\ \text{ounces} \times \frac{1\ \text{pound}}{16\ \text{ounces}} \times \frac{6.00}{1\ \text{pound}} = \$9.75$$

58. Kurt is trying to eat a lower-fat diet than usual. He finds a store that sells 1% ground white-meat turkey breast at $4.00 per pound. He buys one packet weighting 18 ounces and another weighing 22 ounces. How much did he pay?

$$40\ \text{ounces} \times \frac{1\ \text{pound}}{16\ \text{ounces}} \times \frac{4.00}{1\ \text{pound}} = \$10.00$$

59. A physically challenged athlete proves that he is not as disabled as the general public thinks by rolling his special three-wheel racing chair 37.7 miles in one day. How many feet is that?

$$37.7\ \text{miles} \times \frac{5280\ \text{feet}}{1\ \text{mile}} = 199{,}056\ \text{feet}$$

60. Mount Whitney in California is approximately 2.745 miles high. How many feet is that? Round your answer to the nearest 10 feet.

$$2.745\ \text{miles} \times \frac{5280\ \text{feet}}{1\ \text{mile}} = 14{,}490\ \text{feet}$$

61. In 1994, Sue Caivun of China broke the women's world record for the pole vault when she defied gravity and found herself clearing the bar at 13 feet $5\frac{3}{4}$ inches. How many inches high was her pole vault?

$161\frac{3}{4}$ inches

62. Jan Zelezny of the Czech Republic threw his javelin 318 feet 10 inches to break the world's record in 1993. How many inches did his javelin fly?

3826 inches

63. A multimillion-dollar shopping complex in Italy is using nothing but marble for the floors and walls of all public areas. The lobby alone is constructed of 267,905,993 pounds of blue-gray marble. How many tons would that be? Round to the nearest ton. 133,953 tons

64. Last week, the Red Cross and the United Nations airlifted 12,560,950 pounds of food to hungry children. How many tons of food were distributed? Round to the nearest ton. 6280 tons

65. A space probe traveled for 3 years to get to the planet Saturn. How many hours did the trip take? (Assume 365 days in 1 year.) 26,280 hours

66. A seedling peach tree grew for 7 years until it produced its first fruit. How many hours would that be if you assume that there are 365 days in a year? 61,320 hours

There are 6080 feet in a nautical mile and 5280 feet in a land mile. On sea and in the air, distance and speed are often measured in nautical miles. The ratio of regular miles to nautical miles is 38 to 33.

67. A windjammer sailboat was used as a floating school. During one school year, the ship traveled 12,800 nautical miles. What would be the equivalent in land miles? Round to the nearest whole mile.

$$12{,}800\ \text{nautical miles} \times \frac{38\ \text{land miles}}{33\ \text{nautical miles}} = 14{,}739\ \text{land miles}$$

68. A merchant ship took some tourists on board for extra income. The ship took them the equivalent of 850 land miles. How many nautical miles did they travel? Round to the nearest whole mile.

$$850\ \text{land miles} \times \frac{33\ \text{nautical miles}}{38\ \text{land miles}} = 738\ \text{nautical miles}$$

6.2 Metric Measurements: Length

MathPro

Video 13

SSM

After studying this section, you will be able to:

1. *Understand prefixes in metric units.*
2. *Convert from one metric unit of length to another.*

1 *Understanding Prefixes in Metric Units*

The metric system of measurement is used in most industrialized nations of the world. As we move toward a global economy, it is important to be familiar with the metric system. The metric system is designed for ease in calculating and in converting from one unit to another.

In the metric system, the basic unit of length measurement is the meter. A meter is just slightly longer than a yard. To be more precise, the meter is approximately 39.37 inches long.

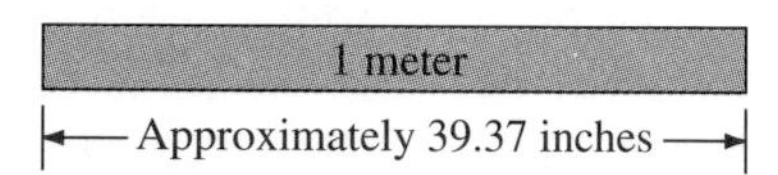

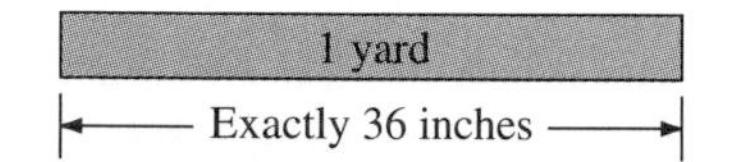

Units that are larger or smaller than the meter are based on the meter and powers of 10. For example, the unit *deka*meter is *ten* meters. The unit *deci*meter is *one-tenth* of a meter. The prefix *deka* means 10. What does the prefix *deci* mean? All the prefixes in the metric system are names for multiples of 10. A list of metric prefixes and their meaning follows.

Prefix	Meaning
kilo-	thousand
hecto-	hundred
deka-	ten
deci-	tenth
centi-	hundredth
milli-	thousandth

The most commonly used prefixes are *kilo-*, *centi-*, and *milli-*. *Kilo-* means thousand, so a *kilo*meter is a thousand meters. Similarly, *centi-* means one hundredth, so a *centi*meter is a one hundredth of a meter. And *milli-* means one thousandth, so a *milli*meter is one thousandth of a meter.

The kilometer is used to measure distances much larger than the meter. How far did you travel in a car? The centimeter is used to measure shorter lengths. What are the dimensions of this textbook? The millimeter is used to measure very small lengths. What is the width of the lead in a pencil?

EXAMPLE 1 Write the prefix that means **(a)** thousand **(b)** tenth

(a) The prefix *kilo-* is used for thousand.

(b) The prefix *deci-* is used for tenth.

Practice Problem 1 Write the prefix that means **(a)** ten **(b)** thousandth ■

2 *Converting from One Metric Unit of Length to Another*

How do we convert from one metric unit to another? For example, how do we change 5 kilometers into an equivalent number of meters?

Recall from Chapter 3 that when we multiply by 10 we move the decimal point one place to the right. When we divide by 10 we move the decimal point one place to the left. Let's see how we use that idea to change from one metric unit to another.

Changing from Larger Metric Units to Smaller Ones

When you change from one metric prefix to another moving to the **right** on this prefix chart, move the decimal point to the **right** the same number of places.

kilometer | *hectometer* | *dekameter* | meter (one) | *decimeter* | *centimeter* | *millimeter*

Thus 1 meter = 100 centimeters because we move two places to the right on the chart of prefixes and we also move the decimal point (1.00) two places to the right.

Now let us examine the four most commonly used metric measurements of length and their abbreviations.

Commonly Used Metric Lengths

1 kilometer (km) = 1000 meters

1 meter (m) (the basic unit of length in the metric system)

1 centimeter (cm) = 0.01 meter

1 millimeter (mm) = 0.001 meter

Now let us see how we can change a measurement stated in a larger unit to an equivalent measurement stated in smaller units.

EXAMPLE 2

(a) Change 5 kilometers to meters. **(b)** Change 20 meters to centimeters.

(a) From *kilometer* to meter we move three places to the right on the prefix chart, so we move the decimal point three places to the right.

5 kilometers = 5.000 meters (move three places) = 5000 meters

(b) To go from meters to *centimeters,* we move two places to the right on the prefix chart. Thus we move the decimal point two places to the right.

20 meters = 20.00 centimeters (move two places) = 2000 centimeters

Practice Problem 2

(a) Change 4 meters to centimeters.
(b) Change 30 centimeters to millimeters. ■

Changing from Smaller Metric Units to Larger Ones

When you change from one metric prefix to another moving to the **left** on this prefix chart, move the decimal point to the **left** the same number of places.

kilometer | hectometer | dekameter | one (meter) | decimeter | centimeter | millimeter

EXAMPLE 3

(a) Change 7 centimeters to meters.

(b) Change 56 millimeters to kilometers.

(a) To go from *centi*meters to meters, we move *two* places to the left on the prefix chart. Thus we move the decimal point two places to the left.

7 centimeters = 0.07 meter (move two places to the left)
= 0.07 meter

(b) To go from *milli*meters to *kilo*meters, we move six places to the left on the prefix chart. Thus we move the decimal point six places to the left.

56 millimeters = 0.000056. kilometer (move six places to the left)
= 0.000056 kilometer

Practice Problem 3

(a) Change 3 *milli*meters to meters.

(b) Change 47 *centi*meters to *kilo*meters. ■

Thinking Metric

Long distances are customarily measured in kilometers. A kilometer is about 0.62 mile. It takes about 1.6 kilometers to make a mile. The following drawing shows the relationship between the kilometer and the mile.

1 kilometer

1 mile (about 1.6 kilometers)

Many small distances are measured in centimeters. A centimeter is about 0.394 inch. It takes about 2.54 centimeters to make an inch. You can get a good idea of their size by looking at a ruler marked in both inches and centimeters.

1 centimeter

1 inch (about 2.54 centimeters)

Try to visualize how many centimeters long or wide this book is. It is 21 centimeters wide and 27.5 centimeters long.

A millimeter is a very small unit of measurement, often used in manufacturing.

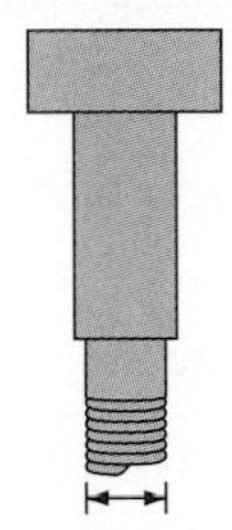

The threaded end of a bolt may be 6 mm wide.

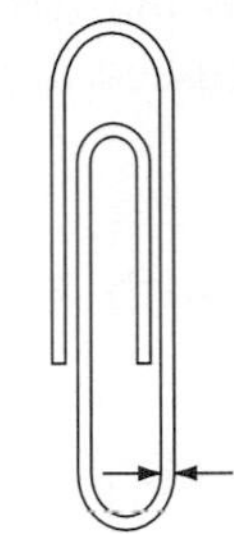

A paper clip is made of wire 0.8 mm thick.

TEACHING TIP It is important to know the relative size of a meter, a centimeter, and a kilometer. Most students will benefit greatly from doing a few problems like Example 4.

Now that you have an understanding of the size of these metric units, let's try to select the most convenient metric unit for measuring the length of an object.

EXAMPLE 4 Bob measured the width of a doorway in his house. He wrote down "73." But what unit of measurement did he use?

(a) 73 kilometers **(b)** 73 meters **(c)** 73 centimeters

The most reasonable choice is **(c)**, 73 centimeters. The other two units would be much too long. A meter is close to a yard, and a doorway would not be 73 yards wide! A kilometer is even larger than a meter.

Practice Problem 4 Joan measured the length of her car. She wrote down 3.8. Which unit of measurement did she use?

(a) 3.8 kilometers **(b)** 3.8 meters **(c)** 3.8 centimeters ■

The most frequent metric conversions in length are done between kilometers, meters, centimeters, and millimeters. Their abbreviations are km, m, cm, and mm, and you should be able to use them correctly.

EXAMPLE 5 Convert each measurement.

(a) 364 m to km **(b)** 982 cm to m **(c)** 16 mm to cm **(d)** 5.2 m to mm

In each of the first three cases, we move the decimal point to the left because we are going from a smaller unit to a larger unit.

(a) 364 m = 0.364 km (three places to left)
= 0.364 km

(b) 982 cm = 9.82 m (two places to left)
= 9.82 m

(c) 16 mm = 1.6 cm (one place to left)
= 1.6 cm

Here we need to move the decimal point to the right because we are going from a larger unit to a smaller unit.

(d) 5.2 m = 5200. mm (three places to right)
= 5200 mm

Practice Problem 5 Convert. **(a)** 1527 m to km **(b)** 375 cm to m

(c) 128 mm to cm **(d)** 46 m to mm ■

The other metric units of length are the hectometer, the dekameter, and the decimeter. These are not used very frequently, but it is good to understand how their lengths relate to the basic unit, the meter. A complete list of the metric lengths we have discussed appears in the following table.

Metric Lengths with Abbreviations

1 kilometer (km) = 1000 meters
1 hectometer (hm) = 100 meters
1 dekameter (dam) = 10 meters
1 meter (m)
1 decimeter (dm) = 0.1 meter
1 centimeter (cm) = 0.01 meter
1 millimeter (mm) = 0.001 meter

TEACHING TIP If students complain that they are getting all the meter prefixes mixed up, remind them that the kilometer, meter, and centimeter are the three most important and most frequently used metric measures. Units like decimeters and hectometers are rarely used and need not be memorized.

EXAMPLE 6 Convert.

(a) 426 decimeters to kilometers **(b)** 9.47 hectometers to meters

(a) We are converting from a smaller unit, dm, to a larger one, km. Therefore, there will be fewer kilometers than decimeters (426 will "get smaller"). We move the decimal point four places to the left.

$$426 \text{ dm} = 0.0426 \text{ km} \quad \text{(four places to left)}$$
$$= 0.0426 \text{ km}$$

(b) We are converting from a larger unit, hm, to a smaller one, m. Therefore, there will be more meters than hectometers (9.47 will "get larger"). We move the decimal point two places to the right.

$$9.47 \text{ hm} = 947. \text{ m} \quad \text{(two places to right)}$$
$$= 947 \text{ m}$$

Practice Problem 6 Convert.

(a) 389 millimeters to dekameters **(b)** 0.48 hectometer to centimeters ■

When several metric measurements are to be added, we change them to a convenient common unit.

EXAMPLE 7 Add. 125 m + 1.8 km + 793 m

First we change the kilometer measurement to a measurement in meters.

$$1.8 \text{ km} = 1800 \text{ m}$$

Then we add.

$$\begin{array}{r} 125 \text{ m} \\ 1800 \text{ m} \\ +\ 793 \text{ m} \\ \hline 2718 \text{ m} \end{array}$$

Practice Problem 7 Add. 782 cm + 2 m + 537 m ■

SIDELIGHT: Extremely Large Metric Distances Is the biggest length in the metric system a kilometer? Is the smallest length a millimeter? No. The system extends to very large units and very small ones. Usually, only scientists use these units.

1 gigameter = 1,000,000,000 meters
1 megameter = 1,000,000 meters
1 kilometer = 1000 meters
1 meter
1 millimeter = 0.001 meter
1 micrometer = 0.000001 meter
1 nanometer = 0.000000001 meter

TEACHING TIP The gigameter is useful in astronomy. Astronomers often refer to a basic measuring length of 1 astronomical unit (AU). The length of 1 astronomical unit is 150 gigameters.

For example, 26 megameters equals a length of 26,000,000 meters. A length of 31 micrometers equals a length of 0.000031 meter.

SIDELIGHT: Metric Measurements for Computers A **byte** is the amount of computer memory needed to store one alphanumeric character. When referring to computers you may hear the following words: **kilobytes, megabytes,** and **gigabytes.** The following chart may help you:

1 gigabyte (GB) = one billion bytes =	1,000,000,000 bytes
1 megabyte (MB) = one million bytes* =	1,000,000 bytes
1 kilobyte (KB) = one thousand bytes† =	1000 bytes

*Sometimes in computer science 1 megabyte is considered to be 1,048,576 bytes.
†Sometimes in computer science 1 kilobyte is considered to be 1024 bytes.

We will consider these measures of computer storage in the Putting Your Skills to Work on page 364.

A Challenge for You

Complete the following:

1. 18 megameters = 18,000 kilometers
2. 26 millimeters = 26,000 micrometers
3. 17 nanometers = 0.000017 millimeter
4. 38 meters = 0.000038 megameter
5. 1.2 gigabytes = 1,200,000,000 bytes
6. 528 megabytes = 528,000,000 bytes
7. 78.9 kilobytes = 78,900 bytes
8. 24.9 gigabytes = 24,900,000,000 bytes

6.2 Exercises

Verbal and Writing Skills *Write the prefix that means*

1. Hundred hecto-

2. Hundredth centi-

3. Tenth deci-

4. Thousandth milli-

5. Thousand kilo-

6. Ten deka-

The following conversions involve metric units that are very commonly used. You should be able to perform each conversion without any notes and without consulting your text.

7. 37 centimeters = 370 millimeters

8. 44 centimeters = 440 millimeters

9. 3.6 kilometers = 3600 meters

10. 7.6 kilometers = 7600 meters

11. 328 millimeters = 0.328 meter

12. 508 millimeters = 0.508 meter

13. 56.3 centimeters = 0.563 meter

14. 96.4 centimeters = 0.964 meter

15. 2 kilometers = 200,000 centimeters

16. 7 kilometers = 700,000 centimeters

17. 78,000 millimeters = 0.078 kilometer

18. 96,000 millimeters = 0.096 kilometer

19. 0.5386 kilometer = 538,600 millimeters

20. 0.3279 kilometer = 327,900 millimeters

Abbreviations are used in the following. Fill in the blanks with the correct value.

21. 35 mm = 3.5 cm = 0.035 m

22. 83 mm = 8.3 cm = 0.083 m

23. 3582 mm = 3.582 m = 0.003582 km

24. 7812 mm = 7.812 m = 0.007812 km

25. 0.32 cm = 0.0032 m = 0.0000032 km

26. 0.81 cm = 0.0081 m = 0.0000081 km

Applications

27. Peter measured the length of his dining room table. Choose the most reasonable measurement.
(a) 2 m **(b)** 2 km **(c)** 2 cm a

28. Wally measured the length of his college identification card. It was which of the following?
(a) 10 km **(b)** 10 m **(c)** 10 cm c

29. Eddie measured the width of a compact disc case. Choose the most appropriate measurement.
(a) 80.5 km **(b)** 80.5 m **(c)** 80.5 mm c

30. The Botanical Gardens are located in the center of the city. If it takes approximately 10 minutes to walk directly from the east entrance to the west entrance, which would be the most likely measurement of the width of the Gardens?
(a) 1.4 km **(b)** 1.4 m **(c)** 1.4 mm a

The following conversions involve some of the metric units that are not used extensively. You should be able to perform each conversion, but it is not necessary to do it from memory.

31. 3 kilometers = 30 hectometers

32. 7 hectometers = 70 dekameters

33. 27 meters = 270 decimeters

34. 35 decimeters = 350 centimeters

35. 198 millimeters = 1.98 decimeters

36. 2.93 decimeters = 0.00293 hectometer

37. 48.2 meters = 0.482 hectometer

38. 435 hectometers = 43.5 kilometers

39. 0.5236 hectometer = 5236 centimeters

40. 0.7175 dekameter = 7175 millimeters

Change to a convenient unit of measure and add.

41. 243 m + 2.7 km + 312 m
243 m + 2700 m + 312 m = 3255 m

42. 845 m + 5.79 km + 701 m
845 m + 5790 m + 701 m = 7336 m

43. 5.2 cm + 361 cm + 968 mm
5.2 cm + 361 cm + 96.8 cm = 463 cm

44. 4.8 cm + 607 cm + 834 mm
4.8 cm + 607 cm + 83.4 cm = 695.2 cm

45. 82 m + 471 cm + 0.32 km
82 m + 4.71 m + 320 m = 406.71 m

46. 46 m + 986 cm + 0.884 km
46 m + 9.86 m + 884 m = 939.86 m

47. The outside casing of a stereo cabinet is built of plywood 0.95 centimeters thick attached to plastic 1.35 centimeters thick and a piece of mahogany veneer, 2.464 millimeters thick. How thick is the stereo casing? 2.5464 cm

48. A plywood board is 2.2 centimeters thick. A layer of tar paper is 3.42 millimeters thick. A layer of false brick siding is 2.7 centimeters thick. A house wall consists of these three layers. How thick is the wall?
2.2 cm + 0.342 cm + 2.7 cm = 5.242 cm *or* 52.42 mm

Mixed Practice

Write true *or* false *for each statement.*

44. 4.8 cm + 607 cm + 834 mm
4.8 cm + 607 cm + 83.4 cm = 695.2 cm

45. 82 m + 471 cm + 0.32 km
82 m + 4.71 m + 320 m = 406.71 m

46. 46 m + 986 cm + 0.884 km
46 m + 9.86 m + 884 m = 939.86 m

13. 56.3 centimeters = <u>0.563</u> meter

14. 96.4 centimeters = <u>0.964</u> meter

49. 1 meter = 0.01 centimeter False

50. 1 kilometer = 0.001 meter False

51. 10 centimeters = 100 millimeters True

52. 10 millimeters = 1 centimeter True

53. An airport runway might be 2 kilometers long. True

54. A man might be 2 meters tall. True

55. A page of this book is about 2 centimeters thick. False

56. The glass in a drinking glass is usually about 2 millimeters thick. False

57. A kilometer is shorter than a mile. True

58. A meter is longer than a yard. True

Applications

59. The longest runway is at Edwards Air Force Base, California. It is 1,192,000 centimeters in length.
(a) How many kilometers long is the runway?
(b) How many meters long is the runway?
a. 11.92 km
b. 11,920 meters

60. The world's longest train run is on the Trans-Siberian line in Russia, from Moscow to Nakhodka on the Sea of Japan. The length of the run, which makes 97 stops, measures 94,380,000 centimeters. The run takes 8 days, 4 hours, and 25 minutes.
(a) How many meters is the run? 943,800 meters
(b) How many kilometers is the run?
943.8 kilometers

61. The highest railroad line in the world is a track on the Morococha branch of the Peruvian State Railways at La Cima. The track is 4818 meters high.
(a) How many centimeters high is the track?
(b) How many kilometers high is the track?
a. 481,800 centimeters
b. 4.818 kilometers

62. The break-up of the former Soviet Union has delayed the completion of the Rogunskaya dam, which will be 335 meters high.
(a) How many kilometers high is the dam supposed to be? 0.335 kilometer
(b) How many centimeters high is the dam supposed to be? 33,500 centimeters

63. A prism with sides measuring 0.004 millimeters (0.0001 inches), barely visible to the naked eye, was created at the National Institute of Standards and Technology in Boulder, Colorado, in 1989. How many kilometers wide is each side of the prism? .000000004 kilometer

64. Type D263 glass, made by Deutsche Spezialglas AG, of Grunenplan, Germany for use in electronic and medical equipment, has a minimum thickness of 0.025 millimeters, and a maximum thickness of 0.035 millimeters. What would be the measurement, in meters of the minimum and maximum thickness of the glass.
0.000025 meters minimum; 0.000035 meters maximum thickness

6.3 Metric Measurements: Volume and Weight

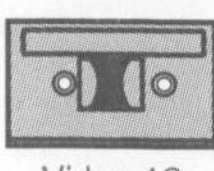

After studying this section, you will be able to:

1 *Convert between metric units of volume.*
2 *Convert between metric units of weight.*

TEACHING TIP It is helpful to emphasize the fact that a liter is just slightly larger than a quart. (1 liter is 1.057 quarts.)

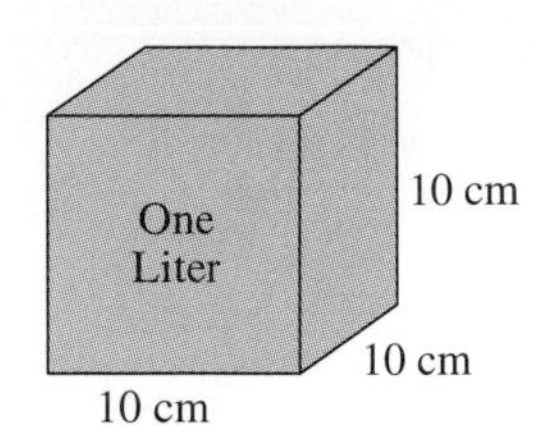

1 *Converting Between Metric Units of Volume*

As products are distributed worldwide, more and more of them are being sold in metric units. Soft drinks come in 1-, 2-, or 3-liter bottles. Labels often print the amount in both American units and metric units. Get used to looking at these labels to gain a sense of the size of metric units of volume.

The basic metric unit for volume is the liter. A liter is defined as the volume of a box 10 cm × 10 cm × 10 cm or 1000 cm^3. A cubic centimeter may be written as cc, so we sometimes see 1000 cc = 1 liter. A liter is slightly larger than a quart; 1 liter of liquid is 1.057 quarts of that liquid.

The most common metric units of volume are the milliliter, the liter, and the kiloliter. Often a capital letter L is used as an abbreviation for *liter.*

> 1 kiloliter (kL) = 1000 liters
> 1 liter (L)
> 1 milliliter (mL) = 0.001 liter

We know that 1000 cc = 1 liter. Dividing each side of that equation by 1000, we get 1 cc = 1 mL.

> When you change one metric prefix to another, move the decimal point in the same direction and the appropriate number of places. Use the prefix chart as a guide.

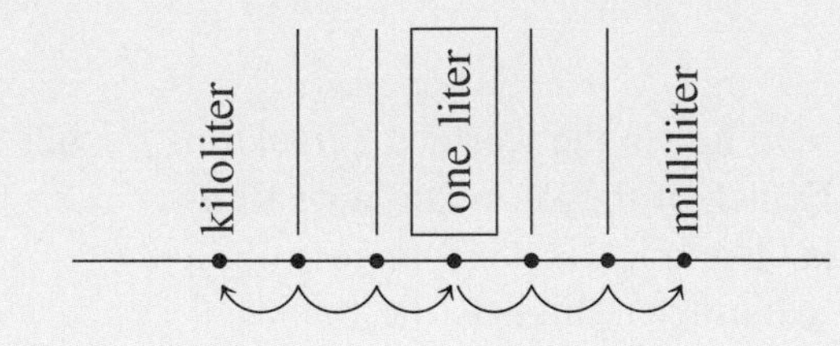

The prefixes for liter follow the pattern we have seen for meter. The *kilo-* is three places to the left of the liter, and the *milli-* is three places to the right.

EXAMPLE 1 Convert.

(a) 3 L = ____ mL **(b)** 24 kL = ____ L **(c)** 0.084 L = ____ mL

The prefix *milli-* is three places to the right. We move the decimal point three places to the right.

(a) 3 L = 3000 mL **(b)** 24 kL = 24,000 L **(c)** 0.084 L = 84 mL

Practice Problem 1 Convert.

(a) 5 L = ____ mL **(b)** 84 kL = ____ L **(c)** 0.732 L = ____ mL ■

EXAMPLE 2 Convert.

(a) 26.4 mL = ____ L **(b)** 5982 mL = ____ L **(c)** 6.7 L = ____ kL

The unit *L* is three places to the left. We move the decimal three places to the left.

(a) 26.4 mL = 0.0264 L **(b)** 5982 mL = 5.982 L **(c)** 6.7 L = 0.0067 kL

TEACHING TIP If your students are confused about the standard abbreviations of some of the units, call their attention to the chart of standard abbreviations at the beginning of Section 6.4. Remind them that it is a good idea to learn the standard abbreviations of these units as they go through the chapter.

Practice Problem 2 Convert.

(a) 15.8 mL = ____ L **(b)** 12,340 mL = ____ L **(c)** 86.3 L = ____ kL ■

The cubic centimeter is often used in medicine. Recall 1 mL = 1 cm^3 or 1 cc.

EXAMPLE 3 Convert.

(a) 26 mL = ____ cm^3 **(b)** 0.82 L = ____ cm^3

A milliliter and a cubic centimeter are equivalent.

(a) 26 mL = 26 cm^3

We use the same rule to convert liters to cubic centimeters as we do to convert liters to milliliters.

(b) 0.82 L = 820 cm^3

TEACHING TIP Students may wonder why there are two names for the same thing. Explain to them that the standard unit is the milliliter (mL) and that this is the one they are expected to learn. The cubic centimeter is used in nursing and a few other technical areas.

Practice Problem 3 Convert.

(a) 396 mL = ____ cm^3 **(b)** 0.096 L = ____ cm^3 ■

EXAMPLE 4 A special cleaning fluid to rinse test tubes in the chemistry lab costs $40.00 per liter. What is the cost per milliliter?

Since the milliliter is only $\frac{1}{1000}$ of a liter, its cost is only $\frac{1}{1000}$ of $40.00.

$$\frac{\$40.00}{1000} = \$0.040$$

Thus each milliliter costs $0.04, which is quite expensive.

Practice Problem 4 A purified acid costs $110 per liter. What does it cost per milliliter? ■

2 *Converting Between Metric Units of Weight*

In the metric system the basic unit of weight is the gram. A *gram* is the weight of the water in a box that is 1 centimeter on each side. To get an idea of how small a gram is, we note that two small paper clips weigh about 1 gram. A gram is only about 0.035 ounce.

One kilogram is 1000 times larger than a gram. A kilogram weighs about 2.2 pounds. The most common measures of weight in the metric system are shown in the following chart.

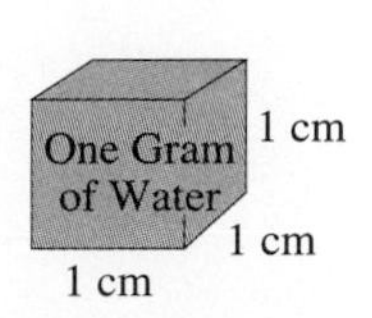

Common Metric Weight Measurements

1 metric ton (t) = 1,000,000 grams

1 kilogram (kg) = 1000 grams

1 gram (g)

1 milligram (mg) = 0.001 gram

When you change one metric prefix to another, move the decimal point in the same direction and the appropriate number of places. Use the prefix chart as a guide.

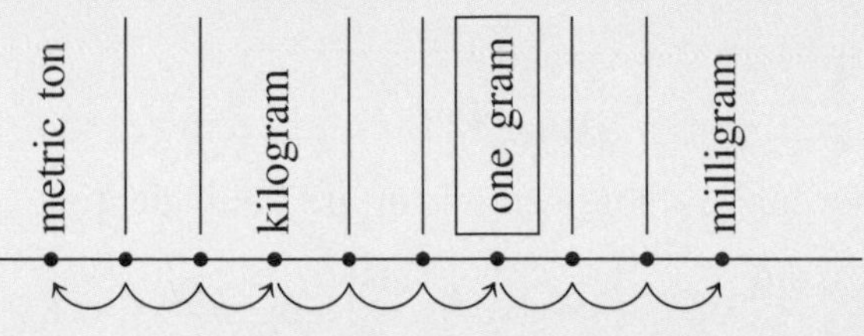

As we convert from a larger unit to the next smaller unit on the chart (for example, from kg to g), we move three places to the right. This moves the decimal point in the measurement three places to the right (4 kg = 4000 g).

EXAMPLE 5 Convert.

(a) 2 t = ____ kg **(b)** 0.42 kg = ____ g **(c)** 19.63 g = ____ mg

(a) 2 t = 2000 kg **(b)** 0.42 kg = 420 g

(c) 19.63 g = 19,630 mg ■

Practice Problem 5 Convert.

(a) 3.2 t = ____ kg **(b)** 7.08 kg = ____ g **(c)** 0.526 g = ____ mg ■

As we convert from a smaller unit to the next larger unit on the chart (for example, from g to kg), we move three places to the left. This moves the decimal point in the measurement three places to the left (6000 g = 6 kg).

EXAMPLE 6 Convert.

(a) 283 kg = ____ t **(b)** 14.628 g = ____ kg **(c)** 7.98 mg = ____ g

(a) 283 kg = 0.283 t **(b)** 14.628 g = 0.014628 kg

(c) 7.98 mg = 0.00798 mg

Practice Problem 6 Convert.

(a) 59 kg = ____ t **(b)** 6.152 g = ____ kg **(c)** 28.3 mg = ____ g ■

EXAMPLE 7 If a chemical costs \$0.03 per gram, what will it cost per kilogram?

Since there are 1000 grams in a kilogram, a chemical that costs \$0.03 per gram would cost 1000 times as much per kilogram.

$$1000 \times \$0.03 = \$30.00$$

The chemical would cost \$30.00 per kilogram.

Practice Problem 7 If coffee costs \$10.00 per kilogram, what will it cost per gram? ■

EXAMPLE 8 Select the most reasonable weight for Tammy's Toyota.

(a) 820 t **(b)** 820 g **(c)** 820 kg **(d)** 820 mg

The most reasonable answer is **(c)** 820 kg. The other weight values are much too large or much too small. Since a kilogram is slightly more than 2 pounds, we see that this weight, 820 kg, more closely tells the weight of a car.

Practice Problem 8 Select the most reasonable weight for Hank, starting linebacker for the college football team.

(a) 120 kg **(b)** 120 g **(c)** 120 mg ■

SIDELIGHT When dealing with very small particles or atomic elements, scientists sometimes use units smaller than a gram.

Small Weight Measurements

1 milligram = 0.001 gram

1 microgram = 0.000001 gram

1 nanogram = 0.000000001 gram

1 picogram = 0.000000000001 gram

We could make the following conversions.

2.6 picograms = 0.0026 nanogram

29.7 micrograms = 0.0297 milligram

58 nanograms = 58,000 picograms

58 nanograms = 0.058 microgram

See Exercises 6.3, problems 75 and 76.

6.3 Exercises

Verbal and Writing Skills

Write the metric unit that represents each of the following.

1. One thousand liters 1 kL

2. One thousandth of a liter 1 mL

3. One thousandth of a gram 1 mg

4. One thousand grams 1 kg

5. 1,000,000 grams 1 t

6. 1000 kilograms 1 t

Perform each conversion.

7. 64 kL = 64,000 L

8. 37 kL = 37,000 L

9. 4.7 L = 4700 mL

10. 8.6 L = 8600 mL

11. 18.9 mL = 0.0189 L

12. 31.5 mL = 0.0315 L

13. 752 L = 0.752 kL

14. 493 L = 0.493 kL

15. 2.43 kL = 2,430,000 mL

16. 1.76 kL = 1,760,000 mL

17. 82 mL = 82 cm^3

18. 152 mL = 152 cm^3

19. 5261 mL = 0.005261 kL

20. 28,156 mL = 0.028156 kL

21. 74 L = 74,000 cm^3

22. 122 L = 122,000 cm^3

23. 162 g = 0.162 kg

24. 294 g = 0.294 kg

25. 35 mg = 0.035 g

26. 13 mg = 0.013 g

27. 6328 mg = 6.328 g

28. 986 mg = 0.986 g

29. 2.92 kg = 2920 g

30. 14.6 kg = 14,600 g

31. 17 t = 17,000 kg

32. 36 t = 36,000 kg

33. 0.32 g = 0.00032 kg

34. 0.78 g = 0.00078 kg

35. 7896 g = 0.007896 t

36. 12,315 g = 0.012315 t

37. 5.9 kg = 5,900,000 mg

38. 0.83 kg = 830,000 mg

Fill in the blanks with the correct values.

39. 7 mL = 0.007 L = 0.000007 kL

40. 18 mL = 0.018 L = 0.000018 kL

41. 128 cm^3 = 0.128 L = 0.000128 kL

42. 199.8 cm^3 = 0.1998 L = 0.0001998 kL

43. 522 mg = 0.522 g = 0.000522 kg

44. 49 mg = 0.049 g = 0.000049 kg

45. 3607 g = 3.607 kg = 0.003607 t

46. 7183 g = 7.183 kg = 0.007183 t

47. Alice bought a jar of apple juice at the store. Choose the most reasonable measurement for its contents.
(a) 0.32 kL
(b) 0.32 L b
(c) 0.32 mL

48. A nurse gave an injection of insulin to a diabetic patient. Choose the most reasonable measurement for the dose.
(a) 4 kL
(b) 4 L
(c) 4 mL c

49. A shipowner purchased a new oil tanker. Choose the most reasonable measurement for its weight.
(a) 4000 t a
(b) 4000 kg
(c) 4000 g

50. Robert bought a new psychology textbook. Choose the most reasonable measurement for its weight.
(a) 0.49 t
(b) 0.49 kg b
(c) 0.49 g

Find the convenient unit of measure and add.

51. 83 L + 822 mL + 30.1 L
83 L + 0.822 L + 30.1 L = 113.922 L

52. 152 L + 473 mL + 77.3 L
152 L + 0.473 L + 77.3 L = 229.773 L

53. 5 t + 3.82 t + 983 kg
5 t + 3.82 t + 0.983 t = 9.803 t

54. 7.8 t + 669 kg + 5.23 t
7.8 t + 0.669 t + 5.23 t = 13.699 t

55. 24 mg + 136 mg + 0.26 kg
24 mg + 136 mg + 260,000 mg = 260,160 mg

56. 78 mg + 221 mg + 0.14 kg
78 mg + 221 mg + 140,000 mg = 140,299 mg

Mixed Practice *Write* true *or* false *for each statement.*

57. 1 milliliter = 0.001 liter
True

58. 1 kiloliter = 1000 liters
True

59. Milk can be purchased in kiloliter jugs at the food store.
False

60. Small amounts of medicine are often measured in liters.
False

61. 1 kilogram = 1000 grams
True

62. 1 gram = 1000 milligrams
True

63. 1 metric ton = 100 grams
False

64. 0.1 gram = 10 milligrams
False

65. A nickel coin weighs about 5 grams.
True

66. A convenient size for a family purchase is 2 kilograms of ground beef.
True

Applications *Convert the units.*

67. A recently developed medication to prevent rejection in heart transplant patients COELS, $358,000 per liter. How much would one milliliter cost?
$358

68. Australia is using a form of natural gas fuel for newly developed cars. If a rancher uses 3.50 kiloliters per year for natural gas and the price is $0.39 per liter, how much will the rancher spend per year? $1365

69. A physician in Mexico is striving to perfect a new anticancer drug. At the moment, he is using a chemical solution that costs $95.50 per gram. How much would it cost to buy 6 kilograms of the solution? $573,000

70. A pharmaceutical firm developed a new vaccine that costs $6.00 per milliliter to produce. How much will it cost the firm to produce 1 liter of the vaccine? $6000

71. A very rare essence of an almost extinct flower found in the Amazon Jungle of South America is extracted by a biogenetic company trying to copy and synthesize it. The company estimates that if the procedure is successful, the product will cost the company $850 per milliliter to produce. How much will it cost the company to produce 0.4 liter of the engineered essence? $340,000

72. In a recent year the price of gold was $9.40 per gram. At that price how much would a kilogram of gold cost? $9400

73. A government rocket is carrying cargo into space for a private company. The charge for the freight shipment is $22,450 per kilogram of the special cargo, and the cargo weighs 0.45 metric ton. How much will the private company have to pay. $10,102,500

74. The individual coal-cutting record is 45.4 metric tons per person in one shift, (6 hours), by five Soviet miners. If the coal is sold for $2.00 per kilogram, what is the value of the coal? $90,800

To Think About *Complete the following.*

★ **75.** 5632 picograms = <u>0.005632</u> micrograms

★ **76.** 0.076182 milligram = <u>76,182</u> nanograms

Cumulative Review Questions

77. 14 out of 70 is what percent?
20%

78. What is 23% of 250?
57.5

79. What is 1.7% of $18,900?
$321.30

80. A salesperson earns a commission of 8%. She sold furniture worth $8960. How much commission will she earn? $716.80

6.4 Conversion of Units (Optional)

MathPro

Video 13

SSM

After studying this section, you will be able to:

1 *Convert units of length, volume, or weight between the metric and American systems.*

2 *Convert temperature readings between Fahrenheit and Celsius degrees.*

1 *Converting Units of Length, Volume, or Weight Between the Metric and American Systems*

So far we've seen how to convert units when working *within* either the American or the metric system. Many people, however, work in *both* the metric and the American systems. If you study such fields as chemistry, electromechanical technology, business, X-ray technology, nursing, or computers, you will probably need to convert measurements between the two systems. We learn that skill in this section.

To convert between American units and metric units, it is helpful to have equivalent values. The most commonly used equivalents are listed below. Most of these equivalents are approximate.

TEACHING TIP College math faculty do not agree on which, if any, of the topics of this section should be required in this course. In many schools the computer science, business, and physical science departments assume that the math department is teaching this material. The majority of math faculty seem to feel it is better to introduce the topic but not require the memorization of all the equivalent measures.

EQUIVALENT MEASURES

	American to Metric	Metric to American
Units of length	1 mile = 1.61 kilometers 1 yard = 0.914 meter 1 foot = 0.305 meter 1 inch = 2.54 centimeters	1 kilometer = 0.62 mile 1 meter = 3.28 feet 1 meter = 1.09 yards 1 centimeter = 0.394 inch
Units of volume	1 gallon = 3.79 liter 1 quart = 0.946 liter	1 liter = 0.264 gallon 1 liter = 1.06 quarts
Units of weight	1 pound = 0.454 kilogram 1 ounce = 28.35 gram	1 kilogram = 2.2 pounds 1 gram = 0.0353 ounce

Remember that to convert from one unit to the other you multiply by a fraction that is equivalent to 1. Create a fraction from the equivalent measures table so that the unit in the denominator cancels the unit you are changing.

To change 5 miles to kilometers, we look in the table and find that 1 mile = 1.61 kilometers. We will use the fraction

$$\frac{1.61 \text{ kilometers}}{1 \text{ mile}}$$

We want to have 1 mile in the denominator.

$$5 \cancel{\text{miles}} \times \frac{1.61 \text{ kilometers}}{1 \cancel{\text{mile}}} = 5 \times 1.61 \text{ kilometers} = 8.05 \text{ kilometers}$$

Thus 5 miles = 8.05 kilometers.

EXAMPLE 1

(a) Convert 3 feet to meters. **(b)** Convert 6 inches to centimeters.

(a) $3 \text{ feet} \times \dfrac{0.305 \text{ meter}}{1 \text{ foot}} = 0.915 \text{ meter}$

(b) $6 \text{ inches} \times \dfrac{2.54 \text{ centimeters}}{1 \text{ inch}} = 15.24 \text{ centimeters}$

Practice Problem 1

(a) Convert 7 feet to meters. **(b)** Convert 4 inches to centimeters. ■

Unit abbreviations are quite common, so we will use them for the remainder of this section. We list them here for your reference.

American Measure (Alphabetical Order)	Standard Abbreviation
feet	ft
gallon	gal
inch	in.
mile	mi
ounce	oz
pound	lb
quart	qt
yard	yd

Metric Measure	Standard Abbreviation
centimeter	cm
gram	g
kilogram	kg
kilometer	km
liter	L
meter	m
millimeter	mm

EXAMPLE 2

(a) Convert 26 m to yd. **(b)** Convert 1.9 km to mi. **(c)** Convert 14 gal to L.
(d) Convert 2.5 L to qt. **(e)** Convert 5.6 lb to kg. **(f)** Convert 152 g to oz.

(a) $26 \text{ m} \times \dfrac{1.09 \text{ yd}}{1 \text{ m}} = 28.34 \text{ yd}$ **(b)** $1.9 \text{ km} \times \dfrac{0.62 \text{ mi}}{1 \text{ km}} = 1.178 \text{ mi}$

(c) $14 \text{ gal} \times \dfrac{3.79 \text{ L}}{1 \text{ gal}} = 53.06 \text{ L}$ **(d)** $2.5 \text{ L} \times \dfrac{1.06 \text{ qt}}{1 \text{ L}} = 2.65 \text{ qt}$

(e) $5.6 \text{ lb} \times \dfrac{0.454 \text{ kg}}{1 \text{ lb}} = 2.5424 \text{ kg}$ **(f)** $152 \text{ g} \times \dfrac{0.0353 \text{ oz}}{1 \text{ g}} = 5.3656 \text{ oz}$

Practice Problem 2

(a) Convert 17 m to yd. **(b)** Convert 29.6 km to mi.
(c) Convert 26 gal to L. **(d)** Convert 6.2 L to qt.
(e) Convert 16 lb to kg. **(f)** Convert 280 g to oz. ■

Some conversions require more than one step.

EXAMPLE 3 Convert 235 cm to ft. Round your answer to the nearest hundredth of a foot.

Our first fraction converts centimeters to inches. Our second fraction converts inches to feet.

$$235\,\cancel{\text{cm}} \times \frac{0.394\,\cancel{\text{in.}}}{1\,\cancel{\text{cm}}} \times \frac{1\text{ ft}}{12\,\cancel{\text{in.}}} = \frac{92.59}{12}\text{ ft}$$

$$= 7.72\text{ ft (rounded to the nearest hundredth)}$$

Practice Problem 3 Convert 180 cm to ft. ■

The same rules can be followed for a rate such as 50 miles per hour.

EXAMPLE 4 Convert 100 km/hr to mi/hr.

TEACHING TIP Remind students that changing rates from feet per second to miles per hour or miles per hour to kilometers per hour does not require a new skill. It is merely using the ''multiply by a unit fraction'' method that we have been using throughout the chapter.

$$\frac{100\,\cancel{\text{km}}}{\text{hr}} \times \frac{0.62\text{ mi}}{1\,\cancel{\text{km}}} = 62\text{ mi/hr}$$

Thus 100 km/hr is approximately equal to 62 mi/hr.

Practice Problem 4 Convert 88 km/hr to mi/hr. (Round to the nearest hundredth.) ■

EXAMPLE 5 A camera film that is 35 mm wide is how many inches wide?

We first convert from millimeters to centimeters by moving the decimal point in the number 35 one place to the left.

$$35\text{ mm} = 3.5\text{ cm}$$

Then we convert to inches using a unit fraction.

$$3.5\,\cancel{\text{cm}} \times \frac{0.394\text{ in.}}{1\,\cancel{\text{cm}}} = 1.379\text{ in.}$$

TEACHING TIP Students often have trouble with square units. They wonder why 1 square yard has 9 square feet instead of 3 square feet. Have them study the figure below to see the comparison. The problem is compounded with metric units. The key to understanding this topic is to always draw a sketch to illustrate the comparison. Encourage students to do this when they get to Exercises 6.4, problems 73 and 74. For classroom activity, ask students how many square centimeters are in one square meter. (Answer is 10,000 square centimeters.) Usually, every student who draws a quick sketch as he or she thinks, gets the right answer to this classroom problem.

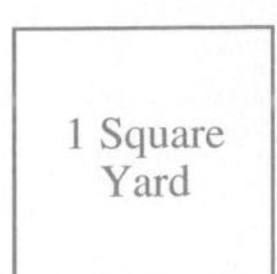

Practice Problem 5 The city police use a 9-mm automatic pistol. If the pistol fires a bullet 9 mm wide, how many inches wide is it? (Round to the nearest hundredth.) ■

SIDELIGHT Suppose we consider a rectangle that measures 2 yards wide by 4 yards long. The area would be 2 yards × 4 yards = 8 square yards. How could you change 8 square yards to square meters? Suppose that we look at 1 square yard. Each side is 1 yard long, which is equivalent to 0.9144 meter.

$$\text{Area} = 1\text{ yard} \times 1\text{ yard} = 0.9144\text{ meter} \times 0.9144\text{ meter}$$

$$\text{Area} = 1\text{ square yard} = 0.8361\text{ square meter}$$

Thus $1\text{ yd}^2 = 0.8361\text{ m}^2$. Therefore

$$8\text{ yd}^2 \times \frac{0.8361\text{ m}^2}{1\text{ yd}^2} = 6.6888\text{ m}^2$$

8 square yards = 6.6888 square meters.

2 *Converting Temperature Readings Between Fahrenheit and Celsius Degrees*

In the metric system temperature is measured on the Celsius scale. Water boils at 100° (100°C) and freezes at 0° (0°C) on the Celsius scale. In the Fahrenheit system water boils at 212° (212°F) and freezes at 32° (32°F).

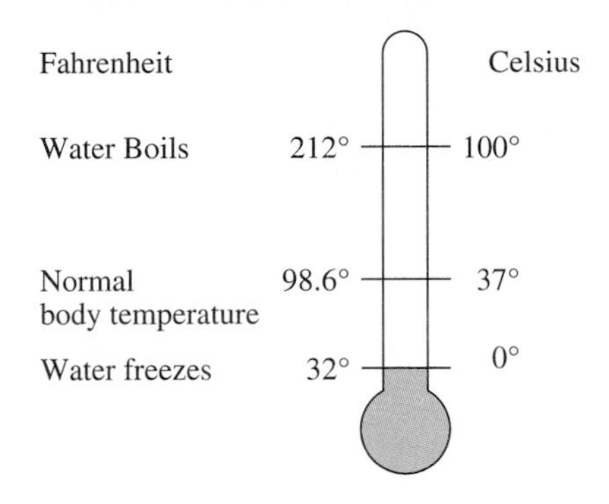

To convert Celsius to Fahrenheit, we can use the formula

$$F = 1.8 \times C + 32$$

EXAMPLE 6 When the temperature is 35°C, what is the Fahrenheit reading?

$$\begin{aligned} F &= 1.8 \times C + 32 \\ &= 1.8 \times 35 + 32 \\ &= 63 + 32 \\ &= 95 \end{aligned}$$

The temperature is 95°F.

Practice Problem 6 Convert 20°C to Fahrenheit temperature. ■

To convert Fahrenheit temperature to Celsius, we can use the formula

$$C = \frac{5 \times F - 160}{9}$$

where F is the number of Fahrenheit degrees and C is the number of Celsius degrees.

EXAMPLE 7 When the temperature is 176°F, what is the Celsius reading?

$$\begin{aligned} C &= \frac{5 \times F - 160}{9} \\ &= \frac{5 \times 176 - 160}{9} \\ &= \frac{880 - 160}{9} = \frac{720}{9} = 80 \end{aligned}$$

The temperature is 80°C.

Practice Problem 7 Convert 181°F to Celsius temperature. ■

CALCULATOR

Converting Temperature

You can use your calculator to convert temperature readings between Fahrenheit and Celsius. To convert 30°C to Fahrenheit temperature enter:

1.8 [×] 30 [+] 32 [=]

Display:

[86]

The temperature is 86°F.

To convert 82.4°F to Celsius temperature enter:

5 [×] 82.4 [−] 160 [=] [÷] 9 [=]

Display:

[28]

The temperature is 28°C.

6.4 Exercises

Perform each conversion. Round the answer to the nearest hundredth when necessary.

1. 7 ft to m
$7\text{ ft} \times \frac{0.305\text{ m}}{1\text{ ft}} = 2.14\text{ m}$

2. 11 ft to m
$11\text{ ft} \times \frac{0.305\text{ m}}{1\text{ ft}} = 3.36\text{ m}$

3. 9 in. to cm
$9\text{ in.} \times \frac{2.54\text{ cm}}{1\text{ in.}} = 22.86\text{ cm}$

4. 13 in. to cm
$13\text{ in.} \times \frac{2.54\text{ cm}}{1\text{ in.}} = 33.02\text{ cm}$

5. 14 m to yd
$14\text{ m} \times \frac{1.09\text{ yd}}{1\text{ m}} = 15.26\text{ yd}$

6. 18 m to yd
$18\text{ m} \times \frac{1.09\text{ yd}}{1\text{ m}} = 19.62\text{ yd}$

7. 26.5 m to yd
$26.5\text{ m} \times \frac{1.09\text{ yd}}{1\text{ m}} = 28.89\text{ yd}$

8. 29.3 m to yd
$29.3\text{ m} \times \frac{1.09\text{ yd}}{1\text{ m}} = 31.94\text{ yd}$

9. 15 km to mi
$15\text{ km} \times \frac{0.62\text{ mi}}{1\text{ km}} = 9.3\text{ mi}$

10. 12 km to mi
$12\text{ km} \times \frac{0.62\text{ mi}}{1\text{ km}} = 7.44\text{ mi}$

11. 24 yd to m
$24\text{ yd} \times \frac{0.914\text{ m}}{1\text{ yd}} = 21.94\text{ m}$

12. 31 yd to m
$31\text{ yd} \times \frac{0.914\text{ m}}{1\text{ yd}} = 28.33\text{ m}$

13. 82 mi to km
$82\text{ mi} \times \frac{1.61\text{ km}}{1\text{ mi}} = 132.02\text{ km}$

14. 68 mi to km
$68\text{ mi} \times \frac{1.61\text{ km}}{1\text{ mi}} = 109.48\text{ km}$

15. 25 m to ft
$25\text{ m} \times \frac{3.28\text{ ft}}{1\text{ m}} = 82\text{ ft}$

16. 35 m to ft
$35\text{ m} \times \frac{3.28\text{ ft}}{1\text{ m}} = 114.8\text{ ft}$

17. 17.5 cm to in.
$17.5\text{ cm} \times \frac{0.394\text{ in.}}{1\text{ cm}} = 6.90\text{ in.}$

18. 19.6 cm to in.
$19.6\text{ cm} \times \frac{0.394\text{ in.}}{1\text{ cm}} = 7.72\text{ in.}$

19. 200 m to yd
$200\text{ m} \times \frac{1.09\text{ yd}}{1\text{ m}} = 218\text{ yd}$

20. 350 m to yd
$350\text{ m} \times \frac{1.09\text{ yd}}{1\text{ m}} = 381.5\text{ yd}$

21. 5 gal to L
$5\text{ gal} \times \frac{3.79\text{ L}}{1\text{ gal}} = 18.95\text{ L}$

22. 5 gal to L
$5\text{ gal} \times \frac{3.79\text{ L}}{1\text{ gal}} = 18.95\text{ L}$

23. 280 gal to L
$280\text{ gal} \times \frac{3.79\text{ L}}{1\text{ gal}} = 1061.2\text{ L}$

24. 140 gal to L
$140\text{ gal} \times \frac{3.79\text{ L}}{1\text{ gal}} = 530.6\text{ L}$

25. 23 qt to L
$23\text{ qt} \times \frac{0.946\text{ L}}{1\text{ qt}} = 21.76\text{ L}$

26. 28 qt to L
$28\text{ qt} \times \frac{0.946\text{ L}}{1\text{ qt}} = 26.49\text{ L}$

27. 19 L to gal
$19\text{ L} \times \frac{0.264\text{ gal}}{1\text{ L}} = 5.02\text{ gal}$

28. 15 L to gal
$15\text{ L} \times \frac{0.264\text{ gal}}{1\text{ L}} = 3.96\text{ gal}$

29. 4.5 L to qt
$4.5\text{ L} \times \frac{1.06\text{ qt}}{1\text{ L}} = 4.77\text{ qt}$

30. 6.5 L to qt
$6.5\text{ L} \times \frac{1.06\text{ qt}}{1\text{ L}} = 6.89\text{ qt}$

31. 32 lb to kg
$32\text{ lb} \times \frac{0.454\text{ kg}}{1\text{ lb}} = 14.53\text{ kg}$

32. 27 lb to kg
$27\text{ lb} \times \frac{0.454\text{ kg}}{1\text{ lb}} = 12.26\text{ kg}$

33. 7 oz to g
$7\text{ oz} \times \frac{28.35\text{ g}}{1\text{ oz}} = 198.45\text{ g}$

34. 9 oz to g
$9\text{ oz} \times \frac{28.35\text{ g}}{1\text{ oz}} = 255.15\text{ g}$

35. 16 kg to lb
$16\text{ kg} \times \frac{2.2\text{ lb}}{1\text{ kg}} = 35.2\text{ lb}$

36. 14 kg to lb
$14\text{ kg} \times \frac{2.2\text{ lb}}{1\text{ kg}} = 30.8\text{ lb}$

37. 126 g to oz
$126\text{ g} \times \frac{0.0353\text{ oz}}{1\text{ g}} = 4.45\text{ oz}$

38. 186 g to oz
$186\text{ g} \times \frac{0.0353\text{ oz}}{1\text{ g}} = 6.57\text{ oz}$

39. 166 cm to ft
$166 \times \frac{0.394}{1} \times \frac{1}{12} = 5.45\text{ ft}$

40. 142 cm to ft
$142 \times 0.394 \times \frac{1}{12} = 4.66\text{ ft}$

41. 16.5 ft to cm
$16.5 \times 12 \times 2.54 = 502.92\text{ cm}$

42. 19.5 ft to cm
$19.5 \times 12 \times 2.54 = 594.36\text{ cm}$

43. 50 km/hr to mi/hr
$50 \times 0.62 = 31\text{ mi/hr}$

44. 60 km/hr to mi/hr
$60 \times 0.62 = 37.2\text{ mi/hr}$

45. 60 mi/hr to km/hr
$60 \times 1.61 = 96.6\text{ km/hr}$

46. 40 mi/hr to km/hr
$40 \times 1.61 = 64.4\text{ km/hr}$

47. A wire that is 13 mm wide is how many inches wide?
$1.3\text{ cm} \times \frac{0.394\text{ in.}}{1\text{ cm}} = 0.51\text{ in.}$

48. A bolt that is 7-mm wide is how many inches wide?
$0.7\text{ cm} \times \frac{0.394\text{ in.}}{1\text{ cm}} = 0.28\text{ in.}$

49. 40°C to Fahrenheit
$1.8 \times 40 + 32 = 104°\text{F}$

50. 60°C to Fahrenheit
$1.8 \times 60 + 32 = 140°\text{F}$

51. 85°C to Fahrenheit
$1.8 \times 85 + 32 = 185°\text{F}$

52. 105°C to Fahrenheit
$1.8 \times 105 + 32 = 221°\text{F}$

53. 12°C to Fahrenheit
$1.8 \times 12 + 32 = 53.6°\text{F}$

54. 21°C to Fahrenheit
$1.8 \times 21 + 32 = 69.8°\text{F}$

55. 68°F to Celsius
$\frac{5 \times 68 - 160}{9} = 20°\text{C}$

56. 131°F to Celsius
$\frac{5 \times 131 - 160}{9} = 55°\text{C}$

57. 168°F to Celsius
$\frac{5 \times 168 - 160}{9} = 75.56°\text{C}$

58. 112°F to Celsius
$\frac{5 \times 112 - 160}{9} = 44.44°\text{C}$

59. 86°F to Celsius
$\frac{5 \times 86 - 160}{9} = 30°\text{C}$

60. 98°F to Celsius
$\frac{5 \times 98 - 160}{9} = 36.67°\text{C}$

Applications

Solve. Round answers to the nearest hundredth when necessary.

61. Mr. and Mrs. Weston have traveled 67 miles on a boat cruise from Seattle, Washington, to Victoria Island, Vancouver, B.C., Canada. They have 36 kilometers until their rendezvous point with another boat. How many kilometers in total will they have traveled? 143.87 km

62. Marcia is traveling from Ixtapa to Zihuatenejo in Mexico. She drives from the center of town to the cliff that makes the descent down to the beaches. The odometer on her American car shows that the first part of her short trip has taken 4 miles. The sign carved into a wooden post says Zihuatenejo 14 KILOMETERS. How many kilometers in total will she have traveled when she arrives at the beach? 20.44 km

63. Pierre had a Jeep imported into France. During a trip from Paris to Lyon, he used 38 liters of gas. The tank, which he filled before starting the trip, holds 15 gallons of gas. How many liters of gas were left in the tank when he arrived, if he had a full tank to start with? 18.85 liters

64. A surgeon is irrigating an abdominal cavity after a cancerous growth is removed. There is a supply of 3 gallons of distilled water in the operating room. The surgeon uses a total of 7 liters of the water during the procedure. How many liters of water are left over after the operation? 4.37 liters

65. One of the heaviest males documented in medical records weighed 635 kg in 1978. What would have been his weight in pounds? 1397 pounds

66. The average weight for a 7-year-old girl is 22.2 kilograms. What is the average weight in pounds?

$22.2 \text{ kg} \times \frac{2.2 \text{ lb}}{1 \text{ kg}} = 48.84 \text{ lb}$

67. In the Australian summer, Ayers Rock in Central Australia is visited by tourists who climb beginning at 4 o'clock in the morning, when the temperature is 19° Celsius, because after 7 o'clock in the morning, the temperature can reach 45°C and people could die of dehydration while climbing. What would be equivalent Fahrenheit temperatures? It is 66.2°F at 4:00 A.M. The temperature may reach 113°F after 7:00 A.M.

68. The holiday turkey in Buenos Aires, Argentina, was roasted at 200° Celsius for 4 hours, (20 minutes per pound). What would have been the temperature in Fahrenheit in Joplin, Missouri? 392°F

69. Metallic tungsten, or wolfram (W), melts at 6188° Fahrenheit. At what Celsius temperature would metallic tungsten melt? 3420°C

70. On top of Old Smoky, water boils at 205°F. What Celsius temperature is this?

$\frac{5 \times 205 - 160}{9} = 96.11°\text{C}$

Round your answer to the nearest thousandth.

71. A pathologist found 0.768 ounce of lead in the liver of a child who died of lead poisoning. How many grams of toxic lead were in the child's liver? 21.773 g

72. A prospector in the Yukon, in Alaska, found a gold nugget that weighed 2.552 ounces while panning in a river. How many grams were in the nugget? 72.349 g

Round your answer to four decimal places.

73. 28 square inches = ? square centimeters

$28 \times 2.54 \times 2.54 = 180.6448$ sq cm

74. 36 square meters = ? square yards

$36 \times 1.09 \times 1.09 = 42.7716$ sq yd

75. Change 48 feet per second to miles per hour.

$48 \times \frac{1}{5280} \times 60 \times 60 = 32.7273$ mi/hr

76. Change 55 miles per hour to feet per second.

$55 \times 5280 \times \frac{1}{60} \times \frac{1}{60} = 80.6667$ ft/sec

Cumulative Review Problems

Do the operations in the correct order.

77. $2^3 \times 6 - 4 + 3$ 47

78. $5 + 2 - 3 + 5 \times 3^2$ 49

79. $2^2 + 3^2 + 4^3 + 2 \times 7$ 91

80. $5^2 + 4^2 + 3^2 + 3 \times 8$ 74

Putting Your Skills to Work

THE MATHEMATICS OF COMPUTER CAPACITY

As we mentioned on page 349 a **byte** is the amount of computer memory needed to store one alphanumeric character. One kilobyte (KB or K) is usually defined as one thousand bytes. One megabyte (MB) is usually defined as one million bytes while one gigabyte (GB) is one billion bytes.

PROBLEMS FOR INDIVIDUAL INVESTIGATION

1. Sharon purchased a new computer with a 1.2-GB hard drive. She installed a version of Windows that uses 16.5 MB of hard drive space. She has stored several personal programs on her computer that use 867 K of hard drive space. What is the maximum amount of space still available on her hard drive? 1,182,633,000 bytes of space

Speed on a computer is often measured in hertz. One **hertz** is a unit of frequency of one cycle per second. One megahertz (MHz) is one million cycles per second. One gigahertz (GHz) is one billion cycles per second. Although different types of tasks on a computer are measured in different ways, it is generally true that the "speed" at which the computer can do a task is measured by megahertz. Theoretically, a 100-MHz computer is twice as fast as a 50-MHz computer.

PROBLEMS FOR GROUP INVESTIGATION AND COOPERATIVE LEARNING

Together with other members of your class, determine the following. Round to the nearest thousandth.

2. On a 40-MHz computer, it took 3.2 seconds for Mary Lou to have her 20-page paper scanned for spelling errors. Approximately how long will it take her using the same software on a 150-MHz computer to do the same task? 0.853 second

3. Engineers and scientists are now researching the possibility of a 1-GHz personal computer in the near future. Theoretically, how long would it take such a computer to scan Mary Lou's paper for spelling errors? 0.128 second

INTERNET CONNECTIONS: Go to ``http://www.prenhall.com/tobey'' to be connected

Site: Windows Station Configurator (Vektron International) or a related site

This site shows some of the options currently available from a computer manufacturer. Answer the following questions.

4. If you choose the largest available hard drive, how much more disk space will you have than you would with the smallest hard drive? Give your answer as a percent.

5. The processor determines the speed of the computer. Suppose it would take 24 seconds for the fastest available computer to complete an operation. How long would the same operation take on the slowest available computer?

6.5 Applied Problems

After studying this section, you will be able to:

1 *Solve applied problems involving metric and American units.*

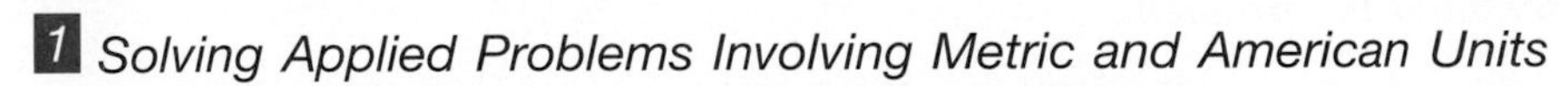

1 *Solving Applied Problems Involving Metric and American Units*

Once again, we will be using the blueprint for solving applied problems that we used in Sections 1.8, 2.9, and 3.7. The following ideas will be helpful.

Read the problem carefully. Draw a picture if this helps you. Fill in the Mathematics Blueprint for Problem Solving. Decide what conversion of units (if any) is needed to help you solve the problem. Perform the calculations. State the answer and the unit of measure. Estimate the answer. Compare the exact answer with the estimate. See if your answer is reasonable.

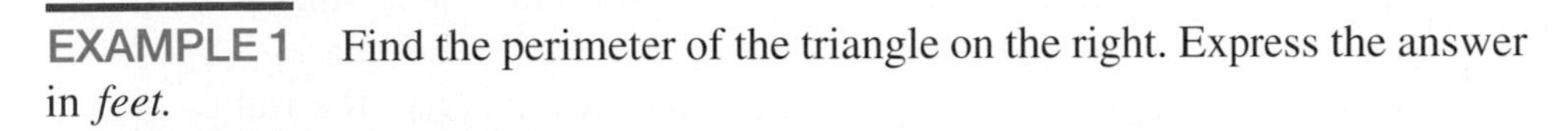

EXAMPLE 1 Find the perimeter of the triangle on the right. Express the answer in *feet.*

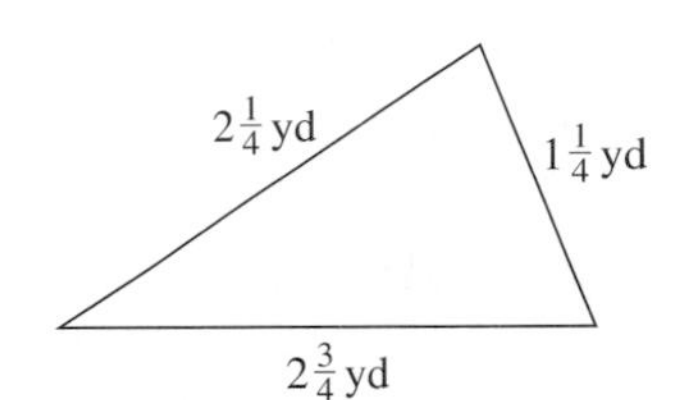

1. *Understand the problem.*
   The perimeter is the sum of the lengths of the sides. We are asked to express the answer in feet. Remember to convert the yards to feet.
2. *Solve and state the answer.*
   Add the three sides.

$$\begin{array}{r} 2\frac{1}{4} \text{ yards} \\ 1\frac{1}{4} \text{ yards} \\ +\ 2\frac{3}{4} \text{ yards} \\ \hline 5\frac{5}{4} \text{ yards} \end{array} = 6\frac{1}{4} \text{ yards}$$

Convert $6\frac{1}{4}$ yards to feet using the fact that 1 yard = 3 feet. To make the calculation easier, we will change $6\frac{1}{4}$ to $\frac{25}{4}$.

$$6\frac{1}{4}\ \cancel{\text{yards}} \times \frac{3 \text{ feet}}{1\ \cancel{\text{yard}}} = \frac{25}{4}\ \cancel{\text{yards}} \times \frac{3 \text{ feet}}{1\ \cancel{\text{yard}}} = \frac{75}{4} \text{ feet} = 18\frac{3}{4} \text{ feet}$$

The perimeter of the triangle is $18\frac{3}{4}$ feet.

3. *Check.*
   We will check by estimating the answer.

$$2\frac{1}{4} \text{ yd} \approx 2 \text{ yd} \qquad 1\frac{1}{4} \text{ yd} \approx 1 \text{ yd} \qquad 2\frac{3}{4} \text{ yd} \approx 3 \text{ yd}$$

Now we add the three sides, using our estimated values.

$$2 + 1 + 3 = 6 \text{ yards} \qquad 6 \text{ yards} = 18 \text{ feet}$$

Our estimated answer, 18 feet, is close to our calculated answer, $18\frac{3}{4}$ feet. Thus our answer seems reasonable. ✓

TEACHING TIP This is a good time to reinforce the concept of estimating your answer. Ask the students to ESTIMATE THE ANSWER FOR PRACTICE PROBLEM 1.

Practice Problem 1 Find the perimeter of the rectangle on the right. Express the answer in *feet.* ■

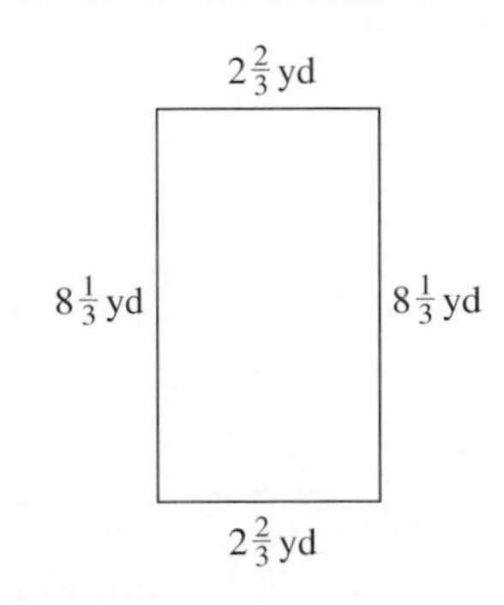

We use the Mathematics Blueprint for Problem Solving when the solution to a problem is not obvious.

TEACHING TIP If students have trouble visualizing Example 2, ask them to visualize how many times they can fill a 1-quart milk bottle from a 1-gallon milk jug. Draw a picture of each on the board. (Answer: 4 times.)

Then ask them if they had 6 1-gallon jugs how much milk could be poured in 24 identical containers. (Answer: 1 quart of milk into each.)

EXAMPLE 2 How many 210-*liter* gasoline barrels can be filled from a tank of 5.04 *kiloliters* of gasoline?

1. *Understand the problem.*

MATHEMATICS BLUEPRINT FOR PROBLEM SOLVING			
Gather the Facts	What Am I Asked to Do?	How Do I Proceed?	Key Points to Remember
We have 5.04 kiloliters of gasoline. We are going to divide the gasoline into smaller barrels that hold 210 liters each.	Find out how many of these smaller 210-liter barrels can be filled.	We need to get all measurements in the same units. We choose to convert 5.04 kiloliters to liters. Then we divide that result by 210 to find out how many barrels will be filled.	To convert 5.04 kiloliters to liters, we move the decimal point 3 places to the right.

2. *Solve and state the answer.*

First we convert 5.04 kiloliters to liters.

$$5.04 \text{ kiloliters} = 5040 \text{ liters}$$

Now we find out how many barrels will be filled. How many 210-liter barrels will 5040 liters fill? Visualize fitting 210-liter barrels into a big barrel that holds 5040 liters.

5040 liters

210 liters | 210 liters | 210 liters

We need to divide

$$\frac{5040 \text{ liters}}{210 \text{ liters}} = 24$$

Thus we can fill 24 of the 210-liter barrels.

3. *Check.*

Estimate each value. 5.04 kiloliters is approximately 5 kiloliters or 5000 liters. 210-liter barrels hold approximately 200 liters. How many times does 200 fit into 5000?

$$\frac{5000}{200} = 25$$

We estimate that 25 barrels can be filled. This is very close to our calculated value of 24 barrels. Thus our answer is reasonable. ✓

Practice Problem 2 A lab assistant must use 18.06 liters of solution to fill 42 jars. How many milliliters of the solution will go into each jar? ■

6.5 Exercises

Applications

Solve each.

1. Find the perimeter of the triangle. Express your answer in *yards*.

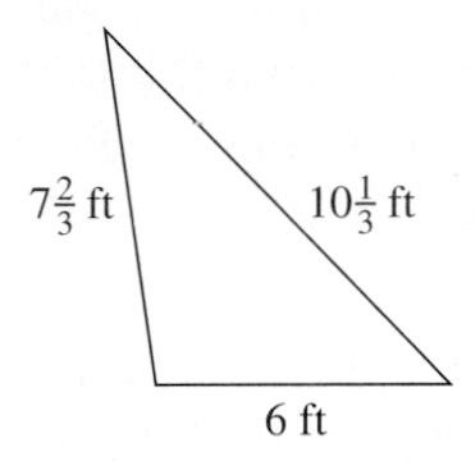

$7\frac{2}{3} + 10\frac{1}{3} + 6 = 24$ ft $= 8$ yd

2. Find the perimeter of the triangle. Express your answer in *yards*.

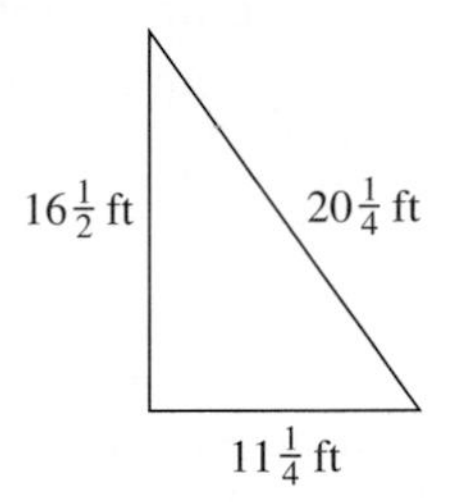

$16\frac{1}{2} + 20\frac{1}{4} + 11\frac{1}{4} = 48$ ft $= 16$ yd

3. The triangular giraffe enclosure at the zoo is being refenced. One side is 86 yards long and a second side is 77 yards long. If the zoo has 522 *feet* of fencing, how much will be left over for the third side? Express your answer in *yards*. 11 yards

4. A farmer has 600 feet of fencing to fence in a triangular region. One side is to be 75 yards and a second 65 yards. How much fencing will the farmer have left for the third side? Express your answer in *yards*.

$75 + 65 = 140$ $\quad \begin{array}{r} 200 \\ -\ 140 \\ \hline 60 \text{ yd} \end{array}$

5. A rectangular picture window measures 87 centimeters × 152 centimeters. Window insulation is applied along all four sides. The insulation costs $7.00 per meter. What will it cost to insulate the window?

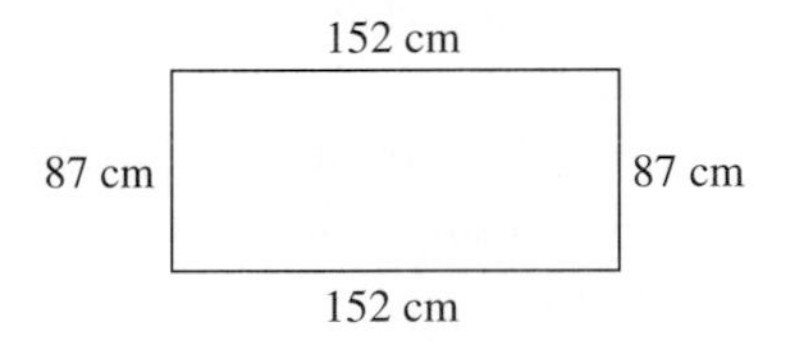

Perimeter = 478 cm $\quad 478 \times 0.07 = \$33.46$

6. A rectangular doorway measures 90 centimeters × 200 centimeters. Weatherstripping is applied on the top and the two sides. The weatherstripping costs $6.00 per meter. What does it cost to weatherstrip the door?

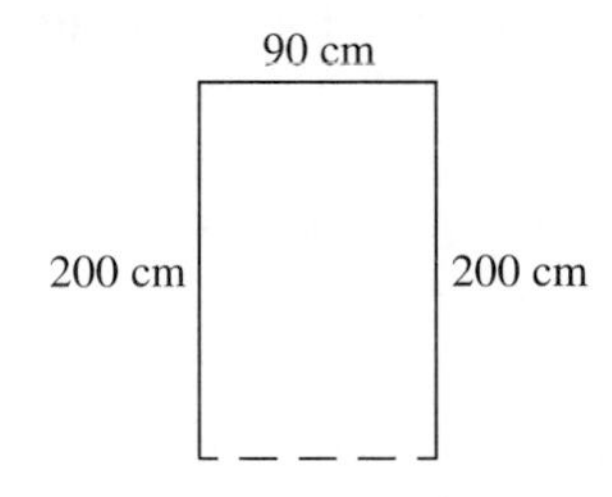

Needed weatherstripping = 490 cm $\quad 490 \times \$0.06 = \$29.40$

7. A beverage company is filling new bottles of strawberry-kiwi fruit juice to be sent to a test market of college students. The packaging division has to put 62 liters into 80 bottles. How many milliliters of juice will be in each bottle?
775 mL per bottle

8. A lab assistant must use 28 *liters* of solvent to fill 40 jars. How many *milliliters* of solvent will be in each jar?

28 L = 28,000 mL $\quad \frac{28{,}000}{40} = 700$ mL per jar

9. A tack supply company, which makes saddles, fittings, bridles, and other equipment for horses and riders, has a length of braided leather 12.4 meters long. The leather must be cut into 4 equal pieces. How many *centimeters* long will each piece be?
310 centimers per piece

10. A stretch of road 1.863 kilometers long has 230 parking spaces of equal length painted in white on the pavement. How many meters long is each parking space? 8.1 meters

11. A hallway outside the kindergarten classrooms in Aalborg, Denmark, has one wall that is 67 meters long, one fifth of the wall is dedicated to paintings made by the children. How many centimeters long is the art portion of the wall? 1340 cm

12. An antenna 186 *meters* tall is on the roof of a building. The top $\frac{1}{8}$ of the antenna has a special rubberized insulation. How many *centimeters* long is the portion of the antenna with this insulation?

186 m = 18,600 cm $\quad \frac{18,600}{8} = 2325$ cm

13. How many 24-*milliliter* samples can be obtained from 3 *liters* of a chemical solution?

3 L = 3000 mL $\quad \frac{3000}{24} = 125$ samples

14. A gourmet balsamic vinegar factory is filling 200-liter oak casks from 8 kiloliters of balsamic vinegar. How many oak casks can be filled?
40 oak casks

15. A bottle of superglue states on its label that it should not be used at temperatures above 85°F. The sign on the display in the store reads NOT TO BE USED AT TEMPERATURES ABOVE 27°C. What is the discrepancy, in degrees Fahrenheit between the two suggested temperatures? The discrepancy is 4.4°F. The sign in the store is 4.9°F less than it should be.

16. The temperature in Dakar, Senegal, today is 33°C. The temperature on this date last year was 94°F. What is the difference, in degrees Fahrenheit, between the temperature in Dakar today and one year ago? The difference is 2.6°F. The temperature last year was 2.6°F greater than today's temperature.

17. A Swedish flight attendant is heating passenger meals for the new American Airline he is working for. He is used to heating meals at 180°C. His co-worker tells him that the food must be heated at 350°F. What is the difference in temperature in degrees Fahrenheit between the two temperatures? Which temperature is hotter? The difference is 6°F. The temperature reading of 180°C is hotter.

18. A French cook wishes to bake potatoes in an oven at 195°C. The American assistant set the oven at 400°F. By how many degrees Fahrenheit was the oven temperature off? Was the oven too hot or too cold?

195°C = 383°F $\quad 400 - 383 = 17$°F too hot

19. Diana and Louisa are traveling in Switzerland at 95 kilometers/hour for 6.2 hours. The total trip will be 560 miles long. How many miles do they still need to travel? Round your answer to the nearest mile. 195 miles further

20. Alex is traveling from Niagara Falls, New York, to Toronto, Canada. He has been driving at 100 kilometers/hour for 4.8 hours. His total trip will be 415 miles. How many miles does Alex have left to travel? Round your answer to the nearest mile. 117 miles

21. Sharon and James Hanson traveled from Arizona to Acapulco, Mexico. The last day of their trip they traveled 520 miles and it took them 8 hours. The maximum speed limit is 110 kilometers/hour.

(a) How many kilometers per hour did they average on the last day of the trip?

(b) Did they break the speed limit?

(a) about 105 kilometers per hour.
(b) Probably not. We cannot be sure, since they could speed for a short time, but we have no evidence to indicate that they broke the speed limit.

22. A small corporate jet travels at 600 kilometers/hour for 1.5 hours. The pilot said the plane will be on time if it travels at 350 miles per hour.

(a) What was the jet's speed in miles per hour?

(b) Will it arrive on time?

(a) 600 km/hr = 372 mi/hr **(b)** Yes

23. A very popular snack bar by the community pool sells quite a bit of soda from the poolside soda fountain. On an average day, 8 pints per minute of various soft drinks are sold. How many gallons per hour of soda are dispensed? 60 gallons per hour

24. A swimming pool is filled at the rate of 6 pints per minute. How many gallons per hour is this?

$6 \times \frac{1}{2} \times \frac{1}{4} = 0.75$ gal/min

$0.75 \times 60 = 45$ gal/hr

25. 3.4 tons of bananas are off-loaded from a ship that has just arrived from Costa Rica. The port taxes imported fruit at $0.015 per pound. What is the tax on the entire shipment? $102

26. Trucks in Sam's home state are taxed each year at $0.03 per pound. Sam's empty truck weighs 1.8 tons. What is his annual tax?
$1.8 \times 2000 \times 0.03 = \$108$

27. A box of raisin bran contains 14 ounces of cereal. Of this, 11 ounces are bran flakes and the rest is raisins. How many *grams* of raisins are in the box?

$14 - 11 = 3 \text{ oz} \quad 3 \text{ oz} \times \frac{28.35 \text{ g}}{1 \text{ oz}} = 85.05 \text{ g}$

28. A can of peaches contains 16 ounces. Fred discovered that 5 ounces were syrup and the rest was fruit. How many *grams* of fruit are in the can?

$16 - 5 = 11 \text{ oz} \quad 11 \text{ oz} \times \frac{28.35 \text{ g}}{1 \text{ oz}} = 311.85 \text{ g}$

★ **29.** Maria bought 18 quarts of motor oil for $1.39 per quart. Her uncle from Mexico is visiting this summer. He said his car will use 12 liters of oil while driving this summer. **(a)** How many extra quarts of oil did Maria buy? **(b)** How much did this extra oil cost her?

(a) $12 \text{ L} \times \frac{1.06 \text{ qt}}{1 \text{ L}} = 12.72 \text{ qt} \quad 18 - 12.72 = 5.28 \text{ qt}$

She bought 5 qt extra.

(b) $5 \times 1.39 = \$6.95$

★ **30.** Carlos is mixing a fruit spray to retard the damage caused by beetles. He has 16 quarts of spray concentrate available. The old recipe he used in Mexico called for 11 liters of spray concentrate. **(a)** How much extra spray concentrate in quarts does he have? **(b)** If the spray costs $2.89 per quart, how much will it cost him to prepare the old recipe?

(a) $11 \text{ L} \times \frac{1.06 \text{ qt}}{1 \text{ L}} = 11.66 \text{ qt} \quad 16 - 11.66 = 4.34 \text{ qt}$

He had 4 quarts extra.

(b) $12 \times 2.89 = \$34.68$

31. Jackson's small motorcycle gets 56 kilometers per liter. If he drives 392 kilometers to Quebec City, Canada, and gas is $0.78 per liter,
(a) How much would the gas used for the trip cost? $5.46
(b) What would be the number of *miles per gallon* Jackson gets on his motorcycle?
132 miles per gallon

32. Ryan's Volvo station wagon runs on diesel fuel. He gets 20 miles per gallon on the highway. If he drives a distance of 275 kilometers to Mexico City, and diesel fuel costs $1.30 per gallon, how much does the fuel used for the trip cost him? Round your answer to the nearest cent. How many kilometers per liter does he get?
$11.08; approx. 8.59 kilometers per liter

Cumulative Review Problems

Solve for n.

33. $\frac{n}{16} = \frac{2}{50}$
$n = 0.64$

34. $\frac{1.7}{n} = \frac{51}{6}$
$n = 0.2$

35. Justin is going to India. While reading his atlas he sees that Bombay is 6 inches from where he would like to travel. The scale shows that 3 inches represent 7.75 miles on the ground. If he travels by train on a straight track, how far will the train go to take him to his destination? 15.5 miles

36. Thompson has a scale model of a famous fishing schooner. Each 2 centimeters on the model represents an actual length of 7.5 yards. The model is 11 centimeters long. How long is the famous fishing schooner?
41.25 yards

Chapter Organizer

Topic	Procedure	Examples
Changing from one American unit to another, p. 338.	1. Find the equality statement that relates what you want to find and what you know. 2. Form a unit fraction. The denominator will contain the units you now have. 3. Simplify and multiply by the unit fraction.	Convert 210 inches to feet. 1. Use 12 inches = 1 foot. 2. Unit fraction = $\frac{1 \text{ foot}}{12 \text{ inches}}$. 3. $210 \cancel{\text{in.}} \times \frac{1 \text{ ft}}{12 \cancel{\text{in.}}} = \frac{210}{12} \text{ ft} = 17.5 \text{ ft}$ Convert 86 yards to feet. 1. Use 1 yard = 3 feet. 2. Unit fraction = $\frac{3 \text{ feet}}{1 \text{ yard}}$. 3. $86 \cancel{\text{yd}} \times \frac{3 \text{ ft}}{1 \cancel{\text{yd}}} = 86 \times 3 \text{ ft} = 258 \text{ ft}$
Changing from one metric unit to another, pp. 345, 352, 353.	When you change from one prefix to another moving to the *left,* move the decimal point to the *left* the same number of places. kilo- = 10^3 hecto- = 10^2 deka- = 10^1 one unit = 1 deci- = 0.1 centi- = 0.01 milli- = 0.001 When you change from one prefix to another moving to the *right,* move the decimal point to the *right* the same number of places.	Change 7.2 meters to kilometers. 1. Move three decimal places to the left. 0.007.2 2. 7.2 m = 0.0072 km. Change 196 centimeters to meters. 1. Move two places to the left. 196. 2. 196 cm = 1.96 m. Change 17.3 liters to milliliters. 1. Move three decimal places to the right. 17.300 2. 17.3 L = 17,300 mL.
Changing from American units to metric units, p. 358.	1. From the list of equivalent measures, pick an equality statement that begins with the unit you now have. 1 mi = 1.61 km 1 yd = 0.914 m 1 ft = 0.305 m 1 in. = 2.54 cm 1 gal = 3.79 L 1 qt = 0.946 L 1 lb = 0.454 kg 1 oz = 28.35 g 2. Multiply by a unit fraction.	Convert 7 gallons to liters. 1. 1 gal = 3.79 L. 2. $7 \cancel{\text{gal}} \times \frac{3.79 \text{ L}}{1 \cancel{\text{gal}}} = 26.53 \text{ L}$ Convert 18 pounds to kilograms. 1. 1 lb = 0.454 kg. 2. $18 \cancel{\text{lb}} \times \frac{0.454 \text{ kg}}{1 \cancel{\text{lb}}} = 8.172 \text{ kg}$

Chapter Organizer

Topic	Procedure	Examples
Changing from metric units to American units, p. 358.	1. From the list of equivalent measures, pick an equality statement that begins with the unit you now have and ends with the unit you want. 1 km = 0.62 mi 1 m = 3.28 ft 1 m = 1.09 yd 1 cm = 0.394 in. 1 L = 0.264 gal 1 L = 1.06 qt 1 kg = 2.2 lb 1 g = 0.0353 oz 2. Multiply by a unit fraction.	Convert 605 grams to ounces. 1. 1 g = 0.0353 oz. 2. $605 \cancel{g} \times \frac{0.0353 \text{ oz}}{1 \cancel{g}} = 21.3565$ oz Convert 80 km/hr to mi/hr. 1. 1 km = 0.62 mi. 2. $80 \frac{\cancel{km}}{hr} \times \frac{0.62 \text{ mi}}{1 \cancel{km}} = 49.6$ mi/hr
Changing from Celsius to Fahrenheit temperature, p. 361.	1. To convert Celsius to Fahrenheit, we use the formula $F = 1.8 \times C + 32$ 2. We replace C by the Celsius temperature. 3. We calculate to find the Fahrenheit temperature.	Convert 65°C to Fahrenheit. 1. $F = 1.8 \times C + 32$ 2. $F = 1.8 \times 65 + 32$ 3. $F = 117 + 32 = 149$ The temperature is 149°F.
Changing from Fahrenheit to Celsius temperature, p. 361.	1. To convert Fahrenheit to Celsius, we use the formula $C = \frac{5 \times F - 160}{9}$ 2. We replace F by the Fahrenheit temperature. 3. We calculate to find the Celsius temperature.	Convert 50°F to Celsius. 1. $C = \frac{5 \times F - 160}{9}$ 2. $C = \frac{5 \times 50 - 160}{9}$ 3. $C = \frac{250 - 160}{9} = \frac{90}{9} = 10$ The temperature is 10°C.

Chapter 6 Review Problems

6.1 *Convert. When necessary, express your answer as a decimal. Round to the nearest hundredth.*

1. 27 ft = 9 yd
2. 33 ft = 11 yd
3. 3 mi = 5280 yd
4. 4 mi = 7040 yd
5. 90 in. = 7.5 ft
6. 78 in. = 6.5 ft
7. 15,840 ft = 3 mi
8. 10,560 ft = 2 mi
9. 4 tons = 8000 lb
10. 7 tons = 14,000 lb
11. 92 oz = 5.75 lb
12. 100 oz = 6.25 lb
13. 15 gal = 60 qt
14. 21 gal = 84 qt
15. 31 pt = 15.5 qt
16. 27 pt = 13.5 qt
17. 2160 sec = 36 min
18. 1380 sec = 23 min
19. 14 hr = 840 min
20. 11 hr = 660 min

6.2 *Convert. Do not round your answer.*

21. 56 cm = 560 mm
22. 29 cm = 290 mm
23. 1763 mm = 176.3 cm
24. 2598 mm = 259.8 cm
25. 9.2 m = 920 cm
26. 7.4 m = 740 cm
27. 5 km = 5000 m
28. 7 km = 7000 m
29. 285 m = 0.285 km
30. 473 m = 0.473 km

Change all units to meters and add.

31. 6.2 m + 121 cm + 0.52 m 7.93 m
32. 9.8 m + 673 cm + 0.48 m 17.01 m
33. 0.024 km + 1.8 m + 983 cm 35.63 m
34. 0.078 km + 5.5 m + 609 cm 89.59 m

6.3 *Convert. Do not round your answer.*

35. 17 kL = 17,000 L
36. 23 kL = 23,000 L
37. 59 mL = 0.059 L
38. 77 mL = 0.077 L
39. 196 kg = 196,000 g
40. 721 kg = 721,000 g
41. 778 mg = 0.778 g
42. 459 mg = 0.459 g
43. 125 kg = 0.125 t
44. 705 kg = 0.705 t
45. 76 kg = 76,000 g
46. 41 kg = 41,000 g
47. 765 cm^3 = 765 mL
48. 423 cm^3 = 423 mL
49. 2.43 L = 2430 cm^3
50. 1.93 L = 1930 cm^3

6.4 *Perform each conversion. Round your answer to the nearest hundredth.*

51. 42 kg = 92.4 lb
52. 33 kg = 72.6 lb
53. 15 ft = 4.58 m
54. 9 ft = 2.75 m
55. 13 oz = 368.55 g
56. 17 oz = 481.95 g
57. 1.8 ft = 54.86 cm
58. 1.3 ft = 39.62 cm
59. 14 cm = 5.52 in.
60. 18 cm = 7.09 in.
61. 20 lb = 9.08 kg
62. 30 lb = 13.62 kg
63. 12 yd = 10.97 m
64. 14 yd = 12.80 m
65. 80 km/hr = 49.6 mi/hr
66. 70 km/hr = 43.4 mi/hr
67. 15°C = 59 °F
68. 25°C = 77 °F
69. 221°F = 105 °C
70. 185°F = 85 °C
71. 32°F = 0 °C

6.5 *Solve each problem. Round your answer to the nearest hundredth when necessary.*

72. A roof is layered with 1.76-cm plywood, 4.32-mm tar paper, and 0.93-cm shingles. How thick is the roof? 3.12 cm

73. Find the perimeter of the triangle.
(a) Express your answer in feet. 17 ft
(b) Express your answer in inches. 204 in.

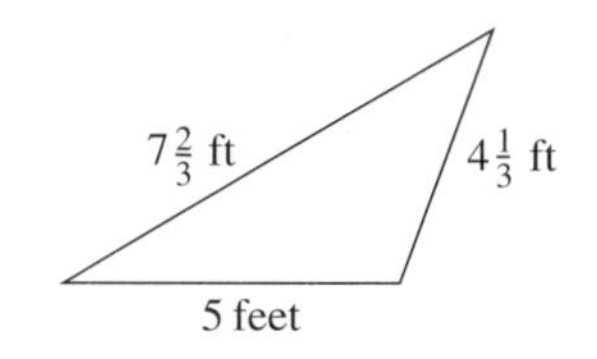

74. Find the perimeter of the rectangle.
(a) Express your answer in meters. 200 m
(b) Express your answer in kilometers. 0.2 km

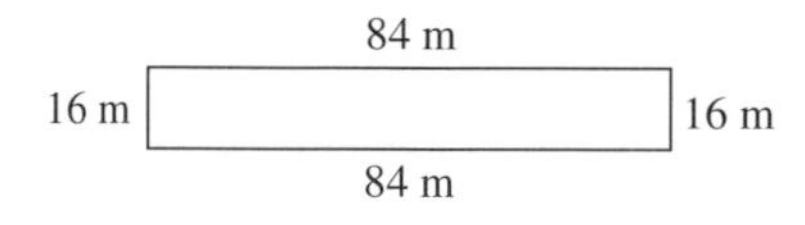

75. The unit price on a box of cereal was \$0.14 per ounce. The net weight was 450 grams. How much did the cereal cost? \$2.22

76. Keshia traveled at 90 km/hr for 3 hours. She needs to travel 200 miles. How much farther does she need to travel? 32.6 miles farther

77. The width of a house in Mexico is 12.6 meters. Kim arrived with enough wooden trim for a house that is 43 feet wide. Did she have enough? How many feet extra or how short was this amount? Yes 1.67 feet extra

78. A chemist has a solution of 4 *liters*. She needs to place it into 24 equal-sized smaller jars. How many *milliliters* will be contained in each jar? 166.67 milliliters per jar

79. A shipment of parts for Europe weighs 2 tons. The cost for overseas shipping is \$0.32 per kilogram. How much will it cost to send the shipment overseas? \$581.12

80. A flagpole is 19 *meters* long. The bottom $\frac{1}{5}$ of it is coated with a special water seal before being placed in the ground. How many *centimeters* long is the portion that has the water seal? 380 centimeters

81. A German cook wishes to bake a cake at 185° Celsius. The oven is set at 390° Fahrenheit. By how many degrees Fahrenheit is the oven temperature different from what is desired? Is the oven too hot or not hot enough? 25°F too hot

Chapter 6 Test

Convert. Express your answer as a decimal rounded to the nearest hundredth when necessary.

1. 1.6 tons = ________ lb

2. 19 ft = ________ in.

3. 21 gal = ________ qt

4. 36,960 ft = ________ mi

5. 3 cups = ________ qt

6. 1800 sec = ________ min

Perform each conversion. Do not round your answer.

7. 27.3 cm = ________ m

8. 9.2 km = ________ m

9. 46 mm = ________ cm

10. 9.88 cm = ________ m

11. 12.7 m = ________ cm

12. 0.936 cm = ________ mm

13. 46 L = ________ kL

14. 127 L = ________ mL

15. 28.9 mg = ________ g

16. 983 g = ________ kg

17. 0.92 L = ________ mL

18. 9.42 g = ________ mg

Perform each conversion. Round to the nearest hundredth when necessary.

19. 42 mi = ________ km

20. 1.78 yd = ________ m

21. 9 cm = ________ in.

22. 38 L = ________ gal

23. 7.3 kg = ________ lb

24. 3 oz = ________ g

1. 3200

2. 228

3. 84

4. 7

5. 0.75

6. 30

7. 0.273

8. 9200

9. 4.6

10. 0.0988

11. 1270

12. 9.36

13. 0.046

14. 127,000

15. 0.0289

16. 0.983

17. 920

18. 9420

19. 67.62

20. 1.63

21. 3.55

22. 10.03

23. 16.06

24. 85.05

Solve each problem. Round your answer to the nearest hundredth when necessary.

25. A rectangular picture frame measures 3 m × 7 m.

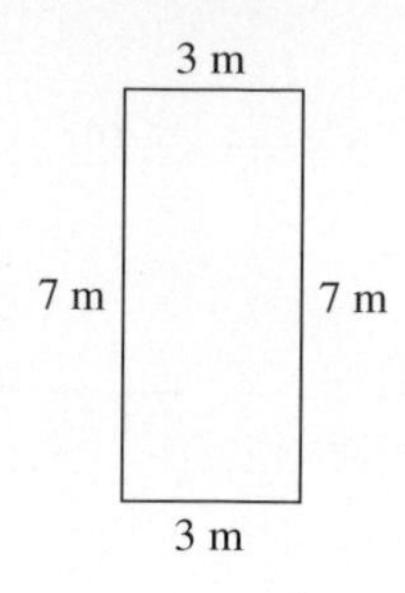

(a) What is the perimeter of the picture frame in meters?

(b) What is the perimeter of the picture frame in yards?

26. The temperature is 80°F today. Kristen's computer contains a warning not to operate above 35°C.

(a) How many degrees Fahrenheit are there between the two temperatures?

(b) Can she use her computer today?

27. A pump is running at 5.5 quarts per minute. How many gallons per hour is this?

28. The speed limit on a Canadian road is 100 km/hr.

(a) How far can Samuel travel at this speed limit in 3 hours?

(b) If Samuel has to travel 200 miles, how much farther will he need to go after 3 hours of driving at 100 km/hr?

25. (a) 20 meters

(b) 21.8 yards

26. (a) 15°F

(b) Yes

27. 82.5 gal/hr

28. (a) 300 km

(b) 14 miles

Cumulative Test for Chapters 1–6

Approximately one half of this test is based on Chapter 6 material. The remainder is based on material covered in Chapters 1–5.

Do each problem. Simplify your answer.

1. Subtract. $\begin{array}{r} 9824 \\ -\ 3796 \\ \hline \end{array}$

2. Multiply. $\begin{array}{r} 608 \\ \times\ 305 \\ \hline \end{array}$

3. Divide. $28\overline{)1932}$

4. Add. $\frac{1}{7} + \frac{3}{14} + \frac{2}{21}$

5. Subtract. $3\frac{1}{8} - 1\frac{3}{4}$

6. Is this proportion true or false? $\frac{21}{35} = \frac{12}{20}$

7. Solve the proportion. $\frac{0.4}{n} = \frac{2}{30}$

8. A piece of wire 6.5 centimeters long weighs 68 grams. What will a 20-centimeter length of the same wire weigh? (Round to the nearest hundredth.)

9. What percent of 66 is 165?

10. Find 18% of 360.

11. 0.5% of what number is 100?

Convert. Express your answer as a decimal rounded to the nearest hundredth when necessary.

12. 38 qt = ________ gal

13. 2.5 tons = ________ lb

14. 7 pt = ________ qt

15. 25 feet = ________ in.

1. 6028

2. 185,440

3. 69

4. $\frac{19}{42}$

5. $1\frac{3}{8}$

6. True

7. $n = 6$

8. 209.23 grams

9. 250%

10. 64.8

11. 20,000

12. 9.5

13. 5000

14. 3.5

15. 300

16.	3700
17.	0.0628
18.	790
19.	0.05
20.	672
21.	106.12
22.	43.58
23.	25.81
24.	14.49
25.	11.88 meters
26.	59°F. The difference is 44°F. The 15°C temperature is higher.
27.	7 miles
28.	1.738 centimeters

Perform each conversion. Do not round your answer.

16. 3.7 km = ____ m **17.** 62.8 g = ____ kg **18.** 0.79 L = ____ mL

19. 5 cm = ____ m **20.** 42 lb = ____ oz

Perform each conversion. Round to the nearest hundredth when necessary.

21. 28 gal = ____ L **22.** 96 lb = ____ kg

23. 7.87 m = ____ ft **24.** 9 mi = ____ km

25. Find the perimeter in *meters* of this triangle.

6 yd
4 yd
3 yd

26. Change 15°C to Fahrenheit temperature. Now find the difference between 15°C and 15°F. Which figure represents the higher temperature?

27. Ricardo traveled on a Mexican highway at 100 km/hr for $1\frac{1}{2}$ hours. He needs to travel a total distance of 100 miles. How far does he still need to travel? (Express your answer in miles.)

28. Two metal sheets are 0.72 centimeter thick and 0.98 centimeter thick, respectively. An insulating foil is placed between them that is 0.38 millimeter thick. When the three layers are placed together, what is the total thickness? Write the exact answer.

CHAPTER 7
Geometry

Experts in forensic science find the areas of footprints left by suspects at the scene of a crime. Using a mathematical formula they determine the approximate weight of the suspect. Do you think you could conduct this type of investigation? Turn to the Putting Your Skills to Work on page 433.

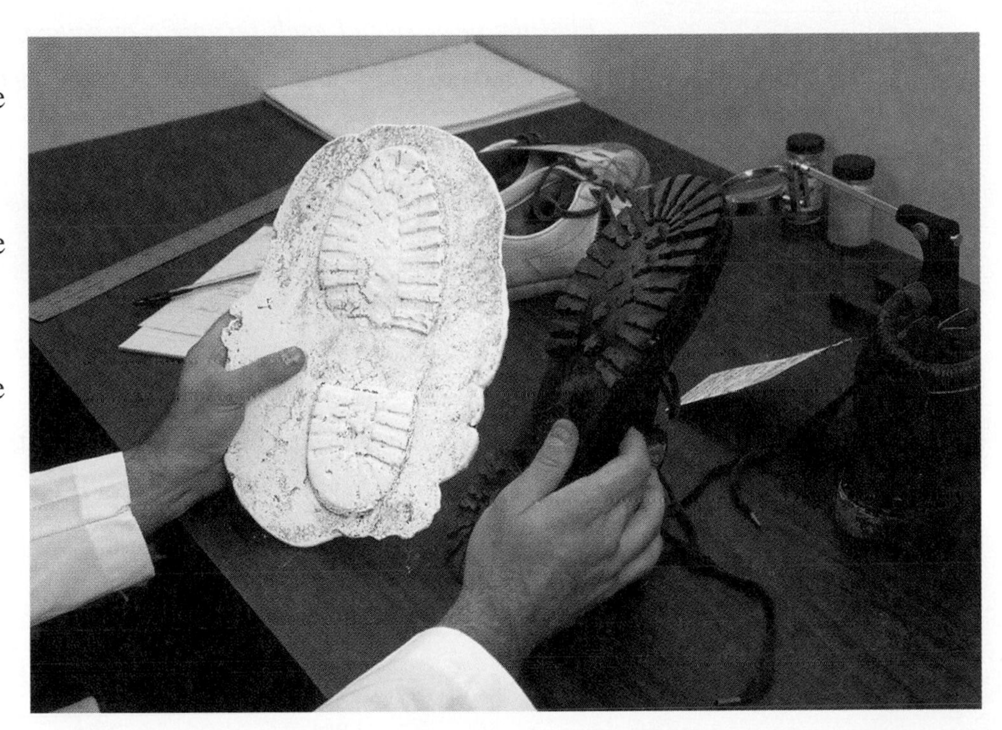

Pretest Chapter 7

1.	18 m
2.	14 m
3.	23 sq cm
4.	2.4 sq cm
5.	25.6 yd
6.	52 ft
7.	135 sq in.
8.	171 sq in.
9.	97 sq m
10.	59°
11.	15.3 m
12.	30 sq m
13.	8
14.	12
15.	6.782

If you are familiar with the topics in this chapter, take this test now. Check your answers with those in the back of the book. If an answer was wrong or you couldn't do a problem, study the appropriate section of the chapter.

If you are not familiar with the topics in this chapter, don't take this test now. Instead, study the examples, work the practice problems, and then take the test.

This test will help you identify which concepts you have mastered and which you need to study further. In each problem round your answer to the nearest tenth when necessary. Use $\pi \approx 3.14$ in your calculation whenever a value of π is necessary.

Section 7.1

Find the perimeter of each rectangle or square.

1. Length = 6.5 m, width = 2.5 m **2.** Length = width = 3.5 m

Find the area of each square or rectangle.

3. Length = width = 4.8 cm **4.** Length = 2.7 cm, width = 0.9 cm

Section 7.2

Find the perimeter of each figure.

5. A parallelogram with one side measuring 9.2 yd and another side measuring 3.6 yd.

6. A trapezoid with sides measuring 17 ft, 5 ft, 25 ft, and 5 ft.

Find the area of each figure.

7. A parallelogram with a base of 27 in. and a height of 5 in.

8. A trapezoid with a height of 9 in. and bases of 16 in. and 22 in.

9.

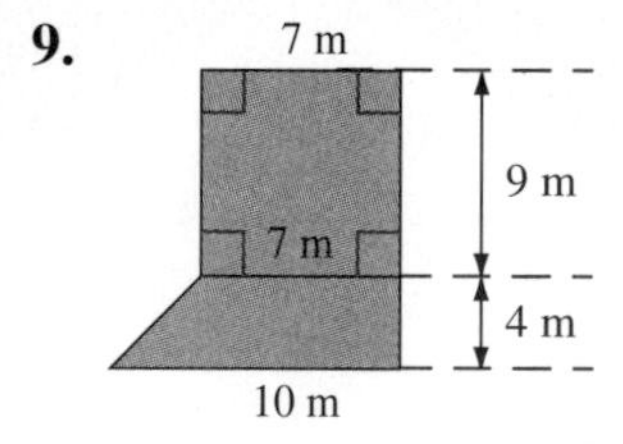

Section 7.3

10. Find the third angle in the triangle if two angles are 42° and 79°.

11. Find the perimeter of the triangle whose sides measure 7.2 m, 4.3 m, and 3.8 m.

12. Find the area of a triangle with a base of 12 m and a height of 5 m.

Section 7.4

Evaluate exactly.

13. $\sqrt{64}$ **14.** $\sqrt{4} + \sqrt{100}$

15. Approximate $\sqrt{46}$ using the square root table on page A–4 or a calculator with a square root key. Round your answer to the nearest thousandth.

Section 7.5

Find the unknown side of the right triangle.

16.

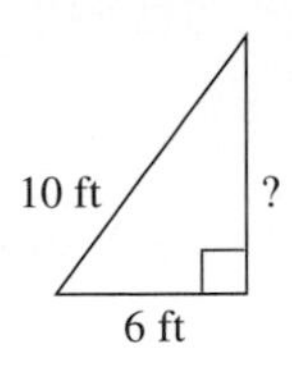

17.

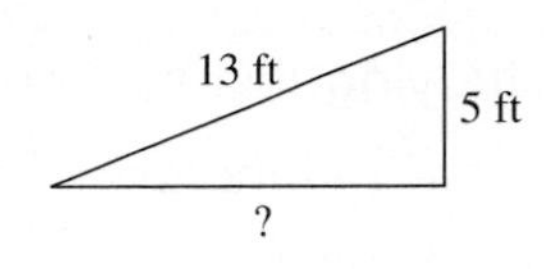

Section 7.6

18. Find the diameter of a circle whose radius is 14 in.

19. Find the circumference of a circle whose diameter is 30 cm.

20. Find the area of a circle whose radius is 7 m.

21. Find the area of the shaded region.

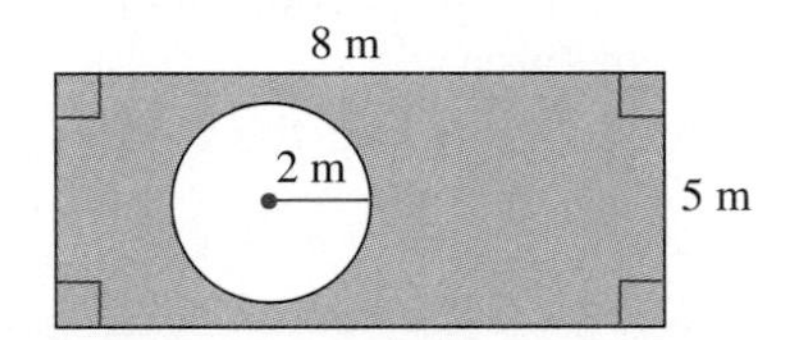

Section 7.7

Find the volume of each object.

22. A rectangular solid with the dimensions of 6 yd by 5 yd by 8 yd.

23. A sphere of radius 3 ft.

24. A cylinder of height 12 in. and radius 7 in.

25. A pyramid of height 21 m with a square base measuring 25 m on a side.

26. A cone of height 30 m and a radius of 6 m.

Section 7.8

Find n in each set of similar triangles.

27.

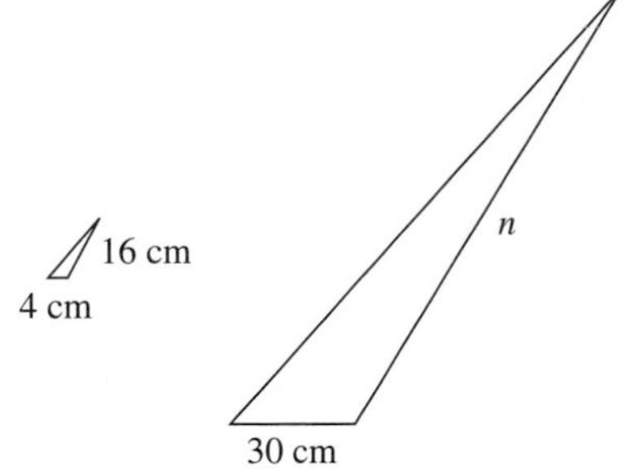

28.

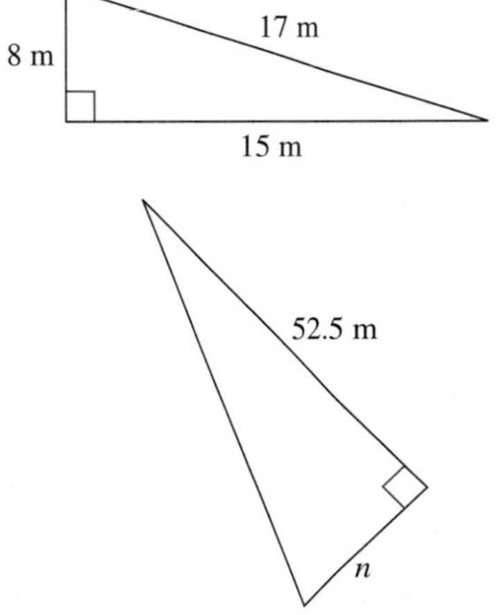

Section 7.9

29. A track field consists of two semicircles and a rectangle.
(a) Find the area of the field.
(b) It costs $0.15 per square yard to fertilize the field. What would it cost to complete this task?

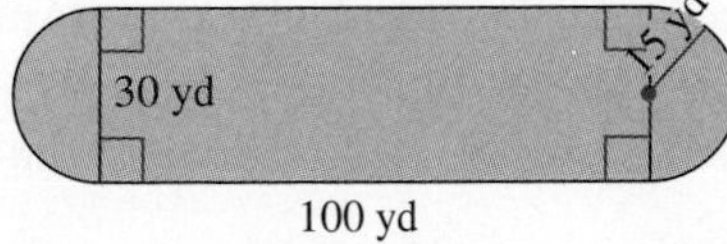

16. 8 ft

17. 12 ft

18. 28 in.

19. 94.2 cm

20. 153.9 sq m

21. 27.4 sq m

22. 240 cu yd

23. 113 cu ft

24. 1846.3 cu in.

25. 4375 cu m

26. 1130.4 cu m

27. $n = 120$ cm

28. $n = 28$ m

29. (a) 3706.5 sq yd

(b) $555.98

7.1 Rectangles and Squares

MathPro

Video 14

SSM

After studying this section, you will be able to:

1. *Find the perimeters of rectangles and squares.*
2. *Find the perimeters of shapes made up of rectangles and squares.*
3. *Find the areas of rectangles and squares.*
4. *Find the areas of shapes made up of rectangles and squares.*

1 *Finding the Perimeters of Rectangles and Squares*

Geometry has a visual aspect that numbers and abstract ideas do not have. We can take pen in hand and draw a picture of a rectangle that represents a room with certain dimensions. We can easily visualize problems such as ''What is the distancc around the outside edges of the room (perimeter)?'' or ''How much carpeting will be needed for the room (area)?''

A rectangle is a four-sided figure like the figures shown below.

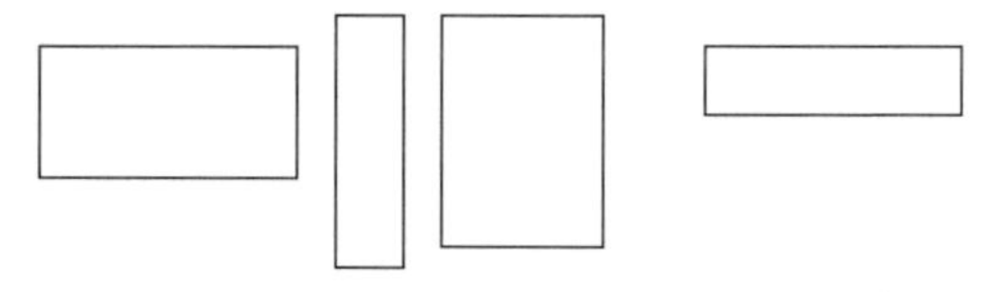

TEACHING TIP The idea of perpendicular sides may be a little vague for students who have absolutely no background in geometry. A good class discussion question is, ''Can you give me some examples from everyday life of perpendicular lines?'' By discussing a couple of the class examples and sketching them on the blackboard, you can clear up any student confusion.

A rectangle has two interesting properties: (1) any two adjoining sides are perpendicular and (2) opposite sides of a rectangle are equal. By ''any two adjoining sides are perpendicular'' we mean that any two sides that meet form an angle that measures 90° (called a *right angle*). These perpendicular sides form one of these shapes: ⌜ ⌝ ⌞ ⌟. When we say that ''opposite sides of a rectangle are equal,'' we mean that the measure of one side is equal to the measure of the side opposite to it. If all four sides are equal (have the same length), then the rectangle is called a *square*.

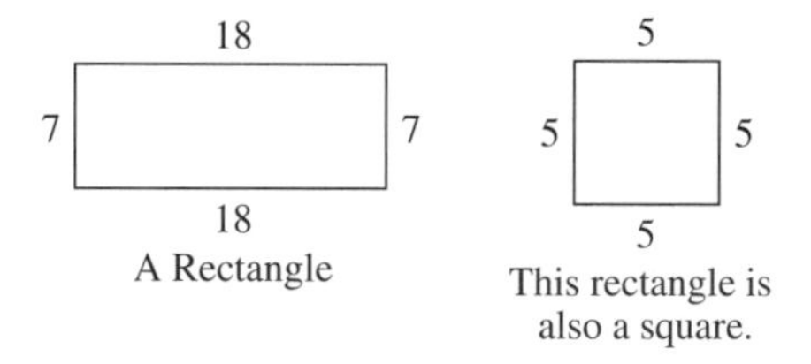

The perimeter of a rectangle is the sum of the lengths of all its sides. To find the perimeter of the rectangle below, we add up the lengths of all the sides of the figure.

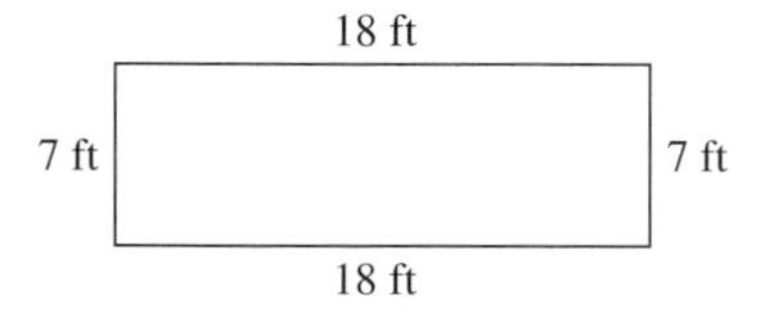

$$\text{Perimeter} = 7 \text{ ft} + 18 \text{ ft} + 7 \text{ ft} + 18 \text{ ft}$$
$$= 50 \text{ ft}$$

Thus the perimeter of the rectangle above is 50 feet.

We could also use a formula to find the perimeter of a rectangle. In the formula we use letters to represent the measurements of the length and width of the rectangle. Let l represent the length, w represent the width, and P represent the perimeter. Note that the length is the longer side and the width is the shorter side. Since the perimeter is found by adding up the measurements all around the rectangle, we see that

$$P = w + l + w + l$$
$$= 2l + 2w$$

When we write $2l$ and $2w$, we mean 2 times l and 2 times w. We can use the formula to find the perimeter of the rectangle.

$$P = 2l + 2w$$
$$= (2)(18 \text{ ft}) + (2)(7 \text{ ft})$$
$$= 36 \text{ ft} + 14 \text{ ft}$$
$$= 50 \text{ ft}$$

Thus the perimeter can be found quickly by using the following formula.

> The **perimeter (P) of a rectangle** is twice the length plus twice the width:
>
> $$P = 2l + 2w$$

EXAMPLE 1 A helicopter has a 3-cm by 5.5-cm insulation pad near the control panel that is rectangular in shape. Find the perimeter of the rectangle.

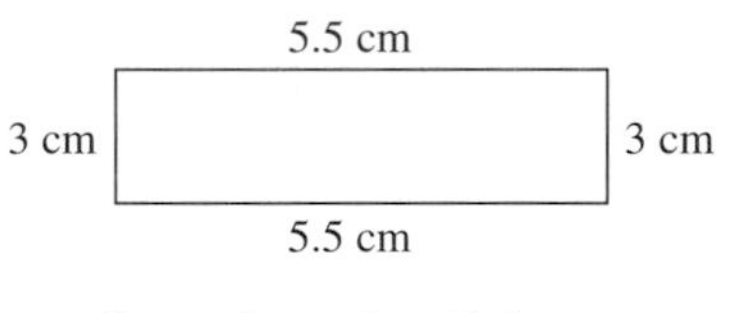

$$\text{Length} = l = 5.5 \text{ cm}$$
$$\text{Width} = w = 3 \text{ cm}$$

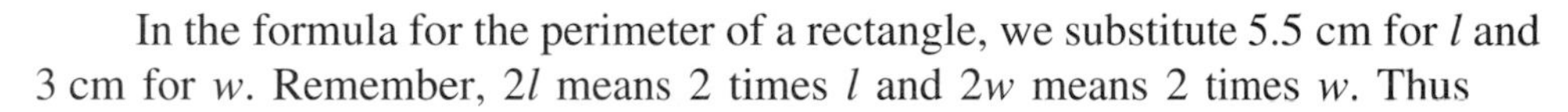

In the formula for the perimeter of a rectangle, we substitute 5.5 cm for l and 3 cm for w. Remember, $2l$ means 2 times l and $2w$ means 2 times w. Thus

$$P = 2l + 2w$$
$$= (2)(5.5 \text{ cm}) + (2)(3 \text{ cm})$$
$$= 11 \text{ cm} + 6 \text{ cm} = 17 \text{ cm}$$

Practice Problem 1 Find the perimeter of the rectangle in the margin. ■

6 m
1.5 m 1.5 m
6 m

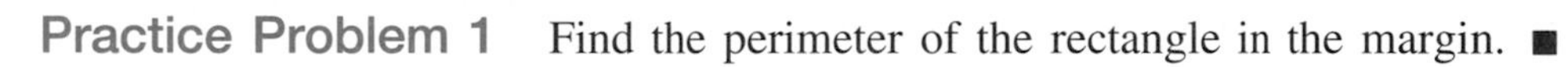

EXAMPLE 2 Find the perimeter of a rectangular wheat field. Its width is 700 yd and its length is 1300 yd.

$$\text{Width} = w = 700 \text{ yd}$$
$$\text{Length} = l = 1300 \text{ yd}$$

$$P = 2l + 2w$$
$$= (2)(1300 \text{ yd}) + (2)(700 \text{ yd})$$
$$= 2600 \text{ yd} + 1400 \text{ yd} = 4000 \text{ yd}$$

Practice Problem 2 Find the perimeter of a corn field. Its width is 860 ft and its length is 1200 ft. ■

A square is a rectangle where all four sides are equal (have the same length). Some examples of squares are shown below.

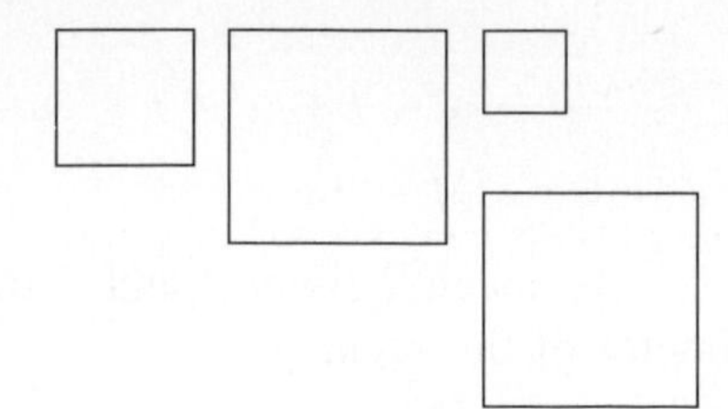

A square, then, is only a special type of rectangle. We can find the perimeter of a square just as we found the perimeter of a rectangle—by adding the measurements of all the sides of the square, but because the lengths of all sides are the same, the formula for the perimeter of a square is very simple. Let s represent the length of one side and P represent the perimeter. So, to find the perimeter, we can write:

> The **perimeter of a square** is four times the length of a side:
>
> $$P = 4s$$

EXAMPLE 3 High Ridge Stables has a new sign at the highway entrance that is in the shape of a square with each side measuring 8.6 yards. Find the perimeter of the sign.

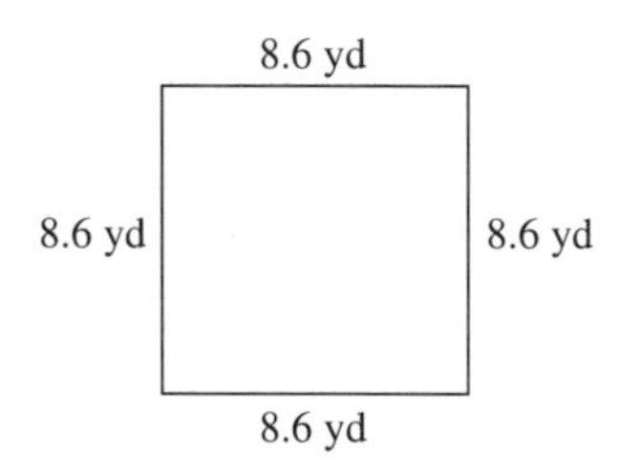

Side = s = 8.6 yd.

$$\begin{aligned} P &= 4s \\ &= (4)(8.6 \text{ yd}) \\ &= 34.4 \text{ yd} \end{aligned}$$

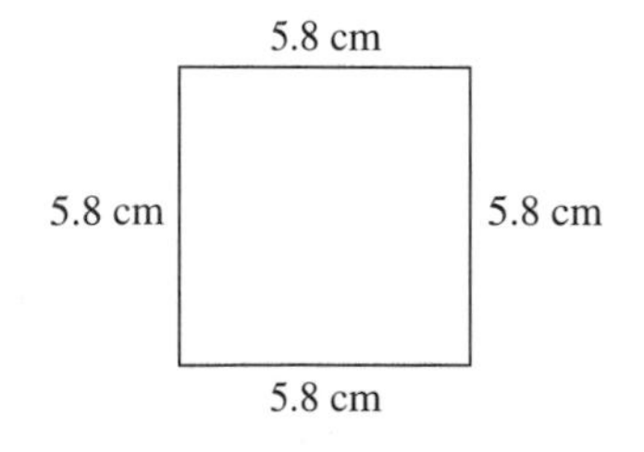

Practice Problem 3 Find the perimeter of the square in the margin. ■

2 *Finding the Perimeters of Shapes Made Up of Rectangles and Squares*

Some figures are a combination of rectangles and squares. To find the perimeter of the total figure, look only at the outside edges.

EXAMPLE 4 Find the perimeter of this figure.

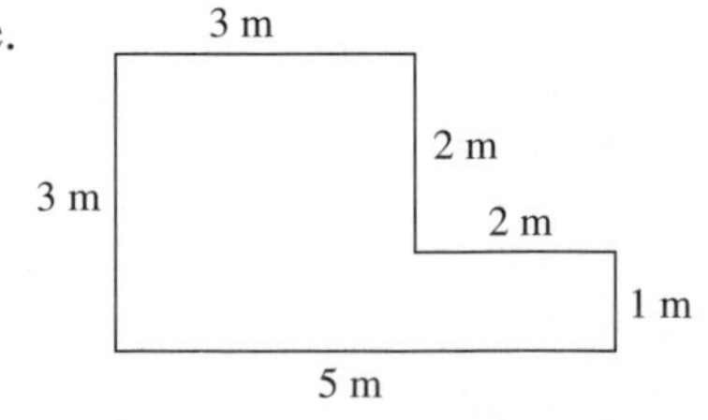

We want to find the distance around the figure. Therefore, we look only at the outside edges. There are six sides to add together.

$$5 + 3 + 3 + 2 + 2 + 1 = 16$$

The perimeter of this shape is 16 m.

4 cm
1 cm
2 cm
1 cm
1 cm
5 cm

Practice Problem 4 Find the perimeter of the figure in the margin. ■

We can apply our knowledge to everyday problems. For example, by knowing how to find the perimeter of a rectangle, we can find out how many feet of picture framing will be needed or how many feet of weather sealing will be used. Consider the following problem.

TEACHING TIP Some students have significant difficulty visualizing an object that is made up of two or more geometric objects. Take the time to show how the object is "created" by putting two shapes together.

EXAMPLE 5 Find the cost of weather stripping needed to seal the edges of the hatch pictured at the right. Weather stripping costs \$0.12 per foot.

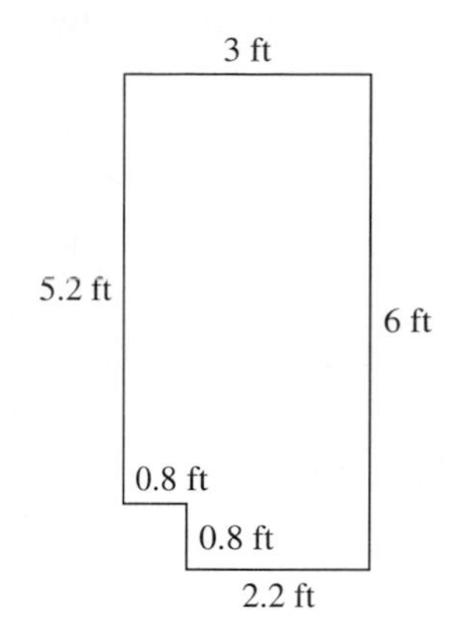

First we need to find the perimeter of the hatch. The perimeter is the sum of all the edges.

$$\begin{array}{r} 3.0 \text{ ft} \\ 6.0 \text{ ft} \\ 2.2 \text{ ft} \\ 0.8 \text{ ft} \\ 0.8 \text{ ft} \\ +\ 5.2\phantom{\text{ ft}} \\ \hline 18.0 \text{ ft} \end{array}$$

The perimeter is 18 ft. Now we calculate the cost.

$$18.0 \cancel{\text{ft}} \times \frac{0.12 \text{ dollar}}{\cancel{\text{ft}}} = \$2.16 \text{ for weather-sealing materials}$$

Practice Problem 5 Find the cost of weather stripping required to seal the edges of this hatch. Weather sealing costs \$0.16 per foot.

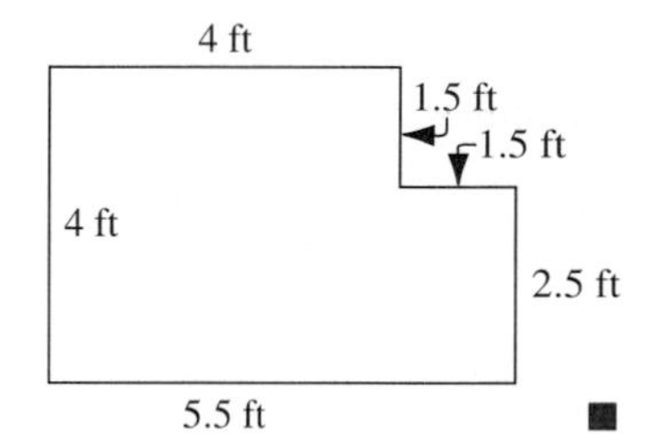

■

3 *Finding the Areas of Rectangles and Squares*

What do we mean by **area?** Area is the measure of the *surface inside* a geometric figure. For example, for a rectangular room, the area is the amount of floor in that room.

One *square meter* is the measure of a square that is 1 m long and 1 m wide.

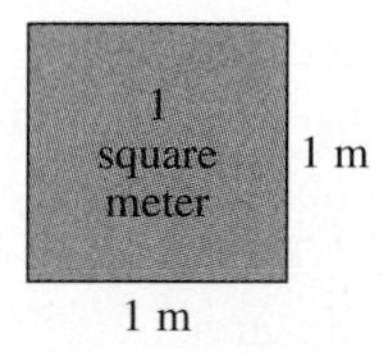

We can abbreviate *square meter* as m^2. In fact, all areas are measured in square meters, square feet, square inches, and so on (written as m^2, ft^2, $in.^2$, and so on).

We can calculate the area of a rectangular region if we know its length and its width. To find the area, *multiply* the length by the width.

> The **area (A) of a rectangle** is the length times the width.
>
> $$A = lw$$

EXAMPLE 6 Find the area of this rectangle.

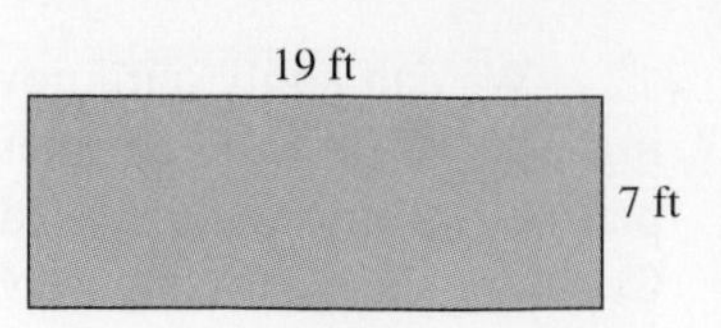

Our answer must be in square feet because we multiply 19 ft by 7 ft.

$$w = 7 \text{ ft} \qquad l = 19 \text{ ft}$$

$$A = (l)(w) = (7 \text{ ft})(19 \text{ ft}) = 133 \text{ ft}^2$$

The area is 133 square feet.

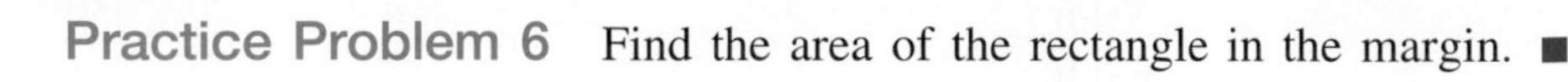

Practice Problem 6 Find the area of the rectangle in the margin. ■

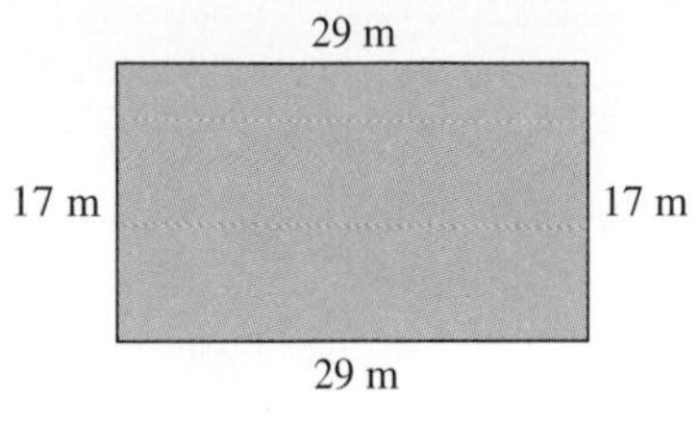

To find the area of a square, we multiply the length of one side by itself.

> The **area of a square** is the square of the length of one side.
>
> $$A = s^2$$

TEACHING TIP You can raise the students' thinking level with the following question: "You have just seen how we multiplied 19 feet by 7 feet to obtain 133 square feet in Example 6. Suppose someone said to you, "Prove it." How could you convince someone that a rectangle that measures 19 feet by 7 feet really has an area of 133 square feet?" Student answers will vary. The most convincing proof to students is to take square tiles exactly 1 foot on each side and show that exactly 133 tiles fit inside the rectangle.

EXAMPLE 7 A square measures 9.6 in. on each side. Find its area.

We know our answer will be measured in square inches. We will write this as in.^{2}.

$$\begin{aligned} A &= s^2 \\ &= (9.6)^2 \\ &= (9.6 \text{ in.})(9.6 \text{ in.}) \\ &= 92.16 \text{ in.}^2 \end{aligned}$$

Practice Problem 7 Find the area of a square computer chip that measures 11.8 mm on each side. ■

4 *Finding the Areas of Shapes Made Up of Rectangles and Squares*

EXAMPLE 8 Consider this shape, which is made up of a rectangle and a square. Find the area of the shaded region.

The shaded region is made up of two separate regions. You can think of each separately, and calculate the area of each one. The total area is just the sum of the two separate areas.

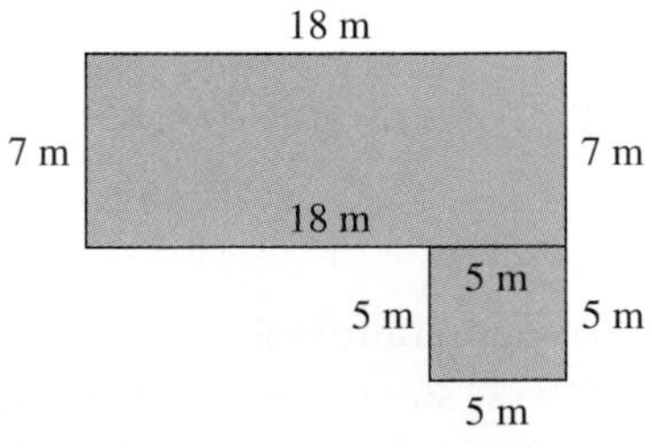

Area of rectangle $= (7)(18) = 126 \text{ m}^2$ Area of square $= 5^2 = 25 \text{ m}^2$

The area of the rectangle	=	126 m^2
+ The area of the square	=	25 m^2
The total area is	=	151 m^2

Practice Problem 8 Find the area of the shaded region in the margin. ■

7.1 Exercises

Verbal and Writing Skills

1. A rectangle has two properties: (1) any two adjoining sides are __perpendicular__ and (2) opposite sides are __equal__.

2. To find the perimeter of a figure, we __add__ the lengths of all of the sides.

3. To find the area of a rectangle, we __multiply__ the length by the width.

4. All area is measured in __square__ units.

Find the perimeter of each rectangle or square.

5. 5.5 mi, 2 mi, 2 mi, 5.5 mi

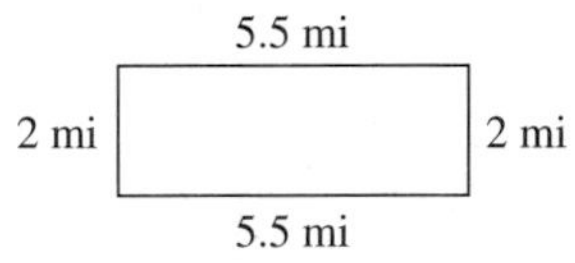

5.5 mi
2.0 mi
5.5 mi
2.0 mi
15 mi

6. 9 cm, 1.5 cm, 1.5 cm, 9 cm

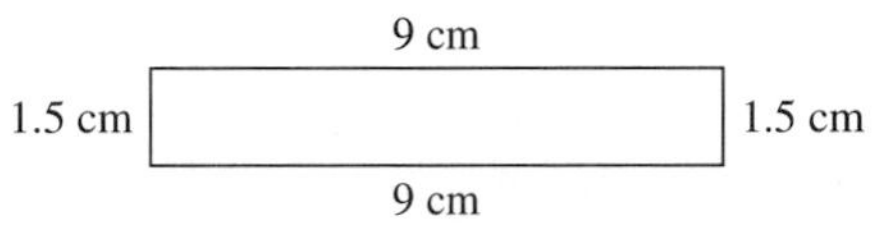

1.5 cm
9.0 cm
1.5 cm
9.0 cm
21 cm

8. 2.5 ft, 9.3 ft, 9.3 ft, 2.5 ft

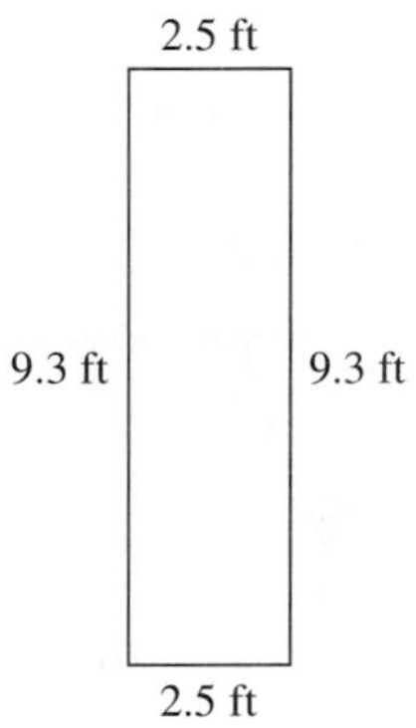

2.5 ft
9.3 ft
2.5 ft
9.3 ft
23.6 ft

7. 5 ft, 6.8 ft, 6.8 ft, 5 ft

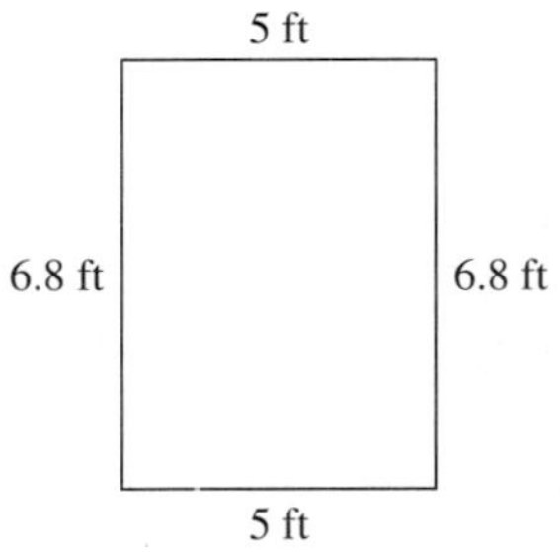

5.0 ft
6.8 ft
5.0 ft
6.8 ft
23.6 ft

9. 12.3 ft, 12.3 ft, 12.3 ft, 12.3 ft

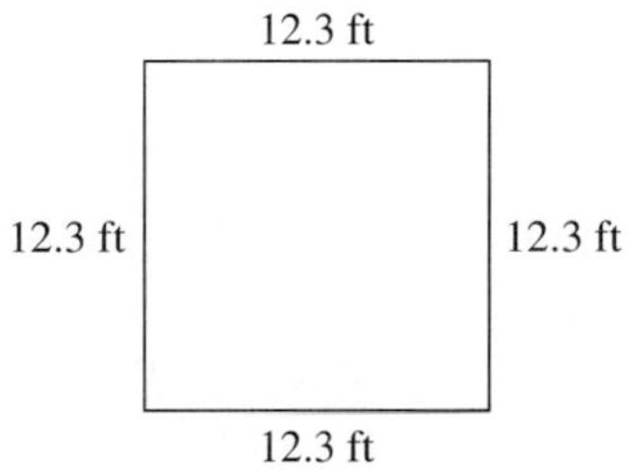

4×12.3 ft $= 49.2$ ft

10. 15.6 ft, 15.6 ft, 15.6 ft, 15.6 ft

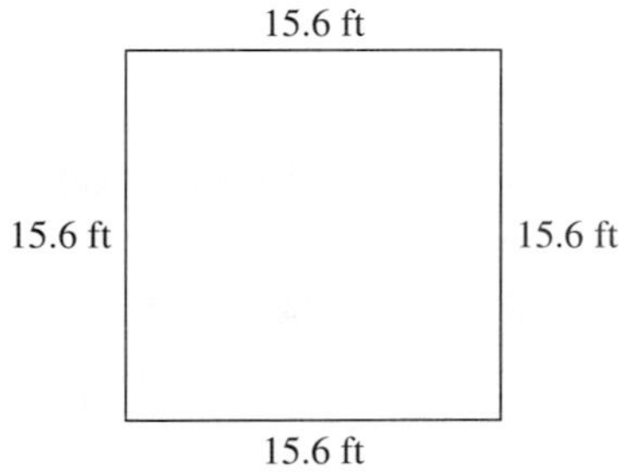

4×15.6 ft $= 62.4$ ft

11. Length = 0.84 mm, width = 0.12 mm

2×0.12 mm $+ 2 \times 0.84$ mm $= 1.92$ mm

12. Length = 6.2 in., width = 1.5 in.

2×1.5 in. $+ 2 \times 6.2$ in. $= 15.4$ in.

13. Length = 7 in., width = 5.73 in.

2×5.73 in. $+ 2 \times 7$ in. $= 25.46$ in.

14. Length = 0.54 km, width = 0.36 km

2×0.36 km $+ 2 \times 0.54$ km $= 1.8$ km

15. Length = width = 4.28 km

4×4.28 km $= 17.12$ km

16. Length = width = 9.63 cm

4×9.63 cm $= 38.52$ cm

Find the perimeter of each rectangle.

17. Length = 0.0089 cm, width = 0.0034 cm

2×0.0089 cm $+ 2 \times 0.0034$ cm $= 0.0246$ cm

18. Length = 0.0093 cm, width = 0.0076 cm

2×0.0093 cm $+ 2 \times 0.0076$ cm $= 0.0338$ cm

19. Length = 15.2 m, width = 6.65 m

2×15.2 m $+ 2 \times 6.65$ m $= 43.7$ m

20. Length = 18.4 m, width = 5.35 m

2×18.4 m $+ 2 \times 5.35$ m $= 47.5$ m

Find the perimeter of each square. The length of the side is given.

21. 13 m
$4 \times 13 \text{ m} = 52 \text{ m}$

22. 14 m
$4 \times 14 \text{ cm} = 56 \text{ cm}$

23. 1.2 mi
$4 \times 1.2 \text{ mi} = 4.8 \text{ mi}$

24. 1.8 m
$4 \times 1.8 \text{ m} = 7.2 \text{ m}$

25. 0.0052 mm
$4 \times 0.0052 \text{ mm} = 0.0208 \text{ mm}$

26. 0.0043 mm
$4 \times 0.0043 \text{ mm} = 0.0172 \text{ mm}$

27. 7.96 cm
$4 \times 7.96 \text{ cm} = 31.84 \text{ cm}$

28. 6.32 cm
$4 \times 6.32 \text{ cm} = 25.28$ c

Find the perimeter of the shapes, which are made up of rectangles and squares.

29.

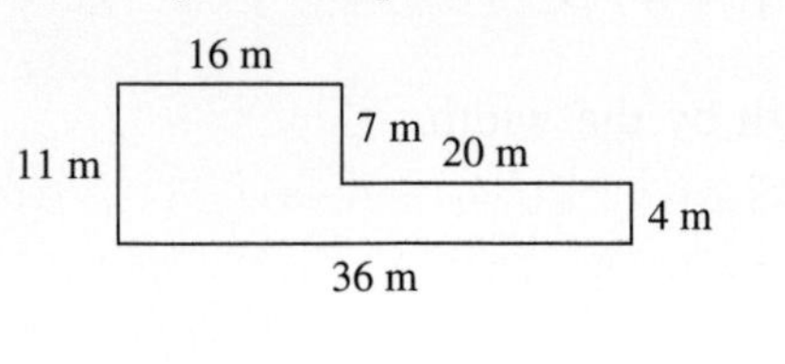

16 m
7 m
20 m
4 m
36 m
11 m
94 m

30.

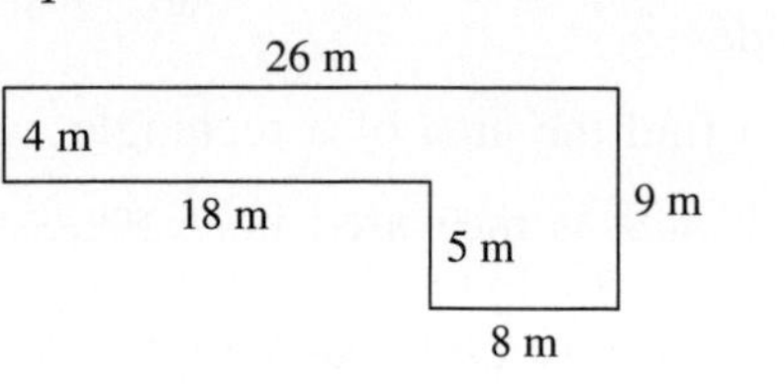

26 m
9 m
8 m
5 m
18 m
4 m
70 m

31.

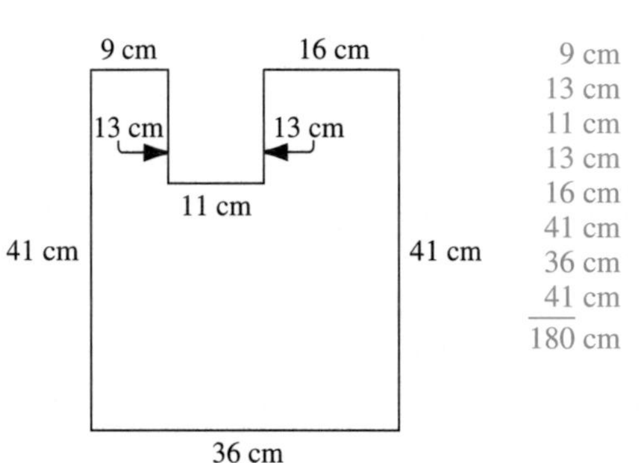

9 cm
13 cm
11 cm
13 cm
16 cm
41 cm
36 cm
41 cm
180 cm

32.

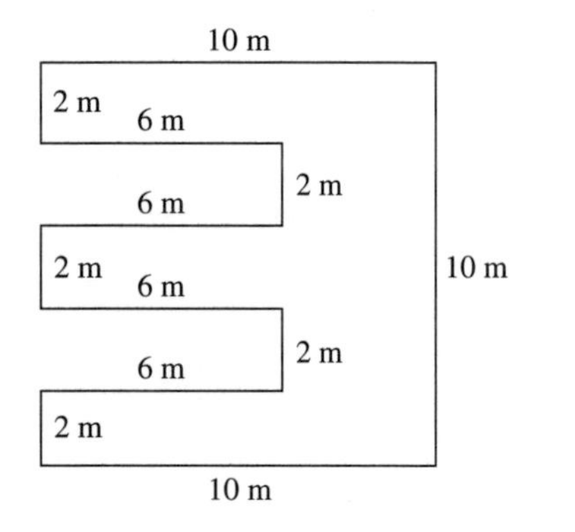

10 m
10 m
10 m
2 m
6 m
2 m
6 m
2 m
6 m
2 m
6 m
2 m
64 m

Find the area of each rectangle or square.

33. Length = 18 in., width = 4 in.
$4 \text{ in.} \times 18 \text{ in.} = 72 \text{ in.}^2$

34. Length = 24 in., width = 7 in.
$7 \text{ in.} \times 24 \text{ in.} = 168 \text{ in.}^2$

35. Length = width = 9.8 ft
$9.8 \text{ ft} \times 9.8 \text{ ft} = 96.04 \text{ ft}^2$

36. Length = width = 7.6 ft
$7.6 \text{ ft} \times 7.6 \text{ ft} = 57.76 \text{ ft}^2$

37. Length = 0.96 m, width = 0.3 m
$0.3 \text{ m} \times 0.96 \text{ m} = 0.288 \text{ m}^2$

38. Length = 0.132 m, width = 0.02 m
$0.02 \text{ m} \times 0.132 \text{ m} = 0.00264 \text{ m}^2$

39. Length = 156 yd, width = 96 yd
$156 \text{ yd} \times 96 \text{ yd} = 14{,}976 \text{ yd}^2$

40. Length = 183 yd, width = 81 yd
$183 \text{ yd} \times 81 \text{ yd} = 14{,}823 \text{ yd}^2$

Find the shaded area.

41.

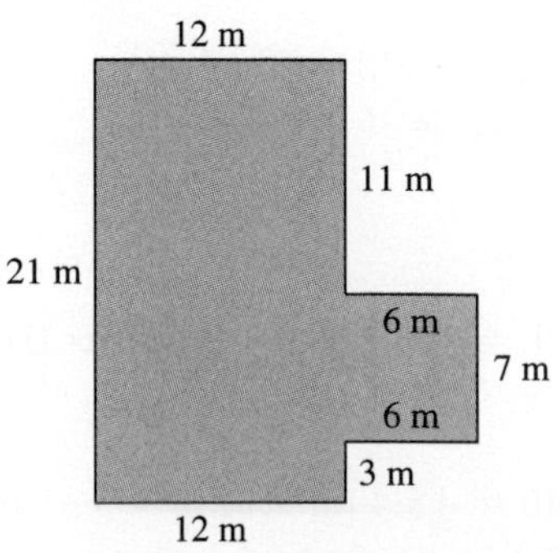

$21 \text{ m} \times 12 \text{ m} + 6 \text{ m} \times 7 \text{ m} = 294 \text{ m}^2$

42.

26 m
10 m
16 m
3 m
3 m
20 m
20 m
52 m

$20 \text{ m} \times 10 \text{ m} + 26 \text{ m} \times 23 \text{ m} + 20 \text{ m} \times 16 \text{ m} = 1118 \text{ m}^2$

Applications

Some of the following problems will require that you find a perimeter. Others will require that you find an area. Read each problem carefully to determine which you are to find.

43. A hotel conference center is building an indoor fitness center area measuring 220 ft × 50 ft. The flooring to cover the space is made of a special 3-layered cushioned tile and costs $12.00 per square foot. How much will it cost for the new flooring? $132,000

44. A beach volleyball area will feature 4 volleyball courts, which will take up an area of 500 ft × 90 ft. To put down an adequate amount of sifted sand, it will cost $0.27 per square foot. How much will it cost to sand the volleyball courts? $12,150

45. The state champions in basketball this year were the Tolland Tigers. The alumni presented them with a giant blown-up photo of the entire team which measures 8.25 ft × 5 feet. A gold metal frame will cost $3.80 per foot. How much will it cost to frame the championship photo? $100.70

46. Sammy's SCUBA Shop is installing a new sign measuring 5.4 ft × 8.1 ft. The sign will be framed in purple neon light, which will cost $32.50 per foot. How much will it cost to frame the sign in purple neon light? $877.50

47. A brand new family water park has built a kiddie pool that looks like a television. The TV pool measures 20.24 meters long and 16.82 meters wide. The pool cover to keep the pool clean at night costs $2.80 per square meter. How much will it cost to purchase the pool cover to keep the pool clean? $953.22

48. A rectangular billboard advertising HARRY'S HEARTHSIDE RESTAURANT measures 32.61 meters long and 23.94 meters wide. The background paint is a "glow-in-the-dark" paint and costs $5.69 per square meter. How much will it cost to paint the "glow-in-the-dark" billboard? $4442.09

A family decides to have custom carpeting installed. It will cost $14.50 per square yard. The binding, which runs along the outside edges of the carpet, will cost $1.50 per yard. Find the cost of carpeting *and* binding *for each room. Note that dimensions are given in feet. (Remember, 1 square yard equals 9 square feet.)*

★49.

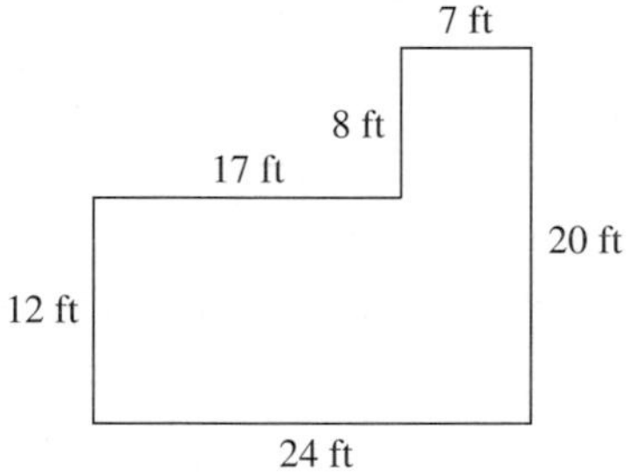

Area = 24 ft × 12 ft + 8 ft × 7 ft = 344 ft²
Perimeter = 12 ft + 17 ft + 8 ft + 7 ft + 20 ft + 24 ft = 88 ft

$$\text{Cost} = 344 \text{ ft}^2 \times \frac{\$14.50}{\text{yd}^2} \times \frac{1 \text{ yd}^2}{9 \text{ ft}^2} + 88 \text{ ft} \times \frac{1 \text{ yd}}{3 \text{ ft}} \times \frac{1.50}{\text{yd}} = \$598.22$$

★50.

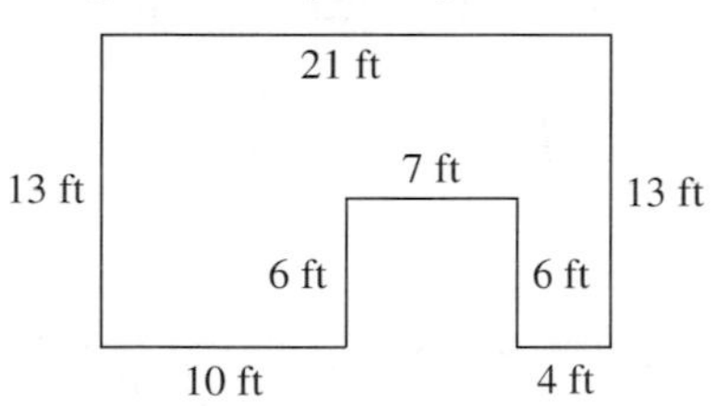

Area = 21 ft × 13 ft − 6 ft × 7 ft = 231 ft²
Perimeter = 2 × 13 ft + 2 × 21 ft + 2 × 6 ft = 80 ft

$$\text{Cost} = 231 \text{ ft}^2 \times \frac{1 \text{ yd}^2}{9 \text{ ft}^2} \times \frac{14.50}{\text{yd}^2} + 80 \text{ ft} \times \frac{1 \text{ yd}}{3 \text{ ft}} \times \frac{1.50}{\text{yd}} = \$412.17$$

Cumulative Review Problems

51. Add.
$$\begin{array}{r} 156.8 \\ 27.2 \\ +\ 39.3 \\ \hline 223.3 \end{array}$$

52. Subtract.
$$\begin{array}{r} 200.57 \\ -\ 193.39 \\ \hline 7.18 \end{array}$$

53. Multiply.
$$\begin{array}{r} 1076 \\ \times\ 20.3 \\ \hline 21{,}842.8 \end{array}$$

54. Divide. $12.3\overline{)19.384}$ 1.58

7.2 Parallelograms and Trapezoids

MathPro

Video 14

SSM

After studying this section, you will be able to:

1 *Find the perimeter and area of a parallelogram.*
2 *Find the perimeter and area of a trapezoid.*

1 *Finding the Perimeter and Area of a Parallelogram*

Parallelograms and trapezoids are figures related to rectangles. Actually, they are in the same ''family,'' the **quadrilaterals** (four-sided figures). For all these figures, the perimeter is the distance around the figure. But each has a different formula for finding area.

TEACHING TIP Students must be clear on the meaning of *parallel lines.* Ask them to define what they mean by parallel lines. Usually, after two or three suggestions they will identify the key property that parallel lines are STRAIGHT LINES that are always the SAME DISTANCE apart.

A **parallelogram** is a four-sided figure in which both pairs of opposite sides are parallel. Parallel lines are two straight lines that are always the same distance apart. The opposite sides of a parallelogram are equal in length.

These figures are parallelograms. Notice that the adjoining sides need not be perpendicular.

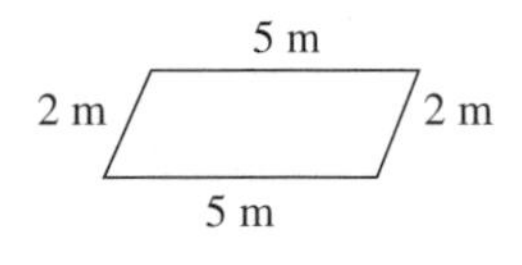

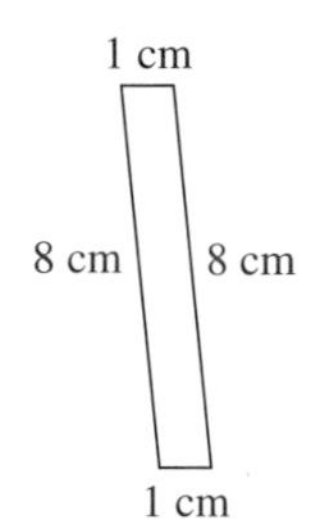

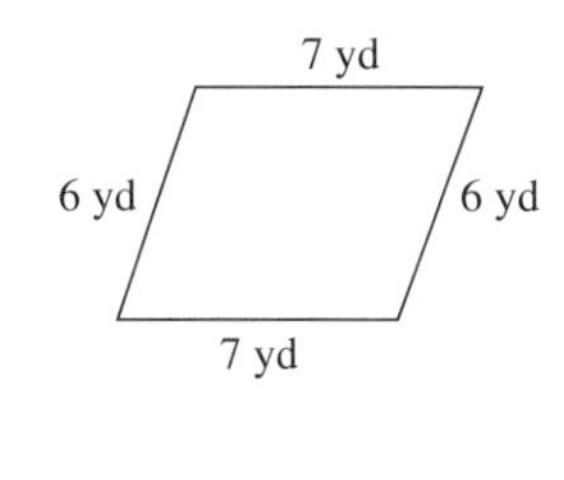

The **perimeter** of a parallelogram is the distance around the parallelogram. It is found by adding the lengths of all the sides of the figure.

EXAMPLE 1 Find the perimeter of the parallelogram.

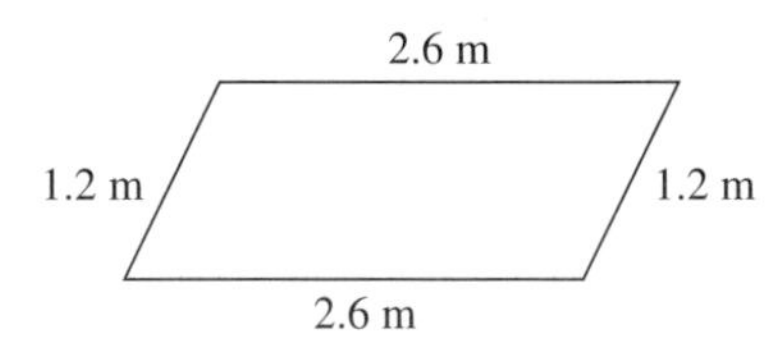

$$P = (2)(1.2) + (2)(2.6)$$
$$= 2.4 + 5.2 = 7.6 \text{ m}$$

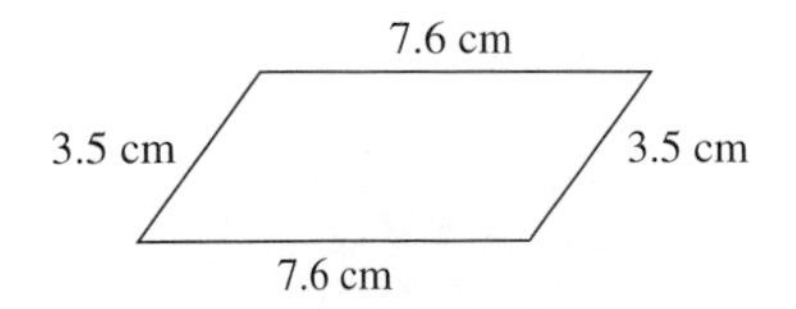

Practice Problem 1 Find the perimeter of the parallelogram in the margin. ■

To find the **area** of a parallelogram, we multiply the base times the height. Any side of a parallelogram can be considered the base. The height is the shortest distance between the base and the side opposite the base. The height is a line segment that is perpendicular to the base. When we write the formula for area, we use the length of the base (b) and the height (h).

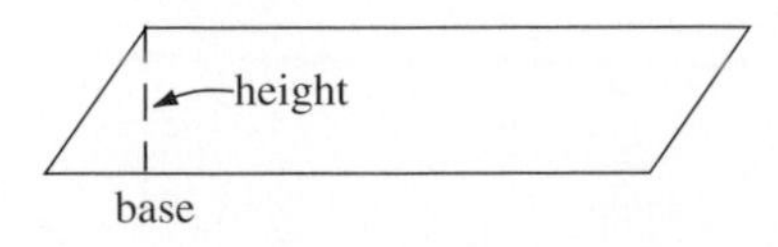

The **area of a parallelogram** is the base (b) times the height (h).

$$A = bh$$

Why is the area of a parallelogram equal to the base times the height? What reasoning leads us to that formula? Suppose that we cut off the triangular region on one side of the parallelogram and move it to the other side.

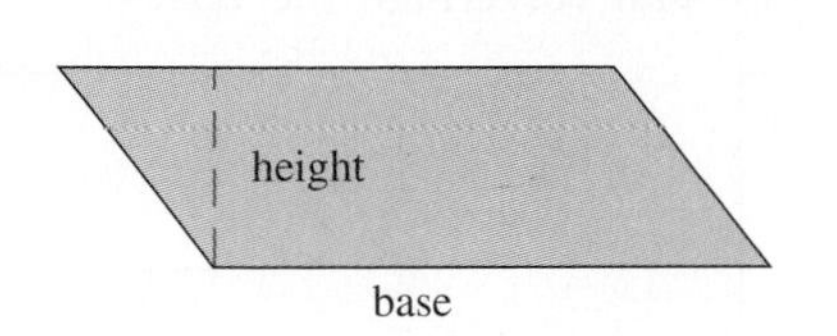

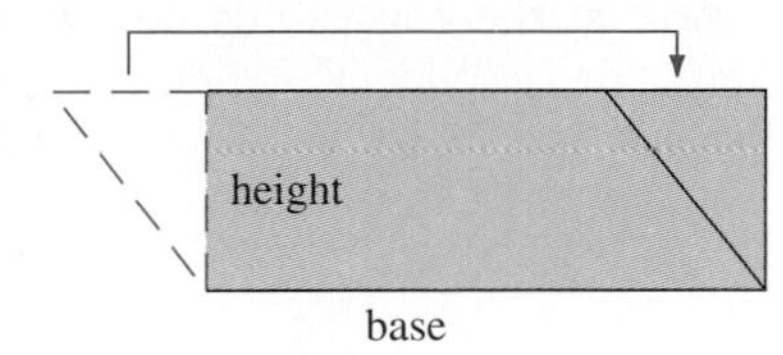

We now have a rectangle.

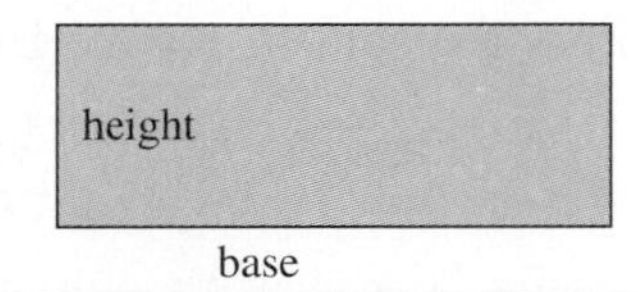

To find the area, we multiply the width by the length. In this case, $A = bh$. Thus finding the area of a parallelogram is like finding the area of a rectangle of length b and width h: $A = bh$. We explore this concept in Exercises 7.2, problems 37 and 38.

EXAMPLE 2 Find the area of the parallelogram with base 7.5 m and height 3.2 m.

$$\begin{aligned} A &= bh \\ &= (7.5 \text{ m})(3.2 \text{ m}) \\ &= 24 \text{ m}^2 \end{aligned}$$

Practice Problem 2 Find the area of a parallelogram with base 10.3 km and height 1.5 km. ■

EXAMPLE 3 Find the perimeter and the area of the parallelogram.

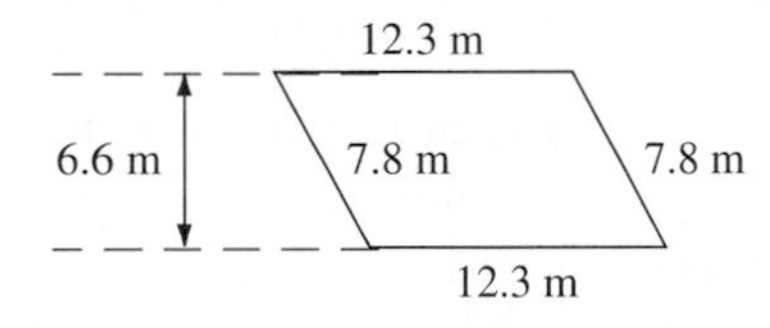

The perimeter is the sum of the lengths of its sides.

$$\begin{aligned} P &= (2)(7.8 \text{ m}) + (2)(12.3 \text{ m}) \\ &= 15.6 \text{ m} + 24.6 \text{ m} = 40.2 \text{ m} \end{aligned}$$

The area formula will use the height, which is labeled in the drawing as 6.6 m. Notice in this parallelogram the height we draw falls outside the figure.

$$\begin{aligned} A &= bh \\ &= (12.3 \text{ m})(6.6 \text{ m}) = 81.18 \text{ m}^2 \end{aligned}$$

Practice Problem 3 Find the perimeter and the area of the parallelogram in the margin. ■

TEACHING TIP Be sure to mention WHY the formula for the area of a parallelogram is logical. This type of reasoning will benefit students throughout this chapter.

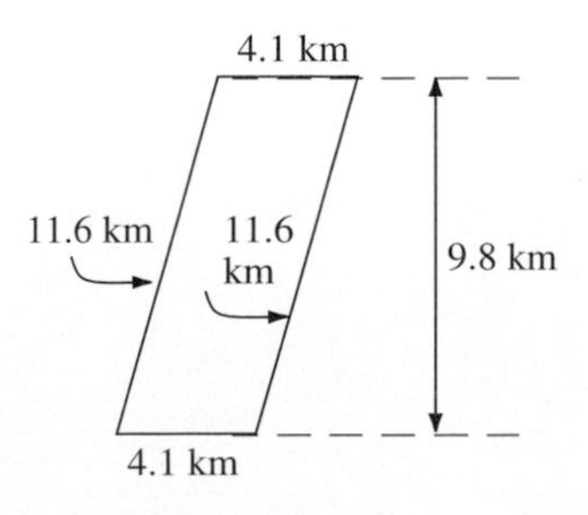

2 *Finding the Perimeter and Area of a Trapezoid*

A trapezoid is a four-sided figure with two parallel sides. The parallel sides are called *bases.* The lengths of the bases do not have to be equal. The adjoining sides do not have to be perpendicular. Sometimes the trapezoid is sitting on a base. Then both bases are horizontal. But be careful. Sometimes the bases are vertical. You can recognize the bases because they are the two parallel sides. This becomes important when you use the formula for finding the area of a trapezoid.

Look at the trapezoids below. See if you can recognize the bases.

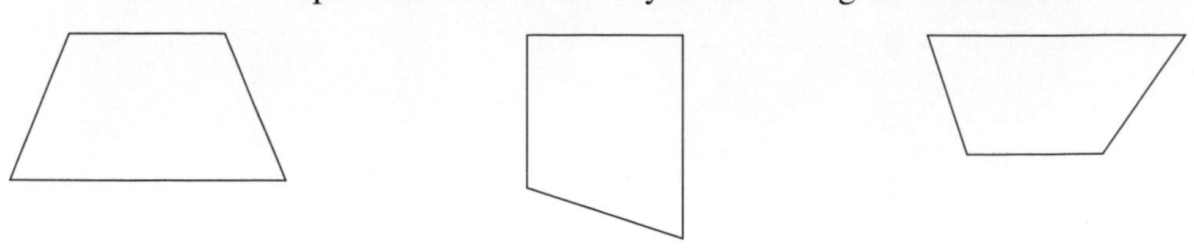

The perimeter of a trapezoid is the sum of the lengths of all of its sides.

EXAMPLE 4 Find the perimeter of the trapezoid on the right.

$$P = 18 \text{ m} + 5 \text{ m} + 12 \text{ m} + 5 \text{ m}$$
$$= 40 \text{ m}$$

12 m
5 m 5 m
18 m

Practice Problem 4 Find the perimeter of a trapezoid with sides of 7 yd, 15 yd, 21 yd, and 13 yd. ■

Remember, we often use parentheses as a way to group numbers together. The numbers inside parentheses should be combined first.

$(5)(7 + 2) = (5)(9)$ *First we add numbers inside the parentheses.*

$= 45$ *Then we multiply.*

The formula for the area of a trapezoid uses parentheses in this way.

The area of a trapezoid is one-half the height times the sum of the bases. (That means you add the bases *first.*)

The **area of a trapezoid** with a shorter base b and a longer base B and height h is

$$A = \frac{h(b + B)}{2}$$

Base = b
height = h
Base = B

EXAMPLE 5 Find the area of the trapezoid on the right.

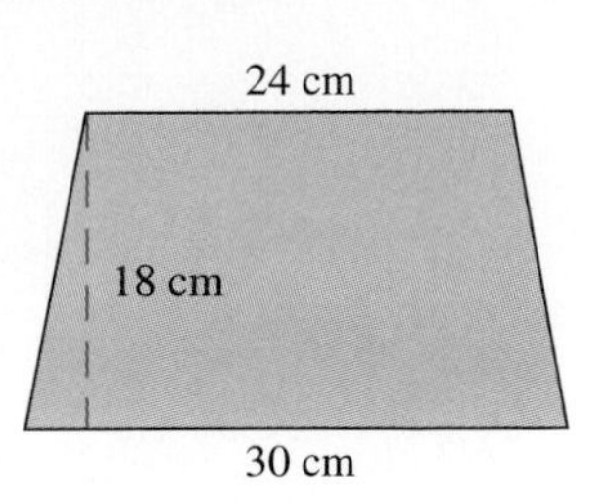

$$A = \frac{h(b + B)}{2}$$

Looking at the diagram, we see that $h = 18$, $b = 24$, and $B = 30$.

$$A = \frac{(18 \text{ cm})(24 \text{ cm} + 30 \text{ cm})}{2}$$

$$= \frac{(18 \text{ cm})(54 \text{ cm})}{2} = \frac{972}{2} \text{ cm}^2 = 486 \text{ cm}^2$$

Practice Problem 5 Find the area of the trapezoid in the margin. ■

1.3 mm
2.6 mm
1.9 mm

EXAMPLE 6 A roadside sign is in the shape of a trapezoid. It has a height of 30 ft and the bases are 60 ft and 75 ft.

(a) What is the area of the sign?

(b) If 1 gallon of paint covers 200 ft^2, how many gallons of paint will be needed to paint the sign?

(a) $A = \frac{h(b + B)}{2}$ where $h = 30$, $b = 60$, and $B = 75$

$$= \frac{(30 \text{ ft})(60 \text{ ft} + 75 \text{ ft})}{2}$$

$$= \frac{(30 \text{ ft})(135 \text{ ft})}{2} = \frac{4050}{2} \text{ ft}^2 = 2025 \text{ ft}^2$$

(b) Each gallon covers 200 ft^2, so we multiply by the fraction $\frac{1 \text{ gal}}{200 \text{ ft}^2}$. This fraction is equivalent to 1.

$$2025 \cancel{\text{ft}^2} \times \frac{1 \text{ gal}}{200 \cancel{\text{ft}^2}} = \frac{2025}{200} \text{ gal}$$

$$= 10.125 \text{ gal}$$

Thus 10.125 gallons of paint would be needed. In real life we would buy 11 gallons of paint to be sure to have enough.

Practice Problem 6 A corner parking lot is shaped like a trapezoid. The trapezoid has a height of 140 yd. The bases measure 180 yd and 130 yd.

(a) Find the area of the parking lot.

(b) If 1 gallon of sealant will cover 100 square yards of the parking lot, how many gallons are needed to cover the entire parking lot? ■

Some area problems involve using two or more separate regions. Remember, areas can be added or subtracted.

EXAMPLE 7 Find the area of the following piece for inlaid woodwork made by a master carpenter. Since this shape is hard to cut, it is made of one trapezoid and one rectangle laid together.

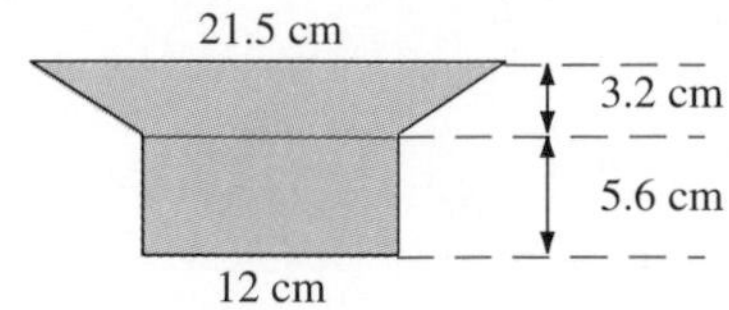

We separate the area into two portions and find the area of each portion separately.

21.5 cm
3.2 mm
12 cm

The area of the trapezoid is

$$A = \frac{h(b + B)}{2}$$

$$= \frac{(3.2\text{ cm})(12\text{ cm} + 21.5\text{ cm})}{2}$$

$$= \frac{(3.2\text{ cm})(33.5\text{ cm})}{2}$$

$$= \frac{107.2}{2}\text{ cm}^2$$

$$= 53.6\text{ cm}^2$$

The area of the rectangle is

$$A = lw$$

$$= (12\text{ cm})(5.6\text{ cm})$$

$$= 67.2\text{ cm}^2$$

We now add each area.

$$\begin{array}{r} 67.2\text{ cm}^2 \\ +\ 53.6\text{ cm}^2 \\ \hline 120.8\text{ cm}^2 \end{array}$$

The total area of the piece for inlaid woodwork is 120.8 cm^2.

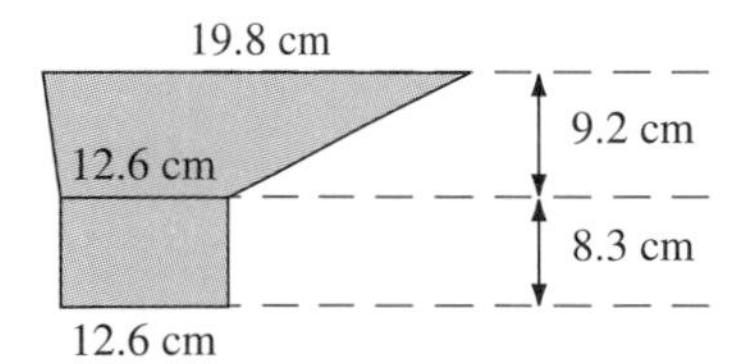

Practice Problem 7 Find the area of the piece for inlaid woodwork shown in the margin. The shape is made of one trapezoid and one rectangle. ■

7.2 Exercises

Verbal and Writing Skills

1. The perimeter of a parallelogram is found by ___adding___ the lengths of all the sides of the figure.

2. To find the area of a parallelogram, multiply the base times the ___height___.

3. The height of a parallelogram is a line segment that is ___perpendicular___ to the base.

4. The area of a trapezoid is one-half the height times the ___sum___ of the bases.

Find the perimeter of each parallelogram.

5. One side measures 2.8 m and a second side measures 17.3 m.
   $2 \times 17.3 \text{ m} + 2 \times 2.8 \text{ m} = 40.2 \text{ m}$

6. One side measures 4.6 m and a second side measures 20.5 m.
   $2 \times 4.6 \text{ m} + 2 \times 20.5 \text{ m} = 50.2 \text{ m}$

7.
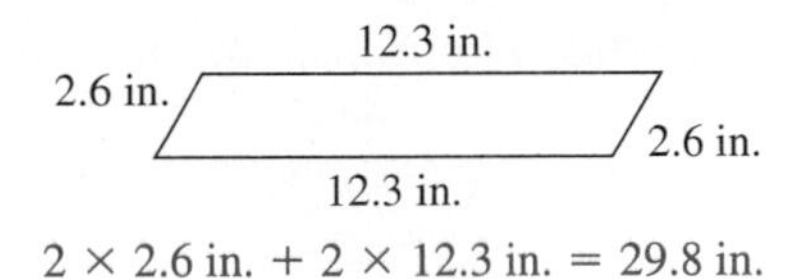

$2 \times 2.6 \text{ in.} + 2 \times 12.3 \text{ in.} = 29.8 \text{ in.}$

8.
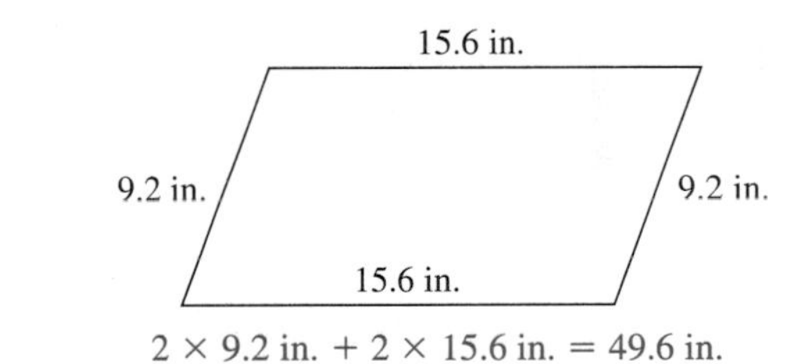

$2 \times 9.2 \text{ in.} + 2 \times 15.6 \text{ in.} = 49.6 \text{ in.}$

Find the area of each parallelogram.

9. The base is 25 m and the height is 42.3 m.
   $25 \text{ m} \times 42.3 \text{ m} = 1057.5 \text{ m}^2$

10. The base is 46 m and the height is 12.3 m.
    $46 \text{ m} \times 12.3 \text{ m} = 565.8 \text{ m}^2$

11. The base is 20.5 m and the height is 21.5 m.
    $20.5 \text{ cm} \times 21.5 \text{ cm} = 440.75 \text{ cm}^2$

12. The base is 36 m and the height is 38.5 m.
    $36 \text{ m} \times 38.5 \text{ m} = 1386 \text{ m}^2$

13. A courtyard is shaped like a parallelogram. Its base is 126 yd and its height is 28 yd. Find its area. $126 \text{ yd} \times 28 \text{ yd} = 3528 \text{ yd}^2$

14. The preferred seating area at the South Shore Music Theatre is in the shape of a parallelogram. Its base is 28 yd and its height is 21.5 yd. Find the area. 602 square yards

Find the perimeter of each trapezoid.

15.
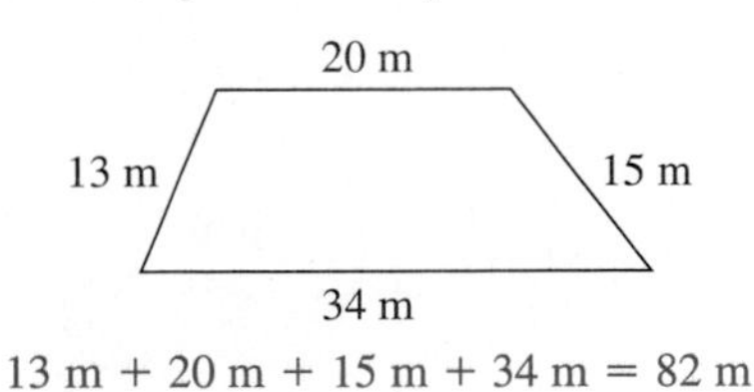

$13 \text{ m} + 20 \text{ m} + 15 \text{ m} + 34 \text{ m} = 82 \text{ m}$

16.
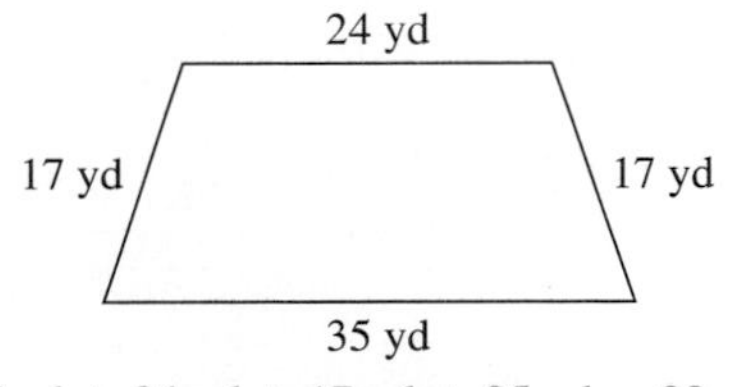

$17 \text{ yd} + 24 \text{ yd} + 17 \text{ yd} + 35 \text{ yd} = 93 \text{ yd}$

17. The four sides are 130 cm, 100 cm, 70 cm, and 80 cm. $80 \text{ cm} + 130 \text{ cm} + 100 \text{ cm} + 70 \text{ cm} = 380 \text{ cm}$

18. The four sides are 160 cm, 65 cm, 185 cm, and 60 cm. $60 \text{ cm} + 160 \text{ cm} + 65 \text{ cm} + 185 \text{ cm} = 470 \text{ cm}$

Find the area of each trapezoid.

19. The height is 7 m and the bases are 3 m and 9 m.

$A = \frac{1}{2} \times 7\text{ m} \times (3\text{ m} + 9\text{ m}) = 42\text{ m}^2$

20. The height is 10 m and the bases are 6 m and 9 m.

$A = \frac{1}{2} \times 10\text{ m} \times (6\text{ m} + 9\text{ m}) = 75\text{ m}^2$

21. The height is 4 cm and the bases are 14 cm and 29 cm.

$A = \frac{1}{2} \times 4\text{ cm} \times (14\text{ cm} + 29\text{ cm}) = 86\text{ cm}^2$

22. The height is 2 cm and the bases are 26 cm and 31 cm.

$A = \frac{1}{2} \times 2\text{ cm} \times (26\text{ cm} + 31\text{ cm}) = 57\text{ cm}^2$

23. The height is 16 yd and the bases are 15 yd and 28 yd.

$A = \frac{1}{2} \times 16\text{ yd} \times (28\text{ yd} + 15\text{ yd}) = 344\text{ yd}^2$

24. The height is 22 yd and the bases are 13 yd and 29 yd.

$A = \frac{1}{2} \times 22\text{ yd} \times (29\text{ yd} + 13\text{ yd}) = 462\text{ yd}^2$

25. A provincial park in Canada is laid out in the shape of a trapezoid. The trapezoid has a height of 20 km. The bases are 24 km and 31 km. Find the area of the park.

$A = \frac{1}{2} \times 20\text{ km} \times (24\text{ km} + 31\text{ km}) = 550\text{ km}^2$

26. An underwater diving area for snorklers and Scuba divers in Key West, Florida, is designated by buoys and ropes, making the diving section into the shape of a trapezoid on the surface of the water. The trapezoid has a height of 265 meters. The bases are 300 meters and 280 meters. Find the area of the designated diving area.

76,850 square meters

Find the areas for the following shapes, which are made of trapezoids, parallelograms, squares, and rectangles.

27.

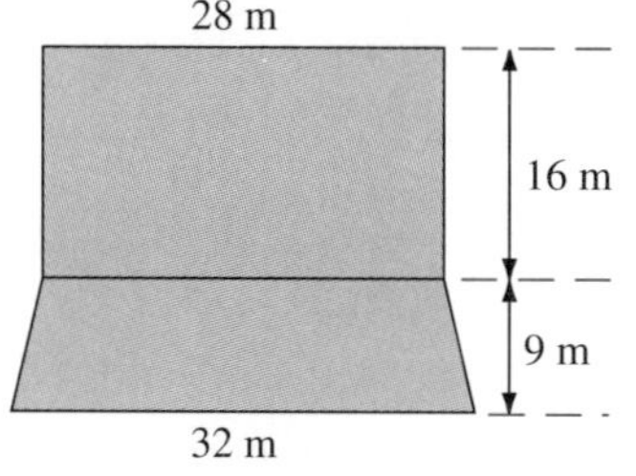

$A = 28\text{ m} \times 16\text{ m} + \frac{1}{2} \times 9\text{ m} \times (28\text{ m} + 32\text{ m}) = 718\text{ m}^2$

28.

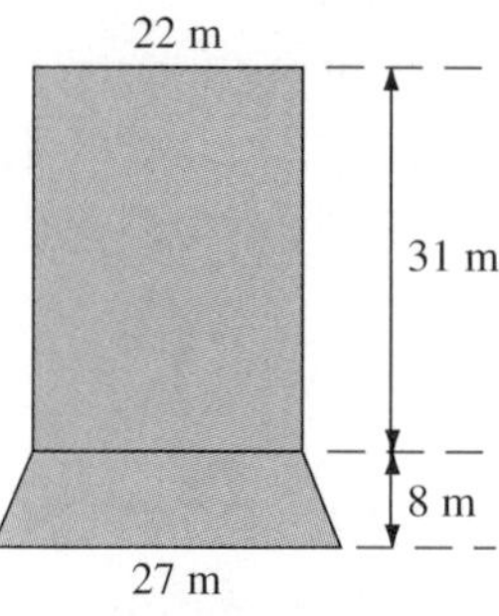

$A = 22\text{ m} \times 31\text{ m} + \frac{1}{2} \times 8\text{ m} \times (22\text{ m} + 27\text{ m}) = 878\text{ m}^2$

29.

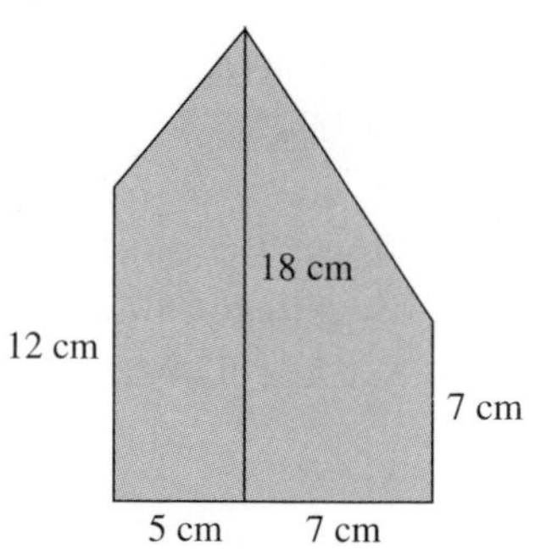

$A = \frac{1}{2} \times 5\text{ cm} \times (12\text{ cm} + 18\text{ cm}) + \frac{1}{2} \times 7\text{ cm} \times (18\text{ cm} + 7\text{ cm}) = 162.5\text{ cm}^2$

30.

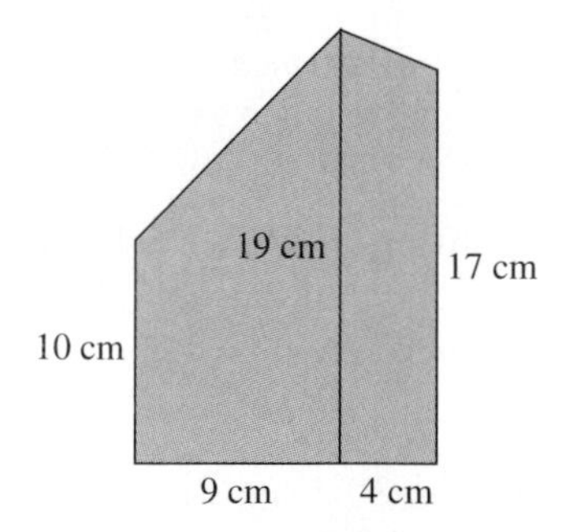

$A = \frac{1}{2} \times 9\text{ cm} \times (10\text{ cm} + 19\text{ cm}) + \frac{1}{2} \times 4\text{ cm} \times (19\text{ cm} + 17\text{ cm}) = 202.5\text{ cm}^2$

31.

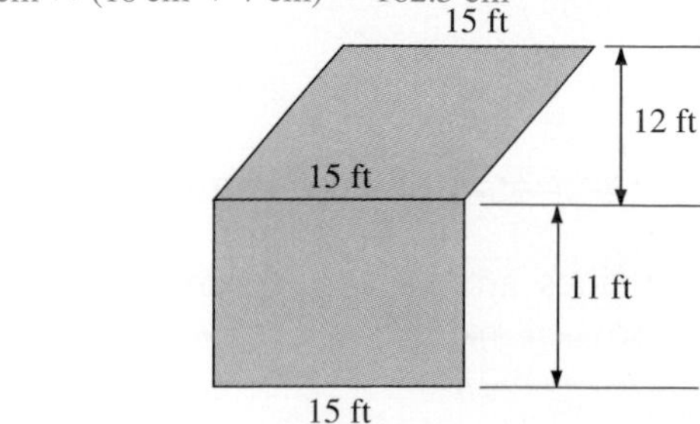

$A = 12\text{ ft} \times 15\text{ ft} + 15\text{ ft} \times 11\text{ ft} = 345\text{ ft}^2$

32.

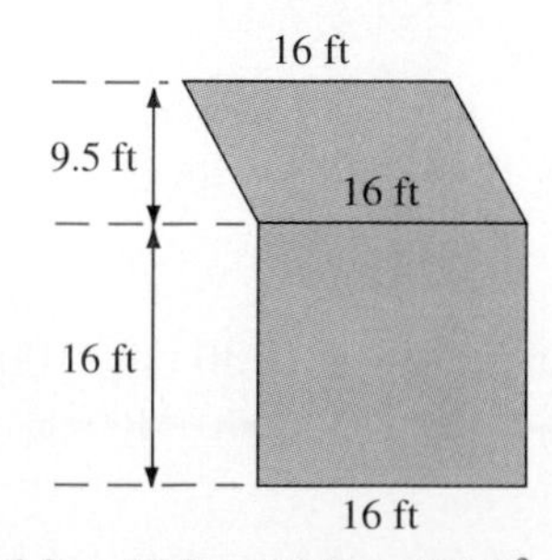

$A = 9.5\text{ ft} \times 16\text{ ft} + 16\text{ ft} \times 16\text{ ft} = 408\text{ ft}^2$

The following shapes represent the lobby of a conference center. The lobby will be carpeted at a cost of $22.00 per square yard. How much will the carpeting cost?

33.

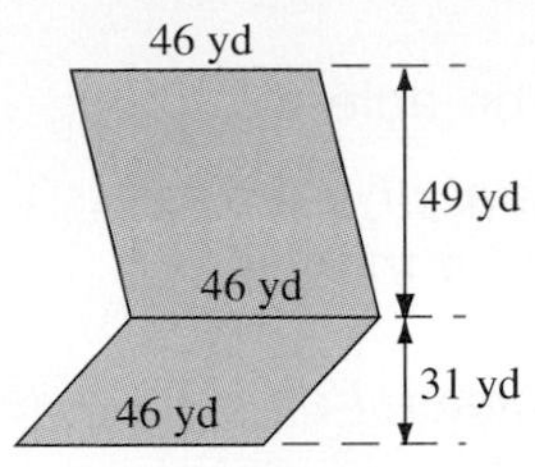

Area = 46 yd × 49 yd + 46 yd × 31 yd = 3680 yd^2

Cost = 3680 yd^2 × $\frac{22.00}{\text{yd}^2}$ = $80,960.00

34.

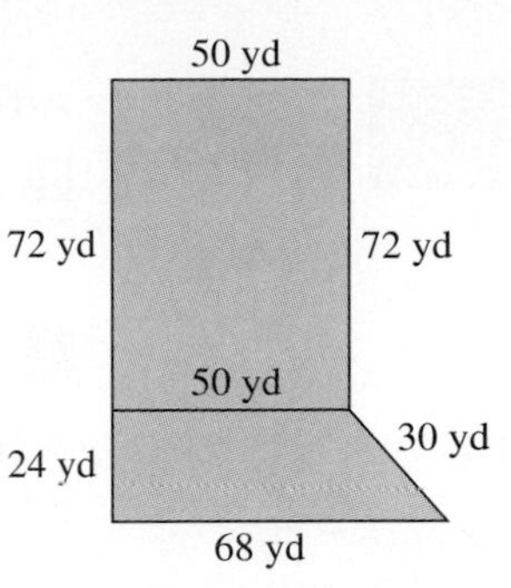

Area = 72 yd × 50 yd + $\frac{1}{2}$ × 24 yd × (50 yd + 68 yd) = 5016 yd^2

Cost = 5016 yd^2 × $\frac{22.00}{\text{yd}^2}$ = $110,352.00

35. A room with a cathedral ceiling has a skylight window in the shape of a trapezoid. Find the area of this trapezoid, whose height is 34.569 in. and whose bases are 17.398 in. and 22.782 in.

$A = \frac{1}{2}$(34.569 in.)(17.398 in. + 22.782 in.) = 694.49 in.^{2} (rounded)

36. Cape Cod Community College has a student union with a cafeteria in the shape of a parallelogram. Find the area of the parallelogram, whose height is 4582.3 cm and whose base is 9946.5 cm.

A = (4582.3 cm)(9946.5 cm) = 45,577,847 cm^2 (rounded)

To Think About

37. See if you can find a formula that would give the area of a regular octagon. (A regular octagon is an eight-sided figure with all sides equal.) The dimensions of the rectangles and trapezoids are labeled on the sketch.

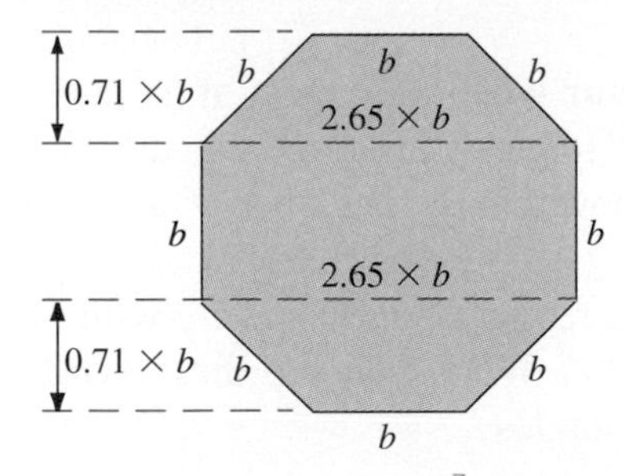

$2 \times \left[\frac{1}{2} \times 0.71 \times b \times (b + 2.65 \times b)\right] + 2.65 \times b \times b$

Area = 5.2415 × b^2 sq units

38. See if you can find a formula for the area of a regular hexagon. Each side is b units long. (A regular hexagon is a six-sided figure with all sides equal.)

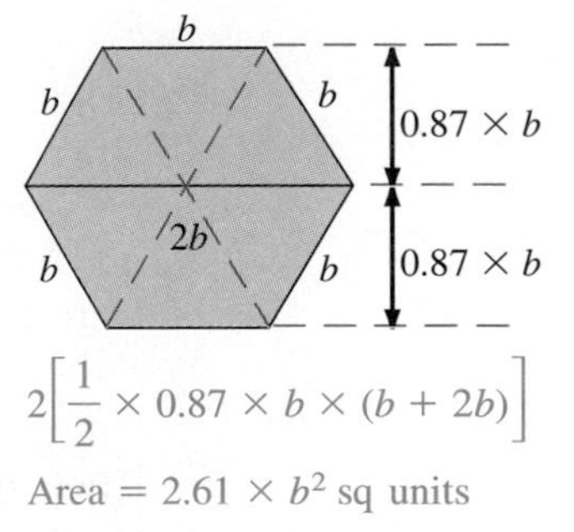

$2\left[\frac{1}{2} \times 0.87 \times b \times (b + 2b)\right]$

Area = 2.61 × b^2 sq units

Cumulative Review Problems

Complete each conversion.

39. 10 yd = 30 ft **40.** 15,840 ft = 3 mi **41.** 18 m = 1800 cm **42.** 26 mm = 2.6 cm

7.3 Angles and Triangles

MathPro

Video 15

SSM

After studying this section, you will be able to:

1. *Understand angles in a quadrilateral or a triangle.*
2. *Find the perimeter and the area of a triangle.*

1 *Understanding Angles in a Quadrilateral or a Triangle*

TEACHING TIP The word *angle* comes from the Latin word *angulus,* which means "corner." The idea that the number of degrees in a complete circle is 360 degrees is attributed by some scholars to ancient civilizations that believed that 60 was a "perfect number" and that 6 of these 60s made a "perfect circle."

A **line** ⟷ extends indefinitely, but a portion of a line, called a **line segment,** has a beginning and an end. An **angle** is formed whenever two lines meet. The two line segments are called the **sides** of the angle. The point at which they meet is called the **vertex** of the angle.

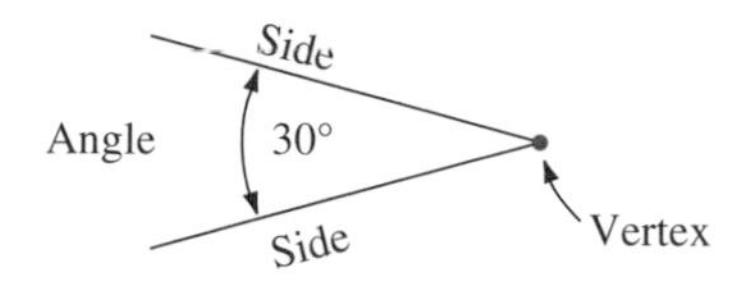

The "amount of opening" of an angle can be measured. Angles are commonly measured in degrees. In this sketch the angle measures 30 degrees, or 30°. The symbol ° indicates degrees. If you fix one side of an angle and keep moving the other side, the angle measure will get larger and larger until eventually you have gone around in one complete revolution.

One complete revolution is 360°.

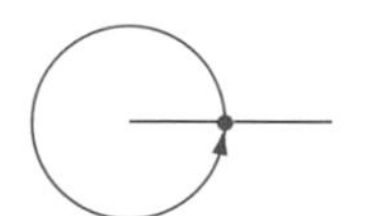

One-half revolution is 180°.

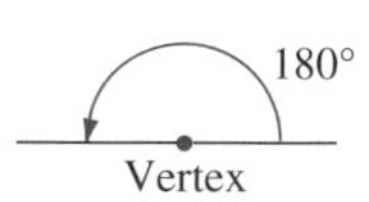

One-fourth revolution is 90°.

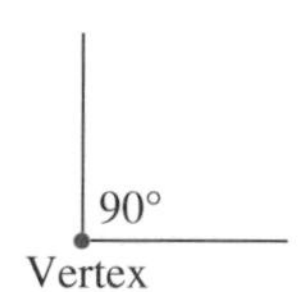

We call two lines **perpendicular** when they meet at an angle of 90°. A 90° angle is called a **right angle.** A 90° angle is often indicated by a small □ at the vertex. Thus when you see ∟ you know that the angle is 90° and also that the sides are perpendicular to each other.

We often label each vertex of a rectangle and a square with the symbol □ to indicate that each angle measures 90°. When you see these symbols, you know that the quadrilaterals are both rectangles.

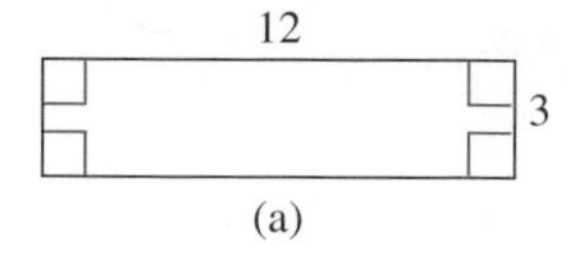

(a)

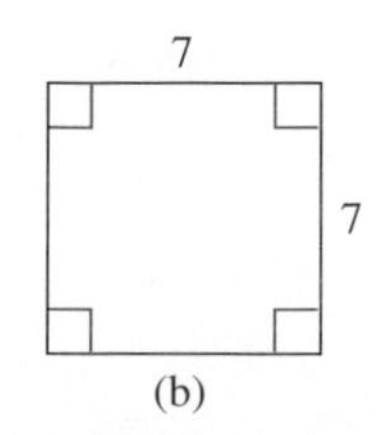

(b)

Recall that the adjacent sides of a rectangle form right angles. Rectangle (b) is also a square. Note that the length of each adjacent side is 7. We can assume that the length of each side is 7. Recall that a square is a rectangle all of whose sides are equal.

The height of any figure is perpendicular to the base. When you see a small □ in a drawing, you know that the line is the height because the line forms a 90° angle. One line is the height and the other line is the base. Look at the drawings of the parallelogram and of the trapezoid below. Identify the height and the base in each figure.

TEACHING TIP It is important to emphasize the use of the small square to indicate that two sides meet at exactly 90 degrees, or a right angle.

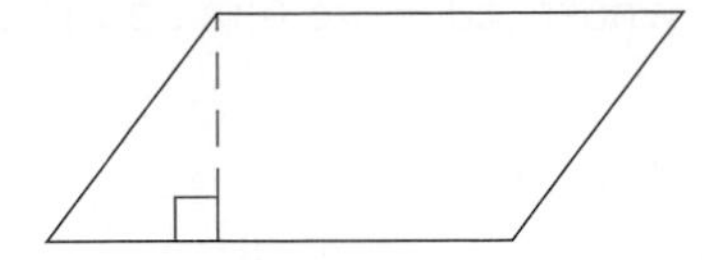
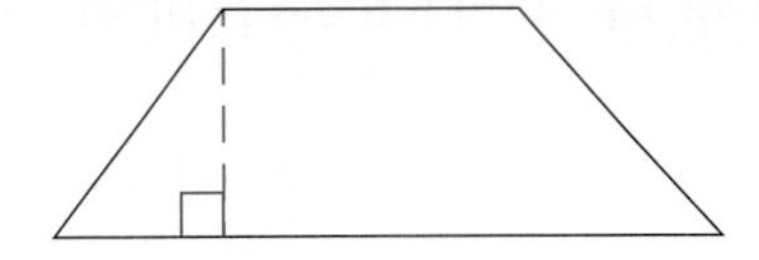

A triangle is a three-sided figure with three angles. The prefix *tri* means "three." Some triangles are shown below.

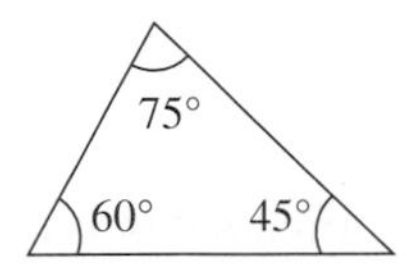

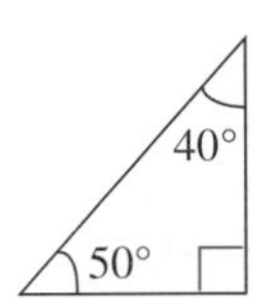

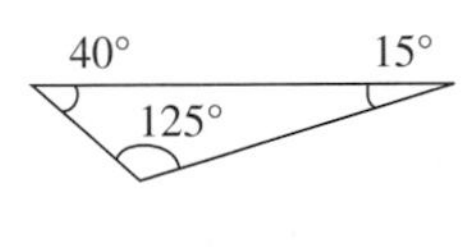

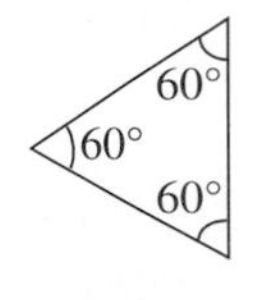

Although all triangles have three sides, not all triangles are shaped the same. The shape of a triangle depends on the size of the angles and the length of the sides.

We will begin our study of triangles by looking at the angles. Although the size of the angles in triangles may be different, the sum of the angle measures of any triangle is always 180°.

> The sum of the measures of the angles in a triangle is 180°.

We can use this fact to find the measure of an unknown angle in a triangle if we know the measures of the other two angles.

EXAMPLE 1 In the triangle to the right, angle A measures 35° and angle B measures 95°. Find the measure of angle C.

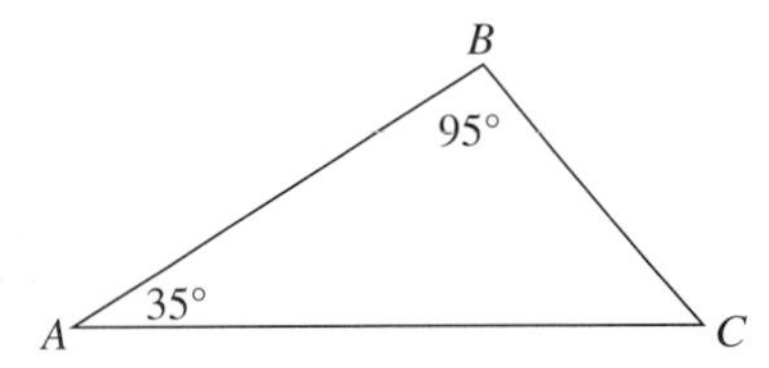

We will use the fact that the sum of the measures of the angles of a triangle is 180°.

$$35 + 95 + x = 180$$
$$130 + x = 180$$

What number x when added to 130 equals 180? Since $130 + 50 = 180$, x must equal 50.

Angle C must equal 50°.

Practice Problem 1 In a triangle, angle B measures 125° and angle C measures 15°. What is the measure of angle A? ■

2 *Finding the Perimeter and the Area of a Triangle*

Recall that the perimeter of any figure is the sum of the lengths of its sides. Thus the perimeter of a triangle is the sum of the lengths of its three sides.

TEACHING TIP Emphasize to students that they must learn the terms *isosceles, equilateral,* and *right triangle.*

EXAMPLE 2 Find the perimeter of a triangle whose sides are 6 in., 5 in., and 8 in.

$$P = 6 \text{ in.} + 5 \text{ in.} + 8 \text{ in.} = 19 \text{ in.}$$

Practice Problem 2 Find the perimeter of a triangle whose sides are 10.5 m, 10.5 m, and 8.5 m. ■

Some triangles have special names. A triangle with two equal sides is called an **isosceles triangle.**

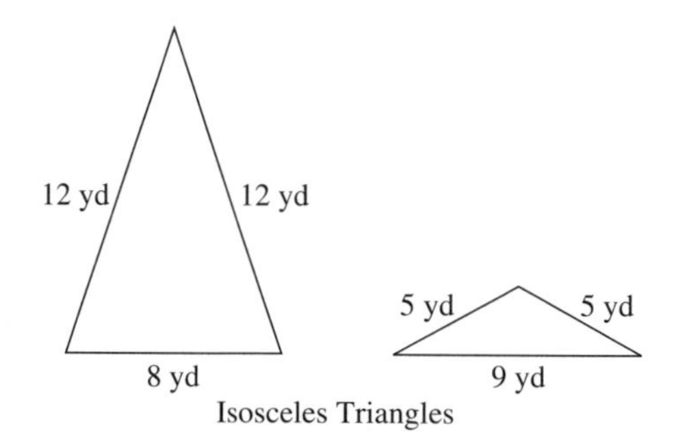

Isosceles Triangles

A triangle with three equal sides is called an **equilateral triangle.** All angles in an equailateral triangle are exactly 60°.

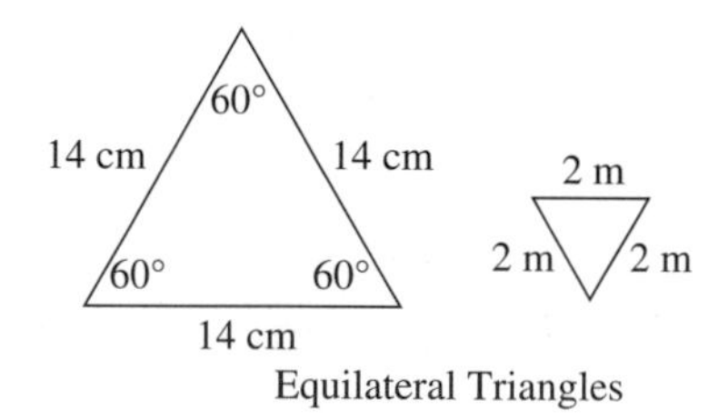

Equilateral Triangles

EXAMPLE 3 Find the perimeter of an equilateral triangle one of whose sides is 12 in.

We know that all the sides of an equilateral triangle are equal. Thus every side measures 12 inches. To find the perimeter, we can add 12 inches three times or we can multiply 3 times 12 inches.

$$P = 12 \text{ in.} + 12 \text{ in.} + 12 \text{ in.} = 36 \text{ in.} \qquad \text{or} \qquad P = (3)(12 \text{ in.}) = 36 \text{ in.}$$

The perimeter of the triangle is 36 in.

Practice Problem 3 The perimeter of an equilateral triangle is 30 cm. Find the length of each side. ■

A triangle with one 90° angle is called a ***right triangle.***

The *height* of any triangle is the distance of a line drawn from a vertex perpendicular to the other side or an extension of the other side. The height may be one of the sides in a right triangle. The height may reach to an extension of one side if one angle of the triangle is greater than 90°. The *base* is perpendicular to the height.

TEACHING TIP Students sometimes have trouble visualizing the height of a triangle if the height lies outside the triangle. Explain that the height lies outside of the triangle if one angle of the triangle is greater than 90°. Point out that height is always the perpendicular line and will be marked with a small square. The base will be the side the height is perpendicular to. If the height is outside the triangle, the base is only that part of the perpendicular that is the side of the triangle and does not include the extension.

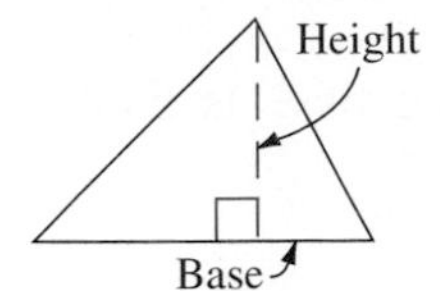

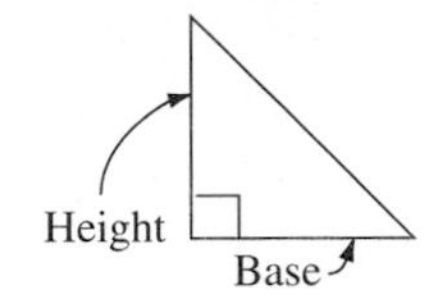

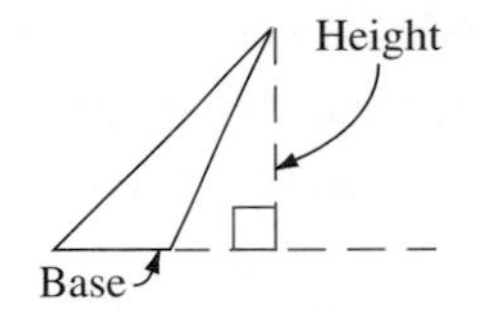

To find the area of a triangle, we need to be able to identify the height and the base of the triangle. The area of any triangle is half of the product of the base times the height of the triangle. The height is measured from the vertex above the base to that base.

The **area of a triangle** is the base times the height divided by 2.

$$A = \frac{bh}{2}$$

h = height

b = base

Where does the 2 come from in the formula $A = \frac{bh}{2}$? Why does this formula for the area of a triangle work? Suppose that we construct a triangle with base b and height h.

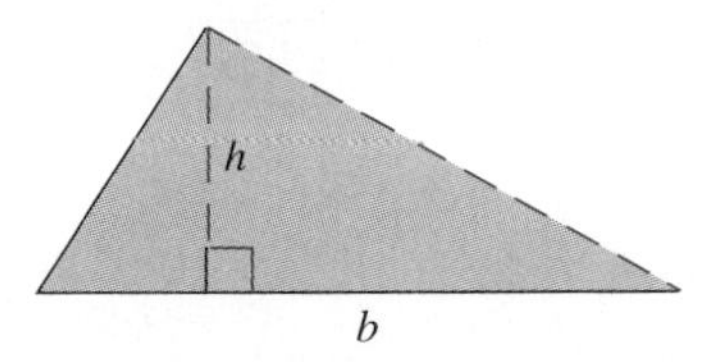

Now let us make an exact copy of the triangle and turn the copy around to the right exactly 180°. Carefully place the two triangles together. We now have a parallelogram of base b and height h. The area of a parallelogram is $A = bh$.

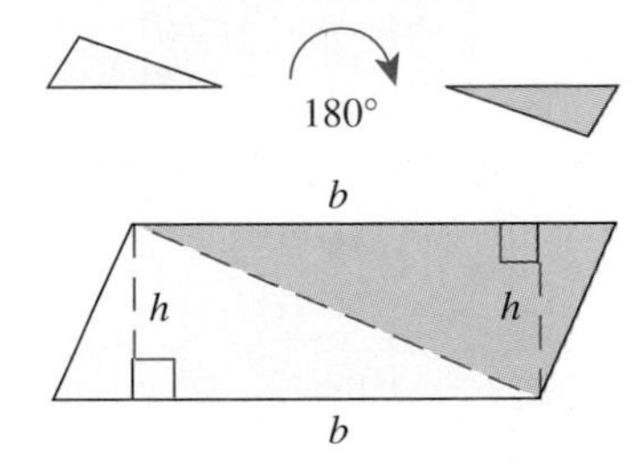

Because the parallelogram has area $A = bh$ and is made up of two triangles of identical shape and area, the area of one of the triangles would be the area of the parallelogram divided by 2. Thus the area of a triangle is $A = \frac{bh}{2}$.

EXAMPLE 4 Find the area of the triangle.

$$A = \frac{bh}{2} = \frac{(23\text{ m})(16\text{ m})}{2} = \frac{368\text{ m}^2}{2} = 184\text{ m}^2$$

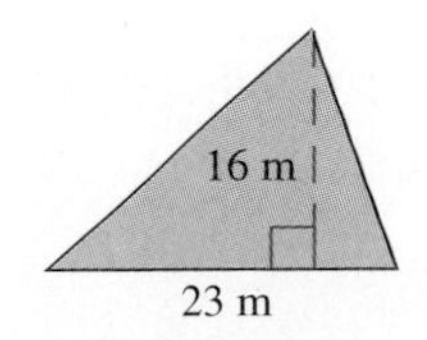

Practice Problem 4 Find the area of the triangle in the margin. ■

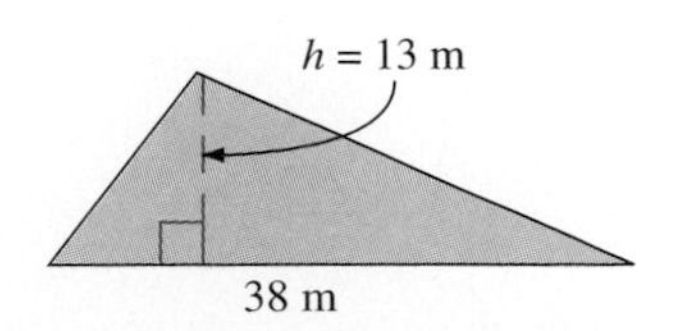

EXAMPLE 5 Find the area of a triangle whose base is 18 mm and whose height is 15 mm.

$$A = \frac{bh}{2} = \frac{(18 \text{ mm})(15 \text{ mm})}{2} = \frac{270 \text{ mm}^2}{2} = 135 \text{ mm}^2$$

Practice Problem 5 Find the area of a triangle whose base is 5 cm and whose height is 16 cm. ■

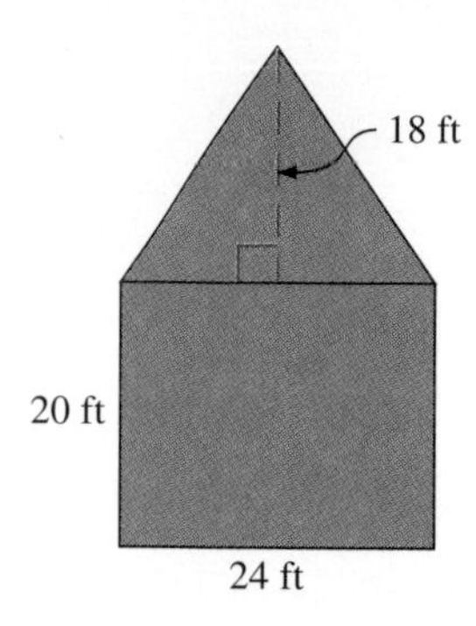

In some geometric shapes, a triangle is combined with rectangles, squares, parallelograms, and trapezoids.

EXAMPLE 6 Find the area of the side of the house shown in the margin.

Because opposite sides of a rectangle are equal, the triangle has a base of 24 ft. Thus

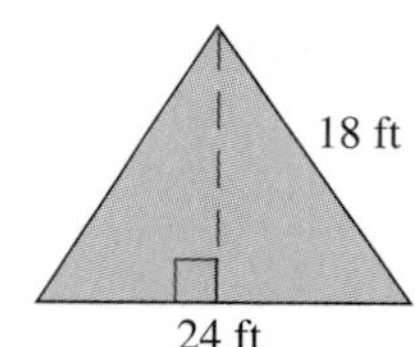

$$A = \frac{bh}{2} = \frac{(24 \text{ ft})(18 \text{ ft})}{2} = \frac{432 \text{ ft}^2}{2} = 216 \text{ ft}^2$$

The area of the rectangle is $A = lw = (24 \text{ ft})(20 \text{ ft}) = 480 \text{ ft}^2$.

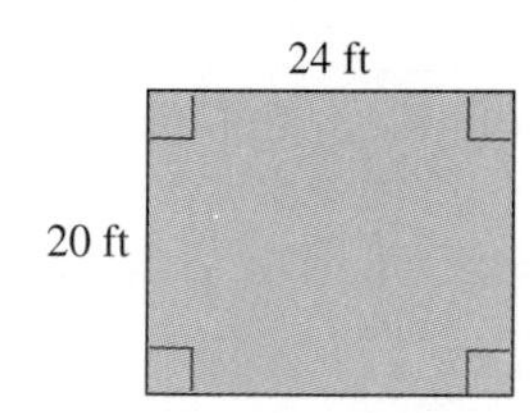

The sum of the two areas is

$$\begin{array}{r} 216 \text{ ft}^2 \\ + \ 480 \text{ ft}^2 \\ \hline 696 \text{ ft}^2 \end{array}$$

Practice Problem 6 Find the area. ■

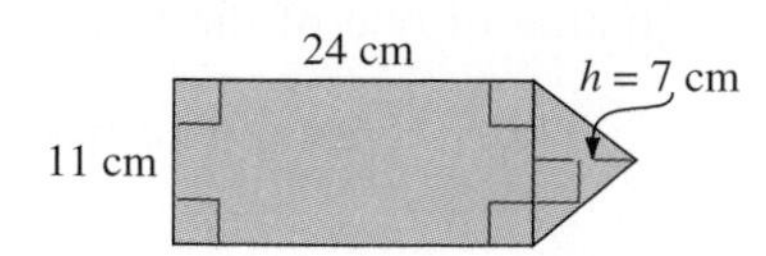

7.3 Exercises

Verbal and Writing Skills

1. A 90° angle is called a ____right____ angle.

2. The sum of the angle measures of a triangle is ____180°____.

3. Explain in your own words how you would find the measure of an unknown angle in a triangle if you knew the measures of the other two angles. Answers will vary. Check students' response for understanding of the procedure.

4. If you were told that a triangle was an isosceles triangle, what could you conclude about the sides of that triangle? You could conclude that two of the sides of the triangle are equal.

5. If you were told that a triangle was an equilateral triangle, what could you conclude about the sides of that triangle? You could conclude that all three sides of the triangle are equal.

6. How do you find the area of a triangle? The area of a triangle is the base times the height divided by 2.

Write true *or* false *for each statement.*

7. Two lines that meet at a 90° angle are perpendicular. True

8. A right triangle has two angles of 90°. False

9. The sum of the angles of a triangle is 180°. True

10. An isosceles triangle has two equal sides. True

11. An equilateral triangle has one angle greater than 90°. False

12. All equilateral triangles are the same size. False

Find the missing angle in each triangle.

13. Two angles are 30° and 90°.
$30° + 90° = 120°$ $180° - 120° = 60°$

14. Two angles are 90° and 45°.
$90° + 45° = 135°$ $180° - 135° = 45°$

15. Two angles are 130° and 20°.
$130° + 20° = 150°$ $180° - 150° = 30°$

16. Two angles are 16° and 18°.
$16° + 18° = 34°$ $180° - 34° = 146°$

17. Two angles are both 45°.
$45° + 45° = 90°$ $180° - 90° = 90°$

18. Two angles are both 60°.
$60° + 60° = 120°$ $180° - 120° = 60°$

19. Two angles are 44° and 8°.
$44° + 8° = 52°$ $180° - 52° = 128°$

20. Two angles are 136° and 18°.
$136° + 18° = 154°$ $180° - 154° = 26°$

Find the perimeter of each triangle.

21. A triangle whose sides are 36 m, 27 m, and 41 m.
$P = 36 \text{ m} + 27 \text{ m} + 41 \text{ m} = 104 \text{ m}$

22. A triangle whose sides are 71 m, 65 m, and 82 m.
$P = 71 \text{ m} + 65 \text{ m} + 82 \text{ m} = 218 \text{ m}$

23. An *isosceles* triangle whose sides are 50 in., 40 in., and 40 in.
$P = 50 \text{ in.} + 40 \text{ in.} + 40 \text{ in.} = 130 \text{ in.}$

24. An *isosceles* triangle whose sides are 36 in., 36 in., and 29 in.
$P = 36 \text{ in.} + 36 \text{ in.} + 29 \text{ in.} = 101 \text{ in.}$

25. An *equilateral* triangle whose side measures 3.5 mi.
$P = 3(3.5 \text{ mi}) = 10.5 \text{ mi}$

26. An *equilateral* triangle whose side measures 4.6 mi.
$P = 3(4.6 \text{ mi}) = 13.8 \text{ mi}$

27. A triangle whose sides are 8.5 in., 7.5 in., and 9.5 in.
$P = 8.5 \text{ in.} + 7.5 \text{ in.} + 9.5 \text{ in.} = 25.5 \text{ in.}$

28. A triangle whose sides are 6.5 in., 10.0 in., and 12.3 in.
$P = 6.5 \text{ in.} + 10.0 \text{ in.} + 12.3 \text{ in.} = 28.8 \text{ in.}$

Find the area of each triangle.

29.

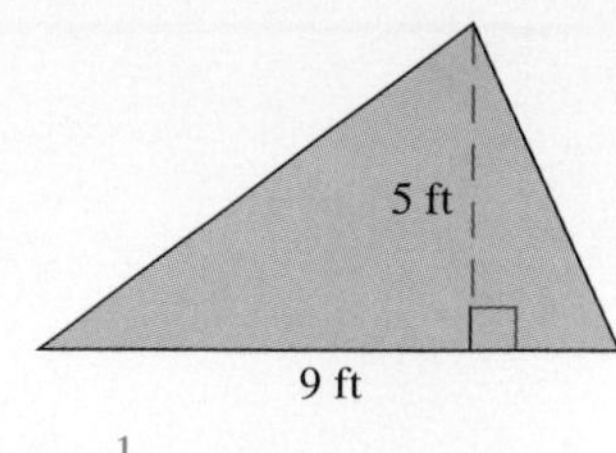

$A = \frac{1}{2} \times 9 \text{ ft} \times 5 \text{ ft} = 22.5 \text{ ft}^2$

30.

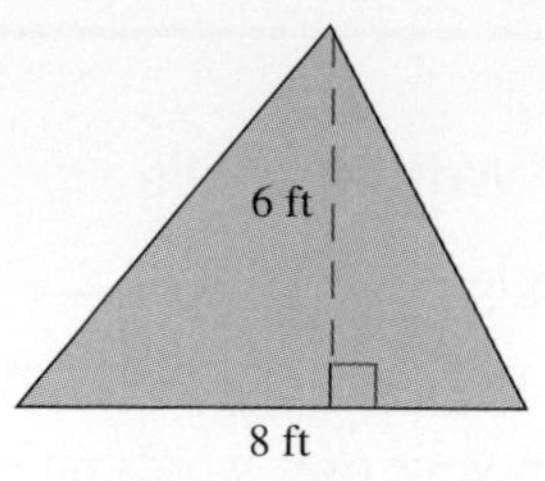

$A = \frac{1}{2} \times 8 \text{ ft} \times 6 \text{ ft} = 24 \text{ ft}^2$

31.

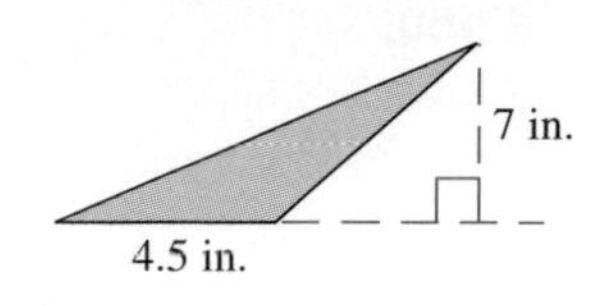

$A = \frac{1}{2} \times 4.5 \text{ in.} \times 7 \text{ in.} = 15.75 \text{ in.}^2$

32.

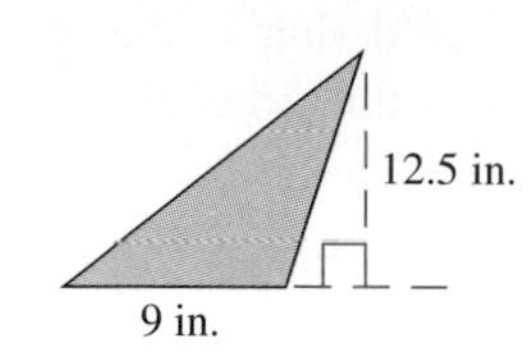

$A = \frac{1}{2} \times 9 \text{ in.} \times 12.5 \text{ in.} = 56.25 \text{ in.}^2$

33. The base is 17.5 cm and the height is 9.5 cm.

$A = \frac{1}{2} \times 9.5 \text{ cm} \times 17.5 \text{ cm} = 83.125 \text{ cm}^2$

34. The base is 3.6 cm and the height is 11.2 cm.

$A = \frac{1}{2} \times 3.6 \text{ cm} \times 11.2 \text{ cm} = 20.16 \text{ cm}^2$

35. The base is 6.7 m and the height is 4.2 m.

$A = \frac{1}{2} \times 6.7 \text{ m} \times 4.2 \text{ m} = 14.07 \text{ m}^2$

36. The base is 8.5 m and the height is 3.6 m.

$A = \frac{1}{2} \times 8.5 \text{ m} \times 3.6 \text{ m} = 15.3 \text{ m}^2$

37. The base is 3.5 yd and the height is 7 yd.

$A = \frac{1}{2} \times 7 \text{ yd} \times 3.5 \text{ yd} = 12.25 \text{ yd}^2$

38. The base is 5.6 yd and the height is 4.8 yd.

$A = \frac{1}{2} \times 5.6 \text{ yd} \times 4.8 \text{ yd} = 13.44 \text{ yd}^2$

Find the perimeter *of each figure. Use* outside edges *only.*

39.

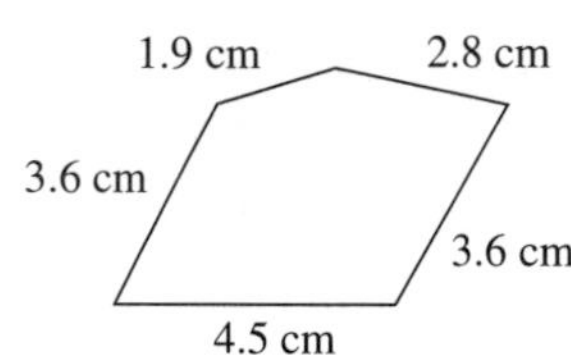

$P = 3.6 \text{ cm} + 4.5 \text{ cm} + 3.6 \text{ cm} + 2.8 \text{ cm} + 1.9 \text{ cm}$
$= 16.4 \text{ cm}$

40.

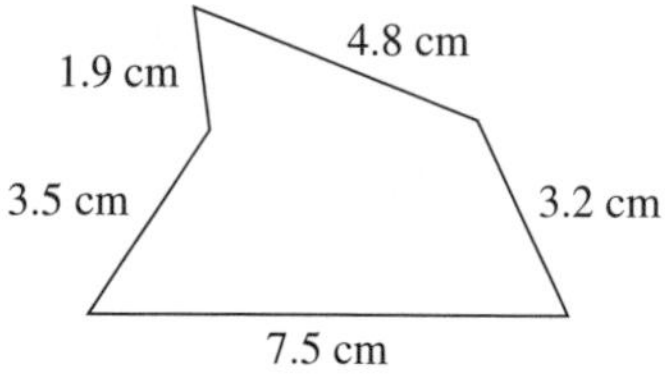

$P = 3.5 \text{ cm} + 7.5 \text{ cm} + 3.2 \text{ cm} + 4.8 \text{ cm} + 1.9 \text{ cm}$
$= 20.9 \text{ cm}$

Find the area of the shaded regions.

41.

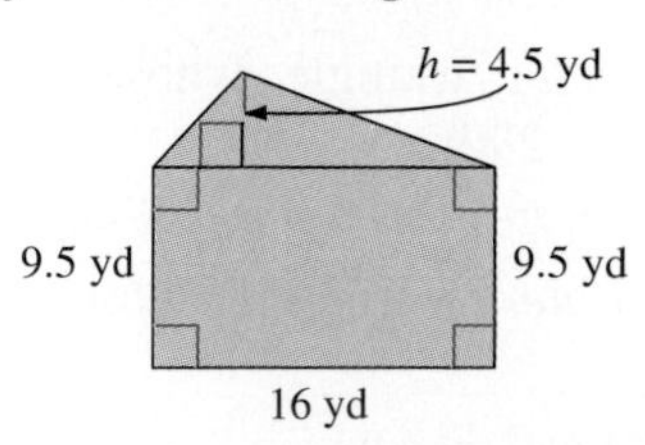

$A = 16 \text{ yd} \times 9.5 \text{ yd} + \frac{1}{2} \times 16 \text{ yd} \times 4.5 \text{ yd} = 188 \text{ yd}^2$

42.

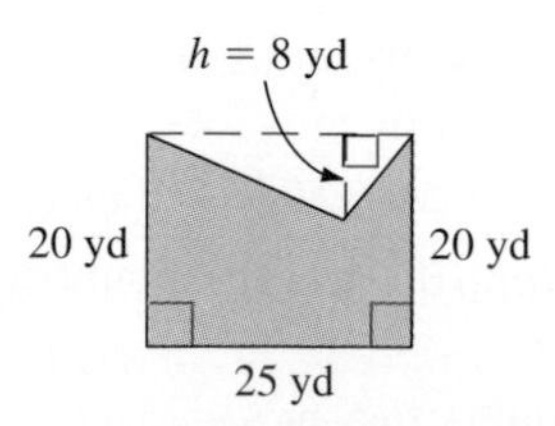

$A = 25 \text{ yd} \times 20 \text{ yd} - \frac{1}{2} \times 25 \text{ yd} \times 8 \text{ yd} = 400 \text{ yd}^2$

Applications

Find the total area of all the vertical sides (front and back) of each building.

43.

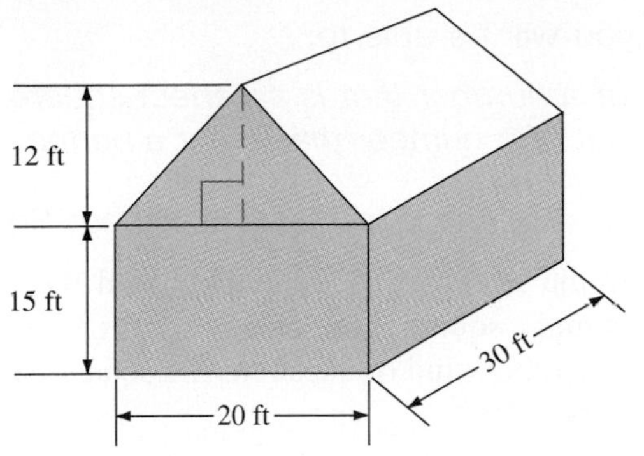

Area of each front and back $= \frac{1}{2}(12 \text{ ft}) \times 20 \text{ ft} + 15 \text{ ft}(20 \text{ ft}) =$ 420 ft^2.
Area of each side $= 15 \text{ ft}(30 \text{ ft}) = 450 \text{ ft}^2$.
Total area $= 2(420 \text{ ft}^2) + 2(450 \text{ ft}^2) = 1740 \text{ ft}^2$.

44.

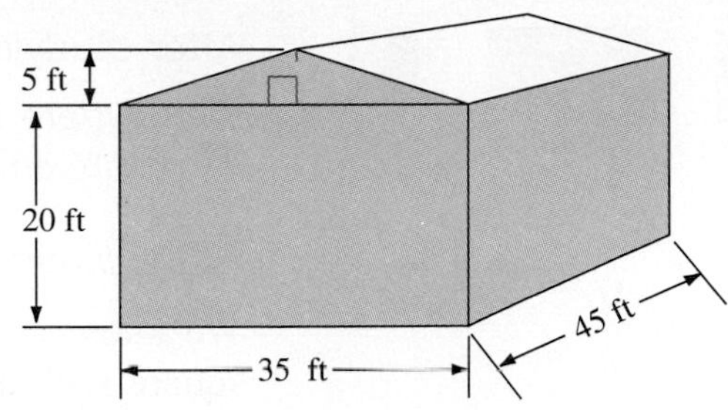

Area of each front and back $= \frac{1}{2}(5 \text{ ft})(35 \text{ ft}) + 20 \text{ ft}(35 \text{ ft}) =$ 787.5 ft^2.
Area of each side $= 20 \text{ ft}(45 \text{ ft}) = 900 \text{ ft}^2$.
Total area $= 2(787.5 \text{ ft}^2) + 2(900 \text{ ft}^2) = 3375 \text{ ft}^2$.

The top surface of the wings of a test plane must be coated with a special lacquer that costs \$90.00 per square yard. Find the cost to coat the shaded wing surface of each plane.

45.

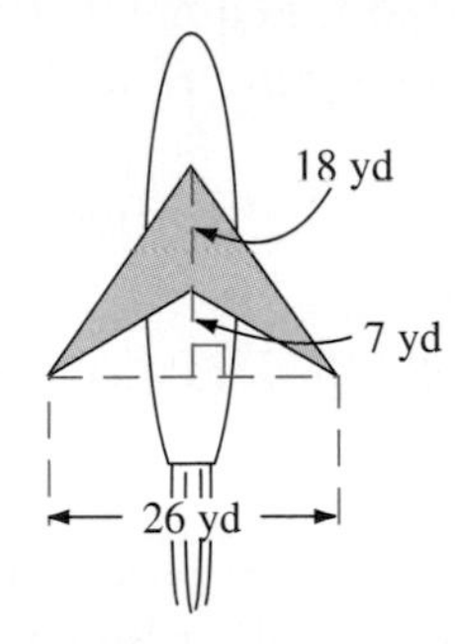

$A = 2 \times \frac{1}{2} \times 18 \text{ yd} \times 13 \text{ yd} = 234 \text{ yd}^2$

$\text{Cost} = 234 \text{ yd}^2 \times \frac{\$90.00}{\text{yd}^2} = \$21{,}060.00$

46.

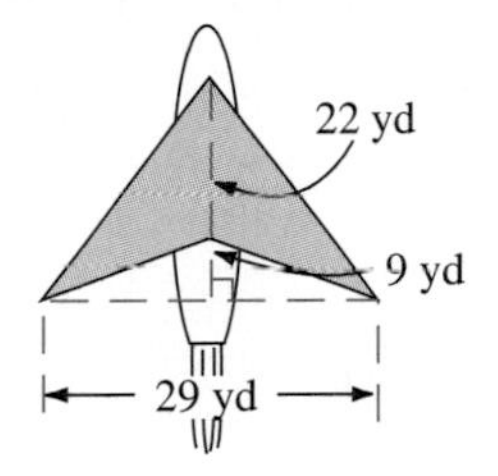

$A = 2 \times \frac{1}{2} \times 22 \text{ yd} \times 14.5 \text{ yd} = 319 \text{ yd}^2$

$\text{Cost} = 319 \text{ yd}^2 \times \frac{\$90.00}{\text{yd}^2} = \$28{,}710.00$

Cumulative Review Problems

Solve for n. Round to the nearest hundredth.

47. $\frac{5}{n} = \frac{7.5}{18}$

$n = 12$

48. $\frac{n}{29} = \frac{7}{3}$

$n = 67.67$

49. On the cruise ship *H.M.S. Salinora,* all restaurants on board must keep the ratio of waitstaff to patrons at 4 to 15. How many waitstaff should be serving if there are 2685 patrons dining at one time? 716

50. Recently, an airline found that after the transatlantic flight to Frankfurt, 68 people out of 300 passengers kept their in-flight magazines, after being encouraged to take the magazines with them to read at their leisure. On a similar flight carrying 425 people, how many in-flight magazines would the airline expect to be taken? Round to the nearest whole number. 96 magazines

7.4 Square Roots

MathPro

Video 15

SSM

After studying this section, you will be able to:

1 *Evaluate the square root of a number that is a perfect square.*
2 *Approximate the square root of a number that is not a perfect square.*

1 *Evaluating the Square Root of a Number That Is a Perfect Square*

We know that by using the formula $A = s^2$ we can quickly find the area of a square with a side of 3 in. We simply square 3 in. That is, $A = (3 \text{ in.})(3 \text{ in.}) = 9 \text{ in.}^2$. Sometimes we want to ask another kind of question. If a square has an area of 64 in.^{2}, what is the length of its sides?

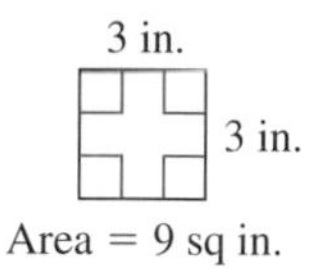

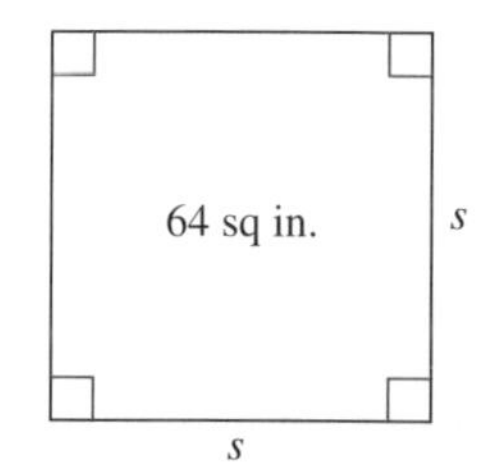

The answer is 8 in. Why? The skill we need to find a number when we are given the square of that number is called *finding the square root.* The square root of 64 is 8.

The **square root** of a number is one of the two identical factors of that number.

The square root of 64 is 8 because $(8)(8) = 64$.

The square root of 9 is 3 because $(3)(3) = 9$.

The symbol for finding the square root of a number is $\sqrt{}$. To write the square root of 64, we write $\sqrt{64} = 8$. Sometimes we speak of finding the square root of a number as *taking* the square root of the number, or we can say that we will *evaluate* the square root of the number. Thus to take the square root of 9, we write $\sqrt{9} = 3$; to evaluate the square root of 9, we write $\sqrt{9} = 3$.

EXAMPLE 1 Find the square roots. **(a)** $\sqrt{25}$ **(b)** $\sqrt{121}$

(a) $\sqrt{25} = 5$ because $(5)(5) = 25$ **(b)** $\sqrt{121} = 11$ because $(11)(11) = 121$

Practice Problem 1 Find the square roots. **(a)** $\sqrt{49}$ **(b)** $\sqrt{169}$ ■

If square roots are added or subtracted, they must be evaluated FIRST, then added or subtracted.

EXAMPLE 2 Find. $\sqrt{25} + \sqrt{36}$

$$\sqrt{25} = 5 \text{ because } (5)(5) = 25$$

$$\sqrt{36} = 6 \text{ because } (6)(6) = 36$$

Thus $\sqrt{25} + \sqrt{36} = 5 + 6 = 11$.

Practice Problem 2 Find. $\sqrt{49} - \sqrt{4}$ ■

When a whole number is multiplied by itself, the number that is obtained is called a **perfect square.**

36 is a perfect square because (6)(6) = 36.

49 is a perfect square (7)(7) = 49.

The numbers 20 or 48 are *not* perfect squares. There is no *whole number* that when squared—multiplied by itself—yields 20 or 48. Consider 20. $4^2 = 16$, which is less than 20. $5^2 = 25$, which is more than 20. We realize, then, that the square root of 20 is between 4 and 5 because 20 is between 16 and 25. Since there is no whole number between 4 and 5, no whole number squared equals 20. Since the square root of a perfect square is a whole number, we can say that 20 is NOT a perfect square.

EXAMPLE 3

(a) Is 81 a perfect square? **(b)** If so, find $\sqrt{81}$.

(a) Yes. 81 is a perfect square because (9)(9) = 81. **(b)** $\sqrt{81} = 9$

Practice Problem 3

(a) Is 144 a perfect square? **(b)** If so, find $\sqrt{144}$. ■

TEACHING TIP Encourage students to fill in the chart of perfect squares and to memorize it. Remind them that if they can find those values quickly, they will find working with square roots much easier.

It is helpful to know the first 15 perfect squares. Take a minute to complete the following table.

Number, n	1	2	3	4	5	6	7	8	9	10	11	12	13	14	15
Number Squared, n^2	1	4	9	16	25	36	49	64	81	100	121	144	169	196	225

2 *Approximating the Square Root of a Number That Is Not a Perfect Square*

If a number is not a perfect square, we can only approximate its square root. This can be done by using a square root table such as the one below. Except for exact values such as $\sqrt{4} = 2.000$, all values are rounded to the nearest thousandth.

Number, n	Square Root of the Number, $\sqrt{n}$	Number, n	Square Root of the Number, $\sqrt{n}$
1	1.000	8	2.828
2	1.414	9	3.000
3	1.732	10	3.162
4	2.000	11	3.317
5	2.236	12	3.464
6	2.449	13	3.606
7	2.646	14	3.742

A square root table is located on page A–4. It gives you the square root of whole numbers up to 200. Square roots can also be found with any calculator that has a square root key. Usually the key looks like this [$\sqrt{\ }$] or this [$\sqrt{x}$]. To find the square root of 8 on a calculator, enter the number 8 and press [$\sqrt{\ }$] or [$\sqrt{x}$]. You will see displayed 2.8284271. (Your calculator may display fewer or more digits.) Remember, no matter how many digits your calculator displays, when we find $\sqrt{8}$, we have only an **approximation.** It is not an exact answer. To emphasize this we use the $\approx$ notation to mean "is approximately equal to." Thus $\sqrt{8} \approx 2.828$.

CALCULATOR

Square Roots

Locate the square root key $\boxed{\sqrt{\ }}$ on your calculator.

1. To find $\sqrt{289}$ enter

   289 $\boxed{\sqrt{\ }}$

   The display should read

   $\boxed{17}$

2. To find $\sqrt{194}$ enter

   194 $\boxed{\sqrt{\ }}$

   The display should read

   $\boxed{13.928388}$

   This is just an approximation of the actual square root. We will round the answer to the nearest thousandth. $\sqrt{194} \approx 13.928$

TEACHING TIP Since today many pocket calculators have a square root key, this might be a good time to encourage students who have a calculator to bring it to class, and to urge those who do not have a calculator to borrow one from a friend or relative for a few days. Square roots always seem less threatening to a student with a calculator.

EXAMPLE 4 Find approximate values using the square root table or a calculator. Round your answer to the nearest thousandth.

(a) $\sqrt{2}$ **(b)** $\sqrt{12}$ **(c)** $\sqrt{7}$

(a) $\sqrt{2} \approx 1.414$ **(b)** $\sqrt{12} \approx 3.464$ **(c)** $\sqrt{7} \approx 2.646$

Practice Problem 4 Approximate the square roots to the nearest thousandth.
(a) $\sqrt{3}$ **(b)** $\sqrt{13}$ **(c)** $\sqrt{5}$ ■

EXAMPLE 5 Find the length of the side of a square that has an area of 81 in.^{2}.

$$\sqrt{81 \text{ in.}^2} = 9 \text{ in.}$$

The side measures 9 in.

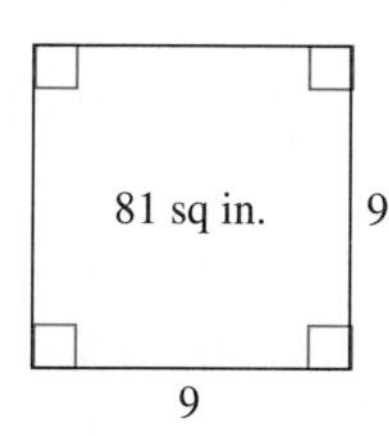

Practice Problem 5 Find the length of the side of a square that has an area of 100 in.^{2}. ■

EXAMPLE 6 Approximate to the nearest thousandth of an inch the length of the side of a square that has an area of 6 in.^{2}.

$$\sqrt{6 \text{ in.}^2} \approx 2.449 \text{ in.}$$

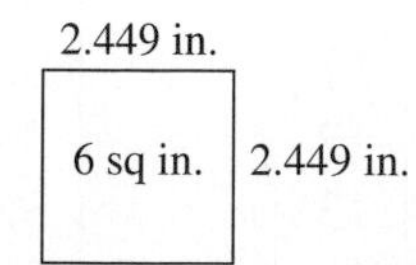

Thus, to the nearest thousandth of an inch, the side measures 2.449 in.

Practice Problem 6 Approximate to the nearest thousandth of a meter the length of the side of a square that has an area of 22 m^2. ■

7.4 Exercises

Verbal and Writing Skills

1. Why is the $\sqrt{25} = 5$? $\sqrt{25} = 5$ because $(5)(5) = 25$.

2. 25 is a perfect square because its square root, 5, is a whole number.

3. Is 32 a perfect square? Why or why not? 32 is not a perfect square because no whole number when multiplied by itself equals 32.

4. How can you approximate the square root of a number that is not a perfect square?

To approximate the square root of a number that is not a perfect square, use the square root table or a calculator.

Find each square root. Do not *use a calculator. Do* not *refer to a table of square roots.*

5. $\sqrt{1}$ 1 **6.** $\sqrt{4}$ 2 **7.** $\sqrt{16}$ 4 **8.** $\sqrt{9}$ 3 **9.** $\sqrt{25}$ 5 **10.** $\sqrt{36}$ 6 **11.** $\sqrt{49}$ 7 **12.** $\sqrt{64}$ 8

13. $\sqrt{100}$ 10 **14.** $\sqrt{81}$ 9 **15.** $\sqrt{121}$ 11 **16.** $\sqrt{144}$ 12 **17.** $\sqrt{169}$ 13 **18.** $\sqrt{196}$ 14 **19.** $\sqrt{0}$ 0 **20.** $\sqrt{225}$ 15

Evaluate the square roots first, then add or subtract the results. Do not *use a calculator or a square root table.*

21. $\sqrt{49} + \sqrt{9}$ $7 + 3 = 10$

22. $\sqrt{25} + \sqrt{64}$ $5 + 8 = 13$

23. $\sqrt{100} + \sqrt{1}$ $10 + 1 = 11$

24. $\sqrt{0} + \sqrt{121}$ $0 + 11 = 11$

25. $\sqrt{36} + \sqrt{64}$ $6 + 8 = 14$

26. $\sqrt{1} + \sqrt{25}$ $1 + 5 = 6$

27. $\sqrt{0} + \sqrt{16}$ $0 + 4 = 4$

28. $\sqrt{36} + \sqrt{81}$ $6 + 9 = 15$

29. $\sqrt{144} - \sqrt{4}$ $12 - 2 = 10$

30. $\sqrt{169} - \sqrt{64}$ $13 - 8 = 5$

31. $\sqrt{121} - \sqrt{36}$ $11 - 6 = 5$

32. $\sqrt{225} - \sqrt{100}$ $15 - 10 = 5$

33. Find an exact value for

$$\sqrt{1} + \sqrt{9} + \sqrt{25} + \sqrt{36} + \sqrt{49}$$

$1 + 3 + 5 + 6 + 7 = 22$

34. Find an exact value for

$$\sqrt{4} + \sqrt{9} + \sqrt{16} + \sqrt{81} + \sqrt{121}$$

$2 + 3 + 4 + 9 + 11 = 29$

35. (a) Is 49 a perfect square? Yes
(b) If so, find $\sqrt{49}$. 7

36. (a) Is 25 a perfect square? Yes
(b) If so, find $\sqrt{25}$. 5

37. (a) Is 256 a perfect square? Yes
(b) If so, find $\sqrt{256}$. 16

38. (a) Is 289 a perfect square? Yes
(b) If so, find $\sqrt{289}$. 17

Use a table of square roots or a calculator with a square root key to approximate the following to the nearest thousandth.

39. $\sqrt{18}$ ≈ 4.243 **40.** $\sqrt{24}$ ≈ 4.899 **41.** $\sqrt{31}$ ≈ 5.568 **42.** $\sqrt{42}$ ≈ 6.481 **43.** $\sqrt{83}$ ≈ 9.110

44. $\sqrt{76}$ ≈ 8.718 **45.** $\sqrt{120}$ ≈ 10.954 **46.** $\sqrt{136}$ ≈ 11.662 **47.** $\sqrt{125}$ ≈ 11.180 **48.** $\sqrt{150}$ ≈ 12.247

Find the length of the side of each square. If the area is not a perfect square, approximate by using a square root table or a calculator with a square root key. Round your answer to the nearest thousandth.

49. A square with area 121 m^2 $\sqrt{121\ m^2} = 11$ m

50. A square with area 169 m^2 $\sqrt{169\ m^2} = 13$ m

51. A square with area 26 m^2 $\sqrt{26\ m^2} \approx 5.099$ m

52. A square with area 34 m^2 $\sqrt{34\ m^2} \approx 5.831$ m

53. A square with area 75 m^2 $\sqrt{75\ m^2} \approx 8.660$ m

54. A square with area 90 m^2 $\sqrt{90\ m^2} \approx 9.487$ m

To Think About

55. Find each square root.

(a) $\sqrt{4}$ 2

(b) $\sqrt{0.04}$ 0.2

(c) $\sqrt{0.0004}$ 0.02

(d) What pattern do you observe? Each answer is obtained from the previous answer by dividing by 10.

(e) Can you find $\sqrt{0.004}$ exactly? Why?
No, because 0.004 isn't a perfect square.

56. Find each square root.

(a) $\sqrt{25}$ 5

(b) $\sqrt{0.25}$ 0.5

(c) $\sqrt{0.0025}$ 0.05

(d) What pattern do you observe? Same as 55d

(e) Can you find $\sqrt{0.025}$ exactly? Why?
No, because 0.025 isn't a perfect square.

Using a calculator with a square root key, evaluate each of the following and round your answer to the nearest thousandth.

57. $\sqrt{42{,}036}$ 205.027

58. $\sqrt{25{,}231}$ 158.843

59. $\sqrt{456} + \sqrt{322}$ 39.299

60. $\sqrt{578} + \sqrt{984}$ 55.410

Cumulative Review Problems

61. The Taronga Zoo in Sydney, Australia, has a viewing tank for its platapi (plural for platapus). The tank is 60 in. high, and 80 in. wide. What is the area of the front of the rectangular tank?
4800 sq. in.

62. Ashrita Furman of Jamaica, New York, holds the world record in ''joggling''—juggling and running at the same time—over 80.5 km. How many meters did he ''joggle''? 80,500 meters

63. The gate that leads from the street to the neighborhood basketball courts measures 92 centimeters wide. How many *meters* wide is the door? 0.92 meters

64. Goliath beetles in Equatorial Africa can weigh up to 98.9 grams. How many *kilograms* can they weigh? 0.0989 kilograms

7.5 The Pythagorean Theorem

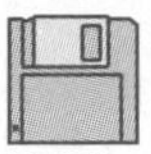

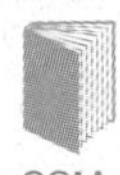

After studying this section, you will be able to:

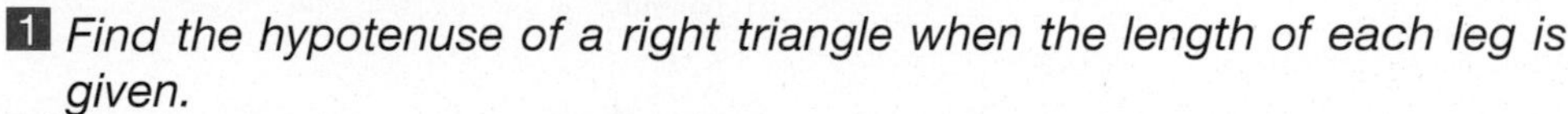

1. *Find the hypotenuse of a right triangle when the length of each leg is given.*

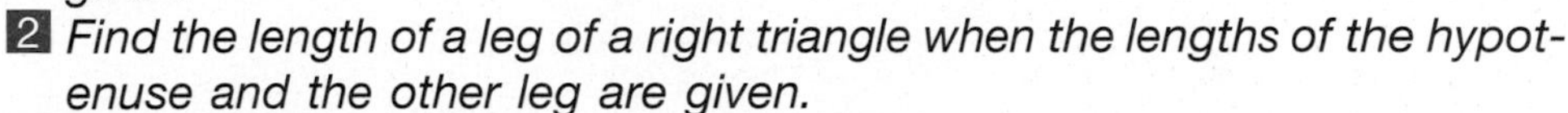

2. *Find the length of a leg of a right triangle when the lengths of the hypotenuse and the other leg are given.*
3. *Solve applied problems using the Pythagorean Theorem.*
4. *Solve to find the missing sides of special right triangles.*

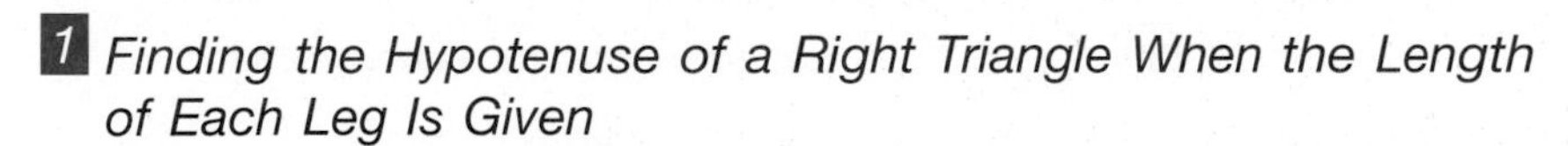

1 *Finding the Hypotenuse of a Right Triangle When the Length of Each Leg Is Given*

The Pythagorean Theorem is a mathematical idea formulated long ago. It is as useful today as it was when it was discovered. The Pythagoreans lived in Italy about 2500 years ago. They studied various mathematical properties. They discovered that for any right triangle, the square of the hypotenuse equals the sum of the squares of the two legs of the triangle. This relationship is known as the **Pythagorean Theorem.** The side opposite the right angle is called the **hypotenuse;** the other two sides are called the **legs** of the right triangle.

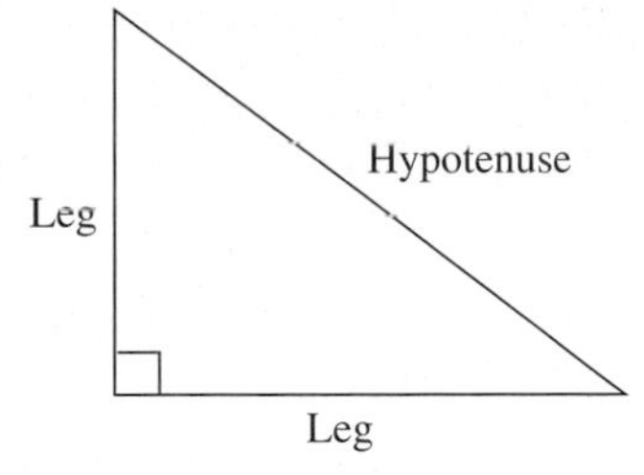

$$(\text{hypotenuse})^2 = (\text{leg})^2 + (\text{leg})^2$$

For example, in the right triangle

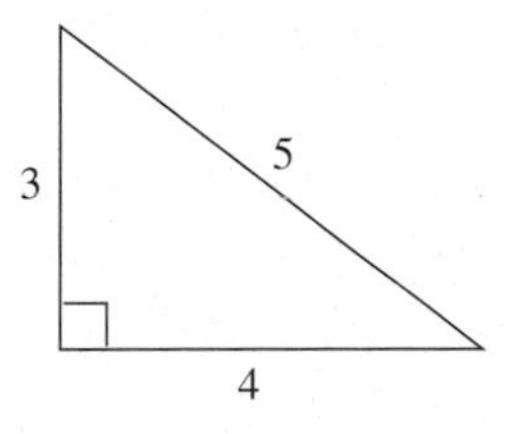

$5^2 = 3^2 + 4^2$

$25 = 9 + 16$

$25 = 25$ ✓

In a right triangle, the hypotenuse is the longest side. It is always opposite the largest angle, the right angle. The legs are the two shorter sides. When we know each leg of a right triangle, we use the property

TEACHING TIP Emphasize that it is important for the students to know what side of a right triangle is the *hypotenuse.* Remind them that they will hear that word many times in mathematics and in other courses.

$$\text{Hypotenuse} = \sqrt{(\text{leg})^2 + (\text{leg})^2}$$

EXAMPLE 1 Find the hypotenuse of a right triangle with legs of 5 in. and 12 in., respectively.

Hypotenuse $= \sqrt{(5)^2 + (12)^2}$

$= \sqrt{25 + 144}$ *Square each value first.*

$= \sqrt{169}$ *Add together the two values.*

$= 13$ in. *Take the square root.*

Practice Problem 1 Find the hypotenuse of a right triangle with legs of 8 m and 6 m, respectively. ■

TEACHING TIP This is a good time to remind students that they should keep reviewing the table of perfect squares they filled out in Section 7.4.

EXAMPLE 2 Find the hypotenuse of a right triangle with legs of 8 m and 15 m, respectively.

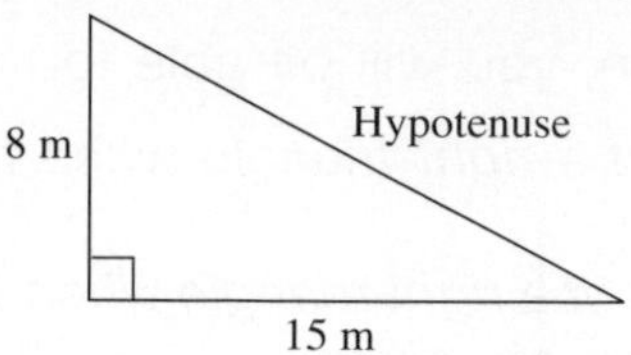

$$\begin{aligned} \text{Hypotenuse} &= \sqrt{(8)^2 + (15)^2} \\ &= \sqrt{64 + 225} && \textit{Square each value first.} \\ &= \sqrt{289} && \textit{Add together the two values.} \\ &= 17 \text{ m} && \textit{Take the square root.} \end{aligned}$$

[Do you see why $\sqrt{289} = 17$? Verify that $(17)(17) = 289$. Finding square roots of large numbers like this sometimes takes a little work.]

Practice Problem 2 Find the hypotenuse of a right triangle with legs of 12 m and 16 m, respectively. ■

Sometimes we cannot find the hypotenuse exactly. In those cases, we often approximate the square root by using a calculator or a square root table.

EXAMPLE 3 Find the hypotenuse of a right triangle with legs of 4 m and 5 m, respectively. Round your answer to the nearest thousandth.

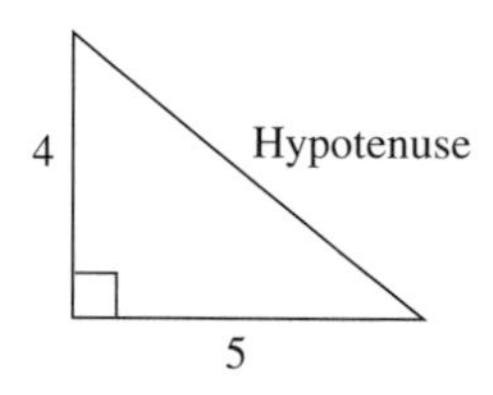

$$\begin{aligned} \text{Hypotenuse} &= \sqrt{(4)^2 + (5)^2} \\ &= \sqrt{16 + 25} && \textit{Square each value first.} \\ &= \sqrt{41} \text{ m} && \textit{Add the two values together.} \end{aligned}$$

Using the square root table or a calculator, we have hypotenuse ≈ 6.403 m

Practice Problem 3 Find to the nearest thousandth the hypotenuse of a right triangle with legs of 3 cm and 7 cm, respectively. ■

2 *Finding the Length of a Leg of a Right Triangle When the Lengths of the Hypotenuse and the Other Leg Are Given*

When we know the hypotenuse and one leg of a right triangle, to find the length of the other leg, we use the property

$$\text{Leg} = \sqrt{(\text{hypotenuse})^2 - (\text{leg})^2}$$

EXAMPLE 4 A right triangle has a hypotenuse of 15 cm and a leg of 12 cm. Find the length of the other leg.

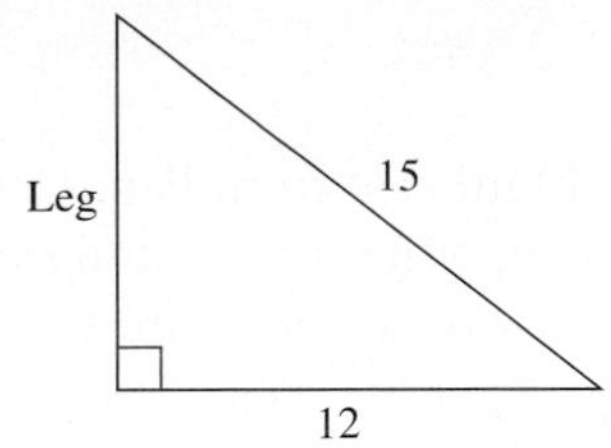

$$\text{Leg} = \sqrt{(15)^2 - (12)^2}$$
$$= \sqrt{225 - 144} \quad \textit{Square each value first.}$$
$$= \sqrt{81} \quad \textit{Subtract.}$$
$$= 9 \text{ cm} \quad \textit{Find the square root.}$$

Practice Problem 4 A right triangle has a hypotenuse of 17 m and a leg of 15 m. Find the length of the other leg. ■

EXAMPLE 5 A right triangle has a hypotenuse of 9 m and a leg of 7 m. Find the length of the other leg correct to the nearest thousandth.

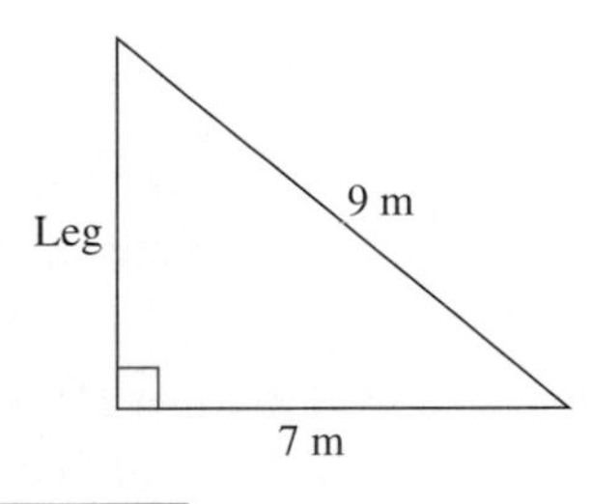

$$\text{Leg} = \sqrt{(9)^2 - (7)^2}$$
$$= \sqrt{81 - 49} \quad \textit{Square each value first.}$$
$$= \sqrt{32} \text{ m} \quad \textit{Subtract the two numbers.}$$

Using a calculator or the square root table, leg $\approx$ 5.657 m.

Practice Problem 5 A right triangle has a hypotenuse of 10 m and a leg of 5 m. Find the length of the other leg correct to the nearest thousandth. ■

3 Solving Applied Problems Using the Pythagorean Theorem

Certain applied problems call for the use of the Pythagorean Theorem in the solution.

EXAMPLE 6 A pilot flies 13 mi east from Pennsville to Salem. She then flies 5 mi south from Salem to Elmer. What is the straight-line distance from Pennsville to Elmer? Round your answer to the nearest tenth of a mile.

1. *Understand the problem.*
   It might help to draw a picture.

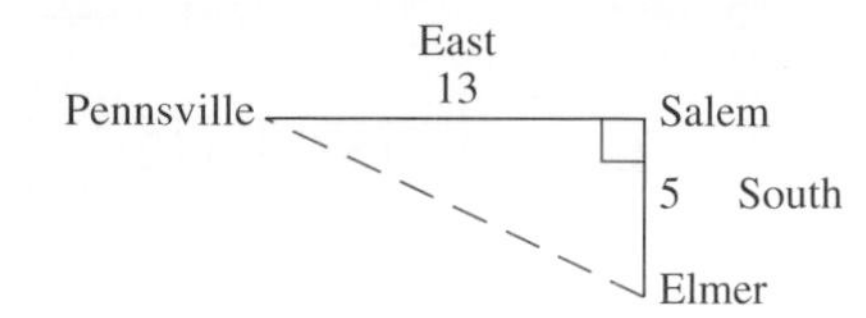

The distance from Pennsville to Elmer is the hypotenuse of the triangle.

2. *Solve and state the answer.*

$$\begin{aligned} \text{Hypotenuse} &= \sqrt{(\text{leg})^2 + (\text{leg})^2} \\ &= \sqrt{(13)^2 + (5)^2} \\ &= \sqrt{169 + 25} \\ &= \sqrt{194} \\ \sqrt{194} &\approx 13.928 \text{ mi} \end{aligned}$$

Rounded to the nearest tenth, the distance is 13.9 mi.

3. *Check.*
   Work backward to check. Use the Pythagorean Theorem. Is

$13.9^2 \approx 13^2 + 5^2$? *(We use $\approx$ because 13.9 is an approximate answer.)*

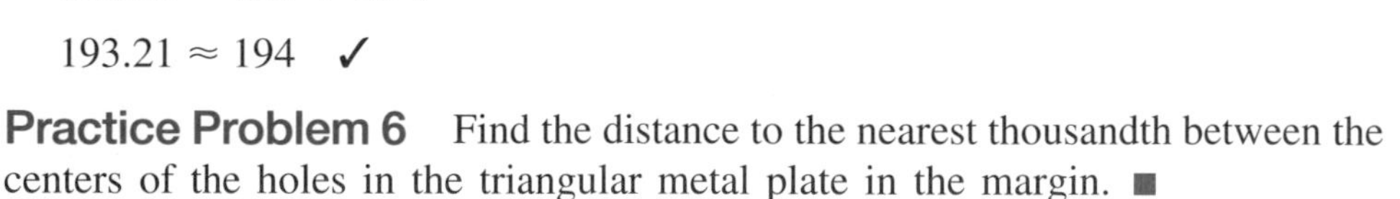

$193.21 \approx 169 + 25$?

$193.21 \approx 194$ ✓

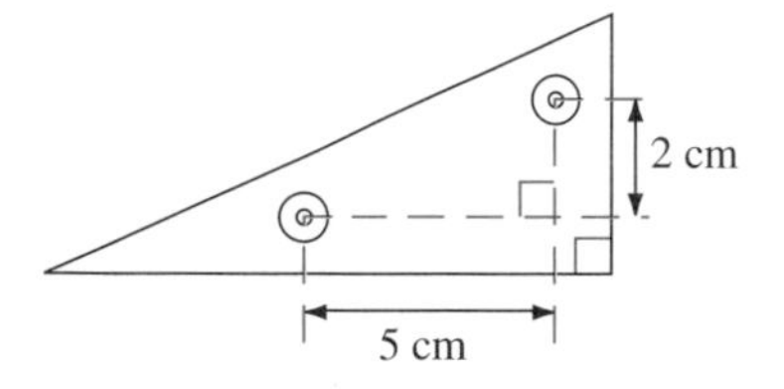

Practice Problem 6 Find the distance to the nearest thousandth between the centers of the holes in the triangular metal plate in the margin. ■

EXAMPLE 7 A 25-ft ladder is placed against a building at a point 22 ft from the ground. What is the distance of the base of the ladder from the building? Round your answer to the nearest tenth.

1. *Understand the problem.*
   Draw a picture.

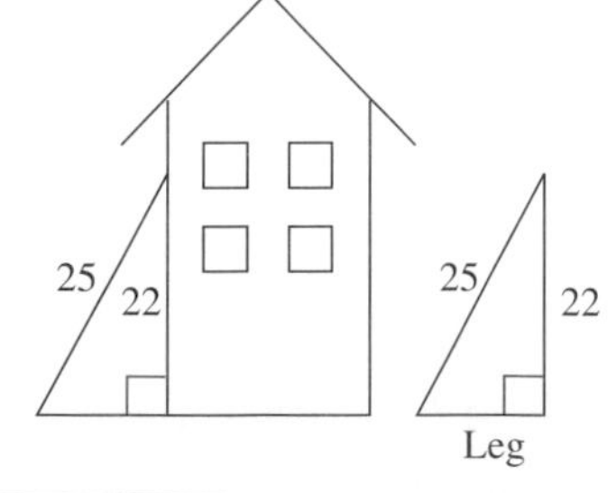

2. *Solve and state the answer.*

$$\begin{aligned} \text{Leg} &= \sqrt{(\text{hypotenuse})^2 - (\text{leg})^2} \\ &= \sqrt{(25)^2 - (22)^2} \\ &= \sqrt{625 - 484} \\ &= \sqrt{141} \\ \sqrt{141} &\approx 11.874 \text{ ft} \end{aligned}$$

If we round to the nearest tenth, the ladder is 11.9 ft from the base of the house.

Practice Problem 7 A kite is out on 30 yd of string. The kite is directly above a rock. The rock is 27 yd from the boy flying the kite. How far above the rock is the kite? Round your answer to the nearest tenth. ■

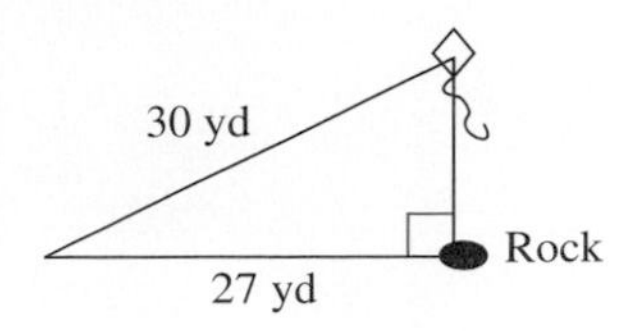

4 Solving to Find the Missing Sides of Special Right Triangles

If we use the Pythagorean Theorem and some other facts from geometry, we can find a relationship among the sides of two special right triangles. The first special right triangle is one that contains an angle that measures 30° and one that measures 60°. We call this the 30°–60°–90° right triangle.

TEACHING TIP Instructors often assume that students have memorized the ratio of the sides of the 30–60–90 right triangle and the 45–45–90 right triangle. But many students have never heard these ratios before or may not remember them from high school. You may need to take extra time to explain this material to the class.

> In a 30°–60°–90° triangle the length of the leg opposite the 30° angle is $\frac{1}{2}$ the length of the hypotenuse.

Notice that the hypotenuse of the first triangle is 10 m and the side opposite the 30° angle is exactly $\frac{1}{2}$ of that, or 5 m. The second triangle has a hypotenuse of 15 yd. The side opposite the 30° angle is exactly $\frac{1}{2}$ of that, or 7.5 yd.

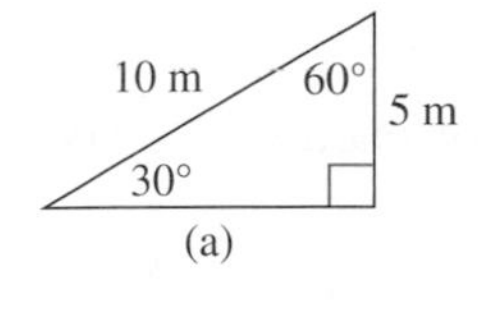

(a)

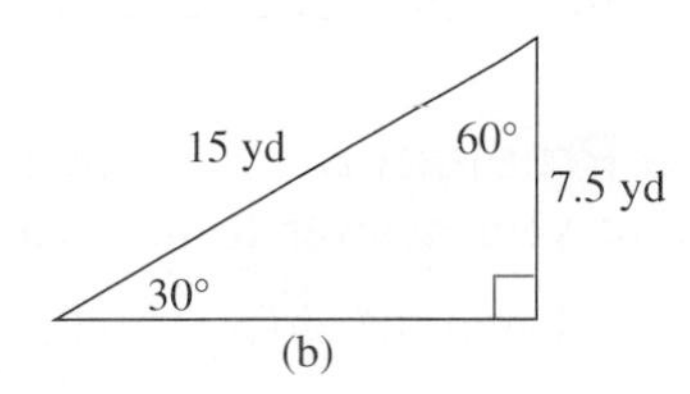

(b)

The second special right triangle is one that contains exactly two angles that each measure 45°. We call this the 45°–45°–90° right triangle.

> In a 45°–45°–90° triangle the sides opposite the 45° angles are equal. The hypotenuse is equal to $\sqrt{2} \times$ the length of either leg.

We will use the decimal approximation $\sqrt{2} \approx 1.414$ with this property.

$$\begin{aligned} \text{Hypotenuse} &= \sqrt{2} \times 7 \\ &\approx 1.414 \times 7 \\ &\approx 9.898 \text{ cm} \end{aligned}$$

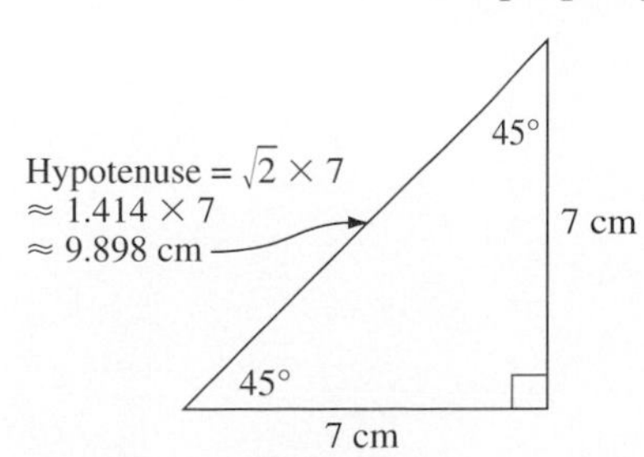

EXAMPLE 8 Find the requested sides for each of the special triangles. Round to the nearest tenth.

(a) Find the length of sides y and x. **(b)** Find the length of hypotenuse z.

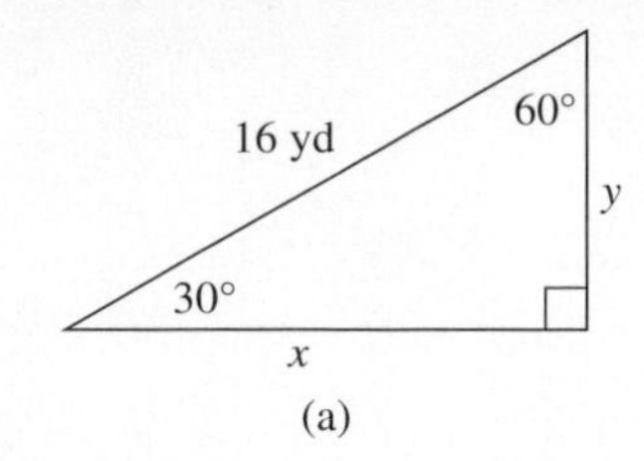

(a)

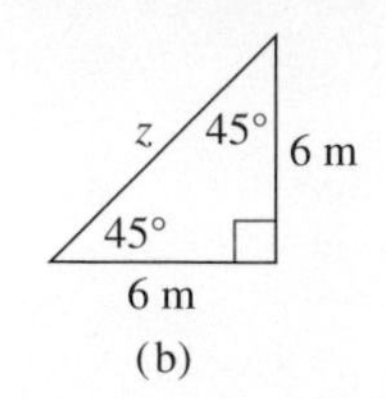

(b)

(a) In a 30°–60°–90° triangle the side opposite the 30° angle is $\frac{1}{2}$ of the hypotenuse.

$$\frac{1}{2} \times 16 = 8$$

Therefore, $y = 8$ yd.

When we know two sides of a right triangle, we find the third side using the Pythagorean Theorem.

$$\begin{aligned} \text{Leg} &= \sqrt{(\text{hypotenuse})^2 - (\text{leg})^2} \\ &= \sqrt{16^2 - 8^2} = \sqrt{256 - 64} \\ &= \sqrt{192} \approx 13.856 \text{ yd} \end{aligned}$$

$x = 13.9$ yd rounded to the nearest tenth.

(b) In a 45°–45°–90° triangle the

$$\begin{aligned} \text{Hypotenuse} &= \sqrt{2} \times \text{leg} \\ &\approx 1.414(6) \\ &= 8.484 \text{ m} \end{aligned}$$

Rounding to the nearest tenth, the hypotenuse = 8.5 m.

Practice Problem 8 Find the requested sides for each of the special triangles. Round your answer to the nearest tenth.

(a) Find the length of sides y and x. **(b)** Find the length of hypotenuse z. ■

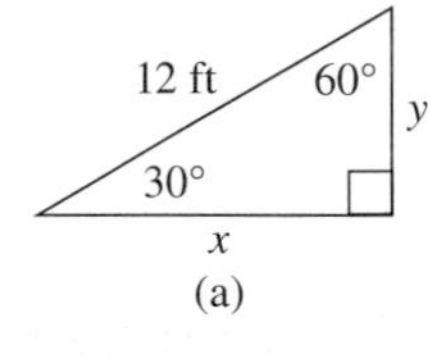

(a)

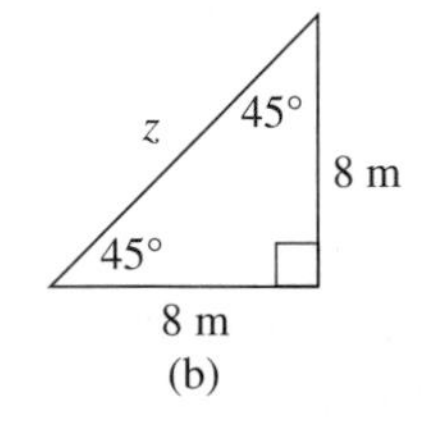

(b)

7.5 Exercises

Find the unknown side of each right triangle. Use a calculator or square root table when necessary and round your answer to the nearest thousandth.

1.

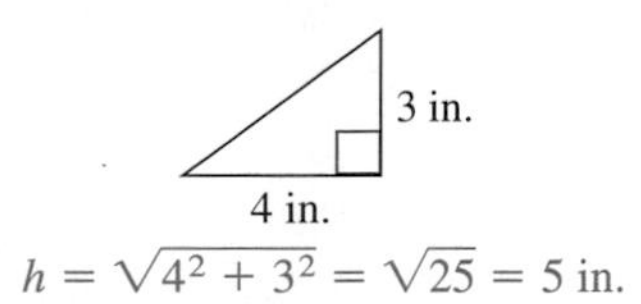

$h = \sqrt{4^2 + 3^2} = \sqrt{25} = 5$ in.

2.

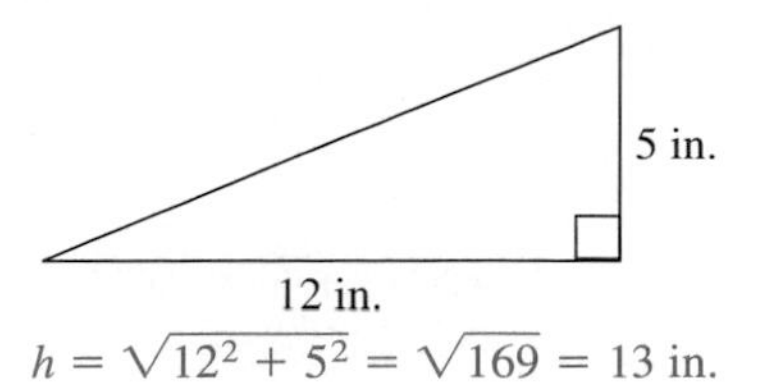

$h = \sqrt{12^2 + 5^2} = \sqrt{169} = 13$ in.

3.

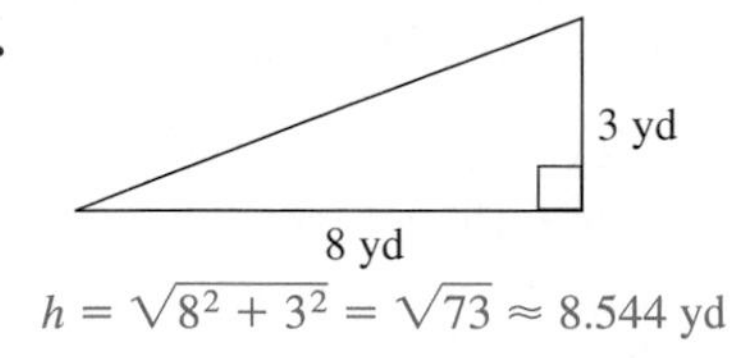

$h = \sqrt{8^2 + 3^2} = \sqrt{73} \approx 8.544$ yd

4.

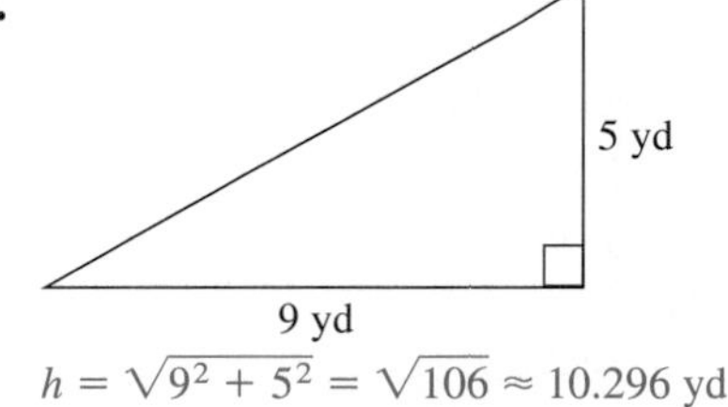

$h = \sqrt{9^2 + 5^2} = \sqrt{106} \approx 10.296$ yd

5.

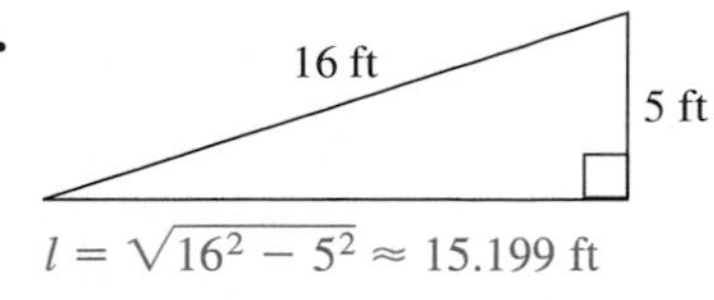

$l = \sqrt{16^2 - 5^2} \approx 15.199$ ft

6.

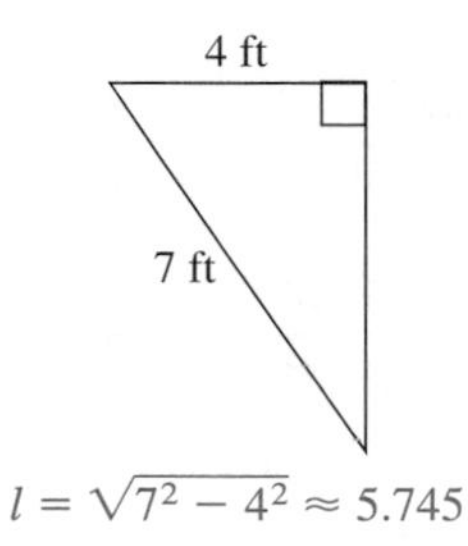

$l = \sqrt{7^2 - 4^2} \approx 5.745$

7.

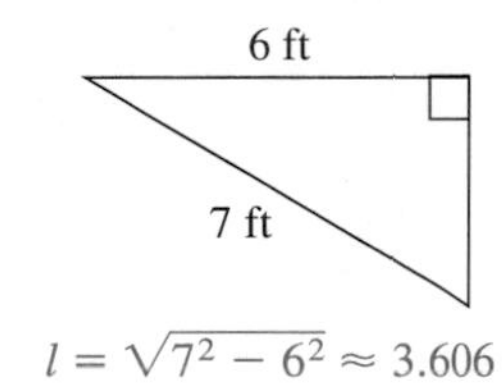

$l = \sqrt{7^2 - 6^2} \approx 3.606$

8.

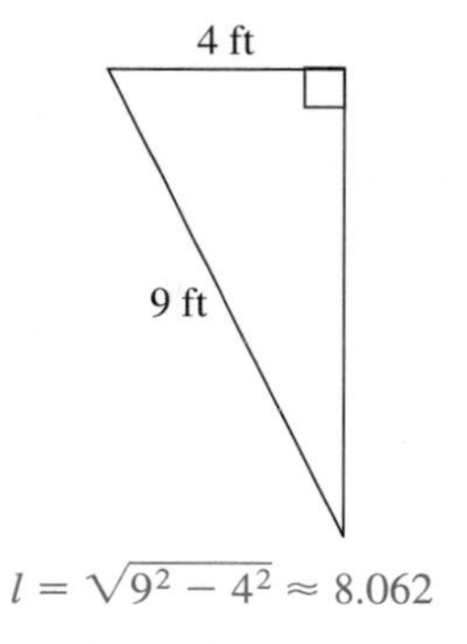

$l = \sqrt{9^2 - 4^2} \approx 8.062$

Find the unknown side of each right triangle to the nearest thousandth using the information given.

9. leg = 8 km, hypotenuse = 13 km
leg $= \sqrt{13^2 - 8^2} = \sqrt{105} \approx 10.247$ km

10. leg = 5 km, hypotenuse = 11 km
leg $= \sqrt{11^2 - 5^2} = \sqrt{96} \approx 9.798$ km

11. leg = 11 m, leg = 3 m
$h = \sqrt{11^2 + 3^2} = \sqrt{130} \approx 11.402$ m

12. leg = 5 m, leg = 4 m
$h = \sqrt{5^2 + 4^2} = \sqrt{41} \approx 6.403$ m

13. leg = 5 m, leg = 5 m
$h = \sqrt{5^2 + 5^2} = \sqrt{50} \approx 7.071$ m

14. leg = 6 m, leg = 6 m
$h = \sqrt{6^2 + 6^2} = \sqrt{72} \approx 8.485$ m

15. hypotenuse = 5 ft, leg = 4 ft
leg $= \sqrt{5^2 - 4^2} = \sqrt{9} = 3$ ft

16. hypotenuse = 10 ft, leg = 6 ft
leg $= \sqrt{10^2 - 6^2} = \sqrt{64} = 8$ ft

17. leg = 9 in., leg = 12 in.
$h = \sqrt{9^2 + 12^2} = 15$ in.

18. leg = 12 in., leg = 16 in.
$h = \sqrt{12^2 + 16^2} = 20$ in.

19. hypotenuse = 13 yd, leg = 11 yd
leg $= \sqrt{13^2 - 11^2} = \sqrt{48} \approx 6.928$ yd

20. hypotenuse = 14 yd, leg = 10 yd
leg $= \sqrt{14^2 - 10^2} = \sqrt{96} \approx 9.798$ yd

Solve each applied problem. Round your answer to the nearest tenth.

21. Find the length of the guy wire supporting the telephone pole.
$h = \sqrt{15^2 + 8^2} = \sqrt{289} = 17$ ft

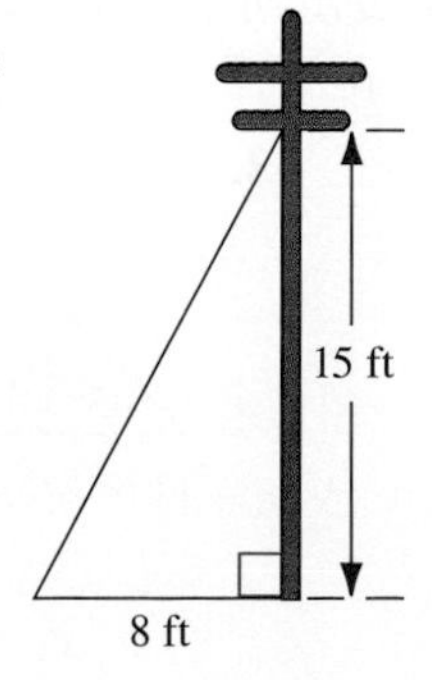

22. Find the length of this ramp to a loading dock.

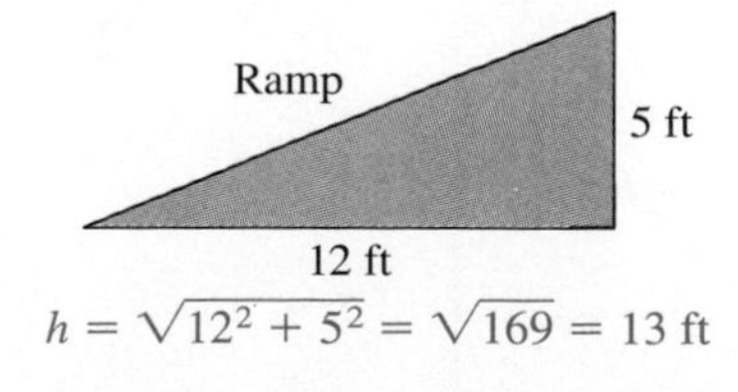

$h = \sqrt{12^2 + 5^2} = \sqrt{169} = 13$ ft

23. Juan runs out of gas in Los Lunas, New Mexico. He walks 4 mi west and then 3 mi south looking for a gas station. How far is he from his starting point?

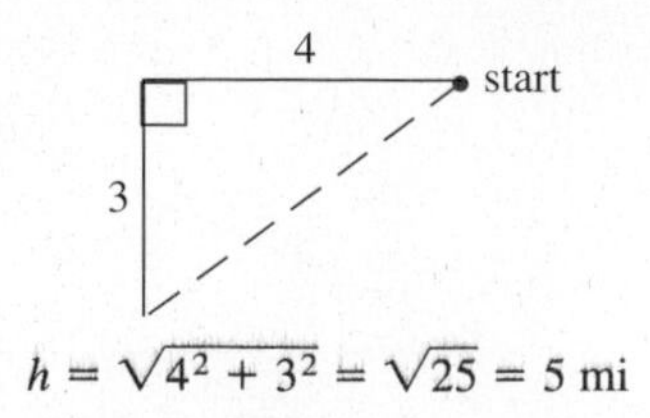

$h = \sqrt{4^2 + 3^2} = \sqrt{25} = 5$ mi

24. A construction project requires a stainless steel plate with holes drilled as shown. Find the distance between the centers of the holes in this triangular plate.

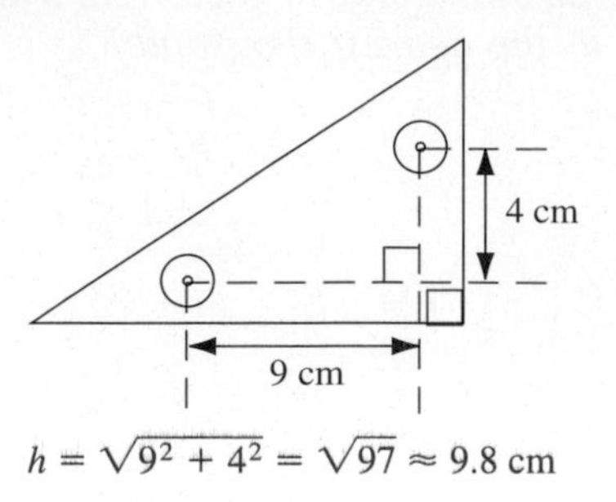

$h = \sqrt{9^2 + 4^2} = \sqrt{97} \approx 9.8$ cm

25. A 20-ft ladder is placed against a college classroom building at a point 18 ft above the ground. What is the distance from the base of the ladder to the building?

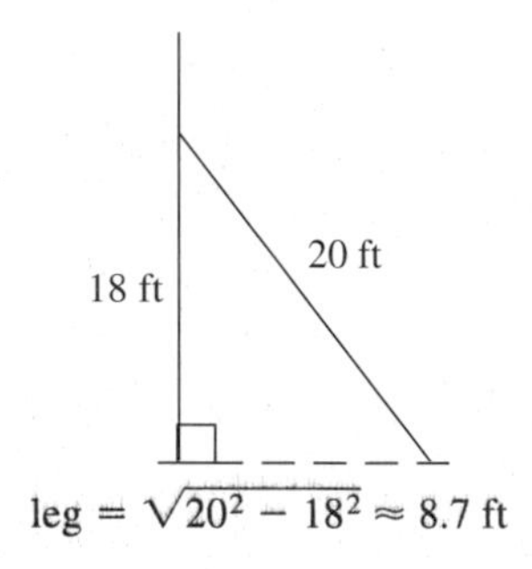

leg $= \sqrt{20^2 - 18^2} \approx 8.7$ ft

26. Barbara is flying her dragon kite on 32 yd of string. The kite is directly above the edge of a pond. The edge of the pond is 30 yd from where the kite is tied to the ground. How far is the kite above the pond?

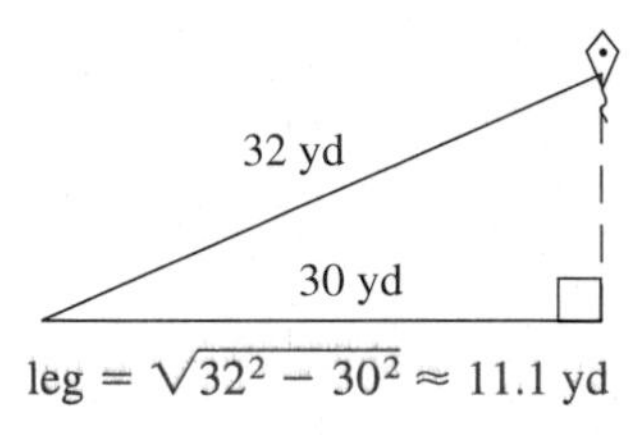

leg $= \sqrt{32^2 - 30^2} \approx 11.1$ yd

Using your knowledge of special right triangles, find the length of each leg. Round your answers to the nearest tenth.

27.

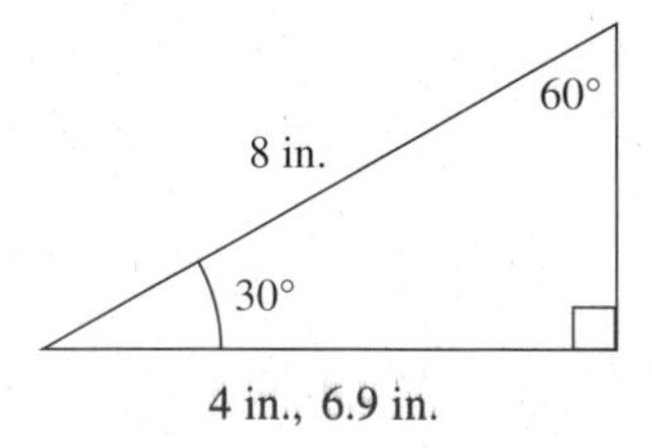

4 in., 6.9 in.

28.

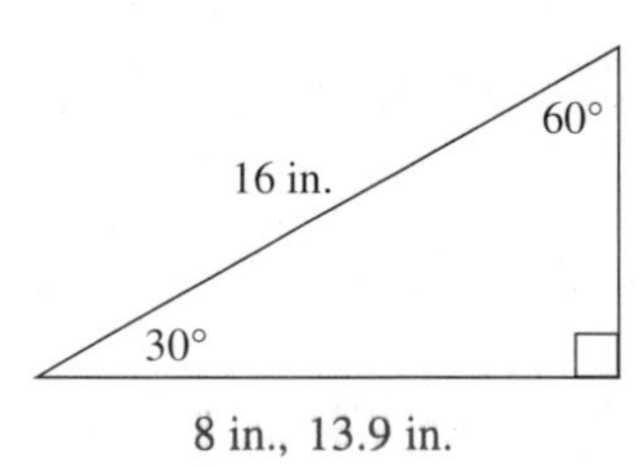

8 in., 13.9 in.

29.

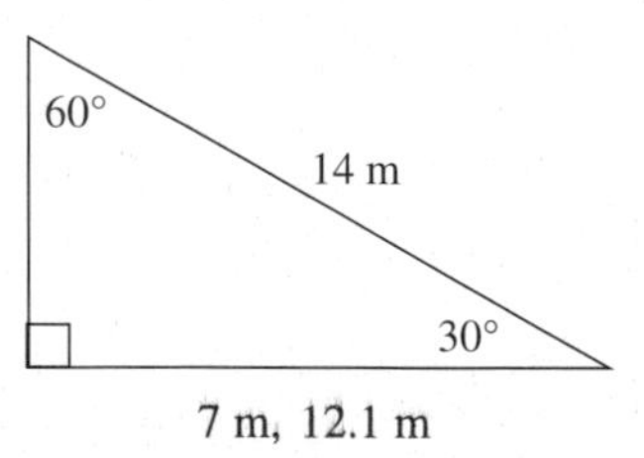

7 m, 12.1 m

30.

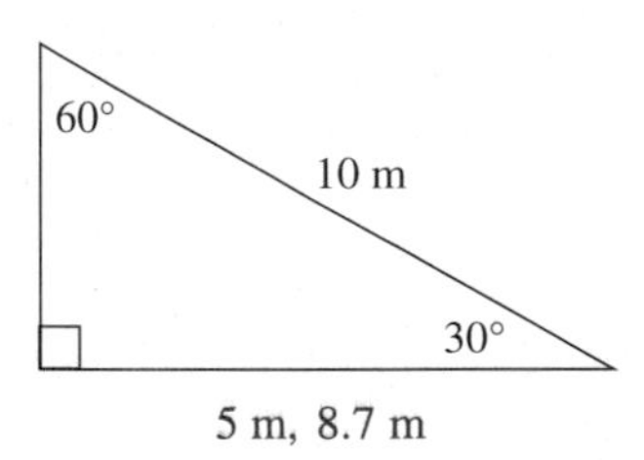

5 m, 8.7 m

Using your knowledge of special right triangles, find the length of the hypotenuse. Round your answer to the nearest tenth.

31.

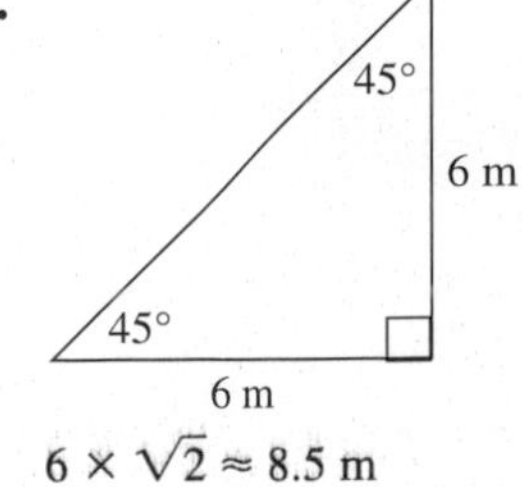

$6 \times \sqrt{2} \approx 8.5$ m

32.

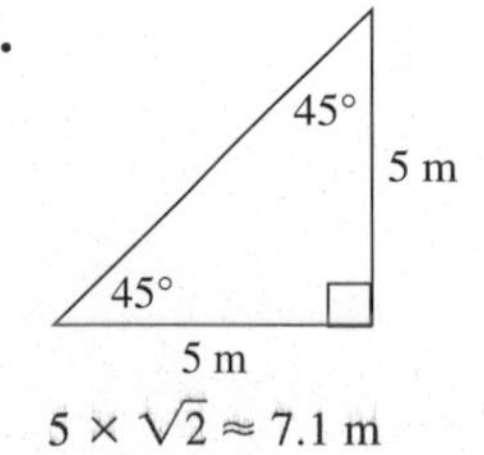

$5 \times \sqrt{2} \approx 7.1$ m

33.

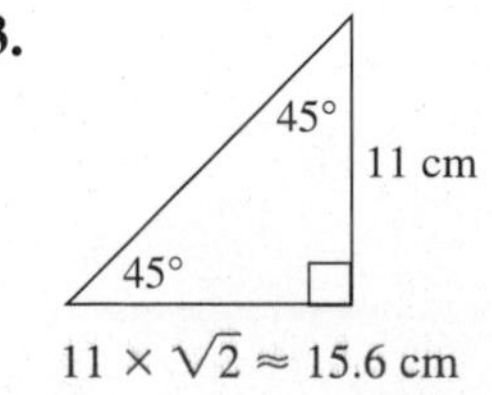

$11 \times \sqrt{2} \approx 15.6$ cm

34.

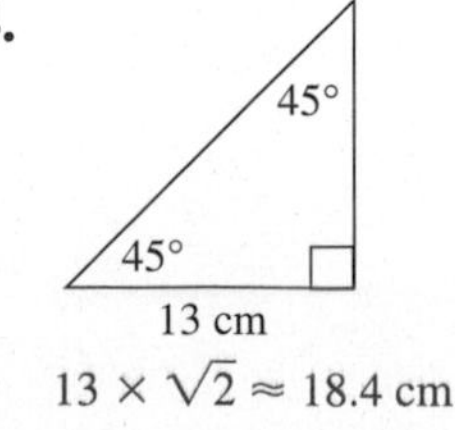

$13 \times \sqrt{2} \approx 18.4$ cm

To Think About

35. A carpenter is going to use a wooden flag pole 10 in. in diameter. He wishes to shape a rectangular base. The base will be 7 in. wide. The carpenter wishes to make the base as tall as possible, minimizing any waste. How tall will the rectangular base be? (Round to the nearest tenth.)

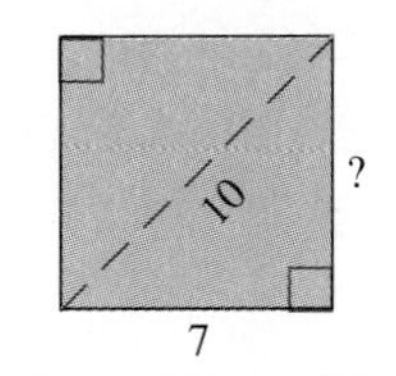

leg $= \sqrt{10^2 - 7^2} = \sqrt{51} \approx 7.1$ in.

36. A 4-m short wave antenna is placed on a garage roof that is 2 m above the lower part of the roof. The base of the garage is 16 m wide. How long is an antenna support from point A to point B?

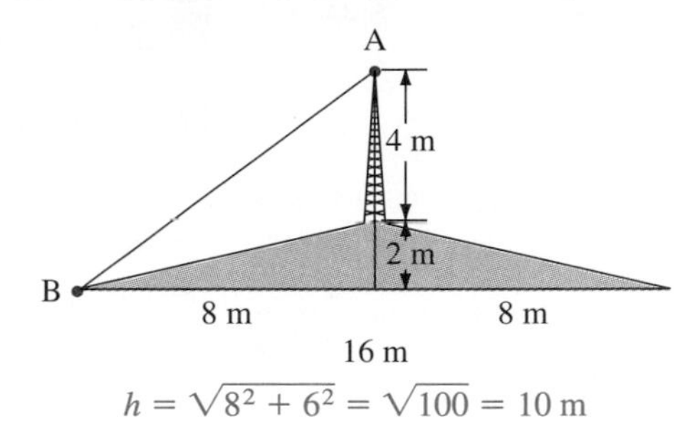

$h = \sqrt{8^2 + 6^2} = \sqrt{100} = 10$ m

Find the exact value of the unknown side of each right triangle.

37. The two legs of the triangle are 48 yd and 20 yd.
$h = \sqrt{48^2 + 20^2} = \sqrt{2704} = 52$ yd

38. The hypotenuse of the triangle is 65 yd and one leg is 33 yd.
leg $= \sqrt{65^2 - 33^2} = \sqrt{3136} = 56$ yd

Cumulative Review Problems

39. Find the area of a triangular piece of land with altitude 22 m and base 31 m.
341 m^2

40. Find the area of a backyard vegetable garden with length 14 m and width 12 m. 168 m^2

41. Find the area of a square window that measures 21 inches on each side. 441 $in.^2$

42. Find the area of the parallelogram-shaped roof of a building with a height 48 yd and base of 88 yd.
4224 yd^2

7.6 Circles

MathPro Video 16 SSM

After studying this section, you will be able to:

1 *Find the area and circumference of a circle.*
2 *Solve area problems containing circles and other geometric shapes.*

1 *Finding the Area and Circumference of a Circle*

TEACHING TIP You may want to point out that no one position on the rim of a circle is "more important" than another. Diplomats have been known to choose a circular table for their more difficult negotiations because every person's position at a circle has equal importance.

Every point on the rim of a circle is the same distance from the center of the circle, so the circle looks the same all around. In geometry we study the relationship between the parts of a circle and we learn how to calculate the distance around a circle as well as the area of a circle.

A **circle** is a figure for which all points are at an equal distance from a given point. This given point is called the **center** of the circle.

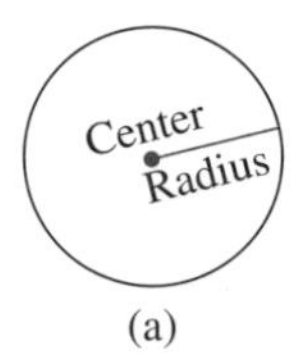

(a)

The **radius** is a line segment from the center to a point on the circle.

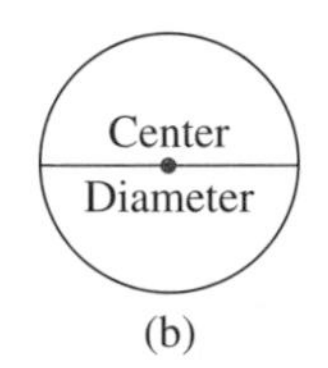

(b)

The **diameter** is a line segment across the circle that passes through the center.

We often use the words **radius** and **diameter** to mean the length of those segments. Note that the plural of radius is radii. Clearly, then,

$$\text{Diameter} = 2 \times \text{radius} \qquad \text{or} \qquad d = 2r$$

We could also say that

$$\text{Radius} = \text{diameter} \div 2 \qquad \text{or} \qquad r = \frac{d}{2}$$

The distance around the rim of the circle is called the **circumference.**

TEACHING TIP Students sometimes wonder why we use 3.14 for pi throughout this book, when they may have used 22/7 as an approximation for pi in high school. Explain that it is more convenient for mathematicians to use a decimal approximation since it can be easily adjusted to any desired level of accuracy by adding more decimal places.

There is a special number called pi, which we denote by the symbol π. π is the number we get when we divide the circumference of a circle by the diameter $\frac{C}{d} = \pi$. π is approximately 3.14159265359. We can approximate π to any number of digits. For all work in this book we will use the following.

> π is approximately 3.14, rounded to the nearest hundredth.

We find the **circumference** C of a circle by multiplying the length of the diameter d times π.

$$C = \pi d$$

EXAMPLE 1 Find the circumference of a quarter if we know the diameter is 2.4 cm. Use $\pi \approx 3.14$. Round your answer to the nearest tenth.

$C = \pi d = (3.14)(2.4 \text{ cm}) = 7.536 \text{ cm} = 7.5 \text{ cm}$ (rounded to the nearest tenth)

Practice Problem 1 Find the circumference of a circle when the diameter is 9 m. Use $\pi \approx 3.14$. Round your answer to the nearest tenth. ■

An alternative formula is $C = 2\pi r$. Remember, $d = 2r$. We can use this formula to find the circumference if we are given the length of the radius.

EXAMPLE 2 Find the circumference of a nickel if we know that it has a radius of 1.05 cm. Use $\pi \approx 3.14$. Round to the nearest tenth.

$$\begin{aligned} C &= 2\pi r \\ &= (2)(3.14)(1.05 \text{ cm}) \\ &= (6.28)(1.05 \text{ cm}) \\ &= 6.594 \text{ cm} \\ &= 6.6 \text{ cm (rounded to the nearest tenth)} \end{aligned}$$

Practice Problem 2 Find the circumference of a circle with radius 14 in. Use $\pi \approx 3.14$. Round to the nearest tenth. ■

When solving word problems, be careful. Ask yourself, "Is the radius given, or is the diameter given?" Then do the calculations accordingly.

EXAMPLE 3 A bicycle tire has a diameter of 24 in. How many feet does the bicycle travel if the wheel makes 1 revolution?

1. *Understand the problem.*
   The distance the wheel travels when it makes 1 revolution is the circumference of the tire. Think of the tire unwinding.

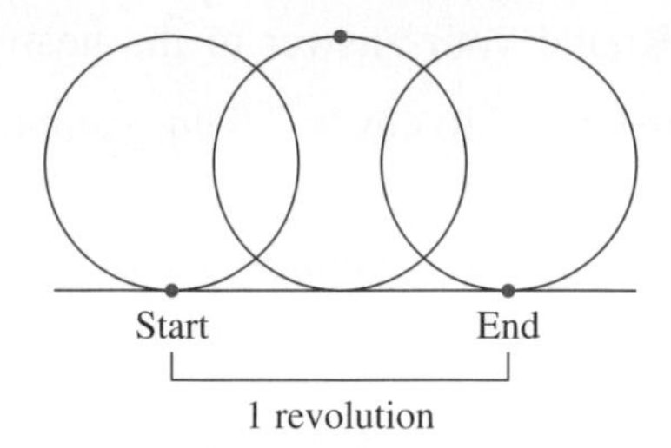

   We are given the *diameter.* The diameter is given in *inches.* The answer should be in *feet.*

2. *Solve and state the answer.*
   Since we are given the diameter, we will use $C = \pi d$. We use 3.14 for π.

$$\begin{aligned} C &= \pi d \\ &= (3.14)(24 \text{ in.}) \\ &= 75.36 \text{ in.} \end{aligned}$$

   We will change 75.36 inches to feet.

$$75.36 \cancel{\text{in.}} \times \frac{1 \text{ ft}}{12 \cancel{\text{in.}}} = 6.28 \text{ ft}$$

   When the wheel makes 1 revolution, the bicycle travels 6.28 ft.

3. *Check.*
   We estimate to check. Since $\pi = 3.14$, we will use 3 for π.

$$C \approx (3)(24 \text{ in.}) \times \frac{1 \text{ ft}}{12 \text{ in.}} \approx 6 \text{ ft} \quad \checkmark$$

Practice Problem 3 A bicycle tire has a diameter of 30 in. How many feet does the bicycle travel if the wheel makes 2 revolutions? ■

The **area of a circle** is the product of π times the radius squared.

$$A = \pi r^2$$

EXAMPLE 4

(a) Estimate the area of a circle whose radius is 6 cm.

(b) Find the exact area of a circle whose radius is 6 cm. Use $\pi \approx 3.14$. Round your answer to the nearest tenth.

(a) Since π is approximately equal to 3.14, we will use 3 for π to estimate the area.

$$\begin{aligned} A &= \pi r^2 \\ &\approx (3)(6\text{ cm})^2 \\ &\approx (3)(6\text{ cm})(6\text{ cm}) \\ &\approx (3)(36\text{ cm}^2) \\ &\approx 108\text{ cm}^2 \end{aligned}$$

Thus our *estimated* area is 108 cm^2.

(b) Now let's compute the exact area.

$$\begin{aligned} A &= \pi r^2 \\ &= (3.14)(6\text{ cm})^2 \\ &= 3.14(6\text{ cm})(6\text{ cm}) \\ &= (3.14)(36\text{ cm}^2) \end{aligned}$$

We *must* square the radius first before multiplying by 3.14.

$$\begin{aligned} &= 113.04\text{ cm}^2 \\ &= 113.0\text{ cm}^2 \text{ (rounded to the nearest tenth)} \end{aligned}$$

Our exact answer is close to the value 108 that we found in part (a).

TEACHING TIP When finding the area, some students try to multiply pi by the value of the radius before they have squared the radius. Remind them of the order of operations that we have used throughout the course. Numbers must be raised to a power first before an indicated multiplication is performed.

Practice Problem 4 Find the area of a circle whose radius is 5 km. Use $\pi = 3.14$. Round your answer to the nearest tenth. Estimate to check. ■

The formula for finding the area uses the length of the radius. If we are given a diameter, we can use the property that $r = \frac{d}{2}$.

EXAMPLE 5 Find the area in square meters of a circle that has a diameter of 18 m. Use $\pi \approx 3.14$. Round your answer to the nearest tenth.

$$r = \frac{d}{2} = \frac{18\text{ m}}{2} = 9\text{ m} \quad \text{The radius is 9 m.}$$

$$\begin{aligned} A &= \pi r^2 \\ &= (3.14)(9\text{ m})^2 \\ &= (3.14)(9\text{ m})(9\text{ m}) \\ &= (3.14)(81\text{ m}^2) \end{aligned}$$

We must square the radius first. Then multiply by 3.14.

$$\begin{aligned} &= 254.34\text{ m}^2 \\ &= 254.3\text{ m}^2 \text{ (rounded to the nearest tenth)} \end{aligned}$$

Practice Problem 5 Find the area of a circle whose diameter is 22 m. Use $\pi \approx 3.14$. Round your answer to the nearest tenth. ■

EXAMPLE 6 Mrs. Heveran wants to buy a circular braided rug that is 8 ft in diameter. Find the cost of the rug at \$35 a square yard.

1. *Understand the problem.*
   We are given the *diameter in feet.* We will need to find the radius. The cost of the rug is in *square yards.* We will need to change square feet to square yards.
2. *Solve and state the answer.*

Find the radius. $r = \frac{d}{2} = \frac{8 \text{ ft}}{2} = 4 \text{ ft}$

Use 3.14 for π.

$$\begin{aligned} A &= \pi r^2 \\ &= (3.14)(4 \text{ ft})^2 \\ &= (3.14)(4 \text{ ft})(4 \text{ ft}) \\ &= (3.14)(16 \text{ ft}^2) \\ &= 50.24 \text{ ft}^2 \end{aligned}$$

Change square feet to square yards. Since 1 yd = 3 ft, $(1 \text{ yd})^2 = (3 \text{ ft})^2$. That is, $1 \text{ yd}^2 = 9 \text{ ft}^2$.

$$50.24 \cancel{\text{ft}^2} \times \frac{1 \text{ yd}^2}{9 \cancel{\text{ft}^2}} = 5.58 \text{ yd}^2$$

Find the cost.

$$\frac{\$35}{1 \cancel{\text{yd}^2}} \times 5.58 \cancel{\text{yd}^2} = \$195.30$$

3. *Check.*
   You may use a calculator to check.

Practice Problem 6 Mr. Lee wants to buy a circular pool cover that is 10 ft in diameter. Find the cost of the pool cover at \$12 a square yard. ■

2 *Solving Area Problems Containing Circles and Other Geometric Shapes*

Several applied area problems have a circular region combined with another region.

EXAMPLE 7 Find the area of the shaded region. Use $\pi \approx 3.14$. Round your answer to the nearest tenth.

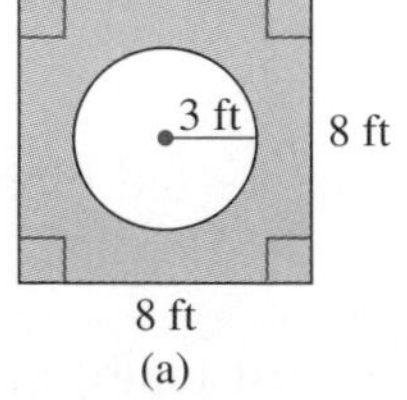

(a)

We will subtract two areas to find the shaded region.

(b)

Area of the square − area of the circle = area of the shaded region

$$A = s^2 \qquad A = \pi r^2$$
$$= (8\text{ ft})^2 \qquad = (3.14)(3\text{ ft})^2$$
$$= 64\text{ ft}^2 \qquad = (3.14)(9\text{ ft}^2)$$
$$= 28.26\text{ ft}^2$$

Area of the square		area of the circle		area of the shaded region
64 ft²	−	28.26 ft²	=	35.74 ft²
			=	35.7 ft² (rounded to nearest tenth)

Practice Problem 7 Find the area of the shaded region in the margin. Use $\pi \approx 3.14$. Round your answer to the nearest tenth. ■

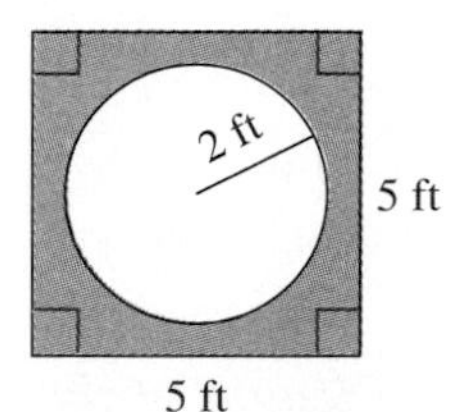

Many geometric shapes involve the semicircle. A **semicircle** is one half of a circle. The area of a semicircle is therefore one-half of the area of a circle.

TEACHING TIP Some students find these area problems with circles and rectangles very difficult. Emphasize the skill of looking at each part separately and then reasoning carefully to determine if two areas should be added or subtracted.

EXAMPLE 8 Find the area of the shaded region. Use $\pi \approx 3.14$. Round to the nearest tenth.

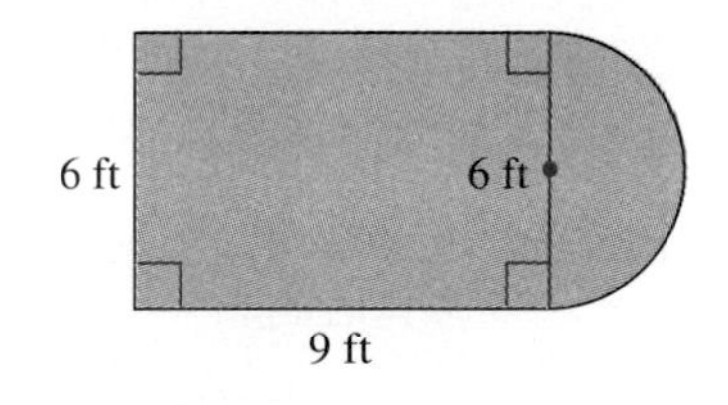

First we will find the area of the semicircle with the diameter of 6 ft.

$$r = \frac{d}{2} = \frac{6\text{ ft}}{2} = 3\text{ ft}$$

The radius is 3 ft. The area of a semicircle with radius 3 ft is

$$A_{\text{semicircle}} = \frac{\pi r^2}{2} = \frac{(3.14)(3\text{ ft})^2}{2} = \frac{(3.14)(9\text{ ft}^2)}{2}$$

$$= \frac{28.26\text{ ft}^2}{2} = 14.13\text{ ft}^2$$

Now we add the area of the rectangle.

$$A = lw = (9\text{ ft})(6\text{ ft}) = 54\text{ ft}^2$$

54.00 ft²	area of rectangle
+ 14.13 ft²	area of semicircle
68.13 ft²	total area

Rounded to the nearest tenth, area = 68.1 ft².

Practice Problem 8 Find the area of the shaded region. Use $\pi \approx 3.14$. Round to the nearest tenth. ■

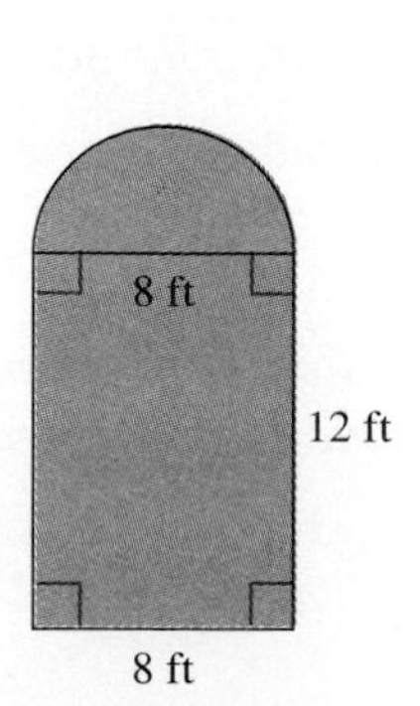

7.6 Exercises

Verbal and Writing Skills

1. The distance around a circle is called the ___circumference___.

2. The radius is a line segment from the ___center___ to a point on the circle.

3. The diameter is two times the ___radius___ of the circle.

4. Explain in your own words how to find the area of a circle if you are given the diameter. You need to divide the diameter by 2, then use the area formula, $A = \pi r^2$.

In all problems use $\pi \approx 3.14$. Round each answer to the nearest tenth.

Find the length of the diameter of a circle if the radius has the value given.

5. $r = 29$ in.
2×29 in. $= 58$ in.

6. $r = 33$ in.
2×33 in. $= 66$ in.

7. $r = 8.5$ mm
2×8.5 mm $= 17$ mm

8. $r = 7.7$ m
2×7.7 m $= 15.4$ m

Find the length of the radius of a circle if the diameter has the value given.

9. $d = 45$ yd
$45 \div 2$ yd $= 22.5$ yd

10. $d = 65$ yd
$65 \div 2$ yd $= 32.5$ yd

11. $d = 3.8$ cm
$3.8 \div 2$ cm $= 1.9$ cm

12. $d = 5.2$ cm
$5.2 \div 2$ cm $= 2.6$ cm

Find the circumference of each circle.

13. Diameter = 32 cm
$C = 3.14 \times 32 \text{ cm} \approx 100.48 \text{ cm}$

14. Diameter = 22 cm
$C = 3.14 \times 22 \text{ cm} \approx 69.08 \text{ cm}$

15. Radius = 11 in.
$C = 2 \times 3.14 \times 11 \text{ in.} \approx 69.1 \text{ in.}$

16. Radius = 15 in.
$C = 2 \times 3.14 \times 15 \text{ in.}$
$= 94.2 \text{ in.}$

A bicycle wheel makes 1 revolution. Determine how far the bicycle travels in inches.

17. The diameter of the wheel is 26 in.
$C \approx (3.14)(26 \text{ in.}) \approx 81.6 \text{ in.}$

18. The diameter of the wheel is 28 in.
$C \approx (3.14)(28 \text{ in.}) \approx 87.9 \text{ in.}$

A bicycle wheel makes 5 revolutions. Determine how far the bicycle travels in feet.

19. The diameter of the wheel is 32 in.
$C \approx (3.14)(32 \text{ in.}) \approx 100.48 \text{ in.}$

$(100.48 \text{ in.})(5) \times \frac{1 \text{ ft}}{12 \text{ in.}} = 41.87 \text{ ft}$

20. The diameter of the wheel is 24 in.
$C \approx (3.14)(24 \text{ in.}) \approx 75.36 \text{ in.}$

$(75.36 \text{ in.})(5) \times \frac{1 \text{ ft}}{12 \text{ in.}} = 31.4 \text{ ft}$

Find the area of each circle.

21. Radius = 5 yd
$A = 3.14 \times (5 \text{ yd})^2 = 78.5 \text{ yd}^2$

22. Radius = 7 yd
$A = 3.14 \times (7 \text{ yd})^2 \approx 153.9 \text{ yd}^2$

23. Radius = 17 m
$A = 3.14 \times (17 \text{ m})^2 \approx 907.5 \text{ m}^2$

24. Radius = 14 m
$A = 3.14 \times (14 \text{ m})^2 \approx 615.4 \text{ m}^2$

25. Diameter = 32 cm
$r = \frac{32}{2} = 16 \quad A = 3.14 \times (16 \text{ cm})^2 \approx 803.8 \text{ cm}^2$

26. Diameter = 44 cm
$r = \frac{44}{2} = 22 \quad A = 3.14 \times (22 \text{ cm})^2 = 1519.8 \text{ cm}^2$

A water sprinkler sends water out in a circular pattern. Determine how large an area is watered.

27. The radius of watering is 8 ft.
$A = 3.14 \times (8 \text{ ft})^2 = 201.0 \text{ ft}^2$

28. The radius of watering is 12 ft.
$A = 3.14 \times (12 \text{ ft})^2 = 452.16 \text{ ft}^2$

A radio station sends out radio waves in all directions from a tower at the center of the circle of broadcast range. Determine how large an area is reached.

29. The diameter is 120 mi.
$r = \frac{120}{2} = 60 \quad A = 3.14 \times (60 \text{ mi})^2 = 11{,}304 \text{ mi}^2$

30. The diameter is 90 mi.
$r = \frac{90}{2} = 45 \quad A = 3.14 \times (45 \text{ mi})^2 = 6358.5 \text{ mi}^2$

Use $\pi \approx 3.14159$ in all calculations for problems 31 and 32.

31. Find the circumference of the circle with a 0.223-m diameter. 0.70057457 m

32. Find the area of the circle whose radius is 1.39 cm. 6.069866 cm^2

Find the area of each semicircle.

33.

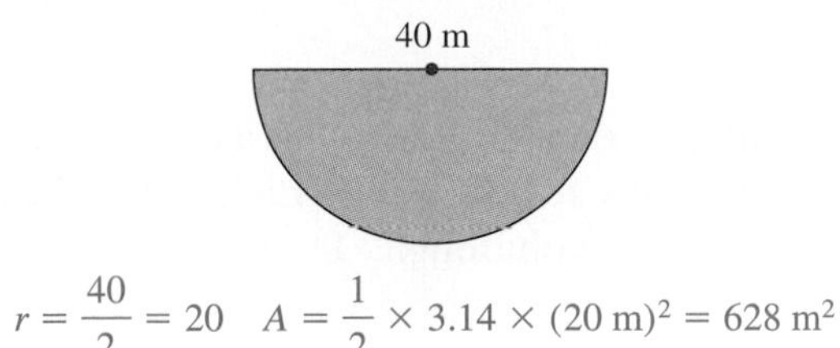

$r = \frac{40}{2} = 20 \quad A = \frac{1}{2} \times 3.14 \times (20\text{ m})^2 = 628\text{ m}^2$

34.

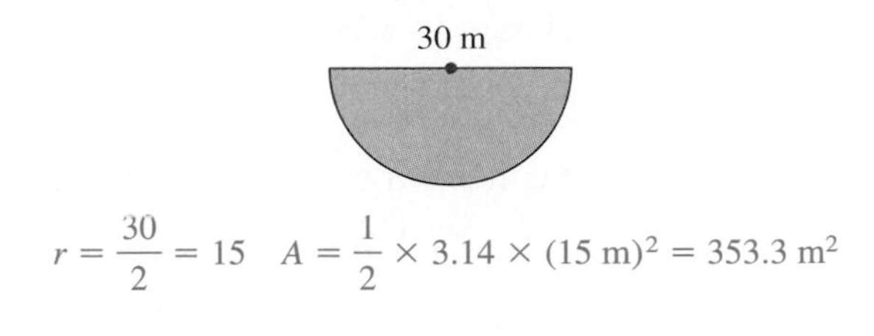

$r = \frac{30}{2} = 15 \quad A = \frac{1}{2} \times 3.14 \times (15\text{ m})^2 = 353.3\text{ m}^2$

Find the area of the shaded regions.

35.

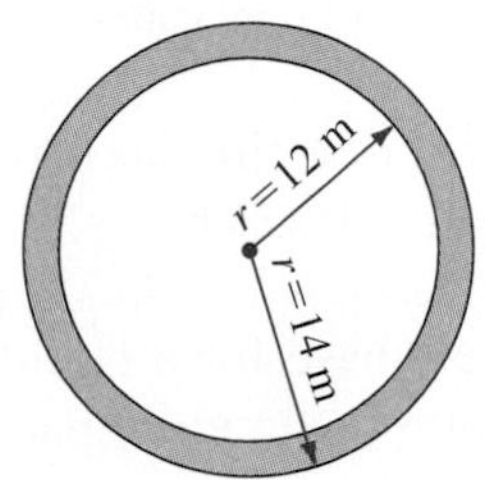

$A = 3.14 \times (14\text{ m})^2 - 3.14 \times (12\text{ m})^2$
$= 615.44\text{ m} - 452.16\text{ m} \approx 163.3\text{ m}^2$

36.

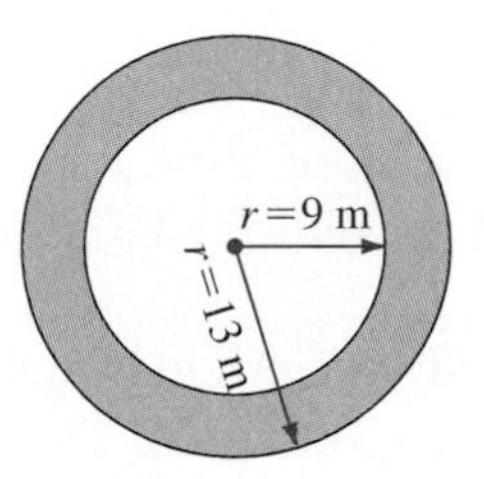

$A = 3.14 \times (13\text{ m})^2 - 3.14 \times (9\text{ m})^2$
$= 530.66\text{ m} - 254.34\text{ m} \approx 276.3\text{ m}^2$

37.

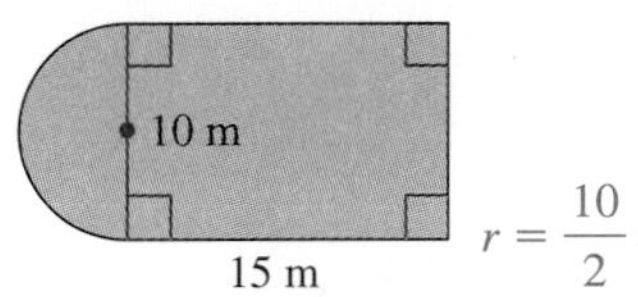

$r = \frac{10}{2} = 5$

$A = \frac{1}{2} \times 3.14 \times (5\text{ m})^2 + 15\text{ m} \times 10\text{ m} = 39.25\text{ m}^2 + 150\text{ m}^2 \approx 189.3\text{ m}^2$

38.

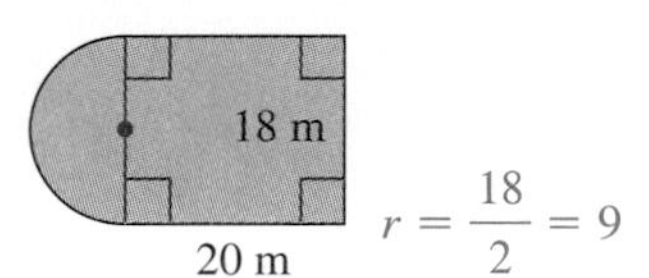

$r = \frac{18}{2} = 9$

$A = \frac{1}{2} \times 3.14 \times (9\text{ m})^2 + 20\text{ m} \times 18\text{ m} = 127.17\text{ m}^2 + 360\text{ m}^2 \approx 487.2\text{ m}^2$

39.

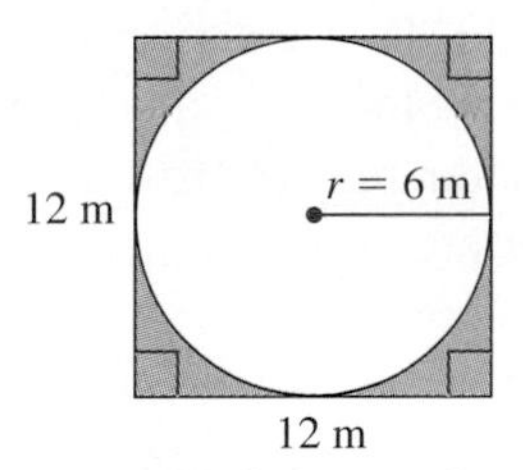

$A = 12\text{ m} \times 12\text{ m} - 3.14 \times (6\text{ m})^2 = 144\text{ m}^2 - 113.04\text{ m}^2 \approx 31.0\text{ m}^2$

40.

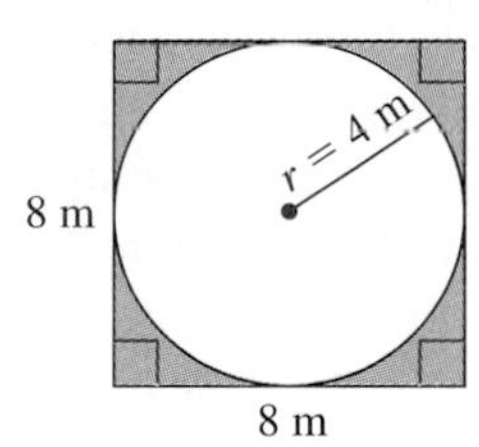

$A = 8\text{ m} \times 8\text{ m} - 3.14 \times (4\text{ m})^2 = 64\text{ m}^2 - 50.24\text{ m}^2 \approx 13.8\text{ m}^2$

Find the cost of fertilizing a playing field at \$0.20 per square yard for the conditions stated.

41. The rectangular part of the field is 120 yd long and the diameter of each semicircle is 40 yd.

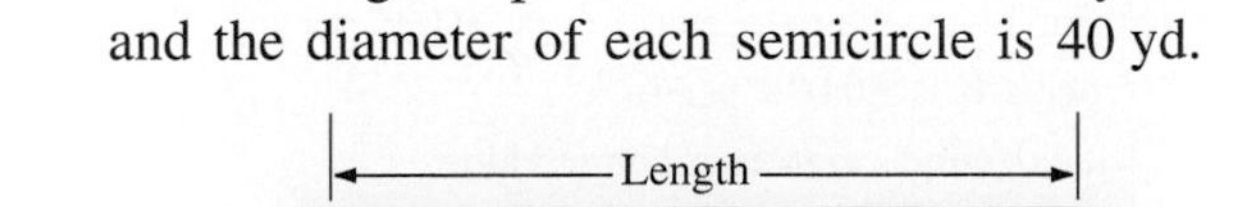
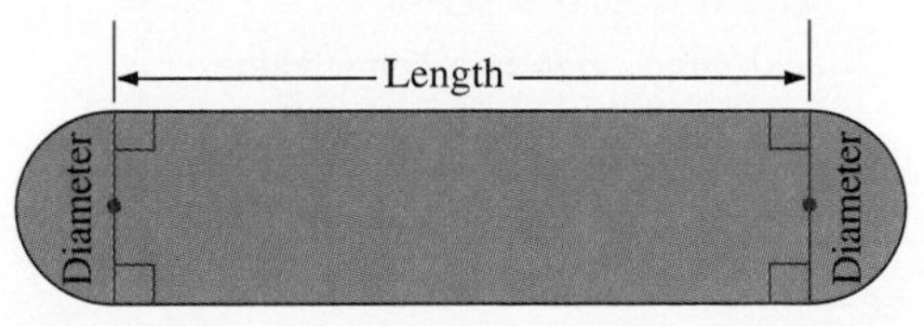

$r = \frac{40}{2} \quad A = 3.14 \times (20\text{ yd})^2 + 120\text{ yd} \times 40\text{ yd} = 1256\text{ yd}^2 + 4800\text{ yd}^2 = 6056\text{ yd}^2$

$\text{Cost} = 6056\text{ yd}^2 \times \frac{0.20}{\text{yd}^2} = \$1211.20$

42. The rectangular part of the field is 110 yd long and the diameter of each semicircle is 50 yd.

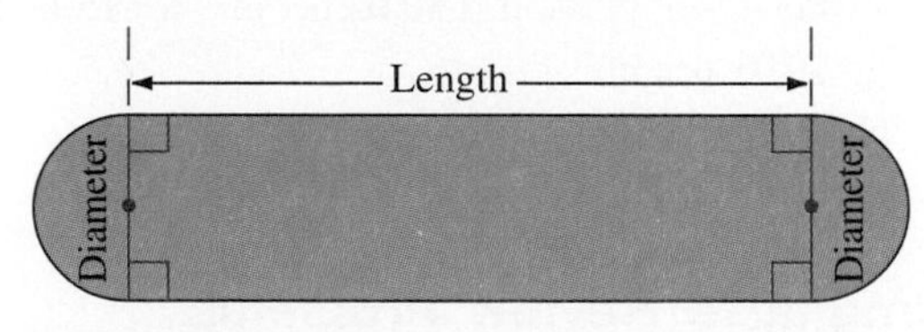

$r = \frac{50}{2} \quad A = 3.14 \times (25\text{ yd})^2 + 110\text{ yd} \times 50\text{ yd} = 7462.5\text{ yd}^2$

$\text{Cost} = 7462.5\text{ yd}^2 \times \frac{0.20}{\text{yd}} = \$1492.50$

Applications

Use $\pi \approx 3.14$ in problems 44–52. Round answers to the nearest hundredth.

43. A porthole window on a freighter ship has a diameter of 2 ft. What is the length of the insulating strip that encircles the window and keeps out wind and moisture? 6.28 ft

44. A manhole cover has a diameter of 3 ft. What is the length of the brass grip-strip that encircles the cover making it easier to manage? 9.42 ft

45. Elena's car has tires with a radius of 14 in. How many feet does her car travel if the wheel makes 35 revolutions? 256.43 feet

46. Jimmy's truck has tires with a radius of 30 inches. How many feet does his truck travel if the wheel makes 9 revolutions? 141.3 feet

47. Carlotta's car has tires with a radius of 16 in. Carlotta moved her car forward 20,096 in. How many complete revolutions did the wheels turn? 200 revolutions

48. Mickey's car has tires with a radius of 15 in. He backed his car up a distance of 9891 in. How many complete revolutions did the wheels turn backing up? 105 revolutions

49. Find the area of a circular Coca-Cola sign with a diameter of 64 in. 3215.36 in.^{2}

50. Find the area of a circular flower bed with a diameter of 14 ft. 153.86 ft^2

51. Sarah bought a circular marble table top to use as her dining room table. The marble is 6 ft in diameter. Find the cost of the table top at $72 per square yard of marble. $226.08

52. Tom made a base for a circular patio by pouring concrete into a circular space 10 ft in diameter. Find the cost at $18 per square yard. $157.00

To Think About

53. A 15-in.-diameter pizza costs $6.00. A 12-in.-diameter pizza costs $4.00. The 12-in.-diameter pizza is cut into six slices. The 15-in.-diameter pizza is cut into eight slices.

(a) What is the cost per slice of the 15-in.-diameter pizza? How many square inches of pizza are in one slice?

(a) $0.75 per slice, 22.1 in.^{2}

(b) What is the cost per slice of the 12-in.-diameter pizza? How many square inches of pizza are in one slice?

(b) $0.67 per slice, 18.8 in.^{2}

(c) If you want more value for your money, which slice of pizza should you buy?

(c) For 12-in. pizza, it is $0.04 per in.^{2}; for 15-in. pizza, it is $0.03 per in.^{2}.
15-inch pizza is a better value.

54. A 14-in.-diameter pizza costs $5.50. It is cut into eight pieces. A 12.5 in. × 12.5 in. square pizza costs $6.00. It is cut into nine pieces.

(a) What is the cost of one piece of the 14-in.-diameter pizza? How many square inches of pizza are in one piece?

(a) $0.69 a piece, 19.2 in.^{2}

(b) What is the cost of one piece of the 12.5 in. × 12.5 in. square pizza? How many square inches of pizza are in one piece?

(b) $0.67 per piece, 17.4 in.^{2}

(c) If you want more value for your money, which piece of pizza should you buy?

(c) For 14-in.-diameter pizza, it is $0.036 per in.^{2}; for square pizza it is $0.039 per in.^{2}.
14-inch round pizza is a better value.

Cumulative Review Problems

55. Find 16% of 87. 13.92

56. What is 0.5% of 60? 0.3

57. 12% of what number is 720? 6000

58. 19% of what number is 570? 3000

7.7 Volume

After studying this section, you will be able to:

MathPro Video 16 SSM

1 *Find the volume of a rectangular solid (box).*
2 *Find the volume of a cylinder.*
3 *Find the volume of a sphere.*
4 *Find the volume of a cone.*
5 *Find the volume of a pyramid.*

1 *Finding the Volume of a Rectangular Solid (Box)*

How much grain can that shed hold? How much water is in the polluted lake? How much air is inside a basketball? These are questions of **volume.** In this section we compute the volume of several three-dimensional geometric figures: the rectangular solid (box), cylinder, sphere, pyramid, and cone.

We can start with a box 1 in. × 1 in. × 1 in.

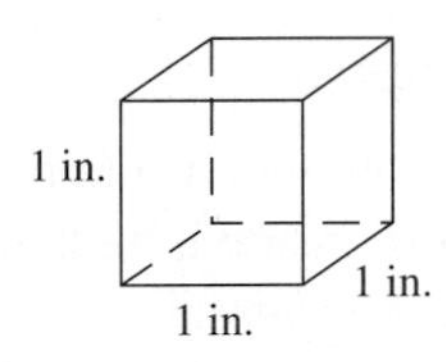

This box has a volume of 1 cubic inch (written 1 in.^{3}). We can use this as a **unit of volume.** Volume is measured in cubic units like cubic meters (abbreviated m^3) or cubic feet (abbreviated ft^3). When we measure volume, we are measuring the space inside an object.

TEACHING TIP Depending on the ability level of your students and the goals of the course, you may find that you prefer to cover only some of the five types of volume problems in this section.

The **volume of a rectangular solid** (box) is the length times the width times the height:

$$V = lwh$$

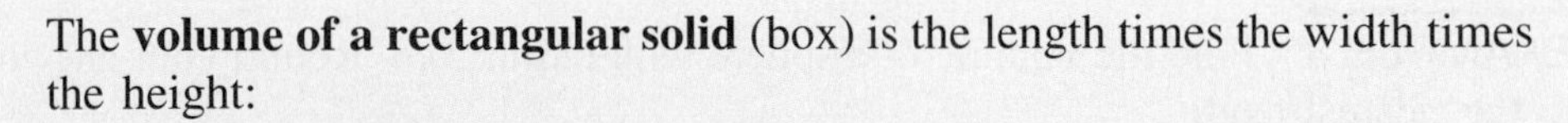

EXAMPLE 1 Find the volume of a box of width 2 ft, length 3 ft, and height 4 ft.

$$V = lwh = (3 \text{ ft})(2 \text{ ft})(4 \text{ ft}) = (6)(4) \text{ ft}^3 = 24 \text{ ft}^3$$

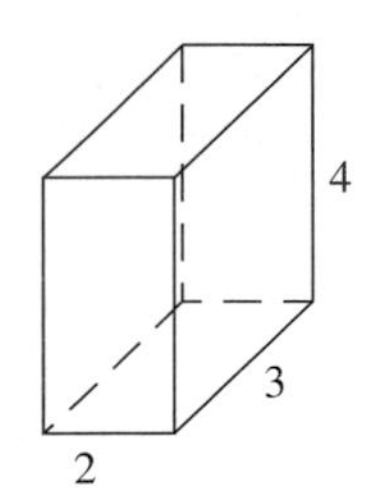

Practice Problem 1 Find the volume of a box of width 5 m, length 6 m, height 2 m. ■

2 *Finding the Volume of a Cylinder*

Cylinders are the shape we observe when we see a tin can or a tube.

The **volume of a cylinder** is the product of the area of a circle πr^2 times the height h.

$$V = \pi r^2 h$$

TEACHING TIP The volume of a lot of unusually shaped solids can be found by using the general concept of "multiply the area of the base by the height." For example, draw a sketch of a solid whose base is a trapezoid or a parallelogram instead of a circle.

We will continue to use π as 3.14, as we did in Section 7.6, on all volume problems requiring the use of π.

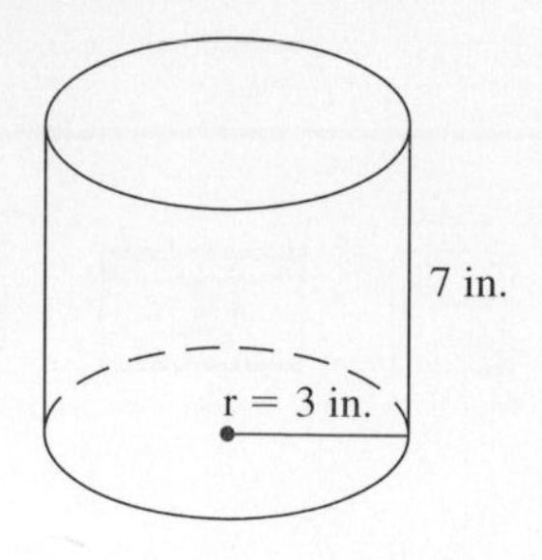

EXAMPLE 2 Find the volume of a cylinder of radius 3 in. and height 7 in. Round the answer to the nearest tenth.

$$V = \pi r^2 h = (3.14)(3 \text{ in.})^2(7 \text{ in.}) = (3.14)(3 \text{ in.})(3 \text{ in.})(7 \text{ in.})$$

Be sure to square the radius before doing any other multiplication.

$$V = (3.14)(9 \text{ in.})^2(7 \text{ in.}) = (28.26 \text{ in.})^2(7 \text{ in.})$$

$$= 197.82 \text{ in.}^3 = 197.8 \text{ in.}^3 \text{ rounded to nearest tenth}$$

Practice Problem 2 Find the volume of a cylinder of radius 2 in. and height 5 in. Round to the nearest tenth. ■

To Think About Look at the formula for the volume of a cylinder. Part of the formula is πr^2. What is πr^2? We could write a formula for the volume of a cylinder in another way. Use the fact that the area of the base of the cylinder is πr^2—that is, $B = \pi r^2$—to write another formula for the volume of a cylinder.

3 *Finding the Volume of a Sphere*

Have you ever considered how you would find the volume of the inside of a ball? How many cubic inches of air are inside a basketball? To answer these questions we need a volume formula for a *sphere.*

The **volume of a sphere** is 4 times π times the radius cubed divided by 3.

$$V = \frac{4\pi r^3}{3}$$

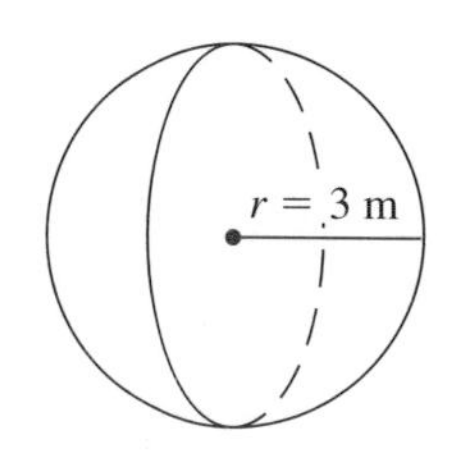

EXAMPLE 3 Find the volume of a sphere with radius 3 m. Round your answer to the nearest tenth.

$$V = \frac{4\pi r^3}{3} = \frac{(4)(3.14)(3 \text{ m})^3}{3}$$

$$= \frac{(4)(3.14)(3)(3)(\cancel{3}) \text{ m}^3}{\cancel{3}}$$

$$V = (4)(3.14)(9) \text{ m}^3 = (12.56)(9) \text{ m}^3 = 113.04 \text{ m}^3$$

$$= 113.0 \text{ m}^3 \text{ rounded to the nearest tenth}$$

Practice Problem 3 Find the volume of a sphere with radius 6 m. Round your answer to the nearest tenth. ■

4 *Finding the Volume of a Cone*

We see the shape of a cone when we look at the sharpened end of a pencil or at an ice cream cone. To find the volume of a cone we use the following formula.

The **volume of a cone** is π times the radius squared times the height divided by 3.

$$V = \frac{\pi r^2 h}{3}$$

h
r

EXAMPLE 4 Find the volume of a cone of radius 7 m and height 9 m.

$$V = \frac{\pi r^2 h}{3}$$

$$= \frac{(3.14)(7\text{ m})^2(9\text{ m})}{3}$$

$$= \frac{(3.14)(7\text{ m})(7\text{ m})(9\text{ m})}{3}$$

$$= (3.14)(49)(3)\text{ m}^3 = (153.86)(3)\text{ m}^3 = 461.58\text{ m}^3$$

$$= 461.6\text{ m}^3 \text{ rounded to the nearest tenth}$$

Practice Problem 4 Find the volume of a cone of radius 5 m and height 12 m. Round to the nearest tenth. ■

5 *Finding the Volume of a Pyramid*

You have seen pictures of the great pyramids of Egypt. These amazing stone structures are over 4000 years old.

The **volume of a pyramid** is obtained by multiplying the area B of the base of the pyramid by the height h and dividing by 3.

$$V = \frac{Bh}{3}$$

EXAMPLE 5 Find the volume of a pyramid with height = 6 m, length of base = 7 m, width of base = 5 m.

The base is a rectangle.

Area of base = (7 m)(5 m) = 35 m^2

Substituting the area of the base 35 m^2 and the height of 6 m, we have

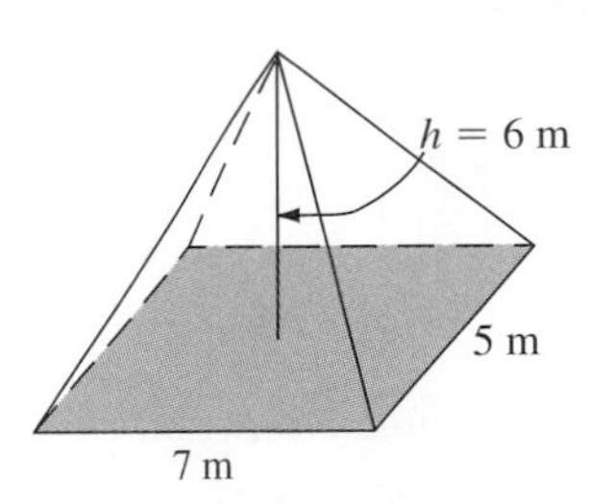

$$V = \frac{Bh}{3} = \frac{(35\text{ m}^2)(6\text{ m})}{3}$$

$$= (35)(2)\text{ m}^3$$

$$= 70\text{ m}^3$$

Practice Problem 5 Find the volume of a pyramid having the dimensions given.

(a) Height 10 m, width 6 m, length 6 m.
(b) Height 15 m, width 7 m, length 8 m. ■

7.7 Exercises

Verbal and Writing Skills

Match the formula for volume with the figure it belongs to on the right.

1. $V = lwh$ box

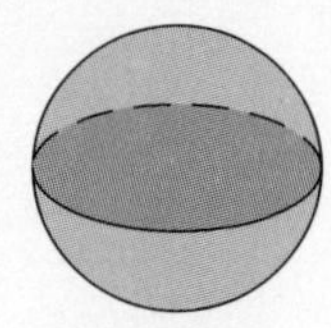

2. $V = \pi r^2 h$ cylinder

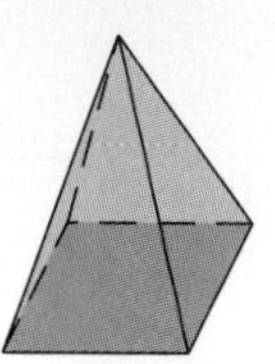

3. $V = \dfrac{4\pi r^3}{3}$ sphere

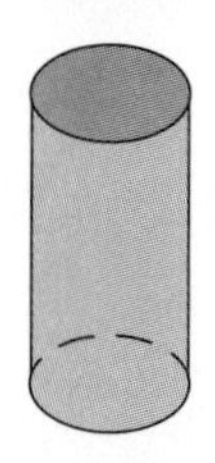

4. $V = \dfrac{\pi r^2 h}{3}$ cone

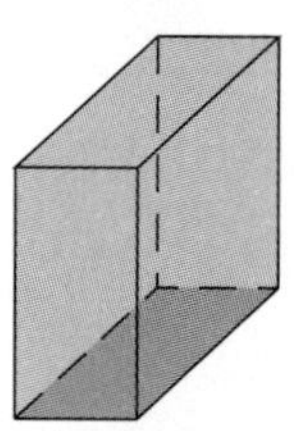

5. $V = \dfrac{Bh}{3}$ pyramid

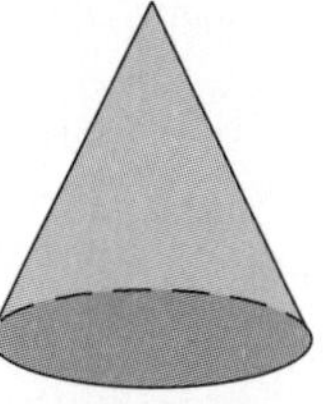

6. $V = s^3$ cube

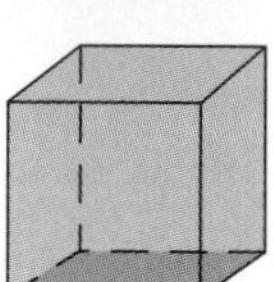

Find each volume. Use $\pi \approx 3.14$. Round each answer to the nearest tenth when necessary.

7. A rectangular solid with width = 2 m, length = 4 m, height = 3 m
$V = 2\text{ m} \times 4\text{ m} \times 3\text{ m} = 24\text{ m}^3$

8. A rectangular solid with width = 5 m, length = 4 m, height = 6 m
$V = 5\text{ m} \times 4\text{ m} \times 6\text{ m} = 120\text{ m}^3$

9. A rectangular solid with width = 14 mm, length = 20 mm, height = 2.5 mm
$V = 20\text{ mm} \times 14\text{ mm} \times 2.5\text{ mm} = 700\text{ mm}^3$

10. A rectangular solid with width = 12 mm, length = 30 mm, height = 1.5 mm
$V = 30\text{ mm} \times 12\text{ mm} \times 1.5\text{ mm} = 540\text{ mm}^3$

11. A cylinder with radius 2 m and height 7 m
$V = 3.14 \times (2\text{ m})^2 \times 7\text{ m} = 87.9\text{ m}^3$

12. A cylinder with radius 3 m and height 8 m
$V = 3.14 \times (3\text{ m})^2 \times 8\text{ m} = 226.1\text{ m}^3$

13. A cylinder with radius 12 m and height 5 m
$V = 3.14 \times (12\text{ m})^2 \times 5\text{ m} = 2260.8\text{ m}^3$

14. A cylinder with radius 9 m and height 6 m
$V = 3.14 \times (9\text{ m})^2 \times 6\text{ m} = 1526.04\text{ m}^3$
$\approx 1526.1\text{ m}^3$

15. A sphere with radius 9 yd

$V = \frac{4 \times 3.14 \times (9 \text{ yd})^3}{3} = 3052.1 \text{ yd}^3$

16. A sphere with radius 12 yd

$V = \frac{4 \times 3.14 \times (12 \text{ yd})^3}{3} = 7234.6 \text{ yd}^3$

17. A sphere with radius 4 m

$V = \frac{4 \times 3.14 \times (4 \text{ m})^3}{3} = 267.9 \text{ m}^3$

18. A sphere with radius 5 m

$V = \frac{4 \times 3.14 \times (5 \text{ m})^3}{3} = 523.3 \text{ m}^3$

Problems 18 and 19 involve hemispheres. A hemisphere is exactly one half of a sphere.

19. Find the volume of a hemisphere with radius = 7 m.

$V = \frac{1}{2} \times \frac{4}{3} \times 3.14 \times (7 \text{ m})^3 \approx 718.0 \text{ m}^3$

20. Find the volume of a hemisphere with radius = 6 m.

$V = \frac{1}{2} \times \frac{4}{3} \times 3.14 \times (6 \text{ m})^3 \approx 452.2 \text{ m}^3$

Find each volume. Use $\pi \approx 3.14$. Round each answer to the nearest tenth.

21. A cone with a height of 12 cm and a radius of 9 cm

$V = \frac{1}{3} \times 3.14 \times (9 \text{ cm})^2 \times 12 \text{ cm} = 1017.4 \text{ cm}^3$

22. A cone with a height of 12 cm and a radius of 9 cm

$V = \frac{1}{3} \times 3.14 \times (9 \text{ cm})^2 \times 12 \text{ cm} \approx 1017.4 \text{ cm}^3$

23. A cone with a height of 10 ft and a radius of 5 ft

$V = \frac{1}{3} \times 3.14 \times (5 \text{ ft})^2 \times 10 \text{ ft} = 261.7 \text{ ft}^3$

24. A cone with a height of 12 ft and a radius of 6 ft

$V = \frac{1}{3} \times 3.14 \times (6 \text{ ft})^2 \times 12 \text{ ft} = 452.2 \text{ ft}^3$

25. A pyramid with a height of 7 m and a square base of 3 m on a side

$V = \frac{1}{3} \times 3 \text{ m} \times 3 \text{ m} \times 7 \text{ m} = 21 \text{ m}^3$

26. A pyramid with a height of 10 m and a square base of 7 m on a side

$V = \frac{1}{3} \times 7 \text{ m} \times 7 \text{ m} \times 10 \text{ m} = 163.3 \text{ m}^3$

27. A pyramid with a height of 5 m and a rectangular base measuring 6 m by 12 m

$V = \frac{1}{3} \times 12 \text{ m} \times 6 \text{ m} \times 5 \text{ m} = 120 \text{ m}^3$

28. A pyramid with a height of 10 m and a rectangular base measuring 8 m by 14 m

$V = \frac{1}{3} \times 14 \text{ m} \times 8 \text{ m} \times 10 \text{ m} = 373.3 \text{ m}^3$

Use $\pi \approx 3.14159$ whenever necessary in solving the following problems.

29. Find the volume of a **box** with dimensions 5.88 m by 4.26 m by 6.82 m.

170.832816 m^3

30. Find the volume of a **cylinder** of radius 3.22 cm and height 8.43 cm.

274.5926 cm^3

31. Find the volume of a **sphere** with radius 5.21 m.

592.3814 m^3

32. Find the volume of a **cone** of radius 21.12 mm and height 32 mm.

14,947.414 mm^3

33. Find the volume of a **pyramid** with height = 9.212 ft, length of rectangular base = 6.22 ft, and length of width = 5.01 ft.

95.6887288 ft^3

34. Find the volume of a **hemisphere** with radius = 8.583 ft.

1324.2671 ft^3

Applications

Use $\pi \approx 3.14$ whenever necessary.

35. The Essex Village neighborhood park has a great sandbox for young children that measures 20 yd × 18 yd. The parks committee will be adding 4 in. of new sand. How many cubic yards of sand will they need?

4 in. $= \frac{1}{9}$ yard $\quad V = 20 \text{ yd} \times 18 \text{ yd} \times \frac{1}{9} \text{ yd} = 40 \text{ yd}^3$

36. Mr. Bledsoe wants to put down a crushed-stone driveway to his summer camp. The driveway is 7 yd wide and 120 yd long. The crushed stone is to be 4 in. thick. How many cubic yards of stone will he need?

4 in. $= \frac{1}{9}$ yd $\quad V = 120 \text{ yd} \times 7 \text{ yd} \times \frac{1}{9} \text{ yd} = 93.3 \text{ yd}^3$

A collar of Styrofoam is made to insulate a pipe. Find the volume of the unshaded region (which represents the collar). The large radius R is to the outer rim. The small radius r is to the edge of the insulation.

37. $r = 3$ in.
$R = 5$ in.
$h = 20$ in.

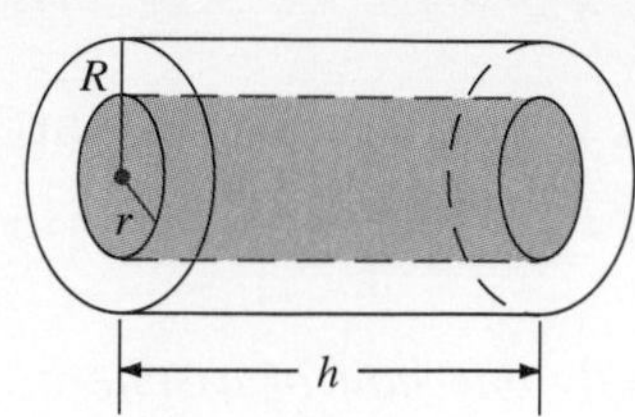

$V = 3.14 \times (5 \text{ in.})^2 \times 20 \text{ in.}$
$- 3.14 \times (3 \text{ in.})^2 \times 20 \text{ in.} = 1004.8 \text{ in.}^3$

38. $r = 4$ in.
$R = 6$ in.
$h = 25$ in.

$V = 3.14 \times (6 \text{ in.})^2 \times 25 \text{ in.} - 3.14 \times (4 \text{ in.})^2 \times 25 \text{ in.} = 1570 \text{ in.}^3$

39. Jupiter has a radius of approximately 45,000 mi. Assuming that it is a sphere, what is its volume?

$V = \frac{4}{3} \times 3.14 \times (45{,}000 \text{ mi})^3$

$V \approx 381{,}510{,}000{,}000{,}000 \text{ mi}^3$

40. A tennis ball has a diameter of 2.5 in. A baseball has a diameter of 2.9 in. What is the difference in *volume* between the baseball and the tennis ball? (Round your answer to nearest tenth.)

$r_b = \frac{2.9 \text{ in.}}{2} \qquad r_t = \frac{2.5 \text{ in.}}{2}$

$D = \frac{4}{3} \times 3.14 \times \left(\frac{2.9 \text{ in.}}{2}\right)^3 - \frac{4}{3} \times 3.14 \times \left(\frac{2.5 \text{ in.}}{2}\right)^3$

$\approx 12.8 \text{ in.}^3 - 8.2 \text{ in.}^3 = 4.6 \text{ in.}^3$

41. The nose cone of a passenger jet is used to receive and send radar. It is made of a special aluminum alloy that costs \$4.00 per cm^3. The cone has a radius of 5 cm and a height of 9 cm. What is the cost of the aluminum needed to make this *solid* nose cone?

$V = \frac{1}{3}(3.14)(5 \text{ cm})^2(9 \text{ cm}) = 235.5 \text{ cm}^3$

$235.5 \text{ cm}^3 \times \frac{\$4.00}{1 \text{ cm}^3} = \$942$

42. A special stainless-steel cone sits on top of a cable television antenna. The cost of the stainless steel is \$3.00 per cm^3. The cone has a radius of 6 cm and a height of 10 cm. What is the cost of the stainless steel needed to make this *solid* steel cone?

$V = \frac{1}{3}(3.14)(6 \text{ cm}^2)(10 \text{ cm}) = 376.8 \text{ cm}^3$

$376.8 \text{ cm}^2 \times \frac{\$3.00}{1 \text{ cm}^3} = \$1130.40$

Suppose that a new pyramid has been found in South America. Find the volume of the pyramid for the conditions given.

43. The stone pyramid has a rectangular base. This rectangular base measures 87 yd by 130 yd. The pyramid has a height of 70 yd.

$V = \frac{1}{3} \times 87 \text{ yd} \times 130 \text{ yd} \times 70 \text{ yd} = 263{,}900 \text{ yd}^3$

44. The pyramid given in problem 43 is made of solid stone. It is not hollow like the pyramids of Egypt. It is composed of layer after layer of cut stone. The stone weighs 422 lb per cubic yard. How many *pounds* will the pyramid weigh? How many tons will the pyramid weigh?

111,365,800 lb or 55,682.9 tons

Cumulative Review Problems

45. Add. $7\frac{1}{3} + 2\frac{1}{4}$

$9\frac{7}{12}$

46. Subtract. $9\frac{1}{8} - 2\frac{3}{4}$

$6\frac{3}{8}$

47. Multiply. $2\frac{1}{4} \times 3\frac{3}{4}$

$\frac{135}{16}$ *or* $8\frac{7}{16}$

48. Divide. $7\frac{1}{2} \div 4\frac{1}{5}$

$\frac{25}{14}$ *or* $1\frac{11}{14}$

Putting Your Skills to Work

THE MATHEMATICAL DETECTIVE

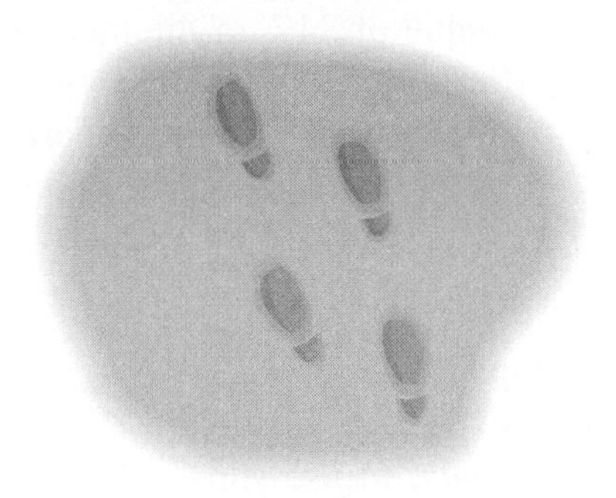

In a number of recent trials, the footprint of the suspect has been a significant aspect of the case. In modern forensic science the use of a footprint to identify a person has become almost as important as the use of a fingerprint. Often the footprint will reveal the exact type of shoe (size, brand name, and model) and the weight of the person who made the footprint.

The approximate weight (W) measured in pounds of the person making a footprint is given by the equation $W = PA$, where P is the pressure made by the person making the footprint and A is the measure of the area of the footprint. For example, suppose a detective at a crime scene discovered that due to the depth of the footprint in soft soil it can be determined that the person making the footprint made the impression at 6 pounds per square inch. Further, the detective determines that the area of the footprint is 20 square inches. Then $W = (6 \text{ lb/in.}^2)(20 \text{ in.}^2) = 120$ lb. The approximate weight of the person making the footprint is 120 pounds.

PROBLEMS FOR INDIVIDUAL INVESTIGATION AND CALCULATION

1. A detective measured a pressure reading of 7 lb/in.^{2} on the ground right next to a footprint whose area was 32 in.^{2}. What was the weight of the person making the footprint? 224 pounds

2. A person who weighs 150 pounds made a footprint at a pressure reading of 5 lb/in.^{2}. What is the area of the footprint? 30 in.^{2}

PROBLEMS FOR GROUP INVESTIGATION AND COOPERATIVE LEARNING

Together with other members of your class determine the answers to the following. Round all answers to the nearest tenth.

3. A footprint is found with the shape as indicated in this drawing. The footprint boundaries of the heel are a rectangle that measures 3.2 inches by 3.5 inches. The bottom of the heel is in the shape of a semicircle with a radius of 1.6 inches. The top of the footprint is shaped like a trapezoid with bases of 5 inches and 6 inches and a height of 3.8 inches. Find the area of the footprint. It is determined that the footprint was made at a pressure reading of 5.5 lb/in.^{2}. Approximately how much did the person making the footprint weigh? 36.1 in.^{2}, 198.7 lb

4. A person whose footprints are 48 inches apart or less (measuring left heel to left heel impressions) is considered to be walking. However a person whose footprints are more than 48 inches apart is considered to be running. A runner's foot hits the ground with greater pressure. To adjust for this in the formula, if a person is running we use an adjusted value of 84% of the measured pressure. If it is determined that person was running because his footprints were 78 inches apart, and his footprints have an area of 27.5 in.^{2} at a pressure of 7.2 lb/in.^{2}, what would be the estimated weight of the person? 166.3 lb

5. Most footprints do not consist of perfect rectangular and trapezoidal shapes, so the area is more difficult to determine. How do you suspect experts in this field determine the area of a footprint if it is not in the exact shape to use some geometrical formula for area? Answers will vary.

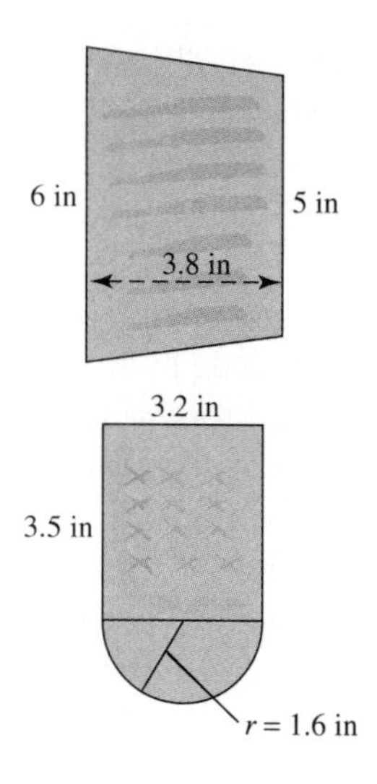

INTERNET CONNECTIONS: Go to ''http://www.prenhall.com/tobey'' to be connected
Site: Kansas Photos (Kansas Bigfoot Center) or a related site

6. This site contains photographs of footprints that may have been made by a bigfoot creature (sasquatch). If these footprints were made at a pressure reading of 6.8 lb/in.^{2}, what was the approximate weight of the creature? (To simplify calculations, you may assume that the footprints are rectangular.)

7.8 Similar Geometric Figures

After studying this section, you will be able to:

1 *Find the corresponding parts of similar triangles.*
2 *Find the corresponding parts of similar geometric figures.*

MathPro Video 17 SSM

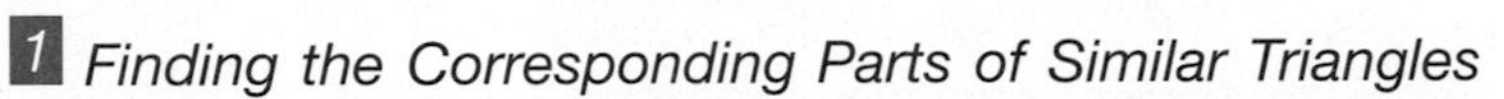

1 *Finding the Corresponding Parts of Similar Triangles*

In English, ''similar'' means that the things being compared are, in general, alike. But in mathematics, ''similar'' means that the things being compared are alike in a special way—they are *alike in shape,* even though they may be different in size. So photographs that are enlarged produce images *similar* to the original; a floor plan of a building is *similar* to the actual building; a model car is *similar* to the actual vehicle.

Two triangles with the same shape but not necessarily the same size are called **similar triangles.** Here are two pairs of similar triangles.

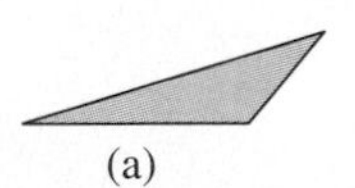
(a)

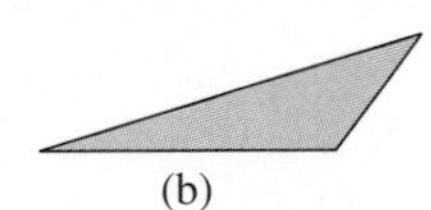
(b)

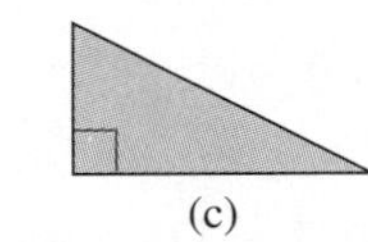
(c)

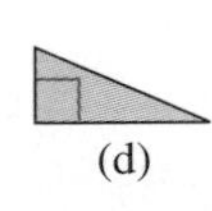
(d)

The two triangles on the right are similar. The smallest angle in the first triangle is angle A. The smallest angle in the second triangle is angle D. Both angles measure 36°. We say that angle A and angle D are corresponding angles in these similar triangles.

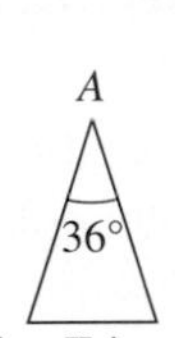

First Triangle

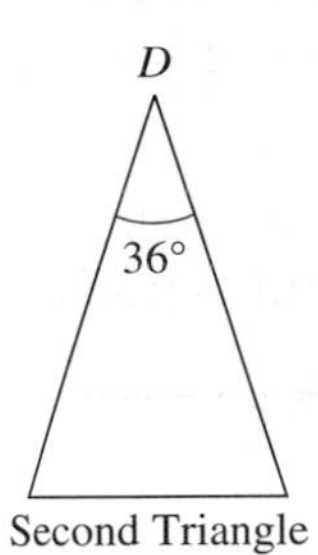

Second Triangle

> The **corresponding angles** of similar triangles are equal.

These two triangles are similar. Notice the corresponding sides.

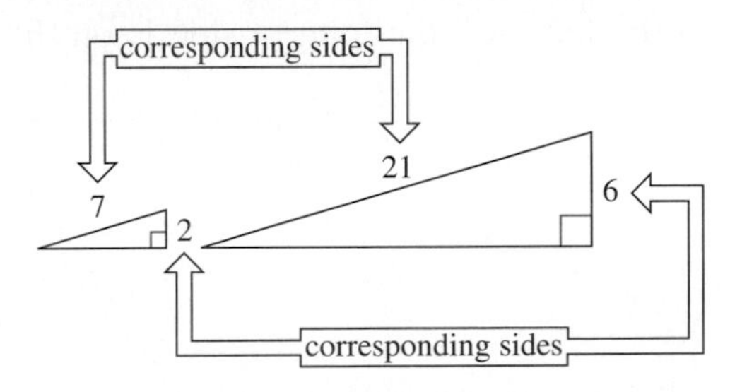

We see that the ratio of 7 to 21 is the same as the ratio of 2 to 6.

$$\frac{7}{21} = \frac{2}{6} \quad \text{is obviously true since} \quad \frac{1}{3} = \frac{1}{3}$$

TEACHING TIP Point out to students that in this section we will be solving a number of proportion problems. You may want to briefly review the steps for solving a proportion.

> The **corresponding sides** of similar triangles have the same ratio.

We can use the fact that corresponding sides of similar triangles have the same ratio to find missing sides of triangles.

EXAMPLE 1 These two triangles are similar. Find the length of side n. Round to the nearest tenth.

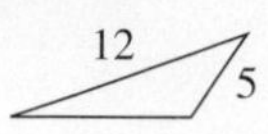

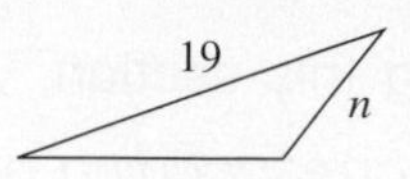

The ratio of 12 to 19 is the same as the ratio of 5 to n.

$$\frac{12}{19} = \frac{5}{n}$$

$12n = (5)(19)$ *Cross-multiply.*

$12n = 95$ *Simplify.*

$\frac{12n}{12} = \frac{95}{12}$ *Divide each side by 12.*

$n = 7.91\overline{6}$

$= 7.9$ *Round to the nearest tenth.*

Side n is of length 7.9.

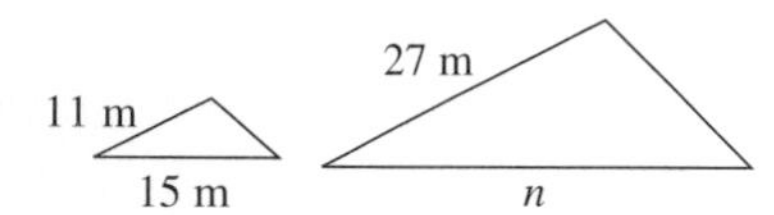

Practice Problem 1 The two triangles in the margin are similar. Find the length of side n. Round to the nearest tenth. ■

Similar triangles are not always oriented the same way. You may find it helpful to rotate one of the triangles so that the similarity is more apparent.

EXAMPLE 2 These two triangles are similar. Name the sides that correspond.

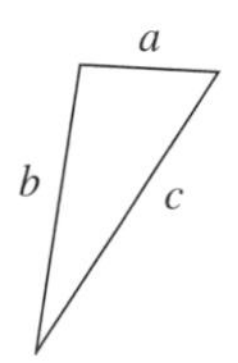

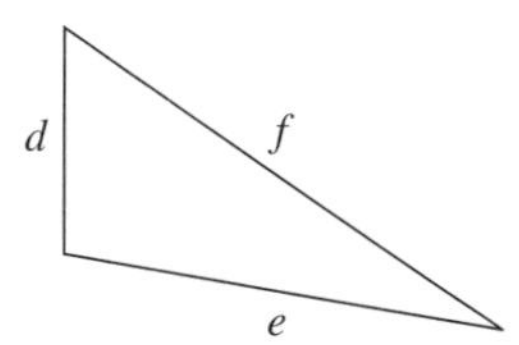

First we turn the second triangle so that the shortest side is on the top, the intermediate side is to the left, and the longest side is on the right.

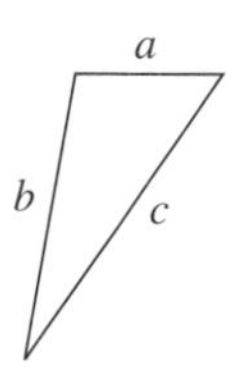

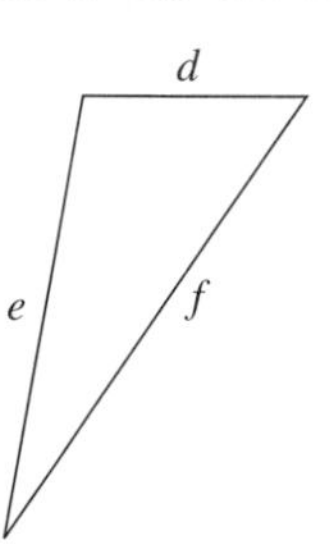

Now the shortest sides of each triangle are on the top, the longest side of each triangle is on the right, and so on. We can see that

a corresponds to d.

b corresponds to e.

c corresponds to f.

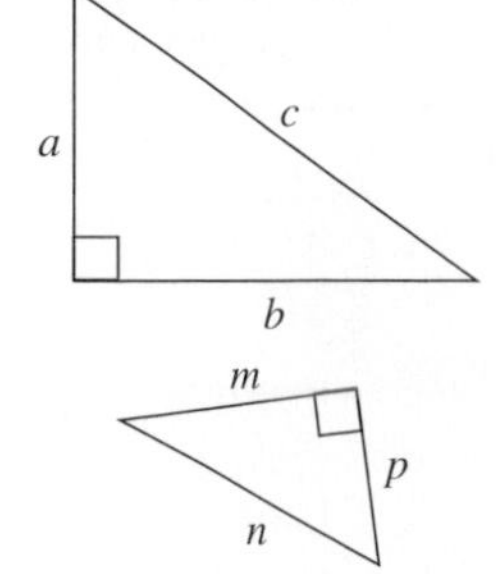

Practice Problem 2 These two triangles in the margin are similar. Name the sides that correspond. ■

> The **perimeters** of similar triangles have the same ratios as the corresponding sides.

EXAMPLE 3 Two triangles are similar. The smaller triangle has sides 5 yd, 7 yd, and 10 yd. The 7-yd side on the smaller triangle corresponds to a side of 21 yd on the larger triangle. What is the perimeter of the larger triangle?

1. *Understand the problem.*
   What are we asked to find? We are asked to find the perimeter of the larger triangle. But we are not given the lengths of all of the sides.
   What are we given? We are given the lengths of all of the sides of a smaller, similar triangle. We are also told that the 7-yd side in the smaller triangle corresponds to the 21-yd side of the larger triangle.
   What do we know? We know that the perimeters of similar triangles have the same ratios as the corresponding sides.
   We can piece this all together and solve.
2. *Solve and state the answer.*
   We begin by finding the perimeter of the smaller triangle.
   $$5 \text{ yd} + 7 \text{ yd} + 10 \text{ yd} = 22 \text{ yd}$$
   We can now write equal ratios. Let $P =$ the unknown perimeter. We will set up our ratios as
   $$\frac{\text{Smaller triangle}}{\text{Larger triangle}}$$
   Remember, once you write the first ratio, be sure to write the terms of the second ratio in the same order. We will write
   $$\begin{matrix}\text{Perimeter of the smaller triangle} \rightarrow \\ \text{Perimeter of the larger triangle} \rightarrow \end{matrix} \frac{22}{P} = \frac{7}{21} \begin{matrix} \leftarrow \text{side of the smaller triangle} \\ \leftarrow \text{side of the larger triangle} \end{matrix}$$
   $$7p = (21)(22)$$
   $$7p = 462$$
   $$\frac{7p}{7} = \frac{462}{7}$$
   $$p = 66$$
   The perimeter of the larger triangle is 66 yards.
3. *Check.*
   Are the ratios the same? Simplify each ratio to check. That is
   $$\frac{7}{21} \stackrel{?}{=} \frac{22}{66}$$
   $$\frac{1 \cdot \cancel{7}}{3 \cdot \cancel{7}} \stackrel{?}{=} \frac{\cancel{2} \cdot \cancel{11}}{\cancel{2} \cdot 3 \cdot \cancel{11}} \quad \textit{Simplify each ratio.}$$
   $$\frac{1}{3} = \frac{1}{3} \quad \checkmark$$

TEACHING TIP Students often wonder where the idea of similar triangles came from. Tell them that the dimensions of the pyramids were determined by measuring and comparing similar triangles. Today models of proposed construction projects are made in which the model is similar in proportion to the final object.

Practice Problem 3 Two triangles are similar. The smaller triangle has sides of 5 yd, 12 yd, and 13 yd. The 12-yd side of the smaller triangle corresponds to a side of 40 yd on the larger triangle. What is the perimeter of the larger triangle? ■

Similar triangles can be used to find distances or lengths that are difficult to measure.

EXAMPLE 4 A flagpole casts a shadow of 36 ft. At the same time a tree that is 3 ft tall has a shadow of 5 ft. How tall is the flagpole?

1. *Understand the problem.*

The shadows cast by the sun shining on vertical objects at the same time of day form similar triangles. We draw a picture.

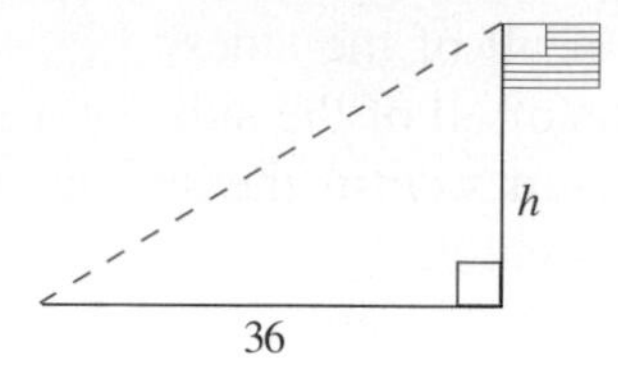

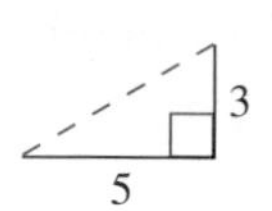

2. *Solve and state the answer.*

Let n = the height of the flagpole. Thus we can say n is to 3 as 36 is to 5.

$$\frac{n}{3} = \frac{36}{5}$$

$$5n = (3)(36)$$

$$5n = 108$$

$$\frac{5n}{5} = \frac{108}{5}$$

$$n = 21.6$$

The flagpole is about 21.6 feet tall.

3. *Check.*

The check is up to you.

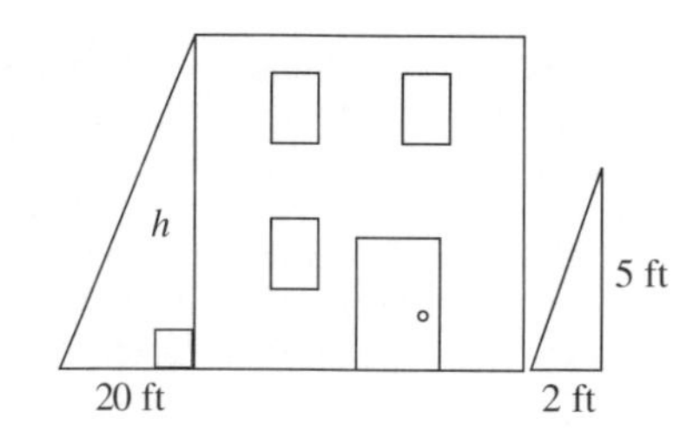

Practice Problem 4 How tall (h) is the building in the margin if the two triangles are similar? ■

2 *Finding the Corresponding Parts of Similar Geometric Figures*

Geometric figures such as rectangles, trapezoids, and circles can be similar figures.

> The **corresponding sides of similar geometric figures** have the same ratio.

EXAMPLE 5 The two rectangles shown here are similar because the corresponding sides of the two rectangles have the same ratio. Find the width of the larger rectangle.

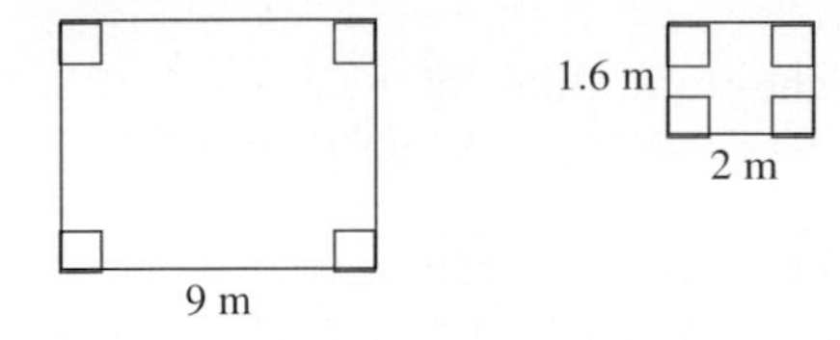

Let w = the width of the larger rectangle.

$$\frac{w}{1.6} = \frac{9}{2}$$

$$2w = (1.6)(9)$$

$$2w = 14.4$$

$$\frac{2w}{2} = \frac{14.4}{2}$$

$$w = 7.2$$

The width of the larger rectangle is 7.2 meters.

Practice Problem 5 The two rectangles in the margin are similar. Find the width of the larger rectangle. ■

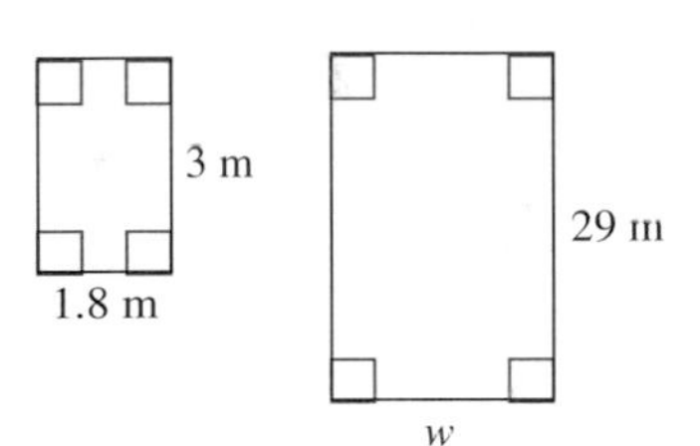

> The perimeters of similar figures—whatever the figures—have the same ratio as their corresponding sides. Circles are special cases. *All* circles are similar. The circumferences of two circles have the same ratio as their radii.

To Think About How would you find the relationship between the areas of two similar geometric figures? Consider the following two similar rectangles?

3 m 7 m 9 m 21 m

The area of the smaller rectangle is (3 m)(7 m) = 21 m^2. The area of the larger rectangle is (9 m)(21 m) = 189 m^2. How could you have predicted this result?

The ratio of small width to large width is $\frac{3}{9} = \frac{1}{3}$. The small rectangle has sides that are $\frac{1}{3}$ as large as the large rectangle. The ratio of the area of the small rectangle to the area of the large rectangle is $\frac{21}{189} = \frac{1}{9}$. Note that $(\frac{1}{3})^2 = \frac{1}{9}$.

Thus we can develop the following principle: The areas of two similar figures are in the same ratio as the square of the ratio of two corresponding sides.

7.8 Exercises

Verbal and Writing Skills

1. Similar figures may be different in ___size___ but they are alike in ___shape___.
2. The corresponding sides of similar triangles have the same ___ratio___.
3. The perimeters of similar figures have the same ratio as their corresponding ___sides___.
4. You are given the lengths of the sides of a large triangle and the length of a corresponding side of a smaller, similar triangle. Explain in your own words how to find the perimeter of the smaller triangle. Answers may vary. Check each answer for understanding.

For each pair of similar triangles, find the missing side n. Round your answer to the nearest tenth when necessary.

5.

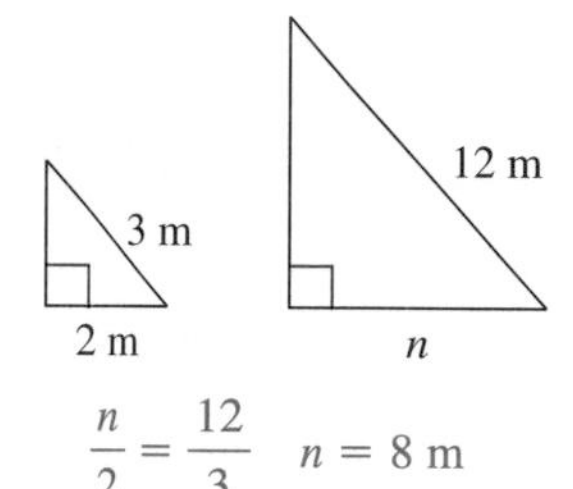

$\frac{n}{2} = \frac{12}{3}$ $n = 8$ m

6.

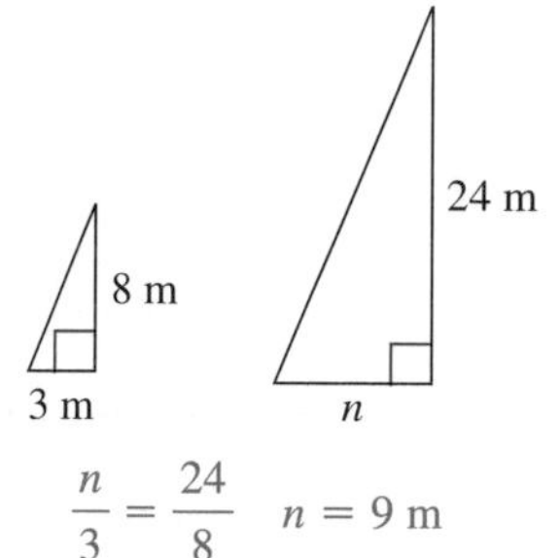

$\frac{n}{3} = \frac{24}{8}$ $n = 9$ m

7.

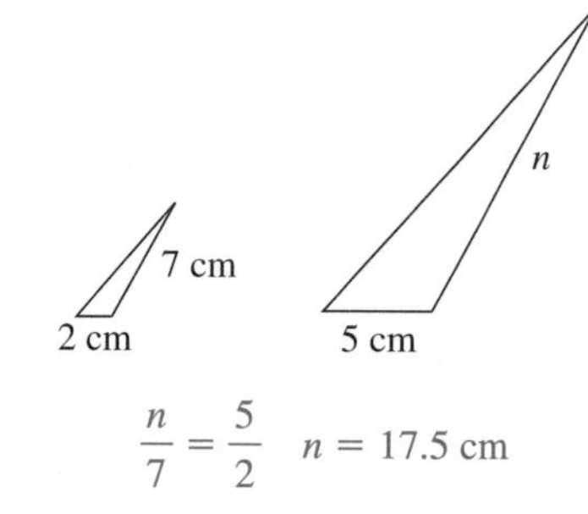

$\frac{n}{7} = \frac{5}{2}$ $n = 17.5$ cm

8.

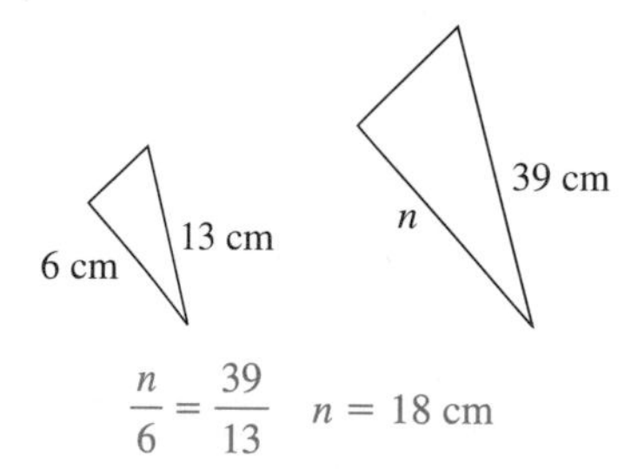

$\frac{n}{6} = \frac{39}{13}$ $n = 18$ cm

9.

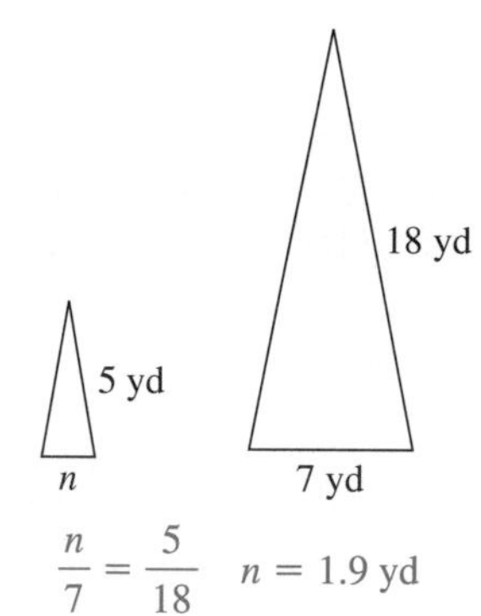

$\frac{n}{7} = \frac{5}{18}$ $n = 1.9$ yd

10.

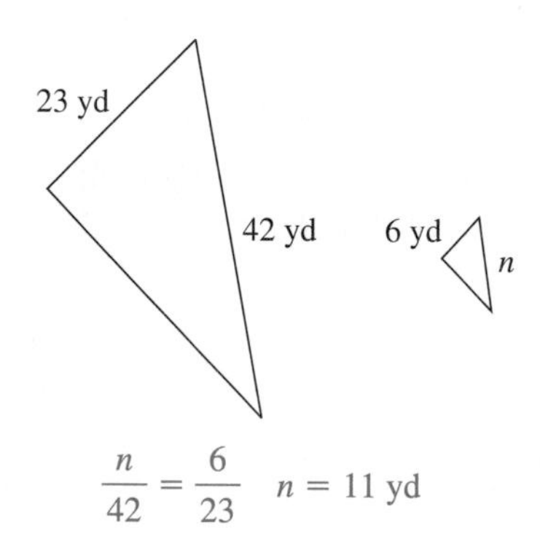

$\frac{n}{42} = \frac{6}{23}$ $n = 11$ yd

Each pair of triangles is similar. Determine which two sides correspond in each case.

11.

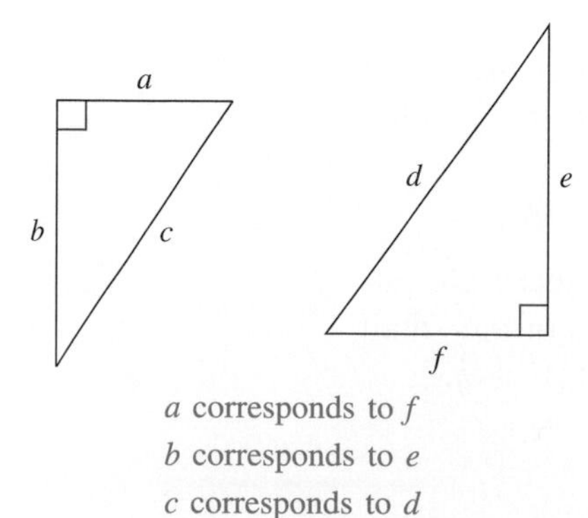

a corresponds to *f*
b corresponds to *e*
c corresponds to *d*

12.

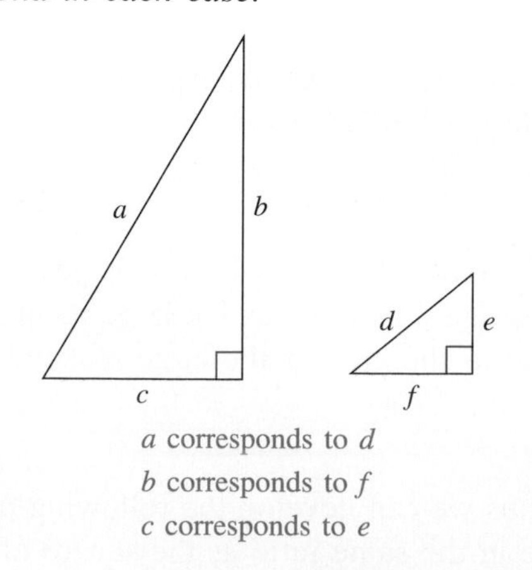

a corresponds to *d*
b corresponds to *f*
c corresponds to *e*

For each pair of similar triangles, find the missing side n. Round your answer to the nearest tenth when necessary.

13.

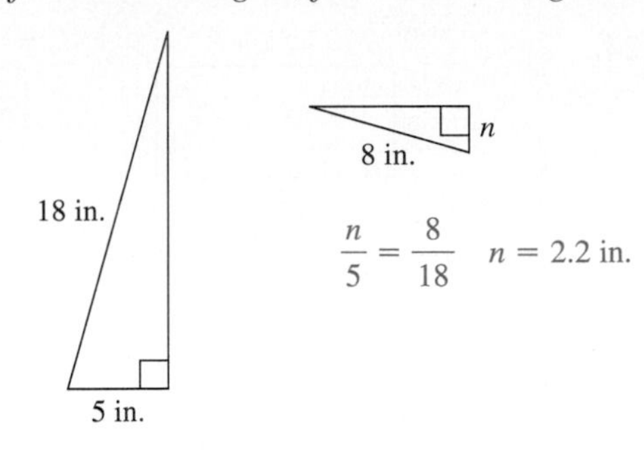

$\frac{n}{5} = \frac{8}{18}$ $\quad n = 2.2$ in.

14.

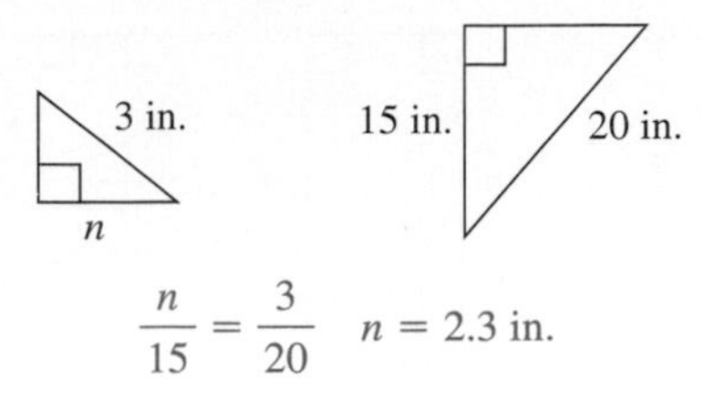

$\frac{n}{15} = \frac{3}{20}$ $\quad n = 2.3$ in.

Applications

15. A sculptor is designing her new triangular masterpiece. Her scale drawing shows that the shortest side of a triangular piece to be made, measures 8 cm. The longest side of the drawing measures 25 cm. The longest side of the actual part to be sculpted must be 10.5 m long. How long will the shortest side of the actual part be? Round to the nearest tenth. 3.36 meters long

16. The zoo has hired a landscape architect to design the triangular lobby of the children's petting zoo. His scale drawing shows that the longest side of the lobby is 9 cm. The shortest side of the lobby is 5 cm The longest side of the actual lobby will be 30 m. How long will the shortest side of the actual lobby be? Round to the nearest tenth.
16.7 meters

17. Janice took a great photo of the entire family at this year's reunion. She takes it to a professional photography studio and asks that the 3-in. by 5-in. photo be blown up to a poster size, which is 3.5 feet tall. What is the smaller dimension (width) of the poster? 2.1 feet wide

18. Jeff and Shelley are planning a new kitchen. The old kitchen measured 9 ft by 12 ft. The new kitchen is similar in shape, but the longest dimension is 19 ft. What is the smaller dimension (width) of the new kitchen?

$\frac{9}{12} = \frac{n}{19}$

$12n = 171$

$n = 14.25$ ft

A flagpole casts a shadow. At the same time a small tree casts a shadow. Use the sketch to find the height n of each flagpole.

19.

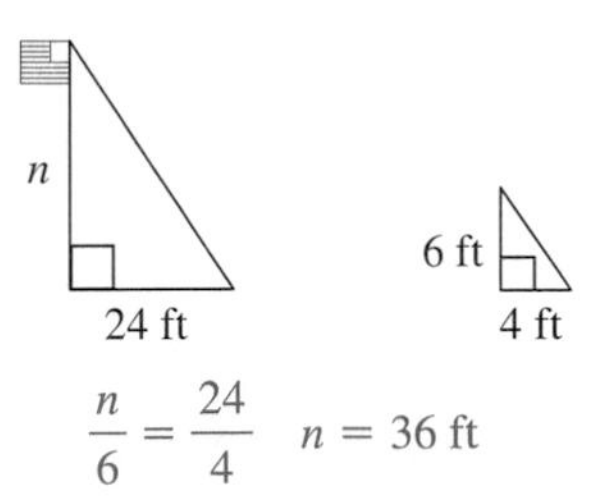

$\frac{n}{6} = \frac{24}{4}$ $\quad n = 36$ ft

20.

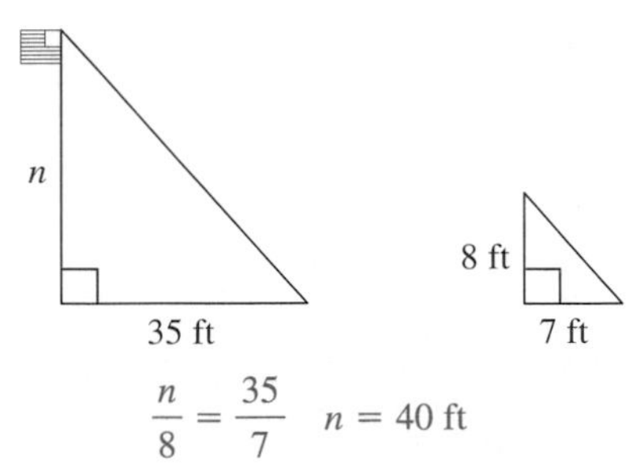

$\frac{n}{8} = \frac{35}{7}$ $\quad n = 40$ ft

21. Lola is standing outside of the shopping mall. She is 5.5 feet tall and her shadow measures 6.5 feet long. The outside of the department store casts a shadow of 96 feet. How tall is the store? Round to the nearest foot. 81.2 feet tall

22. Thomas is rock climbing in Utah. He is 6 feet tall and his shadow measures 8 feet long. The rock he wants to climb casts a shadow of 610 feet. How tall is the rock he is about to climb? 457.5 feet tall

Each pair of figures is similar. Find the missing side. Round to the nearest tenth when necessary.

23.

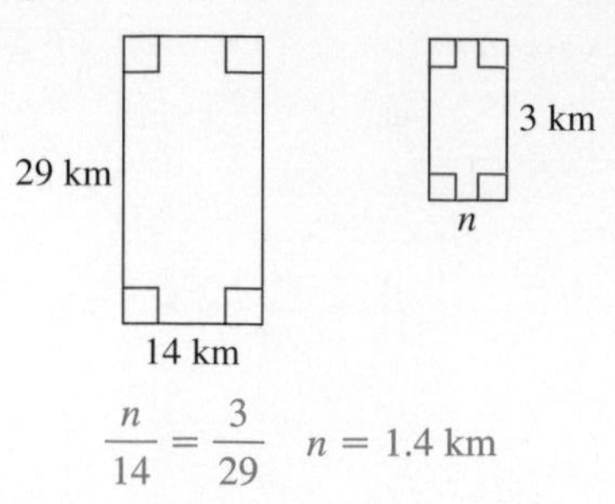

24.

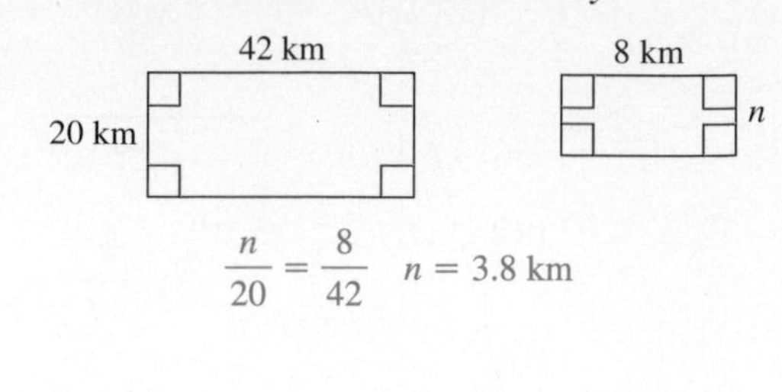

25.

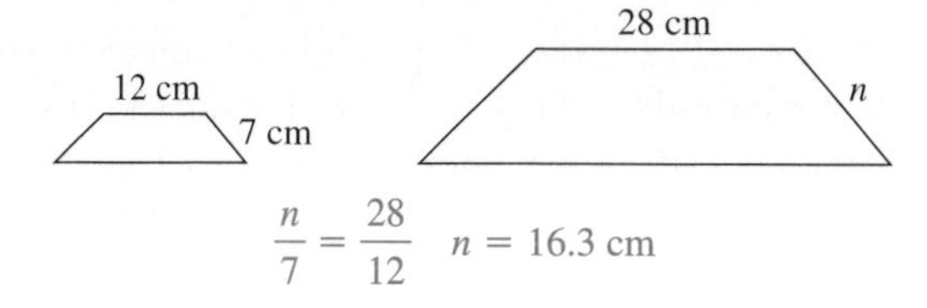

26.

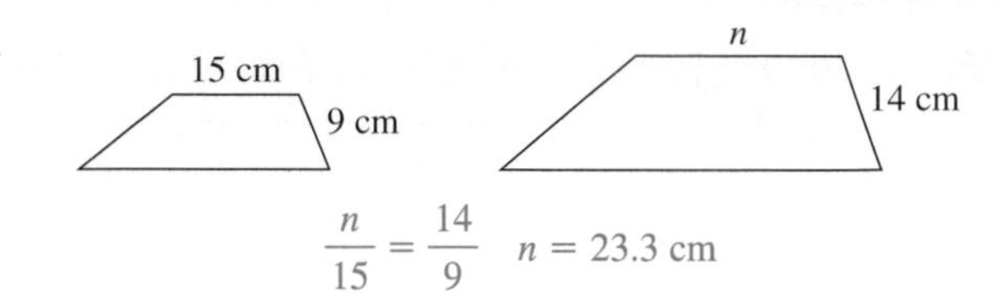

Each pair of figures is similar. Find the missing perimeter.

27.

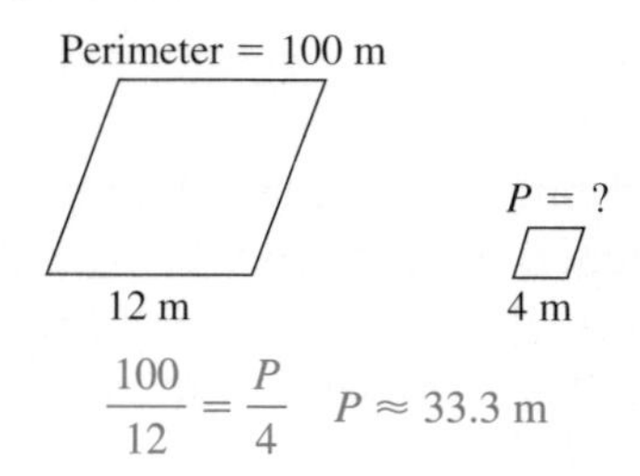

28.

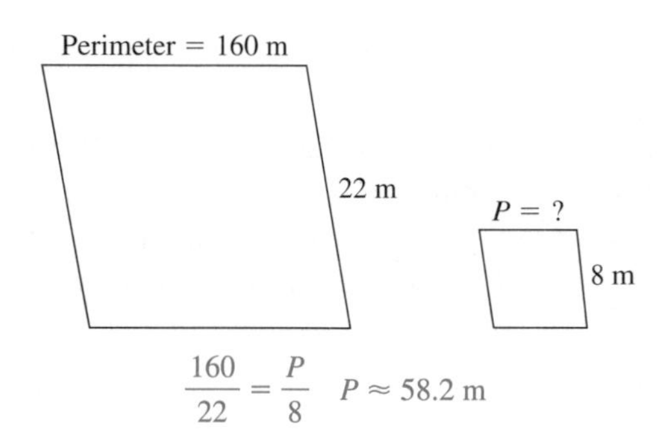

To Think About

Each pair of geometric figures is similar. Find the unknown area.

29.

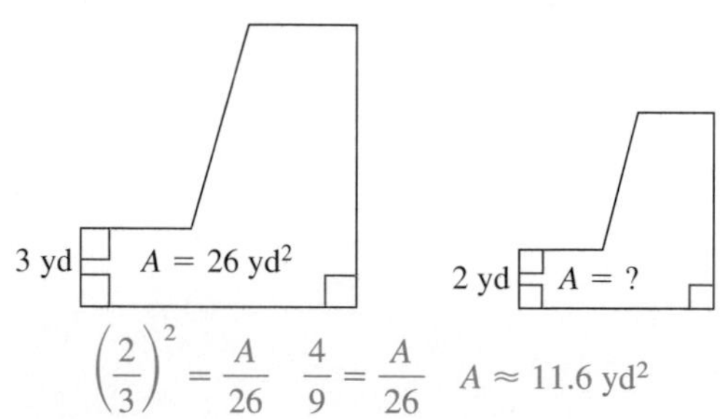

30.

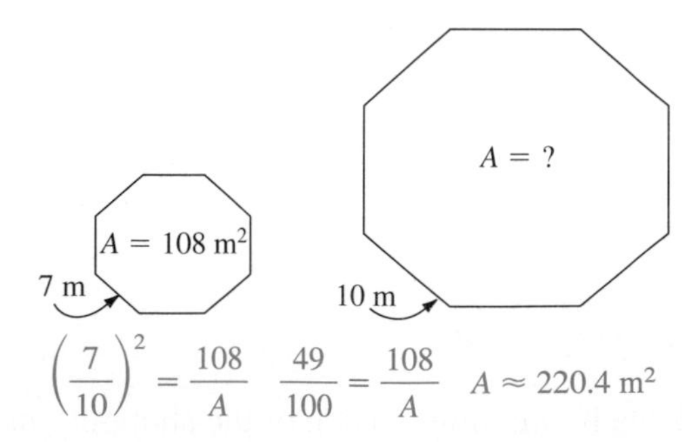

Cumulative Review Problems

Perform the calculation. Use the correct order of operations.

31. $2 \times 3^2 + 4 - 2 \times 5$ 12

32. $300 \div (85 - 65) + 2^4$ 31

33. $(5)(9) - (21 + 3) \div 8$ 42

34. $108 \div 6 + 51 \div 3 + 3^3$ 62

7.9 Applications of Geometry

MathPro

Video 17

SSM

After studying this section, you will be able to:

1 *Solve application problems involving geometric shapes.*

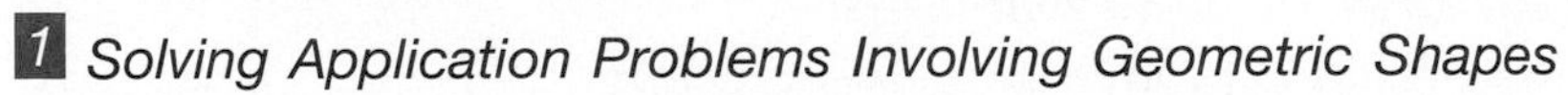

1 *Solving Application Problems Involving Geometric Shapes*

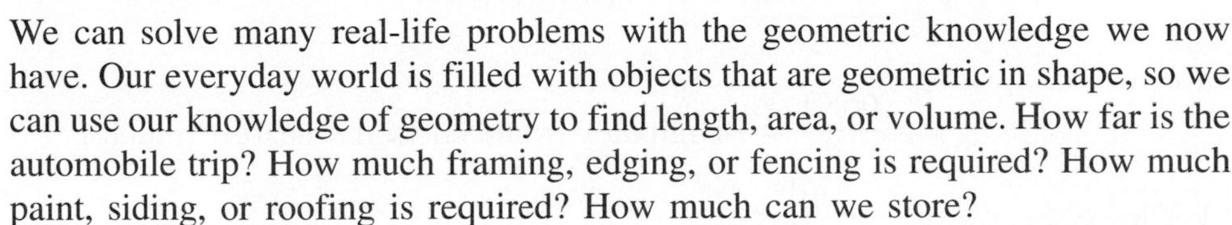

We can solve many real-life problems with the geometric knowledge we now have. Our everyday world is filled with objects that are geometric in shape, so we can use our knowledge of geometry to find length, area, or volume. How far is the automobile trip? How much framing, edging, or fencing is required? How much paint, siding, or roofing is required? How much can we store?

If it is helpful to you, use the Mathematics Blueprint for Problem Solving to organize a plan to solve these applied problems.

EXAMPLE 1 A professional painter can paint 90 ft^2 of wall space in 20 minutes. How long will it take the painter to paint these four walls: 14 ft × 8 ft, 12 ft × 8 ft, 10 ft × 7 ft, and 8 ft × 7 ft?

1. *Understand the problem.*

MATHEMATICS BLUEPRINT FOR PROBLEM SOLVING

Gather the Facts	What Am I Asked to Do?	How Do I Proceed?	Key Points to Remember
Painter paints four walls: 14 ft × 8 ft 12 ft × 8 ft 10 ft × 7 ft 8 ft × 7 ft Painter can paint 90 ft^2 in 20 minutes.	Find out how long it will take the painter to paint the four walls.	(a) Find the total area to be painted. (b) Then find how long it will take to paint the total area.	The area of each rectangular wall is obtained by multiplying the length by the width. To get the time, we multiply by the fraction $1 = \dfrac{20 \text{ min}}{90 \text{ ft}^2}$

2. *Solve and state the answer.*

(a) Find the total area of the four walls.

Each wall is a rectangle. The first one is 8 ft wide and 14 ft long.

$$A = lw$$
$$= (14 \text{ ft})(8 \text{ ft}) = 112 \text{ ft}^2$$

We find the area of the other three walls.

$$(12 \text{ ft})(8 \text{ ft}) = 96 \text{ ft}^2 \qquad (10 \text{ ft})(7 \text{ ft}) = 70 \text{ ft}^2 \qquad (8 \text{ ft})(7 \text{ ft}) = 56 \text{ ft}^2$$

The total area is obtained by adding.

$$\begin{array}{r} 112 \text{ ft}^2 \\ 96 \text{ ft}^2 \\ 70 \text{ ft}^2 \\ +\ 56 \text{ ft}^2 \\ \hline 334 \text{ ft}^2 \end{array}$$

TEACHING TIP Sometimes students want to do Example 1 by solving a proportion. They usually set up the proportion 90 square feet can be done in 20 minutes as 334 square feet can be done in t minutes. Solve the equation for t.

(b) Determine how long it will take to paint the four walls.
Now we multiply the total area by a fraction that tells us that 90 ft^2 can be painted in 20 minutes. We are looking for time, so we want the unit for time in the numerator.

$$334 \text{ ft}^2 \times 1 = 334 \cancel{\text{ft}^2} \times \frac{20 \text{ mi}}{90 \cancel{\text{ft}^2}}$$

$$= \frac{(334)(20)}{90} \text{ min}$$

$$= \frac{6680}{90} = 74 \text{ minutes (rounded to the nearest minute)}$$

3. *Check.*
Estimate the answer.

$$14 \times 8 \approx 10 \times 8 = 80 \text{ ft}^2$$
$$12 \times 8 \approx 10 \times 8 = 80 \text{ ft}^2$$
$$10 \times 7 = 70 \text{ ft}^2$$
$$8 \times 7 = 56 \text{ ft}^2$$

If we estimate the sum of the number of square feet, we will have

$$80 + 80 + 70 + 56 = 286 \text{ ft}^2$$

Now since 60 minutes = 1 hour, we know that if you can paint 90 ft^2 in 20 minutes, you can paint 270 ft^2 in one hour. Our estimate of 286 ft^2 is slightly more than 270 ft^2, so we would expect that the answer would be slightly more than one hour.

Thus our calculated value of 74 minutes (1 hour 14 minutes) seems reasonable. ✓

Practice Problem 1 Mike rented an electric floor sander. It will sand 80 ft^2 of hardwood floor in 15 minutes. He needs to sand the floors in three rooms. The floor dimensions are 24 ft × 13 ft, 12 ft × 9 ft, and 16 ft × 3 ft. How long will it take him to sand the floors in all three rooms? ■

EXAMPLE 2 Carlos and Rosetta want to put vinyl siding on the front of their home in West Chicago. The house dimensions are shown. The door dimensions are 6 ft × 3 ft. The windows measure 2 ft × 4 ft.

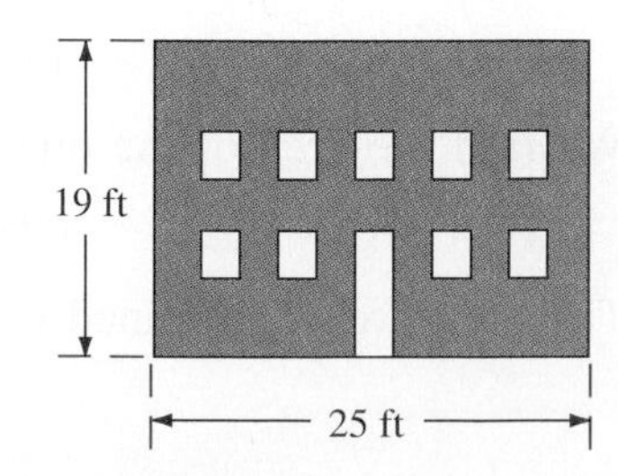

(a) Excluding windows and doors, how many square feet of siding will be needed?

(b) If the siding costs $2.25 per square foot, how much will it cost to side the front of the house?

1. *Understand the problem.*

MATHEMATICS BLUEPRINT FOR PROBLEM SOLVING			
Gather the Facts	**What Am I Asked to Do?**	**How Do I Proceed?**	**Key Points to Remember**
House measures 19 ft × 25 ft. Windows measure 2 ft × 4 ft. Door measures 6 ft × 3 ft. Siding costs \$2.25 per square foot.	Find the cost to put siding on the front of the house.	(a) Find the area of the entire front of the house by multiplying 19 ft by 25 ft. Find the area of one window by multiplying 2 ft by 4 ft. Find the area of the door by multiplying 6 ft by 3 ft. (b) Multiply shaded area by \$2.25.	(a) To obtain the shaded area we must subtract the area of nine windows and one door from the area of the entire front. (b) We must multiply the resulting area by cost of the siding per foot.

2. *Solve and state the answer.*

(a) Find the area of the front of the house (shaded area).
We will find the area of the large rectangle representing the front of the house. Then we will subtract the area of the windows and the door.

$$\text{Area each window} = (2\text{ ft})(4\text{ ft}) = 8\text{ ft}^2$$

$$\text{Area of 9 windows} = (9)(8\text{ ft})^2 = 72\text{ ft}^2$$

$$\text{Area of 1 door} = (6\text{ ft})(3\text{ ft}) = 18\text{ ft}^2$$

$$\text{Area of 9 windows} + 1\text{ door} = 90\text{ ft}^2$$

$$\text{Area of large rectangle} = (19\text{ ft})(25\text{ ft}) = 475\text{ ft}^2$$

Total area of front of house	475 ft^2
− Area of 9 windows and 1 door	$-\ 90\text{ ft}^2$
= Total area to be covered	385 ft^2

We see that 385 ft^2 of siding will be needed.

(b) Find the cost of the siding.

$$\text{Cost} = 385\ \cancel{\text{ft}^2} \times \frac{\$2.25}{1\ \cancel{\text{ft}^2}} = \$866.25$$

The cost to put up siding on the front of the house is \$866.25.

3. *Check.*
We leave the check up to you.

TEACHING TIP After covering a couple of examples in Section 7.9, you may want to ask the students, "Can any of you think of some practical geometry problem like this that you or your friends have encountered in everyday life?" Often they will think of problems like the cost of paving a driveway, building a deck on the house, or finding the volume of the cylinders of a car to determine compression. Solving a real-life example from students' lives tends to impress on them that this section is really "relevant."

Practice Problem 2 In the margin is a sketch of a roof.

(a) What is the area of the roof?

(b) How much would it cost to install new roofing on the roof area shown if the roofing costs \$2.75 per square yard? (*Hint:* 9 square feet = 1 square yard.) ■

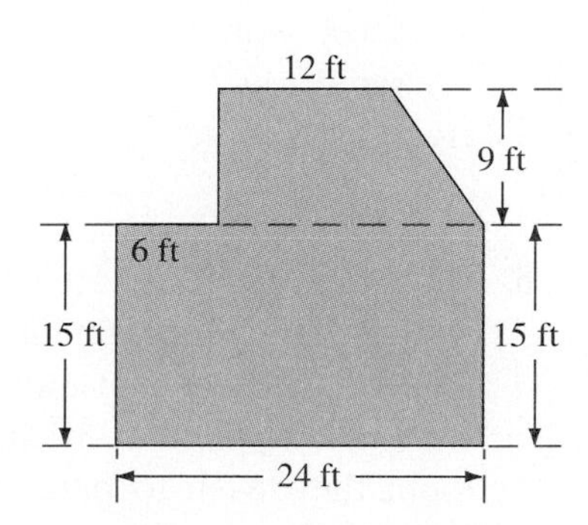

7.9 Exercises

Applications

Round all answers to the nearest tenth unless otherwise directed.

1. Jim and Linda have installed a new garage door for their one car garage in Duluth, Minnesota. The rectangular door measures 6 feet, 3 inches wide and is 6 feet, 9 inches tall. It is a well-built garage with a solid concrete floor, but the winter snow storms often cause blowing snow to come in between the door and the garage. Therefore, Jim and Linda are installing weather stripping on all four sides of the door. The weather stripping costs \$0.85 per foot. How much will it cost them for the required amount of weather stripping? \$22.10

2. The Kim family started a flower garden. The rectangular garden measures 12 ft long and 7 ft wide. To keep out small animals, they had to surround it with fencing, which costs \$0.85 per foot. How much did the fencing cost?

$\frac{\$0.85}{\text{ft}} \times (2 \times 12 \text{ ft} + 2 \times 7 \text{ ft}) = \$32.30$

3. Monica drives to work each day from Bethel to Bridgeton. The sketch shows the two possible routes.

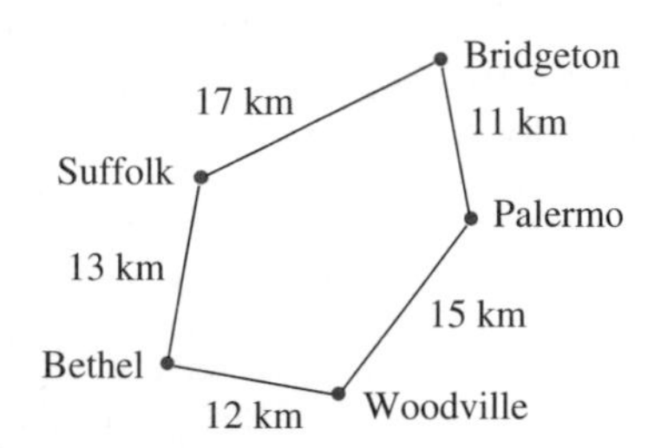

(a) How many kilometers is the trip if she drives through Suffolk? What is her average speed if this trip takes her 0.4 hour?

$13 \text{ km} + 17 \text{ km} = 30 \text{ km} \quad \frac{30 \text{ km}}{0.4 \text{ hr}} = 75 \text{ km/hr}$

(b) How many kilometers is the trip if she drives through Woodville and Palermo? What is her average speed if this trip takes her 0.5 hour?

$15 \text{ km} + 12 \text{ km} + 11 \text{ km} = 38 \text{ km} \quad \frac{38 \text{ km}}{0.5 \text{ hr}} = 76 \text{ km/hr}$

(c) Over which route does she travel at a more rapid rate?

Through Woodville and Palermo

4. Robert drives from work to either a convenience food store and then home or to a food supermarket and then home. The sketch shows the distances.

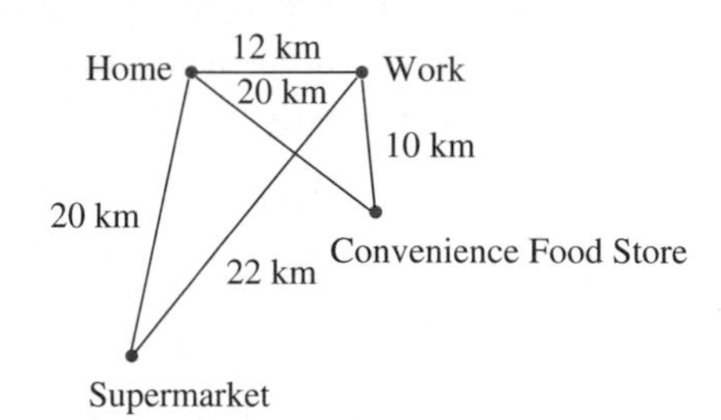

(a) How far does he travel if he goes from work to the supermarket and then home? How fast does he travel if the trip takes 0.6 hour?

$22 \text{ km} + 20 \text{ km} = 42 \text{ km} \quad \text{Ave. sp.} \ \frac{42 \text{ km}}{0.6 \text{ hr}} = 70 \text{ km/hr}$

(b) How far does he travel if he goes from work to the convenience food store and then home? How fast does he travel if the trip takes 0.5 hour?

$10 \text{ km} + 20 \text{ km} = 30 \text{ km} \quad \text{Ave. sp.} \ \frac{30 \text{ km}}{0.5 \text{ hr}} = 60 \text{ km/hr}$

(c) Over which route does he travel at a more rapid rate?

Via the supermarket

5. Dave and Linda McCormick are painting the walls of their apartment in Seattle. When they worked together at Linda's mother's house they were able to paint 80 ft^2 in 25 minutes. The living room they wish to paint has one wall that measures 16 feet by 7 feet, one wall that measures 14 feet by 7 feet, and two walls that measure 12 feet by 7 feet. How long will it take Dave and Linda together to paint the living room of their apartment? Round to the nearest minute.

118 minutes *or* 1 hour, 58 minutes

6. A professional wallpaper hanger can wallpaper 120 ft^2 in 15 minutes. She will be papering four walls in a house. They measure 7 ft × 10 ft, 7 ft × 14 ft, 6 ft × 10 ft, and 6 ft × 8 ft. How many minutes will it take her to paper all four walls?

$(7 \text{ ft} \times 10 \text{ ft}) + (7 \text{ ft} \times 14 \text{ ft}) + (6 \text{ ft} \times 10 \text{ ft}) + (6 \text{ ft} \times 8 \text{ ft}) = 276 \text{ ft}^2$

$276 \text{ ft}^2 \times \frac{15 \text{ min}}{120 \text{ ft}^2} = 34.5 \text{ min}$

7. The floor area of the recreation room at Yvonne's house is shown in the drawing. How much will it cost to carpet the room if the carpet costs $15 per *square yard*?

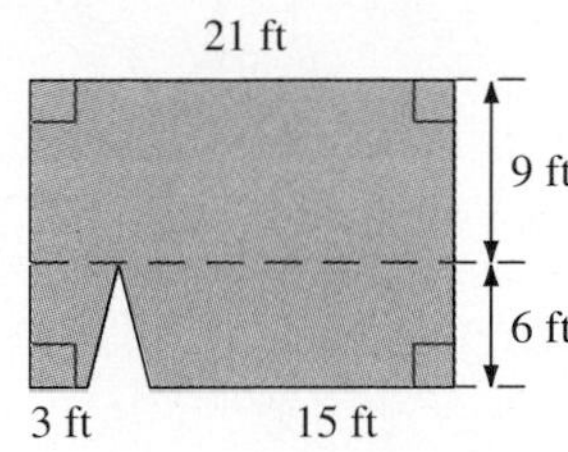

$A = 21 \text{ ft} \times 15 \text{ ft} - \frac{1}{2} \times 3 \text{ ft} \times 6 \text{ ft} = 306 \text{ ft}^2$

$\text{Cost} = 306 \text{ ft}^2 \times \frac{1 \text{ yd}^2}{9 \text{ ft}^2} \times \frac{15.00}{\text{yd}^2} = \$510.00$

8. The side view of a barn is shown in the diagram. The cost of aluminum siding is $18 per *square yard.* How much will it cost to put siding on this side of the barn?

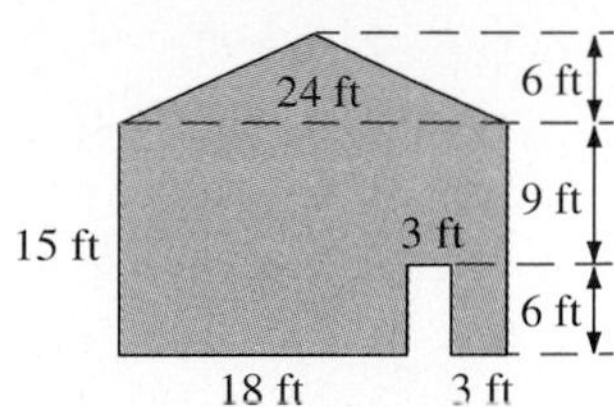

$A = \frac{1}{2} \times 24 \text{ ft} \times 6 \text{ ft} + 15 \text{ ft} \times 24 \text{ ft} - 3 \text{ ft} \times 6 \text{ ft} = 414 \text{ ft}^2$

$\text{Cost} = 414 \text{ ft}^2 \times \frac{1 \text{ yd}^2}{9 \text{ ft}^2} \times \frac{18.00}{\text{yd}^2} = \$828.00$

9. George Washington Thomas drives a diesel truck with two fuel tanks. Each tank is cylindrical in shape. The tanks each have a diameter of 3 feet and a height of 9 feet. How many gallons of diesel fuel can the truck hold if there are 7.5 gal in 1 ft^3? Round your answer to the nearest tenth.

$V = (1.5) \times (1.5) \times (3.14) \times (9) \times (2)$
$= 127.17$ cubic feet in 2 tanks.
$G = (7.5) \times (127.17) \approx 953.8$ gallons

10. A conical pile of grain is 18 ft in diameter and 12 ft high. How many bushels of grain are in the pile if 1 cubic foot = 0.8 bushel?

$r = \frac{18}{2} \text{ ft} \quad V = \frac{1}{3} \times 3.14 \times (9 \text{ ft})^2 \times 12 \text{ ft} = 1017.36 \text{ ft}^3$

$B = 1017.36 \text{ ft}^3 \times \frac{0.8 \text{ bu}}{1 \text{ ft}^3} \approx 813.9 \text{ bushels}$

11. Find the volume of a concrete connector for the city sewer system. A diagram of the connector is shown. It is shaped like a box with a hole of diameter 2 m. If it is formed using concrete that costs $1.20 per cubic meter, how much will the necessary concrete cost?

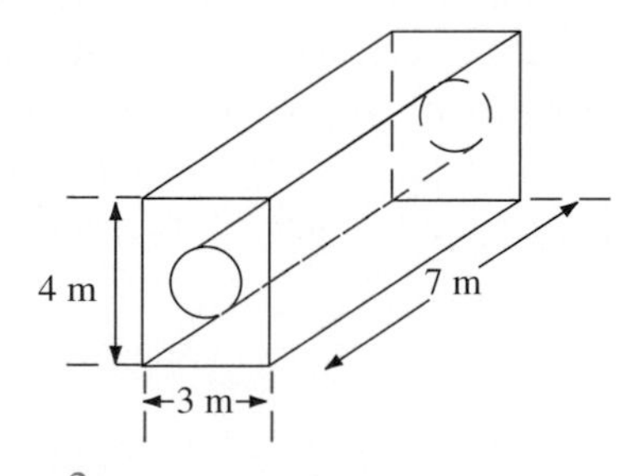

$r = \frac{2}{2} \text{ m}$

$V = 7 \text{ m} \times 3 \text{ m} \times 4 \text{ m}$
$-3.14 \times (1 \text{ m})^2 \times 7 \text{ m} = 62.02 \text{ m}^3$

$\text{Cost} = 62.02 \text{ m}^3 \times \frac{1.20}{\text{m}^3} = \$74.42$

12. A dentist places a gold filling in a tooth in the shape of a cylinder with a hemispherical top. The radius r of the filling is 1 mm. The height is 2 mm. Find the volume of the filling. If dental gold costs $95.00 per cubic millimeter, how much did the gold cost for the filling?

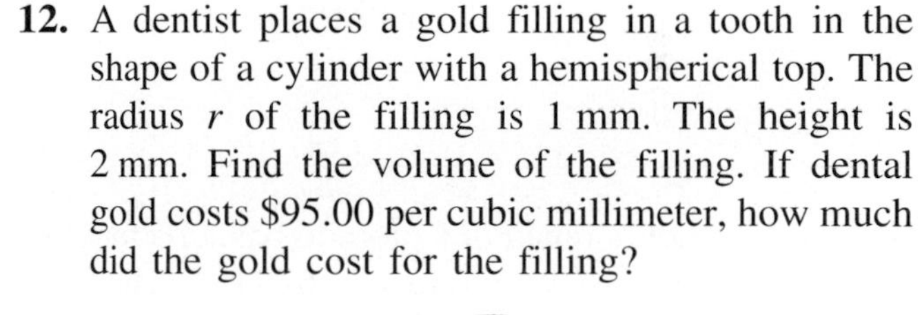

$V = \frac{1}{2} \times \frac{4}{3} \times 3.14 \times (1 \text{ mm})^3$
$+ 3.14 \times 2 \text{ mm} \times (1 \text{ mm})^2 \quad V \approx 8.37 \text{ mm}^3$

$C = 8.37 \text{ mm}^3 \times \frac{95.00}{\text{mm}^3} = \$795.15$

13. The *Landstat* satellite orbits the earth in an almost circular pattern. Assume that the radius of orbit (distance from center of earth to the satellite) is 6500 km.

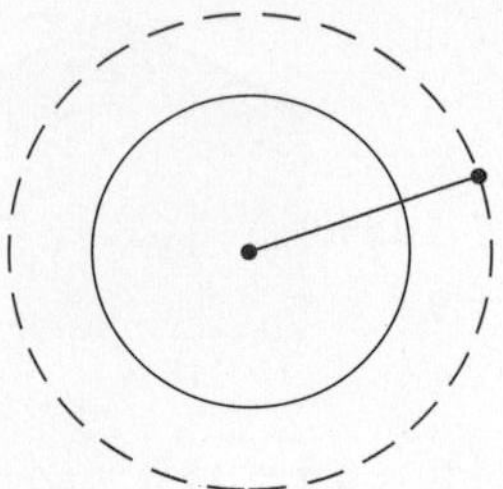

(a) How many kilometers long is one orbit of the satellite? (That is, what is the circumference of the orbit path?)

$C = 2 \times 3.14 \times 6500 \text{ km} = 40{,}820 \text{ km}$

(b) If the satellite goes through one orbit around the earth in 2 hours, what is its speed in kilometers per hour?

$S = \dfrac{40{,}820 \text{ km}}{2 \text{ hr}} \quad S = 20{,}410 \text{ km/hr}$

14. The North City Park is constructed in a shape that includes one fourth of a circle. It is shaded red in this sketch.

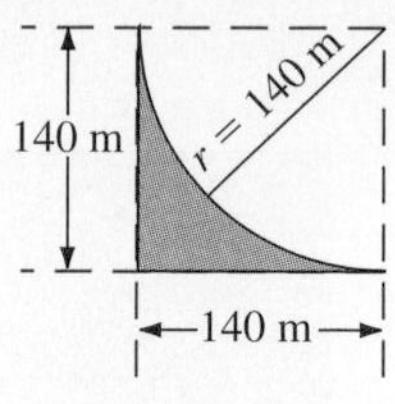

(a) What is the area of the park?

$A = \dfrac{1}{4} \times 3.14 \times (140 \text{ m})^2 = 15{,}386 \text{ m}^2$

(b) How much will it cost to resod the park with new grass at $4.00 per square meter?

$C = 15{,}386 \text{ m}^2 \times \dfrac{4.00}{\text{m}^2} = \$61{,}544.00$

15. Dr. Dobson's ranch in Colorado has a rectangular field that measures 456 feet by 625 feet. The fence surrounding the field needs to be replaced. If the fencing costs $4.57 per *yard,* what will it cost for materials to replace the fence? Round your answer to the nearest cent. $9880.34

16. Boston Sunoco Dealerships owns a large gasoline storage tank in South Boston. The diameter of the tank is 122.9 feet. The height of the tank is 55.8 feet. How many gallons of gasoline will it hold if there are 7.5 gal in 1 ft^3? Round your answer to the nearest tenth. 4,962,138.5 gallons

Cumulative Review Problems

Divide.

17. $16\overline{)2048}$ 128

18. $27\overline{)24{,}705}$ 915

19. $1.3\overline{)0.325}$ 0.25

20. $0.52\overline{)2.5324}$ 4.87

Chapter Organizer

Topic	Procedure	Examples
Perimeter of a rectangle, p. 380.	$P = 2l + 2w$	Find the perimeter of a rectangle with width = 3 m, length = 8 m. $P = (2)(8 \text{ m}) + (2)(3 \text{ m}) = 16 \text{ m} + 6 \text{ m} = 22 \text{ m}$
Perimeter of a square, p. 382.	$P = 4s$	Find the perimeter of a square with side length $s = 6$ m. $P = (4)(6 \text{ m}) = 24 \text{ m}$
Area of a rectangle, p. 383.	$A = lw$	Find the area of a rectangle with width = 2 m, length = 7 m. $A = (7 \text{ m})(2 \text{ m}) = 14 \text{ m}^2$
Area of a square, p. 384.	$A = s^2$	Find the area of a square with a side of 4 m. $A = s^2 = (4 \text{ m})^2 = 16 \text{ m}^2$
Perimeter of parallelograms, trapezoids, and triangles, pp. 388, 390, 398.	Add up the lengths of all sides.	Find the perimeter of a triangle with sides of 3 m, 6 m, and 4 m. $3 \text{ m} + 6 \text{ m} + 4 \text{ m} = 13 \text{ m}$
Area of a parallelogram, p. 389.	$A = bh$ b = length of base h = height 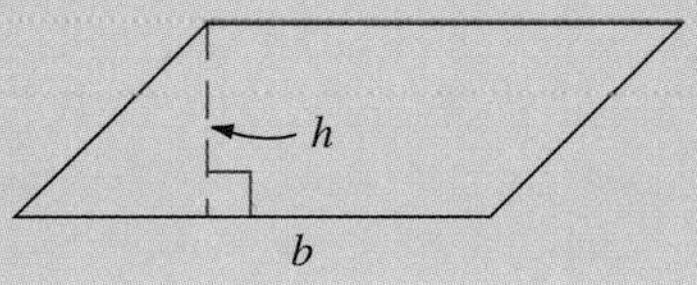	Find the area of a parallelogram with a base of 12 m and a height of 7 m. $A = bh = (12 \text{ m})(7 \text{ m}) = 84 \text{ m}^2$
Area of a trapezoid, p. 390.	$A = \frac{h(b + B)}{2}$ b = length of shorter base B = length of longer base h = height 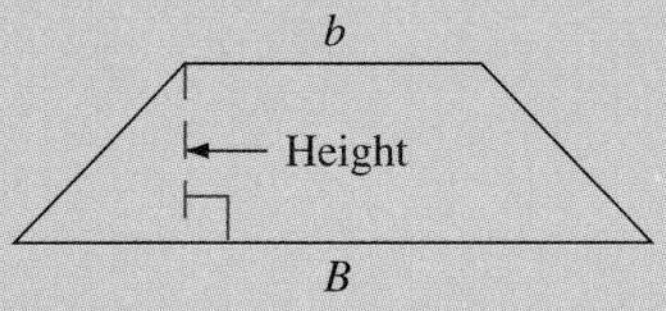	Find the area of a trapezoid whose height is 12 m and whose bases are 17 m and 25 m. $A = \frac{(12 \text{ m})(17 \text{ m} + 25 \text{ m})}{2} = \frac{(12 \text{ m})(42 \text{ m})}{2}$ $= \frac{504 \text{ m}^2}{2} = 252 \text{ m}^2$
The sum of the measures of the three interior angles of a triangle is 180°, p. 397.	In a triangle, to find one missing angle if two are given: **1.** Add up the two known angles. **2.** Subtract the sum from 180°.	Find the missing angle if two known angles in a triangle are 60° and 70°. **1.** $60° + 70° = 130°$ **2.** $180° - 130° = 50°$ The missing angle is 50°.
Area of a triangle, p. 399.	$A = \frac{bh}{2}$ b = base h = height 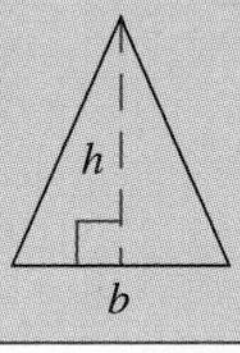	Find the area of the triangle whose base is 1.5 m and whose height is 3 m. $A = \frac{bh}{2} = \frac{(1.5 \text{ m})(3 \text{ m})}{2} = \frac{4.5 \text{ m}^2}{2} = 2.25 \text{ m}^2$

Chapter Organizer

Topic	Procedure	Examples
Evaluating square roots of numbers that are perfect squares, p. 404.	The square root of a number is one of two identical factors of that number.	$\sqrt{0} = 0$ because $(0)(0) = 0$ $\sqrt{4} = 2$ because $(2)(2) = 4$ $\sqrt{100} = 10$ because $(10)(10) = 100$ $\sqrt{169} = 13$ because $(13)(13) = 169$
Approximating the square root of a number that is not a perfect square, p. 405.	**1.** If a calculator with a square root key is available, enter the number and then press the $\boxed{\sqrt{x}}$ or $\boxed{\sqrt{\ }}$ key. The approximate value will be displayed. **2.** If using a square root table, find the number n, then look for the square root of that number. The approximate value will be correct to the nearest thousandth. Number, n — Square Root of That Number, $\sqrt{n}$ 31 — 5.568 32 — 5.657 33 — 5.745 34 — 5.831	**1.** Find on a calculator. **(a)** $\sqrt{13}$ **(b)** $\sqrt{182}$ Round to the nearest thousandth. **(a)** 13 $\boxed{\sqrt{x}}$ 3.60555127 rounds to 3.606 **(b)** 182 $\boxed{\sqrt{x}}$ 13.49073756 rounds to 13.491 **2.** Find from a square root table. **(a)** $\sqrt{31}$ **(b)** $\sqrt{33}$ **(c)** $\sqrt{34}$ To the nearest thousandth, the approximate values are. **(a)** $\sqrt{31} = 5.568$ **(b)** $\sqrt{33} = 5.745$ **(c)** $\sqrt{34} = 5.831$
Finding the hypotenuse of a right triangle when given the length of each leg, p. 409.	Hypotenuse $= \sqrt{(\text{leg})^2 + (\text{leg})^2}$ Leg Leg Hypotenuse	Find the hypotenuse of a triangle with sides of 9 m and 12 m, respectively. Hypotenuse $= \sqrt{(12)^2 + (9)^2}$ $= \sqrt{144 + 81} = \sqrt{225}$ $= 15$ m
Finding the leg of a right triangle when given the length of the other leg and the hypotenuse, p. 410.	Leg $= \sqrt{(\text{hypotenuse})^2 - (\text{leg})^2}$	Find the leg of a right triangle. The hypotenuse is 14 in. and one leg is 12 in. Round your answer to nearest thousandth. Leg $= \sqrt{(14)^2 - (12)^2} = \sqrt{196 - 144} = \sqrt{52}$ Using a calculator or a square root table, leg ≈ 7.211 in.
Solving applied problems involving the Pythagorean Theorem, p. 412.	**1.** Read the problem carefully. **2.** Draw a sketch. **3.** Label the two sides that are given. **4.** If the hypotenuse is unknown, use Hypotenuse $= \sqrt{(\text{leg})^2 + (\text{leg})^2}$ **5.** If one leg is unknown, use Leg $= \sqrt{(\text{hypotenuse})^2 - (\text{leg})^2}$	A boat travels 5 mi south and then 3 mi east. How far is it from the starting point? Round your answer to the nearest tenth. 5 miles south 3 miles east Hypotenuse $= \sqrt{(5)^2 + (3)^2}$ $= \sqrt{25 + 9} = \sqrt{34}$ Using a calculator or a square root table, distance is approximately 5.8 mi.

Chapter Organizer

Topic	Procedure	Examples
The special 30°–60°–90° right triangle, p. 413.	The length of the leg opposite the 30° angle is $\frac{1}{2}\times$ the length of the hypotenuse.	Find y. $y = \frac{1}{2}(26) = 13$ m 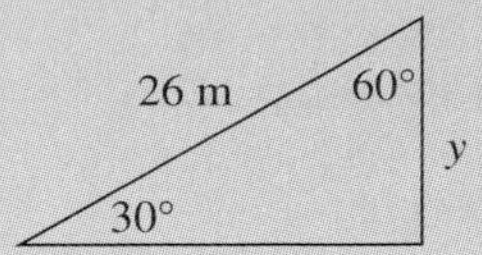
The special 45°–45°–90° right triangle, p. 413.	The sides opposite the 45° angles are equal. The hypotenuse is $\sqrt{2}\times$ the length of either leg.	Find z. $z = \sqrt{2}(13) \approx (1.414)(13) \approx 18.382$ m 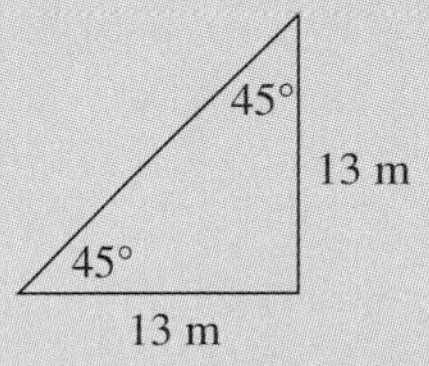
Radius and diameter of a circle, p. 418.	$r =$ radius $d =$ diameter $r = \frac{d}{2}$ $d = 2r$	What is the radius of a circle with diameter 50 in.? $r = \frac{50 \text{ in.}}{2} = 25$ in. What is the diameter of a circle with radius 16 in.? $d = (2)(16 \text{ in.}) = 32$ in.
Pi, p. 418.	Pi is a decimal that goes on forever. It can be approximated by as many decimal places as needed. $\pi = \frac{\text{circumference of a circle}}{\text{diameter of same circle}}$	Use $\pi \approx 3.14$ for all calculations. Unless otherwise directed, round your final answer to the nearest tenth when any calculation involves π.
Circumference of a circle, p. 418.	$C = \pi d$	Find the circumference of a circle with a diameter of 12 ft. $C = \pi d = (3.14)(12 \text{ ft}) = 37.68$ ft $= 37.7$ ft (to nearest tenth)
Area of a circle, p. 420.	$A = \pi r^2$ **1.** Square the radius first. **2.** Then multiply the result by 3.14.	Find the area of a circle with radius 7 ft. $A = \pi r^2 = (3.14)(7 \text{ ft})^2 = (3.14)(49 \text{ ft}^2)$ $= 153.86 \text{ ft}^2$ $= 153.9 \text{ ft}^2$ (to nearest tenth)
Volume of a rectangular solid (box), p. 427.	$V = lwh$	Find the volume of a box whose dimensions are 5 m by 8 m by 2 m. $V = (5 \text{ m})(8 \text{ m})(2 \text{ m}) = (40)(2) \text{ m}^3 = 80 \text{ m}^3$
Volume of a cylinder, p. 427.	$r =$ radius $h =$ height $V = \pi r^2 h$ **1.** Square the radius first. **2.** Then multiply the result by 3.14 and by the height. 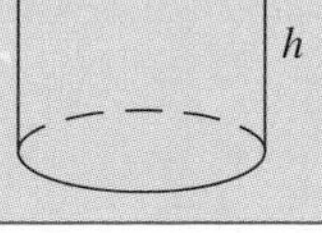	Find the volume of a cylinder with a radius of 7 m and a height of 3 m. $V = \pi r^2 h = (3.14)(7 \text{ m})^2(3 \text{ m})$ $= (3.14)(49)(3) \text{ m}^3$ $= (153.86)(3) \text{ m}^3 = 461.58 \text{ m}^3$ $= 461.6 \text{ m}^3$ (rounded to nearest tenth)

Chapter Organizer

Topic	Procedure	Examples
Volume of a sphere, p. 428.	$V = \frac{4\pi r^3}{3}$ 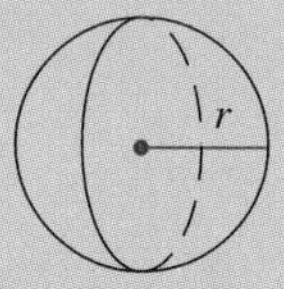 r = radius	Find the volume of a sphere of radius 3 m. $V = \frac{4\pi r^3}{3} = \frac{(4)(3.14)(3\text{ m})^3}{3}$ $= \frac{(4)(3.14)(\overset{9}{\cancel{27}})\text{ m}^3}{\underset{1}{\cancel{3}}} = (4)(3.14)(9)\text{ m}^3$ $= (12.56)(9)\text{ m}^3 = 113.04\text{ m}^3$ $= 113.0\text{ m}^3$ (rounded to nearest tenth)
Volume of a cone, p. 428.	$V = \frac{\pi r^2 h}{3}$ r = radius h = height 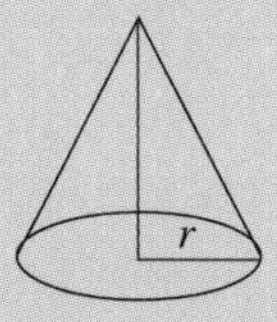 	Find the volume of a cone of height 9 m and radius 7 m. $V = \pi r^2 h = \frac{(3.14)(7\text{ m})^2(9\text{ m})}{3}$ $= \frac{(3.14)(7^2)(\overset{3}{\cancel{9}})\text{ m}^3}{\underset{1}{\cancel{3}}} = (3.14)(49)(3)\text{ m}^3$ $= (153.86)(3)\text{ m}^3 = 461.58\text{ m}^3$ $= 461.6\text{ m}^3$ (rounded to the nearest tenth)
Volume of a pyramid, p. 429.	$V = \frac{Bh}{3}$ 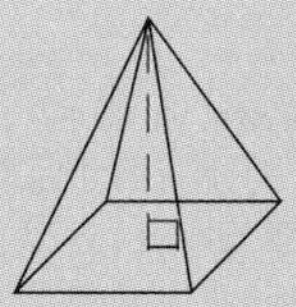 B = area of the base h = height **1.** Find the area of the base. **2.** Multiply this area by the height and divide the result by 3.	Find the volume of a pyramid whose height is 6 m and whose rectangular base is 10 m by 12 m. **1.** $B = (12\text{ m})(10\text{ m}) = 120\text{ m}^2$ **2.** $V = \frac{(120)(\overset{2}{\cancel{6}})\text{ m}^3}{\underset{1}{\cancel{3}}} = (120)(2)\text{ m}^3 = 240\text{ m}^3$
Similar figures, corresponding sides, p. 435.	The corresponding sides of similar figures have the same ratio.	Find n in the following similar figures. $\frac{n}{4} = \frac{9}{3}$ $3n = 36$ $n = 12$ m 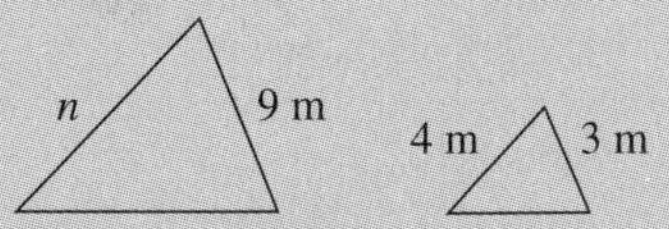
Similar figures, corresponding perimeters, p. 439.	The perimeters of similar figures have the same ratio as the corresponding sides. For reasons of space, the procedure for the areas of similar figures is not given here but may be found in the text (see p. 439).	These two figures are similar. Find the perimeter of the larger figure. $\frac{6}{12} = \frac{29}{p}$ $6p = (12)(29)$ 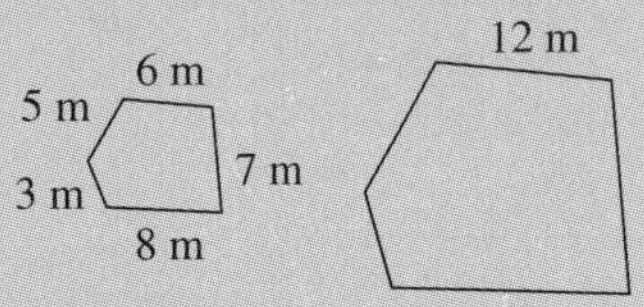 $6p = 348$ $\frac{6p}{6} = \frac{348}{6}$ $p = 58$ The perimeter of the larger figure is 58 m.

Chapter 7 Review Problems

In each problem in the chapter review, round your answer to the nearest tenth when necessary. Use $\pi \approx 3.14$ in all calculations requiring the use of π.

7.1 *Find the perimeter of each square and rectangle.*

1. Length = 8.3 m, width = 1.6 m

$P = 19.8$ m

2. Length = 9.6 cm, width = 2.8 cm

$P = 24.8$ cm

3. Length = width = 5.8 yd

$P = 23.2$ yd

4. Length = width = 2.4 yd

$P = 9.6$ yd

Find the area of each square and rectangle.

5. Length = 5.9 cm, width = 2.8 cm

$A = 16.5\text{ cm}^2$

6. Length = 9.3 m, width = 7.9 m

$A = 73.5\text{ m}^2$

7. Length = width = 4.3 in.

$A = 18.5\text{ in.}^2$

8. Length = width = 7.2 in.

$A = 51.8\text{ in.}^2$

Find the perimeter of each object, made up of rectangles and squares.

9.

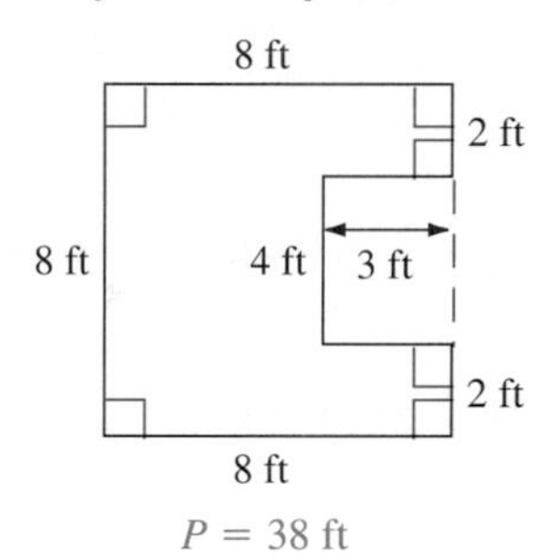

$P = 38$ ft

10.

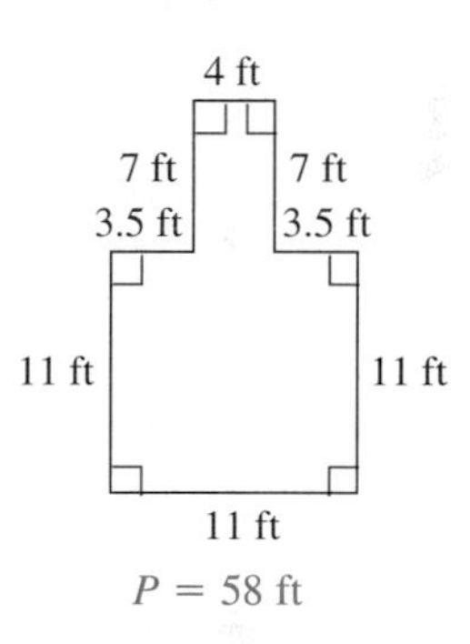

$P = 58$ ft

Find the shaded area of each object, made up of rectangles and squares.

11.

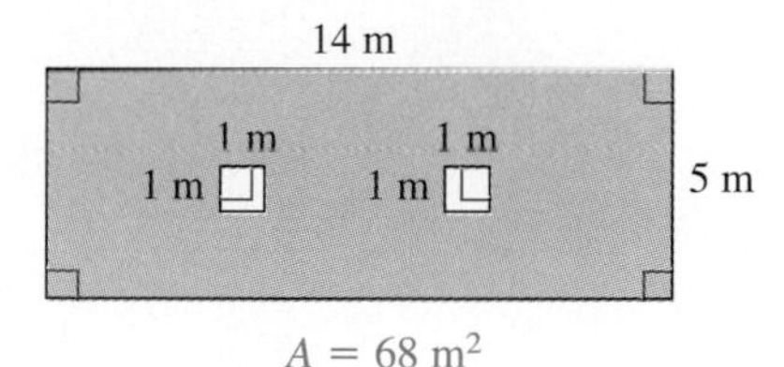

$A = 68\text{ m}^2$

12.

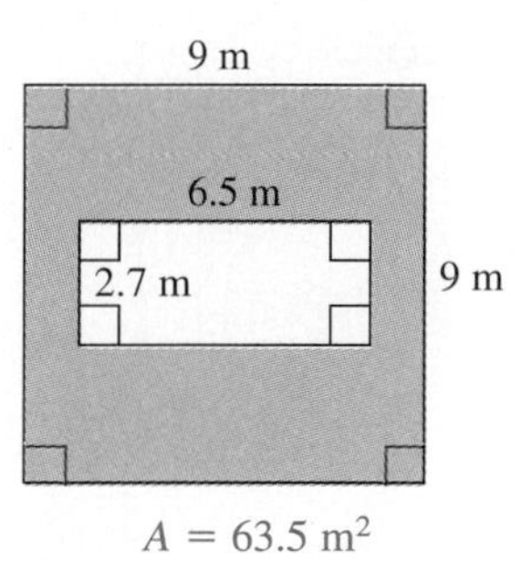

$A = 63.5\text{ m}^2$

7.2 *Find the perimeter of each parallelogram or trapezoid.*

13. Two sides of the parallelogram are 43 m and 7.2 m.

$P = 100.4$ m

14. Two sides of the parallelogram are 26 cm and 12.2 cm.

$P = 76.4$ cm

15. The sides of the trapezoid are 22 in., 13 in., 32 in., and 13 in.

$P = 80$ in.

16. The sides of the trapezoid are 5 mi, 22 mi, 5 mi, and 30 mi.

$P = 62$ mi

Find the area of each parallelogram or trapezoid.

17. The parallelogram has a base of 90 m and a height of 30 m.

$A = 2700\text{ m}^2$

18. The parallelogram has a base of 82 m and a height of 25 m.

$A = 2050\text{ m}^2$

19. The trapezoid has a height of 14 yd and bases of 20 yd and 28 yd.
$A = 336 \text{ yd}^2$

20. The trapezoid has a height of 36 yd and bases of 17 yd and 23 yd.
$A = 720 \text{ yd}^2$

Find the shaded area of each region, made up of parallelograms, trapezoids, and rectangles.

21.

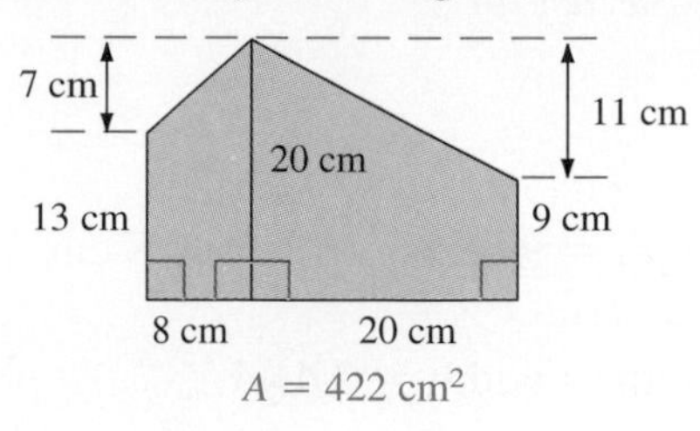

$A = 422 \text{ cm}^2$

22.

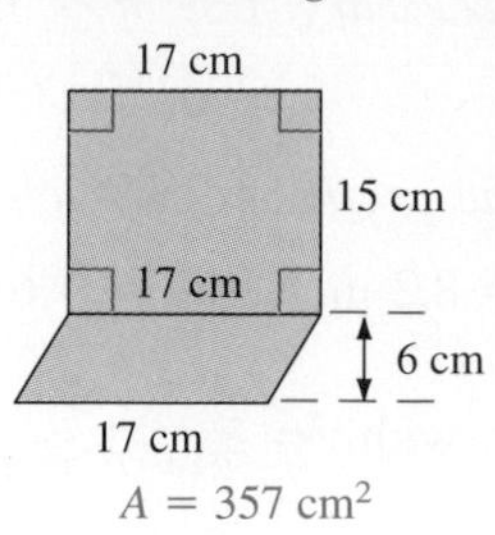

$A = 357 \text{ cm}^2$

7.3 *Find the perimeter of each triangle.*

23. The sides of the triangle are 10 ft, 5 ft, and 7 ft.
$P = 22$ ft

24. The sides of the triangle are 5.5 ft, 3 ft, and 5.5 ft.
$P = 14$ ft

Find the measure of the third angle in each triangle.

25. Two known angles are 15° and 12°.
153°

26. Two known angles are 62° and 78°.
40°

Find the area of each triangle.

27. Base = 9.6 m, height = 5.1 m
$A = 24.5 \text{ m}^2$

28. Base = 12.5 m, height = 9.5 m
$A = 59.4 \text{ m}^2$

29. Base = 12 cm, height = 7.6 cm
$A = 45.6 \text{ cm}^2$

30. Base = 18.2 m, height = 4.4 m
$A = 40.0 \text{ m}^2$

Find the area of each region, made up of triangles and rectangles.

31.

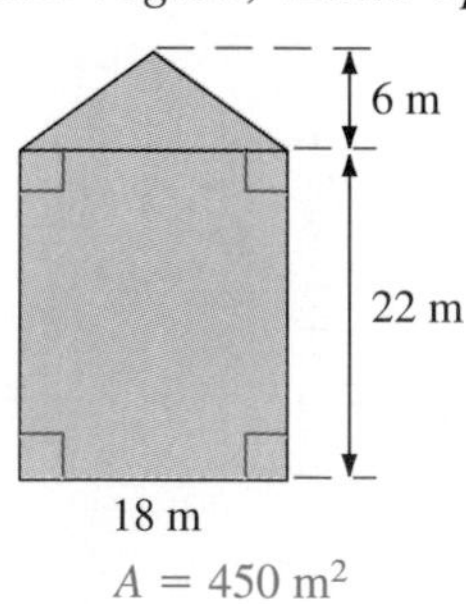

$A = 450 \text{ m}^2$

32.

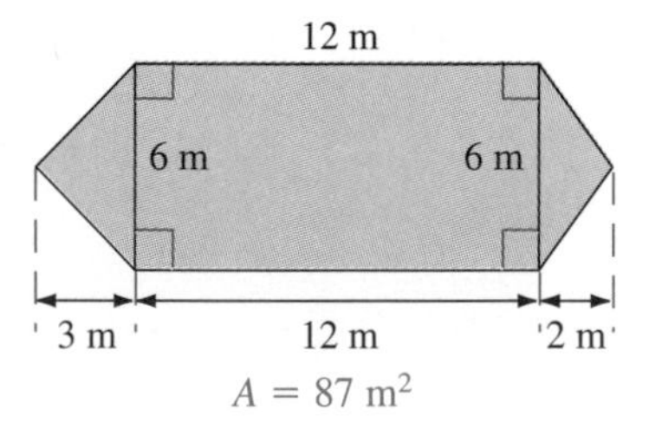

$A = 87 \text{ m}^2$

7.4

Evaluate exactly.

33. $\sqrt{81}$
9

34. $\sqrt{64}$
8

35. $\sqrt{100}$
10

36. $\sqrt{121}$
11

37. $\sqrt{144}$
12

38. $\sqrt{225}$
15

39. $\sqrt{36} + \sqrt{0}$
6

40. $\sqrt{25} + \sqrt{1}$
6

41. $\sqrt{9} + \sqrt{4}$
5

42. $\sqrt{25} + \sqrt{49}$
12

Approximate using a square root table or a calculator with a square root key. Round your answer to the nearest thousandth when necessary.

43. $\sqrt{35}$
≈ 5.916

44. $\sqrt{45}$
≈ 6.708

45. $\sqrt{88}$
≈ 9.381

46. $\sqrt{76}$
≈ 8.718

47. $\sqrt{171}$
≈ 13.077

48. $\sqrt{180}$
≈ 13.416

7.5

Find the unknown side of the right triangle. If the answer cannot be obtained exactly, use a square root table or a calculator with a square root key. Round your answer to the nearest hundredth when necessary.

49.

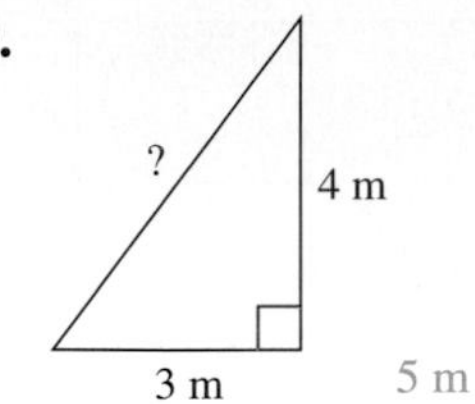

5 m

50.

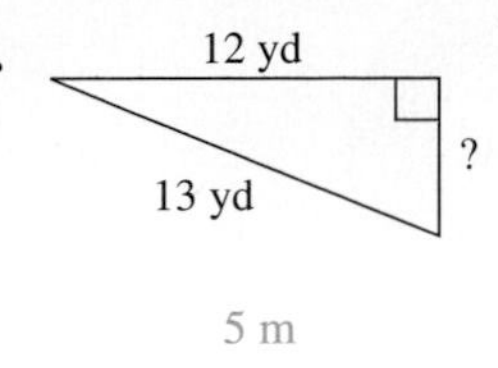

5 m

51.

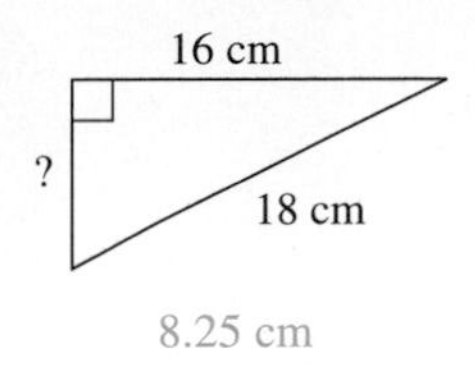

8.25 cm

52.

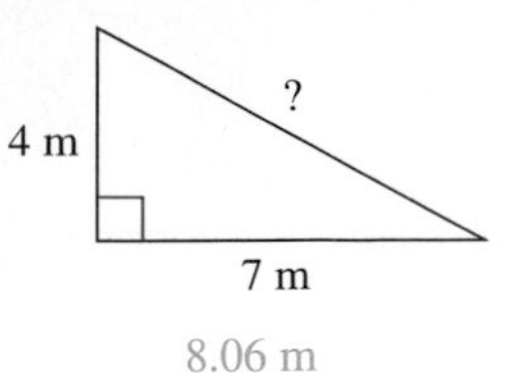

8.06 m

In problems 53–56, round your answer to the nearest tenth.

53. Find the distance between the centers of the holes of a metal plate with the dimensions labeled in this sketch. 6.4 cm

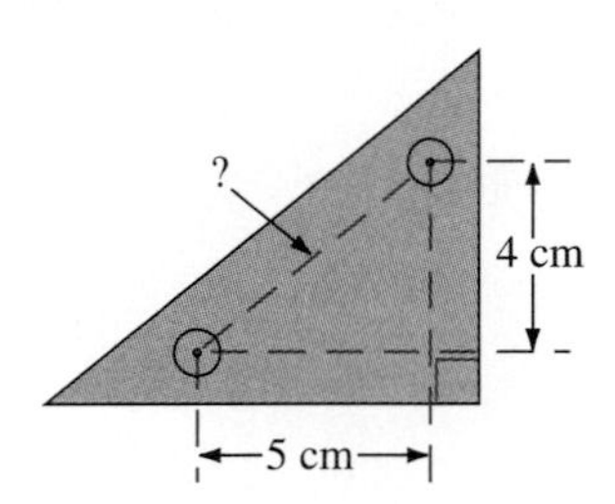

54. A building ramp has the following dimensions. Find the length of the ramp. 9.2 ft

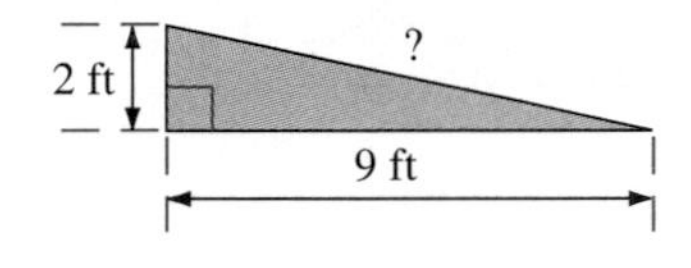

55. A shed is built with the following dimensions. Find the distance from the peak of the roof to the horizontal support brace. 6.3 ft

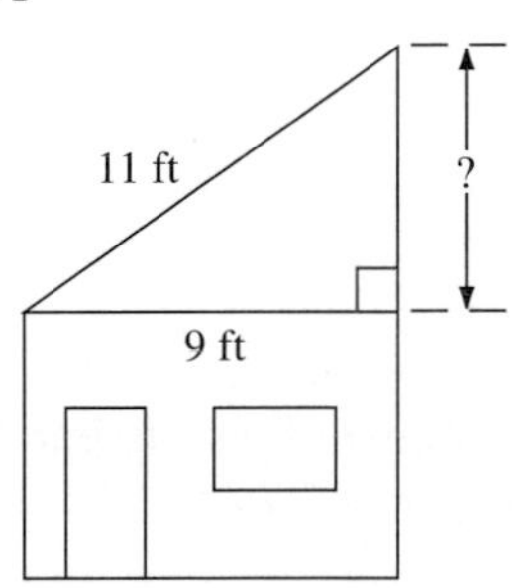

56. Find the width of a door if it is 6 ft tall and the diagonal line measures 7 ft. 3.6 ft

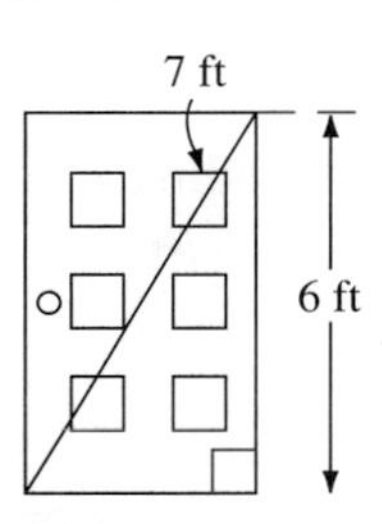

7.6

57. What is the *diameter* of a circle whose radius is 53 cm? 106 cm

58. What is the *radius* of a circle whose diameter is 48 cm? 24 cm

59. Find the *circumference* of a circle with diameter 12 in. 37.7 in.

60. Find the *circumference* of a circle with radius 7 in. 44.0 in.

Find the area of each circle.

61. Radius = 6 m
$A = 113.0 \text{ m}^2$

62. Radius = 9 m
$A = 254.3 \text{ m}^2$

63. Diameter = 16 ft
$A = 201.0 \text{ ft}^2$

64. Diameter = 14 ft
$A = 153.9 \text{ ft}^2$

Find the area of each shaded region, made up of circles, semicircles, rectangles, trapezoids, and parallelograms.

65.

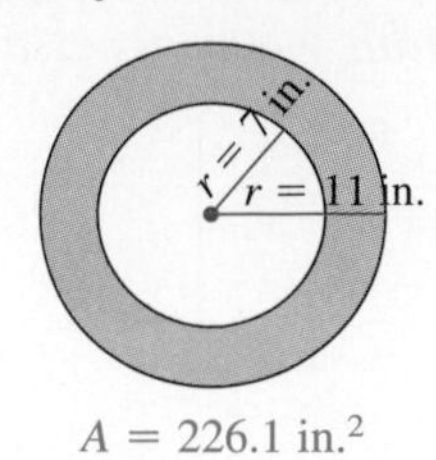

$A = 226.1 \text{ in.}^2$

66.

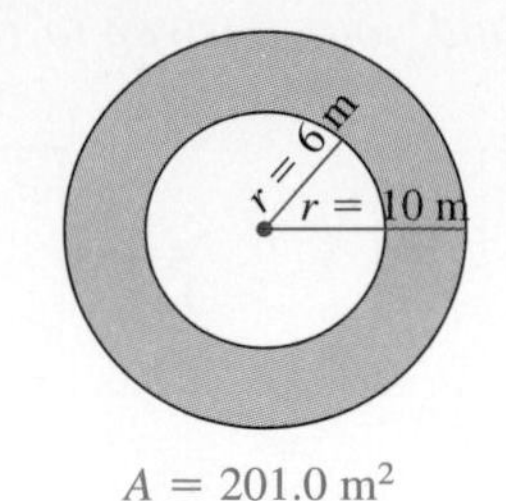

$A = 201.0 \text{ m}^2$

67.

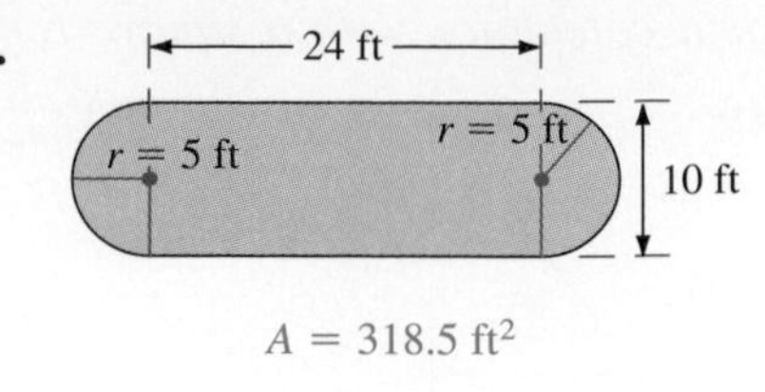

$A = 318.5 \text{ ft}^2$

68.

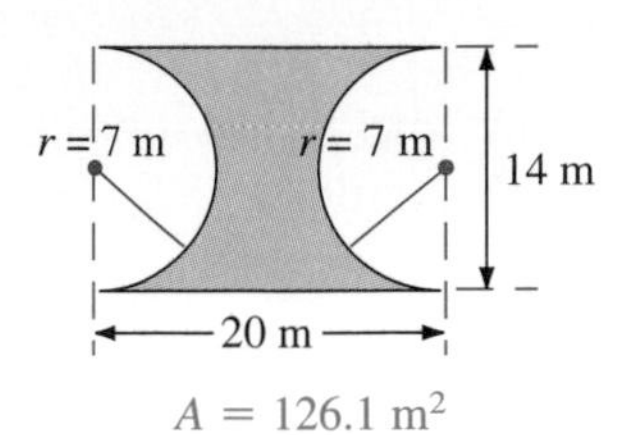

$A = 126.1 \text{ m}^2$

69.

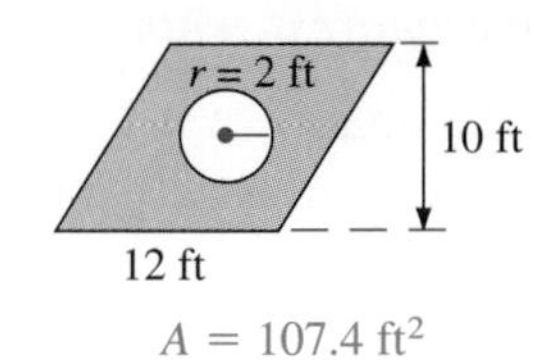

$A = 107.4 \text{ ft}^2$

70.

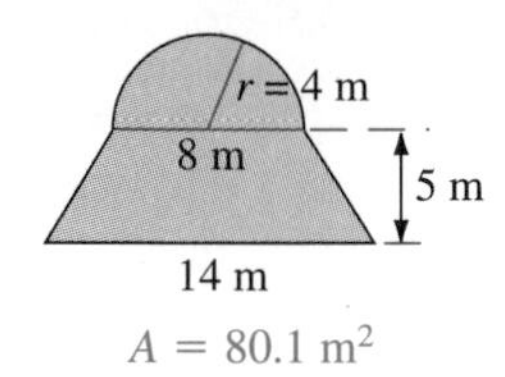

$A = 80.1 \text{ m}^2$

7.7 *Find the volume of each object.*

71. A rectangular box measuring 3 ft by 6 ft by 2.5 ft.
$V = 45 \text{ ft}^3$

72. A rectangular box measuring 20 m by 15 m by 1.5 m.
$V = 450 \text{ m}^3$

73. A sphere with radius 15 ft.
$V = 14{,}130 \text{ ft}^3$

74. A sphere with radius 1.2 ft.
$V = 7.2 \text{ ft}^3$

75. Find the volume of a storage can that is 2 m tall and has a radius of 7 m.
$V = 307.7 \text{ m}^3$

76. Find the volume of a soup can 18 cm tall and having a radius of 9 cm.
$V = 4578.1 \text{ cm}^3$

77. A pyramid that is 18 m high and whose rectangular base measures 16 m by 18 m.
$V = 1728 \text{ m}^3$

78. A pyramid that is 12 m high and whose square base measures 5 m by 5 m.
$V = 100 \text{ m}^3$

79. Find the volume of a cone of sand 9 ft tall with a radius of 20 ft.
$V = 3768 \text{ ft}^3$

80. A chemical has polluted a volume of ground in a cone shape. The depth of the cone is 30 yd. The radius of the cone is 17 yd. Find the volume of polluted ground.
$V = 9074.6 \text{ yd}^3$

7.8 *Find n in each set of similar triangles.*

81.

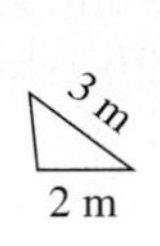

45 m

n

$n = 30$ m

82.

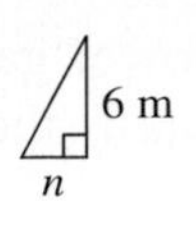

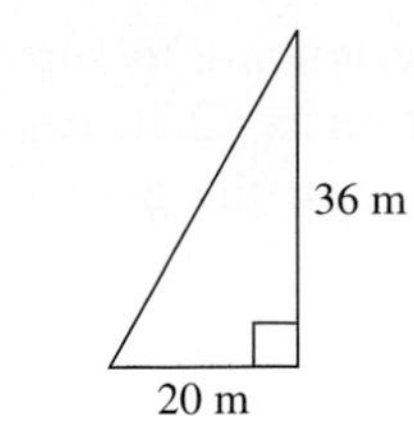

$n = 3.3$ m

83.

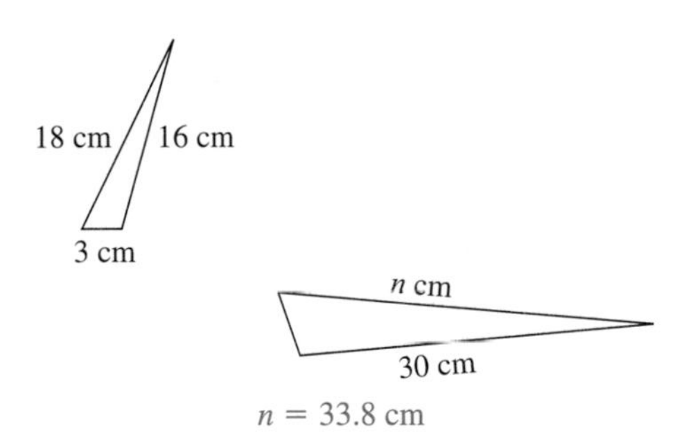

$n = 33.8$ cm

84.

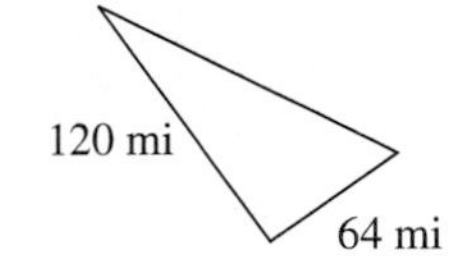

$n = 9.6$ mi

Find the perimeter of the larger of the two similar figures.

85.

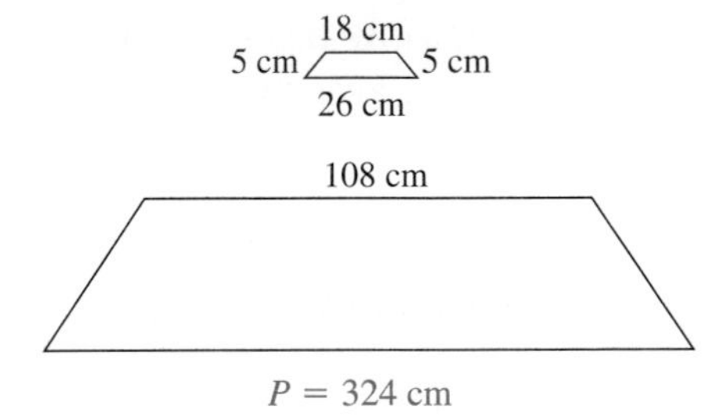

$P = 324$ cm

86.

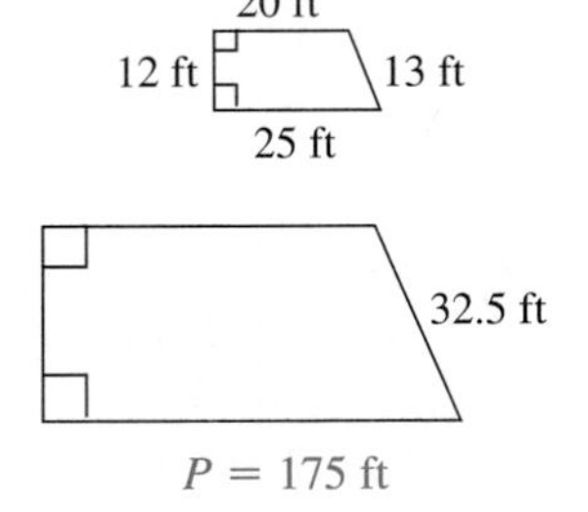

$P = 175$ ft

87. The fencing to surround a flower garden 14 ft by 8 ft costs $2.10 per foot. What is the total cost to fence in the garden? $92.40

88. A picture measures 3 cm by 5 cm. A photographer wishes to enlarge it to a width (shorter distance) of 126 cm.

(a) What will be the length of the new picture? 210 cm

(b) How much larger is the *area* of the new picture compared to the old picture? 1764 times larger

89. The Wilsons are carpeting a recreation room with the dimensions shown. Carpeting costs $8.00 per square yard. How much will the carpeting cost?

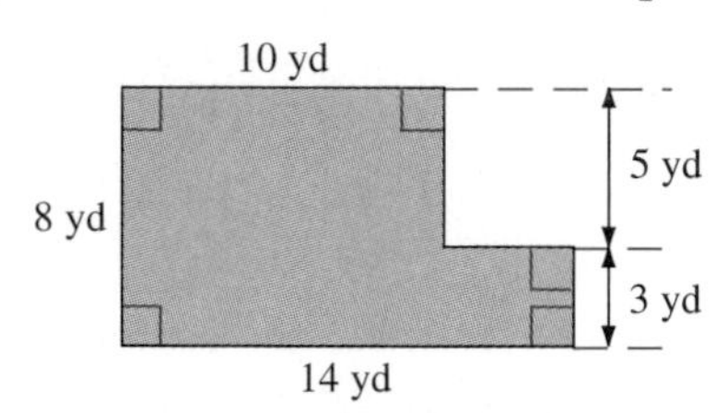

$736.00

90. A conical tank holds acid in a chemistry lab. The tank has a radius of 9 in. and a height of 24 in. How many cubic inches does the tank hold? The acid weighs 16 g per cubic inch. What is the weight of the acid if the tank is full? $V = 2034.7 \text{ in.}^3$ $W = 32{,}555.2$ g

91. A silo has a cylindrical shape with a hemisphere dome. It has dimensions as shown.

(a) What is its volume in cubic feet? $V = 21{,}873.2 \text{ ft}^3$

(b) If 1 cubic foot = 0.8 bushel, how many bushels of grain will it hold? $B = 17{,}498.6$ bushels

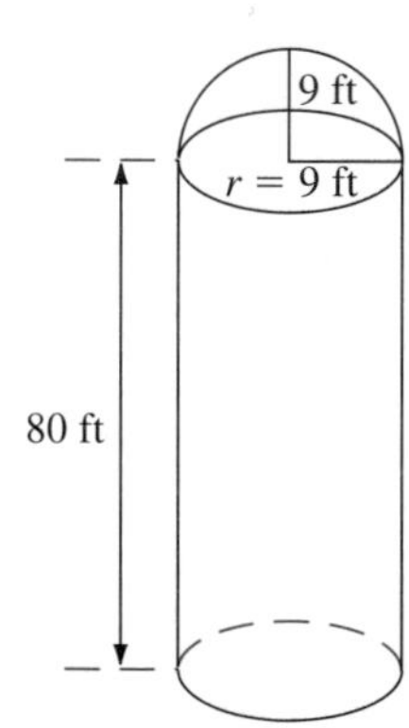

92. (a) How many kilometers is it from Homeville to Seaview if you drive through Ipswich? How fast do you travel if it takes 0.5 hour to travel that way? 50 km 100 km/h

(b) How many kilometers is it from Homeville to Seaview if you drive through Acton and Westville? How fast do you travel if it takes 0.8 hour to travel that way? 56 km 70 km/hr

(c) Over which route do you travel at a more rapid rate? Through Ipswich

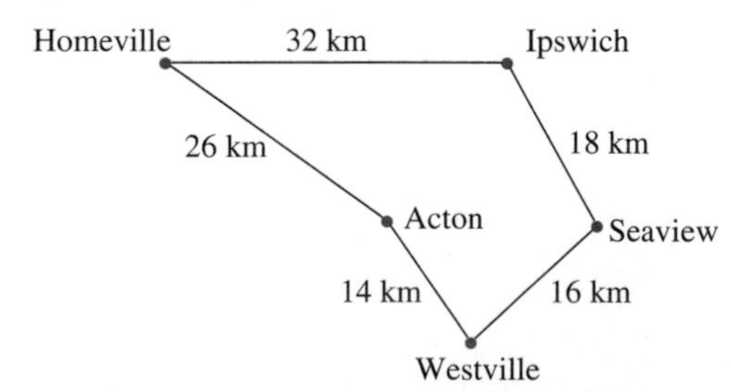

Chapter 7 Test

Find the perimeter for each.

1. A rectangle that measures 9 yd × 11 yd

2. A square with side 8.5 ft

3. A parallelogram with sides measuring 6.5 m and 3.5 m

4. A trapezoid with sides measuring 22 m, 13 m, 32 m, and 13 m

5. A triangle with sides measuring 4.4 m, 10.8 m, and 9.6 m

Find the area for each. Round each area to the nearest tenth.

6. A rectangle that measures 10 yd × 18 yd

7. A square 10.2 m on a side

8. A parallelogram with a height of 6 m and a base of 13 m

9. A trapezoid with a height of 9 m and bases of 7 m and 25 m

10. A triangle with a base of 4 cm and a height of 6 cm

11. A triangle with a base of 15 m and a height of 7 m

Evaluate exactly.

12. $\sqrt{81}$ **13.** $\sqrt{64}$ **14.** $\sqrt{1} + \sqrt{121}$ **15.** $\sqrt{0} + \sqrt{49}$

Approximate using a square root table or a calculator with a square root key. Round your answer to the nearest thousandth when necessary.

16. $\sqrt{54}$ **17.** $\sqrt{120}$ **18.** $\sqrt{187}$

Find the unknown side of the right triangle. Use a calculator or a square root table to approximate square roots to the nearest thousandth.

19.

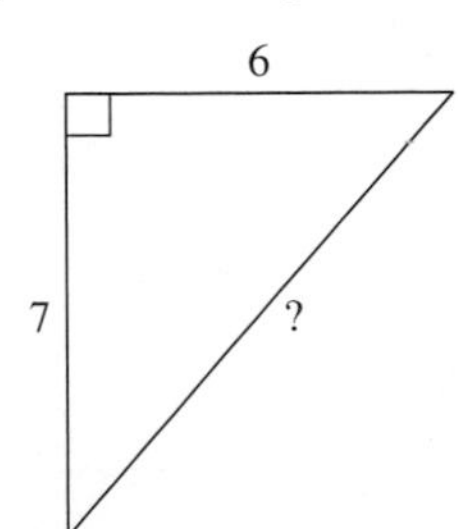

20.

1.	$P = 40$ yd
2.	$P = 34$ ft
3.	$P = 20$ m
4.	$P = 80$ m
5.	$P = 24.8$ m
6.	$A = 180$ yd^2
7.	$A = 104.0$ m^2
8.	$A = 78$ m^2
9.	$A = 144$ m^2
10.	$A = 12$ cm^2
11.	$A = 52.5$ m^2
12.	9
13.	8
14.	12
15.	7
16.	≈ 7.348
17.	≈ 10.954
18.	≈ 13.675
19.	9.22
20.	10

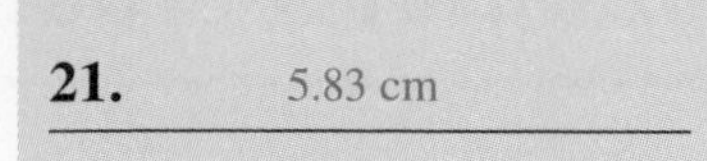

21. 5.83 cm

22. 9 ft

23. $C = 37.7$ in.

24. $A = 254.3\ \text{ft}^2$

25. $A = 107.4\ \text{in.}^2$

26. $A = 144.3\ \text{in.}^2$

27. $V = 840\ \text{m}^3$

28. $V = 803.8\ \text{m}^3$

29. $V = 33.5\ \text{m}^3$

30. $V = 508.7\ \text{ft}^3$

31. $V = 56\ \text{m}^3$

32. $n = 43.2$ cm

33. $n = 46.7$ ft

34. $A = 6456\ \text{yd}^2$ $2582.40

21. Find the distance between the centers of the holes drilled in a rectangular metal plate with the dimensions labeled in this sketch.

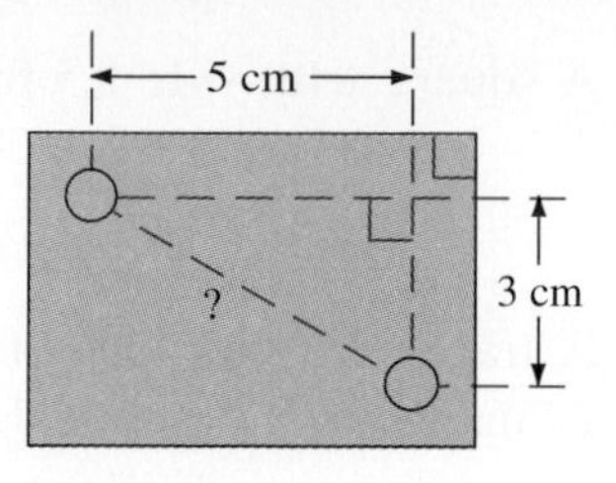

22. A 15-ft-tall ladder is placed so that it reaches 12 ft up on the wall of a house. How far is the base of the ladder from the wall of the house?

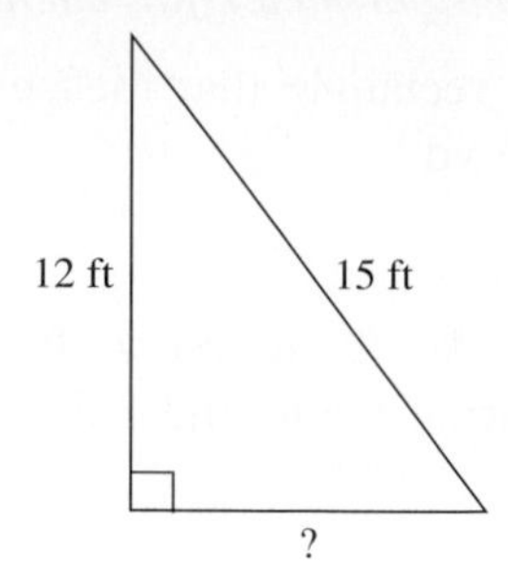

23. Find the circumference of a circle with *radius* 6 in.

24. Find the area of a circle with *diameter* 18 ft.

Find the shaded area of each region, made up of circles, semicircles, rectangles, squares, trapezoids, or parallelograms.

25.

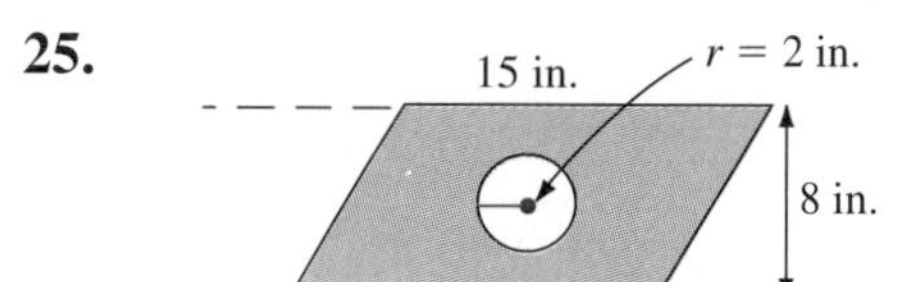

26.

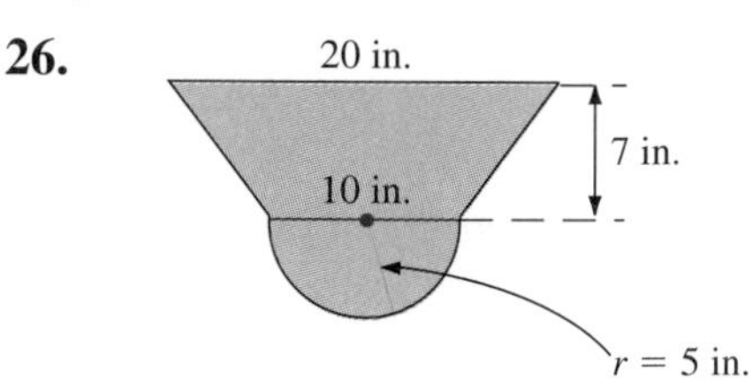

Find the volume of each.

27. A rectangular box measuring 7 m by 12 m by 10 m

28. A cone with a height of 12 m and a radius of 8 m

29. A sphere of radius 2 m

30. A cylinder of height 2 ft and radius 9 ft

31. A pyramid of height 14 m and whose rectangular base measures 4 m by 3 m.

Each pair of triangles is similar. Find the missing side n.

32.

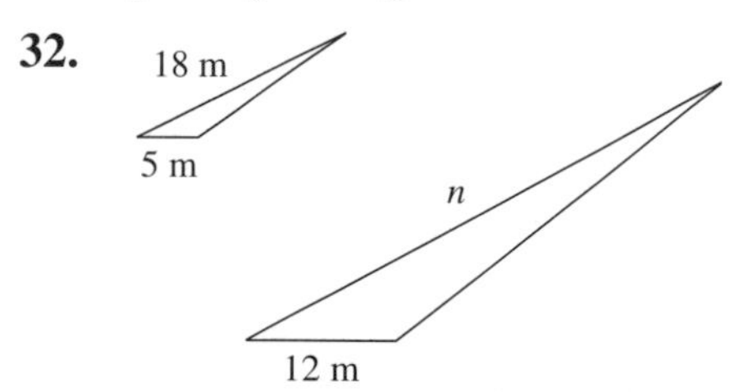

33.

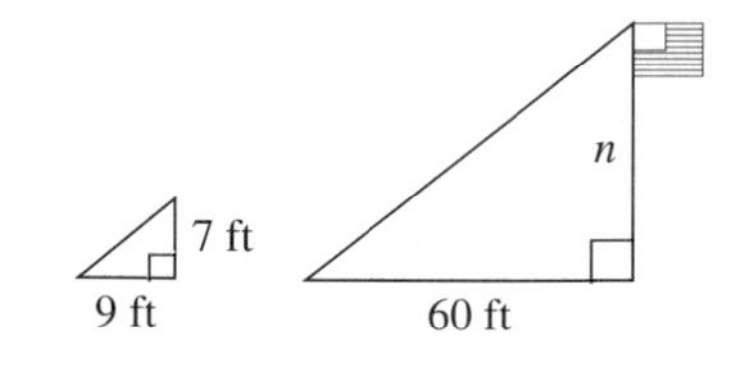

Solve each applied problem.

34. An athletic field has the dimensions shown.

(a) What is the area of the athletic field?

(b) How much will it cost to fertilize it at $0.40 per square yard?

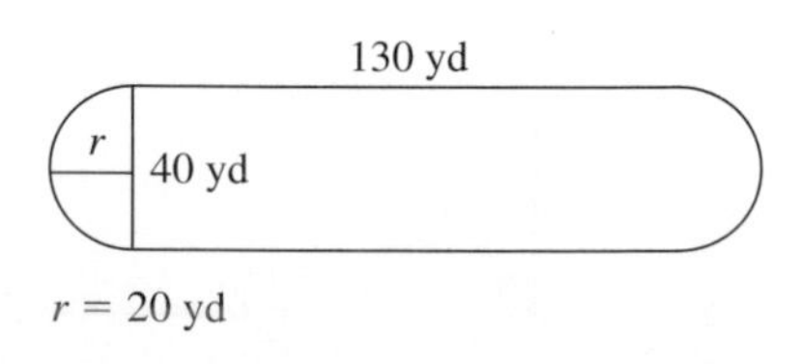

Cumulative Test for Chapters 1–7

Approximately one-half of this test is based on Chapter 7 material. The remainder is based on material covered in Chapters 1–6.

Do each problem. Simplify your answer.

1. Add. $\begin{array}{r} 126{,}350 \\ 278{,}120 \\ +\ 531{,}290 \\ \hline \end{array}$

2. Multiply. $\begin{array}{r} 163 \\ \times\ 205 \\ \hline \end{array}$

3. Subtract. $\dfrac{17}{18} - \dfrac{11}{30}$

4. Divide. $\dfrac{3}{7} \div 2\dfrac{1}{4}$

5. Round to the nearest hundredth. 56.1279

6. Multiply. $\begin{array}{r} 9.034 \\ \times\ \ \ 0.8 \\ \hline \end{array}$

7. Divide. $0.021\overline{)1.743}$

8. Find n. $\dfrac{3}{n} = \dfrac{2}{18}$

9. There are seven teachers for every 100 students. If there are 56 teachers at the university, how many students would you expect?

10. Michael scored 18 baskets out of 24 shots on the court. What percent of his shots went into the basket?

11. 0.8% of what number is 16?

12. What is 18.5% of 220?

13. Convert 586 cm to m.

14. Convert 42 yd to in.

15. Ben traveled 88 km. How many miles did he travel? (1 kilometer = 0.62 mile; 1 mile = 1.61 kilometers.)

In problems 16–30, round your answer to the nearest tenth when necessary. Use $\pi \approx 3.14$ in calculations involving π.

Find the perimeter of each object.

16. A rectangle of length 17 m and width 8 m

17. A trapezoid with sides of 86 cm, 13 cm, 96 cm, and 13 cm

18. Find the circumference of a circle with diameter 18 yd.

Find the area of each object.

19. A triangle with base 1.2 cm and height 2.4 cm

20. A trapezoid with height 18 m and bases of 26 m and 34 m

21.

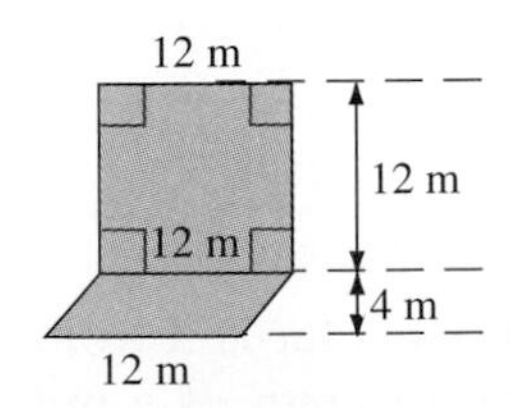

22.

35 yd
20 yd
6 yd
6 yd
20 yd
5 yd
24 yd

23. A circle with radius 4 m.

1. 935,760

2. 33,415

3. $\dfrac{26}{45}$

4. $\dfrac{4}{21}$

5. 56.13

6. 7.2272

7. 83

8. $n = 27$

9. 800 students

10. 75%

11. 2000

12. 40.7

13. 5.86 m

14. 1512 in.

15. 54.56 mi

16. $P = 50$ m

17. $P = 208$ cm

18. 56.5 yd

19. $A = 1.4\ \text{cm}^2$

20. $A = 540\ \text{m}^2$

21. $A = 192\ \text{m}^2$

22. $A = 664\ \text{yd}^2$

23. $A = 50.2\ \text{m}^2$

24.	$V = 2411.5\text{ m}^3$
25.	$V = 3052.1\text{ cm}^3$
26.	$V = 3136\text{ cm}^3$
27.	$V = 2713.0\text{ m}^3$
28.	$n = 33.4$ m
29.	$n = 4.1$ ft
30.	124 yd^2 $992.00
31.	11
32.	≈ 7.550
33.	10.440 in.
34.	4.899 m
35.	13.9 mi

Find the volume of each object.

24. A cylinder with a height of 12 m and a radius of 8 m

25. A sphere with a radius of 9 cm

26. A pyramid with height 32 cm and a rectangular base 14 cm by 21 cm

27. A cone of height 18 m and a radius of 12 m

Find the value of n in each pair of similar figures.

28.

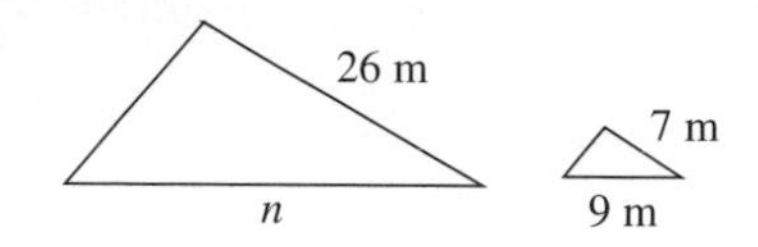

29.

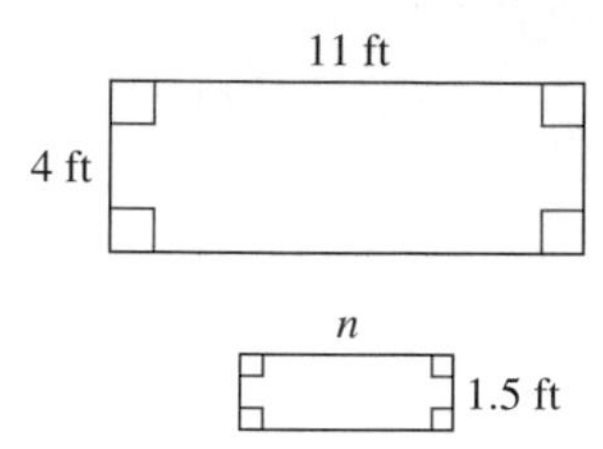

Solve each problem.

30. Mary Ann and Wong Twan have a recreation room with the dimensions shown in the figure. They wish to carpet it at a cost of $8.00 per square yard.

(a) How many square yards of carpet are needed?

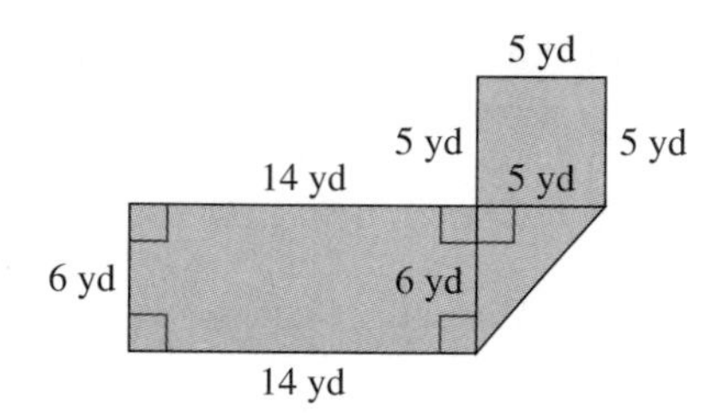

(b) How much will it cost?

31. Evaluate exactly. $\sqrt{36} + \sqrt{25}$

32. Approximate to the nearest thousandth using a table or a calculator. $\sqrt{57}$

Find the unknown side of the right triangle. Use a calculator or square root table if necessary. Round answers to the nearest thousandth.

33.

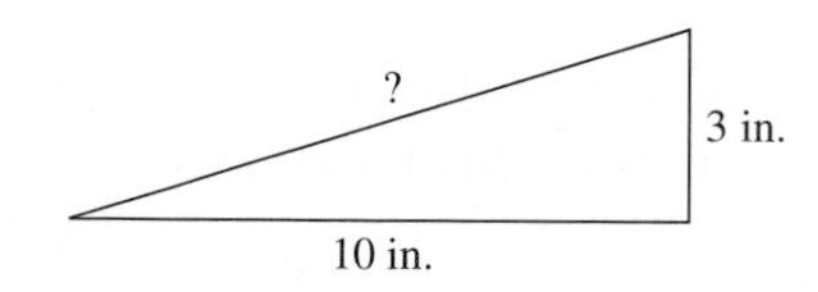

34.

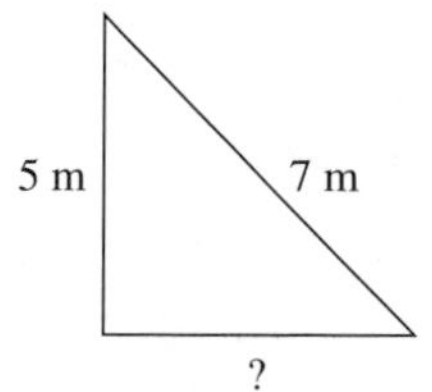

35. A boat travels 12 mi south and then 7 mi east. How far is it from its original starting point? Round your answer to the nearest tenth of a mile.

CHAPTER 8
Statistics

As you look at trucks speeding down the highway do you ever wonder how many years these trucks have been on the road? Or do you wonder how many miles these trucks have been driven? As we build newer and more modern trucks, is the nation's fleet of trucks getting younger or getting older? If you would like to use your knowledge of statistics and graphs to help you answer these questions, please turn to the Putting Your Skills to Work on page 477.

Pretest Chapter 8

If you are familiar with the topics in this chapter, take this test now. Check your answers with those in the back of the book. If an answer was wrong or you couldn't do a problem, study the appropriate section of the chapter.

If you are not familiar with the topics in this chapter, don't take this test now. Instead, study the examples, work the practice problems, and then take the test.

This test will help you identify which concepts you have mastered and which you need to study further.

1. Under age 18
2. 43%
3. 17%
4. 1650 students
5. 350 students
6. 300 people
7. 600 people
8. 3rd quarter of both 1993 and 1994
9. 4th quarter of 1993
10. 100 more people
11. 100 more people

Section 8.1

The 5000 students on campus were recorded by age. The circle graph depicts their distribution by age.

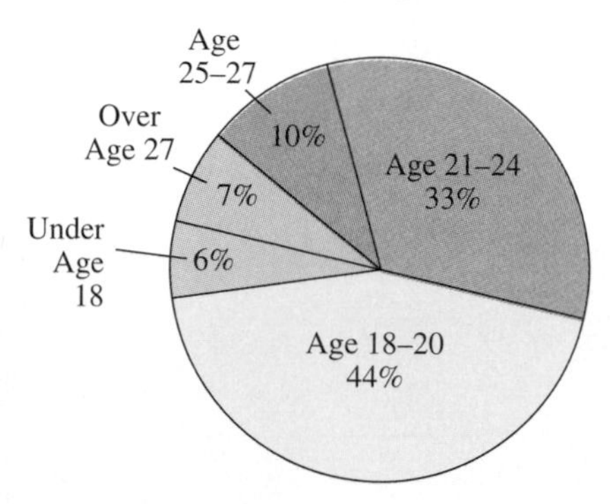

1. What age group has the smallest percent of the student body?

2. What percent of the students are between 21 and 27?

3. What percent of the students are 25 or older?

4. If 5000 students are at the university, *how many students* are between 21 and 24 years of age?

5. *How many students* are over age 27?

Section 8.2

The double-bar graph indicates the number of people unemployed in Pacerville during each quarter of 1993 and 1994.

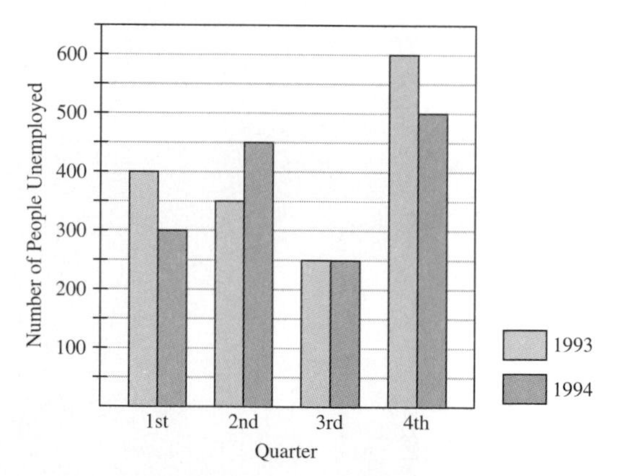

6. How many people were unemployed in Pacerville during the first quarter of 1994?

7. How many people were unemployed in Pacerville during the fourth quarter of 1993?

8. When were the fewest number of people unemployed in Pacerville?

9. When were the greatest number of people unemployed in Pacerville?

10. How many more people were unemployed in the fourth quarter of 1993 than the fourth quarter of 1994?

11. How many more people were unemployed during the second quarter of 1994 than the second quarter of 1993?

The line graph indicates sales and production of color television sets by a major manufacturer during the specified months.

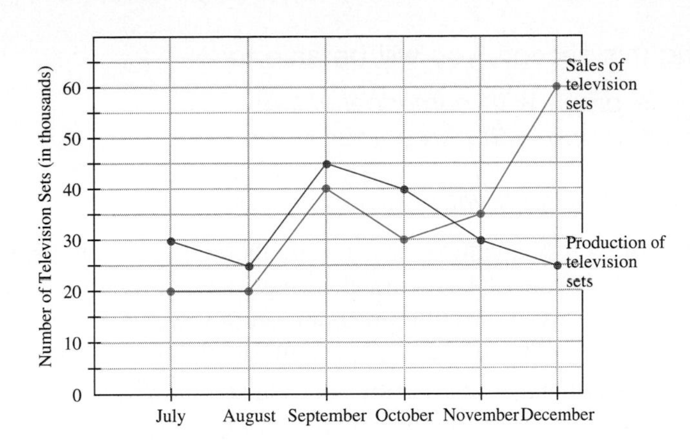

12. During what months was the production of television sets the lowest?
13. During what month was the sales of television sets the highest?
14. What was the first month in which the production of television sets was lower than the sales of television sets?
15. How many television sets were sold in August?
16. How many television sets were produced in November?

12. August and December

13. December

14. November

15. 20,000 sets

16. 30,000 sets

Section 8.3

The histogram tells us the number of miles a car was driven before the car was discarded or sold to a junk dealer.

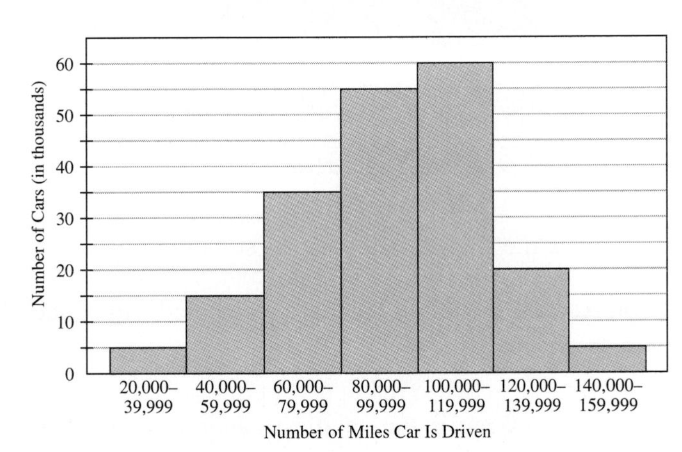

17. How many discarded or junked cars had been driven 80,000 to 99,999 mi?
18. How many discarded or junked cars had been driven 100,000 to 119,999 mi?
19. How many discarded or junked cars had been driven 120,000 mi or more?
20. How many discarded or junked cars had been driven less than 60,000 mi?

17. 55,000 cars

18. 60,000 cars

19. 25,000 cars

20. 20,000 cars

Section 8.4

A typist produced the following number of pages.

Mon.	Tues.	Wed.	Thurs.	Fri.
40	42	44	27	32

21. Find the *mean* number of pages typed per day.
22. Find the *median* number of pages typed per day.

21. 37 pages per day

22. 40 pages per day

8.1 Circle Graphs

MathPro

Video 17

SSM

After studying this section, you will be able to:

1. *Read a circle graph with numerical values.*
2. *Read a circle graph with percentage values.*

1 *Reading a Circle Graph with Numerical Values*

Statistics is that branch of mathematics that collects and studies data. Once the data are collected, they must be organized so that the information is easily readable. We use graphs to give a visual representation of the data that is easy to read. Graphs appeal to the eye. Their visual nature allows them to communicate information about the complicated relationships among statistical data. For this reason, newspapers often use graphs to help their readers quickly grasp information.

Circle graphs are especially helpful for showing the relationship of parts to a whole. The entire circle represents 100%; the pie-shaped pieces represent the subcategories. The following circle graph divides the 10,000 students at Westline College into five categories.

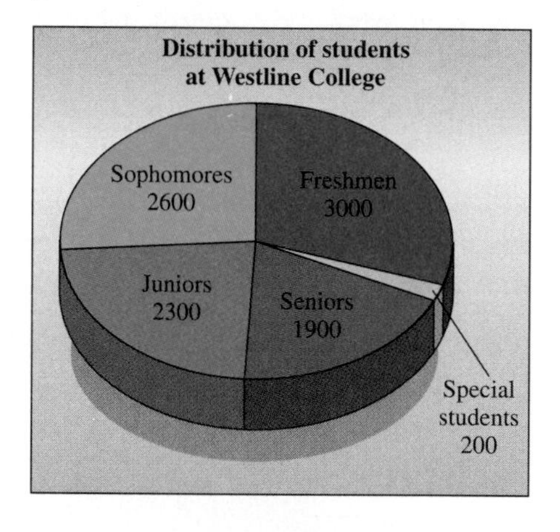

TEACHING TIP With the advent of computer-generated graphics, students rarely have to construct circle graphs. Still, they often have to interpret them. This section of the text deals with interpretation only.

EXAMPLE 1 What is the largest category of students?

The largest pie-shaped section of the circle is labeled ''Freshmen.'' Thus the largest category is *freshmen students.*

Practice Problem 1 What is the smallest category of students? ■

EXAMPLE 2 How many students are either sophomores or juniors?

There are 2600 sophomores and 2300 juniors. If we add these two numbers, we have $2600 + 2300 = 4900$. Thus we see that there are *4900 students who are either sophomores or juniors.*

Practice Problem 2 How many students are either freshmen or special students? ■

EXAMPLE 3 What is the ratio of freshmen to seniors?

Number of freshmen ⟶ 3000
Number of seniors ⟶ 1900

Thus $\frac{3000}{1900} = \frac{30}{19}$

The ratio of freshmen to seniors is $\frac{30}{19}$.

TEACHING TIP Emphasize that the order of the ratio is important. The ratio of freshmen to sophomores is not the same as the ratio of sophomores to freshmen. Remind students that the word following "to" is always the denominator of the ratio fraction.

Practice Problem 3 What is the ratio of freshmen to sophomores? ■

EXAMPLE 4 What is the ratio of seniors to the total number of students?

There are 1900 seniors. We find the total of all the students by adding the number of students in each section of the graph. There are 10,000 students. The ratio of seniors to the total number of students is

$$\frac{1900}{10{,}000} = \frac{19}{100}$$

Practice Problem 4 What is the ratio of freshmen to the total number of students? ■

2 *Reading a Circle Graph with Percentage Values*

Together, the Great Lakes form the largest body of fresh water in the world. The total area of these five lakes is about 290,000 mi^2. The percentage of this total area taken up by each of the Great Lakes is shown in the circle graph at the right.

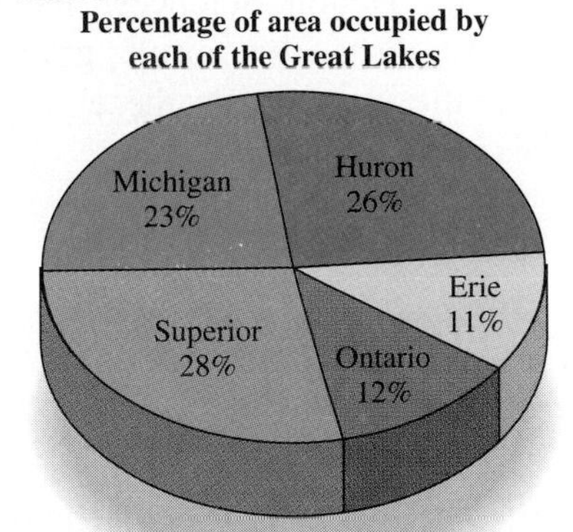

TEACHING TIP Students sometimes get confused about when they can add percents. You can only add percents that are of the same value. Tell students that since these are all percents of the total area taken up by the five Great Lakes, it is all right to add them together.

EXAMPLE 5 Which of the Great Lakes occupies the second-largest area in square miles?

The largest percent corresponds to the biggest area, which is occupied by Lake Superior. The second-largest percent corresponds to Lake Huron's area. Therefore, *Lake Huron has the second-largest area in square miles.*

Practice Problem 5 Which of the Great Lakes occupies the third-largest area in square miles? ■

EXAMPLE 6 What percent of the total area is occupied either by Lake Erie or Lake Ontario?

If we add 11% for Lake Erie and 12% for Lake Ontario, we get

$$11\% + 12\% = 23\%$$

Thus 23% of the area is occupied by either Lake Erie or Lake Ontario.

Practice Problem 6 What percent of the total area is occupied by either Lake Superior or Lake Michigan? ■

EXAMPLE 7 How many of the total 290,000 mi^2 are occupied by Lake Michigan?

Remember that we multiply percent times base to obtain amount. Here 23% of 290,000 is what is occupied by Lake Michigan.

$$(0.23)(290{,}000) = n$$
$$= 66{,}700$$

Thus about 66,700 mi^2 are occupied by Lake Michigan.

Practice Problem 7 How many of the total 290,000 mi^2 are occupied by Lake Superior? ■

Sometimes a circle graph is used to investigate the distribution of one part of a larger group. For example, there were 690,000 bachelor's degrees awarded in the United States in 1993 to students majoring in the six most popular subject areas: Business, Social Sciences, Engineering, Health Sciences, Psychology, and English. This circle graph shows how the degrees in these six subject areas are distributed.

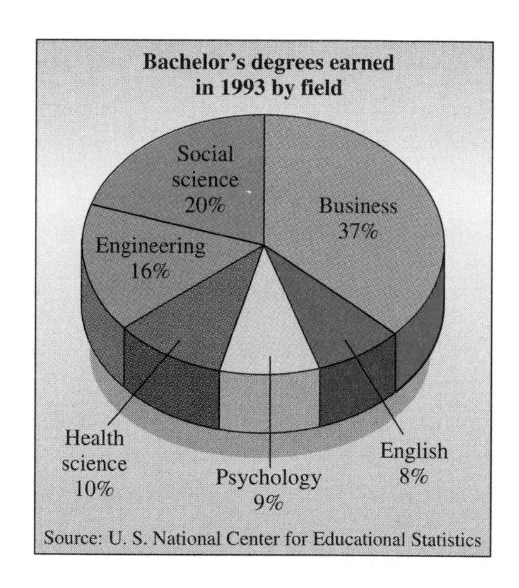

EXAMPLE 8

(a) What percent of the bachelor's degrees represented in this circle graph are in the fields of psychology or business?

(b) Of the approximately 690,000 degrees awarded in these six fields, how many are awarded in the field of Engineering?

(a) We add 37% to 9% to obtain 46%. Thus 46% of the bachelor's degrees represented by this circle are in the fields of psychology or business.

(b) We take 16% of the 690,000 people who obtained degrees in these six areas. Thus we have $(0.16)(690{,}000) = 110{,}400$. There were 110,400 degrees in Engineering awarded in 1993.

Practice Problem 8

(a) What percent of the bachelor's degrees represented in this circle graph are in the fields of Health Science or English?

(b) How many bachelor's degrees in Social Science were awarded in 1993? ■

8.1 Exercises

The following circle graph displays Bob and Linda McDonald's monthly $2000 family budget. Use the circle graph to answer questions 1–10.

1. What category takes the largest amount of the budget? Rent
2. What category takes the least amount of the budget? Utilities
3. How much money is allotted each month for contributions? $200
4. How much money is allotted each month for transportation? $350
5. How much money in total is allotted each month for utilities and transportation?
   $350 + $150 = $500
6. How much money in total is allotted for either food or rent?
   $500 + $600 = $1100

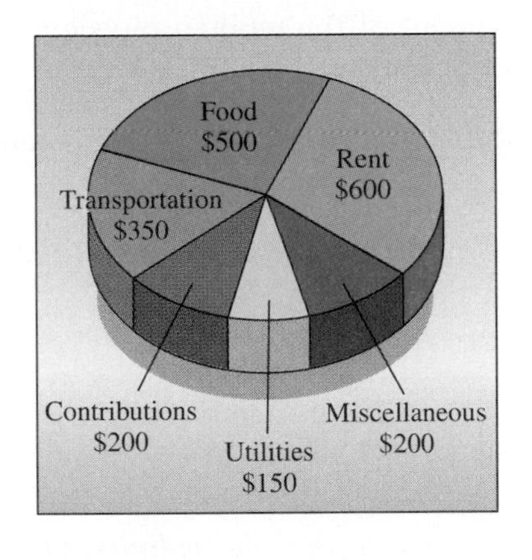

7. What is the ratio of money spent for food to the amount of money spent for transportation?
   $\frac{\$500}{\$350} = \frac{10}{7}$
8. What is the ratio of money spent for rent to the amount of money spent for utilities?
   $\frac{\$600}{\$150} = \frac{4}{1}$
9. What is the ratio of money spent for rent to the total amount of the monthly budget?
   $\frac{\$600}{\$2000} = \frac{3}{10}$
10. What is the ratio of money spent for food to the total amount of the monthly budget?
   $\frac{\$500}{\$2000} = \frac{1}{4}$

A major league pitcher has thrown 650 pitches during the first part of the baseball season. The following circle graph shows the results of his pitches. Use the circle graph to answer questions 11–20.

11. What category had the least number of pitches?
   Hit batters
12. What category had the second-highest number of pitches? Strikes
13. How many pitches were balls? 294
14. How many pitches were strikes? 144
15. How many pitches were either hits or balls?
   85 + 294 = 379 pitches
16. How many pitches were either strikes or fly outs and ground outs? 144 + 124 = 268 pitches

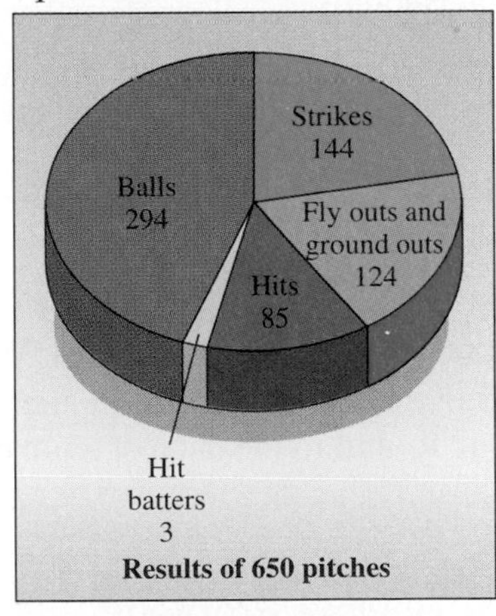

Results of 650 pitches

17. What is the ratio of the number of strikes to the total number of pitches?
   $\frac{144}{650} = \frac{72}{325}$
18. What is the ratio of the number of balls to the total number of pitches?
   $\frac{294}{650} = \frac{147}{325}$
19. What is the ratio of the number of balls to the number of strikes?
   $\frac{294}{144} = \frac{49}{24}$
20. What is the ratio of the number of fly outs and ground outs to the number of hits?
   $\frac{124}{85}$

In 1992 there were approximately 118,000 women physicians in the United States. The following circle graph divides them by age. Use the circle graph to answer questions 21–26.

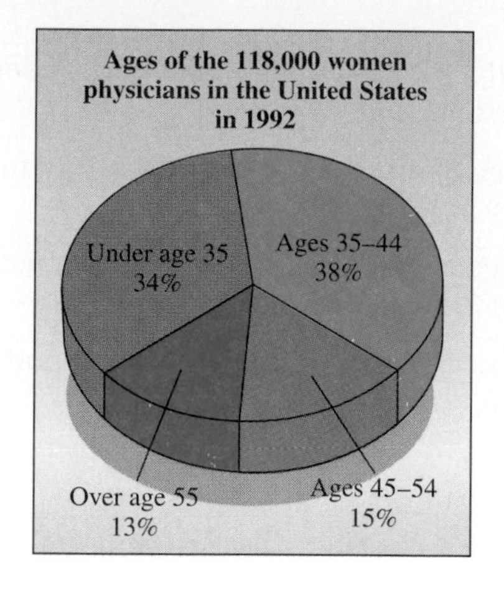

21. What percent of the women physicians are between the ages of 45 and 54? 15%

22. What percent of the women physicians are between the ages of 35 and 44? 38%

23. What percent of the women physicians are under the age of 45? 38% + 34% = 72%

24. What percent of the women physicians are over the age of 45? 13% + 15% = 28%

25. How many of the 118,000 women physicians are between the ages of 35 and 44?
$0.38 \times 118{,}000 = 44{,}840$

26. How many of the 118,000 women physicians are over age 55? $0.13 \times 118{,}000 = 15{,}340$

Researchers estimate that the religious faith distribution of the 5,000,000,000 people in the world in 1988 was approximately that displayed in the following circle graph. Use the graph to answer questions 27–32.

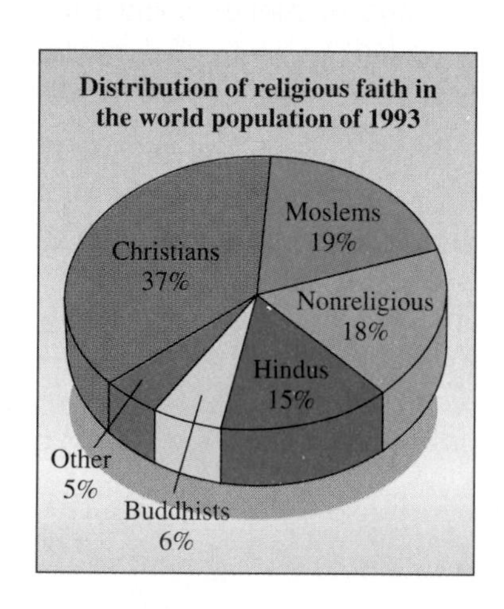

27. Approximately how many of the 5,000,000,000 people are Christians?
$0.37 \times 5{,}000{,}000{,}000 = 1{,}850{,}000{,}000$

28. Approximately how many of the 5,000,000,000 people are Moslems?
$0.19 \times 5{,}000{,}000{,}000 = 950{,}000{,}000$

29. What percent of the world's population is either Moslem or nonreligious? 19% + 18% = 37%

30. What percent of the world's population is either Hindu or Buddhist? 15% + 6% = 21%

31. What percent of the world's population is *not* Moslem? 100% − 19% = 81%

32. What percent of the world's population is *not* Christian? 100% − 37% = 63%

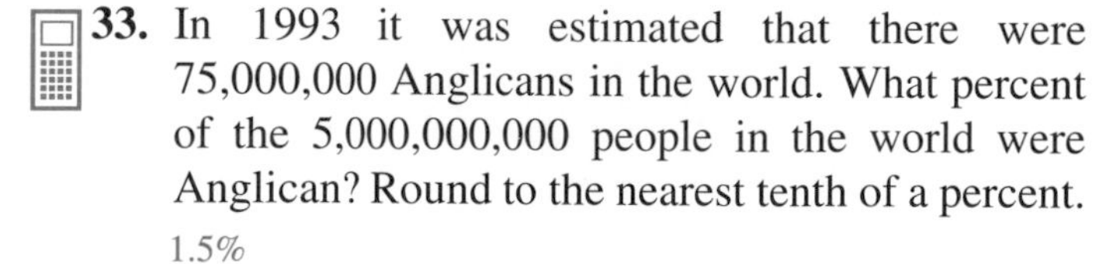

33. In 1993 it was estimated that there were 75,000,000 Anglicans in the world. What percent of the 5,000,000,000 people in the world were Anglican? Round to the nearest tenth of a percent.
1.5%

34. The circle contains a sector labeled Christians. What percent of the Christian sector represents Anglicans? Round to the nearest tenth of a percent. 4.1%

Cumulative Review Problems

35. Find the area of a triangle with base 6 in. and height 14 in. 42 in.^{2}

36. Find the area of a parallelogram with base of 17 in. and height 12 in. 204 in.^{2}

37. How many gallons of paint will it take to cover the four sides of a barn. Two sides measure 7 yd by 12 yd and two sides measure 7 yd by 20 yd. Assume a gallon of paint covers 28 square yards.
16 gal

38. A circular reflector has a radius of 8 cm. How many grams of silver will it take to cover the reflector if each gram will cover 64 sq cm? Assume that the area of the reflector is covered on one side only. (Use $\pi \approx 3.14$). About 3 g

8.2 Bar Graphs and Line Graphs

MathPro

Video 18

SSM

After studying this section, you will be able to:

1 *Read and interpret a bar graph.*
2 *Read and interpret a double-bar graph.*
3 *Read and interpret a line graph.*
4 *Read and interpret a comparison line graph.*

1 *Reading and Interpreting a Bar Graph*

Bar graphs are helpful for seeing changes over a period of time. Bar graphs or line graphs are especially helpful when the same type of data is repeatedly studied. The following bar graph shows the population of California from 1940 to 1990.

TEACHING TIP Explain to students that the scale of these bar graphs is not always as simple as this example. Some will have a scale of 12 million, 12.2 million, 12.4 million, etc. Students will encounter such examples as they read *USA Today*, *Time*, or *Newsweek*.

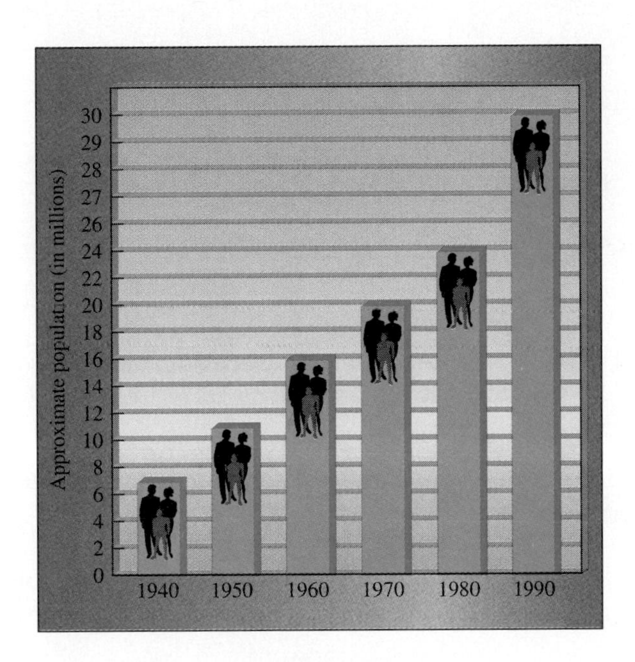

EXAMPLE 1 What was the approximate population of California in 1970?

The bar for 1970 rises to 20. This represents 20 million; thus the approximate population is 20,000,000.

Practice Problem 1 What was the approximate population of California in 1980? ■

EXAMPLE 2 What was the increase in population from 1980 to 1990?

The bar for 1980 rises to 24. Thus the approximate population is 24,000,000. The bar for 1990 rises to 30. Thus the approximate population is 30,000,000. To find the increase in population from 1980 to 1990, we subtract.

$$30{,}000{,}000 - 24{,}000{,}000 = 6{,}000{,}000$$

Practice Problem 2 What was the increase in population from 1940 to 1960? ■

2 *Reading and Interpreting a Double-Bar Graph*

Double-bar graphs are useful for making comparisons. For example, when a company is analyzing its sales, it may want to compare different years or different quarters. The following double-bar graph illustrates the sales of new cars at a local Ford dealership for two different years, 1995 and 1996. The sales are recorded for each quarter of the year.

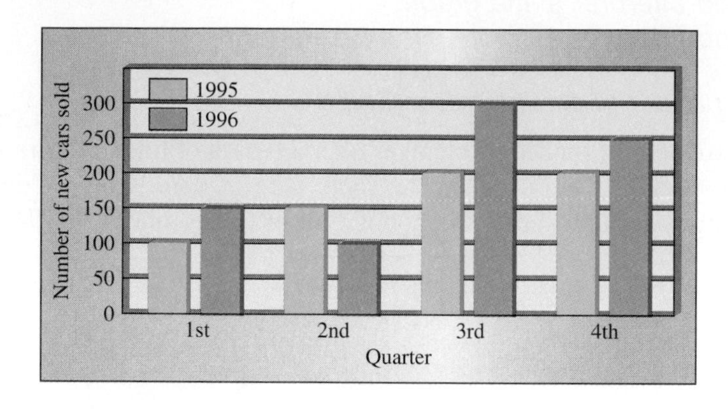

EXAMPLE 3 How many cars were sold in the second quarter of 1995?

The bar rises to 150 for the second quarter of 1995. Therefore, 150 cars were sold.

Practice Problem 3 How many cars were sold in the fourth quarter of 1996? ■

EXAMPLE 4 How many more cars were sold in the third quarter of 1996 than in the third quarter of 1995?

From the double-bar graph, we see that 300 cars were sold in the third quarter of 1996 and that 200 cars were sold in the third quarter of 1995.

$$\begin{array}{r} 300 \\ -\ 200 \\ \hline 100 \end{array}$$

Thus 100 more cars were sold.

Practice Problem 4 How many fewer cars were sold in the second quarter of 1996 than in the second quarter of 1995? ■

3 *Reading and Interpreting a Line Graph*

A line graph is useful for showing trends over a period of time. In a line graph only a few points are actually plotted from measured values. The points are then connected by straight lines to show a trend. The intervening values between points may not lie exactly on the line. The following line graph shows the number of customers per month coming into a restaurant in a vacation community.

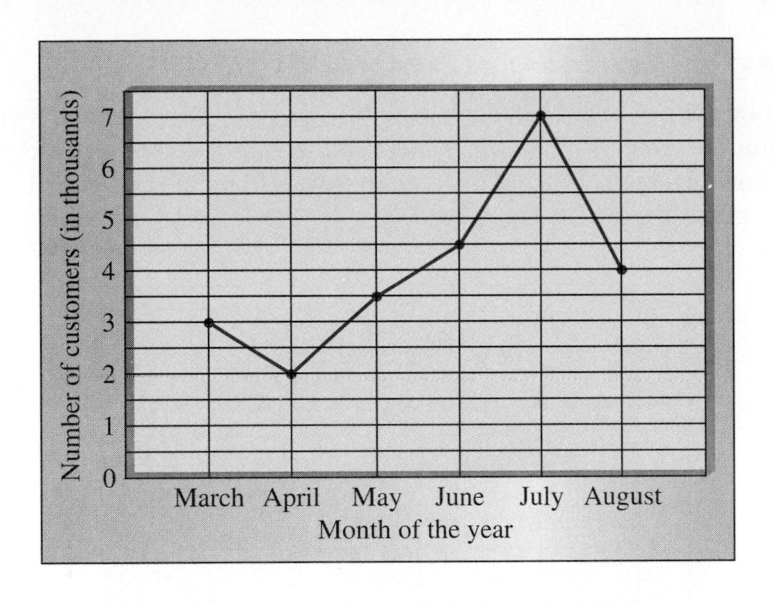

TEACHING TIP Sometimes students try to interpret in between the points on a line graph. In the line graph for Example 5, they may say, "The number of customers was decreasing from the middle of March to the end of March." This, of course, is *not* valid. The graph is only valid for one point plotted in March. Remind students that the purpose of the lines between the plotted points is to show the *direction of trends,* not to show intermediate values between the points.

EXAMPLE 5 In which month did the fewest number of customers come into the restaurant?

The lowest point on the graph occurs in the month of April. *Thus the fewest number of customers came in April.*

Practice Problem 5 In which month did the greatest number of customers come into the restaurant? ■

EXAMPLE 6

(a) Approximately how many customers per month came during the month of June?

(b) From May to June did the number of customers increase or decrease?

(a) Notice that the dot is halfway between 4 and 5. This represents a value halfway between 4000 and 5000 customers. Thus we would estimate *4500 customers* came during the month of June.

(b) Between May and June the line goes up, so *the number of customers increased.*

Practice Problem 6

(a) Approximately how many customers per month came during the month of May?

(b) From March to April did the number of customers increase or decrease? ■

EXAMPLE 7 Between what two months is the *increase* in attendance the largest?

The line from June to July goes upward at the steepest angle. This represents the largest increase. (You can check this by reading the numbers from the left axis.) Thus the greatest increase in attendance is *between June and July.*

Practice Problem 7 Between what two months did the biggest *decrease* occur? ■

4 *Reading and Interpreting a Comparison Line Graph*

Two or more sets of data can be compared by using a *comparison line graph.* A comparison line graph shows two or more line graphs together. A different style for each line distinguishes them. Note that using a blue line and a red line in the following graph makes it easy to read.

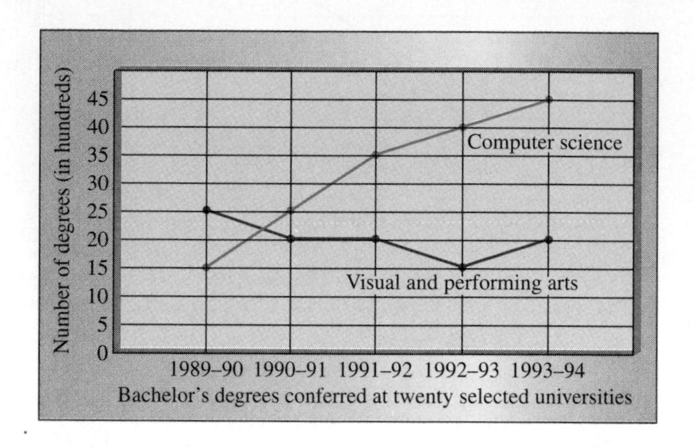

TEACHING TIP When discussing the comparison line graph example, ask students if they remember seeing a similar graph in a newspaper or magazine in the last year. Students often have a wealth of interesting examples that they remember. Class discussion of this topic helps them to see how relevant this type of graph is to everyday life.

EXAMPLE 8 How many bachelor's degrees in computer science were awarded in the academic year 1991–92?

Because the dot corresponding to 1991–92 is opposite 35 and the scale is in hundreds, we have $35 \times 100 = 3500$. Thus *3500 degrees were awarded in computer science in 1991–92.*

Practice Problem 8 How many bachelor's degrees in visual and performing arts were awarded in the academic year 1992–93? ■

EXAMPLE 9 In what academic year were more degrees awarded in the visual and performing arts than in computer science?

The only year when more bachelor's degrees were awarded in the visual and performing arts was *the academic year 1989–90.*

Practice Problem 9 What was the first academic year in which more degrees were awarded in computer science than in the visual and performing arts? ■

8.2 Exercises

The bar graph below shows the approximate population of Texas from 1940 to 1990. Use the graph to answer questions 1–6.

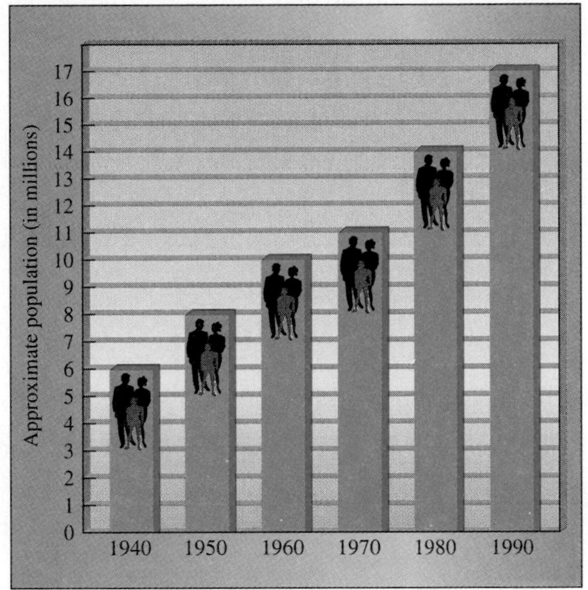

1. What was the approximate population in 1960? 10 million people

2. What was the approximate population in 1950? 8 million people

3. What was the approximate population in 1980? 14 million people

4. What was the approximate population in 1990? 17 million people

5. According to this bar graph, between what years did the population of Texas increase by the smallest amount? 1960–1970

6. According to this bar graph, between what years did the population of Texas increase by the largest amount? 1970–1980 and 1980–1990

The following double-bar graph illustrates the number of students at 10 selected universities at 20-year intervals. The student population is divided into men and women. Use the graph to answer questions 7–14.

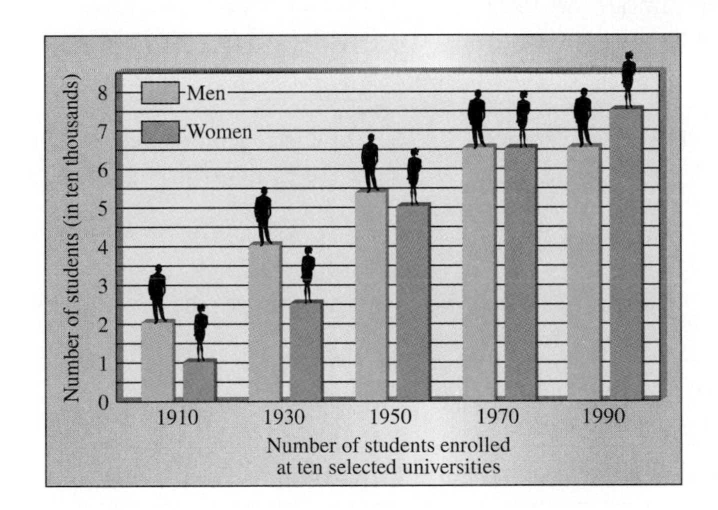

7. How many women students were enrolled in 1950? 50,000 women

8. How many men students were enrolled in 1930? 40,000 men

9. In what year was the number of women students enrolled greater than the number of men students enrolled? 1990

10. In what year was the number of women students enrolled equal to the number of men students enrolled? 1970

11. How many more women students were enrolled in 1950 than in 1930? 25,000 more

12. How many more men students were enrolled in 1970 than in 1950? 10,000 more

13. In what 20-year period did the greatest increase in the enrollment of women take place? 1930 to 1950

14. In what 20-year period did the greatest increase in the enrollment of men take place? 1910 to 1930

The following line graph shows Wentworth Construction Company's profits during the last six years. Use the graph to answer questions 15–20.

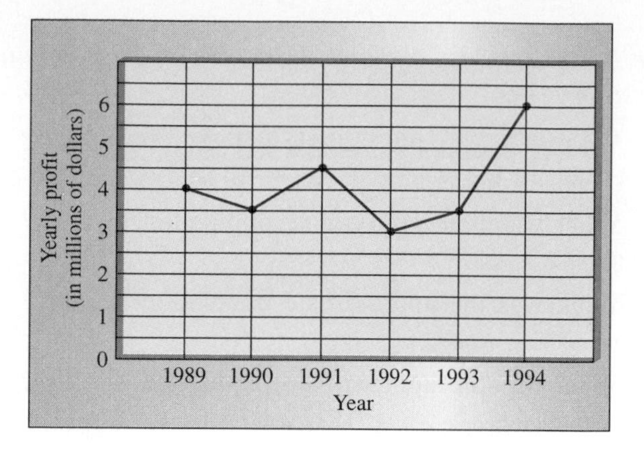

15. What was the profit in 1993? 3.5 million dollars

16. What was the profit in 1989? 4 million dollars

17. What year had the lowest profit? 1992

18. What year had the highest profit? 1994

19. How much greater was the profit in 1991 than in 1990? one million dollars

20. How much greater was the profit in 1994 than in 1989? 2 million dollars

The following comparison line graph indicates the rainfall for the last six months of two different years in Springfield. Use the graph to answer questions 21–26.

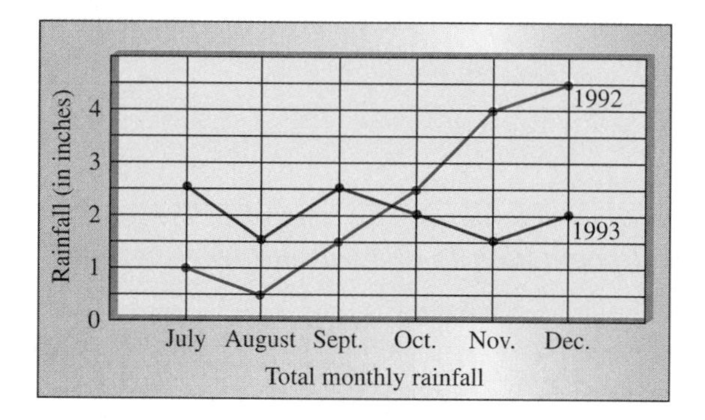

21. In September 1993, how many inches of rain were recorded? 2.5 in.

22. In October of 1992, how many inches of rain were recorded? 2.5 in.

23. During what months was the rainfall of 1993 less than the rainfall of 1992? October, November, and December

24. During what months was the rainfall of 1993 greater than the rainfall of 1992? July, August, and September

25. How many more inches of rain fell in November of 1992 than in October 1992? 1.5 in. more

26. How many more inches of rain fell in September of 1992 than in August 1992? 1 in. more

To Think About

The table below shows the number of pizzas sold at a pizza place near a college campus.

Number of Pizzas Sold by Alfredo's Pizza Parlor	300	400	100	200	600
Month of the Year	Jan.	Feb.	Mar.	Apr.	May

27. Use the graph paper on the right to construct a line graph of the information in the table. Let the vertical scale (height) represent the number of pizzas sold. Let the horizontal scale (width) represent the month.

28. Is the biggest change on the graph an increase or a decrease in the number of pizzas sold per month? increase

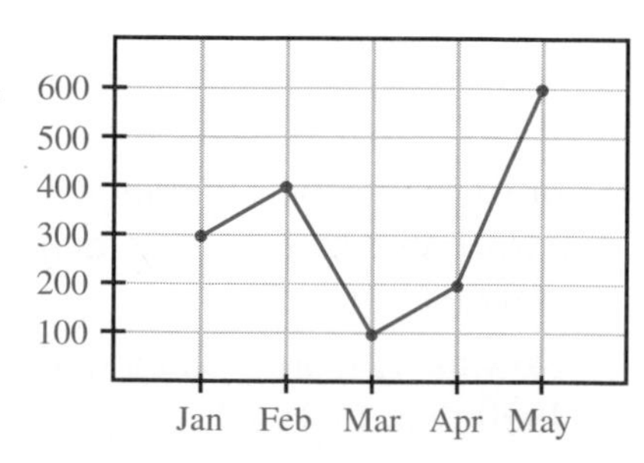

Cumulative Review Problems

Do each calculation in the proper order.

29. $7 \times 6 + 3 - 5 \times 2$ 35

30. $(5 + 6)^2 - 18 \div 9 \times 3$ 115

Putting Your Skills to Work

THE AGING OF AMERICA'S TRUCKS

Does it seem like the trucks on our nation's roads are getting older? Statistics kept on truck registrations show that more and more trucks are staying on the road for a greater number of years.

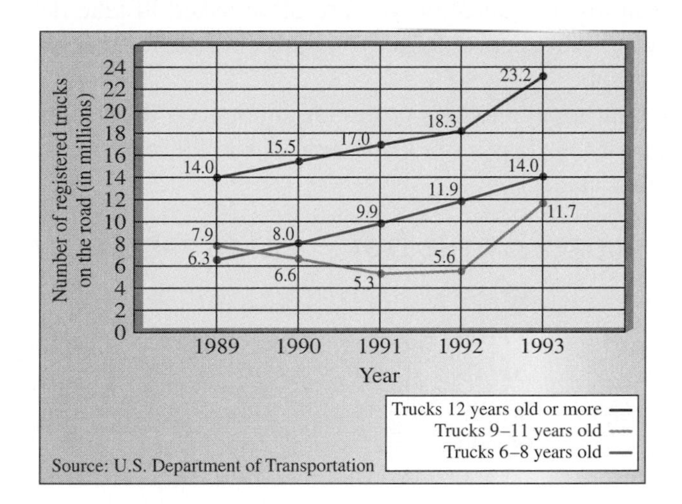

Based on the graph above please determine the following.

PROBLEMS FOR INDIVIDUAL INVESTIGATION AND ANALYSIS

1. During what year were more registered trucks on the road that were 9–11 years old than there were trucks that were 6–8 years old? 1989

2. Which age category saw the greatest increase in the number of registered trucks during the period 1989 to 1993? Trucks that are 12 years old or more

PROBLEMS FOR GROUP INVESTIGATION AND COOPERATIVE LEARNING

Together with some other members of your class determine the answers to the following.

3. From 1989 to 1993 the mean age of a registered truck increased from 7.9 years to 8.6 years. If there were 52.8 million trucks on the road in 1989, how many million trucks would be expected in 1993 due to this increase in age of the trucks?
approximately 57.5 million trucks

4. There were 53.2 million trucks registered in 1989 and 82.5 million trucks registered in 1993. What percent of the registered trucks in 1989 were 6 years old or older? What percent of the registered trucks in 1993 were 6 years old or older? What can you conclude from these statistics?
53.0% were 6 years old or older in 1989.
59.3% were 6 years old or older in 1993.
A greater percentage of the trucks were older in 1993.

INTERNET CONNECTIONS: Go to `http://www.prenhall.com/tobey` to be connected

Site: Truck Inventory and Use Survey (Bureau of Transportation Statistics/U.S. Department of Transportation) or a related site. *Note: This site requires the Adobe Acrobat Reader (which is free) to view the documents.*

View the Truck Inventory and Use Survey for your state. This document includes many data tables relating to the trucks in your state. Locate the information in Table 1 that gives the percent of trucks of varying ages over the past 20 years.

5. What percentage of the trucks used in 1987 were over 4 years old? in 1992?

6. Which age category saw the greatest increase in the percent of trucks in use from 1982 to 1987?

8.3 Histograms

MathPro

Video 18

SSM

After studying this section, you will be able to:

1 *Understand and interpret a histogram.*
2 *Construct a histogram from raw data.*

1 *Understanding and Interpreting a Histogram*

In business or in higher education you are often asked to take data and organize them in some way. This section shows you the technique for making a *histogram*—a type of bar graph.

Suppose that a mathematics professor announced the results of a class test. The 40 students in the class scored between 50 and 99 on the test. The results are displayed on this chart.

SCORES ON THE TEST	CLASS FREQUENCY
50–59	4
60–69	6
70–79	16
80–89	8
90–99	6

The results in the table can be organized in a special type of bar graph known as a ***histogram.*** In a histogram the width of each bar is the same. The width represents the range of scores on the test. This is called a ***class interval.*** The height of each bar gives the class frequency of each class interval. The ***class frequency*** is the number of times a score occurs in a particular class interval.

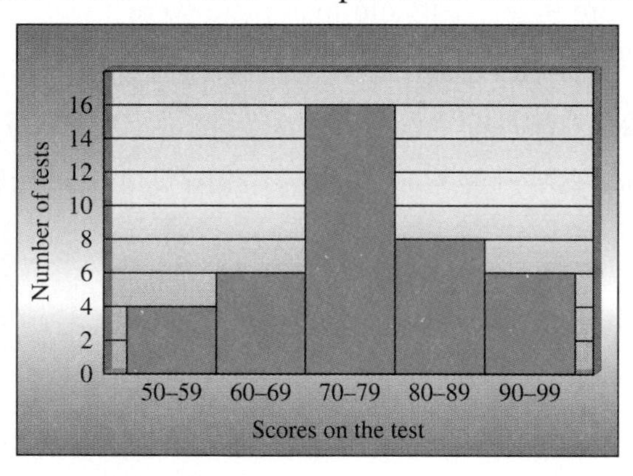

TEACHING TIP Students will sometimes encounter bar graphs that have an abbreviated scale on the side. This indicates that the bar graph has been cut off. Remind them that the effect of the histogram may be exaggerated by a cut-off scale. This applies to all the types of graphs covered in Section 8.2 as well. If the left-hand scale does not start at zero at the bottom of the graph, the graph is said to have a cut-off scale.

EXAMPLE 1 How many students scored a B on the test if the professor considers a test score of 80–89 as a B?

Since the 80–89 bar rises to a height of 8, there were *eight students* who scored a B on the test.

Practice Problem 1 How many students scored a D on the test if the professor considers a test score of 60–69 as a D? ■

EXAMPLE 2 How many students scored less than 80 on the test?

From the histogram, we see that there are three different bar heights to be included. Four tests were scored 50–59, six tests were scored 60–69, and 16 tests were scored 70–79. When we combine $4 + 6 + 16 = 26$, we can see that *26 students scored less than 80* on the test.

Practice Problem 2 How many students scored greater than 69 on the test? ■

The following histogram tells us about the length of life of 110 new light bulbs tested at a research center. The number of hours the bulbs lasted is indicated on the horizontal scale. The frequency of bulbs lasting that long is indicated on the vertical scale.

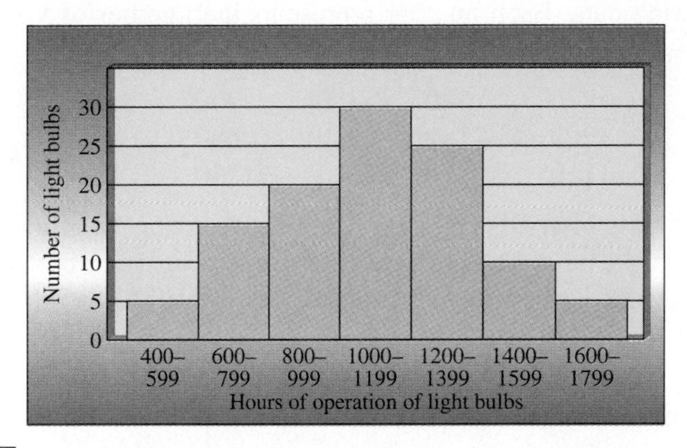

TEACHING TIP When reviewing Examples 3 and 4 with the class, bring up the topic of quality control and the use of graphs to track quality maintenance and improvements in manufactured products.

EXAMPLE 3 How many light bulbs lasted between 1400 and 1599 hours?

The bar with a range of 1400–1599 hours rises to 10. Thus 10 light bulbs lasted that long.

Practice Problem 3 How many light bulbs lasted between 800 and 999 hours? ■

EXAMPLE 4 How many light bulbs lasted fewer than 1000 hours?

We see that there are three different bar heights to be included. Five bulbs lasted 400–599 hours, 15 bulbs lasted 600–799 hours, and 20 bulbs lasted 800–999 hours. We add $5 + 15 + 20 = 40$. Thus 40 light bulbs lasted less than 1000 hours.

Practice Problem 4 How many light bulbs lasted more than 1199 hours? ■

2 *Constructing a Histogram from Raw Data*

To construct a histogram, we start with *raw data,* data that have not yet been organized or interpreted. We perform the following steps.

1. Select data class intervals of equal width for the data.
2. Make a table with class intervals and a *tally* (count) of how many numbers occur in each interval. Add up the tally to find the class frequency for each class interval.
3. Draw the histogram.

First we will practice making the table. Later we will use the table to draw the histogram.

EXAMPLE 5 Complete a table to determine the frequency of each class interval for the following data. Each number represents the number of kilowatt-hours of electricity used in a home during a one-month period.

770	520	850	900	1100
1200	1150	730	680	900
1160	590	670	1230	980

1. We select class intervals of equal width for the data. We choose intervals of 200. We might have chosen smaller or larger intervals, but we choose 200 because it gives us a convenient number of intervals to work with, as we will see.
2. We make a table. We write down the class intervals, then count (tally) how many numbers occur within each interval. Then we write the total. This is the class frequency.

KILOWATT-HOURS USED CLASS INTERVAL	TALLY	FREQUENCY
500–699	||||	4
700–899	|||	3
900–1099	|||	3
1100–1299	~~||||~~	5

TEACHING TIP You may find students will ask for an additional example after you have reviewed the procedure in Examples 5 and 6. You may use the following: "Thirty-one students in a math class recorded their height to the nearest inch. The following results were turned in: 55, 60, 66, 72, 79, 52, 75, 70, 60, 59, 54, 65, 71, 76, 67, 64, 56, 63, 68, 69, 60, 60, 65, 64, 63, 69, 68, 67, 71, 66, 70. These raw data can be arranged as follows:

Height Range in Inches	*Frequency*
48–53	1
54–59	4
60–65	10
66–71	12
72–77	3
78–83	1

Practice Problem 5 Each number below represents the weight in pounds of a new car.

2250	1760	2000	2100	1900
1640	1820	2300	2210	2390
2150	1930	2060	2350	1890

Complete the following table to determine the frequency of each class interval for the data above.

WEIGHT IN POUNDS CLASS INTERVAL	TALLY	FREQUENCY
1600–1799		
1800–1999		
2000–2199		
2200–2399		

■

EXAMPLE 6 Draw a histogram from the table in Example 5.

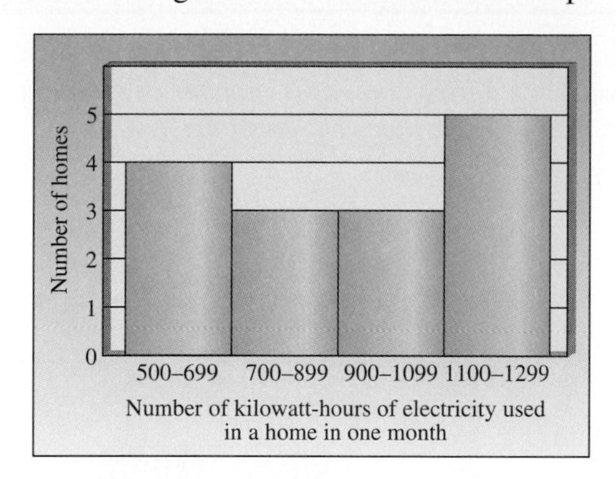

Practice Problem 6 Draw a histogram using the data from the table in Practice Problem 5. ■

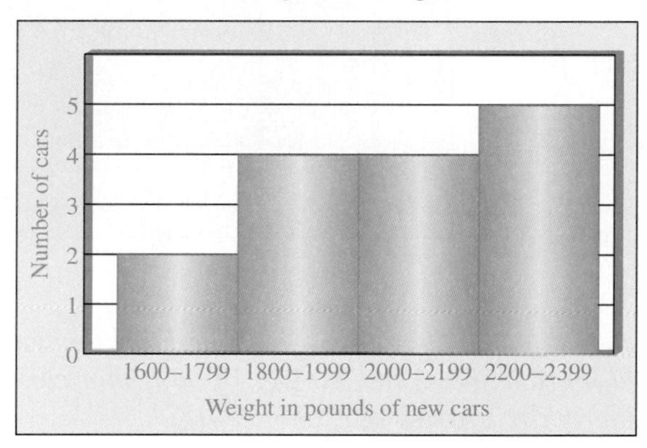

EXAMPLE 7 Draw a histogram from the following table of data concerning books that were purchased in 1994 in the United States.

Age Category of People Purchasing Books	Approximate Number of Books Purchased in the U.S. in 1994
Under 25	90,000,000
25–34	350,000,000
35–44	580,000,000
45–54	450,000,000
55–64	310,000,000
65 and Older	350,000,000
Source: U.S. Bureau of Economic Analysis	

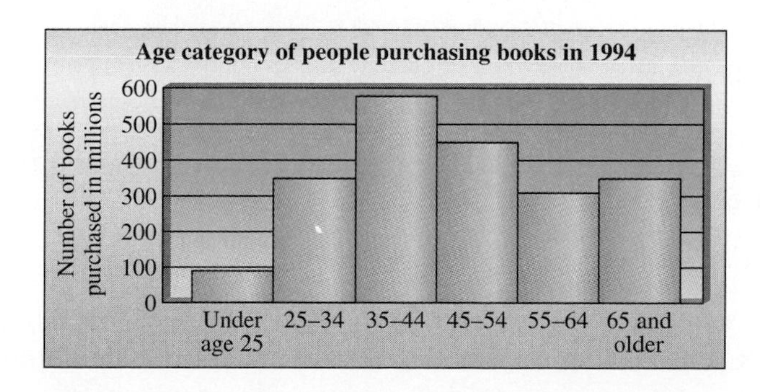

Practice Problem 7 Based on the histogram, between what two age categories is the greatest difference in the number of books purchased? ■

8.3 Exercises

A total of 190 rental cars were studied. The following histogram shows the number of rental cars fitting a certain rating classification in miles per gallon. Use the histogram to answer questions 1–8.

1. How many cars achieve between 28 and 30.9 mi per gallon? 10 cars

2. How many cars achieve between 19 and 21.9 mi per gallon? 50 cars

3. How many cars achieve between 25 and 27.9 mi per gallon? 35 cars

4. How many cars achieve between 16 and 18.9 mi per gallon? 35 cars

5. How many cars achieve more than 24.9 mi per gallon? $10 + 35 = 45$ cars

6. How many cars achieve less than 22 mi per gallon? $35 + 50 = 85$ cars

7. How many cars achieve between 19 and 27.9 mi per gallon? $35 + 50 + 60 = 145$ cars

8. How many cars achieve between 16 and 24.9 mi per gallon? $35 + 50 + 60 = 145$ cars

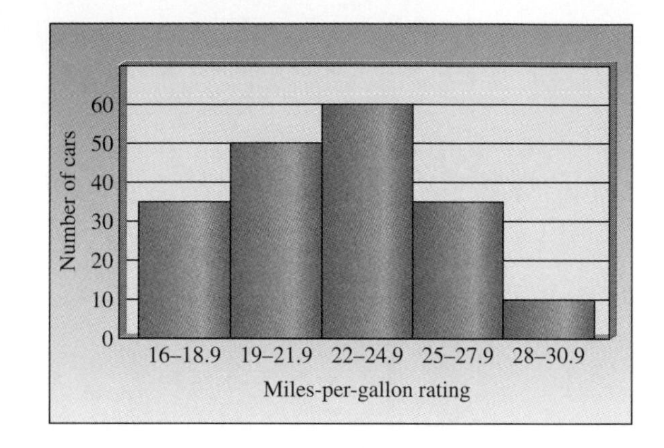

In an air-pollution study a large factory was studied for 47 operating days. Each day the number of tons of sulfur oxides emitted was measured. The following histogram shows the number of days that emissions of sulfur oxide reached a certain class interval. Use the histogram to answer questions 9–18.

9. For how many days did the factory emit between 13 and 16.9 tons of sulfur oxides? 12 days

10. For how many days did the factory emit between 17 and 20.9 tons of sulfur oxides? 16 days

11. For how many days did the factory emit between 25 and 28.9 tons of sulfur oxides? 3 days

12. For how many days did the factory emit between 21 and 24.9 tons of sulfur oxides? 9 days

13. For how many days did the factory emit less than 25 tons of sulfur oxides?
$5 + 12 + 16 + 9 = 42$ days

14. For how many days did the factory emit more than 16.9 tons of sulfur oxides?
$16 + 9 + 3 + 2 = 30$ days

15. For how many days did the factory emit between 13 and 24.9 tons of sulfur oxides?
$12 + 16 + 9 = 37$ days

16. For how many days did the factory emit between 17 and 28.9 tons of sulfur oxides?
$16 + 9 + 3 = 28$ days

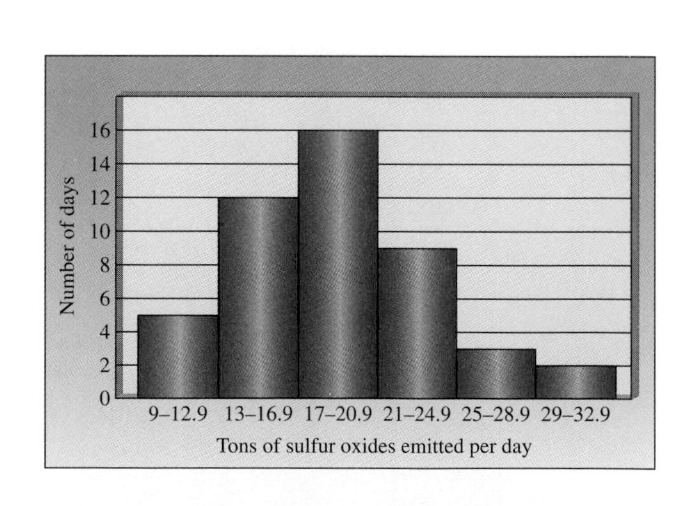

17. On what percent of the 47 operating days were 17 or more tons of sulfur oxides emitted? Round to the nearest tenth of a percent.
$\frac{30}{47} \approx 63.8\%$

18. On what percent of the 47 operating days were less than 21 tons of sulfur oxides emitted? Round to the nearest tenth of a percent.
$\frac{33}{47} \approx 70.2\%$

Complete a table to determine the frequency of each class interval of the following data. Each number represents the highest temperature in degrees Fahrenheit for the day in Boston for February.

23°	26°	30°	18°	42°	17°	19°
51°	42°	38°	36°	12°	18°	14°
20°	24°	26°	30°	18°	17°	16°
35°	38°	40°	33°	19°	22°	26°

<table>
<tr><th></th><th>TEMPERATURE CLASS INTERVAL</th><th>TALLY</th><th>FREQUENCY</th><th></th><th>TEMPERATURE CLASS INTERVAL</th><th>TALLY</th><th>FREQUENCY</th></tr>
<tr><td>19.</td><td>12°–16°</td><td>|||</td><td>3</td><td>20.</td><td>17°–21°</td><td>𝍸|||</td><td>8</td></tr>
<tr><td>21.</td><td>22°–26°</td><td>𝍸|</td><td>6</td><td>22.</td><td>27°–31°</td><td>||</td><td>2</td></tr>
<tr><td>23.</td><td>32°–36°</td><td>|||</td><td>3</td><td>24.</td><td>37°–41°</td><td>|||</td><td>3</td></tr>
<tr><td>25.</td><td>42°–46°</td><td>||</td><td>2</td><td>26.</td><td>47°–51°</td><td>|</td><td>1</td></tr>
</table>

27. Construct a histogram using the table prepared in problems 19–26.

Complete a table to determine the frequency of each class interval of the following data. Each number represents the cost of a prescription purchased by the Lin family this year.

\$28.50	\$16.00	\$32.90	\$46.20	\$ 9.85
\$27.30	\$16.00	\$41.95	\$36.00	\$24.20
\$ 7.65	\$ 8.95	\$ 4.50	\$11.35	\$ 7.75
\$12.30	\$21.85	\$46.20	\$15.50	\$ 4.50

<table>
<tr><th></th><th>PURCHASE PRICE CLASS INTERVAL</th><th>TALLY</th><th>FREQUENCY</th><th></th><th>PURCHASE PRICE CLASS INTERVAL</th><th>TALLY</th><th>FREQUENCY</th></tr>
<tr><td>28.</td><td>$ 4.00–$ 9.99</td><td>𝍸|</td><td>6</td><td>29.</td><td>$10.00–$15.99</td><td>|||</td><td>3</td></tr>
<tr><td>30.</td><td>$16.00–$21.99</td><td>|||</td><td>3</td><td>31.</td><td>$22.00–$27.99</td><td>||</td><td>2</td></tr>
<tr><td>32.</td><td>$28.00–$33.99</td><td>||</td><td>2</td><td>33.</td><td>$34.00–$39.99</td><td>|</td><td>1</td></tr>
<tr><td>34.</td><td>$40.00–$45.99</td><td>|</td><td>1</td><td>35.</td><td>$46.00–$51.99</td><td>||</td><td>2</td></tr>
</table>

36. Construct a histogram using the table constructed in problems 28–35.

Cumulative Review Problems

37. Solve for n. $\frac{126}{n} = \frac{36}{17}$ $n = 59.5$

38. Solve for n. $\frac{n}{18} = \frac{3.5}{9}$ $n = 7$

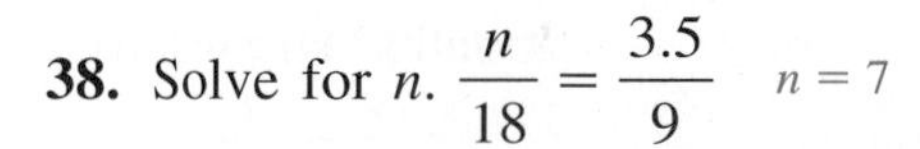

39. Ben and Trish Hale are going to have a Texas barbeque next week. According to Trish's grandmother, the recipe calls for 3 pounds of chicken for every 5 people. The Hales are expecting 20 people. How much chicken should they buy? 12 pounds

40. Tim and Judy Newitt worked as scientists on Mount Washington last year. They found that every 23 in. of snow corresponded to 2 in. of water. During the month of January they measured 150 in. of snow at the mountain weather observatory. How many inches of water does this correspond to? Round your answer to the nearest tenth of an inch. 13 in.

27.

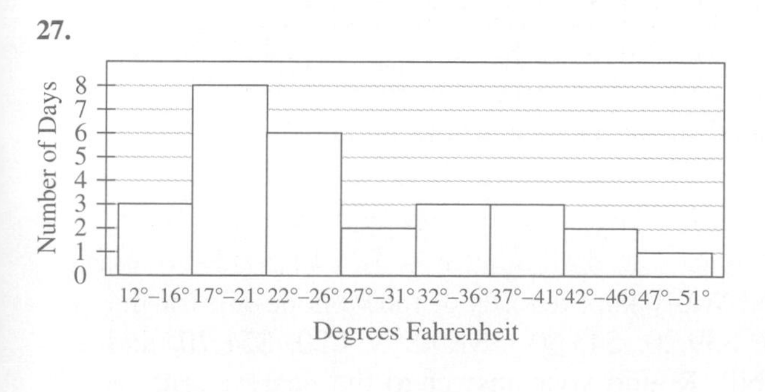

36.

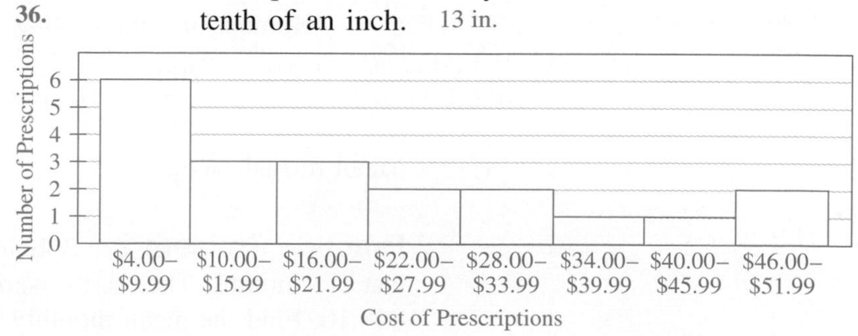

8.4 Mean and Median

MathPro

Video 18

SSM

After studying this section, you will be able to:

1 *Find the mean of a set of numbers.*
2 *Find the median value of a set of numbers.*

1 *Finding the Mean of a Set of Numbers*

We often want to know the ''middle value'' of a group of numbers. In this section we learn that, in statistics, there is more than one way of describing this middle value: there is the *mean* of the group of numbers, and there is the *median* of the group of numbers. In some situations it's more helpful to look at the mean, and in others it's more helpful to look at the median. We'll learn to tell which situations lend themselves to one or the other.

The **mean** of a set of values is the sum of the values divided by the number of values. The mean is often called the **average.**

EXAMPLE 1 Find the average or mean test score of a student who has test grades of 71, 83, 87, 99, 80, and 90.

We take the sum of the six tests and divide the sum by 6.

$$\begin{array}{l} \text{Sum of tests} \longrightarrow \\ \text{Number of tests} \longrightarrow \end{array} \frac{71 + 83 + 87 + 99 + 80 + 90}{6} = \frac{510}{6} = 85$$

The mean is 85.

Practice Problem 1 Find the average or mean of the following test scores: 88, 77, 84, 97, 89. ■

The mean value is often rounded to a certain decimal-place accuracy.

TEACHING TIP In cases like Example 2, discuss with students what happens when we try to write answers like ''The average is 23.45872 miles per gallon.'' Remind them that the accuracy of the answer is dependent on the accuracy of the data that are being averaged. Usually, we round our answer to the same number of significant digits that are in the items being averaged. In Example 2 we have two-digit accuracy in the miles per gallon ratings, so it is logical to round our answer to two significant digits.

EXAMPLE 2 Carl recorded the mi/gal achieved by his car for the last two months. His results were:

Week	1	2	3	4	5	6	7	8
Miles per Gallon	26	24	28	29	27	25	24	23

What is the mean mi/gal figure for the last 8 weeks? Round your answer to the nearest mile per gallon.

$$\begin{array}{l} \text{Sum of values} \longrightarrow \\ \text{Number of values} \longrightarrow \end{array} \frac{26 + 24 + 28 + 29 + 27 + 25 + 24 + 23}{8}$$

$$= \frac{206}{8} = 26 \quad \text{rounded to the nearest whole number}$$

Mean mi/gal rating is 26.

Practice Problem 2 Bob and Wally kept records of their phone bill for the last six months. Their bills were \$39.20, \$43.50, \$81.90, \$34.20, \$51.70, and \$48.10. Find the mean monthly bill. Round your answer to the nearest cent. ■

2 *Finding the Median Value of a Set of Numbers*

If a set of numbers is arranged in order from smallest to largest, the **median** is that value that has the same number of values above it as below it.

EXAMPLE 3 Find the median value of the following package weights: 6 lb, 7 lb, 9 lb, 12 lb, 13 lb, 16 lb, 17 lb.

The numbers are arranged in order.

6, 7, 9	12	13, 16, 17
three numbers	↑ middle number	three numbers

There are three numbers smaller than 12 and three numbers larger than 12. Thus 12 is the median.

Practice Problem 3 Find the median value of the following lengths of telephone calls: 2 minutes, 3 minutes, 7 minutes, 12 minutes, 13 minutes, 15 minutes, 18 minutes, 20 minutes, 21 minutes. ■

If the numbers are not arranged in order, then the first step is to put them in order.

EXAMPLE 4 Find the median cost of the following costs for a microwave oven: \$100, \$60, \$120, \$200, \$190, \$130, \$320, \$290, \$180.

We must arrange the numbers in order from smallest to largest (or largest to smallest).

\$60, \$100, \$120, \$130	\$180	\$190, \$200, \$290, \$320
four numbers	↑ middle number	four numbers

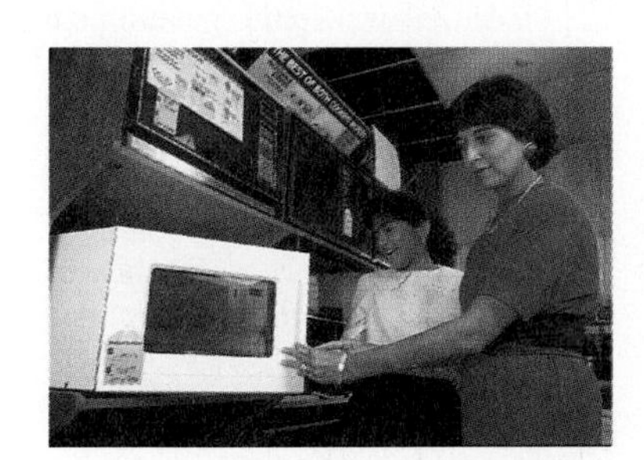

Thus \$180 is the median cost.

Practice Problem 4 Find the median weekly salary of the following salaries: \$320, \$150, \$400, \$600, \$290, \$180, \$450. ■

If a list of numbers contains an even number of items, then of course there is no one middle number. In this situation we obtain the median by taking the average of the two middle numbers.

EXAMPLE 5 Find the median of the following numbers: 13, 16, 18, 26, 31, 33, 38, 39.

13, 16, 18	26, 31	33, 38, 39
three numbers	two middle numbers	three numbers

The average (mean) of 26 and 31 is

$$\frac{26 + 31}{2} = \frac{57}{2} = 28.5$$

Thus the median value is 28.5.

Practice Problem 5 Find the median value of the following numbers: 88, 90, 100, 105, 118, 126. ■

TEACHING TIP After discussing the SIDELIGHT for a few minutes, ask students if they can think of some examples where the mean is used as an average for data but the figure is misleading. Usually, students offer several good suggestions. Among the best are the following:

A. The mean salary for a family in poor and developing countries is often listed as higher than it should be because of a small class of well-paid government officials and leaders.

B. A certain major university bragged that its ''experienced'' faculty had an average of 15 years of university teaching. However, the records showed that most of the faculty had taught only 1–3 years. In fact, about 35% of the faculty were near retirement and had taught (in most cases) over 35 years.

SIDELIGHT When would someone want to use the mean, and when would someone want to use the median? Which is more helpful? The mean, or average, is used more frequently. It is most helpful when the data are distributed fairly evenly, that is, when no one value is ''much larger'' or ''much smaller'' than the rest. For example, a company had employees with annual salaries of \$9000, \$11,000, \$14,000, \$15,000, \$17,000, and \$20,000. All the salaries fall within a fairly limited range. The mean salary

$$\frac{9000 + 11{,}000 + 14{,}000 + 15{,}000 + 17{,}000 + 20{,}000}{6} = \$14{,}333.33$$

gives us a reasonable idea of the ''average'' salary. However, suppose the company had six employees with salaries of \$9000, \$11,000, \$14,000, \$15,000, \$17,000, and \$90,000. Talking about the mean salary, which is \$26,000, is deceptive. No one earns a salary very close to the mean salary. The ''average'' worker in that company does not earn around \$26,000. In this case, the median value is more appropriate. Here the median is \$14,500. Some problems of this type are included in Exercises 8.4, problems 35 and 36.

EXPLORING MATHEMATICAL TOPICS: THE MODE

Another value that is sometimes used to describe a set of data is the mode. The **mode** of a set of data is the number or numbers that occur most often. For example, if a student had test scores of 89, 94, 96, 89, and 90, we would say that the mode is 89. If two values occur most often, we say that the data have two modes (or are bimodal). For example, if the ages of students in a calculus class were 33, 27, 28, 28, 21, 19, 18, 25, 26, and 33, we would say that the modes were 28 and 33. The mode is not used as frequently in statistics as the mean and the median. We will practice finding the mode in Exercises 37–42.

8.4 Exercises

In problems 1–14, find the mean. Round to the nearest tenth when necessary.

1. A student received grades of 89, 92, 83, 96, and 99 on math quizzes. Mean = 91.8

2. A student received grades of 77, 88, 90, 92, 83, and 84 on six history quizzes. Mean = 85.7

3. The Windy City Passport Photo Center received the following number of telephone calls over the last six days: 23, 45, 63, 34, 21, and 42? Mean = 38

4. The local Hertz rental car office received the following number of inquiries over the last seven days: 34, 57, 61, 22, 43, 80, and 39. Mean = 48

5. The last five houses built in town sold for the following prices: $89,000, $93,000, $62,000, $102,000, $89,000. Mean = $87,000

6. Luis priced a sofa at six local stores. The prices were $499, $359, $600, $450, $529, $629. Mean = $511

7. Sam watched television last week for the following number of hours: Mean = 5 hrs

Mon.	Tues.	Wed.	Thurs.	Fri.	Sat.	Sun.
8	3	2	5	6.5	7.5	3

8. Steve's car got the following miles per gallon results during the last 6 months: Mean = 26.2 mi/gal

Jan.	Feb.	Mar.	Apr.	May	June
23	22	25	28	29	30

9. The population on the island of Guam has increased significantly over the last 30 years. Find an approximate value for the mean population for this 30-year period from the following population chart: Mean population = 98,500

1960 Population	1970 Population	1980 Population	1990 Population
67,000	86,000	107,000	134,000

10. Desmond has taken four three-credit courses at college each of four semesters. His semester grade averages are: Mean = 2.64

Semester 1	Semester 2	Semester 3	Semester 4
2.90	2.66	2.52	2.48

11. The captain of the college baseball team achieved the following results:

	Game 1	Game 2	Game 3	Game 4	Game 5
Hits	0	2	3	2	2
Times at Bat	5	4	6	5	4

Find his batting average by dividing his total number of hits by the total times at bat.
$9 \div 24 = 0.375$

12. The captain of the college bowling team had the following results after practice:

	Practice 1	Practice 2	Practice 3	Practice 4
Score (Pins)	541	561	840	422
Number of Games	3	3	4	2

Find her bowling average by dividing the total number of pins scored by the total number of games. $2364 \div 12 = 197$ pins

13. Frank and Wally traveled to the West Coast during the summer. The number of miles they drove and the number of gallons of gas they used are recorded below.

	Day 1	Day 2	Day 3	Day 4
Miles Driven	276	350	391	336
Gallons of Gas	12	14	17	14

Find the average miles per gallon achieved by the car on the trip by dividing the total number of miles driven by the total number of gallons used.
1353 ÷ 57 = 23.7 mi/gal

14. Cindy and Andrea traveled to Boston this fall. The number of miles they drove and the number of gallons of gas they used are recorded below.

	Day 1	Day 2	Day 3	Day 4
Miles Driven	260	375	408	416
Gallons of Gas	10	15	17	16

Find the average miles per gallon achieved by the car on the trip by dividing the total number of miles driven by the total number of gallons used.
1459 ÷ 58 = 25.2 mi/gal

In problems 15–30, find the median value.

15. 22, 36, 45, 47, 48, 50, 58 median = 47

16. 37, 39, 46, 53, 57, 60, 63 median = 53

17. 1052, 968, 1023, 999, 865, 1152 median = 1011

18. 1400, 1329, 1200, 1386, 1427, 1350 median = 1368

19. 0.52, 0.69, 0.71, 0.34, 0.58 median = 0.58

20. 0.26, 0.12, 0.35, 0.43, 0.28 median = 0.28

21. The annual salaries of the employees of a local cable television office are \$17,000, \$11,600, \$23,500, \$15,700, \$26,700, \$31,500.
median = \$20,250

22. The costs of six cars recently purchased by the Weston Company were \$18,270, \$11,300, \$16,400, \$9100, \$12,450, \$13,800. median = \$13,125

23. The number of minutes spent on the phone per day by each of eight San Diego teenagers living on the same block is 40 minutes, 108 minutes, 62 minutes, 12 minutes, 24 minutes, and 31 minutes.
median = 35.5 minutes

24. The ages of 10 people swimming laps at the YMCA pool each morning: 60, 18, 24, 36, 39, 32, 70, 12, 15, and 85. median = 34 years

25. The phone bill for Dr. Price's cellular phone over the last seven months: \$109, \$207, \$420, \$218, \$97, \$330, and \$185. median = \$207

26. The price of the same compact disc sold at several different music stores or by mail-order: \$15.99, \$11.99, \$5.99, \$12.99, \$14.99, \$9.99, \$13.99, \$7.99, and \$10.99. median = \$11.99

27. The numbers of potential actors who tried out for the school play at Hamilton-Wenham Regional High School over the last ten years: 36, 48, 44, 64, 60, 71, 22, 36, 53, and 37. median = 46 actors

28. The number of injuries during the high school football season for the Badgers over the last eight years: 10, 17, 14, 29, 30, 19, 25, and 21.
median = 20 injuries

Find the mean. Round to the nearest cent when necessary.

29. The salaries of eight small business owners in Big Rapids: \$30,000, \$74,500, \$47,890, \$89,000, \$57,645, \$78,090, \$110,370, and \$65,800.
mean = \$69,161.88

30. The prices of nine laptop computers with a Pentium chip: \$5679, \$6902, \$1530, \$2738, \$2999, \$4105, \$3655, \$5980, and \$4430. mean = \$4224.22

Find the median value.

31. 2576, 8764, 3700, 5000, 7200, 4700, 9365, 1987
4850

32. 15.276, 21.375, 18.90, 29.2, 14.77, 19.02 18.96

Find the mean.

33. Oil is imported into the United States in great quantities. The following chart is provided by the U.S. Energy Information Administration.

Year	Amount of Oil Imported (in millions of barrels)
1991	2151
1992	2110
1993	2220
1994	2578
1995	2643

(a) Find the mean number of barrels of oil imported into the United States per year.
2,340,400,000 barrels of oil per year

(b) If one barrel contains 42 gallons of oil, find the mean number of gallons of oil imported into the United States per year.
98,296,800,000 gallons of oil per year

34. The capacity of the United States to produce electricity continues to increase as the demand for electricity continues to grow. The following chart is provided by the U.S. Energy Information Administration.

Year	Summer Capacity (in millions of kilowatts)
1990	685.1
1991	690.9
1992	695.4
1993	694.3
1994	703.0

(a) Find the mean capacity of the United States to produce electricity in the summer measured in number of kilowatts.
693,740,000 kilowatts capacity

(b) A kilowatt is 1000 watts. Find the mean capacity to produce electricity in the summer measured in watts.
693,740,000,000 watts capacity

To Think About

35. A local travel office has 10 employees. Their monthly salaries are $1500, $1700, $1650, $1300, $1440, $1580, $1820, $1380, $2900, $6300.

(a) Find the mean. $2157

(b) Find the median. $1615

(c) Which of these numbers best represents ''what the average person earns''? Why?
The median because the mean is affected by the high amount $6300.

36. A college track star in California ran the 100 meter event in eight track meets. Her times were 11.7 seconds, 11.6 seconds, 12.0 seconds, 12.1 seconds, 11.9 seconds, 18 seconds, 11.5 seconds, 12.4 seconds.

(a) Find the mean. 12.65 sec

(b) Find the median. 11.95 sec

(c) Which of these numbers best represents ''her average running time''? Why?
The median because the large time, 18 seconds, affects the mean.

Exploring Mathematical Topics: The Mode

Find the mode for each of the following.

37. 60, 65, 68, 60, 72, 59, 80
The mode is 60.

38. 86, 84, 82, 87, 84, 88, 90
The mode is 84.

39. 121, 150, 116, 150, 121, 181, 117, 123
The two modes are 121 and 150.

40. 144, 143, 140, 141, 149, 144, 141, 150
The two modes are 141 and 144.

41. The last six bicycles sold at the Skol Bike shop cost: $249, $649, $439, $259, $269, $249
The mode is $249.

42. The last six color televisions sold at the local Circuit City cost: $315, $430, $515, $330, $430, $615
The mode is $430.

Chapter Organizer

Topic	Procedure	Examples
Circle graphs, p. 466.	The percentage of the 200 police officers within a given age range is illustrated. The following circle graph describes the age of the 200 men and women of the Glover City police force. 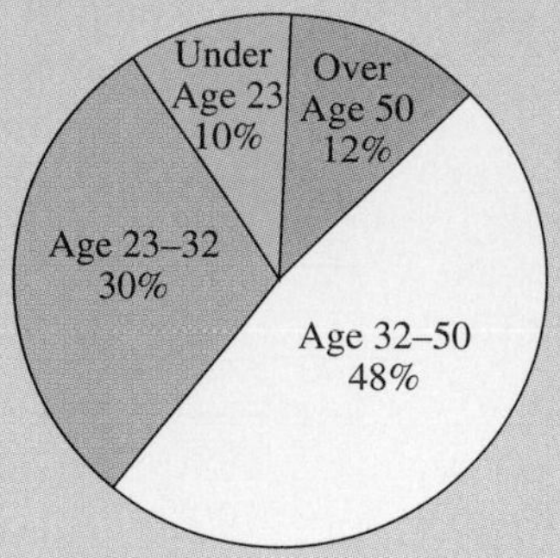 Under Age 23 10%; Over Age 50 12%; Age 23–32 30%; Age 32–50 48%	1. What percent of the police force is between 23 and 32 years old? 30% 2. How many men and women in the police force are over 50 years old? 12% of 200 = (0.12)(200) = 24 people
Bar graphs and double-bar graphs, p. 471.	The following double-bar graph illustrates the sales of color television sets by a major store chain for 1989 and 1990 in three regions of the country. 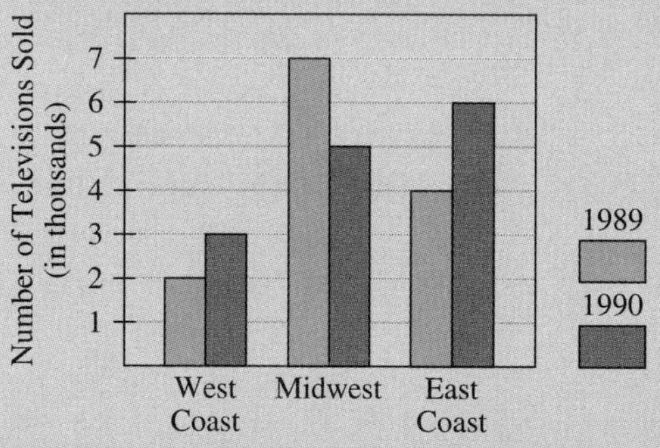 Number of Televisions Sold (in thousands): 1–7; West Coast, Midwest, East Coast; 1989, 1990	1. How many color television sets were sold by the chain on the East Coast in 1990? 6000 sets 2. How many *more* color television sets were sold in 1990 than in 1989 on the West Coast? 3000 sets were sold in 1990; 2000 sets were sold in 1989. $\begin{array}{r} 3000 \\ -\ 2000 \\ \hline 1000 \end{array}$ sets more in 1990
Line graphs and comparison line graphs, pp. 472–74.	The following line graph indicates the number of visitors to Wetlands State Park during a four-month period in 1989 and 1990. 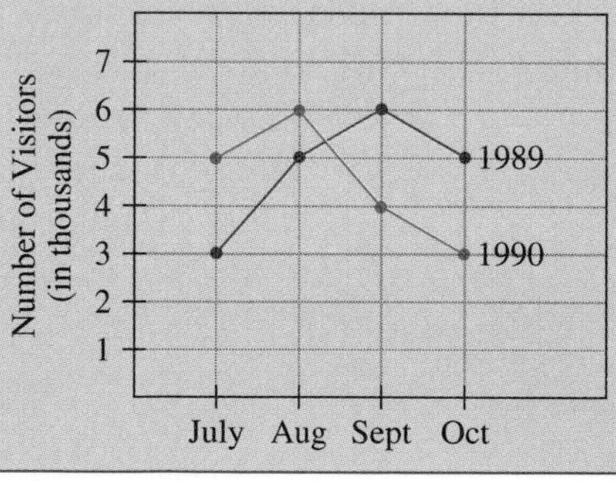 Number of Visitors (in thousands): 1–7; July, Aug, Sept, Oct; 1989, 1990	1. How many visitors came to the park in July of 1989? 3000 visitors 2. In what months were there more visitors in 1989 than in 1990? September and October 3. The sharpest *decrease* in attendance took place between what two months? Between August 1990 and September 1990

Chapter Organizer

Topic	Procedure	Examples
Histograms, p. 478.	The following histogram indicates the number of quizzes in a math class that were scored within each interval on a 15-point quiz. Number of Quizzes with That Score 0 2 4 6 8 10 12 14 16 18 20 0–3 4–7 8–11 12–15 Score on the Quiz	**1.** How many quizzes had a score between 8 and 11? 20 quizzes **2.** How many quizzes had a score of less than 8? $12 + 6 = 18$ quizzes
Finding the mean, p. 484.	The *mean* of a set of values is the sum of the values divided by the number of values. The mean is often called the *average.*	**1.** Find the mean of 19, 13, 15, 25, and 18. $\frac{19 + 13 + 15 + 25 + 18}{5} = \frac{90}{5} = 18$ The mean is 18.
Finding the median, p. 485.	**1.** Arrange the numbers in order from smallest to largest. **2.** If there is an odd number of values, the middle value is the median. **3.** If there is an even number of values, the average of the two middle values is the median.	**1.** Find the median of 19, 29, 36, 15, and 20. First we arrange in order from smallest to largest: 15, 19, 20, 29, 36. 15, 19 (two numbers) 20 ↑ (middle number) 29, 36 (two numbers) The median is 20. **2.** Find the median of 67, 28, 92, 37, 81, and 75. First we arrange in order from smallest to largest: 28, 37, 67, 75, 81, 92. There is an even number of values. 28, 37 67, 75 81, 92 ↑ two middle numbers $\frac{67 + 75}{2} = \frac{142}{2} = 71$ The median is 71.

Chapter 8 Review Problems

A student found that there were a total of 120 personal computers owned by students in the dormitory. The following circle graph displays the distribution of manufacturers of these computers. Use the graph to answer questions 1–8.

1. How many personal computers were manufactured by IBM? 36 computers

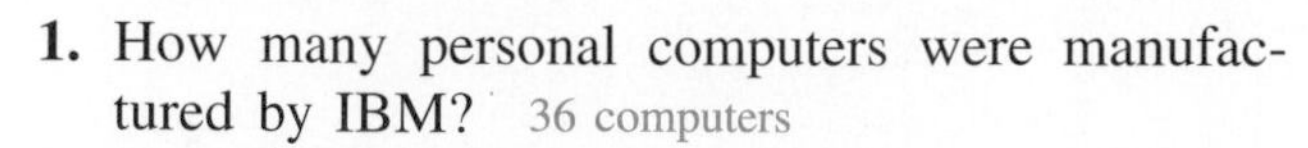

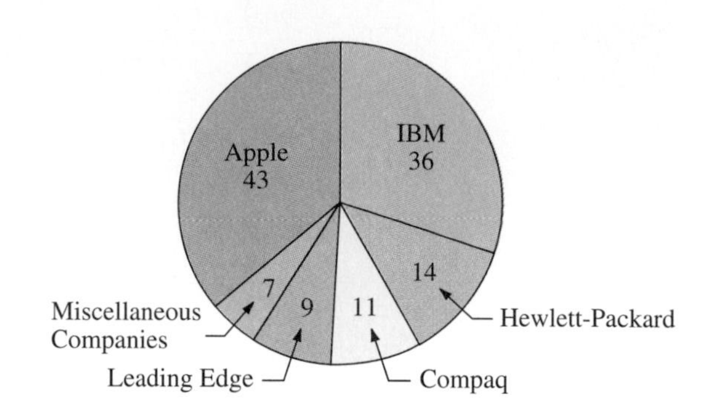

2. How many personal computers were manufactured by Apple? 43 computers

3. How many personal computers were manufactured by Leading Edge or Compaq? 20 computers

4. How many personal computers were manufactured by Hewlett-Packard or miscellaneous companies? 21 computers

5. What is the *ratio* of the number of computers manufactured by IBM to the number of computers manufactured by Leading Edge? $\frac{4}{1}$

6. What is the *ratio* of the number of computers manufactured by Hewlett-Packard to the number of computers manufactured by miscellaneous companies? $\frac{2}{1}$

7. What *percent* of the 120 computers is manufactured by Leading Edge? 7.5%

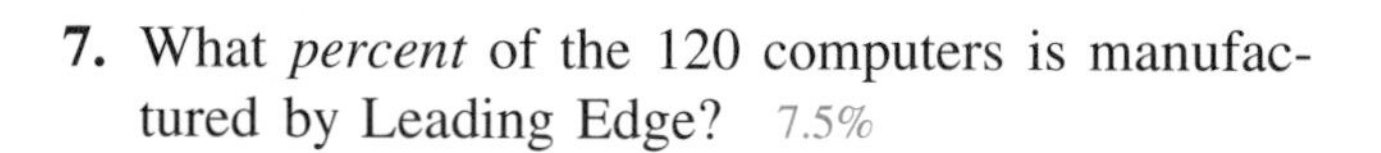

8. What *percent* of the computers is manufactured by IBM? 30%

Nancy and Wally Worzowski's family monthly budget of $2400 is displayed in the following circle graph. Use the graph to answer questions 9–16.

9. What percent of the budget is allotted for transportation? 15%

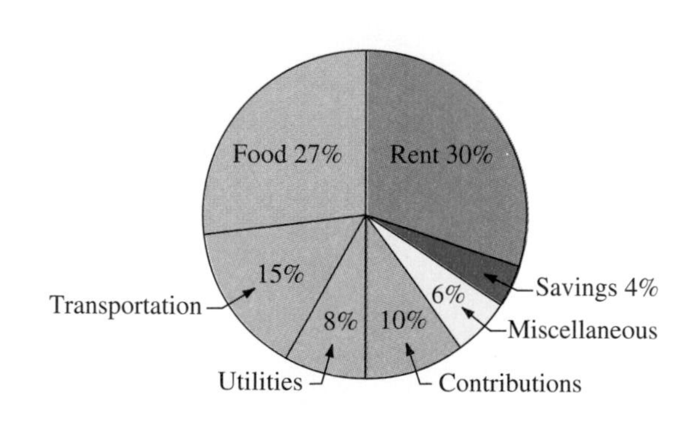

10. What percent of the budget is allotted for savings? 4%

11. What percent of the budget is used up by the food and rent categories? 57%

12. What percent of the budget is used up by the transportation, utilities, and savings categories? 27%

13. Of the total $2400, how much money per month is budgeted for utilities? $192

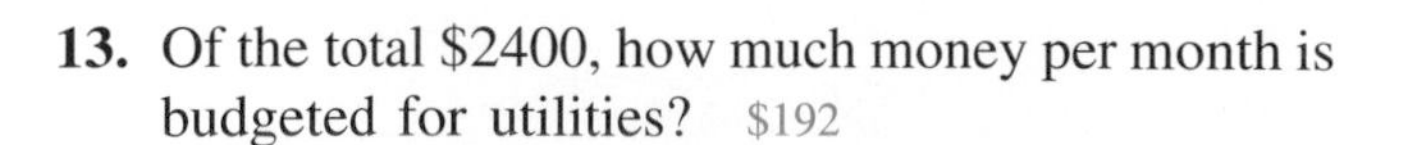

14. Of the total $2400, how much money per month is budgeted for transportation? $360

15. Of the total $2400, how much money per month is budgeted for the rent and savings categories? $816

16. Of the total $2400, how much money per month is budgeted for the transportation and food categories? $1008

The following double-bar graph illustrates the number of customers at Reid's Steak House for each quarter for the years 1993 and 1994. Use the graph to answer questions 17–24.

17. How many customers came to the restaurant in the second quarter of 1993? 6000 customers

18. How many customers came to the restaurant in the third quarter of 1994? 4000 customers

19. When did the restaurant have the greatest number of customers? 4th quarter 1994

20. When did the restaurant have the fewest number of customers? 3rd quarter 1993

21. By how much did the number of customers increase from the first quarter of 1993 to the first quarter of 1994? 1000 customers

22. By how much did the number of customers increase from the fourth quarter of 1993 to the fourth quarter of 1994? 1500 customers

23. During the third quarter (July, August, September) there was road construction in front of the restaurant in both 1993 and 1994. Does the graph indicate the possibility that this might have caused a drop in the number of customers? Yes

24. In the second quarter (April, May, June) the owner spent more on advertising in 1993 than in 1994. Does the graph suggest that this change might have caused a drop in the number of customers? Yes

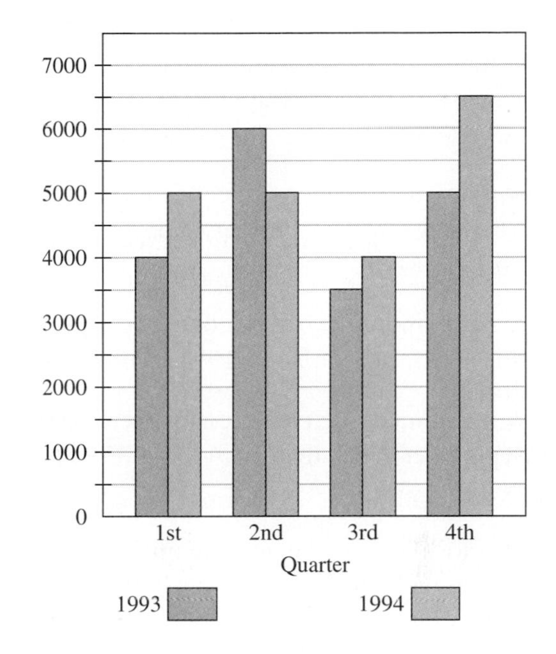

The following line graph shows the number of graduates of Williamston University during the last 6 years. Use the graph to answer questions 25–32.

25. How many Williamston University students graduated in 1993? 400 students

26. How many Williamston University students graduated in 1992? 500 students

27. How many Williamston University students graduated in 1994? 650 students

28. How many Williamston University students graduated in 1991? 450 students

29. How many more Williamston University students graduated in 1990 than in 1989? 100 students more

30. How many more Williamston University students graduated in 1992 than in 1991? 50 students more

31. Between what two years did the number of graduates decline? 1992–1993

32. Between what two years did the number of graduates increase by the greatest amount? 1993–1994

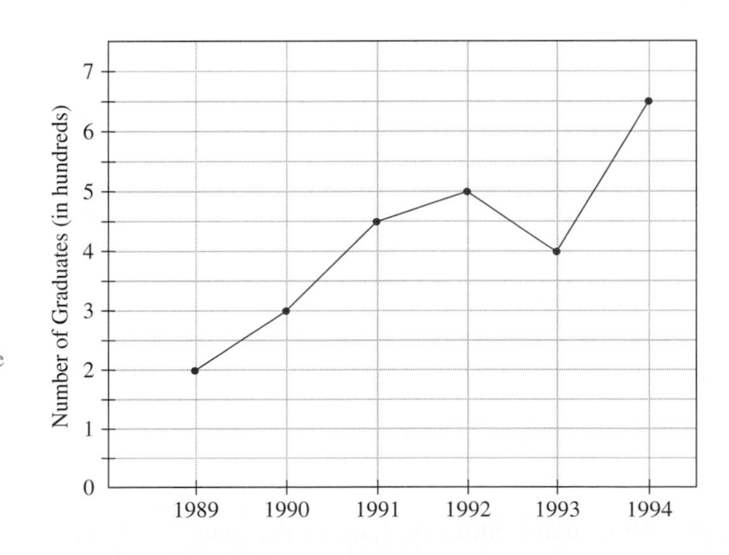

The following comparison line graph shows the number of ice cream cones purchased at the Junction Ice Cream Stand during a five-month period in 1991 and in 1992. Use this graph to answer questions 33–40.

33. How many ice cream cones were purchased in July 1992? 45,000 cones

34. How many ice cream cones were purchased in August 1991? 30,000 cones

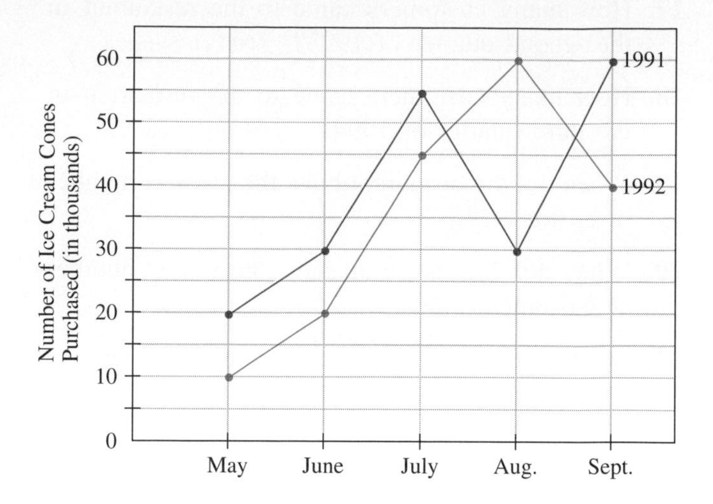

35. How many more ice cream cones were purchased in May 1991 than in May 1992?
10,000 cones more

36. How many more ice cream cones were purchased in August 1992 than in August 1991?
30,000 cones more

37. How many more ice cream cones were purchased in July 1991 than in June 1991? 25,000 cones more

38. How many more ice cream cones were purchased in September 1991 than in August 1991?
30,000 cones more

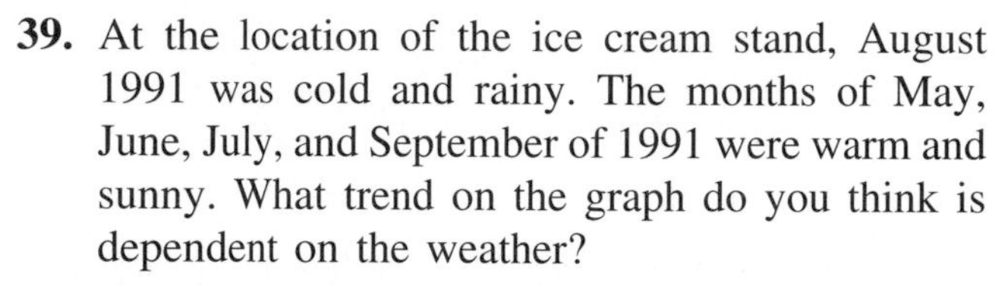

39. At the location of the ice cream stand, August 1991 was cold and rainy. The months of May, June, July, and September of 1991 were warm and sunny. What trend on the graph do you think is dependent on the weather?
The cooler the temperature, the fewer cones sold.

40. July and August of 1992 were warm and very sunny, while May and June of 1992 were cloudy at the location of the ice cream stand. What trend on the graph do you think is dependent on the weather?
The warmer the temperature, the more cones sold.

The state highway department made a survey of the bridges over all of its county and state highways. The number of bridges of each age interval is displayed in the following histogram. Use the histogram to answer questions 41–46.

41. How many bridges are between 40 and 59 years old? 50 bridges

42. How many bridges are between 60 and 79 years old? 25 bridges

43. The greatest number of bridges in the state are between __20__ and __39__ years old.

44. The highway commissioner has ordered an immediate inspection of all bridges older than 79 years old. How many bridges will be inspected?
25 bridges

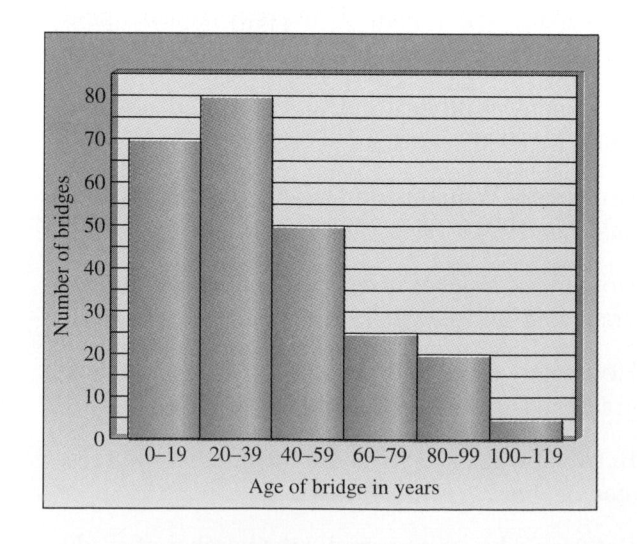

45. All bridges less than 40 years old have had a recent inspection by the highway department. How many bridges were inspected recently?
150 bridges

46. How many more bridges in the state are in the category 60–79 years old than are in the category 80–99 years old? 5 bridges more

Complete a table to determine the frequency of each class interval of the following data. Each number tells how many defective televisions were identified daily in the production of 400 new color television sets by a major manufacturer during the last 28 days.

3	5	8	2	0	1	7
13	6	3	4	1	8	12
0	2	16	5	7	13	14
12	10	17	5	4	0	3

	Number of Defects Class Interval	Tally	Frequency
47.	0–3	𝍸𝍸	10
48.	4–7	𝍸 |||	8
49.	8–11	|||	3
50.	12–15	𝍸	5
51.	16–19	||	2

52. Construct a histogram using the table prepared in problems 47–51.

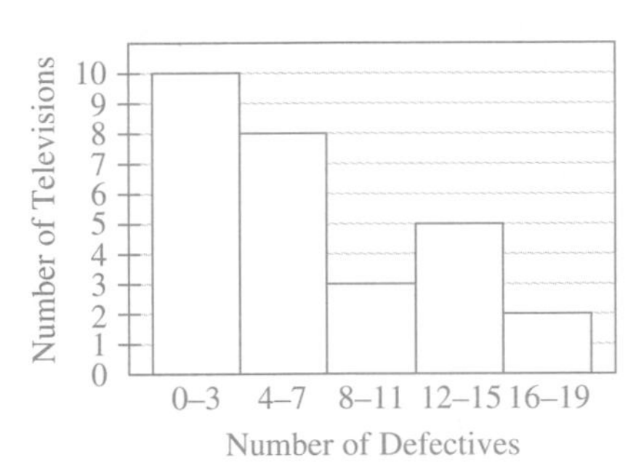

53. Based on the data of problems 47–51, how often were between 0 and 7 defective television sets identified in the production? 18 times

Find the mean value (average) for each of the following items in problems 54–59.

54. The last seven-day maximum temperature readings in Los Angeles in July were: 86°, 83°, 88°, 95°, 97°, 100°, 81°. 90°

55. The amount of groceries purchased by the Michael Stallard family each week for the last seven weeks was $87, $105, $89, $120, $139, $160, $98. $114

56. The number of college textbooks purchased by each of eight men living at Jenkins House during his four years of college was 76, 20, 91, 57, 42, 21, 75, 82. 58 textbooks

57. The number of women students enrolled in the school of engineering at Westwood University during each of the last 10 years was 151, 140, 148, 156, 183, 201, 205, 228, 231, 237. 188 women

58. The number of cars parked in the Central City garage for each of the last five days was 1327, 1561, 1429, 1307, 1481. 1421 cars

59. The number of employees throughout the nation employed by Freedom Rent a Car for the last six years was 882, 913, 1017, 1592, 1778, 1936. 1353 employees

Find the median value for each of the following in problems 60–65.

60. The scores on a recent mathematics exam: 69, 57, 100, 87, 93, 65, 77, 82, 88. 82

61. The number of students taking Abnormal Psychology for the fall semester for the last nine years at Elmson College: 77, 83, 91, 104, 87, 58, 79, 81, 88. 83 students

62. The costs of eight trucks purchased by the Highway Department: $28,500, $29,300, $21,690, $35,000, $37,000, $43,600, $45,300, $38,600. $36,000

63. The costs of 10 houses recently purchased at Stillwater: $98,000, $150,000, $120,000, $139,000, $170,000, $156,000, $135,000, $144,000, $154,000, $126,000. $141,500

64. The number of cups of coffee consumed by each of the students of the 7:00 A.M. Biology III class during the last semester: 38, 19, 22, 4, 0, 1, 5, 9, 18, 36, 43, 27, 21, 19, 25, 20. 19.5 cups

65. The number of deliveries made each day by the Northfield House of Pizza: 21, 16, 0, 3, 19, 24, 13, 18, 9, 31, 36, 25, 28, 14, 15, 26. 18.5 deliveries

66. The scores on eight tests taken by Wong Yin in calculus last semester were 96, 98, 88, 100, 31, 89, 94, 98. Which is a better measure of his usual score, the *mean* or the *median?* Why?
The median, because of one low score, 31.

67. The ten salespersons of People's Dodge sold the following number of cars last month: 13, 16, 8, 4, 5, 19, 15, 18, 39, 12. Which is a better measure of the usual sales of these salespersons, the *mean* or the *median?* Why?
The median, because of the one high data item, 39.

Chapter 8 Test

A state Highway Safety Commission recently reported the results of inspecting 300,000 automobiles. The following circle graph depicts the percent of automobiles that passed and the percent that had one or more safety violations. Use this graph to answer questions 1–5.

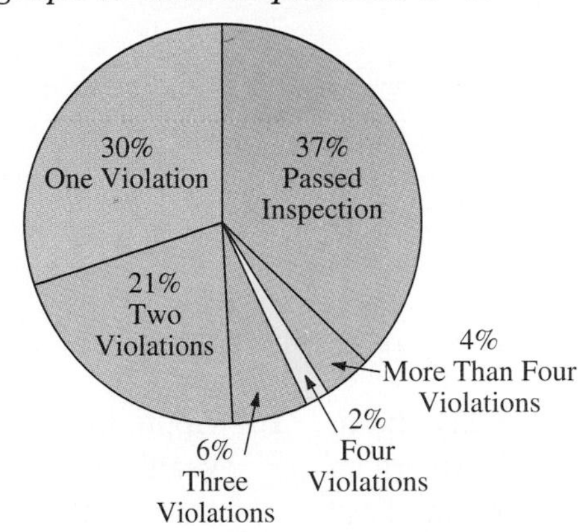

1. What percent of the automobiles passed inspection?

2. What percent of the automobiles had two safety violations?

3. What percent of the automobiles had more than two safety violations?

4. If 300,000 automobiles were inspected, *how many* of them had one safety violation?

5. If 300,000 automobiles were inspected, *how many* of them had either two violations or three violations?

The following double-bar graph indicates the number of cars sold at Boley's Chrysler during each quarter of 1992 and 1993. Use the graph to answer questions 6–11.

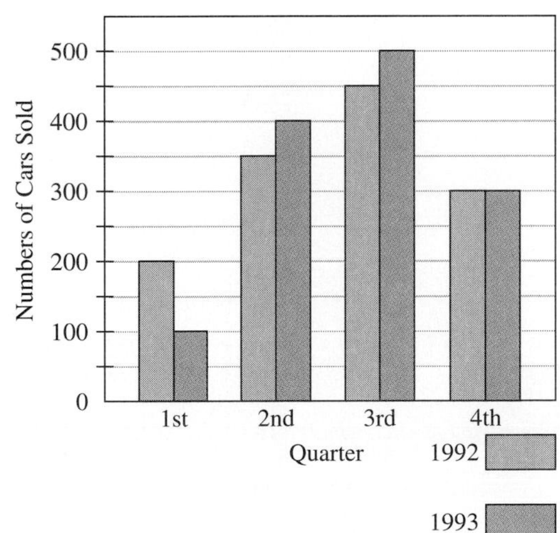

6. How many cars were sold in the second quarter of 1992?

7. How many cars were sold in the third quarter of 1993?

8. When was the greatest number of cars sold?

9. During which quarter were more cars sold in 1992 than in 1993?

10. How many more cars were sold in the second quarter of 1993 than in the second quarter of 1992?

11. How many more cars were sold in the third quarter of 1992 than in the fourth quarter of 1992?

1. 37%

2. 21%

3. 12%

4. 90,000 automobiles

5. 81,000 automobiles

6. 350 cars

7. 500 cars

8. 3rd quarter 1993

9. 1st quarter

10. 50 cars more

11. 150 cars more

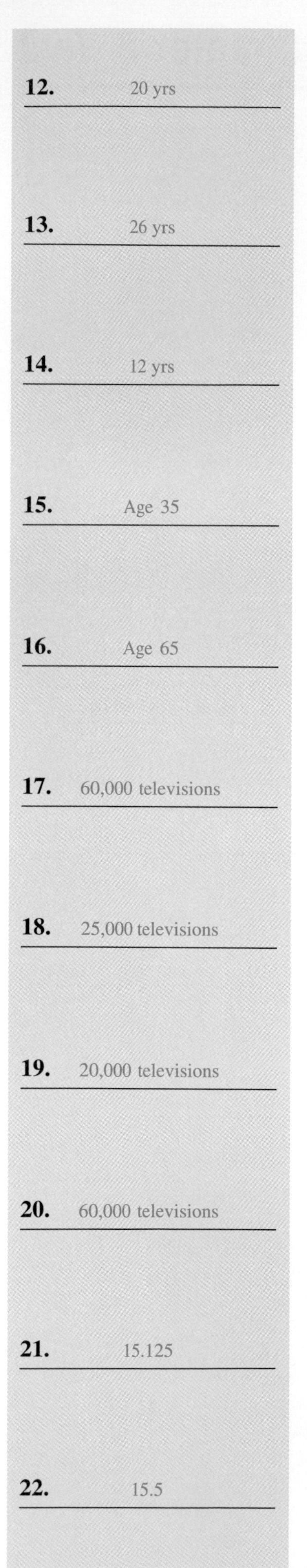

A research study by 10 midwestern universities produced the following line graph. Use the graph to answer questions 12–16.

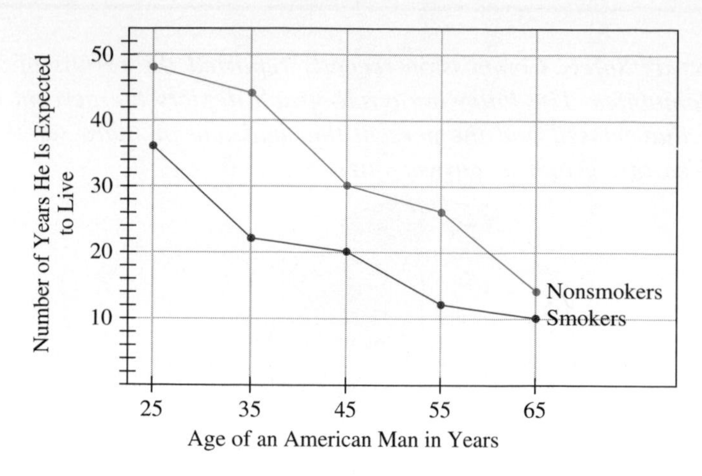

12. Approximately how many more years is a 45-year-old American man expected to live if he smokes?

13. Approximately how many more years is a 55-year-old American man expected to live if he does not smoke?

14. According to this graph, approximately how much longer is a 25-year-old nonsmoker expected to live than a 25-year-old smoker?

15. According to this graph, at what age is the difference between the life expectancy of a smoker and a nonsmoker the greatest?

16. According to this graph, at what age is the difference between the life expectancy of a smoker and a nonsmoker the smallest?

The following histogram was prepared by a consumer research group. Use the histogram to answer questions 17–20.

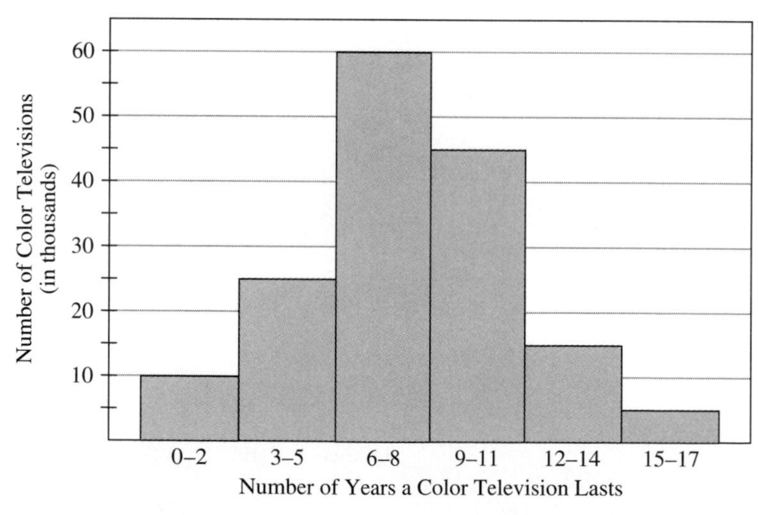

17. How many color television sets lasted 6–8 years?

18. How many color television sets lasted 3–5 years?

19. How many color television sets lasted more than 11 years?

20. How many color television sets lasted 9–14 years?

A chemistry student had the following scores on eight quizzes in her chemistry class: 19, 16, 15, 12, 18, 17, 14, 10

21. Find the *mean* quiz score.

22. Find the *median* quiz score.

Cumulative Test for Chapters 1–8

Approximately one-half of this test is based on Chapter 8 material. The remainder is based on material covered in Chapters 1–7.

1. Add.
1376 + 2804 + 9003 + 7642

2. Multiply. 3004×26

3. Subtract. $7\frac{1}{5} - 3\frac{3}{8}$

4. Divide. $10\frac{4}{5} \div 3\frac{1}{2}$

5. Round to the nearest hundredth.
2864.3719

6. Subtract.
$$\begin{array}{r} 200.58 \\ -\ 127.93 \\ \hline \end{array}$$

7. Divide. $52.0056 \div 0.72$

8. Find n. $\frac{7}{n} = \frac{35}{3}$

9. Of every 2030 cars manufactured, three have major engine defects. If the total number of these cars manufactured was 26,390, approximately how many would have major engine defects?

10. What is 1.3% of 25?

11. 72% of what number is 252?

12. Convert 198 cm to m.

13. Convert 18 yd to ft.

14. Find the area of a circle with radius of 3 in. Round your answer to the nearest tenth.

15. Find the perimeter of a square with a side of 17 in.

The 12,000-member student body of Mason University consists of five groups: freshmen, sophomores, juniors, seniors, and graduate students. The distribution by category is displayed on the following graph.

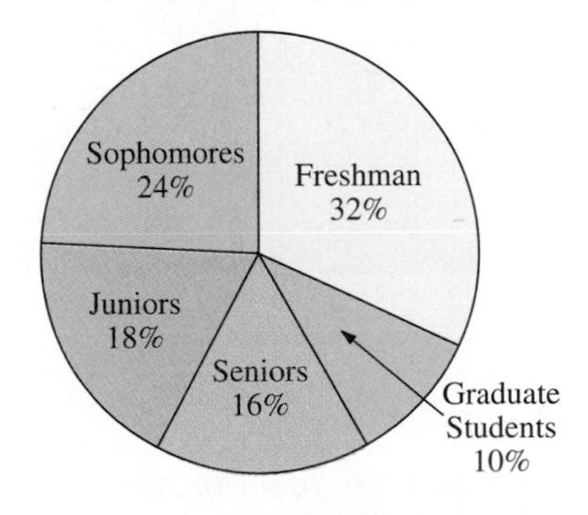

16. What *percent* are either juniors or seniors?

17. How *many* of the 12,000 students are freshmen?

1. 20,825

2. 78,104

3. $\frac{153}{40}$ *or* $3\frac{33}{40}$

4. $\frac{108}{35}$ *or* $3\frac{3}{35}$

5. 2864.37

6. 72.65

7. 72.23

8. $n = 0.6$

9. 39 cars

10. 0.325

11. 350

12. 1.98 m

13. 54 ft

14. 28.3 in.^{2}

15. 68 in.

16. 34%

17. 3840 students

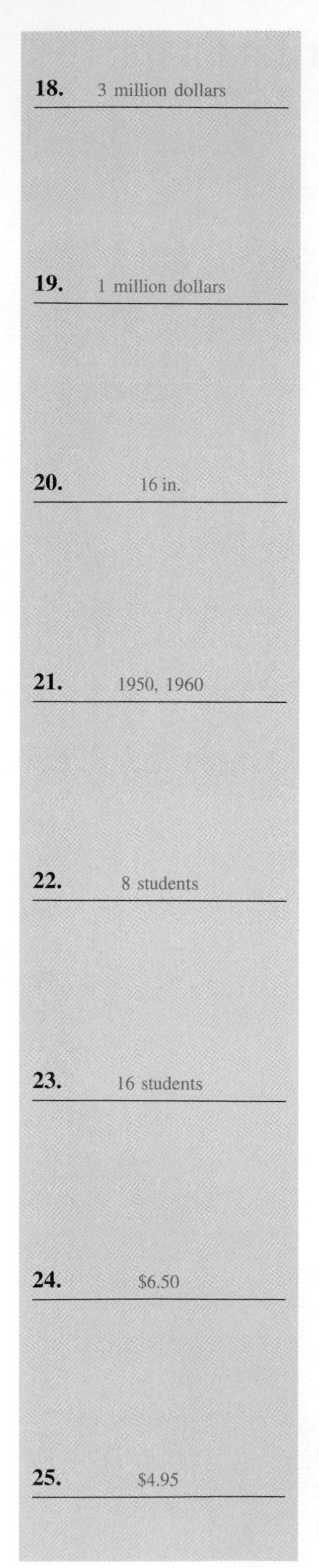

The following double-bar graph indicates the quarterly profits for Dedalon Corporation for 1993 and 1994.

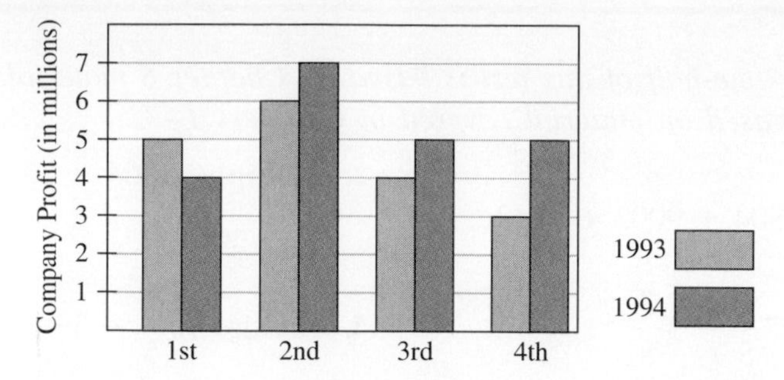

18. What was the quarterly profit for Dedalon Corporation in the fourth quarter of 1993?

19. How much greater was the profit of Dedalon Corporation in the second quarter of 1994 than in the second quarter of 1993?

The following comparison line graph depicts the annual rainfall in Dixville compared to the annual rainfall in Weston for five specific years.

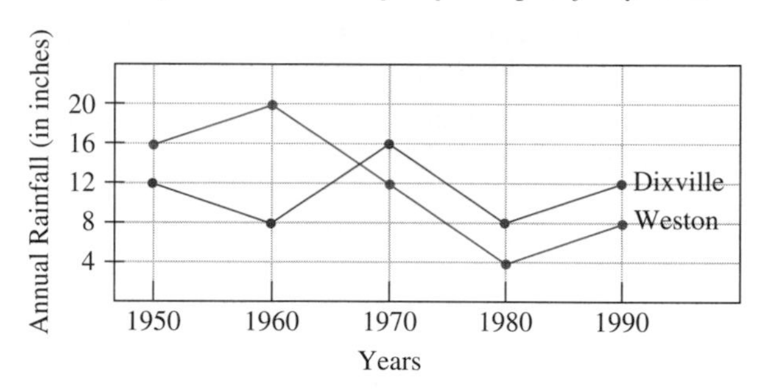

20. How many inches of rain fell in Dixville in 1970?

21. In what years was the annual rainfall in Weston greater than the annual rainfall in Dixville?

The following histogram depicts the number of students in the math course Basic Mathematics who fall into various age groups.

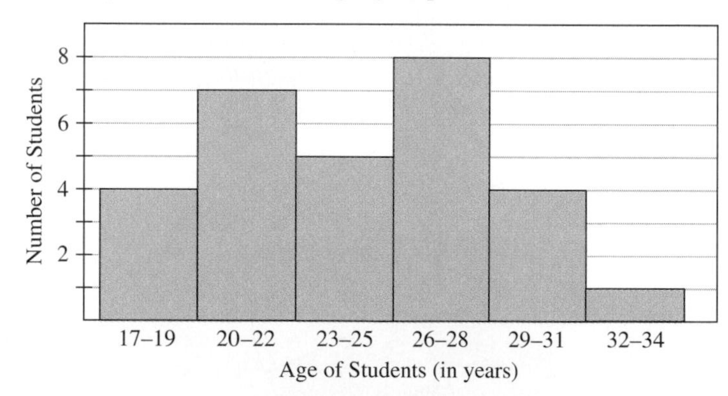

22. How many students are 26–28 years of age?

23. How many students are under 26 years of age?

The following are the hourly wages of six employees of the Hamilton House of Pizza: $5.00, $4.50, $3.95, $4.90, $7.00, $13.65.

24. Find the *mean* hourly wage.

25. Find the *median* hourly wage.

CHAPTER 9

Signed Numbers

The use of signed numbers is very helpful when determining the time in different parts of the world. If you took a six-hour plane flight that left Bangkok, Thailand, at 7:30 A.M. and landed in Bucharest, Romania, do you know at what time you would expect to land? To find out if you can solve problems like this, turn to the Putting Your Skills to Work on page 530.

Pretest Chapter 9

If you are familiar with the topics in this chapter, take this test now. Check your answers with those in the back of the book. If an answer was wrong or you couldn't do a problem, study the appropriate section of the chapter.

If you are not familiar with the topics in this chapter, don't take this test now. Instead, study the examples, work the practice problems, and then take the test.

This test will help you identify which concepts you have mastered and which you need to study further.

1. -19
2. -9
3. 4.5
4. 0
5. $-\frac{4}{12}$ *or* $-\frac{1}{3}$
6. $-\frac{7}{6}$
7. -7
8. 1.7
9. -8
10. -31
11. $\frac{14}{17}$
12. -12
13. 1.4
14. -2.8
15. 42
16. $\frac{19}{15}$ *or* $1\frac{4}{15}$

Section 9.1

Add.

1. $-7 + (-12)$ **2.** $-15 + 6$

3. $7.6 + (-3.1)$ **4.** $8 + (-5) + 6 + (-9)$

5. $\frac{5}{12} + \left(-\frac{3}{4}\right)$ **6.** $-\frac{5}{6} + \left(-\frac{1}{3}\right)$

7. $-2.8 + (-4.2)$ **8.** $-3.7 + 5.4$

Section 9.2

Subtract.

9. $13 - 21$ **10.** $-13 - 18$

11. $\frac{5}{17} - \left(-\frac{9}{17}\right)$ **12.** $-19 - (-7)$

13. $-4.9 - (-6.3)$ **14.** $2.8 - 5.6$

15. $21 - (-21)$ **16.** $\frac{2}{3} - \left(-\frac{3}{5}\right)$

Section 9.3

Multiply or divide.

17. $(-3)(-8)$ **18.** $-48 \div (-12)$

19. $-72 \div 9$ **20.** $(5)(-4)(2)(-1)\left(-\frac{1}{4}\right)$

21. $\frac{72}{-3}$ **22.** $\frac{-\frac{3}{4}}{-\frac{4}{5}}$

23. $(-8)(-2)(-4)$ **24.** $120 \div (-12)$

Section 9.4

Perform each operation in the proper order.

25. $24 \div (-4) + 28 \div (-7)$ **26.** $18 \div 3(5) + (-5) \div (-5)$

27. $7 + (-9) + 2(-5)$ **28.** $8(-6) \div (-10)$

29. $5 - (-6) + 18 \div (-3)$ **30.** $9(-3) + 4(-2) - (-6)$

31. $\frac{12 + 8 - 4}{(-4)(3)(4)}$ **32.** $\frac{56 \div (-7) - 1}{3(-9) + 6(-3)}$

Section 9.5

Write in scientific notation.

33. 80,000 **34.** 0.0005

Write in standard notation.

35. 6.7×10^{-3} **36.** 1.32×10^{6}

17. 24

18. 4

19. -8

20. -10

21. -24

22. $\frac{15}{16}$

23. -64

24. -10

25. -10

26. 31

27. -12

28. 4.8

29. 5

30. -29

31. $-\frac{1}{3}$

32. $\frac{1}{5}$

33. 8×10^{4}

34. 5×10^{-4}

35. 0.0067

36. 1,320,000

9.1 Addition of Signed Numbers

After studying this section, you will be able to:

1 *Add two signed numbers with the same sign.*
2 *Add two signed numbers with different signs.*
3 *Add three or more signed numbers.*

1 *Adding Two Signed Numbers with the Same Sign*

In Chapters 1–8 we worked with whole numbers, fractions, and decimals. In this chapter we enlarge the set of numbers we work with to include numbers that are less than zero. Many real-life situations require using numbers that are less than zero. A debt that is owed, a financial loss, temperatures that fall below zero, and elevations that are below sea level can only be expressed in numbers that are less than zero, or negative numbers.

The following is a graph of the financial reports of four small airlines. It shows positive numbers—those numbers that fall above zero—and negative numbers—those numbers that fall below zero. The positive numbers represent money gained. The negative numbers represent money lost.

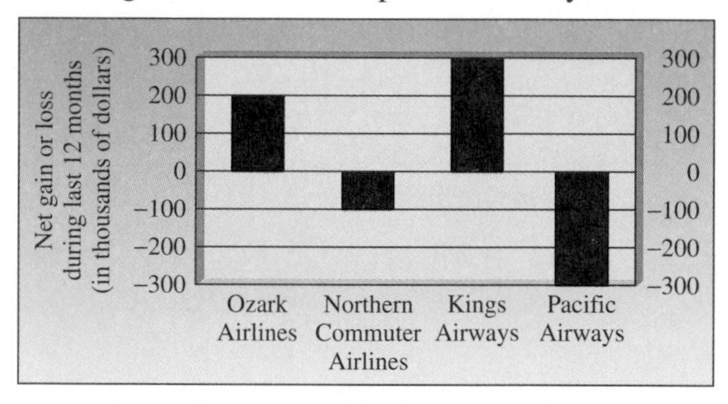

In the last 12 months a value of $-100{,}000$ is shown for Northern Commuter Airlines. This means that Northern Commuter Airlines lost \$100,000 during the year. A value of $-300{,}000$ is recorded for Pacific Airlines. What does this mean?

Another way to picture positive and negative numbers is on a number line. A number line is a line on which each point is associated with a number. The numbers may be positive or negative, whole numbers, fractions, or decimals. Positive numbers are to the right of zero on the number line. Negative numbers are to the left of zero on the number line. Zero is neither positive nor negative.

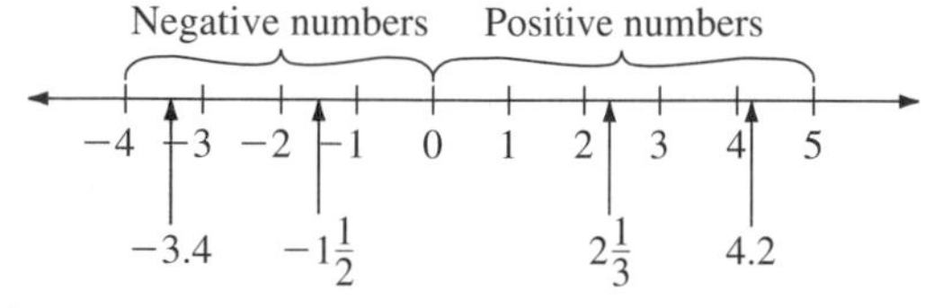

Positive numbers can be written with a plus sign—for example, $+2$—but this is not usually done. Positive 2 is usually written as 2. It is understood that the *sign* of the number is positive although it is not written. Negative numbers must always have the negative sign so that we know they are negative numbers. Negative 2 is written as -2. The *sign* of the number is negative.

Absolute Value

Sometimes we are only interested in the distance a number is from zero. For example, the distance from 0 to $+3$ is 3. The distance from 0 to -3 is also 3. Notice that distance is always a positive number, regardless of which direction we travel on the number line. This distance is called the *absolute value.*

> The **absolute value** of a number is the distance between that number and zero on the number line.

The symbol for absolute value is $| \ |$. When we write $|5|$, we are looking for the distance from 0 to 5 on the number line. Thus $|5| = 5$. This is read, "The absolute value of 5 is 5." $|-5|$ is the distance from 0 to -5 on the number line. Thus $|-5| = 5$. This is read, "The absolute value of -5 is 5."

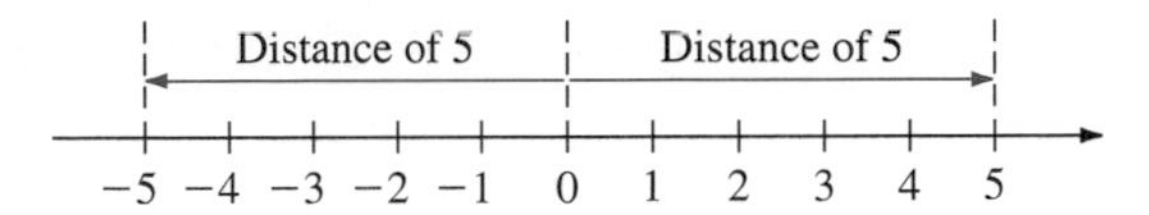

Other examples of absolute value are shown below.

$$|6| = 6 \qquad |-3| = 3$$

$$|7.2| = 7.2 \qquad \left|-\frac{1}{5}\right| = \frac{1}{5}$$

$$|0| = 0 \qquad |-26| = 26$$

We use the concept of absolute value to develop rules for adding signed numbers. We begin by looking at addition of numbers with the same sign. Although you are already familiar with the addition of positive numbers, we will look at an example.

Suppose that we earn \$52 one day and earn \$38 the next day. To learn what our 2-day total is, we add the positive numbers. We earn

$$(+\$52) + (+\$38) = +\$90$$

The money is coming in, and the plus sign records a gain. Notice that we added the numbers and that the sign of the sum was the same as the sign of the addends.

Now let's consider an example of addition of two negative numbers.

Suppose that we consider money spent as negative dollars. If we spend \$52 one day ($-\$52$) and we spend \$38 the next day ($-\$38$), we must add two negative numbers. What is our financial position?

$$(-\$52) + (-\$38) = -\$90$$

We have spent \$90. The negative sign tells us the direction of the money: out! Notice that we added the numbers and that the sign of the sum was the same as the sign of the addends.

These examples suggest the addition rule for two numbers with the same sign.

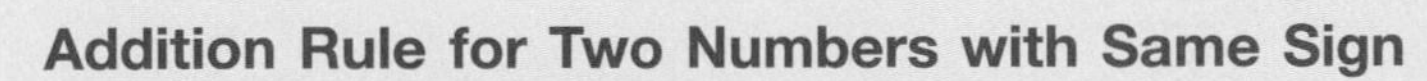

Addition Rule for Two Numbers with Same Sign

To add two numbers with the same sign:

1. Add the absolute value of the numbers.
2. Use the common sign in the answer.

EXAMPLE 1 Add. **(a)** $7 + 5$ **(b)** $-6 + (-2)$ **(c)** $-3.2 + (-5.6)$

(a)
$$\begin{array}{r} 7 \\ +5 \\ \hline 12 \end{array}$$
We add the absolute value of the numbers 7 and 5. The positive sign, although not written, is common to both numbers. The answer is a positive 12. (The + sign is not written.)

(b)
$$\begin{array}{r} -6 \\ + -2 \\ \hline -8 \end{array}$$
We add the absolute value of the numbers 6 and 2 without regard to sign.

We use a negative sign in our answer because the negative sign is common to both numbers.

(c)
$$\begin{array}{r} -3.2 \\ + -5.6 \\ \hline -8.8 \end{array}$$
We add the absolute value of the numbers 3.2 and 5.6.

We use a negative sign in our answer because we added two negative numbers.

Take some time to look over Example 1. Does it make sense to you? These sign rules are very important. Be sure you can obtain the correct answers for Practice Problem 1.

Practice Problem 1 Add. **(a)** $9 + 14$
(b) $-7 + (-3)$ **(c)** $-4.5 + (-1.9)$ ■

These rules can be applied to fractions as well.

EXAMPLE 2 Add. **(a)** $-\frac{3}{13} + \left(-\frac{5}{13}\right)$ **(b)** $\frac{5}{18} + \frac{1}{3}$ **(c)** $-\frac{1}{7} + \left(-\frac{3}{5}\right)$

(a)
$$\begin{array}{r} -\frac{3}{13} \\ + -\frac{5}{13} \\ \hline -\frac{8}{13} \end{array}$$
The common sign here is a negative sign.

(b) The LCD = 18. The first fraction already has the LCD.

$$\begin{array}{rcr} \frac{5}{18} & = & \frac{5}{18} \\ +\frac{1}{3} \cdot \frac{6}{6} & = & +\frac{6}{18} \\ \hline & & \frac{11}{18} \end{array}$$

We add two positive numbers, so the answer is positive.

(c) The LCD = 35.

$$\frac{1}{7} \cdot \frac{5}{5} = \frac{5}{35}$$

Because $\frac{1}{7} = \frac{5}{35}$ it follows that $-\frac{1}{7} = -\frac{5}{35}$.

$$\frac{3}{5} \cdot \frac{7}{7} = \frac{21}{35}$$

Because $\frac{3}{5} = \frac{21}{35}$ it follows that $-\frac{3}{5} = -\frac{21}{35}$. Thus

$$\begin{array}{r} -\frac{1}{7} \\ + -\frac{3}{5} \\ \hline \end{array} \quad \text{is equivalent to} \quad \begin{array}{r} -\frac{5}{35} \\ + -\frac{21}{35} \\ \hline -\frac{26}{35} \end{array}$$

We add two negative numbers, so the answer is negative.

TEACHING TIP Some students devote all their attention to finding the LCD and forget the sign of the fraction. Suggest that it is good practice to double-check the sign of any addition problem involving fractions.

Practice Problem 2 Add.

(a) $-\frac{2}{15} + \left(-\frac{12}{15}\right)$ **(b)** $\frac{5}{12} + \frac{1}{4}$ **(c)** $-\frac{1}{6} + \left(-\frac{2}{7}\right)$ ■

It is interesting to see how often negative numbers appear in statements of the Federal Budget. Observe the data in the following bar graph.

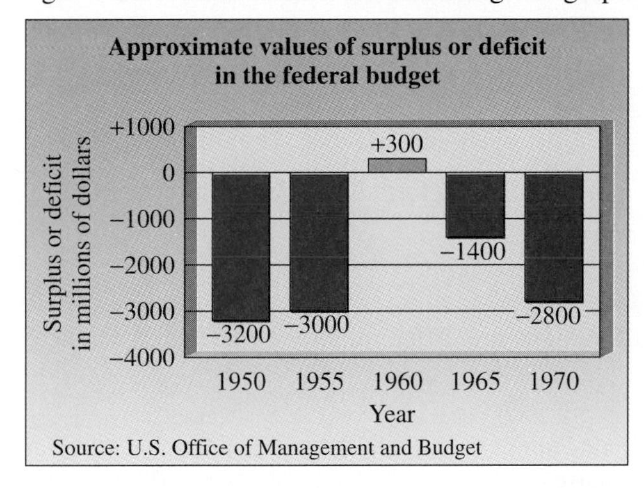

EXAMPLE 3 Find the total value of surplus or deficit for the two years 1950 and 1955.

We add (−\$3200 million) + (−\$3000 million) to obtain −\$6200 million.
This is a total of −\$6.2 billion dollars.
The total debt for these two years is \$6,200,000,000.

Practice Problem 3 Find the total value of surplus or deficit for the two years 1965 and 1970. ■

2 *Adding Two Signed Numbers with Different Signs*

Let's look at some real-life situations involving addition of signed numbers with different signs. Suppose that we earn \$52 one day and we spend \$38 the next day. If we combine (add) the two transactions, it would look like

$$(+\$52) + (-\$38) = +\$14$$

This is a situation with which we are familiar. What we actually do is subtract. That is, we take the difference between \$52 and \$38. Notice that the sign of the larger number is positive and that the sign of the answer is also positive.

Let's look at another situation. Suppose we earn \$38 and we spend \$52. The situation would look like this.

$$(-\$52) + (+\$38) = -\$14$$

We end up owing \$14, which is represented by a negative number. To find the sum, we actually find the difference between \$52 and \$38. Notice that if we do not account for sign, the larger number is 52. The sign of that number is negative and the sign of the answer is also negative. This suggests the addition rule for two numbers with different signs.

Addition Rule for Two Numbers with Different Signs

To add two numbers with different signs:

1. Subtract the absolute value of the numbers.
2. Use the sign of the number with the larger absolute value.

TEACHING TIP After covering the addition of two numbers with the same sign and with different signs, ask students to do the following additional class exercises at their seats.

A. $-12 + (-3) =$
B. $45 + (-25) =$
C. $36 + 10 =$
D. $-18 + 24 =$
E. $-30 + (-40) =$

After allowing enough time for the average student to do these, list the 5 correct answers (−15, 20, 46, 6, −70) and remind students of the rule involved in each case. This will help them to quickly think of which rule applies when they are doing several problems involving the addition of two signed numbers.

EXAMPLE 4 Add. **(a)** $8 + (-10)$ **(b)** $-8 + 10$ **(c)** $-16.6 + 12.3$ **(d)** $16.6 + (-12.3)$

(a)
$$\begin{array}{r} 8 \\ + -10 \\ \hline -2 \end{array}$$
The signs are different, so we find the difference: $10 - 8 = 2$.

The sign of the number with the larger absolute value is negative, so the answer is negative.

(b)
$$\begin{array}{r} -8 \\ +\ 10 \\ \hline 2 \end{array}$$
The signs are different, so we find the difference: $10 - 8 = 2$.

The sign of the number with the larger absolute value is positive, so the answer is positive.

(c)
$$\begin{array}{r} -16.6 \\ +\ 12.3 \\ \hline -4.3 \end{array}$$
The signs are different, so we find the difference.

The sign of the number with the larger absolute value is negative, so the answer is negative.

(d) $\begin{array}{r} 16.6 \\ + -12.3 \\ \hline 4.3 \end{array}$ The signs are different, so we find the difference.

The sign of the number with the larger absolute value is positive, so the answer is positive.

Practice Problem 4 Add.

(a) $7 + (-12)$ **(b)** $-7 + 12$ **(c)** $-20.8 + 15.2$ **(d)** $20.8 + (-15.2)$ ■

Notice that in **(c)** and **(d)** of Example 4 the number with the larger absolute value is on top. This makes the numbers easier to subtract. Because addition is commutative, we could have written **(a)** and **(b)** as

$$\begin{array}{r} -10 \\ + \quad 8 \\ \hline \end{array} \qquad \begin{array}{r} 10 \\ + -8 \\ \hline \end{array}$$

which makes the computation easier. If you are adding two numbers with different signs, place the number with the larger absolute value on top so that you can find the difference easily.

As noted, the commutative property of addition holds for signed numbers.

Commutative Property of Addition

For any real numbers a, b

$$a + b = b + a$$

EXAMPLE 5 Last night the temperature dropped to $-14°$F. From that low, today the temperature rose 33°F. What was the highest temperature today?

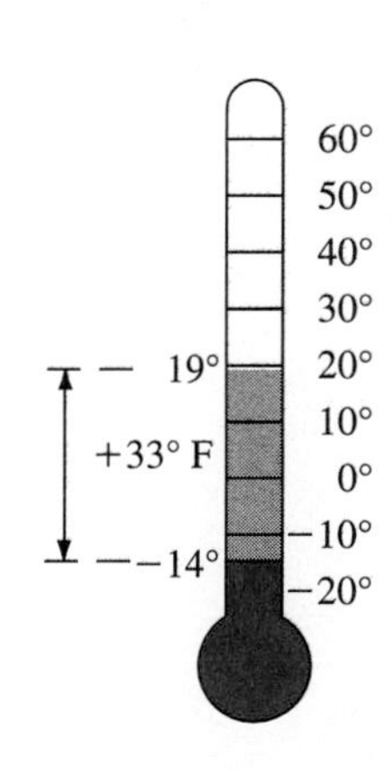

We want to add $-14°$F and 33°F. Because addition is commutative, it does not matter if we add $-14 + 33$ or $33 + (-14)$.

$$\begin{array}{r} 33°\text{F} \\ + -14°\text{F} \\ \hline 19°\text{F} \end{array}$$

The 33 is larger than 14. The difference between 33 and 14 is 19. The number with the larger absolute value is positive, so the answer is positive.

Practice Problem 5 Last night the temperature dropped to $-19°$F. From that low, today the temperature rose 28°F. What was the highest temperature today? ■

3 *Adding Three or More Signed Numbers*

We can add three or more numbers using these rules. Since addition is associative, we may group the numbers to be added in any way. That is, it does not matter which two numbers are added first.

EXAMPLE 6 Add. $24 + (-16) + (-10)$

We can go from left to right and start with 24, or we can start with -16.

Step 1 $\begin{array}{r} 24 \\ + -16 \\ \hline 8 \end{array}$ *or* **Step 1** $\begin{array}{r} -16 \\ + -10 \\ \hline -26 \end{array}$

Step 2 $\begin{array}{r} 8 \\ + -10 \\ \hline -2 \end{array}$ **Step 2** $\begin{array}{r} -26 \\ + \quad 24 \\ \hline -2 \end{array}$

Practice Problem 6 Add. $36 + (-21) + (-18)$ ■

> **Associative Property of Addition**
> For any three real numbers a, b, c
> $$(a + b) + c = a + (b + c)$$

If there are many numbers to add, it may be easier to add the positive numbers and the negative numbers separately and then combine the results.

EXAMPLE 7 Add. $-8 + 3 + 5 + (-7) + (-4) + 6$

Let us add separately the positive numbers and the negative numbers.

$$\begin{array}{r} -8 \\ -7 \\ + -4 \\ \hline -19 \end{array} \qquad \begin{array}{r} 3 \\ 5 \\ +6 \\ \hline 14 \end{array}$$

Now we add together $-19 + 14$, which have opposite signs. The number with the larger absolute value is -19.

$$\begin{array}{r} -19 \\ + \quad 14 \\ \hline -5 \end{array}$$

The difference between 19 and 14 is 5.
19 is larger than 14, so the answer is negative.

Practice Problem 7 Add. $-5 + 3 + (-2) + 8 + (-7) + 9$ ■

EXAMPLE 8 The results of a new company's operations over five months are listed in this table. What is the company's overall profit or loss over the five-month period?

NET OPERATIONS Profit/Loss Statement in Dollars		
Month	Profit	Loss
January	30,000	
February		−50,000
March		−10,000
April	20,000	
May	15,000	

First we will add separately the positive numbers and the negative numbers.

$$\begin{array}{r} 30{,}000 \\ 20{,}000 \\ +\ 15{,}000 \\ \hline 65{,}000 \end{array} \qquad \begin{array}{r} -50{,}000 \\ +\ -10{,}000 \\ \hline -60{,}000 \end{array}$$

Now we add the positive number \$65,000 and the negative number −\$60,000.

$$\begin{array}{r} \$65{,}000 \\ +\ -\$60{,}000 \\ \hline \$\ 5{,}000 \end{array}$$

The company had an overall profit of \$5000 for the five-month period.

Practice Problem 8 The results of the next five months of operations for the same company are listed in the table. What is the overall profit or loss over this five-month period?

NET OPERATIONS Profit/Loss Statement in Dollars		
Month	Profit	Loss
June		−20,000
July	30,000	
August	40,000	
September		−5,000
October		−35,000

■

TEACHING TIP Students are often interested to learn that negative numbers were used by the Chinese to some extent around 300 B.C. However, this idea did not spread rapidly to other parts of the world. It was not until the 1700s that most mathematics textbooks included operations with signed numbers. Some scientists feel that certain properties of electricity, magnetism, and astronomy would have been discovered earlier in the history of civilization if negative numbers had gained earlier acceptance.

9.1 Exercises

Add each pair of signed numbers, which have the same sign.

1. $9 + 15$ 24

2. $8 + 7$ 15

3. $-6 + (-11)$ -17

4. $-5 + (-13)$ -18

5. $-4.9 + (-2.1)$ -7

6. $-8.3 + (-3.7)$ -12

7. $8.9 + 7.6$ 16.5

8. $5.8 + 2.7$ 8.5

9. $\frac{1}{5} + \frac{2}{7}$ $\frac{17}{35}$

10. $\frac{2}{3} + \frac{1}{4}$ $\frac{11}{12}$

11. $-\frac{1}{12} + \left(-\frac{5}{6}\right)$ $-\frac{11}{12}$

12. $-\frac{1}{16} + \left(-\frac{3}{4}\right)$ $-\frac{13}{16}$

Add each pair of signed numbers, which have different signs.

13. $14 + (-5)$ 9

14. $15 + (-6)$ 9

15. $-16 + 9$ -7

16. $-9 + 6$ -3

17. $-17 + 12$ -5

18. $-21 + 15$ -6

19. $-36 + 58$ 22

20. $-42 + 57$ 15

21. $-9.3 + 6.5$ -2.8

22. $-7.2 + 4.4$ -2.8

23. $\frac{1}{12} + \left(-\frac{3}{4}\right)$ $-\frac{8}{12} = -\frac{2}{3}$

24. $\frac{1}{15} + \left(-\frac{3}{5}\right)$ $-\frac{8}{15}$

Mixed Practice

Add.

25. $\frac{7}{9} + \left(-\frac{2}{9}\right)$ $\frac{5}{9}$

26. $\frac{5}{12} + \left(-\frac{7}{12}\right)$ $-\frac{2}{12}$ *or* $-\frac{1}{6}$

27. $-18 + (-4)$ -22

28. $-34 + (-2)$ -36

29. $1.5 + (-2.2)$ -0.7

30. $3.6 + (-4.1)$ -0.5

31. $\frac{3}{14} + \frac{2}{7}$ $\frac{7}{14} = \frac{1}{2}$

32. $\frac{7}{15} + \frac{3}{10}$ $\frac{23}{30}$

33. $-15 + (-23)$ -38

34. $-21 + (-16)$ -37

35. $-5.5 + (-2.1)$ -7.6

36. $-7.7 + (-3.2)$ -10.9

37. $13 + (-9)$ 4

38. $-18 + 7$ -11

39. $\frac{1}{2} + \left(-\frac{2}{3}\right)$ $-\frac{1}{6}$

40. $-\frac{3}{4} + \left(-\frac{5}{8}\right)$ $-\frac{11}{8}$ *or* $-1\frac{3}{8}$

41. $8.6 + 9.5$ 18.1

42. $-3.1 + 10$ 6.9

43. $-5 + \left(-\frac{1}{2}\right)$ $-5\frac{1}{2}$

44. $\frac{5}{6} + (-3)$ $-\frac{13}{6}$ *or* $-2\frac{1}{6}$

Add.

45. $6 + (-14) + 4$
$-8 + 4 = -4$

46. $15 + (-2) + 4$
$13 + 4 = 17$

47. $-6 + 3 + (-12) + 4$
$-3 - 8 = -11$

48. $-2 + 15 + (-5) + 4$
$13 + (-1) = 12$

49. $-7 + 6 + (-2) + 5 + (-3) + (-5)$
$-1 + 3 - 8 = -6$

50. $-2 + 1 + (-12) + 7 + (-4) + (-1)$
$-1 + (-5) + (-5) = -11$

51. $5 + (-8) + 7 + 20 + (-8) + (-6)$
$-3 + 27 + (-14) = 10$

52. $2 + (-10) + 6 + 13 + (-10) + (-4)$
$-8 + 19 + (-14) = -3$

★**53.** $\left(-\frac{1}{5}\right) + \left(-\frac{2}{3}\right) + \left(\frac{4}{25}\right) + \left(-\frac{1}{9}\right)$
$-\frac{45}{225} + \left(-\frac{150}{225}\right) + \frac{36}{225} + \left(-\frac{25}{225}\right) = -\frac{184}{225}$

★**54.** $\left(-\frac{1}{7}\right) + \left(-\frac{5}{21}\right) + \left(\frac{3}{14}\right) + \left(-\frac{3}{7}\right)$
$\left(-\frac{6}{42}\right) + \left(-\frac{10}{42}\right) + \left(\frac{9}{42}\right) + \left(-\frac{18}{42}\right) = -\frac{25}{42}$

Applications

Find the profit or loss situation for a company after the following reports.

55. A \$43,000 loss in February followed by a \$51,000 loss in March. $-\$94,000$

56. A \$16,000 loss in May followed by a \$25,000 loss in June. $-\$41,000$

57. A \$28,000 profit in July followed by a \$19,000 loss in August. \$9,000

58. An \$89,000 profit in April followed by a \$33,000 loss in May. \$56,000

59. A \$35,000 loss in April, a \$17,000 profit in May, and a \$20,000 loss in June. $-\$38,000$

60. A \$34,000 loss in October, a \$12,000 loss in November, and a \$15,500 profit in December. $-\$30,500$

Find the new temperature.

61. Last night, the temperature was -1°F. In the morning, the temperature dropped 17°F. What was the new temperature? -18°F

62. This morning the temperature was -12°F. This afternoon, the temperature rose 19°F. What was the new temperature? 7°F

63. This morning, the temperature was -5°F. This evening, the temperature rose 4°F. What was the new temperature? -1°F

64. Last night the temperature was -7°F. The temperature dropped 15°F this afternoon. What was the new temperature? -22°F

Cumulative Review Problems

65. Use $V = \frac{4\pi r^3}{3}$ to find the volume of a sphere of radius 6 ft. Use $\pi = 3.14$ and round your answer to the nearest tenth. 904.3 ft^3

66. Use $V = \frac{Bh}{3}$ to find the volume of a pyramid whose rectangular base measures 9 m by 7 m and whose height is 10 m. 210 m^3

9.2 Subtraction of Signed Numbers

MathPro

Video 19

SSM

After studying this section, you will be able to:

1. *Subtract one signed number from another.*
2. *Solve problems involving both addition and subtraction of signed numbers.*
3. *Solve simple applied problems that involve the subtraction of signed numbers.*

1 *Subtracting One Signed Number from Another*

We begin our discussion by defining the word **opposite.** The opposite of a positive number is a negative number with the same absolute value. For example, the opposite of 7 is -7. The opposite of a negative number is a positive number with the same absolute value. For example, the opposite of -9 is 9. If a number is the opposite of another number, these two numbers are at an equal distance from zero on the number line.

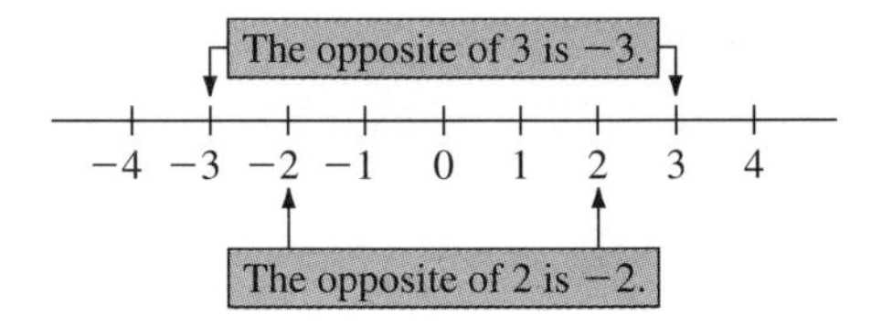

The sum of a number and its opposite is zero.

$$-3 + 3 = 0 \qquad 2 + (-2) = 0$$

We will use the concept of opposite to develop a way to subtract integers.

Let's think about how a checking account works. Suppose that you deposit \$25 and the bank adds a service charge of \$5 for a new checkbook. Your account looks like this:

$$\$25 + (-\$5) = \$20$$

Suppose instead that you deposit \$25 and the bank adds no charge. The next day, you write a check for \$5. The result of these two transactions is

$$\$25 - \$5 = \$20$$

Note that your account has grown by the same amount of money (\$20) in both cases.

We see that adding a negative 5 to 25 is the same as subtracting a positive 5 from 25. That is, $25 + (-5) = 20$ and $25 - 5 = 20$.

Subtracting is equivalent to adding the opposite.

SUBTRACTING	ADDING THE OPPOSITE
$25 - 5 = 20$	$25 + (-5) = 20$
$19 - 6 = 13$	$19 + (-6) = 13$
$7 - 3 = 4$	$7 + (-3) = 4$
$15 - 5 = 10$	$15 + (-5) = 10$

We define a rule for the subtraction of signed numbers.

Subtraction of Signed Numbers

To subtract signed numbers, add the opposite of the second number to the first number.

TEACHING TIP Some students will ask, "Why do we have to change subtraction to adding the opposite?" The most satisfying answer seems to be that if we get used to thinking of subtraction as adding the opposite, then we will be able to consider problems with indicated subtraction and addition such as $-5 - 3 - (-2)$ as addition problems like $-5 + (-3) + 2$. When we are able to do this, we can use the commutative property, which makes many problems easier to solve.

Thus to do a subtraction problem, we first change it to an equivalent addition problem in which the first number does not change but the second number is replaced by its opposite. Then we follow the rules of *addition* for signed numbers.

EXAMPLE 1 Subtract.

$-8 - (-2)$

$-8 + 2$ ← Form the opposite of -2, which is 2.

↑ Change subtraction to addition.

Using the rules of addition, for two numbers with opposite signs, we add

$$-8 + 2 = -6$$

Practice Problem 1 Subtract. $-10 - (-5)$ ■

EXAMPLE 2 Subtract. **(a)** $5 - (-9)$ **(b)** $7 - 8$ **(c)** $-12 - 16$

(a) $5 - (-9)$

$5 + 9$ ← Form the opposite of -9, which is 9.

↑ Change subtraction to addition.

Using the rules of addition, for two numbers with the same sign, we have

$$5 + 9 = 14$$

(b) $7 - 8$

$7 + (-8)$ ← Form the opposite of 8, which is -8.

↑ Change subtraction to addition.

Now we use the rules of addition for two numbers with opposite signs.

$$7 + (-8) = -1$$

(c) $-12 - 16$

$-12 + (-16)$ ← Form the opposite of 16, which is -16.

↑ Change subtraction to addition.

Now we follow the rules of addition of two numbers with the same sign.

$$-12 + (-16) = -28$$

Practice Problem 2 **(a)** $5 - 12$ **(b)** $-11 - 17$ **(c)** $4 - (-17)$ ■

Sometimes the numbers we subtract are fractions or decimals.

EXAMPLE 3 Subtract. **(a)** $5.6 - (-8.1)$ **(b)** $-\frac{6}{11} - \left(-\frac{1}{22}\right)$

(a) Change the subtraction to adding the opposite. Then add.

$$5.6 - (-8.1) = 5.6 + 8.1$$
$$= 13.7$$

(b) Change the subtraction to adding the opposite. Then add.

$$-\frac{6}{11} - \left(-\frac{1}{22}\right) = -\frac{6}{11} + \frac{1}{22}$$

$$= -\frac{6}{11} \cdot \frac{2}{2} + \frac{1}{22}$$ *We see that the LCD = 22. We change $\frac{6}{11}$ to a fraction with a denominator of 22.*

$$= -\frac{12}{22} + \frac{1}{22}$$ *Add.*

$$= -\frac{11}{22} \text{ or } -\frac{1}{2}$$

Practice Problem 3 Subtract. **(a)** $3.6 - (-9.5)$ **(b)** $-\frac{5}{8} - \left(-\frac{5}{24}\right)$ ■

SCIENTIFIC CALCULATOR

Negative Numbers

To enter a negative number on a scientific calculator, find the key marked [+/−]. To enter the number −3, press the key 3 and then the key +/−. The display should read

[−3]

To find (−32) + (−46), enter

32 [+/−] [+] 46 [+/−] [=]

The display should read

[−78]

Try
(a) (−756) + 184
(b) 92 + (−51)
(c) (−618) − (−824)
(d) (−36) + (−10) − (−15)

Note: The [+/−] key changes the sign of a number from + to − or − to +.

Remember that in performing subtraction of two signed numbers,

1. The first number does not change.
2. The subtraction sign is changed to addition.
3. We write the opposite of the second number.
4. We find the result of this addition problem.

Think of each subtraction problem as a problem of adding the opposite.

If you see $7 - 10$, think $7 + (-10)$.
If you see $-3 - 19$, think $-3 + (-19)$.
If you see $6 - (-3)$, think $6 + (+3)$.

EXAMPLE 4 Subtract.

(a) $7 - 10$ **(b)** $-3 - 19$ **(c)** $6 - (-3)$

(d) $-4 - (-2)$ **(e)** $-\frac{1}{2} - \left(-\frac{1}{3}\right)$ **(f)** $2.7 - (-5.2)$

(a) $7 - 10 = 7 + (-10) = -3$ **(b)** $-3 - 19 = -3 + (-19) = -22$

(c) $6 - (-3) = 6 + 3 = 9$ **(d)** $-4 - (-2) = -4 + 2 = -2$

(e) $-\frac{1}{2} - \left(-\frac{1}{3}\right) = -\frac{1}{2} + \frac{1}{3} = -\frac{3}{6} + \frac{2}{6} = -\frac{1}{6}$

(f) $2.7 - (-5.2) = 2.7 + 5.2 = 7.9$

Practice Problem 4 Subtract.

(a) $8 - 14$ **(b)** $-6 - 13$ **(c)** $20 - (-5)$

(d) $-9 - (-3)$ **(e)** $-\frac{1}{5} - \left(-\frac{1}{2}\right)$ **(f)** $3.6 - (-5.5)$ ■

2 Solving Problems Involving Both Addition and Subtraction of Signed Numbers

If a problem involves both addition and subtraction, work from left to right and perform each operation in order.

EXAMPLE 5 Perform this set of operations working from left to right.

$-8 - (-3) + (-5) = -8 + 3 + (-5)$ *First we change subtracting a -3 to adding a 3.*

$= -5 + (-5)$ *The sum of $-8 + 3 = -5$.*

$= -10$ *The sum of $-5 + (-5) = -10$.*

Practice Problem 5 Perform the following set of operations.

$$-5 - (-9) + (-14)$$ ■

3 Solving Simple Applied Problems That Involve the Subtraction of Signed Numbers

When we want to find the difference in altitude between two mountains, we subtract. We subtract the lower altitude from the higher altitude. Look at the illustration at the right. The difference in altitude between A and B is 3480 ft − 1260 ft = 2220 ft.

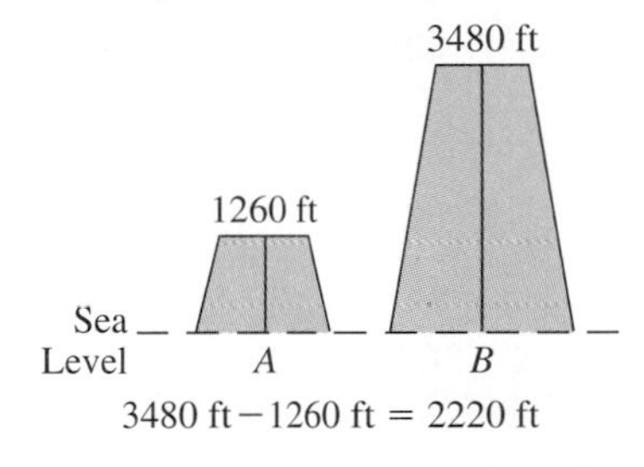

3480 ft − 1260 ft = 2220 ft

Land that is below sea level is considered to have a negative altitude. The Dead Sea is 1286 ft below sea level. Look at the illustration below. The difference in altitude between C and D is 2590 ft − (−1286 ft) = 3876 ft.

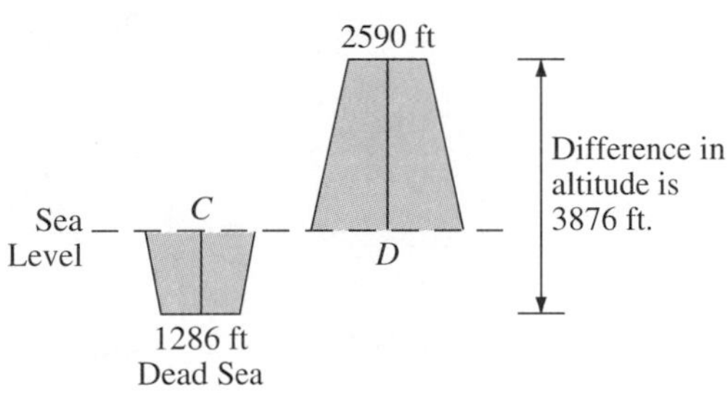

2590 ft − (−1286 ft) = 2590 ft + 1286 ft = 3876 ft

TEACHING TIP Students find this example helpful. During the spring flooding season the Esconti River continued to rise for four days. The following data were recorded:

Day	Height of River (number of feet above normal)
Monday	−4
Tuesday	−2
Wednesday	+5
Thursday	+8

A. What was the difference in height from Monday to Tuesday?

$-2 - (-4) = +2$ ft

B. What was the difference in height from Tuesday to Wednesday?

$+5 - (-2) = +7$ ft

C. What was the difference in height from Wednesday to Thursday?

$+8 - (+5) = +3$ ft

The three subtraction examples here seem to make sense to students because they know the result should be positive in each case, since the river is rising.

EXAMPLE 6 Find the difference in temperature between 38°F in the day in Anchorage, Alaska, and −26°F at night.

We subtract the lower temperature from the higher temperature.

$$38 - (-26) = 38 + 26 = 64$$

The difference is 64°F.

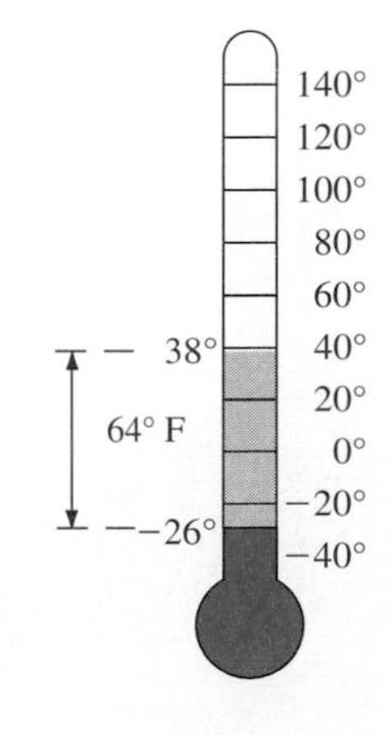

Practice Problem 6 Find the difference in temperature between 31°F in the day in Fairbanks, Alaska, and −37° at night. ■

9.2 Exercises

In each case, subtract signed numbers by adding the opposite of the second number to the first number.

1. $3 - 9$
$3 + (-9) = -6$

2. $5 - 12$
$5 + (-12) = -7$

3. $-5 - 9$
$-5 + (-9) = -14$

4. $-8 - 5$
$-8 + (-5) = -13$

5. $-12 - (-10)$
$-12 + 10 = -2$

6. $-27 - (-12)$
$-27 + 12 = -15$

7. $3 - (-21)$
$3 + 21 = 24$

8. $10 - (-23)$
$10 + 23 = 33$

9. $12 - 30$
$12 + (-30) = -18$

10. $10 - 14$
$10 + (-14) = -4$

11. $-12 - (-15)$
$-12 + 15 = 3$

12. $-17 - (-30)$
$-17 + 30 = 13$

13. $150 - 210$
$150 + (-210) = -60$

14. $500 - 150$
$500 + (-150) = 350$

15. $300 - (-256)$
$300 + 256 = 556$

16. $420 - (-300)$
$420 + 300 = 720$

17. $-58 - 32$
$-58 + (-32) = -90$

18. $-75 - 25$
$-75 + (-25) = -100$

19. $-45 - (-85)$
$-45 + 85 = 40$

20. $-48 - (-60)$
$-48 + 60 = 12$

21. $-2.5 - 4.2$
$(-2.5) + (-4.2) = -6.7$

22. $-4.1 - 3.9$
$-4.1 + (-3.9) = -8$

23. $10.6 - 3.5$
$10.6 + (-3.5) = 7.1$

24. $6.5 - 3.7$
$6.5 + (-3.7) = 2.8$

25. $-10.9 - (-2.3)$
$-10.9 + 2.3 = -8.6$

26. $-6.8 - (-2.9)$
$-6.8 + 2.9 = -3.9$

27. $4.8 - (-2.1)$
$4.8 + 2.1 = 6.9$

28. $6.3 - (-5.8)$
$6.3 + 5.8 = 12.1$

29. $\frac{1}{4} - \left(-\frac{3}{4}\right)$
$\frac{1}{4} + \frac{3}{4} = 1$

30. $\frac{5}{7} - \left(-\frac{6}{7}\right)$
$\frac{5}{7} + \frac{6}{7} = \frac{11}{7}$ *or* $1\frac{4}{7}$

31. $-\frac{5}{6} - \frac{1}{3}$
$-\frac{5}{6} + \left(-\frac{2}{6}\right) = -\frac{7}{6}$ *or* $-1\frac{1}{6}$

32. $-\frac{2}{8} - \frac{1}{4}$
$-\frac{2}{8} + \left(-\frac{2}{8}\right) = -\frac{1}{2}$

33. $-\frac{5}{12} - \left(-\frac{1}{4}\right)$
$-\frac{5}{12} + \frac{3}{12} = -\frac{1}{6}$

34. $-\frac{7}{14} - \left(-\frac{1}{7}\right)$
$-\frac{7}{14} + \frac{2}{14} = -\frac{5}{14}$

35. $\frac{7}{11} - \frac{1}{2}$
$\frac{14}{22} + \left(-\frac{11}{22}\right) = \frac{3}{22}$

36. $\frac{5}{13} - \frac{1}{3}$
$\frac{15}{39} + \left(-\frac{13}{39}\right) = \frac{2}{39}$

Perform each set of operations working from left to right.

37. $2 - (-8) + 5$
$10 + 5 = 15$

38. $7 - (-3) + 9$
$10 + 9 = 19$

39. $-5 - 6 - (-11)$
$-11 + 11 = 0$

40. $-3 - 12 - (-5)$
$-15 + 5 = -10$

41. $7 - (-2) - (-8)$
$9 + 8 = 17$

42. $6 - (-3) - (-5)$
$9 + 5 = 14$

43. $-16 - (-2) - 4$
-18

44. $-12 - (-3) - 5$
-14

45. $9 - 3 - 2 - 6$
$6 - 2 - 6 = 4 - 6 = -2$

46. $12 - 5 - 4 - 8$
$7 - 4 - 8 = 3 - 8 = -5$

47. $-16 + 9 - (-2) - 8$
$-7 + 2 - 8 = -13$

48. $-12 + 8 - (-16) - 9$
$-4 + 16 - 9 = 3$

49. $\frac{1}{4} + \left(-\frac{1}{12}\right) - \left(-\frac{2}{3}\right) - \frac{5}{6} - \frac{1}{2}$
$-\frac{1}{2}$

50. $-\frac{1}{30} + \left(-\frac{2}{3}\right) - \left(-\frac{5}{6}\right) - \frac{7}{10} + \frac{7}{15}$
$-\frac{1}{10}$

Applications

Use your knowledge of signed numbers to answer the following questions.

51. Find the difference in altitude between a mountain 5277 ft high and a canyon 844 ft below sea level.
$5277 - (-844) = 5277 + 844 = 6121$ ft

52. Find the difference in altitude between a mountain 4128 ft high and a gorge 312 ft below sea level.
$4128 - (-312) = 4128 + 312 = 4440$ ft

53. Find the difference in temperature in Alta, Utah, between 23°F during the day and −19°F at night.
$23 - (-19) = 23 + 19 = 42$°F

54. Find the difference in temperature in Fairbanks, Alaska, between 27°F in the day and −33°F at night. $27° - (-33°) = 60$°F

A company's profit and loss statement in dollars for the last five months is shown below.

Month	Profit	Loss
January	18,700	
February		−32,800
March		−6,300
April	43,500	
May		−12,400

55. What is the difference between the loss in March and the loss in February?
$-32{,}800 - (-6300) = -26{,}500$

56. What is the difference between the loss in March and the loss in May?
$-12{,}400 - (-6300) = -6100$

57. What is the difference between the profit in January and the loss in March?
$18{,}700 - (-6300) = 25{,}000$

58. What is the difference between the profit in April and the loss in May?
$43{,}500 - (-12{,}400) = 55{,}900$

To Think About

59. Give an example of how a bank might want to do the calculation $50 - (-80)$ and would accomplish this by adding $50 + 80$.
If the bank wishes to remove an erroneous debit of $80 from a customer's account having a balance of $50, then it would add

Bal.	$50.00
Credit	$80.00
New Bal.	$130.00

60. Give an example of how a bank might want to do the calculation $-100 - 50$ and would actually accomplish this by adding $-100 + (-50)$.

Bal.	0.00
Ck #1	−$100.00
Ck #2	−$50.00
New Bal.	−$150.00

Cumulative Review Problems

Perform in order.

61. $20 \times 2 \div 10 + 4 - 3$
5

62. $5 \times 7 + 6 \times 3 - 8$
45

9.3 Multiplication and Division of Signed Numbers

MathPro Video 19 SSM

After studying this section, you will be able to:

1 *Multiply or divide two signed numbers.*
2 *Multiply three or more signed numbers.*

1 *Multiplying and Dividing Two Signed Numbers*

We are familiar with multiplying two positive numbers. For example, $(2)(8) = 16$ gives us no problem. But how do we handle negative numbers in multiplication? What would a negative number *mean* in multiplication? Well, if you work for 2 hours at 8 dollars an hour, $(2)(8) = 16$ tells you what you have gained. But if you pay someone for working 2 hours at 8 dollars an hour, $(2)(-8) = -16$ tells you that you have \$16 less than what you started with.

TEACHING TIP Explain to students that multiplication of signed numbers and multiplication of other types of expressions in algebra are indicated by the use of the parentheses. They will see this notation in any math courses they take in the future.

Recall the different ways we can indicate multiplication.

$$3 \times 5 \qquad 3 \cdot 5 \qquad (3)(5) \qquad 3(5)$$

It is common to use parentheses to mean multiplication.

EXAMPLE 1 Evaluate.

(a) $(7)(8)$ **(b)** $3(12)$ **(c)** $2(6)(3)$ **(d)** $(4)(8)(2)$

(a) $(7)(8) = 56$ **(b)** $3(12) = 36$

(c) $2(6)(3) = 12(3) = 36$ **(d)** $(4)(8)(2) = 32(2) = 64$

Practice Problem 1 Evaluate.

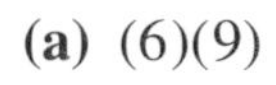

(a) $(6)(9)$ **(b)** $7(12)$ **(c)** $3(5)(8)$ **(d)** $(6)(2)(7)$ ■

Let us look at another real-life application of multiplication of signed numbers. Suppose that water is flowing *into* a tank at the rate of 3 gallons a minute. 2 minutes from *now* there will be 6 gallons more in the tank. That is

$$(2)(3) = 6$$

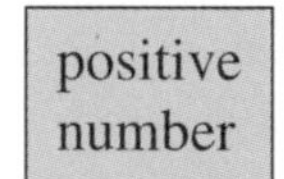

positive number × positive number = positive number

Now suppose that water is flowing *out* of the tank at the rate of 3 gallons per minute. 2 minutes *ago* there were 6 gallons more in the tank than there are now. 3 gallons of water flowing *out* of the tank is represented by the number (-3). 2 minutes *ago* is represented by the number (-2). Thus we have

$$(-2)(-3) = +6$$

negative number × negative number = positive number

These examples suggest the following rule.

> **Multiplication and Division Rule for Two Numbers with the Same Sign**
>
> To multiply or divide two numbers with the same sign, multiply or divide the absolute values. The sign of the result is positive.

EXAMPLE 2 Multiply.

(a) $(-8)(-3)$ **(b)** $-5(-6)$ **(c)** $\left(-\frac{1}{2}\right)\left(-\frac{3}{5}\right)$ **(d)** $(0.6)(1.3)$

In each case, we are multiplying two numbers with the same sign. We will always obtain a positive number.

(a) $(-8)(-3) = 24$ **(b)** $-5(-6) = 30$

(c) $\left(-\frac{1}{2}\right)\left(-\frac{3}{5}\right) = \frac{3}{10}$ **(d)** $(0.6)(1.3) = 0.78$

Practice Problem 2 Multiply.

(a) $(-8)(-8)$ **(b)** $-10(-6)$ **(c)** $\left(-\frac{1}{3}\right)\left(-\frac{2}{7}\right)$ **(d)** $(1.2)(0.4)$ ■

Because division is related to multiplication, we find, just as in multiplication, whenever we divide two numbers with the same sign, the result is a positive number.

EXAMPLE 3 Divide.

(a) $\frac{-20}{-10}$ **(b)** $(-50) \div (-2)$

(c) $(-9.9) \div (-3.0)$ **(d)** $\frac{\frac{2}{3}}{\frac{1}{4}}$

(a) $\frac{-20}{-10} = 2$ **(b)** $(-50) \div (-2) = 25$

(c) $(-9.9) \div (-3.0) = 3.3$ **(d)** $\frac{2}{3} \div \frac{1}{4} = \left(\frac{2}{3}\right)\left(\frac{4}{1}\right) = \frac{8}{3}$ *or* $2\frac{2}{3}$

Practice Problem 3 Divide.

(a) $\frac{-16}{-8}$ **(b)** $-78 \div (-2)$

(c) $(-1.2) \div (-0.5)$ **(d)** $\frac{\frac{3}{5}}{\frac{1}{10}}$ ■

Let's take another look at our real-life example of water flowing into and out of a tank. Suppose that water is flowing into a tank at the rate of 3 gallons a minute. 2 minutes ago there were 6 gallons less in the tank. Remember, we use -2 to represent 2 minutes ago. We have

$$(-2)(3) = -6$$

negative number × positive number = negative number

TEACHING TIP Students often find it helpful to consider a few examples from everyday life that illustrate multiplying a positive number by a negative number to obtain a negative number. Students seem to find the following illustrations helpful.

A. Repeated plays where yardage is lost in a football game. Four plays losing 3 yards each is a loss of 12 yards. $4(-3) = -12$

B. Repeated sales made by a company at a loss. Four sales made losing 9 dollars each is a loss of 36 dollars. $4(-9) = -36$

C. Repeated dives by a submarine going farther and farther beneath the surface. The submarine captain gave 4 commands for the sub to dive 50 feet farther down. The total depth dropped was 200 feet. $4(-50) = -200$. Repeatedly interjecting practical examples of different types tends to convince students that the sign rules are logical and not "artificial."

Now suppose water is flowing out of the tank at the rate of 3 gallons per minute. 2 minutes from now there will be 6 gallons less in the tank. Remember, we use -3 to represent the water flowing out of the tank. We have

$$(2)(-3) = -6$$

positive number	$\times$	negative number	$=$	negative number

These examples suggest the following rule for multiplying numbers when the signs are not the same.

Multiplication Rule for Two Numbers with Different Signs

To multiply two numbers with different signs, multiply the absolute values. The result is negative.

EXAMPLE 4 Multiply.

(a) $2(-8)$ **(b)** $(-3)(25)$ **(c)** $\left(-\frac{1}{3}\right)\left(\frac{2}{5}\right)$ **(d)** $0.6(-2.4)$

In each case, we are multiplying two signed numbers with different signs. We will always get a negative number for an answer.

(a) $2(-8) = -16$ **(b)** $(-3)(25) = -75$

(c) $\left(-\frac{1}{3}\right)\left(\frac{2}{5}\right) = -\frac{2}{15}$ **(d)** $0.6(-2.4) = -1.44$

Practice Problem 4 Multiply.

(a) $(-8)(5)$ **(b)** $3(-60)$ **(c)** $\left(-\frac{1}{5}\right)\left(\frac{3}{7}\right)$ **(d)** $0.5(-6.7)$ ■

A similar rule applies to division of two numbers when the signs are not the same.

TEACHING TIP An interesting example of division of two numbers that are different in sign is the following. Mount Washington recorded the following Fahrenheit temperatures in January for a four-day period: $-30°$, $-38°$, $-40°$, and $-32°$. What was the average temperature over this four-day period? $-30 + (-38) + (-40) + (-32) = -140$. Now -140 divided by 4 yields -35. The average temperature was -35 degrees Fahrenheit.

Division Rule for Two Numbers with Different Signs

To divide two numbers with different signs, divide the absolute values. The result is negative.

EXAMPLE 5 Divide.

(a) $-20 \div 5$ **(b)** $36 \div (-18)$ **(c)** $\frac{-20.8}{2.6}$ **(d)** $\frac{\frac{3}{5}}{-\frac{9}{13}}$

(a) $-20 \div 5 = -4$ **(b)** $36 \div (-18) = -2$ **(c)** $\frac{-20.8}{2.6} = -8$

(d) $\frac{3}{5} \div \left(-\frac{9}{13}\right) = \frac{\overset{1}{\cancel{3}}}{5}\left(-\frac{13}{\underset{3}{\cancel{9}}}\right) = -\frac{13}{15}$

Practice Problem 5 Divide.

(a) $-50 \div 25$ **(b)** $49 \div (-7)$ **(c)** $\frac{21.12}{-3.2}$ **(d)** $\frac{-\frac{2}{7}}{\frac{4}{13}}$ ■

2 *Multiplying Three or More Signed Numbers*

When multiplying more than two numbers, multiply any two numbers first, then multiply the result by another number. Continue until each factor has been used.

TEACHING TIP Remind students that multiplication is commutative. Thus to evaluate an expression like $12(-3)(5)(-\frac{1}{4})$, they may prefer—although it is not necessary—to write the problem in the form $12(-\frac{1}{4})(-3)(5)$ in order to facilitate cancellation.

EXAMPLE 6 Multiply. $5(-2)(-3)$

$5(-2)(-3) = -10(-3)$ First multiply $5(-2) = -10$.

$= 30$ Then multiply $-10(-3) = 30$.

Practice Problem 6 Multiply. $(-6)(3)(-4)$ ■

EXAMPLE 7 Multiply. $6\left(-\frac{1}{2}\right)(-4)(-2)$

First we note that in multiplying the first two numbers the number 6 and the denominator of the fraction have a common factor, 2. We divide out the common factor.

$$\frac{6}{1}\left(-\frac{1}{2}\right) = \frac{\overset{3}{\cancel{6}}}{1}\left(-\frac{1}{\underset{1}{\cancel{2}}}\right) = -\frac{3}{1} = -3$$

Then we continue the multiplication.

$$(-3)(-4)(-2) = 12(-2) = -24$$

TEACHING TIP Students usually enjoy finding the pattern of signs if you ask them to do the following four problems and determine a pattern.

A. $(-3)(-2)(-4)(-6) = 144$
B. $(-3)(-2)(-4)(6) = -144$
C. $(-3)(-2)(4)(6) = 144$
D. $(-3)(2)(4)(6) = -144$

Usually, they will come up with the rule that if there is an odd number of negative factors, the answer is negative, but if there is an even number of negative factors, the answer is positive.

Practice Problem 7 Multiply. $4(-8)\left(-\frac{1}{4}\right)(-3)$ ■

EXAMPLE 8 A phosphate ion has an electrical charge of -3. If 10 phosphate ions are removed from a substance, what is the change in the charge of the remaining substance?

Removing 10 ions from a substance can be represented by the number -10. We can use multiplication to determine the result of removing 10 ions with an electrical charge of -3.

$$(-3)(-10) = 30$$

Thus the change in charge would be $+30$.

Practice Problem 8 An oxide ion has an electrical charge of -2. If 6 oxide ions are added to a substance, what is the change in the charge of the new substance? ■

9.3 Exercises

Multiply.

1. $9(-3)$
-27

2. $7(-5)$
-35

3. $(-6)(-4)$
24

4. $(-3)(-10)$
30

5. $(-7)(9)$
-63

6. $(-9)(6)$
-54

7. $(3)(9)$
27

8. $(6)(5)$
30

9. $(-20)(-3)$
60

10. $(-30)(-6)$
180

11. $(-20)(8)$
-160

12. $(-15)(12)$
-180

13. $(2.5)(-0.6)$
-1.5

14. $(8.5)(-0.3)$
-2.55

15. $(-1.5)(-2.5)$
3.75

16. $(-4.0)(-3.8)$
15.2

17. $\left(-\frac{2}{5}\right)\left(\frac{3}{7}\right)$
$-\frac{6}{35}$

18. $\left(-\frac{5}{12}\right)\left(-\frac{2}{3}\right)$
$\frac{5}{18}$

19. $\left(-\frac{4}{12}\right)\left(-\frac{3}{23}\right)$
$\frac{1}{23}$

20. $\left(-\frac{7}{9}\right)\left(\frac{3}{28}\right)$
$-\frac{1}{12}$

Divide.

21. $60 \div (-6)$
-10

22. $64 \div (-8)$
-8

23. $121 \div 11$
11

24. $169 \div 13$
13

25. $-70 \div (-5)$
14

26. $-60 \div (-3)$
20

27. $-16 \div 8$
-2

28. $-63 \div 7$
-9

29. $\frac{48}{-6}$
-8

30. $\frac{52}{-13}$
-4

31. $\frac{-72}{-8}$
9

32. $\frac{-56}{-7}$
8

33. $\frac{1}{2} \div \left(-\frac{3}{5}\right)$
$-\frac{5}{6}$

34. $-\frac{2}{7} \div \frac{3}{4}$
$-\frac{8}{21}$

35. $\frac{-\frac{4}{5}}{-\frac{7}{10}}$
$1\frac{1}{7}$ *or* $\frac{8}{7}$

36. $\frac{-\frac{26}{15}}{-\frac{13}{7}}$
$\frac{14}{15}$

37. $49.2 \div (-6)$
-8.2

38. $30.4 \div (-4)$
-7.6

39. $\frac{-55.8}{-9}$
6.2

40. $\frac{-27.3}{-7}$
3.9

Multiply.

41. $3(-6)(-4)$
$(-18)(-4) = 72$

42. $3(-5)(-9)$
$(-15)(-9) = 135$

43. $(-8)(4)(-6)$
$(-32)(-6) = 192$

44. $(-7)(-2)(-3)$
$14(-3) = -42$

45. $2(-8)(3)\left(-\frac{1}{3}\right)$
$(-48)\left(-\frac{1}{3}\right) = 16$

46. $7(-2)(-5)\left(\frac{1}{7}\right)$
$70\left(\frac{1}{7}\right) = 10$

47. $(-5)(2)(-1)(-3)$
$10(-3) = -30$

48. $(-8)(-2)(3)(-1)$
$48(-1) = -48$

49. $8(-3)(-5)(0)(-2)$
0

50. $9(-6)(-4)(-3)(0)$
0

Perform each operation. Simplify your answer.

★**51.** $\left(-\frac{2}{3}\right)\left(-\frac{3}{4}\right)\left(-\frac{5}{6}\right)\left(-\frac{7}{8}\right)\left(-\frac{9}{10}\right)$

$-\frac{21}{64}$

★**52.** $\left(-\frac{2}{3}\right) \div \left(-\frac{2}{3}\right)\left(\frac{3}{5}\right) \div \left(\frac{3}{10}\right)\left(-\frac{1}{2}\right)$

-1

Calculate.

53. $(2.76)(-3.21)(1.09)$

-9.656964

54. $(-0.024)(0.001)(-0.61)$

0.00001464

55. $(-3288) \div (-0.213)$

$15{,}436.61972$

56. $(-62.8712) \div (0.0763)$

-824

Applications

Ions are atoms or groups of atoms with positive or negative electrical charges. The charges of some ions are given below.

aluminum	+3	chloride	−1	magnesium	+2
oxide	−2	phosphate	−3	silver	+1

Find the total charge.

57. 17 magnesium ions $17(2) = 34$

58. 8 phosphate ions $8(-3) = -24$

59. 4 silver ions and 2 chloride ions

$4(1) + 2(-1) = 2$

60. 3 oxide ions and 9 chloride ions

$3(-2) + 9(-1) = -15$

61. 5 magnesium, 6 oxide, and 12 aluminum ions

$5(2) + 6(-2) + 12(3) = 34$

62. 7 aluminum, 4 chloride, and 9 oxide ions

$7(3) + 4(-1) + 9(-2) = -1$

63. Eight chloride ions are removed from a substance. What is the change in the charge of the remaining substance? +8

64. Six oxide ions are removed from a substance. What is the change in the charge of the remaining substance? +12

To Think About

65. Bob multiplied −12 times a mystery number that was not negative and he did not get a negative answer. What was the mystery number? 0

66. Fred divided a mystery number that was not negative by −3 and did not get a negative answer. What was the mystery number? 0

In problems 67 and 68, the letters a, b, c, and d represent numbers that may be either positive or negative. The first three numbers are multiplied to obtain the fourth number. This can be stated by the equation $(a)(b)(c) = d$.

67. In the case where a, c, and d are all negative numbers, is the number b positive or negative?

b is negative

68. In the case where a is positive but b and d are negative, is the number c positive or negative?

c is positive

Cumulative Review Problems

69. Find the area of a parallelogram with height of 6 in. and base of 15 in. 90 in.^{2}

70. Find the area of a trapezoid with height of 12 m and bases of 18 m and 26 m, respectively. 264 m^2

9.4 Order of Operations with Signed Numbers

MathPro Video 20 SSM

After studying this section, you will be able to:

1 *Calculate with signed numbers using more than one operation.*

1 *Calculating with Signed Numbers Using More Than One Operation*

If we have several computations to do, such as

$$(-48) - (-8) + 20 - (-10)$$

we will become confused unless we perform the operations in the proper order. The order of operations with signed numbers is similar to the order of operations with whole numbers, which we discussed in Section 1.7. If there are no grouping symbols or exponents, we need to remember that

> **1.** Multiplication and division are done first, in whichever order these operations first present themselves, working from left to right.
>
> **2.** Then addition and subtraction are done from left to right.

EXAMPLE 1 Perform the indicated operations in the proper order.

$$-6 \div (-2)(5)$$

Multiplication and division are of equal priority. So we work from left to right and divide first.

$$\underbrace{-6 \div (-2)}_{3}(5)$$
$$3 \quad (5) = 15$$

Practice Problem 1 Perform the indicated operations in the proper order.

$$20 \div (-5)(-3) \blacksquare$$

TEACHING TIP Remind students that keeping extra parentheses in the problem after the first step of operations, as in Example 2, will usually increase their accuracy as they do problems of this type.

EXAMPLE 2 Perform the indicated operations in the proper order.

(a) $7 + 6(-2)$ **(b)** $9 \div 3 - 16 \div (-2)$ **(c)** $4 - 2(7) - 5$

(a) Multiplication and division must be done first. We begin with $6(-2)$.

$$7 + \underbrace{6(-2)}$$
$$7 + (-12) = -5$$

TEACHING TIP Encourage students to show the step of changing $3 - (-8)$ to $3 + 8$ when working out the problem.

(b) There is no multiplication, but there is division, and we do that first.

$$\underbrace{9 \div 3}_{3} - \underbrace{16 \div (-2)}_{(-8)}$$
$$3 - (-8) = 3 + 8 \quad \textit{Transform subtraction to adding the opposite.}$$
$$= 11$$

(c) $4 - \underbrace{2(7)} - 5$

$$4 - 14 - 5 = 4 + (-14) + (-5) \quad \textit{Transform subtraction to adding the opposite.}$$
$$= -15$$

Practice Problem 2 Perform the indicated operations.

(a) $25 \div (-5) + 16 \div (-8)$ **(b)** $9 + 20 \div (-4)$ **(c)** $13 + 7(-6) - 8$ ■

> If a fraction has operations written in the numerator or in the denominator or both, these operations must be done first. Then the fraction may be simplified or the division carried out.

EXAMPLE 3 Perform the indicated operation in the proper order.

(a) $\dfrac{5+6-2}{-8+3-7}$ **(b)** $\dfrac{7(-2)+4}{8 \div (-2)(5)}$

(a) Begin by evaluating the numerator and then the denominator.

$$\frac{5+6-2}{-8+3-7} = \frac{9}{-12}$$ *Simplify the fraction.*

$$= -\frac{3}{4}$$ *We place the negative sign in front of the fraction in the answer.*

(b)

$$\frac{7(-2)+4}{8 \div (-2)(5)} = \frac{-14+4}{(-4)(5)}$$ *We perform the multiplication and division first in the numerator and the denominator respectively.*

$$= \frac{-10}{-20}$$ *Simplify the fraction.*

$$= \frac{1}{2}$$

Note that the answer is positive since a negative number divided by a negative number gives a positive result.

TEACHING TIP Emphasize to students that problems like Example 3 can have the answer written three ways. The negative sign can actually be written before the fraction, before the 3, or before the 4. You may want to mention, however, that it is generally preferred not to have the negative sign in the denominator of the fraction.

Practice Problem 3 Perform the indicated operations.

(a) $\dfrac{12-2-3}{21+3-10}$ **(b)** $\dfrac{9(-3)-5}{2(-4) \div (-2)}$ ■

EXAMPLE 4 The Fahrenheit temperature for the last five days of January was $-18°$, $-13°$, $-3°$, $9°$, $-15°$. What was the average temperature?

First we add the temperatures, then divide by 5.

$$\frac{-18+(-13)+(-3)+9+(-15)}{5} = \frac{-40}{5} = -8$$

The average temperature was $-8°$F.

Practice Problem 4 A company had profit/loss reports as follows: January $-\$10{,}000$, February $+\$30{,}000$, March $-\$18{,}000$, April $-\$6000$. Find the average monthly profit or loss for this time period. ■

9.4 Exercises

Perform the operations in the proper order.

1. $8 + (-3) + (-6)$
-1

2. $9 + (-2) + (-5)$
2

3. $6 + 5(-4)$
$6 + (-20) = -14$

4. $-34 + 4(3)$
$-34 + 12 = -22$

5. $16 + 32 \div (-4)$
$16 + (-8) = 8$

6. $15 - (-18) \div 3$
$15 - (-6) = 21$

7. $24 \div (-3) + 16 \div (-4)$
$-8 + (-4) = -12$

8. $(-56) \div (-7) + 30 \div (-15)$
$8 + (-2) = 6$

9. $3(-4) + 5(-2) - (-3)$
$-12 + (-10) + 3 = -19$

10. $6(-3) + 8(-1) - (-2)$
$-18 + (-8) + 2 = -24$

11. $-18 \div 2 + 9$
$-9 + 9 = 0$

12. $-20 \div 2 + 10$
$-10 + 10 = 0$

13. $5 - 30 \div 3$
$5 - 10 = -5$

14. $3 - 12 \div 4$
$3 - 3 = 0$

15. $36 \div 12(-2)$
$3(-2) = -6$

16. $20 \div 4(8)$
$5(8) = 40$

17. $5(5) - 5$
$25 - 5 = 20$

18. $9(-6) + 6$
$-54 + 6 = -48$

19. $3(-4) + 6(-2) - 3$
$-12 + (-12) - 3 = -27$

20. $-6(7) + 8(-3) + 5$
$-42 + (-24) + 5 = -61$

21. $8(-7) - 2(-3)$
$-56 + 6 = -50$

22. $10(-5) - 4(12)$
$-50 + (-48) = -98$

23. $16 - 4(8) + 18 \div (-9)$
$16 - 32 + (-2) = -18$

24. $20 - 3(-2) + (-20) \div (-5)$
$20 + 6 + 4 = 30$

25. $16 \div (-2)(3) + 5 - 6$
$-8(3) + 5 - 6 = -24 + 5 - 6 = -25$

26. $-60 \div (-3)(4) - 8 + 4$
$20(4) - 8 + 4 = 80 + (-8) + 4 = 76$

In problems 27–36 simplify the numerator and denominator first, using the proper order of operations. Then reduce the fraction if possible.

27. $\dfrac{8 + 6 - 12}{3 - 6 + 5}$
$\dfrac{2}{2} = 1$

28. $\dfrac{10 - 5 - 7}{8 + 3 - 9}$
$\dfrac{-2}{2} = -1$

29. $\dfrac{6(-2) + 4}{6 - 3 - 5}$
$\dfrac{-12 + 4}{-2} = 4$

30. $\dfrac{8 + 2 - 6}{3(-5) + 13}$
$\dfrac{4}{-15 + 13} = -2$

31. $\dfrac{-16 \div (-2)}{3(-4) - 4}$
$\dfrac{8}{-12 - 4} = -\dfrac{1}{2}$

32. $\dfrac{2(-3) - 5 + 1}{10 \div (-5)}$
$\dfrac{-6 - 5 + 1}{-2} = 5$

33. $\dfrac{4 - (-3) - 5}{20 \div (-10)}$
$\dfrac{2}{-2} = -1$

34. $\dfrac{10 + (-3) - (-5)}{32 \div (-8)}$
$\dfrac{12}{-4} = -3$

35. $\dfrac{7 - (-1)}{9 - 9 \div (-3)}$
$\dfrac{8}{9 + 3} = \dfrac{2}{3}$

36. $\dfrac{4 - (-3) + 1}{2 - 2 \div (-2)}$
$\dfrac{8}{3}$ or $2\dfrac{2}{3}$

Perform the operations in the proper order.

37. $\dfrac{1}{2} \div \left(-\dfrac{2}{3}\right)\left(-\dfrac{3}{7}\right) + \left(-\dfrac{5}{14}\right)$
$\dfrac{1}{2}\left(-\dfrac{3}{2}\right)\left(-\dfrac{3}{7}\right) + \left(-\dfrac{5}{14}\right) = \dfrac{9}{28} + \left(-\dfrac{10}{28}\right) = -\dfrac{1}{28}$

38. $(0.3)(-2.9)(-3.5) + (50.6) \div (-2.0)$
$(-0.87)(-3.5) + (-25.3) = (3.045) + (-25.3) = -22.255$

Applications

The following chart gives the average monthly temperatures in Fahrenheit for two cities in Alaska.

	Jan.	Feb.	Mar.	Apr.	May	June	July	Aug.	Sept.	Oct.	Nov.	Dec.
Barrow	−14	−20	−16	−2	19	33	39	38	31	14	−1	−13
Fairbanks	−13	−4	9	30	48	59	62	57	45	25	4	−10

Use the chart to solve each problem. Round your answer to the nearest tenth of a degree.

39. What is the average temperature in Barrow during the three months of December, January, and February?

$\frac{-13 + (-14) + (-20)}{3} = -15.7°\text{F}$

40. What is the average temperature in Fairbanks during the three winter months of January, February, and March?

$\frac{-13 + (-4) + 9}{3} = -2.7°\text{F}$

41. What is the average temperature in Barrow from October through March?

$\frac{14 + (-1) + (-13) + (-14) + (-20) + (-16)}{6} = -8.3°\text{F}$

42. What is the average temperature in Fairbanks from September through March?

$\frac{45 + 25 + 4 + (-10) + (-13) + (-4) + 9}{7} = 8°\text{F}$

★ **43.** What is the average yearly temperature in Barrow?

$\frac{108}{12} = 9°\text{F}$

★ **44.** What is the average yearly temperature in Fairbanks?

$\frac{312}{12} = 26°\text{F}$

Cumulative Review Problems

45. A telephone wire that is 3840 m long would be how long measured in kilometers? 3.84 km

46. A container with 36.8 g of protein would be equivalent to how many milligrams of protein? 36,800 mg

47. Find the area of a trapezoid with the height of 14 in. and bases of 23 in. and 37 in., respectively. 420 in.^{2}

48. Find the area of a parallelogram with a height of 15 m and a base of 22 m. 330 m^2

Putting Your Skills to Work

MATHEMATICS OF UNIVERSAL TIME

The basis of time for almost all locations in the world is considered the mean solar time for the meridian that passes through Greenwich, England. This time is called *Universal Time* or *Greenwich Time*. To determine the time in another location in the world a time chart similar to the one at the bottom of this page is used. By adding the appropriate signed number the new time can be determined.

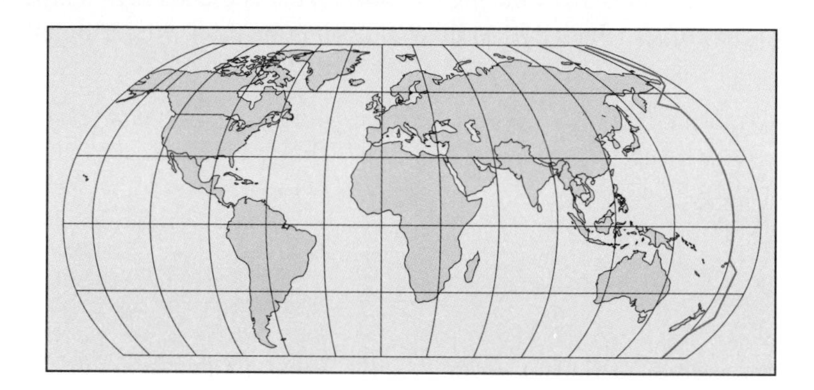

For example if it is 3 P.M. in London to find the time in Beijing, China, we merely add: $3 + (+8) = 11$. Thus at that hour, it is 11 P.M. in Beijing. Of course adjustments have to be made for changing from A.M. to P.M. and P.M. to A.M. If it is 4 P.M. in London, to find the time in Denver, Colorado, we add: $4 + (-7) = -3$. This is 3 hours earlier than noon, so the time in Denver would be 9 A.M.

Location	Difference from Universal Time	Location	Difference from Universal Time
Algiers, Algeria	+1	Karachi, Pakistan	+5
Bangkok, Thailand	+7	London, England	0
Beijing, China	+8	Melbourne, Australia	+10
Bombay, India	+5.5	Mexico City, Mexico	−6
Boston, MA, USA	−5	Montevideo, Uruguay	−3
Bucharest, Romania	+2	Moscow, Russia	+3
Darwin, Australia	+8.5	Netherland Antilles	−4
Denver, CO, USA	−7	Saint John's, Newfoundland	−3.5
Dubai, United Arab Emirates	+4	San Diego, CA, USA	−8
Fiji Islands	+12	Seoul, Korea	+9
French Polynesia	−10	Taipei, Taiwan	+8
Kabul, Afghanistan	+4.5		

PROBLEMS FOR INDIVIDUAL INVESTIGATION AND ANALYSIS

1. It is 1 P.M. in London. What time is it in Seoul, Korea? 10 P.M.

2. If it is 9 P.M. in London, what time is it in the Netherland Antilles? 5 P.M.

3. If it is 6 A.M. in Taipei, Taiwan, what time is it in Moscow, Russia? 1 A.M.

4. If it is 11 A.M. in San Diego, California, what time is it in the Fiji Islands? At that moment in time, is it the same day in both locations?
No. It is 7 A.M. the next day in the Fiji Islands

PROBLEMS FOR GROUP INVESTIGATION AND COOPERATIVE LEARNING

Together with other members of your class see if you can determine the following.

5. If you fly from Bangkok, Thailand, to Bucharest, Romania, and it is 7:30 A.M. when you leave for an airplane flight that lasts 6 hours, what time would you expect to land? 8:30 A.M. in Bucharest

6. You leave Darwin, Australia, and take an airplane flight to Karachi, Pakistan. The flight lasts 5 hours. You left Darwin at 10:45 A.M. What time would you expect to land? 12:15 P.M. in Karachi

7. You are staying in Saint John's, Newfoundland. You wish to place a call to the U.S. Embassy at Dubai, United Arab Emirates. It is 11:00 P.M. in Saint John's. You know you need to call the embassy between 9:00 A.M. and 4:00 P.M. local time. Is it appropriate to make the call at this time? If not, how long do you have to wait (how many more hours will it be) until you can make the call?
No. You will need to wait $2\frac{1}{2}$ hours.

8. Dr. and Mrs. Martinez took a plane flight from Mexico City that left at 6:15 A.M. It was supposed to arrive in Bombay, India (with one stop for fuel), 17 hours later. They actually arrived in Bombay at 2:00 P.M. local time the next day. How many hours late was the plane?
The plane was $3\frac{1}{4}$ hours late

INTERNET CONNECTIONS: Go to ''http://www.prenhall.com/tobey'' to be connected
Site: World Time Zones (U.S. Naval Observatory) or a related site

Use the information at this site to answer the following questions. Assume that Daylight Savings Time is not in effect anywhere.

9. If it is 3 P.M. in Aruba, what time is it in Bangladesh?

10. If it is 5 A.M. on Tuesday in Japan, what time and day is it in Jamaica?

11. If it is 3 P.M. on Friday in French Polynesia, what time and day is it in France?

12. Alan lives in Hawaii. He wants to call his friend in Pakistan when it is between 8 A.M. and 4 P.M. on Monday in Pakistan. What local time and day should he call?

9.5 Scientific Notation

MathPro

Video 21

SSM

After studying this section, you will be able to:

1 *Change numbers in standard notation to scientific notation.*
2 *Change numbers in scientific notation to standard notation.*
3 *Add or subtract numbers in scientific notation.*

1 *Changing Numbers in Standard Notation to Scientific Notation*

Scientists who frequently work with very large or very small measurements use a certain way to write numbers, called **scientific notation.** Our usual way of writing a number, which we call "standard notation" or "ordinary form," expresses the distance to the nearest star, Proxima Centauri as 24,800,000,000,000 miles. In scientific notation, we more conveniently write this as

$$2.48 \times 10^{13} \text{ miles}$$

For a very small number, like two millionths, the standard notation is

$$0.000002$$

The same quantity in scientific notation is

$$2 \times 10^{-6}$$

Notice that each number in scientific notation has two parts: (1) a number that is 1 or greater but less than 10, which is multiplied by (2) a power of 10. That power is called an *integer.* In this section we learn how to go back and forth between standard and scientific notation.

> A positive number is in scientific notation if it is in the form $a \times 10^n$, where a is a number greater than (or equal to) 1 and less than 10, and n is an integer.

We begin our investigation of scientific notation by looking at the large numbers. We recall from Section 1.6 that

$$10 = 10^1$$

$$100 = 10^2$$

$$1000 = 10^3$$

The number of zeros tells us the number for the exponent. To write a number in scientific notation, we want the first number to be greater than 1 and less than 10, and the second number to be a power of 10.

Let's see how this can be done. Consider the following.

$$6700 = \underbrace{6.7}_{\substack{\textit{greater than 1} \\ \textit{and less than 10}}} \times 1000 = 6.7 \times \underset{\substack{\uparrow \\ \textit{a power} \\ \textit{of 10}}}{10^3}$$

Let us look at two more cases.

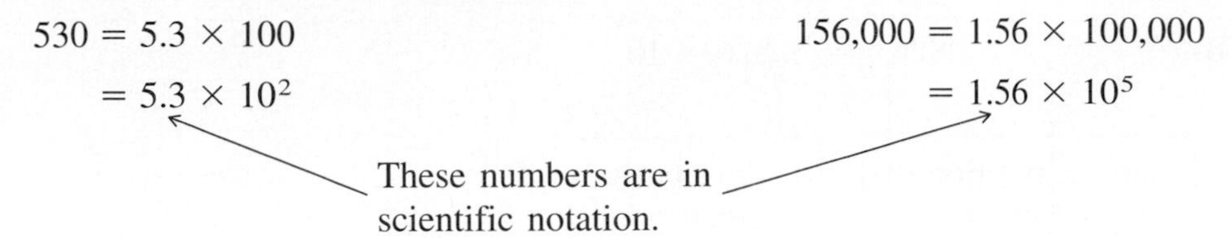

530 = 5.3 × 100
= 5.3 × 10²
156,000 = 1.56 × 100,000
= 1.56 × 10⁵
These numbers are in scientific notation.

Now that we have seen some examples of numbers in scientific notation, let us think through the steps in the next example.

EXAMPLE 1 Write 156 in scientific notation.

We first change the given number to a number greater than or equal to 1 and less than 10. To determine the power of 10, we look at the shift of the decimal point.

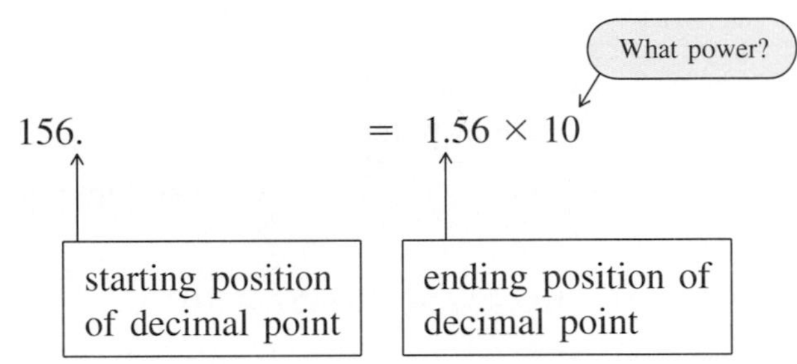

The decimal point moves 2 places to the left. We therefore put 2 for the power of 10.

$$156 = 1.56 \times 10^2$$

Practice Problem 1 Write 896 in scientific notation. ■

It is important to remember that all numbers *greater than* 10 always have a positive exponent when expressed in scientific notation.

EXAMPLE 2 Write in scientific notation.

(a) 57 **(b)** 9826 **(c)** 163,457

(a)

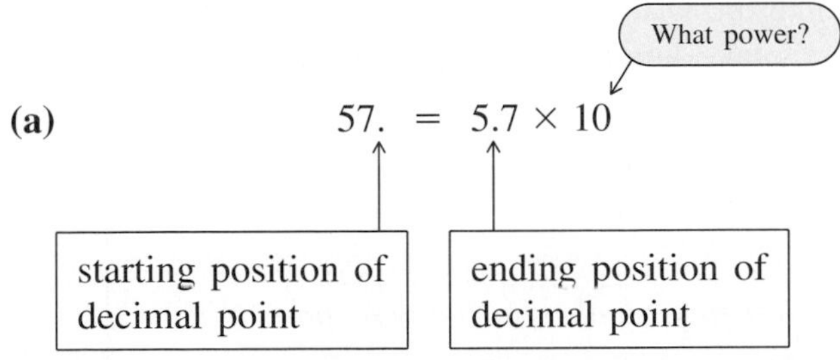

The decimal point moves 1 place to the left. We therefore put 1 for the power of 10.

$$57 = 5.7 \times 10^1$$

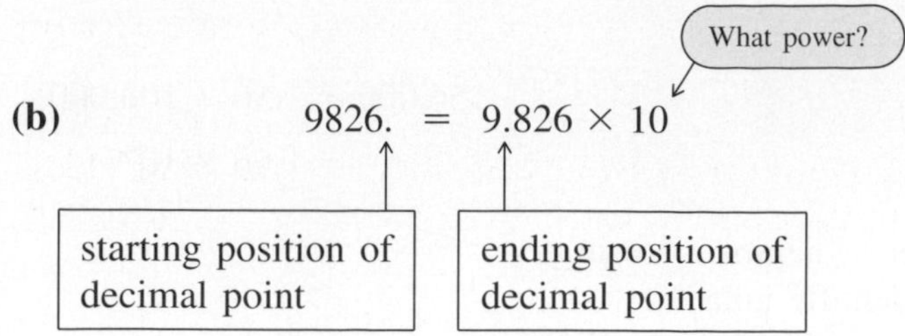

TEACHING TIP To demonstrate the usefulness of scientific notation, tell students that scientists and astronomers often refer to a measurement of one solar mass. This is 1.99×10^{30} kilograms. Ask students to try writing that number out without using scientific notation!

The decimal point moves 3 places to the left. We therefore put 3 for the power of 10.

$$9826 = 9.826 \times 10^3$$

What power?

(c) 163,457. = 1.63457 × 10

starting position of decimal point | ending position of decimal point

The decimal point moves 5 places to the left. We therefore put 5 for the power of 10.

$$163{,}457 = 1.63457 \times 10^5$$

SCIENTIFIC CALCULATOR

Standard to Scientific Notation

Many calculators have a setting that will display numbers in scientific notation. Consult your manual on how to do. To convert 154.32 into scientific notation first change your setting to display in scientific notation. Often SCI is displayed on the calculator. Then enter:

154.32 [=]

Display:

[1.5432 02]

1.5432 02 means 1.5432×10^2.

Note that your calculator display may show the power of 10 in a different manner. Be sure to change your setting back to the regular display when you are done.

Practice Problem 2 Write in scientific notation.

(a) 48 **(b)** 3729 **(c)** 506,936 ■

When changing to scientific notation, all zeros to the *right* of the final nonzero digit may be eliminated. This does not change the value of your answer.

EXAMPLE 3 Write in scientific notation.

(a) 3100 **(b)** 700 **(c)** 4,500,000

(a) $3100 = 3.100 \times 10^3 = 3.1 \times 10^3$ **(b)** $700 = 7.00 \times 10^2 = 7 \times 10^2$
(c) $4{,}500{,}000 = 4.500000 \times 10^6 = 4.5 \times 10^6$

Practice Problem 3 Write in scientific notation.

(a) 4600 **(b)** 900 **(c)** 3,800,000 ■

Now we will look at numbers that are *less than* 1. In the introduction to this section we saw that 0.000002 can be written as 2×10^{-6}. How is it possible to have a negative exponent? Let's take another look at the powers of 10.

$10^3 = 1000$

$10^2 = 100$

$10^1 = 10$

$10^0 = 1$ *Recall that any number to the zero power is* 1.

Now, following this pattern, what would you expect the next number on the left to be? What would be the next number on the right? Each number on the right is one-tenth the number above it. We continue the pattern.

$10^{-1} = 0.1$

$10^{-2} = 0.01$

$10^{-3} = 0.001$

Thus you can see how it is possible to have negative exponents. We use negative exponents to write numbers in scientific notation that are less than 1.

Let's look at 0.76. This number is less than 1. Recall that in our definition for scientific notation the first number must be greater than or equal to 1 and less than 1. To change 0.76 to such a number, we will have to move the decimal point.

$$0.76 = 7.6 \times 10^{-1}$$

The decimal point moves 1 place to the right. We therefore put a -1 for the power of 10. We use a negative exponent because the original number is less than 1. Let's look at two more cases.

$0.0025 = 2.5 \times 10^{-3}$ $\qquad$ $0.00088 = 8.8 \times 10^{-4}$

These numbers are in scientific notation.

When we start with a number that is less than 1 and write it in scientific notation, we will get a result with 10 to a negative power. Think carefully through the steps of the following example.

TEACHING TIP Whether changing from, or to, scientific notation, it is important to remember that all numbers less than 1 always have a negative exponent when expressed in scientific notation.

EXAMPLE 4 Write in scientific notation.

(a) 0.036 **(b)** 0.72 **(c)** 0.000589 **(d)** 0.008

(a) We change the given number to a number greater than or equal to 1 and less than 10. Thus we change 0.036 to 3.6.

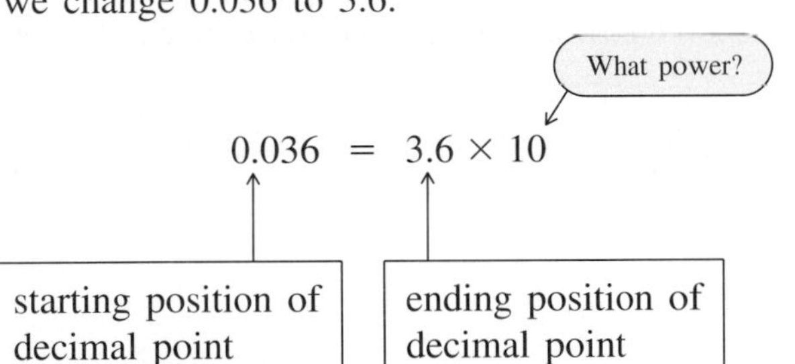

The decimal point moved 2 places to the right. Since the original number is less than 1, we use -2 for the power of 10.

$$0.036 = 3.6 \times 10^{-2}$$

(b)

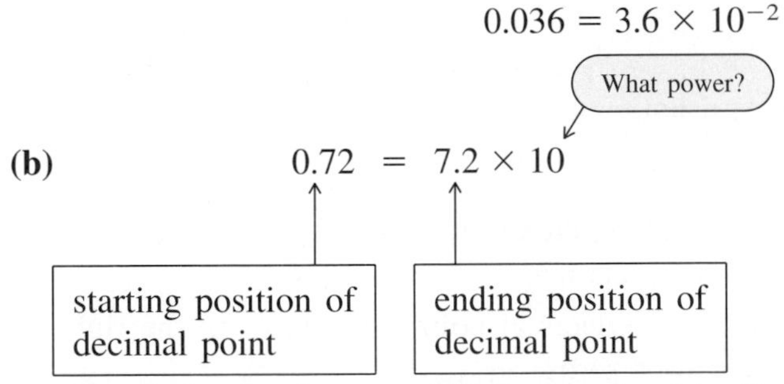

The decimal point moved 1 place to the right. Because the original number is less than 1, we use -1 for the power of 10.

$$0.72 = 7.2 \times 10^{-1}$$

Similarly, we find that

(c) $0.000589 = 5.89 \times 10^{-4}$ $\quad$ Move the decimal point four places to the right.

(d) $0.008 = 8 \times 10^{-3}$ $\quad$ Move the decimal point three places to the right.

Practice Problem 4 Write in scientific notation.

(a) 0.076 **(b)** 0.982 **(c)** 0.000312 **(d)** 0.006 ■

It is very helpful when writing numbers in scientific notation to remember these two concepts.

1. A number that is larger than 10 will always have a positive exponent as the power of 10 when it is written in scientific notation.
2. A number that is smaller than 1 will always have a negative exponent as a power of 10 when it is written in scientific notation.

2 *Changing Numbers in Scientific Notation to Standard Notation*

Often we are given a number in scientific notation and want to write it in standard notation—that is, we want to write it in what we consider ''ordinary form.'' To do this, we reverse the process we've just used. If the number in scientific notation has a positive power of 10, we know the number is greater than 10. Therefore, we move the decimal point to the right to convert to standard notation.

EXAMPLE 5 Write in standard notation.

(a) 8.32×10^2 **(b)** 5.8671×10^4

(a) $8.32 \times 10^2 = 832$ Move the decimal point two places to the right.

(b) $5.8671 \times 10^4 = 58{,}671$ Move the decimal point four places to the right.

Practice Problem 5 Write in standard notation.

(a) 3.08×10^2 **(b)** 6.543×10^3 ■

In some cases, we will need to add zeros as we move the decimal point to the right.

EXAMPLE 6 Write in standard notation.

(a) 5.6×10^2 **(b)** 9.8×10^5 **(c)** 3×10^3

(a) $5.6 \times 10^2 = 560$ Move the decimal point two places to the right. Add one zero.

(b) $9.8 \times 10^5 = 980{,}000$ Move the decimal point five places to the right. Add four zeros.

(c) $3 \times 10^3 = 3000$ Move the decimal point three places to the right. Add three zeros.

Practice Problem 6 Write in standard notation.

(a) 8.5×10^2 **(b)** 4.3×10^5 **(c)** 6×10^4 ■

If the number in scientific notation has a negative power of 10, we know the number is less than 1. Therefore, we move the decimal point to the left to convert to standard notation. In some cases, we need to add zeros as we move the decimal point to the left.

EXAMPLE 7 Write in standard notation.

(a) 7.16×10^{-1} **(b)** 2.48×10^{-3} **(c)** 1.2×10^{-4}

(a) $7.16 \times 10^{-1} = 0.716$ Move the decimal point one place to the left.

(b) $2.48 \times 10^{-3} = 0.00248$ Move the decimal point three places to the left. Add two zeros between the decimal point and 2.

(c) $1.2 \times 10^{-4} = 0.00012$ Move the decimal point four places to the left. Add three zeros between the decimal point and 1.

Practice Problem 7 Write in standard notation.

(a) 8.56×10^{-1} **(b)** 7.72×10^{-3} **(c)** 2.6×10^{-5} ■

3 *Adding or Subtracting Numbers in Scientific Notation*

Scientists calculate numbers in scientific notation when they study galaxies (the huge) and when they study microbes (the tiny). To add or subtract numbers in scientific notation, the numbers must have the same power of 10.

> Numbers in scientific notation may be added or subtracted if they have the same power of 10. We add or subtract the decimal part and leave the power of 10 unchanged.

EXAMPLE 8

(a) Add. 5.89×10^{20} miles + 3.04×10^{20} miles

(b) Subtract. 9.63×10^{17} pounds − 2.98×10^{17} pounds

(a)
$$\begin{array}{r} 5.89 \times 10^{20} \text{ miles} \\ +\ 3.04 \times 10^{20} \text{ miles} \\ \hline 8.93 \times 10^{20} \text{ miles} \end{array}$$

(b)
$$\begin{array}{r} 9.63 \times 10^{17} \text{ pounds} \\ -\ 2.98 \times 10^{17} \text{ pounds} \\ \hline 6.65 \times 10^{17} \text{ pounds} \end{array}$$

TEACHING TIP This concept of adding numbers in scientific notation is very important. Students often miss the fact that each number must have the same power of 10 before the addition can be carried out.

Practice Problem 8

(a) Add. 6.85×10^{22} kilograms + 2.09×10^{22} kilograms

(b) Subtract. 8.04×10^{30} tons − 6.98×10^{30} tons ■

EXAMPLE 9 Add. $7.2 \times 10^{6} + 5.2 \times 10^{5}$

$$5.2 \times 10^{5} = 520{,}000$$

But 520,000 can be written as 0.52×10^{6}. Now we can add.

$$\begin{array}{r} 7.2 \times 10^{6} \\ +\ 0.52 \times 10^{6} \\ \hline 7.72 \times 10^{6} \end{array}$$

Practice Problem 9 Subtract. $4.36 \times 10^{5} - 3.1 \times 10^{4}$ ■

9.5 Exercises

Write in scientific notation.

1. 35
3.5×10^1

2. 68
6.8×10^1

3. 137
1.37×10^2

4. 542
5.42×10^2

5. 2148
2.148×10^3

6. 6825
6.825×10^3

7. 120
1.2×10^2

8. 340
3.4×10^2

9. 500
5×10^2

10. 100
1×10^2

11. 26,300
2.63×10^4

12. 78,100
7.81×10^4

13. 288,000
2.88×10^5

14. 238,000
2.38×10^5

15. 4,632,000
4.632×10^6

16. 2,034,000
2.034×10^6

17. 12,000,000
1.2×10^7

18. 28,000,000
2.8×10^7

19. 0.67
6.7×10^{-1}

20. 0.42
4.2×10^{-1}

21. 0.398
3.98×10^{-1}

22. 0.512
5.12×10^{-1}

23. 0.00279
2.79×10^{-3}

24. 0.00613
6.13×10^{-3}

25. 0.4
4×10^{-1}

26. 0.9
9×10^{-1}

27. 0.0015
1.5×10^{-3}

28. 0.0087
8.7×10^{-3}

29. 0.000016
1.6×10^{-5}

30. 0.000072
7.2×10^{-5}

31. 0.00000531
5.31×10^{-6}

32. 0.00000198
1.98×10^{-6}

33. 0.0007
7×10^{-4}

34. 0.00005
5×10^{-5}

35. 0.0000001
1×10^{-7}

36. 0.000000009
9×10^{-9}

Write in standard notation.

37. 1.6×10^1
16

38. 2.8×10^2
280

39. 5.36×10^4
53,600

40. 2.19×10^4
21,900

41. 6.2×10^{-2}
0.062

42. 3.5×10^{-2}
0.035

43. 5.6×10^{-5}
0.000056

44. 3.2×10^{-5}
0.000032

45. 6.3×10^{-4}
0.00063

46. 7.8×10^{-3}
0.0078

47. 9×10^{11}
900,000,000,000

48. 2×10^{12}
2,000,000,000,000

49. 3×10^{-7}
0.0000003

50. 5×10^{-8}
0.00000005

51. 3.862×10^{-8}
0.00000003862

52. 8.139×10^{-9}
0.000000008139

53. 4.6×10^{12}
4,600,000,000,000

54. 3.8×10^{11}
380,000,000,000

55. 6.721×10^{10}
67,210,000,000

56. 4.039×10^9
4,039,000,000

Applications

Write each number in scientific notation.

57. In 1 year, light will travel 5,878,000,000,000 miles. 5.878×10^{12} miles

58. The world's forests total 2,700,000,000 acres of wooded area. 2.7×10^9 acres

59. Yellow light has a wavelength of 0.00000059 meter. 5.9×10^{-7} meter

60. An electron has a charge of 0.00000000048 electrostatic unit. 4.8×10^{-10} electrostatic unit

61. The population of the United States is about 264,000,000 people. 2.64×10^8 people

62. The flowing duckweed of Australia has a small fruit that weighs 0.0000024 oz. 2.4×10^{-6} oz

Write each number in standard notation.

63. During the period August 15–25, 1875, a huge swarm of Rocky Mountain locusts invaded Nebraska. The swarm contained an estimated 1.25×10^{13} insects.

12,500,000,000,000 insects

64. The gross national product of the United States in 1992 was 5.9619×10^{15} dollars.

$5,961,900,000,000,000

65. The diameter of a red corpuscle of human blood is about 7.5×10^{-5} centimeter.

0.000075 centimeter

66. Recently a tiny hole with an approximate diameter of 3.16×10^{-10} meter was produced on the surface of molybdenum disulfide by Doctors Heckl and Maddocks using a chemical method involving a mercury drill. 0.000 000 000 316 meter

67. During the eruption of the Taupo volcano in New Zealand in 130 A.D., approximately 1.4×10^{10} tons of pumice were carried up in the air.

14,000,000,000 tons

68. The approximate mass of a water molecule is 3×10^{-22} gram. 0.0000000000000000000003 gram

Add.

69. 3.38×10^{7} dollars + 5.63×10^{7} dollars

9.01×10^{7} dollars

70. 8.17×10^{9} atoms + 2.76×10^{9} atoms

$10.93 \times 10^{9} = 1.093 \times 10^{10}$ atoms

71. 4.52×10^{9} dollars + 3.41×10^{9} dollars

7.93×10^{9} dollars

72. 5.62×10^{12} pounds + 2.18×10^{12} pounds

7.8×10^{12} pounds

Subtract.

73. 7.18×10^{15} miles − 2.79×10^{15} miles

4.39×10^{15} miles

74. 5.29×10^{12} acres − 1.99×10^{12} acres

3.3×10^{12} acres

75. 4×10^{8} feet − 3.76×10^{7} feet

$40 \times 10^{7} - 3.76 \times 10^{7} = 36.24 \times 10^{7} = 3.624 \times 10^{8}$ feet

76. 9×10^{10} meters − 1.26×10^{9} meters

$90 \times 10^{9} - 1.26 \times 10^{9} = 88.74 \times 10^{9} = 8.874 \times 10^{10}$ meters

Cumulative Review Problems

Calculate.

77. 7.63×2.18

16.6334

78. 5.92×1.98

11.7216

79. $0.53\overline{)0.13674}$ 0.258

80. $0.42\overline{)0.15204}$ 0.362

Chapter Organizer

Topic	Procedure	Examples
Absolute value, p. 504.	The absolute value of a number is the distance between the number on the number line and zero.	$\lvert -6 \rvert = 6 \quad \lvert 3 \rvert = 3 \quad \lvert 0 \rvert = 0$
Adding signed numbers with the same sign, p. 505.	To add two numbers with the same sign: **1.** Add the absolute value of the numbers. **2.** Use the common sign in the answer.	$12 + 5 = 17$ $-6 + (-8) = -14$ $-5.2 + (-3.5) = -8.7$ $-\frac{1}{7} + \left(-\frac{3}{7}\right) = -\frac{4}{7}$
Adding signed numbers with different signs, p. 508.	To add two numbers with different signs: **1.** Subtract the absolute value of the numbers. **2.** Use the sign of the number with the larger absolute value.	$14 + (-8) = 6$ $-14 + 8 = -6$ $-3.2 + 7.1 = 3.9$ $\frac{5}{13} + \left(-\frac{8}{13}\right) = -\frac{3}{13}$
Subtracting signed numbers, p. 514.	To subtract signed numbers, add the opposite of the second number to the first number.	$-9 - (-3) = -9 + 3 = -6$ $5 - (-7) = 5 + 7 = 12$ $8 - 12 = 8 + (-12) = -4$ $-4 - 13 = -4 + (-13) = -17$ $-\frac{1}{12} - \left(-\frac{5}{12}\right) = -\frac{1}{12} + \frac{5}{12} = \frac{4}{12} = \frac{1}{3}$
Multiplying or dividing signed numbers with the same sign, p. 520.	To multiply or divide two numbers with the same sign, multiply or divide the absolute values. The sign of the result is positive.	$(0.5)(0.3) = 0.15$ $(-6)(-2) = 12$ $\frac{-20}{-2} = 10$ $\frac{-\frac{1}{3}}{-\frac{1}{7}} = \left(-\frac{1}{3}\right)\left(-\frac{7}{1}\right) = \frac{7}{3}$
Multiplying or dividing signed numbers with different signs, p. 522.	To multiply or divide two numbers with different signs, multiply or divide the absolute values. The result is negative.	$7(-3) = -21$ $(-6)(4) = -24$ $(-36) \div (2) = -18$ $\frac{41.6}{-8} = -5.2$
Combined operations, with addition, subtraction, multiplication, and division, p. 526.	**1.** First multiply and divide from left to right. **2.** Then add and subtract from left to right.	Perform the operations in the proper order. **(a)** $-12 + 30 \div (-5) = -12 + (-6) = -18$ **(b)** $-21 \div 3 - 26 \div (-2) = -7 - (-13)$ $= -7 + 13 = 6$
Simplifying fractions with combined operations in numerator and denominator, p. 527.	**1.** Perform the operations in the numerator. **2.** Perform the operations in the denominator. **3.** Simplify the fraction.	Perform the operations in the proper order. $\frac{7(-4) - (-2)}{8 - (-5)}$ The numerator is $7(-4) - (-2) = -28 - (-2) = -26$ The denominator is $8 - (-5) = 8 + 5 = 13$ Thus the fraction becomes $-\frac{26}{13} = -2$.

Chapter Organizer

Topic	Procedure	Examples
Writing a number in scientific notation, p. 532.	**1.** Move the decimal point to the position immediately to the right of the first nonzero digit. **2.** Count the number of decimal places you moved the decimal point. **3.** Multiply the result by a power of 10 equal to the number of places moved. If the number you started with is larger than 10, you use a positive exponent. If the number you started with is less than 1, you use a negative exponent.	Write in scientific notation. **(a)** 178 **(b)** 25,000,000 **(c)** 0.006 **(d)** 0.00001732 **(a)** $178 = 1.78 \times 10^2$ **(b)** $25{,}000{,}000 = 2.5 \times 10^7$ **(c)** $0.006 = 6 \times 10^{-3}$ **(d)** $0.00001732 = 1.732 \times 10^{-5}$
Changing from scientific notation to standard notation, p. 536.	**1.** If the exponent of 10 is *positive,* move the decimal point to the *right* as many places as the exponent shows. Insert extra zeros as necessary. **2.** If the exponent of 10 is *negative,* move the decimal point to the *left* as many places as the exponent shows. *Note:* Remember that numbers in scientific notation that have positive exponents are always greater than 10. Numbers in scientific notation that have negative exponents are always less than 1.	Write in standard notation. **(a)** 8×10^6 **(b)** 1.23×10^4 **(c)** 7×10^{-2} **(d)** 8.45×10^{-5} **(a)** $8 \times 10^6 = 8{,}000{,}000$ **(b)** $1.23 \times 10^4 = 12{,}300$ **(c)** $7 \times 10^{-2} = 0.07$ **(d)** $8.45 \times 10^{-5} = 0.0000845$

Chapter 9 Review Problems

9.1 *Add.*

1. $21 + (-7)$ 14

2. $-18 + (-2)$ −20

3. $-15 + (-7)$ −22

4. $13 + (-5)$ 8

5. $-20 + 5$ −15

6. $-18 + 4$ −14

7. $-3.6 + (-5.2)$ −8.8

8. $-7.6 + (-1.2)$ −8.8

9. $-\frac{1}{5} + \left(-\frac{1}{3}\right)$ $-\frac{8}{15}$

10. $-\frac{2}{7} + \frac{5}{14}$ $\frac{1}{14}$

11. $20 + (-14)$ 6

12. $-80 + 60$ −20

13. $7 + (-2) + 9 + (-3)$ 11

14. $6 + (-3) + 8 + (-9)$ 2

15. $8 + (-7) + (-6) + 3 + (-2) + 8$ 4

16. $4 + (-10) + 6 + (-3) + (-8) + 7$ −4

9.2 *Subtract.*

17. $12 - 16$ −4

18. $18 - 36$ −18

19. $-2 - 7$ −9

20. $-7 - 6$ −13

21. $-36 - (-21)$ −15

22. $-21 - (-28)$ 7

23. $12 - (-7)$ 19

24. $14 - (-3)$ 17

25. $1.6 - 3.2$ −1.6

26. $-5.2 - 7.1$ −12.3

27. $-\frac{2}{5} - \left(-\frac{1}{3}\right)$ $-\frac{1}{15}$

28. $\frac{1}{6} - \left(-\frac{5}{12}\right)$ $\frac{7}{12}$

Perform the operations from left to right.

29. $5 - (-2) - (-6)$ 13

30. $-15 - (-3) + 9$ −3

31. $9 - 8 - 6 - 4$ −9

32. $-7 - 8 - (-3)$ −12

9.3 *Multiply or divide.*

33. $7(-2)$ -14

34. $8(-7)$ -56

35. $(-10)(-5)$ 50

36. $(-16)(-3)$ 48

37. $\left(-\frac{2}{7}\right)\left(-\frac{1}{5}\right)$ $\frac{2}{35}$

38. $\left(-\frac{6}{15}\right)\left(\frac{5}{12}\right)$ $-\frac{1}{6}$

39. $(5.2)(-1.5)$ -7.8

40. $(-3.6)(-1.2)$ 4.32

41. $-60 \div (-20)$ 3

42. $-18 \div (-3)$ 6

43. $\frac{-36}{4}$ -9

44. $\frac{-60}{12}$ -5

45. $\frac{-13.2}{-2.2}$ 6

46. $\frac{48}{-3.2}$ -15

47. $\frac{-\frac{2}{5}}{\frac{4}{7}}$ $-\frac{7}{10}$

48. $\frac{-\frac{1}{3}}{-\frac{7}{9}}$ $\frac{3}{7}$

49. $3(-5)(-2)$ 30

50. $6(-8)(-1)$ 48

51. $4(-7)(-8)\left(-\frac{1}{2}\right)$ -112

52. $3(-6)(-5)\left(-\frac{1}{5}\right)$ -18

9.4 *Perform the operations in the proper order.*

53. $8(-5) - (-6)$ -34

54. $12 \div (-3) + 5$ 1

55. $7 - 7(-1)$ 14

56. $7(-8) - (-2)$ -54

57. $8 - (-30) \div 6$ 13

58. $26 + (-28) \div 4$ 19

59. $2(-6) + 3(-4) - (-13)$ -11

60. $-49 \div (-7) + 3(-2)$ 1

61. $36 \div (-12) + 50 \div (-25)$ -5

62. $21 - (-30) \div 15$ 23

63. $50 \div 25(-4)$ -8

64. $-80 \div 20(-3)$ 12

65. $9(-9) + 9$ -72

In problems 66–69 simplify the numerator and denominator first, using the proper order of operations. Then reduce the fraction if possible.

66. $\frac{5 - 9 + 2}{3 - 5}$ 1

67. $\frac{4(-6) + 8 - 2}{15 - 7 + 2}$ $-\frac{9}{5}$ or $-1\frac{4}{5}$

68. $\frac{20 \div (-5) - (-6)}{(2)(-2)(-5)}$ $\frac{1}{10}$

69. $\frac{6 - (-3) - 2}{5 - 2 \div (-1)}$ 1

9.5 *Write in scientific notation.*

70. 4160 4.16×10^3

71. 3,700,000 3.7×10^6

72. 218,000 2.18×10^5

73. 47,320,000,000 4.732×10^{10}

74. 0.004 4.0×10^{-3}

75. 0.007 7.0×10^{-3}

76. 0.0000218 2.18×10^{-5}

77. 0.00000763 7.63×10^{-6}

78. 0.5136 5.136×10^{-1}

79. 0.02173 2.173×10^{-2}

Write in standard notation.

80. 9×10^6 9,000,000

81. 7×10^5 700,000

82. 1.89×10^4 18,900

83. 3.76×10^3 3760

84. 7.52×10^{-2} 0.0752

85. 6.61×10^{-3} 0.00661

86. 9×10^{-7} 0.0000009

87. 8×10^{-8} 0.00000008

88. 5.36×10^{-4} 0.000536

89. 1.98×10^{-5} 0.0000198

Add or subtract. Express your answer in scientific notation.

90. $5.26 \times 10^{11} + 3.18 \times 10^{11}$ 8.44×10^{11}

91. $7.79 \times 10^{15} + 1.93 \times 10^{15}$ 9.72×10^{15}

92. $7.79 \times 10^{30} + 5.63 \times 10^{30}$ 1.342×10^{31}

93. $5.04 \times 10^{26} + 9.39 \times 10^{26}$ 1.443×10^{27}

94. $3.42 \times 10^{14} - 1.98 \times 10^{14}$ 1.44×10^{14}

95. $1.76 \times 10^{26} - 1.08 \times 10^{26}$ 6.8×10^{25}

Chapter 9 Test

Add.

1. $-13 + 8$
2. $-24 + (-8)$
3. $6.7 + (-2.9)$
4. $-3 + (-6) + 7 + (-4)$
5. $10 + (-7) + 3 + (-9)$
6. $-\frac{1}{4} + \left(-\frac{5}{8}\right)$

Subtract.

7. $-32 - 6$
8. $23 - 18$
9. $\frac{4}{5} - \left(-\frac{1}{3}\right)$
10. $-50 - (-7)$
11. $-2.5 - (-6.5)$
12. $4.8 - 2.7$
13. $\frac{1}{12} - \left(-\frac{5}{6}\right)$
14. $23 - (-23)$

Multiply or divide.

15. $(-20)(-6)$
16. $64 \div (-8)$
17. $-40 \div (-4)$

1. -5

2. -32

3. 3.8

4. -6

5. -3

6. $-\frac{7}{8}$

7. -38

8. 5

9. $\frac{17}{15}$ *or* $1\frac{2}{15}$

10. -43

11. 4

12. 2.1

13. $\frac{11}{12}$

14. 46

15. 120

16. -8

17. 10

18.	-18
19.	3
20.	$-\frac{7}{10}$
21.	56
22.	-32
23.	17
24.	14
25.	-8
26.	-88
27.	-32
28.	22
29.	$-\frac{1}{7}$
30.	$-\frac{1}{6}$
31.	8.054×10^4
32.	7×10^{-6}
33.	0.0000936
34.	$72{,}000$

Multiply or divide.

18. $(-9)(-1)(-2)(4)\left(\frac{1}{4}\right)$

19. $\dfrac{-39}{-13}$

20. $\dfrac{-\frac{3}{5}}{\frac{6}{7}}$

21. $(-7)(-2)(4)$

22. $96 \div (-3)$

Perform the operations in the proper order.

23. $7 - 2(-5)$

24. $(-42) \div (-7) + 8$

25. $18 \div (-3) + 24 \div (-12)$

26. $8(-5) + 6(-8)$

27. $3 - 9 - (-4) + 6(-5)$

28. $-48 \div (-6) - 7(-2)$

29. $\dfrac{3 + 8 - 5}{(-4)(6) + (-6)(3)}$

30. $\dfrac{5 + 28 \div (-4)}{7 - (-5)}$

Write in scientific notation.

31. 80,540

32. 0.000007

Write in standard notation.

33. 9.36×10^{-5}

34. 7.2×10^4

Cumulative Test for Chapters 1–9

Approximately one half of this test is based on Chapter 9 material. The remainder is based on material covered in Chapters 1–8.

Do each problem. Simplify your answers.

1. Subtract. $\begin{array}{r} 28{,}981 \\ -\ 16{,}598 \\ \hline \end{array}$

2. Divide. $36\overline{)4572}$

3. Add. $3\frac{1}{4} + 8\frac{2}{3}$

4. Multiply. $1\frac{5}{6} \times 2\frac{1}{2}$

5. Round to the nearest thousandth. 9.812456

6. Add. $5.82 + 38.964 + 0.571 + 9.305 + 8.8$

7. Multiply. 12.89×5.12

8. Find n. $\frac{n}{8} = \frac{56}{7}$

9. For every 156 parts manufactured, there are seven defects. If 2808 parts are manufactured, how many defects would you expect?

10. What is 0.8% of 38?

11. 12% of what number is 480?

12. Convert 94 km to m.

13. Convert 180 in. to yd.

14. Find the area of a circle with radius 5 m. Round your answer to the nearest tenth.

1. 12,383

2. 127

3. $\frac{143}{12}$ *or* $11\frac{11}{12}$

4. $\frac{55}{12}$ *or* $4\frac{7}{12}$

5. 9.812

6. 63.46

7. 65.9968

8. $n = 64$

9. 126 defective parts

10. 0.304

11. 4000

12. 94,000 m

13. 5 yd

14. 78.5 m^2

15. (a) 300 students

(b) 1100 students

16. 13

17. −4.7

18. $\frac{5}{12}$

19. −11

20. −5

21. −60

22. $\frac{6}{5}$ *or* $1\frac{1}{5}$

23. 18

24. 4

25. −2

26. $\frac{4}{15}$

27. 5.79863×10^5

28. 7.8×10^{-4}

29. 38,500,000

30. 0.00007

15. The following histogram depicts the ages of students at Wolfville College.

(a) How many students are between ages 23 and 25?

(b) How many students are older than 19 years?

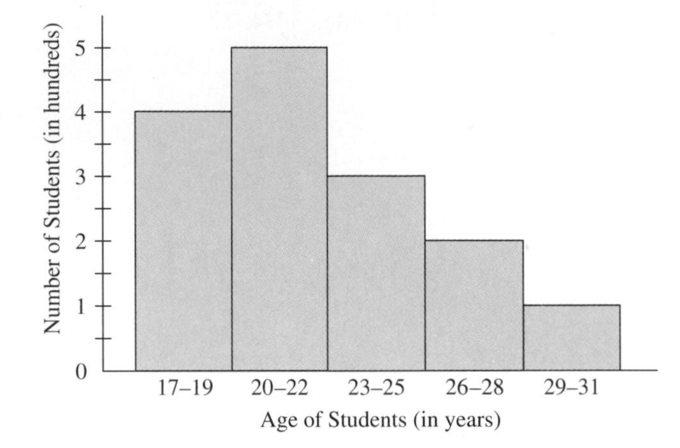

16. Evaluate. $\sqrt{36} + \sqrt{49}$

Add.

17. $-1.2 + (-3.5)$

18. $-\frac{1}{4} + \frac{2}{3}$

Subtract.

19. $7 - 18$

20. $-8 - (-3)$

Multiply or divide.

21. $(5)(-3)(-1)(-2)(2)$

22. $\dfrac{-\frac{3}{7}}{-\frac{5}{14}}$

Perform the operations in the proper order.

23. $6 - 3(-4)$

24. $(-20) \div (-2) + (-6)$

25. $\dfrac{(-2)(-1) + (-4)(-3)}{1 + (-4)(2)}$

26. $\dfrac{(-11)(-8) \div 22}{1 - 7(-2)}$

Write in scientific notation.

27. 579,863

28. 0.00078

Write in standard notation.

29. 3.85×10^7

30. 7×10^{-5}

CHAPTER 10

Introduction to Algebra

If you needed to prepare the water supply for a city in the year 2040 could you make a reasonable projection of how many people will need water based on the projected size of the town population that year? This becomes a critical skill for city planners and other individuals helping towns, cities, and entire states prepare for future needs in serving the population. If you think you can solve these types of problems please turn to the Putting Your Skills to Work on page 575.

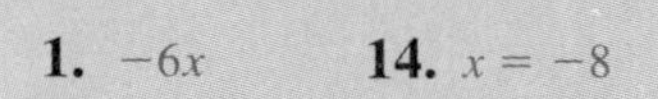

Pretest Chapter 10

1. $-6x$
2. y
3. $-3a + 2b$
4. $-12x - 4y + 10$
5. $16x - 2y - 6$
6. $4a - 12b + 3c$
7. $42x - 18y$
8. $-3a - 15b + 3$
9. $-3a - 6b + 12c + 10$
10. $x - 8y$
11. $x = 37$
12. $x = 3.5$
13. $x = \frac{5}{4}$ or $1\frac{1}{4}$
14. $x = -8$
15. $x = 5$
16. $x = \frac{3}{2}$ or $1\frac{1}{2}$
17. $x = 7$
18. $x = 1.6$
19. $x = 3$
20. $x = 2.25$
21. $c = 9 + p$
22. $l = 2w + 5$
23. a = height of Mt. Ararat; $a - 1758$ = height of Mt. Hood
24. width = 19 m; length = 35 m
25. 7.75 ft 10.25 ft

If you are familiar with the topics in this chapter, take this test now. Check your answers with those in the back of the book. If an answer was wrong or you couldn't do a problem, study the appropriate section of the chapter.

If you are not familiar with the topics in this chapter, don't take this test now. Instead, study the examples, work the practice problems, and then take the test.

This test will help you identify which concepts you have mastered and which you need to study further.

Section 10.1

Combine like terms.

1. $9x - 15x$

2. $-8y + 12y - 3y$

3. $6a - 5b - 9a + 7b$

4. $5x - y + 2 - 17x - 3y + 8$

5. $7x - 14 + 5y + 8 - 7y + 9x$

6. $4a - 7b + 3c - 5b$

Section 10.2

Simplify.

7. $(6)(7x - 3y)$

8. $(-3)(a + 5b - 1)$

9. $(-2)(1.5a + 3b - 6c - 5)$

10. $(5)(2x - y) - (3)(3x + y)$

Section 10.3–Section 10.5

Solve for the variable.

11. $5 + x = 42$

12. $x + 2.5 = 6$

13. $x - \frac{3}{4} = \frac{1}{2}$

14. $7x = -56$

15. $5.4x = 27$

16. $\frac{3}{5}x = \frac{9}{10}$

17. $5x - 9 = 26$

18. $12 - 3x = 7x - 4$

19. $(5)(x - 1) = 7 - (3)(x - 4)$

20. $3x + 7 = (5)(5 - x)$

Section 10.6

Translate the English sentence into an equation.

21. The computer weighs 9 kg more than the printer. Use c to represent the weight of the computer and p to represent the weight of the printer.

22. The length of the rectangle is 5 m longer than double the width. Use l to represent the length and w the width of the rectangle.

Write algebraic expressions for each of the specified quantities using the same variable.

23. Mount Hood is 1758 m shorter than Mount Ararat in Turkey. Use the variable a to describe the height of each mountain in meters.

Section 10.7

Solve each problem by using an equation.

24. A rectangular field has a perimeter of 108 m. The length is 3 m less than double the width. Find the dimensions of the field.

25. An 18-ft board is cut into two pieces. One piece is 2.5 ft longer than the other. Find the length of each piece.

10.1 Variables and Like Terms

After studying this section, you will be able to:

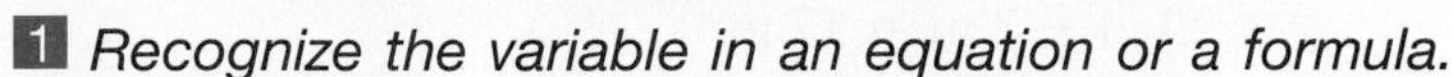
1 *Recognize the variable in an equation or a formula.*
2 *Combine like terms containing a variable.*

MathPro
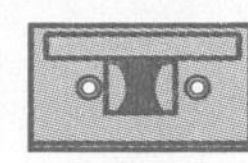
Video 21

SSM

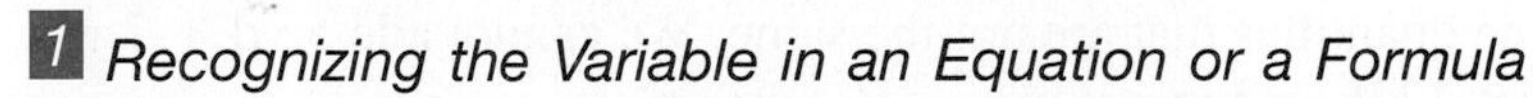

1 *Recognizing the Variable in an Equation or a Formula*

In algebra we reason and solve problems by means of symbols. A **variable** is a symbol, usually a letter of the alphabet, that stands for a number. We can use the variable even though we may not know what number the variable stands for. We can find that number by following a logical order of steps. These are the rules of algebra.

We begin by taking a closer look at variables. In the formula for the area of a circle, the equation $A = \pi r^2$ contains two variables. r represents the value of the radius. A represents the value of the area. π is a known value. We often use the decimal approximation 3.14 for π.

EXAMPLE 1 Name the variables in the following equations.

(a) $A = lw$ **(b)** $p = 2w + 2l$ **(c)** $V = \dfrac{4\pi r^3}{3}$

(a) $A = lw$ The variables are A, l, and w.

(b) $p = 2w + 2l$ The variables are p, w, and l.

(c) $V = \dfrac{4\pi r^3}{3}$ The variables are V and r.

Practice Problem 1 Name the variables.

(a) $A = \dfrac{bh}{2}$ **(b)** $C = \pi d$ **(c)** $V = lwh$ ■

We have seen various ways to write multiplication. For example, three times n can be written as $3 \times n$. In algebra we usually do not write the multiplication symbol. We can simply write $3n$ to mean three times n. A number just to the left of a variable indicates that the number is multiplied by the variable. Thus $4ab$ means 4 times a times b. A number just to the left of a parentheses also means multiplication. Thus $3(n + 8)$ means $3 \times (n + 8)$. So a formula can be written with a multiplication sign or without a multiplication sign.

EXAMPLE 2 Write the formula without a multiplication sign.

(a) $V = \dfrac{B \times h}{3}$ **(b)** $A = \dfrac{h \times (B + b)}{2}$

(a) $V = \dfrac{Bh}{3}$ **(b)** $A = \dfrac{h(B + b)}{2}$

TEACHING TIP Example 2 is helpful because students may have seen formulas written both ways. It is important for them to know that the formula $A = b \times h$ is the same as the formula $A = bh$. You may want to give some additional examples, such as $F = 1.8 \times C + 32$ is equivalent to $F = 1.8C + 32$, and $I = P \times R \times T$ is equivalent to $I = PRT$.

Practice Problem 2 Write the formula without a multiplication sign.

(a) $p = 2 \times w + 2 \times l$ **(b)** $A = \pi \times r^2$ ■

2 Combining Like Terms Containing a Variable

Recall that when we work with measurements we combine like quantities. A carpenter, for example, might perform the following calculations.

$$20 \text{ m} - 3 \text{ m} = 17 \text{ m}$$

$$7 \text{ in.} + 9 \text{ in.} = 16 \text{ in.}$$

We cannot combine quantities that are not the same. We cannot add 5 yd + 7 gal. We cannot subtract 12 lb − 3 in.

Similarly, when using variables, we can add or subtract only when the same variable is used. For example, we can add $4a + 5a = 9a$, but we cannot add $4a + 5b$.

TEACHING TIP Remind students that the word *term* is used extensively in mathematics, so it is best to make sure they grasp its definition now.

A **term** is a number, a variable, or a product of a number and one or more variables separated in an expression by a + sign or a − sign. In the expression $2x + 4y + (-1)$ there are three terms, $2x$, $4y$, and -1. **Like terms** have identical variables and identical exponents, so in the expression $3x + 4y + (-2x)$, the two terms $3x$ and $(-2x)$ are called *like terms*. The terms $2x^2y$ and $5xy$ are not like terms, since the exponents for the variable x are not the same. To combine like terms, you combine the numbers, called the numerical **coefficients,** that are directly in front of the terms by using the rules for adding signed numbers.

EXAMPLE 3 Add. $5x + 7x$

We add $5 + 7 = 12$. Thus

$$5x + 7x = 12x$$

Practice Problem 3 Add. $9x + 2x$ ■

When we combine signed numbers, we try to combine them mentally. Thus to combine $7 - 9$, we think $7 + (-9)$ and we write -2. In a similar way, to combine $7x - 9x$, we think $\boxed{7x + (-9x)}$ and we write $-2x$. Your instructor may ask you to write out this "think" step as part of your work. In the following example we show this extra step inside a $\boxed{}$ box. You should determine from your instructor whether he or she feels this step is necessary.

EXAMPLE 4 Combine.

(a) $14x - 20x$ **(b)** $3x + 7x - 15x$ **(c)** $9x - 12x - 11x$

(a) $14x - 20x = \boxed{14x + (-20x)} = -6x$

(b) $3x + 7x - 15x = \boxed{3x + 7x + (-15x)} = 10x + (-15x) = -5x$

(c) $9x - 12x - 11x = \boxed{9x + (-12x) + (-11x)} = -3x + (-11x) = -14x$

Practice Problem 4 Combine.

(a) $26x - 18x$ **(b)** $8x - 22x + 5x$ **(c)** $19x - 7x - 12x$ ■

A variable without a numerical coefficient is understood to have a coefficient of 1.

$$7x + y \text{ means } 7x + 1y. \qquad 5a - b \text{ means } 5a - 1b.$$

EXAMPLE 5 Combine.

(a) $3x - 8x + x$ **(b)** $12x - x - 20.5x$

(a) $3x - 8x + x = 3x - 8x + 1x = -5x + 1x = -4x$

(b) $12x - x - 20.5x = 12x - 1x - 20.5x = 11.0x - 20.5x = -9.5x$

TEACHING TIP Students often forget that a variable without a numerical coefficient is understood to have a coefficient of 1. In addition to Example 5, you may want to show them the following illustrations.
A. $8w + w - 6w = 3w$
B. $-13z - 18z - z = -32z$

Practice Problem 5 Combine.

(a) $9x - 12x + x$ **(b)** $5.6x - 8x + 10x$ ■

Numbers cannot be combined with variable terms.

EXAMPLE 6 Combine.

(a) $3 - 10x + 5 - 18x$ **(b)** $7.8 - 2.3x + 9.6x - 10.8$

In each case, we combine the numbers separately and the variable terms separately. It may help to use the commutative and associative properties first.

(a) $3 - 10x + 5 - 18x$
$= 3 + 5 - 10x - 18x$
$= 8 - 28x$

(b) $7.8 - 2.3x + 9.6x - 10.8$
$= 7.8 - 10.8 - 2.3x + 9.6x$
$= -3 + 7.3x$

Practice Problem 6 Combine.

(a) $-8 + 2x - 9 - 15x$ **(b)** $17.5 - 6.3x - 8.2x + 10.5$ ■

There may be more than one letter in a problem. Keep in mind, however, that only like terms may be combined.

EXAMPLE 7 Combine.

(a) $5x + 2y + 8 - 6x + 3y - 4$ **(b)** $2n + p + q - 8n + 5p - 6q$

For convenience, we will rearrange the problem to place like terms next to each other. This is an optional step; you do not need to do this.

(a) $5x - 6x + 2y + 3y + 8 - 4 = -1x + 5y + 4 = -x + 5y + 4$

(b) $2n - 8n + 1p + 5p + 1q - 6q = -6n + 6p - 5q$

The order of the terms in an answer is not important in this type of problem. The answer to Example 7(a) could have been $5y + 4 - x$ or $4 + 5y - x$. Often we give the answer with the letters in alphabetical order.

Practice Problem 7 Combine.

(a) $2w + 3z - 12 - 5w - z - 16$ **(b)** $-3a + 2b + c - 7a - 5b - 2c$ ■

10.1 Exercises

Name the variables in each equation.

1. $G = 5xy$
G, x, y

2. $N = 3bh$
N, b, h

3. $S = 4\pi r^2$
S, r

4. $S = 3\pi r^3$
S, r

5. $p = \dfrac{4ab}{3}$
p, a, b

6. $p = \dfrac{7ab}{4}$
p, a, b

Write each equation without a multiplication sign.

7. $p = 4 \times s^2$
$p = 4s^2$

8. $p = 2 \times w + 2 \times l$
$p = 2w + 2l$

9. $A = \dfrac{b \times h}{2}$
$A = \dfrac{bh}{2}$

10. $A = b \times h$
$A = bh$

11. $V = \dfrac{4 \times \pi \times r^3}{3}$
$V = \dfrac{4\pi r^3}{3}$

12. $V = l \times w \times h$
$V = lwh$

13. $H = 2 \times a - 3 \times b$
$H = 2a - 3b$

14. $F = 4 \times p - 2 \times q$
$F = 4p - 2q$

Combine like terms.

15. $9x + 8x$
$17x$

16. $7x + 4x$
$11x$

17. $-16x + 26x$
$10x$

18. $-12x + 40x$
$28x$

19. $2x - 8x + 5x$
$-x$

20. $9x + 3x - 7x$
$5x$

21. $6x - 12x - 15x$
$-21x$

22. $8x - 3x - 20x$
$-15x$

23. $x + 3x + 8 - 7$
$4x + 1$

24. $5x - x + 9 + 16$
$4x + 25$

25. $1.3x + 10 - 2.4x - 3.6$
$-1.1x + 6.4$

26. $3.8x + 2 - 1.9x - 3.5$
$1.9x - 1.5$

27. $-8.2x + 6.1x + 4 - 5.3$
$-2.1x - 1.3$

28. $4x - 12.5x - 6.5 + 2.7$
$-8.5x - 3.8$

29. $-13x + 7 - 19x - 10$
$-13x - 19x + 7 - 10 = -32x - 3$

30. $6 - 20x - 8 + 39x$
$-20x + 39x + 6 - 8 = 19x - 2$

31. $4x + 3y - 8 - 2x - 6y + 4$
$4x - 2x + 3y - 6y - 8 + 4$
$= 2x - 3y - 4$

32. $9x + 5y - 6 - 7x + 8y - 5$
$9x - 7x + 5y + 8y - 6 - 5$
$= 2x + 13y - 11$

33. $5a - 3b - c + 8a - 2b - 6c$
$5a + 8a - 3b - 2b - c - 6c$
$= 13a - 5b - 7c$

34. $10a + 5b - 7c - a - 8b - 3c$
$10a - a + 5b - 8b - 7c - 3c$
$= 9a - 3b - 10c$

35. $-7s - 2r + 3 - s - r - 8$
$-3r - 8s - 5$

36. $-15s + 9r - 7 - 3s + 4r - 12$
$-18s + 13r - 19$

37. $8n - 12p + q + 3q - 8p - 9p$
$8n - 29p + 4q$

38. $17n + 3p - 5n - 7q + 5p + 4n$
$16n + 8p - 7q$

39. $7x - 9.5x + 3.6 - 12x - 14x - 8$
$-28.5x - 4.4$

40. $2x - 6.5x + 5.3 - 10x - 22x - 11$
$-36.5x - 5.7$

41. $7x - 8y + 3 + 8y - 3 - 7x$
$7x - 7x - 8y + 8y + 3 - 3 = 0$

42. $14x - 3y + 12 - 14x - 12 + 3y$
$14x - 14x - 3y + 3y + 12 - 12 = 0$

43. $5.2x - 3.4y + 6.8 - 7.2x - 9.5y$
$-2x - 12.9y + 6.8$

44. $3x - 8.6y + 5.5 - 9.6x + 3.2y$
$-6.6x - 5.4y + 5.5$

Cumulative Review Problems

Solve for n. Round to the nearest hundredth when necessary.

45. $\dfrac{n}{6} = \dfrac{12}{15}$
$n = 4.8$

46. $\dfrac{n}{9} = \dfrac{36}{40}$
$n = 8.1$

47. $6n = 18$
$n = 3$

48. $5.5n = 46.75$
$n = 8.5$

10.2 The Distributive Property

After studying this section, you will be able to:

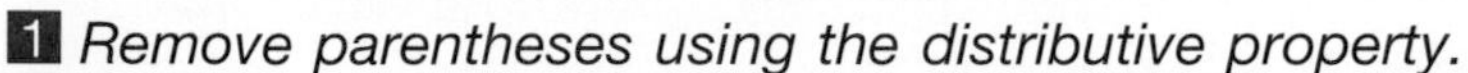

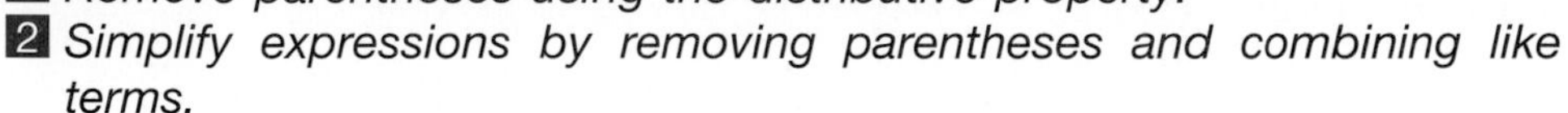

1 *Remove parentheses using the distributive property.*
2 *Simplify expressions by removing parentheses and combining like terms.*

1 *Removing Parentheses Using the Distributive Property*

What do we mean by the word *property* in mathematics? A **property** is an essential characteristic. A property of addition is an essential characteristic of addition. In this section we learn about the distributive property and how to use this property to simplify expressions.

Sometimes we encounter expressions like $4(x + 3)$. We'd like to be able to simplify the expression so that no parentheses appear. Notice that this expression contains two operations, multiplication and addition. We will use the distributive property to distribute the 4 to both terms in the addition. That is,

$$4(x + 3) \qquad \text{is equal to} \qquad 4(x) + 4(3)$$

The multiplication of the numerical coefficient 4 can be ''distributed over'' the expression $x + 3$ by multiplying the 4 by each of the terms in the parentheses. The expression $4(x) + 4(3)$ can be simplified to $4x + 12$, which has no parentheses. Thus we can use the distributive property to ''remove the parentheses.''

Using variables, we can write the distributive property two ways.

> **Distributive Properties of Multiplication over Addition**
>
> If a, b, and c are signed numbers, then
>
> $$a(b + c) = ab + ac \qquad \text{and} \qquad (b + c)a = ba + ca$$

A numerical example shows that the distributive property works.

$$(7)(4 + 6) = (7)(4) + (7)(6)$$
$$(7)(10) = 28 + 42$$
$$70 = 70$$

When the number is to the left of the parentheses, we use

$$a(b + c) = ab + ac.$$

EXAMPLE 1 Simplify. **(a)** $(4)(x + 3)$ **(b)** $(-3)(x + 3y)$ **(c)** $6(3a + 7b)$

(a) $(4)(x + 3) = 4x + (4)(3) = 4x + 12$

(b) $(-3)(x + 3y) = -3x + (-3)(3)(y) = -3x + (-9y) = -3x - 9y$

(c) $6(3a + 7b) = 6(3a) + (6)(7b) = 18a + 42b$

Practice Problem 1 Simplify

(a) $(7)(x + 5)$ **(b)** $(-4)(x + 2y)$ ■

Sometimes the number is to the right of the parentheses, so we use

$$(b + c)a = ba + ca.$$

TEACHING TIP You may want to warn students that although they need to be able to simplify both $5(x + 2y)$ and $(x + 2y)5$, it is the first type of problem that they will see most often.

EXAMPLE 2 Simplify. **(a)** $(5x + y)(2)$ **(b)** $(3x - y)(3)$

(a) $(5x + y)(2) = (5x)(2) + (y)(2)$
$= 10x + 2y$

(b) $(3x - y)(3) = (3x)(3) + (-y)(3)$
$= 9x + (-3y)$
$= 9x - 3y$

Notice in Example 2(a) that we write our final answer with the numerical coefficient to the left of the variable. We would not leave $(y)(2)$ as an answer but would write $2y$.

Practice Problem 2 Simplify. **(a)** $(x + 3y)(8)$ **(b)** $(2x - 7)(2)$ ■

Sometimes the distributive property is used with three terms within the parentheses. When a parentheses is used inside a parentheses, the outside ( ) is often changed to a bracket [ ] notation.

TEACHING TIP Warn students that it is very easy to make a sign error when doing problems of this type. After looking over Example 3 ask students to do the following exercises.
A. $-9(-3 - 9x + 5y)$
B. $12(2 - 7w + z)$
Ask them to check their work carefully. Answers are $27 + 81x - 45y$ and $24 - 84w + 12z$.

EXAMPLE 3 Simplify. **(a)** $(7)(2x - y + 7)$
(b) $(-5)(x + 2y - 8)$ **(c)** $(1.5x + 3y + 7)(2)$

(a) $(7)(2x - y + 7) = (7)[2x + (-1y) + 7] = 7(2x) + 7(-1y) + 7(7)$
$= 14x + (-7y) + 49$
$= 14x - 7y + 49$

(b) $(-5)(x + 2y - 8) = (-5)[x + 2y + (-8)] = -5(x) + (-5)(2y) + (-5)(-8)$
$= -5x + (-10y) + 40$
$= -5x - 10y + 40$

(c) $(1.5x + 3y + 7)(2) = (1.5x)(2) + (3y)(2) + (7)(2) = 3x + 6y + 14$

In this example every step is shown in detail. You may find that you do not need to write so many steps.

Practice Problem 3 Simplify. **(a)** $(8)(3x - y - 6)$
(b) $(-5)(x + 4y + 5)$ **(c)** $(3)(2.2x + 5.5y + 6)$ ■

TEACHING TIP Caution students to be most careful of + and − signs in this type of problem. Ask them to simplify the following problems at their seats.
A. Simplify.
$-3(2x - 3y) - 2(x + y)$
B. Simplify.
$2(-5a + 2b) - 5(-3a - 3b)$
Students are often amazed at how easy it is to make a sign error and not obtain the correct answers of $-8x + 7y$ for problem A and $5a + 19b$ for problem B.

The parentheses may contain four terms. The coefficients may be decimals or fractions.

EXAMPLE 4 Simplify. **(a)** $(3)(4.5x + 3.2y - z - 2.5)$
(b) $(-7)(x - 5y + 10z - 10)$ **(c)** $\frac{2}{3}\left(x + \frac{1}{2}y - \frac{1}{4}z + \frac{1}{5}\right)$

(a) $(3)[4.5x + 3.2y + (-1z) + (-2.5)] = 13.5x + 9.6y + (-3z) + (-7.5)$
$= 13.5x + 9.6y - 3z - 7.5$

(b) $(-7)[1x + (-5y) + 10z + (-10)] = -7x + 35y + (-70z) + 70$
$= -7x + 35y - 70z + 70$

(c) $\frac{2}{3}\left(x + \frac{1}{2}y - \frac{1}{4}z + \frac{1}{5}\right)$

$= \left(\frac{2}{3}\right)(x) + \left(\frac{2}{3}\right)\left(\frac{1}{2}y\right) + \left(\frac{2}{3}\right)\left(-\frac{1}{4}z\right) + \left(\frac{2}{3}\right)\left(\frac{1}{5}\right)$

$= \frac{2}{3}x + \frac{1}{3}y - \frac{1}{6}z + \frac{2}{15}$

Practice Problem 4 Simplify. **(a)** $(8)(1.2x + 2.4y - 3z - 1.5)$

(b) $(-9)(x + 3y - 4z + 5)$ **(c)** $\frac{3}{2}\left(\frac{1}{2}x - \frac{1}{3}y + 4z - \frac{1}{2}\right)$ ■

2 *Simplifying Expressions by Removing Parentheses and Combining Like Terms*

After removing parentheses we may have a chance to combine like terms. The direction "simplify" means remove parentheses, combine like terms, and leave the answer in as simple, and correct, a form as possible.

EXAMPLE 5 Simplify. $(2)(x + 3y) + (3)(4x + 2y)$

$(2)(x + 3y) + (3)(4x + 2y) = 2x + 6y + 12x + 6y$ *Use the distributive property.*

$= 14x + 12y$ *Combine like terms.*

Practice Problem 5 Simplify. $(3)(2x + 4y) + (2)(5x + y)$ ■

EXAMPLE 6 Simplify. **(a)** $(-3)(2x + 4) + (2)(-3 + 5x)$
(b) $(2)(x - 3y) + (5)(2x + 6)$

(a) $(-3)(2x + 4) + (2)(-3 + 5x) = -6x - 12 - 6 + 10x$ *Use the distributive property.*
Combine like terms.

$= 4x - 18$

(b) $(2)(x - 3y) + (5)(2x + 6) = 2x - 6y + 10x + 30$ *Use the distributive property.*

$= 12x - 6y + 30$ *Combine like terms.*

Notice that in the final step of Example 6(b) only the x terms could be combined. There are no other like terms.

Practice Problem 6 Simplify.

(a) $(-4)(x - 5) + (3)(-1 + 2x)$ **(b)** $(2)(x - 3y) + (5)(2x + 6)$ ■

10.2 Exercises

Verbal and Writing Skills

1. A <u>variable</u> is a symbol, usually a letter of the alphabet, that stands for a number.

2. What is the variable in the expression $5x + 9$? x

3. Identify the like terms in the expression $3x + 2y - 1 + x - 3y$. $3x$ and x, $2y$ and $-3y$

4. Explain the distributive property in your own words. Give an example.
Answers will vary. Check students' answers for understanding.

Simplify.

5. $(5)(x + 7)$
$5x + 5(7) = 5x + 35$

6. $(7)(3x + 1)$
$7(3x) + 7(1) = 21x + 7$

7. $4(2x + 7)$
$4(2x) + 4(7) = 8x + 28$

8. $6(3x + 8)$
$6(3x) + 6(8) = 18x + 48$

9. $(-2)(x + y)$
$-2x - 2y$

10. $(-5)(x + y)$
$-5x - 5y$

11. $(-7)(1.5x - 3y)$
$-7(1.5x) - 7(-3y)$
$= -10.5x + 21y$

12. $(-4)(3.5x - 6y)$
$-4(3.5x) - 4(-6y)$
$= -14x + 24y$

13. $(-3x + 7y)(-10)$
$(-3x)(-10) + (7y)(-10)$
$= 30x - 70y$

14. $(-2x + 8y)(-12)$
$(-2x)(-12) + (8y)(-12)$
$= 24x - 96y$

15. $(2)(x + 3y - 5)$
$2(x) + 2(3y) - (2)(5)$
$= 2x + 6y - 10$

16. $(3)(2x + y - 6)$
$3(2x) + 3(y) - 3(6)$
$= 6x + 3y - 18$

17. $-3(a + b + 4c)$
$(-3)a + (-3)b + (-3)(4c)$
$= -3a - 3b - 12c$

18. $-4(7a - b + c)$
$(-4)(7a) + (-4)(-b) + (-4)(c)$
$= -28a + 4b - 4c$

19. $(2a - 5b + c)(-8)$
$-16a + 40b - 8c$

20. $(-3a - b + 2c)(-7)$
$21a + 7b - 14c$

21. $(15)(-12a + 2.2b + 6.7)$
$-180a + 33b + 100.5$

22. $(14)(-10a + 3.2b + 4.5)$
$-140a + 44.8b + 63$

23. $(4)(a - 3b + 5c + 7)$
$4a - 12b + 20c + 28$

24. $(7)(2a - b + 3c - 8)$
$14a - 7b + 21c - 56$

25. $(-2)(1.3x - 8.5y - 5z + 12)$
$-2.6x + 17y + 10z - 24$

26. $(-3)(1.4x - 7.6y - 9z - 4)$
$-4.2x + 22.8y + 27z + 12$

27. $\frac{1}{2}\left(2x - 3y + 4z - \frac{1}{2}\right)$
$x - \frac{3}{2}y + 2z - \frac{1}{4}$

28. $\frac{1}{3}\left(-3x + \frac{1}{2}y + 2z - 3\right)$
$-x + \frac{1}{6}y + \frac{2}{3}z - 1$

Applications

29. The area of a trapezoid is $A = \dfrac{h(B + b)}{2}$. Write this formula without parentheses and without a multiplication sign.

$A = \dfrac{hB + hb}{2}$

30. The surface area of a cylinder is $S = 2\pi r(h + r)$. Write this formula without parentheses and without a multiplication sign.

$S = 2\pi rh + 2\pi r^2$

Simplify. Be sure to combine like terms.

31. $(3)(2x + y) + (2)(x - y)$

$6x + 3y + 2x - 2y = 8x + y$

32. $(3)(3x + 2y) + (4)(x - y)$

$9x + 6y + 4x - 4y = 13x + 2y$

33. $4(5x - 1) + 7(x - 5)$

$20x - 4 + 7x - 35 = 27x - 39$

34. $6(5x - 1) + 3(x - 4)$

$30x - 6 + 3x - 12 = 33x - 18$

35. $(-2)(a - 3b) + (3)(-a + 4b)$

$-2a + 6b - 3a + 12b = -5a + 18b$

36. $(-2)(3a - b) + (4)(-a + 3b)$

$-6a + 2b - 4a + 12b = -10a + 14b$

37. $(6)(3x + 2y) - (4)(x + 7)$

$18x + 12y - 4x - 28 = 14x + 12y - 28$

38. $(5)(4x + 3y) - (3)(x + 8)$

$20x + 15y - 3x - 24 = 17x + 15y - 24$

39. $(1.5)(x + 2.2y) + (3)(2.2x + 1.6y)$

$1.5x + 3.3y + 6.6x + 4.8y = 8.1x + 8.1y$

40. $(2.4)(x + 3.5y) + (2)(1.4x + 1.9y)$

$2.4x + 8.4y + 2.8x + 3.8y = 5.2x + 12.2y$

41. $(3)(a + b + 2c) - (4)(3a - b + 2c)$

$3a + 3b + 6c - 12a + 4b - 8c = -9a + 7b - 2c$

42. $(2)(2a + b + c) - (5)(2a - b + c)$

$4a + 2b + 2c - 10a + 5b - 5c = -6a + 7b - 3c$

To Think About

43. Illustrate the distributive property by using the area of two rectangles.

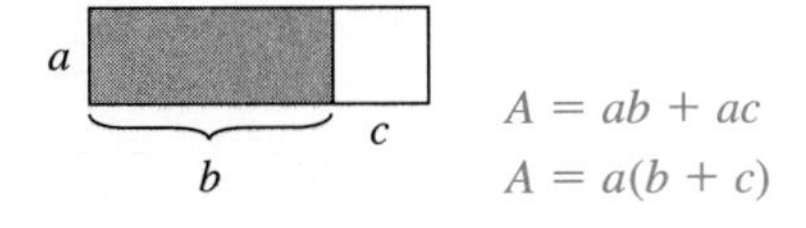

$A = ab + ac$
$A = a(b + c)$

44. Show that multiplication is distributive over subtraction by using the area of two rectangles.

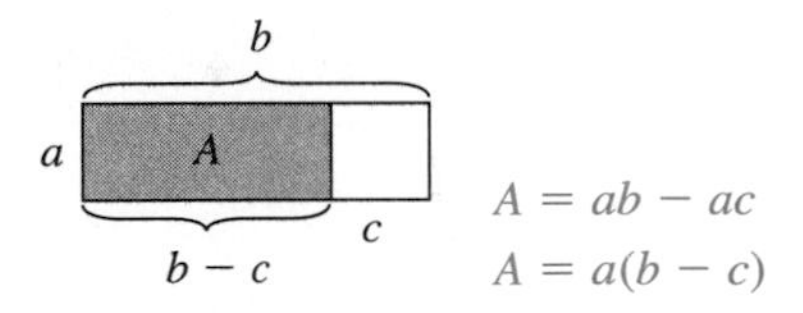

$A = ab - ac$
$A = a(b - c)$

Cumulative Review Problems

Round all answers to the nearest tenth. Use $\pi = 3.14$ when necessary.

45. Find the circumference of a circular fan with a diameter of 12 inches.

37.68 in.

46. Find the area of a circular security mirror in a convenience store with a radius of 8 inches.

200.96 in.^{2}

47. Find the area of the shaded figure.

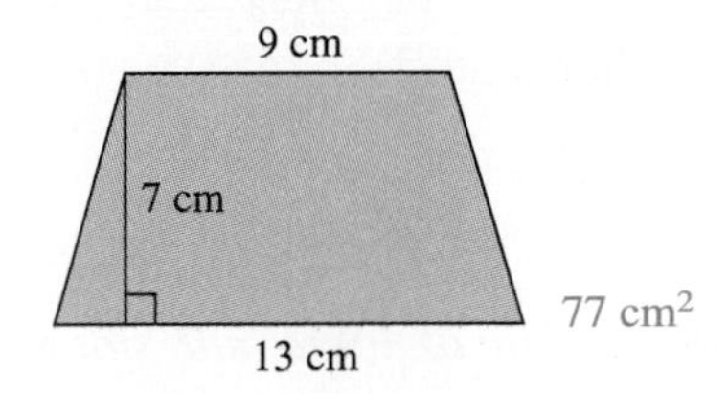

77 cm^2

48. Find the volume of a cylinder with radius 4 in. and height 6 in.

301.4 in.^{3}

10.3 Solve Equations Using the Addition Property

MathPro

Video 22

SSM

After studying this section, you will be able to:

1 *Solve equations using the addition property.*

1 *Solving Equations Using the Addition Property.*

One of the most important skills in algebra is that of "solving an equation." Starting from an equation with a variable whose value is unknown, we transform the equation into a simpler, equivalent equation by choosing a logical step. In the following sections we'll learn some of the "logical steps" to choose in solving an equation successfully.

We begin with the concept of an equation. An **equation** is a mathematical statement that says that two expressions are equal. The statement $x + 5 = 13$ means that some number (x) plus 5 equals 13. Experience with the basic addition facts tells us that $x = 8$ since $8 + 5 = 13$. Therefore, 8 is the solution to the equation. The **solution** of an equation is that number which makes the equation true. What if we did not know that $8 + 5 = 13$? How could we find the value of x? Let's look at the first "logical step" in solving this equation.

The important thing to remember when using "logical steps" is that an equation is like a balance scale. Whatever you do to one side of the equation, you must do the exact same thing to the other side of the equation to maintain the balance.

> **Addition Property of Equations**
>
> You may add the same number to each side of an equation to obtain an equivalent equation.

We want to make $x + 5 = 13$ into a simpler equation, $x =$ some number. What can we add to each side of the equation so that $x + 5$ becomes simply x? We can add the opposite of $+5$. Let's see what happens when we do.

$$x + 5 = 13$$

$$x + 5 + (-5) = 13 + (-5)$$ *Remember to add a (-5) to both sides of the equation.*

$$x + 0 = 8$$

$$x = 8$$

We found the solution to the equation.

Notice that in the second step above the number 5 "moved" to the other side of the equation. We know that we may add a number to both sides of the equation. But what number should we choose to add? We always add the opposite of the number we want to "move to the other side."

EXAMPLE 1 Solve. $x - 9 = 3$

$$x - 9 = 3$$ *Think: "Add the opposite of -9 to both sides of the equation."*

$$x - 9 + 9 = 3 + 9$$

$$x + 0 = 12$$

$$x = 12$$

To check the solution, we substitute 12 for x in the original equation.

$$x - 9 = 3$$
$$12 - 9 \stackrel{?}{=} 3$$
$$3 = 3 \quad \checkmark$$

It is always a good idea to check your solution, especially in more complicated equations.

Practice Problem 1 Solve. $x + 7 = -8$ ■

Equations can contain integers, decimals, or fractions. We use the addition property to solve equations where only one operation is involved, addition or subtraction.

EXAMPLE 2 Solve each equation. Check your solution.

(a) $y - 11 = -6$ **(b)** $x + 1.5 = 4$ **(c)** $x - \frac{3}{4} = \frac{3}{8}$

We use the addition property to solve these equations since each equation involves only addition or subtraction.

(a)
$$y - 11 = -6$$
Think: "Add the opposite of -11 *to both sides of the equation."*
$$y - 11 + 11 = -6 + 11$$
$$y + 0 = 5$$
$$y = 5$$

Check.
$$y - 11 = -6$$
Substitute 5 *for* y *in the original equation.*
$$5 - 11 \stackrel{?}{=} -6$$
$$-6 = -6 \quad \checkmark$$

(b)
$$x + 1.5 = 4$$
Think: "Add the opposite of 1.5 *to both sides of the equation."*
$$x + 1.5 + (-1.5) = 4 + (-1.5)$$
$$x = 2.5$$

Check.
$$x + 1.5 = 4$$
Substitute 2.5 *for* x *in the original equation.*
$$2.5 + 1.5 \stackrel{?}{=} 4$$
$$4 = 4 \quad \checkmark$$

(c)
$$y - \frac{3}{4} = \frac{3}{8}$$
Think: "Add the opposite of $-\frac{3}{4}$ *to both sides of the equation."*
$$y - \frac{3}{4} + \frac{3}{4} = \frac{3}{8} + \frac{3}{4}$$
We need to change $\frac{3}{4}$ *to* $\frac{6}{8}$.
$$y + 0 = \frac{3}{8} + \frac{6}{8}$$
$$y = \frac{9}{8} \text{ or } 1\frac{1}{8}$$

Check

$$y - \frac{3}{4} = \frac{3}{8} \qquad \textit{Substitute } \frac{9}{8} \textit{ for } y \textit{ in the original equation.}$$

$$\frac{9}{8} - \frac{3}{4} \stackrel{?}{=} \frac{3}{8}$$

$$\frac{9}{8} - \frac{6}{8} \stackrel{?}{=} \frac{3}{8}$$

$$\frac{3}{8} = \frac{3}{8} \quad \checkmark$$

Practice Problem 2 Solve each equation. Check your solution.

(a) $x + 8 = 2$ **(b)** $y - 3.2 = 9$ **(c)** $x + \frac{1}{6} = \frac{2}{3}$ ■

Sometimes you will need to use the addition property twice. If variables and numbers appear on both sides of the equation, you will want to get all of the variables on one side of the equation and all of the numbers on the other side of the equation.

EXAMPLE 3 Solve. Check your solution. $2x + 7 = x + 9$

We begin by moving $+7$ to the right-hand side of the equation.

$$2x + 7 = x + 9$$

$$2x + 7 - 7 = x + 9 - 7 \qquad \textit{Add } -7 \textit{ to both sides of the equation.}$$

$$2x = x + 2$$

Now we need to move x to the left-hand side of the equation.

$$2x = x + 2$$

$$2x - x = x - x + 2 \qquad \textit{Add } -x \textit{ to both sides of the equation.}$$

$$x = 2$$

We would certainly want to check this solution. Solving the equation took several steps and we might have made a mistake along the way. To check, we substitute 2 for x in the original equation.

Check.

$$2x + 7 \stackrel{?}{=} x + 9$$

$$2(2) + 7 \stackrel{?}{=} 2 + 9$$

$$4 + 7 \stackrel{?}{=} 2 + 9$$

$$11 = 11 \quad \checkmark$$

Practice Problem 3 Solve. Check your solution. $3x - 5 = 2x + 1$ ■

10.3 Exercises

Verbal and Writing Skills

1. An equation is an mathematical statement that says that two expressions are equal.
2. The solution of an equation is that number which makes the equation true.
3. To use the addition property, we add the opposite of the number we want to "move" to the other side of the equation.
4. To use the addition property to solve the equation $x - 8 = 9$, we add $+8$ to both sides of the equation.

Solve for the variable.

5. $y - 5 = 9$
   $y - 5 + 5 = 9 + 5$
   $y = 14$
6. $y - 7 = 6$
   $y - 7 + 7 = 6 + 7$
   $y = 13$
7. $x + 6 = 15$
   $x + 6 + (-6) = 15 + (-6)$
   $x = 9$
8. $x + 8 = 12$
   $x + 8 + (-8) = 12 + (-8)$
   $x = 4$
9. $x + 16 = -2$
   $x + 16 + (-16) = -2 + (-16)$
   $x = -18$
10. $y + 12 = -8$
    $y + 12 + (-12) = -8 + (-12)$
    $y = -20$
11. $x - 3 = -11$
    $x - 3 + 3 = -11 + 3$
    $x = -8$
12. $y - 15 = -8$
    $y - 15 + 15 = -8 + 15$
    $y = 7$
13. $4 + x = 13$
    $4 + (-4) + x = 13 + (-4)$
    $x = 9$
14. $7 + y = 11$
    $7 + (-7) + y = 11 + (-7)$
    $y = 4$
15. $-12 + x = 7$
    $-12 + 12 + x = 7 + 12$
    $x = 19$
16. $-20 + y = 10$
    $-20 + 20 + y = 10 + 20$
    $y = 30$
17. $x + 2.4 = 5.6$
    $x + 2.4 + (-2.4) = 5.6 + (-2.4)$
    $x = 3.2$
18. $y + 5.2 = 8.9$
    $y + 5.2 + (-5.2) = 8.9 + (-5.2)$
    $y = 3.7$
19. $x - 4.6 = 5.2$
    $x - 4.6 + 4.6 = 5.2 + 4.6$
    $x = 9.8$
20. $y - 5.4 = 3.1$
    $y - 5.4 + 5.4 = 3.1 + 5.4$
    $y = 8.5$
21. $x + 3.7 = -5$
    $x + 3.7 + (-3.7) = -5 + (-3.7)$
    $x = -8.7$
22. $y + 6.2 = -3.1$
    $y + 6.2 + (-6.2) = -3.1 + (-6.2)$
    $y = -9.3$
23. $x - 7.6 = -8$
    $x - 7.6 + 7.6 = -8 + 7.6$
    $x = -0.4$
24. $y - 8.3 = -6$
    $y - 8.3 + 8.3 = -6 + 8.3$
    $y = 2.3$
25. $x + \frac{1}{4} = \frac{3}{4}$
    $x + \frac{1}{4} + \left(-\frac{1}{4}\right) = \frac{3}{4} + \left(-\frac{1}{4}\right)$
    $x = \frac{2}{4}$ *or* $\frac{1}{2}$
26. $y + \frac{3}{8} = \frac{5}{8}$
    $y + \frac{3}{8} + \left(-\frac{3}{8}\right) = \frac{5}{8} + \left(-\frac{3}{8}\right)$
    $y = \frac{2}{8}$ *or* $\frac{1}{4}$
27. $x - \frac{3}{5} = \frac{2}{5}$
    $x - \frac{3}{5} + \frac{3}{5} = \frac{2}{5} + \frac{3}{5}$
    $x = \frac{5}{5}$ *or* 1
28. $y - \frac{2}{7} = \frac{6}{7}$
    $y - \frac{2}{7} + \frac{2}{7} = \frac{6}{7} + \frac{2}{7}$
    $y = \frac{8}{7}$ *or* $1\frac{1}{7}$
29. $x + \frac{2}{3} = -\frac{5}{6}$
    $x + \frac{2}{3} + \left(-\frac{2}{3}\right) = -\frac{5}{6} + \left(-\frac{2}{3}\right)$
    $x = -\frac{9}{6}$ *or* $-1\frac{1}{2}$
30. $y + \frac{1}{2} = -\frac{3}{4}$
    $y + \frac{1}{2} + \left(-\frac{1}{2}\right) = -\frac{3}{4} + \left(-\frac{1}{2}\right)$
    $y = -\frac{5}{4}$ *or* $-1\frac{1}{4}$
31. $x - \frac{2}{5} = -\frac{3}{4}$
    $x - \frac{2}{5} + \frac{2}{5} = -\frac{3}{4} + \frac{2}{5}$
    $x = -\frac{7}{20}$
32. $x - \frac{1}{4} = -\frac{2}{3}$
    $x - \frac{1}{4} + \frac{1}{4} = -\frac{2}{3} + \frac{1}{4}$
    $x = -\frac{5}{12}$

Solve for the variable. You may need to use the addition property twice.

33. $3x - 5 = 2x + 9$
$3x - 2x = 9 + 5$
$x = 14$

34. $5x + 1 = 4x - 3$
$5x - 4x = -3 - 1$
$x = -4$

35. $2x - 7 = x - 19$
$2x - x = -19 + 7$
$x = -12$

36. $4x + 6 = 3x + 5$
$4x - 3x = 5 - 6$
$x = -1$

37. $7x - 9 = 6x - 7$
$7x - 6x = -7 + 9$
$x = 2$

38. $4x - 8 = 3x + 12$
$4x - 3x = 12 + 8$
$x = 20$

39. $5x + 15 = 4x + 8$
$5x - 4x = 8 - 15$
$x = -7$

40. $8x - 13 = 7x + 9$
$8x - 7x = 9 + 13$
$x = 22$

Mixed Practice

Solve for the variable.

41. $y - \frac{1}{2} = 6$
$y - \frac{1}{2} + \frac{1}{2} = 6 + \frac{1}{2}$
$y = 6\frac{1}{2}$

42. $x + 1.2 = -3.8$
$x + 1.2 - 1.2 = -3.8 - 1.2$
$x = -5$

43. $z + 13 = 5$
$z + 13 - 13 = 5 - 13$
$z = -8$

44. $-6 + x = -15$
$-6 + 6 + x = -15 + 6$
$x = -9$

45. $5.9 + y = -5$
$5.9 - 5.9 + y = -5 - 5.9$
$y = -10.9$

46. $z + \frac{2}{3} = \frac{7}{12}$
$z + \frac{2}{3} - \frac{2}{3} = \frac{7}{12} - \frac{2}{3}$
$z = -\frac{1}{12}$

47. $2x - 1 = x + 5$
$2x - x = 5 + 1$
$x = 6$

48. $\frac{3}{4} + y = \frac{5}{8}$
$\frac{3}{4} - \frac{3}{4} + y = \frac{5}{8} - \frac{3}{4}$
$y = -\frac{1}{8}$

To Think About

49. In the equation $x + 9 = 12$, we add -9 to both sides of the equation to solve. What would you do to solve the equation $3x = 12$? Why?
To solve the equation $3x = 12$, divide both sides of the equation by 3 so that x stands alone.

Cumulative Review

Combine like terms.

50. $3x - 5x + x$
$-x$

51. $2a - 3b + 4a - b$
$6a - 4b$

52. $5x - y + 3 - 2x + 4y$
$3x + 3y + 3$

Simplify.

53. $7(2x + 3y) - 3(5x - 1)$
$-x + 21y + 3$

54. In 1995 there were 132.3 million people in the total labor force of the United States. Of this number 7.4 million people were unemployed. What percent of the total labor force was unemployed in 1995? Round your answer to the nearest tenth of a percent. 5.6%

10.4 Solve Equations Using the Division or Multiplication Property

After studying this section, you will be able to:

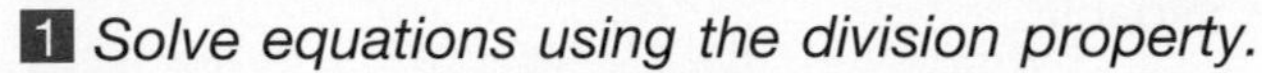
1 *Solve equations using the division property.*

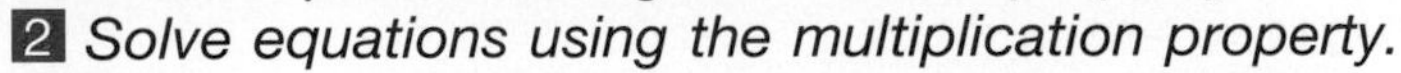
2 *Solve equations using the multiplication property.*

MathPro

Video 22

SSM

1 *Solving Equations Using the Division Property*

Recall that when we solve an equation using the addition property, we transform the equation to a simpler equation where $x =$ some number. We use the same idea to solve the equation $3n = 75$. Think: "What can we do to the left side of the equation so that n stands alone?" If we divide the left side of the equation by 3, n will stand alone. Remember, however, whatever you do to one side of the equation, you must do to the other side of the equation.

$$3n = 75$$

$$\frac{3n}{3} = \frac{75}{3}$$

$$n = 25$$

This is one of two important procedures used to solve equations.

> **Division Property of Equations**
>
> You may divide each side of an equation by the same number to obtain an equivalent equation.

EXAMPLE 1 Solve for n.

(a) $6n = 72$ **(b)** $12n = 30$

(a) $6n = 72$ *The variable n is multiplied by* 6.

$\frac{6n}{6} = \frac{72}{6}$ *Divide each side by* 6.

$n = 12$ $(72 \div 6 = 12)$

(b) $12n = 30$ *The variable n is multiplied by* 12.

$\frac{12n}{12} = \frac{30}{12}$ *Divide each side by* 12.

$n = 2.5$ $(30 \div 12 = 2.5)$

TEACHING TIP It has been a while since students have done proportion equations. You may want to give them a few exercises of this type to do at their seats. Solve for n for each of the following.

A. $8n = 72$

B. $5.5n = 55$

C. $1.7n = 5.1$

(The answers are $n = 9$, $n = 10$, and $n = 3$.)

Practice Problem 1 Solve for n.

(a) $8n = 104$ **(b)** $5n = 36$ ■

Sometimes the coefficient of the variable is a negative number. In solving problems of this type, we therefore need to divide each side of the equation by that negative number.

EXAMPLE 2 Solve for the variable.

(a) $-3n = 51$ **(b)** $-11x = -55$

(a) $-3n = 51$ *The coefficient of n is* -3.

$\frac{-3n}{-3} = \frac{51}{-3}$ *Divide each side of the equation by* -3.

$n = -17$ *Watch your signs!* $[51 \div (-3) = -17]$

(b) $-11x = -55$ *The coefficient of x is* -11.

$\frac{-11x}{-11} = \frac{-55}{-11}$ *Divide each side of the equation by* -11.

$n = 5$ *Watch your signs!* $[-55 \div (-11) = 5]$

Practice Problem 2 Solve for the variable.

(a) $-7n = 35$ **(b)** $-9n = -108$ ■

Sometimes the coefficient of the variable is a decimal. In solving problems of this type, we need to divide each side of the equation by that decimal number.

EXAMPLE 3 Solve for the variable.

(a) $2.5y = 20$ **(b)** $0.3x = 2.7$

(a) $2.5y = 20$ *The coefficient of y is* 2.5.

$\frac{2.5y}{2.5} = \frac{20}{2.5}$ *Divide each side of the equation by* 2.5.

$$2.5_{\wedge}\overline{)20.0_{\wedge}} = 8$$

$y = 8$

To check, substitute 8 for y in the original equation.

$$2.5(8) \stackrel{?}{=} 20$$

$$20 = 20 \checkmark$$

(b) $0.3x = 2.7$ *The coefficient of x is* 0.3.

$\frac{0.3x}{0.3} = \frac{2.7}{0.3}$ *Divide each side of the equation by* 0.3.

$$0.3_{\wedge}\overline{)2.7_{\wedge}} = 9$$

$x = 9$

It is always best to check the solution to equations involving decimals or fractions. Does $0.3(9) = 2.7$? Multiply to check.

Practice Problem 3 Solve for the variable. Check your solution.

(a) $3.2x = 16$ **(b)** $0.5y = 2.9$ ■

2 *Solving Equations Using the Multiplication Property*

Sometimes the coefficient of the variable is a fraction such as $\frac{3}{4}x = 6$. Think: "What can we do to the left side of the equation so that x will stand alone?" Recall that when you multiply a fraction by its reciprocal, the product is 1. That is, $\frac{4}{3} \cdot \frac{3}{4} = 1$. We will use this idea to solve the equation. But remember, whatever you do to one side of the equation, you must do to the other side of the equation.

$$\frac{3}{4}x = 6$$

$$\frac{4}{3} \cdot \frac{3}{4}x = 6 \cdot \frac{4}{3}$$

$$1x = \frac{\overset{2}{\cancel{6}}}{1} \cdot \frac{4}{\cancel{3}}$$

$$x = 8$$

> **Multiplication Property of Equations**
>
> You may multiply each side of an equation by the same number to obtain an equivalent equation.

EXAMPLE 4 Solve for the variable and check your solution.

(a) $\frac{2}{3}x = \frac{1}{2}$ **(b)** $\frac{5}{8}y = 1\frac{1}{4}$ **(c)** $1\frac{1}{2}z = 3$

(a) Think: "What can we multiply the left-hand side of the equation by to obtain a 1 multiplied by the variable?"

$$\frac{2}{3}x = \frac{1}{2}$$

$$\frac{3}{2} \cdot \frac{2}{3}x = \frac{1}{2} \cdot \frac{3}{2}$$ *Multiply both sides of the equation by $\frac{3}{2}$ because $\frac{3}{2} \cdot \frac{2}{3} = 1$.*

$$x = \frac{3}{4}$$

Check.

$$\frac{2}{3}x = \frac{1}{2}$$

$$\frac{2}{3} \cdot \frac{3}{4} \stackrel{?}{=} \frac{1}{2}$$ *Substitute $\frac{3}{4}$ for x in the original equation.*

$$\frac{1}{2} = \frac{1}{2} \quad \checkmark$$

(b)

$$\frac{5}{8}y = 1\frac{1}{4}$$

$$\frac{5}{8}y = \frac{5}{4}$$ *Change the mixed number to a fraction. It will be easier to work with.*

$$\frac{8}{5} \cdot \frac{5}{8}y = \frac{5}{4} \cdot \frac{8}{5}$$ *Multiply both sides of the equation by $\frac{8}{5}$ because $\frac{8}{5} \cdot \frac{5}{8} = 1$.*

$$y = 2$$

Check.

$$\frac{5}{8}y = 1\frac{1}{4}$$

$$\frac{5}{8}(2) \stackrel{?}{=} 1\frac{1}{4}$$ *Substitute 2 for y in the original equation.*

$$\frac{10}{8} \stackrel{?}{=} 1\frac{1}{4}$$

$$1\frac{2}{8} \stackrel{?}{=} 1\frac{1}{4}$$

$$1\frac{1}{4} = 1\frac{1}{4} \quad ✓$$

(c)

$$1\frac{1}{2}z = 3$$

$$\frac{3}{2}z = 3$$ *Change the mixed number to a fraction.*

$$\frac{2}{3} \cdot \frac{3}{2}z = 3 \cdot \frac{2}{3}$$ *Multiply both sides of the equation by $\frac{2}{3}$. Why?*

$$z = 2$$

It is always a good idea to check the solution to an equation involving fractions. We leave the check for this solution up to you.

Practice Problem 4 Solve for the variable and check your solution.

(a) $\frac{3}{5}x = \frac{1}{4}$ **(b)** $\frac{1}{6}y = 2\frac{2}{3}$ **(c)** $3\frac{1}{5}z = 4$ ■

Hint: Remember to write all mixed numbers as improper fractions before performing other algebraic operations to solve linear equations. An alternate method that may also be used is to convert the mixed number to decimal form. Thus equations like $4\frac{1}{8}x = 12$ can also first be written as $4.125x = 12$.

10.4 Exercises

Verbal and Writing Skills

1. How is an equation similar to a balance scale? A sample answer is: To maintain the balance, whatever you do to one side of the scale, you need to do the exact same thing to the other side of the scale.

2. The division property states that we may divide each side of an equation by the same number to obtain an equivalent equation.

3. To change $\frac{3}{4}x = 5$ to a simpler equation, multiply both sides of the equation by $\frac{4}{3}$.

4. Given the equation $1\frac{3}{5}y = 2$, we multiply both sides of the equation by $\frac{5}{8}$ to solve for y. Why?

The coefficient of the variable $1\frac{3}{5} = \frac{8}{5}$. We multiply both sides of the equation by $\frac{5}{8}$ because $\frac{8}{5} \cdot \frac{5}{8} = 1$.

Solve for the variable. Use the division property of equations.

5. $4x = 36$

$\frac{4x}{4} = \frac{36}{4}$

$x = 9$

6. $8x = 56$

$\frac{8x}{8} = \frac{56}{8}$

$x = 7$

7. $7y = -28$

$\frac{7y}{7} = \frac{-28}{7}$

$y = -4$

8. $5y = -45$

$\frac{5y}{5} = \frac{-45}{5}$

$y = -9$

9. $-9x = 18$

$\frac{-9x}{-9} = \frac{18}{-9}$

$x = -2$

10. $-7y = 49$

$\frac{-7y}{-7} = \frac{49}{-7}$

$y = -7$

11. $-5x = -40$

$\frac{-5x}{-5} = \frac{-40}{-5}$

$x = 8$

12. $-4y = -28$

$\frac{-4y}{-4} = \frac{-28}{-4}$

$y = 7$

13. $72 = 8n$

$\frac{72}{8} = \frac{8n}{8}$

$9 = n$

14. $48 = 6n$

$\frac{48}{6} = \frac{6n}{6}$

$8 = n$

15. $27 = -3m$

$\frac{27}{-3} = \frac{-3m}{-3}$

$-9 = m$

16. $64 = -8m$

$\frac{64}{-8} = \frac{-8m}{-8}$

$-8 = m$

17. $1.5x = 9$

$\frac{1.5x}{1.5} = \frac{9}{1.5}$

$x = 6$

18. $3.2y = 16$

$\frac{3.2y}{3.2} = \frac{16}{3.2}$

$y = 5$

19. $0.6x = 6$

$\frac{0.6x}{0.6} = \frac{6}{0.6}$

$x = 10$

20. $0.5y = 50$

$\frac{0.5y}{0.5} = \frac{50}{0.5}$

$y = 100$

21. $5.5z = 9.9$

$\frac{5.5z}{5.5} = \frac{9.9}{5.5}$

$z = 1.8$

22. $3.5n = 7.7$

$\frac{3.5n}{3.5} = \frac{7.7}{3.5}$

$n = 2.2$

23. $-0.6x = 2.7$

$\frac{-0.6x}{-0.6} = \frac{2.7}{-0.6}$

$x = -4.5$

24. $-0.5y = 0.8$

$\frac{-0.5y}{-0.5} = \frac{0.8}{-0.5}$

$y = -1.6$

25. $\frac{2}{3}x = 6$

$\frac{3}{2} \cdot \frac{2}{3}x = \frac{6}{1} \cdot \frac{3}{2}$

$x = 9$

26. $\frac{3}{4}x = 3$

$\frac{4}{3} \cdot \frac{3}{4}x = \frac{3}{1} \cdot \frac{4}{3}$

$x = 4$

27. $\frac{2}{5}y = 4$

$\frac{5}{2} \cdot \frac{2}{5}y = \frac{4}{1} \cdot \frac{5}{2}$

$y = 10$

28. $\frac{5}{6}y = 10$

$\frac{6}{5} \cdot \frac{5}{6}y = \frac{10}{1} \cdot \frac{6}{5}$

$y = 12$

29. $\frac{3}{5}n = \frac{3}{4}$

$\frac{5}{3} \cdot \frac{3}{5}n = \frac{3}{4} \cdot \frac{5}{3}$

$n = \frac{5}{4}$ *or* $1\frac{1}{4}$

30. $\frac{2}{3}z = \frac{1}{3}$

$\frac{3}{2} \cdot \frac{2}{3}z = \frac{1}{3} \cdot \frac{3}{2}$

$z = \frac{1}{2}$

31. $\frac{3}{10}x = \frac{1}{5}$

$\frac{10}{3} \cdot \frac{3}{10}x = \frac{1}{5} \cdot \frac{10}{3}$

$x = \frac{2}{3}$

32. $\frac{4}{7}y = \frac{4}{5}$

$\frac{7}{4} \cdot \frac{4}{7}y = \frac{4}{5} \cdot \frac{7}{4}$

$y = \frac{7}{5}$ *or* $1\frac{2}{5}$

33. $\frac{1}{2}x = 1\frac{1}{3}$

$\frac{2}{1} \cdot \frac{1}{2}x = \frac{4}{3} \cdot \frac{2}{1}$

$x = \frac{8}{3}$ *or* $2\frac{2}{3}$

34. $\frac{9}{10}y = 3\frac{3}{5}$

$\frac{10}{9} \cdot \frac{9}{10}y = \frac{18}{5} \cdot \frac{10}{9}$

$y = 4$

35. $1\frac{1}{4}z = 10$

$\frac{4}{5} \cdot \frac{5}{4}z = \frac{10}{1} \cdot \frac{4}{5}$

$z = 8$

36. $3\frac{1}{2}n = 21$

$\frac{2}{7} \cdot \frac{7}{2}n = \frac{21}{1} \cdot \frac{2}{7}$

$n = 6$

To Think About

37. Mical solves an equation by multiplying both sides of the equation by 0.5 when he should have divided. He gets an answer of 4.5. What should the correct answer be?

Undo the multiplication. $4.5 \div 0.5 = 9$

Now divide. $9 \div 0.5 = 18$

18 is the correct answer.

38. Alexis solves an equation by multiplying both sides of the equation by $\frac{2}{3}$ when she should have multiplied both sides by the reciprocal. She gets an answer of 4. What should the correct answer be?

Undo the multiplication. $4\left(\frac{3}{2}\right) = 6$

Now multiply by the reciprocal. $6\left(\frac{3}{2}\right) = 9$

9 is the correct answer.

Cumulative Review

Perform the operation in the proper order.

39. $5(-2) + 3(-6) + 8$

-20

40. $10 - 4(-3) + (-10) - (-2)$

14

Combine like terms.

41. $6 - 3x + 5y + 7x - 12y$

$4x - 7y + 6$

42. $4b - 2a + 5 - 6a + 3 - b$

$-8a + 3b + 8$

43. The amount of electricity that could be produced for $26.10 in 1970 cost $130.90 to produce in 1995. What is the percent of increase in the cost to produce electricity from 1970 to 1995? Round to the nearest tenth. Source: U.S. Bureau of Labor Statistics

401.5%

44. In 1995 a total of 62.3 million barrels of oil were produced in the world. In that year the United Stated produced 6.5 million barrels of oil. What percent of the world's oil production was produced in the United States in 1995? Round to the nearest tenth. Source: U.S. Energy Administration

10.4%

10.5 Solve Equations Using Two Properties

After studying this section, you will be able to:

1 *Use two properties to solve an equation.*
2 *Solve equations where the variable is on both sides of the equals sign.*
3 *Solve equations with parentheses.*

MathPro

Video 23

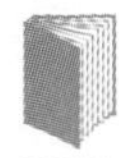
SSM

1 *Using Two Properties to Solve an Equation*

To solve an equation, we take a logical step to change the equation to a simpler equivalent equation. The simpler equivalent equation is $x =$ some number. To do this, we use the addition property, the division property, or the multiplication property. In this section you will use more than one property to solve complex equations. Each time you use a property you take a step toward solving the equation. At each step you try to isolate the variable. That is, you try to get the variable to stand alone.

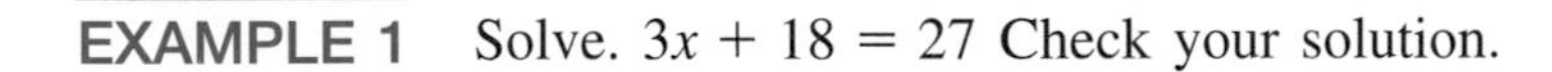

EXAMPLE 1 Solve. $3x + 18 = 27$ Check your solution.

TEACHING TIP It is important to stress determining the correct number to add to each side of the equation. As an oral exercise after discussing Example 1 or a similar problem, ask students to decide, "What number should be added to each side of the equation in each of the following problems?"
A. $-7x + 2 = 51$
B. $-12x - 13 = 47$
C. $20x + 3 = -103$
(The answers are -2 for A, $+13$ for B, and -3 for C.)

We want only x terms on the left and only numbers on the right. We begin by "moving" 18 to the other side of the equals sign.

$3x + 18 + (-18) = 27 + (-18)$ *Add the opposite of 18 to both sides of the equation so that 3x stands alone.*

$$3x = 9$$

$$\frac{3x}{3} = \frac{9}{3}$$ *Divide both sides of the equation by 3 so that x stands alone.*

$$x = 3$$ *The solution to the equation is $x = 3$.*

Check.

$3(3) + 18 \stackrel{?}{=} 27$ *Substitute 3 for x in the original equation.*

$$9 + 18 \stackrel{?}{=} 27$$

$$27 = 27 \quad ✓$$

Practice Problem 1 Solve. Check your solution. $5x + 13 = 33$ ■

EXAMPLE 2 Solve. $9x - 5 = -41$

We begin by "moving" -5 to the right-hand side of the equals sign.

$9x - 5 + 5 = -41 + 5$ *Add the opposite of -5 to both sides of the equation.*

$$9x = -36$$

$$\frac{9x}{9} = \frac{-36}{9}$$ *Divide both sides of the equation by 9.*

$$x = -4$$

The check is left up to you.

Practice Problem 2 Solve. Check your solution. $7x - 8 = -50$ ■

2 *Solving Equations Where the Variable Is on Both Sides of the Equals Sign*

Sometimes variables appear on both sides of the equation. When this occurs, we need to move the variables.

EXAMPLE 3 Solve. $8x = 5x - 21$

We want to "move" $5x$ to the left-hand side of the equation so that all of the variables are on one side of the equation and all of the numbers are on the other side of the equation.

$8x + (-5x) = 5x + (-5x) - 21$ *Add the opposite of 5x to both sides of the equation.*

$$3x = -21$$

$$\frac{3x}{3} = \frac{-21}{3}$$ *Divide both sides of the equation by 3.*

$$x = -7$$

Practice Problem 3 Solve. $4x = -8x + 42$ ■

Sometimes you will have to move the variables to one side of the equation and the numbers to the other side of the equation. To do this, you will have to use the addition property twice.

EXAMPLE 4 Solve. $2x + 9 = 5x - 3$

We begin by moving the numbers to the right-hand side of the equation.

$2x + 9 + (-9) = 5x - 3 + (-9)$ *Add −9 to both sides of the equation.*

$$2x = 5x - 12$$

Now we want to move the variable to the left-hand side of the equation.

$2x + (-5x) = 5x + (-5x) - 12$ *Add −5x to both sides of the equation.*

$$-3x = -12$$

$$\frac{-3x}{-3} = \frac{-12}{-3}$$ *Divide both sides of the equation by 3.*

$$x = 4$$

Check

$2(4) + 9 \stackrel{?}{=} 5(4) - 3$ *Substitute 4 for x in the original equation.*

$$8 + 9 \stackrel{?}{=} 20 - 3$$

$$17 = 17 \checkmark$$

Practice Problem 4 Solve. $4x - 7 = 9x + 13$ ■

If there are like terms on one side of the equation, these should be combined first. Then proceed as above.

EXAMPLE 5 Solve for the variable.

(a) $-5 + 2y + 8 = 7y + 23$ **(b)** $9z + 3 - 2z = 18 + 4z - 4$

(a) We begin by combining the like terms on the left-hand side of the equation.

$2y + 3 = 7y + 23$ *We combined $-5 + 8$.*

$2y + (-7y) + 3 = 7y + (-7y) + 23$ *"Move" the variables to the left-hand side of the equation.*

$-5y + 3 = 23$

$-5y + 3 + (-3) = 23 + (-3)$ *"Move" the 3 to the right-hand side of the equation.*

$-5y = 20$

$\frac{-5y}{-5} = \frac{20}{-5}$ *Divide both sides of the equation by -5.*

$y = -4$

Check.

$$-5 + (2)(-4) + 8 \stackrel{?}{=} (7)(-4) + 23$$

$$-5 + (-8) + 8 \stackrel{?}{=} -28 + 23$$

$$-5 = -5 \checkmark$$

(b) We begin by combining the like terms on both sides of the equation.

$7z + 3 = 4z + 14$ *We combined $9z - 2z$ on the left. We combined $18 - 4$ on the right.*

$7z + (-4z) + 3 = 4z + (-4z) + 14$ *"Move" the variables to the left.*

$3z + 3 + (-3) = 14 + (-3)$ *"Move" 3 to the right.*

$3z = 11$

$\frac{3z}{3} = \frac{11}{3}$ *Divide both sides by 3.*

$z = \frac{11}{3}$ *We leave the solution as a fraction.*

The check is up to you.

Practice Problem 5 Solve for the variable.

(a) $4x - 23 = 3x + 7 - 2x$ **(b)** $7y - 5 + 2y = 2 + 3y + 7$ ■

It is wise to check your answer when solving this type of linear equation. The chance of making a simple error with signs is quite high. Checking gives you a chance to detect this type of error.

TEACHING TIP Usually, about a quarter of the students will not take the time to collect like terms on one side of the equation before they add a value to each side of the equation. Emphasize that this step is very important because it makes the problem much shorter and reduces the chance of making an error.

3 *Solving Equations with Parentheses*

If a problem contains one or more parentheses, remove them using the distributive property. Then collect like terms on each side of the equation. Then solve.

EXAMPLE 6 Solve for x.

$$(2)(x + 3) = 5x - 8 + 10x$$

$2x + 6 = 5x - 8 + 10x$	*Remove parentheses by using the distributive property.*
$2x + 6 = 15x - 8$	*Add like terms on the right-hand side of the equation.*
$2x + 6 + (-6) = 15x + (-8) + (-6)$	*Add -6 to each side.*
$2x = 15x + (-14)$	
$2x + (-15x) = 15x + (-15x) + (-14)$	*Add $-15x$ to each side.*
$-13x = -14$	
$\dfrac{-13x}{-13} = \dfrac{-14}{-13}$	*Divide each side by -13.*
$x = \dfrac{14}{13}$	*Leave the answer in fractional form. (Note that the answer is positive!)*

Practice Problem 6 Solve for x.

$$(3)(x - 2) + 5x = 7x + 10 \quad ■$$

We now list a procedure that may be used to help you remember all the steps we are using to solve equations.

Procedure to Solve Equations

1. Remove any parentheses.
2. Collect like terms on each side of the equation.
3. Add the appropriate value to both sides of the equation to get all numbers on one side.
4. Add the appropriate value to both sides of the equation to get all variable terms on the other side.
5. Divide both sides of the equation by the numerical coefficient of the variable.
6. Check by substituting back into the original equation.

You have probably noticed that steps 3 and 4 are interchangeable. You can do step 3 and then step 4, or you can do step 4 and then step 3.

10.5 Exercises

Check to see if the given answer is a solution to the equation.

1. Is $x = -1$ a solution to $-4 - 8x = -2 - 6x$?
$-4 + 8 \stackrel{?}{=} -2 + 6 \quad 4 = 4$ Yes

2. Is $x = -5$ a solution to $-2x + 7 = -4x - 3$?
$10 + 7 \stackrel{?}{=} 20 - 3 \quad 17 = 17$ Yes

3. Is $x = 2$ a solution to $3 - 4x = 5 - 3x$?
$3 - 8 \stackrel{?}{=} 5 - 6 \quad -5 \neq -1$ No

4. Is $x = 5$ a solution to $3x + 2 = -2x - 23$?
$15 + 2 \stackrel{?}{=} -10 - 23 \quad 17 \neq -33$ No

Solve.

5. $4x + 9 = 5$
$4x = -4 \quad x = -1$

6. $7x + 16 = 2$
$7x = -14 \quad x = -2$

7. $12x - 30 = 6$
$12x = 36 \quad x = 3$

8. $15x - 10 = 35$
$15x = 45 \quad x = 3$

9. $9x - 3 = -7$
$9x = -4 \quad x = -\frac{4}{9}$

10. $6x - 9 = -12$
$6x = -3 \quad x = -\frac{1}{2}$

11. $2x = 7x + 25$
$-5x = 25 \quad x = -5$

12. $4x = 6x + 14$
$-2x = 14 \quad x = -7$

13. $-9x = 3x - 10$
$-12x = -10 \quad x = \frac{5}{6}$

14. $-7x = 2x + 11$
$-9x = 11 \quad x = -\frac{11}{9}$

15. $18 - 2x = 4x + 6$
$-6x = -12 \quad x = 2$

16. $3x + 4 = 7x - 12$
$-4x = -16 \quad x = 4$

17. $8 + x = 3x - 6$
$-2x = -14 \quad x = 7$

18. $9 - 8x = 3 - 2x$
$-6x = -6 \quad x = 1$

19. $7 + 3x = 6x - 8$
$-3x = -15 \quad x = 5$

20. $2x - 7 = 3x + 9$
$-x = 16 \quad x = -16$

21. $5 + 2y = 7 + 5y$
$-3y = 2 \quad y = -\frac{2}{3}$

22. $12 + 5y = 9 - 3y$
$8y = -3 \quad y = -\frac{3}{8}$

23. $4y + 5 = 2y - 9$
$2y = -14 \quad y = -7$

24. $4y + 7 = 6y - 7$
$-2y = -14 \quad y = 7$

25. $-10 + 6y + 2 = 3y - 26$
$3y = -18 \quad y = -6$

26. $6 - 5x + 2 = 4x + 35$
$-9x = 27 \quad x = -3$

27. $12 + 4y - 7 = 6y - 9$
$-2y = -14 \quad y = 7$

28. $6 - 8x - 12 = -13x - 21$
$5x = -15 \quad x = -3$

29. $5z + 7 - 3z = 17 - 9z + 12$
$11z = 22 \quad z = 2$

30. $8x - 6 + 3x = 5 + 7x - 19$
$4x = -8 \quad x = -2$

31. $9 - 3y + 12 = 7y - 15 + 2y$
$36 = 12y \quad y = 3$

32. $13 + 5x - 7 = 3x - 8 - 12x$
$14x = -14 \quad x = -1$

33. $(5)(x + 4) = 4x + 15$
$5x + 20 = 4x + 15 \quad x = -5$

34. $(4)(x + 2) = 8x + 12$
$4x + 8 = 8x + 12 \quad x = -1$

35. $(3)(x + 4) + 2x = 7$
$3x + 12 + 2x = 7 \quad x = -1$

36. $2x + (3)(x - 4) = 3$
$5x = 15 \quad x = 3$

37. $9 + (6)(x - 3) = 3x$
$3x = 9 \quad x = 3$

38. $8 + (2)(x + 9) = 4x$
$-2x = -26 \quad x = 13$

39. $(2)(y + 3) = (4)(g + 5) - 7$
$2y + 6 = 4y + 13 \quad y = -\frac{7}{2}$

40. $13 + (7)(2y - 1) = (5)(y + 6)$
$14y + 6 = 5y + 30 \quad y = \frac{8}{3}$

41. $3(x - 6) = 4(x - 2) - 10$
$3x - 18 = 4x - 18 \quad x = 0$

42. $5x - 6(x + 1) = 2x - 6$
$-x - 6 = 2x - 6 \quad x = 0$

Solve for x.

★ **43.** $(3)(x + 0.2) - (2)(x + 0.25) = (2)(x + 0.3) - 0.5$
$x + 0.1 = 2x + 0.1 \quad -x = 0 \quad x = 0$

★ **44.** $(0.2)(x + 3) - (2)(0.5x - 0.75) = (0.3)(x + 2) - 2.9$
$-0.8x + 2.1 = 0.3x - 2.3 \quad -1.1x = -4.4 \quad x = 4$

To Think About

45. **(a)** Solve $7 + 3x = 6x - 8$ by collecting x terms on the left. $-3x = -15 \quad x = 5$

(b) Solve by collecting x terms on the right.
$15 = 3x \quad 5 = x$

(c) Which method is easier? Why?
Answers will vary

46. **(a)** Solve $8 + 5x = 2x - 6 + x$ by collecting x terms on the left.
$8 + 5x = 3x - 6 \quad 2x = -14 \quad x = -7$

(b) Solve by collecting x terms on the right.
$14 = -2x \quad -7 = x$

(c) Which method is easier? Why?
Answers will vary

Cumulative Review Problems

Use $\pi = 3.14$. Round your answer to the nearest tenth.

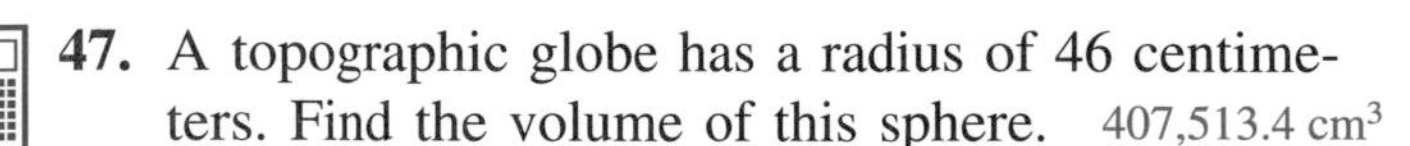

47. A topographic globe has a radius of 46 centimeters. Find the volume of this sphere. 407,513.4 cm^3

48. Find the area of the shaded region in the figure on the right.
23.4 $in.^2$

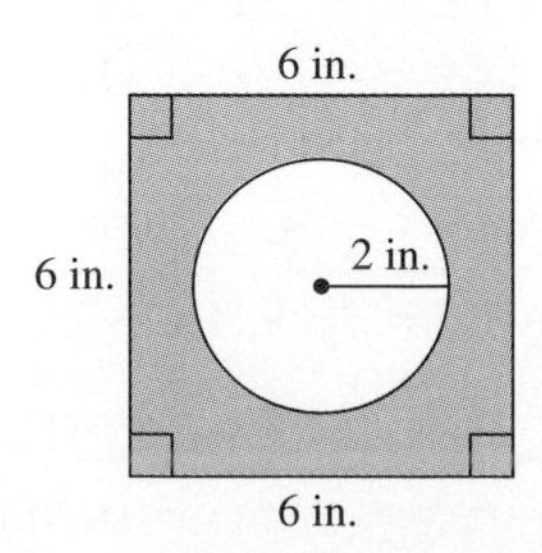

Putting Your Skills to Work

USING MATHEMATICS TO PREDICT POPULATION INCREASES

The U.S. Bureau of the Census not only counts the population of the United States through a census, it makes projections about the number of people likely to be living in this country in the future. The following chart represents the estimate of how the population totals of our country are projected to change through the year 2030.

Of course several factors are hard to predict. These are economic growth, health advances and the spread of diseases, immigration patterns and policies of the United States, and the fertility rate of the population. The table shows a most likely rate of increase figure as well as both a lower and higher rate. Which rate proves to be the best predictor would be influenced by these factors we have listed

UNITED STATES RESIDENT POPULATION PROJECTION (MEASURED IN MILLIONS)

Source: U.S. Bureau of the Census

Year	Most Likely Population	Lower Rate of Population Increase	Higher Rate of Population Increase
2000	274.6	271.2	278.1
2010	297.7	281.5	314.6
2020	322.7	288.8	357.7
2030	346.9	291.1	405.1
2040	370.0	287.7	458.4

The population of the United States in 1996 was 265.3 million people. If we want the most likely population for the United States in the year 2020 we would use the table to find the value 322.7 million people. Then we could make the following calculation:

$\frac{322.7}{265.3} \approx 1.216 = 121.6\%$. Thus we could say that the likely projected pop-population in the year 2020 is 121.6% of the population in 1996. If we form the Population Increase Equation

$$N = (P)(C)$$

where

N = the expected new population

P = the percent that the new population is of the original population

C = the current population in 1996

Then we could write the equation $N = (1.216)(C)$. If a city, town, or entire state experienced growth at approximately the national rate of population growth, then a city or town could approximate its population for the year 2020. For example if a city of population 52,340 in 1996 wanted to determine an estimated population for the year 2020 in order to prepare for the appropriate level of fire and police protection, they could use the equation to perform this calculation:

$$N = (1.216)(52{,}340) = 63{,}645.44$$

Rounded to the nearest person, the estimated population of the city in the year 2020 would be 63,645.

PROBLEMS FOR INDIVIDUAL INVESTIGATION AND ANALYSIS

In each of the following problems determine an equation in the form of $N = (P)(C)$ where P is measured to the nearest thousandth.

1. Find the equation appropriate for a city that experiences the same level of growth from 1996 to 2040 as the entire country in terms of the ''Most Likely Population.'' $N = (1.395)(C)$

2. Find the equation appropriate for a city that experiences the same level of growth from 1996 to 2010 as the entire country if both increase at the ''Higher Rate of Population Increase.'' $N = (1.186)(C)$

3. Find the equation appropriate for a city that experiences the same level of growth from 1996 to 2000 as the entire country if both increase at the ''Lower Rate of Population Increase.'' $N = (1.022)(C)$

PROBLEMS FOR GROUP INVESTIGATION AND COOPERATIVE LEARNING

Round the value of P to the nearest thousandth, then calculate.

4. A city with a population of 346,00 in 1996 generally increases at the same rate as the population of the country. What would be the most likely population in the year 2030? 452,568

5. A town with a population of 12,680 people in 1996 is planning for a new school system for the year 2040. If the population increased at the ''Higher Rate of Population Increase,'' what would be the maximum population of the town? If the population increased at the ''Lower Rate of Population Increase,'' what would be the minimum population of the town?
Maximum population: 21,911
Minimum population: 13,745

INTERNET CONNECTIONS: Go to ''`http://www.prenhall.com/tobey`'' to be connected

Site: Population Growth Remains High (United Nations Population Fund) or a related site

This site gives the mid-1996 world population as well as high and low world population projections for the years 2015 and 2050. Use this information to answer the following questions.

6. Find the equation appropriate for a country that experiences the same level of growth from 1996 to 2015 as the entire world if both increase according to the low projection.

7. Find the equation appropriate for a country that experiences the same level of growth from 1996 to 2050 as the entire world if both increase according to the high projection.

10.6 Translating English to Algebra

After studying this section, you will be able to:

1 *Translate simple comparisons in English into mathematical equations using two given variables.*
2 *Write algebraic expressions for several quantities using one given variable.*

MathPro

Video 23

SSM

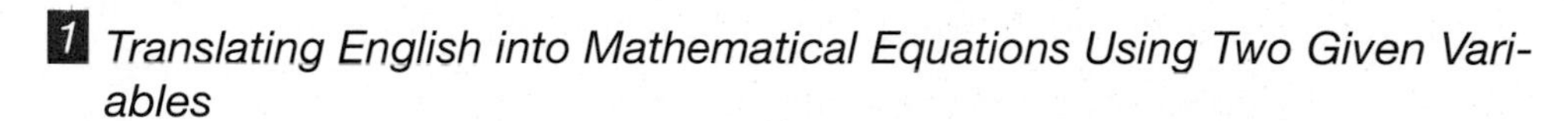

1 *Translating English into Mathematical Equations Using Two Given Variables*

In the preceding section you learned how to solve an equation. We can use equations to solve applied problems, but before we can do that, we need to know how to write an equation that will represent the situation in a word problem.

In this section we practice *translating* to help you to write your own mathematical equations. That is, we translate English expressions into algebraic expressions. In the next section we'll apply this translation skill to a variety of word problems.

The following chart presents the mathematical symbols generally used in translating English phrases into equations.

The English Phrase:	Is Usually Represented by the Symbol:
greater than increased by more than added to taller than total of sum of	+
less than decreased by smaller than fewer than shorter than difference of	−
double	2×
product of times	×
triple	3×
ratio of quotient of	÷
is was has has the value of costs weighs equals represents amounts to	=

The relationship between two or more objects described in an English sentence can often be conveniently expressed as a short equation using variables. For example, if we say in English "Bob's salary is $1000 greater than Fred's salary," we can express the mathematical relationship by the equation

$$b = 1000 + f$$

where b represents Bob's salary and f represents Fred's salary.

EXAMPLE 1 Translate the English sentence into an equation using variables. Use r to represent Roberto's weight and j to represent Juan's weight.

Roberto's weight is 42 lb more than Juan's weight.

Roberto's weight	is	42 lb	more than	Juan's weight
↓	↓	↓	↓	↓
r	$=$	42	$+$	j

The equation $r = 42 + j$ could also be written as

$$r = j + 42$$

Both are correct translations because addition is commutative.

Practice Problem 1 Translate the English sentence into an equation using variables. Use t to represent Tom's height and a to represent Abdul's height.

Tom's height is 7 in. more than Abdul's height. ■

When translating the phrases "less than" or "fewer than," it is important to watch out that the number is subtracted *from* the variable. Words like "costs" "weighs," or "has the value of" act like an equality sign (=).

EXAMPLE 2 Translate the English sentence into an equation using variables. Use c to represent the cost of the chair in dollars and s to represent the cost of the sofa in dollars.

The chair costs $200 less than the sofa.

The chair	costs		$200 less than
↓	↓		↘
c	$=$	s	$-$ 200
		↑	
		the sofa	

Note the order of the equation. We subtract 200 from s, so we have $s - 200$. It would be incorrect to write $200 - s$. Do you see why?

TEACHING TIP After you have explained a few examples, have the class do the following exercises at their seats. In each case, let $b =$ the number of articles Bob has and $m =$ the number of articles Mike has.

A. Bob has 12 more articles than Mike has.

B. Mike has 7 less articles than Bob has.

C. Mike has 5 more than double the number of articles Bob has.

D. Bob has 15 less than triple the number of articles Mike has.

(Answers: A: $b = 12 + m$; B: $m = b - 7$; C: $m = 5 + 2b$; D: $b = 3m - 15$)

Practice Problem 2 Translate the English sentence into an equation using variables. Use n to represent the number of students in the noon class and m to represent the number of students in the morning class.

The noon class has 24 students less than the morning class. ■

In a similar way we can translate the phrases "more than" or "greater than," but since addition is commutative, we find it easier to write the mathematical symbols in the same order as the words in the English sentence. Note that words such as "drove" or "carried" sometimes are translated with an equality symbol.

EXAMPLE 3 Translate the English sentence into an equation using variables. Use t to represent the number of miles driven on Tuesday and m to represent the number of miles driven on Monday.

On Tuesday Yvonne drove 80 mi more than she did on Monday.

On Tuesday	Yvonne drove	80 mi	more than	on Monday
t	$=$	80	$+$	m

Practice Problem 3 Translate the English sentence into an equation with variables. Use t to represent the number of boxes carried on Thursday and f to represent the number of boxes carried on Friday.

On Thursday Adrianne carried five more boxes into the dorm than she did on Friday. ■

EXAMPLE 4 Translate the English sentence into an equation with variables. Use l to represent the length and w to represent with width.

The length of the rectangle is 3 inches more than double the width.

The length	is	3	more than	double the width
l	$=$	3	$+$	$2w$

Note here that if the width is w, then to double the width is to multiply the width by 2. We may write this as $2 \times w$ or $2(w)$ or, more simply, as $2w$.

Practice Problem 4 Translate the English sentence into an equation with variables. Use l to represent the length and w to represent the width.

The length of the rectangle is 7 m more than triple the width. ■

2 *Writing Algebraic Expressions for Several Quantities Using One Given Variable*

In each of the examples so far we have used two *different variables*. Now we'll learn how to write algebraic expressions for several quantities using the *same variable*. In the next section we'll use this skill to write and solve equations.

A mathematical expression that contains a variable is often called an **algebraic expression.**

EXAMPLE 5 Write algebraic expressions for Bob's salary and Fred's salary. Fred's salary is $150 more than Bob's salary. Use the letter b.

Let b = Bob's salary.

Let $\underbrace{b + 150}_{\text{\$150 more than Bob's salary}}$ = Fred's salary.

Notice that Fred's salary is compared to Bob's salary. Thus it is logical to let Bob's salary be b and then express Fred's salary as $150 more than Bob's.

Practice Problem 5 Write algebraic expressions for Sally's trip and Melinda's trip. Melinda's trip is 380 mi longer than Sally's trip. Use the letter s. ■

TEACHING TIP Present to the students this thought-stimulating problem. Let Bob's salary last year be x dollars. Describe in symbols: **A.** His salary was increased by $400 and then he received a 10% raise. **B.** He received a 10% raise, and then he received a further increase of $400. **C.** Which salary would be higher for Bob?
(Answers: A: $1.10\ (x + 400)$; B: $1.10x + 400$; C: His salary is higher if he gets the $400 raise first.)

TEACHING TIP After you have covered Examples 7 and 8, or similar class examples, have the students try the following exercises at their seats. In each case, express each quantity in terms of x.

A. The Dodge truck weighs 300 lb more than the Ford truck.

B. The psychology class has 20 more than double the number of students in the music appreciation class.

C. The first side of a triangle is 7 inches longer than the second. The third side is 2 inches shorter than double the second.

(Answers: A: Dodge truck $= x + 300$, Ford truck $= x$; B: psychology class $= 20 + 2x$, music appreciation class $= x$; C: length of first side $= x + 7$, length of second side $= x$, length of third side $= 2x - 2$.)

EXAMPLE 6 Write algebraic expressions for the size of each of two angles of a triangle. Angle B of the triangle is 34° less than angle A. Use the letter A.

Let A = the number of degrees in angle A.

Let $\underbrace{A - 34}_{34° \text{ less than angle } A}$ = the number of degrees in angle B.

Practice Problem 6 Write algebraic expressions for the height of each of two buildings. Larson Center is 126 ft shorter than McCormick Hall. Use the letter m. ■

Often in algebra when we write expressions for one or two unknown quantities, we use the letter x.

EXAMPLE 7 Write algebraic expressions for the number of students in each of two classes. The Monday class has 17 more students than the Tuesday class. Use the letter x.

Let x = the number of students in the Tuesday class.

Let $x + 17$ = the number of students in the Monday class.

Practice Problem 7 Write an algebraic expression for the number of dollars it costs to make each of two purchases. A sofa costs $215 more than a chair. Use the letter x. ■

EXAMPLE 8 Write algebraic expressions for the length of each of three sides of a triangle. The second side is 4 in. longer than the first. The third side is 7 in. shorter than triple the length of the first side. Use the letter x.

Since the other two sides are compared to the first side, we start by writing an expression for the first side.

Let x = the length of the first side.

Let $x + 4$ = the length of the second side.

Let $3x - 7$ = the length of the third side.

Practice Problem 8 Write an algebraic expression for the length of each of three sides of a triangle. The second side is double the length of the first side. The third side is 6 in. longer than the first side. Use the letter x. ■

10.6 Exercises

Verbal and Writing Skills

Translate the English sentence into an equation using the variables indicated for each problem.

1. Emilio's height is 7 in. more than Martin's height. Use e for Emilio's height and m for Martin's height.
$e = 7 + m$

2. Kathy's tomato plants are 3 ft higher than Sylvia's tomato plants. Use k for Kathy's tomato plants and s for Sylvia's tomato plants.
$k = 3 + s$

3. The college trunk for Henry's books weighs 42 lb more than the college trunk for Fred's books. Use h for the weight of Henry's books and f for the weight of Fred's books.
$h = 42 + f$

4. The large cereal box contains 7 oz more than the small box of cereal. Use l for the number of ounces in the large cereal box and s for the number of ounces in the small cereal box.
$l = s + 7$

5. The bracelet costs $107 less than the necklace. Use b for the bracelet and n for the necklace.
$b = n - 107$

6. The doctor has 20 fewer patients this week than last week. Use t to represent the number of patients this week and l to represent the number of patients last week.
$t = l - 20$

7. The movie had 111 fewer people on Friday than it did on Saturday. Use f for the number of people who saw the movie on Friday and s for the number of people who saw the movie on Saturday.
$f = s - 111$

8. The car dealership sold 42 fewer used cars than new cars. Use u for the number of used cars and n for the number of new cars.
$u = n - 42$

9. Dwight threw the javelin 5 meters less than Evan did. Use d for the distance Dwight threw the javelin and e for the distance that Evan threw the javelin.
$d = e - 5$

10. The Swiss chocolate bar is 4 cm thicker than the Belgian chocolate bar. Use s for the thickness of the Swiss chocolate bar and b for the thickness of the Belgian chocolate bar.
$s = 4 + b$

11. The length of the rectangle is 7 m longer than double the width. Use l for the length and w for the width of the rectangle.
$l = 2w + 7$

12. The length of the rectangle is 5 m longer than triple the width. Use l for the length and w for the width of the rectangle.
$l = 3w + 5$

13. The length of the second side of a triangle is 2 in. shorter than triple the length of the first side of the triangle. Use s to represent the dimension of the second side and f to represent the dimension of the first side.
$s = 3f - 2$

14. The length of the second side of a triangle is 4 in. shorter than double the length of the first side of the triangle. Use s to represent the dimension of the second side and f to represent the dimension of the first side.
$s = 2f - 4$

15. The cost of the average home in the 1990s is $3000 more than triple the cost of the average home in the 1960s. Use s to represent the cost of the average home in the 1960s and n to represent the cost of the average home in the 1990s.
$n = 3000 + 3s$

16. The cost of buying an average bag of groceries in the 1980s was $40 more than double the cost of an average bag of groceries in the 1950s. Use e to represent the average bag of groceries in the 1980s and f to represent the average bag of groceries in the 1950s.
$e = 40 + 2f$

17. The combined number of hours Jenny and Sue watch television per week is 26 hours. Use j for the number of hours Jenny watches television and s for the number of hours Sue watches television.

$j + s = 26$

18. The attendance at the Giants game was greater than the attendance at the Jets game. The difference in attendance between the two games was 7234 fans. Use g for the number of fans at the Giants game and j for the number of fans at the Jets game.

$g - j = 7234$

19. The product of hourly wages by the time worked results in \$500. Let h = number of dollars per hour paid for hourly wages and t = number of hours worked on time records.

$ht = 500$

20. The ratio of men to women at Central College is 5 to 3. Let m = the number of men and w = the number of women.

$\frac{m}{w} = \frac{5}{3}$ *or* $3m = 5w$

Write algebraic expressions for each quantity using the given variable.

21. Jim's salary is \$600 more than Charlie's salary. Use the letter c.

c = Charlie's salary $c + 600$ = Jim's salary

22. The truck's weight is 1500 lb more than the car's weight. Use the letter c.

c = car's weight $c + 1500$ = truck's weight

23. Barbara's car trip was 386 mi longer than Julie's car trip. Use the letter j.

j = Julie's mileage $j + 386$ = Barbara's mileage

24. The cost of the television set was \$212 more than the cost of the compact disc player. Use the letter p.

p = cost of player $p + 212$ = cost of television

25. Angle A of the triangle is 46° less than angle B. Use the letter b.

b = number of degrees in angle B $b - 46°$ = number of degrees in angle A

26. The top of the box is 38 cm shorter than the side of the box. Use the letter s.

s = length of the side of the box in cm $s - 38$ = length of the top of the box in cm

27. Mount Everest is 4430 m taller than Mount Whitney. Use the letter w.

w = height of Mt. Whitney in meters $w + 4430$ = height of Mt. Everest in meters

28. Mount McKinley is 1802 m taller than Mount Ranier. Use the letter r.

r = height of Mt. Ranier in meters $r + 1802$ = height of Mt. McKinley in meters

29. Wally has taken twice as many biology classes as Ernesto. Use the letter e.

e = classes taken by Ernesto $2e$ = classes taken by Wally

30. Juanita has worked for the company three times as long as Charo. Use the letter c.

c = Charo's time $3c$ = Juanita's time

31. The length of the rectangle is 12 m longer than triple the width. Use the letter x.

x = width $3x + 12$ = length

32. The length of the rectangle is 7 m longer than double the width. Use the letter x.

x = width $2x + 7$ = length

33. The second angle of a triangle is double the first. The third angle of the triangle is 14° smaller than the first. Use the letter x.

x = 1st angle $2x$ = second angle $x - 14$ = third angle

34. The second angle of a triangle is triple the first. The third angle of the triangle is 36° larger than the first. Use the letter x.

x = 1st angle $3x$ = 2nd angle $x + 36$ = 3rd angle

Cumulative Review Problems

Perform the operations in the proper order.

35. $-6 - (-7)(2)$

8

36. $24 \div (-6) + 5$

1

37. $5 - 5 + 8 - (-4) + 2 - 15$

-1

38. $(2)(-3)(-1)(3)(-1)$

-18

10.7 Applications

After studying this section, you will be able to:

1 *Solve problems involving comparisons.*
2 *Solve problems involving geometric formulas.*
3 *Solve problems involving rates and percents.*

MathPro  Video 23  SSM

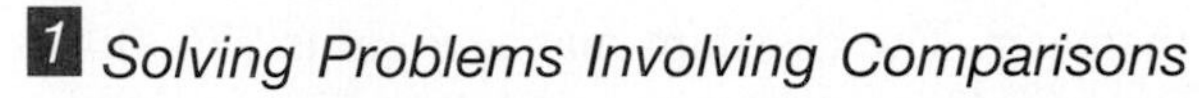

1 *Solving Problems Involving Comparisons*

To solve the following problem, we use the three steps for problem solving with which you are familiar, plus another step: *Write an equation.*

EXAMPLE 1 A 12-ft board is cut into two pieces. The longer piece is 3.5 ft longer than the shorter piece. What is the length of each piece?

1. *Understand the problem.*
Draw a diagram.

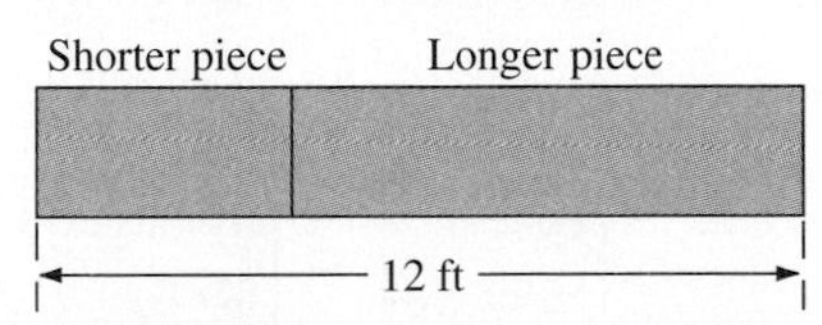

Since the longer piece is *compared to* the shorter piece, we let the variable represent the shorter piece. Let x = the length of the shorter piece. The longer piece is 3.5 ft longer than the shorter piece. Let $x + 3.5$ = the length of the longer piece. The sum of the two pieces is 12 ft. We write an equation.

2. *Write an equation.*

$$x + x + 3.5 = 12$$

3. *Solve the equation and state the answer.*

$$x + x + 3.5 = 12$$

$$2x + 3.5 = 12 \qquad \textit{Collect like terms.}$$

$$2x + 3.5 + (-3.5) = 12 + (-3.5) \qquad \textit{Add } -3.5 \textit{ to each side.}$$

$$2x = 8.5 \qquad \textit{Collect like terms.}$$

$$\frac{2x}{2} = \frac{8.5}{2} \qquad \textit{Divide each side by 2.}$$

$$x = 4.25$$

The shorter piece is 4.25 ft long.

$$x + 3.5 = \text{the longer piece}$$

$$4.25 + 3.5 = 7.75$$

The longer piece is 7.75 ft long.

4. *Check.*
We verify solutions to word problems by making sure that all the calculated values satisfy the original conditions. Do the two pieces add up to 12 ft?

$$4.25 + 7.75 \stackrel{?}{=} 12$$

$$12 = 12 \quad ✓ \quad \text{Yes}$$

Is one piece 3.5 ft longer than the other?

$$7.75 \stackrel{?}{=} 3.5 + 4.25$$

$$7.75 = 7.75 \quad ✓ \quad \text{Yes}$$

Practice Problem 1 An 18-ft board is cut into two pieces. The longer piece is 4.5 ft longer than the shorter piece. What is the length of each piece? ■

If one item is less than another, we use subtraction. We usually let the variable represent the larger quantity.

TEACHING TIP Students tend to want to avoid writing down the initial steps of "Let x = José's annual salary" and "Let $x - 7600$ = the assistant manager's annual salary." Encourage them not to skip these steps by telling them that taking the time to write these steps will help them organize the facts of the problem, determine which answer is which, and check the answers to the problem.

EXAMPLE 2 José is a store manager. The assistant manager earns \$7600 less annually than José does. The sum of José's annual salary and the assistant manager's annual salary is \$42,000. How much does each earn?

1. *Understand the problem.*

Let x = José's annual salary.

Let $x - 7600$ = the assistant manager's annual salary.

José's salary x	+	assistant manager's salary $x - 7600$	=	\$42,000 total annual salary of the two people.

2. *Write an equation.*

$$x + x - 7600 = 42{,}000$$

3. *Solve the equation and state the answer.*

$$x + x - 7600 = 42{,}000$$

$$2x - 7600 = 42{,}000 \quad \textit{Collect like terms.}$$

$$2x + (-7600) + 7600 = 42{,}000 + 7600 \quad \textit{Add 7600 to each side.}$$

$$2x = 49{,}600 \quad \textit{Collect like terms.}$$

$$\frac{2x}{2} = \frac{49{,}600}{2} \quad \textit{Divide each side by 2.}$$

$$x = 24{,}800$$

José earns \$24,800 annually. The assistant manager earns \$7600 less.

$$x - 7600 = 24{,}800 - 7600 = 17{,}200$$

The assistant manager earns \$17,200 annually.

4. *Check.*

Does the assistant manager earn \$7600 less than José?

$$24{,}800 - 7600 \stackrel{?}{=} 17{,}200$$

$$17{,}200 = 17{,}200 \ \checkmark \ \text{Yes}$$

Is the sum of the two salaries 42,000?

$$24{,}800 + 17{,}200 \stackrel{?}{=} 42{,}000$$

$$42{,}000 = 42{,}000 \ \checkmark \ \text{Yes}$$

Practice Problem 2 Mike and Linda bought a sofa. They also bought an easy chair that cost \$186 less than the sofa. The total cost of the two items was \$649. How much did each item cost? ■

Sometimes three items are compared. Let a variable represent the quantity to which things are compared. Then write an expression for the other two quantities.

EXAMPLE 3 Professor Jones is teaching 332 students in three sections of general psychology this semester. His noon class has 23 students more than his 8:00 A.M. class. His 2:00 P.M. class has 36 students fewer than his 8:00 A.M. class. How many students are in each class?

1. *Understand the problem.*
   Each class enrollment is compared to the enrollment in the 8:00 A.M. class.

   Let $x =$ the number of students in the 8:00 A.M. class.

   The noon class has 23 more students than the 8:00 A.M. class.

   Let $x + 23 =$ the number of students in the noon class.

   The 2:00 P.M. class has 36 fewer students than the 8:00 A.M. class.

   Let $x - 36 =$ the number of students in the 2:00 P.M. class.

   The total enrollment for the three sections is 332.
   You can draw a diagram.

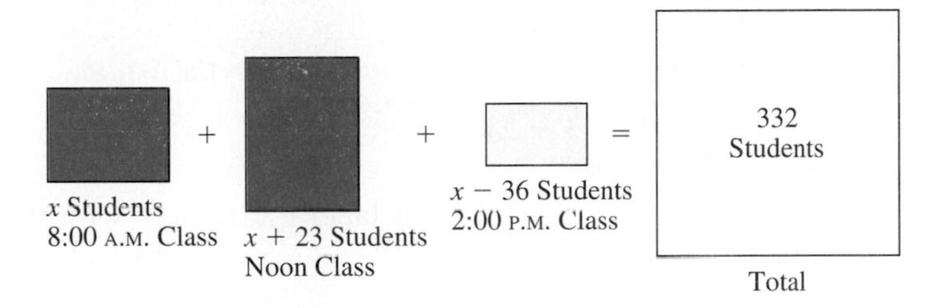

2. *Write an equation.* $x + x + 23 + x - 36 = 332$

3. *Solve the equation and state the answer.*

$$x + x + 23 + x - 36 = 332$$

$$3x - 13 = 332 \qquad \textit{Collect like terms.}$$

$$3x + (-13) + 13 = 332 + 13 \qquad \textit{Add 13 to each side.}$$

$$\frac{3x}{3} = \frac{345}{3} \qquad \textit{Divide each side by 3.}$$

$$x = 115 \quad \text{8:00 A.M. class}$$

$$x + 23 = 115 + 23 = 138 \quad \text{noon class}$$

$$x - 36 = 115 - 36 = 79 \quad \text{2:00 P.M. class}$$

   Thus there are 115 students in the 8:00 A.M. class, 138 students in the noon class, and 79 students in the 2:00 P.M. class.

4. *Check.*
   Does the number of students in each class total 332?

$$115 + 138 + 79 \stackrel{?}{=} 332$$

$$332 = 332 \ \checkmark \ \text{Yes}$$

   Does his noon class have 23 students more than his 8:00 A.M. class?

$$138 \stackrel{?}{=} 23 + 115$$

$$138 = 138 \ \checkmark \ \text{Yes}$$

   Does his 2:00 P.M. class have 36 students fewer than his 8:00 A.M. class?

$$79 \stackrel{?}{=} 115 - 36$$

$$79 = 79 \ \checkmark \ \text{Yes}$$

TEACHING TIP Have the students check the following problem answers. Central Freight dispatched 5 fewer trucks on Monday than on Tuesday. The company dispatched 9 more trucks on Wednesday than on Tuesday. The total number of trucks dispatched on the three days was 173 trucks. A student obtained the following answers: 53 trucks dispatched on Monday, 58 trucks on Tuesday, and 62 trucks on Wednesday. Was the student's answer correct? (Answer: No. The sum of the trucks does total 173. The number of trucks Monday is 5 fewer than on Tuesday. However, the number of trucks on Wednesday is not 9 more than the number on Tuesday.) Examples like this help students see the need for checking all three parts of word problems to see if the answers they obtained are correct.

Practice Problem 3 The city airport had a total of 349 departures on Monday, Tuesday, and Wednesday. There were 29 more departures on Tuesday than on Monday. There were 16 fewer departures on Wednesday than on Monday. How many departures occurred on each day? ■

2 *Solving Problems Involving Geometric Formulas*

Some applied problems have to do with the geometric properties of two-dimensional figures. The problems may involve perimeter or the measure of the angles in a triangle.

Recall that when we double something, we are multiplying by 2. That is, if something is x units, then double that value is $2x$. Triple that value is $3x$.

EXAMPLE 4 A farmer wishes to fence in a rectangular field with 804 ft of fence. The length is to be 3 ft longer than *double the width.* How long and how wide is the field?

1. *Understand the problem.*
   The perimeter of a rectangle is given by $P = 2w + 2l$.
   Let w = the width.

   The length is 3 ft *longer than double the width.*

   $$\text{Length} = 3 + 2w$$

   Thus $2w + 3$ = the length.
   You may wish to draw a diagram and label the figures with the given facts.

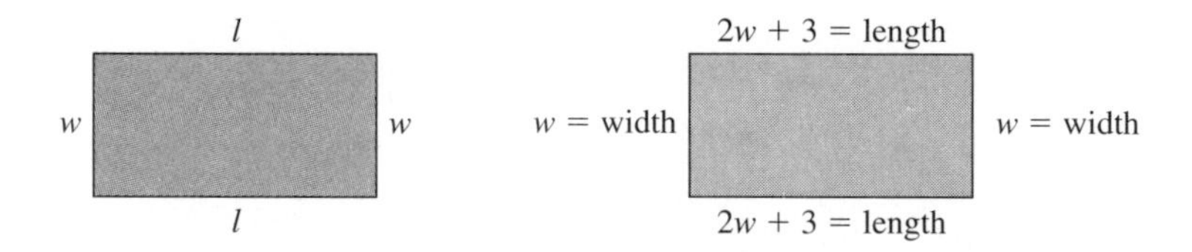

2. *Write an equation.*
   Substitute the given facts into the perimeter formula.

   $$2w + 2l = P$$
   $$2w + 2(2w + 3) = 804$$

3. *Solve the equation and state the answers.*

   $$2w + 2(2w + 3) = 804$$
   $$2w + 4w + 6 = 804 \qquad \textit{Use the distributive property.}$$
   $$6w + 6 = 804 \qquad \textit{Collect like terms.}$$
   $$6w + 6 + (-6) = 804 + (-6) \qquad \textit{Add } -6 \textit{ to each side.}$$
   $$6w = 798 \qquad \textit{Simplify.}$$
   $$\frac{6w}{6} = \frac{798}{6} \qquad \textit{Divide each side by 6.}$$
   $$w = 133$$

   The width is 133 ft.
   The length $= 2w + 3$. When $w = 133$ we have

   $$(2)(133) + 3 = 266 + 3 = 269$$

   Thus the length is 269 ft.

4. *Check.*

Is the length 3 ft longer than double the width?

$$269 \stackrel{?}{=} 3 + (2)(133)$$
$$269 \stackrel{?}{=} 3 + 266$$
$$269 \stackrel{?}{=} 269 \quad ✓ \quad \text{Yes}$$

Is the perimeter 804 ft?

$$(2)(133) + (2)(269) \stackrel{?}{=} 804$$
$$266 + 538 \stackrel{?}{=} 804$$
$$804 = 804 \quad ✓ \quad \text{Yes}$$

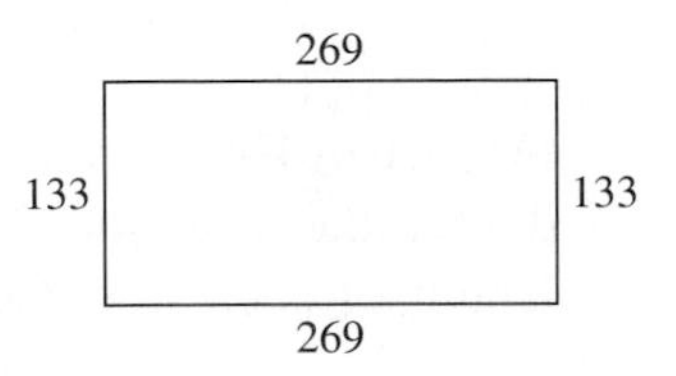

TEACHING TIP Students sometimes forget that in perimeter problems there is an essential difference between triangle problems and rectangle problems. They often write $P = w + l$ for a rectangle instead of the correct formula $P = 2w + 2l$.

Practice Problem 4 What are the length and width of a rectangular field that has a perimeter of 772 ft and a length that is 8 ft longer than double the width? ■

EXAMPLE 5 The perimeter for a triangular rug section is 21 ft. The second side is double the length of the first side. The third side is 3 ft longer than the first side. Find the length of the three sides of the rug.

Let x = the length of the first side.

Let $2x$ = the length of the second side.

Let $x + 3$ = the length of the third side.

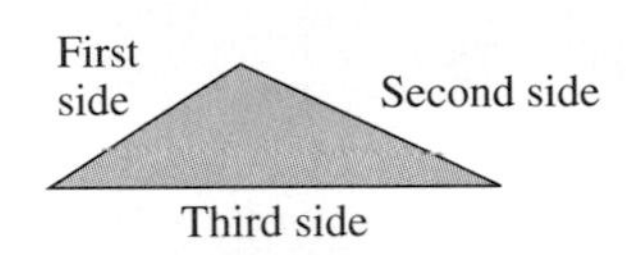

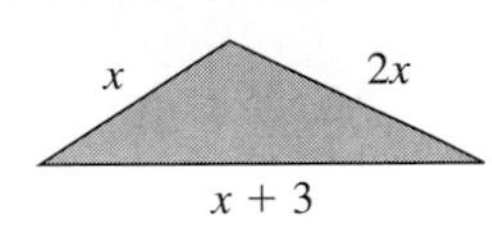

The distance around the three sides totals 21 ft.
Thus

$$x + 2x + x + 3 = 21 \quad \textit{The perimeter equation.}$$
$$4x + 3 = 21 \quad \textit{Collect like terms.}$$
$$4x + 3 + (-3) = 21 + (-3) \quad \textit{Add } -3 \textit{ to each side.}$$
$$4x = 18 \quad \textit{Simplify.}$$
$$\frac{4x}{4} = \frac{18}{4} \quad \textit{Divide each side by 4.}$$
$$x = 4.5$$

The first side is 4.5 ft long.

$$2x = (2)(4.5) = 9 \text{ ft}$$

The second side is 9 ft long.

$$x + 3 = 4.5 + 3 = 7.5 \text{ ft}$$

The third side is 7.5 ft long.

Check.

Do the three sides add up to a perimeter of 21 ft?

$$4.5 + 9 + 7.5 \stackrel{?}{=} 21$$
$$21 = 21 \quad ✓ \quad \text{Yes}$$

Is the second side double the length of the first side?

$$9 \stackrel{?}{=} (2)(4.5)$$

$$9 = 9 \quad ✓ \quad \text{Yes}$$

Is the third side 3 ft longer than the first side?

$$7.5 \stackrel{?}{=} 3 + 4.5$$

$$7.5 = 7.5 \quad ✓ \quad \text{Yes}$$

Practice Problem 5 The perimeter of a triangle is 36 m. The second side is double the first side. The third side is 10 m longer than the first side. Find the length of each side. Check your solutions. ■

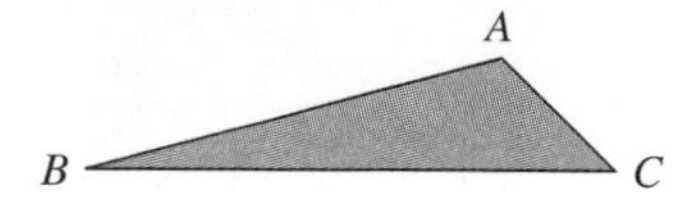

EXAMPLE 6 A triangle has three angles, A, B, and C. Angle C is triple angle B. Angle A is 105° larger than angle B. Find the measure of each angle. Check your answer.

Let x = the number of degrees in angle B.

Let $3x$ = the number of degrees in angle C.

Let $x + 105$ = the number of degrees in angle A.

The sum of the interior angles of a triangle is 180°. Thus we can write

$$x + 3x + x + 105 = 180$$

$$5x + 105 = 180$$

$$5x + 105 + (-105) = 180 + (-105)$$

$$5x = 75$$

$$\frac{5x}{5} = \frac{75}{5}$$

$$x = 15$$

Angle B measures 15°.

$$3x = (3)(15) = 45$$

Angle C measures 45°.

$$x + 105 = 15 + 105 = 120$$

Angle A measures 120°.

Check.
Do the angles total 180°?

$$15 + 45 + 120 \stackrel{?}{=} 180$$

$$180 = 180 \quad ✓ \quad \text{Yes}$$

Is angle C triple angle B?

$$45 \stackrel{?}{=} (3)(15)$$

$$45 = 45 \quad ✓ \quad \text{Yes}$$

Is angle A 105° larger than angle B? $\quad 120 \stackrel{?}{=} 105 + 15$

$$120 = 120 \quad ✓ \quad \text{Yes}$$

Practice Problem 6 Angle C of a triangle is triple angle A. Angle B is 30° less than angle A. Find the measure of each angle. ■

3 *Solving Problems Involving Rates and Percents*

You can use equations to solve problems that involve rates and percents. Recall that the commission a salesperson earns is based on the total sales made. For example, a saleswoman earns \$40 if she gets a 4% commission and she sells \$1000 worth of products. That is, 4% of \$1000 = \$40. Sometimes a salesperson earns a base salary. The commission will then be added to the base salary to determine the total salary. You can find the total salary if you know the amount of the sales. How would you find the amount of sales if the salary were known? We will use an equation.

EXAMPLE 7 This month's salary for an encyclopedia saleswoman was \$3000. This includes her base monthly salary of \$1800 plus a 5% commission on total sales. Find the total sales for the month.

Total salary \$3000	=	base salary of \$1800	+	5% commission on total sales

Let s = the amount of total sales.
Then $0.05s$ = the amount of commission earned from the sales.

$$3000 = 1800 + 0.05s$$

$$1200 = 0.05s$$

$$\frac{1200}{0.05} = \frac{0.05s}{0.05}$$

$$24{,}000 = s$$

She sold \$24,000 worth of encylopedias.

Check.
Does 5% of \$24,000 added to \$1800 yield a salary of \$3000?

$$(0.05)(24{,}000) + 1800 \stackrel{?}{=} 3000$$

$$1200 + 1800 \stackrel{?}{=} 3000$$

$$3000 = 3000 \quad \checkmark \quad \text{Yes}$$

TEACHING TIP An additional example, similar to Example 7, that you may want to show your students is: Roberta earned \$1960 last month. Her salary is based on a monthly salary of \$1000 plus 8% commission on her monthly sales. Find the amount of her monthly sales last month. (Answer: Her sales were \$12,000 for the month.)

Practice Problem 7 A salesperson at a boat dealership earns \$1000 a month plus a 3% commission on the total sales price of the boats he sells. Last month he earned \$3250. What was the total sales price of the boats he sold? ■

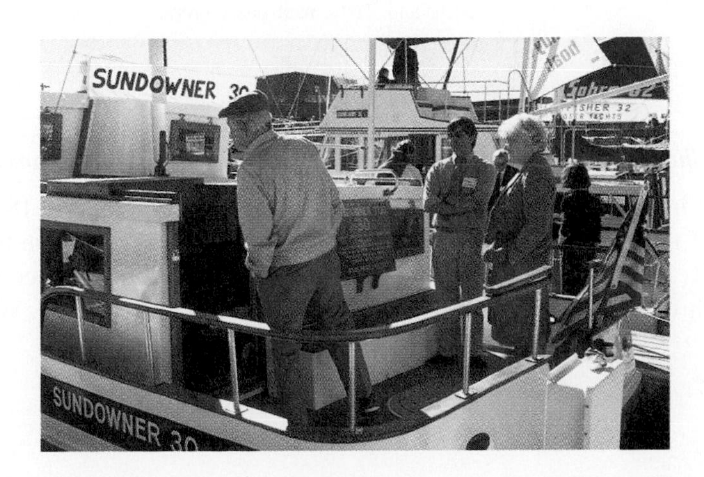

10.7 Exercises

Applications

Solve each problem by using an equation. Show what you set the variable equal to.

1. A 16-ft board is cut into two pieces. The longer piece is 5.5 ft longer than the shorter piece. What is the length of each piece?

x = length of shorter piece

$x + 5.5$ = length of longer piece

$x + 5.5 + x = 16$

$x = 5.25$ ft $x + 5.5 = 10.75$ ft

2. A 20-ft board is cut into two pieces. The longer piece is 4.5 ft longer than the shorter piece. What is the length of each piece?

x = length of shorter piece

$x + 4.5$ = length of longer piece

$x + x + 4.5 = 20$

$x = 7.75$ ft $x + 4.5 = 12.25$ ft

3. In two days Ramone Sánchez drove 560 mi to a vacation campground. He drove 97 more miles the first day than the second day. How far did he drive each day?

x = no. of miles driven 2nd day

$x + 97$ = no. of miles driven 1st day

$x + 97 + x = 560$

$x = 231.5$ mi $x + 97 = 328.5$ mi

4. In two days Felicia drove 325 mi to visit an aunt. She drove 55 more miles the first day than the second day. How far did she drive each day?

x = no. of miles driven 2nd day

$x + 55$ = no. of miles driven 1st day

$x + 55 + x = 325$ $x = 135$ mi

$x + 55 = 190$ mi

5. At the gym, Jodie can leg-press 140 lb less than Jack. Together they are able to leg-press 670 lb of weights. How much can each leg-press separately?

Jodie can leg-press 265 lb.
Jack can leg-press 405 lb.

6. Sandra raised \$271 less than Marlena for their favorite charity. Together they raised \$1977. How much did each woman raise for charity?

Sandra raised \$853.
Marlena raised \$1124.

7. The Business Club's thrice-yearly car wash serviced 398 cars this year. A total of 84 more cars participated in May than in November. A total of 43 fewer cars were washed in July than in November. How many cars were washed during each month?

119 cars in November, 203 cars in May, 76 cars in July

8. An American Airlines jet traveling to Austin, Texas, is carrying 274 passengers. The plane is carrying three times as many business class passengers as first class passengers. The plane is carrying two more than thirteen times as many coach passengers as first class passengers. How many passengers of each type are on the plane?

There are 16 first class passengers, 48 business class passengers, and 210 coach passengers.

Solve each problem by using an equation. Show what you set the variable equal to. Check your answers.

9. A 12-ft solid cherry wood tabletop is cut into two pieces to allow for an insert later on. Of the two original pieces, the shorter piece is 4.7 ft shorter than the longer piece. What is the length of each piece?

The shorter piece is 3.65 feet long.
The longer piece is 8.35 feet long.

10. An artist had painted a huge painting 18 feet long. Her goal is to cut the canvas and have two pieces of the same painting. The longer piece of canvas is 6.5 ft longer than the shorter piece. What is the length of each piece?

The shorter piece is 5.75 feet long.
The longer piece is 12.25 feet long.

11. The perimeter of a rectangle is 52 cm. The length is 2 cm less than triple the width. What are the dimensions of the rectangle?

x = width of rectangle

$3x - 2$ = length of rectangle

$2x + 2(3x - 2) = 52$

$x = 7$ cm $\quad 3x - 2 = 19$ cm

12. The perimeter of a rectangle is 64 cm. The length is 4 cm less than triple the width. What are the dimensions of the rectangle?

x = width of rectangle

$3x - 4$ = length of rectangle

$2x + 2(3x - 4) = 64$

$x = 9$ cm $\quad 3x - 4 = 23$ cm

13. A geometric puzzle has a triangular playing piece with a perimeter of 44 cm. The length of the second side is double the first side. The length of the third side is 12 cm longer than the first side. Find the length of each side.

The sides are 8 cm, 16 cm, and 20 cm.

14. There is a triangular piece of land adjoining an oil field in Texas, with a perimeter of 271 meters. The length of the second side is double the first side. The length of the third side is 15 m longer than the first side. Find the length of each side.

The sides are 64 meters, 128 meters, and 79 meters.

15. A triangular desk flag at the United Nations has a perimeter of 199 mm. The second side is 20 mm longer than the first side. The third side is 4 mm shorter than the first side. Find the length of each side. The sides are 61 mm, 81 mm, and 57 mm.

16. A triangular part of an earring has a perimeter of 66 mm. The second side is 2 mm longer than the first side. The third side is 5 mm shorter then the first side. Find the length of each side. The sides are 23 mm, 25 mm, and 18 mm.

17. A triangle has three angles, A, B, and C. Angle B is triple angle A. Angle C is 40° larger than angle A. Find the measure of each angle.

x = no. of degrees in angle A $\quad 3x$ = no. of degrees in angle B

$x + 40$ = no. of degrees in angle C

$x + 3x + x + 40 = 180$

$x = 28°$ $\quad 3x = 84°$ $\quad x + 40 = 68°$

18. A triangle has three angles, A, B, and C. Angle B is triple angle A. Angle C is 15° smaller than angle A. Find the measure of each angle.

x = no. of degrees in angle A $\quad 3x$ = no. of degrees in angle B

$x - 15$ = no. of degrees in angle C

$x + 3x + x - 15 = 180$

$x = 39°$ $\quad 3x = 117°$ $\quad x - 15 = 24°$

19. A saleswoman earns a base monthly salary of \$2000 plus a 2% commission on total sales. Last month her total salary was \$2800. What was the amount of her total sales?

x = total sales $\quad 2000 + 0.02x = \$2800$ $\quad x = \$40{,}000$

20. A salesman earns a base monthly salary of \$1500 plus a 3% commission on total sales. Last month his total salary was \$2400. What was the amount of his total sales?

x = total sales $\quad 1500 + 0.03x = 2400$ $\quad x = \$30{,}000$

21. A realtor charges \$100 to place a rental listing plus 12% of the yearly rent. An apartment in Central City was rented by the realtor for one year. She charged the landowner \$820. How much does the apartment cost to rent for one year?
x = yearly rent $100 + 0.12x = 820$
$0.12x = 720$ $x = \$6000$

22. A realtor charges \$50 to place a rental listing plus 9% of the yearly rent. An apartment in the town of West Longmeadow was rented by the realtor for one year. He charged the landowner \$482. How much does the apartment cost to rent for one year?
x = yearly rent $50 + 0.09x = 482$ $x = \$4800$

23. The population of Melbourne this year is 2,895,300. This was a 2% increase over the population last year. What was the population last year? Round to the nearest hundred. 2,838,500

24. The population of Monroe this year is 12,580. This was a 3% increase over the population last year. What was the population last year? Round to the nearest ten. 12,210

25. A rectangular field is 8.2066 yd longer than double the width. The perimeter is 2876.39 yd. What are the dimensions of the field?
Width = 476.6628 yd Length = 961.5322 yd

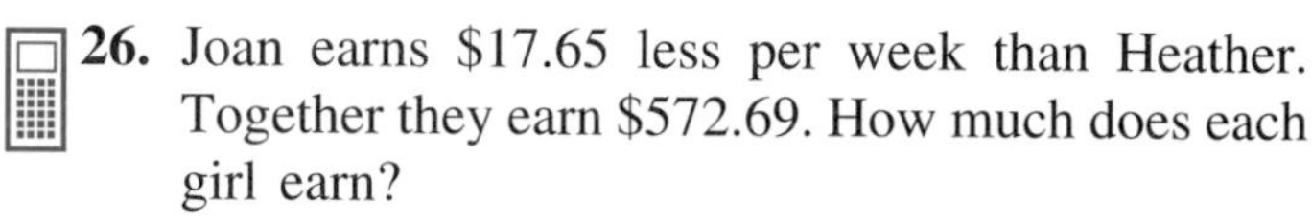

26. Joan earns \$17.65 less per week than Heather. Together they earn \$572.69. How much does each girl earn?
\$295.17 Heather \$277.52 Joan

To Think About

27. The cost of three computer programs at a computer discount store is \$570.33. The cost of the second program is \$20 less than double the cost of the first program. The cost of the third program is \$17 more than triple the cost of the second program. How much does each program cost?
first program = \$70.37 third program = \$379.22
second program = \$120.74

28. The cost of three vintage cars at a recent antique car club auction is \$51,000. The cost of the Model T is \$6000 less than double the cost of an Edsel. The cost of a GTO is \$3000 less than triple the cost of the Model T. How much does each vehicle cost?
The Edsel costs \$8000.
The Model T costs \$10,000.
The GTO costs \$27,000.

Cumulative Review Problems

29. What is 16.5% of 350?
57.75

30. What percent of 20 is 12?
60%

31. 38% of what number is 190?
500

32. What is 0.52% of 4000?
20.8

Chapter Organizer

Topic	Procedure	Examples
Combining like terms, p. 550.	If the terms are like terms, you combine the numerical coefficients directly in front of the variables.	Combine like terms. **(a)** $7x - 8x + 2x = -1x + 2x = x$ **(b)** $3a - 2b - 6a - 5b = -3a - 7b$ **(c)** $a - 2b + 3 - 5a = -4a - 2b + 3$
The distributive properties, p. 553.	$a(b + c) = ab + ac$ and $(b + c)a = ba + ca$	$(5)(x - 4y) = 5x - 20y$ $(3)(a + 2b - 6) = 3a + 6b - 18$ $(-2x + y)(7) = -14x + 7y$
Problems involving parentheses and like terms, p. 555.	**1.** Remove the parentheses using the distributive property. **2.** Combine like terms.	Simplify. $(2)(4x - y) - (3)(-2x + y) = 8x - 2y + 6x - 3y$ $= 14x - 5y$
Solving equations using the addition property, p. 558.	**1.** Add the appropriate value to both sides of the equation so that the variable is on one side and a number is on the other side of the equal sign. **2.** Check by substituting your answer back into the original equation.	Solve for x. $x - 2.5 = 7$ $x - 2.5 + 2.5 = 7 + 2.5$ $x + 0 = 9.5$ $x = 9.5$ Check. $x - 2.5 = 7$ $9.5 - 2.5 \stackrel{?}{=} 7$ $7 = 7$
Solving equations using the division property, p. 563.	**1.** Divide both sides of the equation by the numerical coefficient of the variable. **2.** Check by substituting your answer back into the original equation.	Solve for x. $-12x = 60$ $\frac{-12x}{-12} = \frac{60}{-12}$ $x = -5$ Check. $-12x = 60$ $(-12)(-5) \stackrel{?}{=} 60$ $60 = 60$
Solving equations using the multiplication property, p. 565.	**1.** Multiply both sides of the equation by the reciprocal of the numerical coefficient of the variable. **2.** Check by substituting your answer back into the original equation.	Solve for x. $\frac{3}{4}x = \frac{5}{8}$ $\frac{4}{3} \cdot \frac{3}{4}x = \frac{5}{8} \cdot \frac{4}{3}$ $x = \frac{5}{6}$ Check. $\frac{3}{4}x = \frac{5}{8}$ $\left(\frac{3}{4}\right)\left(\frac{5}{6}\right) \stackrel{?}{=} \frac{5}{8}$ $\frac{5}{8} = \frac{5}{8}$ ✓

Chapter Organizer

Topic	Procedure	Examples
Solving equations using more than one step, p. 569.	1. Remove any parentheses. 2. Collect like terms on each side of the equation. 3. Add the appropriate value to both sides of the equation to get all numbers on one side. 4. Add the appropriate value to both sides of the equation to get all variable terms on the other side. 5. Divide both sides of the equation by the numerical coefficient of the variable. 6. Check by substituting back into the original equation.	Solve for x. $5x - (2)(6x - 1) = (3)(1 + 2x) + 12$ $5x - 12x + 2 = 3 + 6x + 12$ $-7x + 2 = 15 + 6x$ $-7x + 2 + (-2) = 15 + (-2) + 6x$ $-7x = 13 + 6x$ $-7x + (-6x) = 13 + 6x + (-6x)$ $-13x = 13$ $\frac{-13x}{-13} = \frac{13}{-13}$ $x = -1$ Check. For $x = -1$ $5x - (2)(6x - 1) = (3)(1 + 2x) + 12$ $(5)(-1) - (2)[6(-1) - 1] \stackrel{?}{=} (3)[1 + (2)(-1)] + 12$ $-5 - (2)[-6 - 1] \stackrel{?}{=} (3)[1 - 2] + 12$ $-5 - (2)[-7] \stackrel{?}{=} (3)[-1] + 12$ $-5 + 14 \stackrel{?}{=} -3 + 12$ $9 = 9$ ✓
Translating an English sentence into an equation, p. 577.	When translating English into an equation: (see table below)	Translate a comparison in English into an equation using two given variables. Use t to represent Thursday's temperature and w to represent Wednesday's temperature. The temperature Thursday was 12 degrees higher than the temperature on Wednesday. Temperature Thursday → t; was → $=$; 12° → 12; higher than → $+$; Temperature Wednesday → w

The English Phrase:	Is Usually Represented by the Symbol:
greater than increased by more than added to taller than total of sum of	+
less than smaller than fewer than shorter than difference of	−
product of times	×
double	2 ×
triple	3 ×
ratio of quotient of	÷
is costs has the value of weighs has was equals represents amounts to	=

Chapter Organizer		
Topic	**Procedure**	**Examples**
Writing algebraic expressions for several quantities, p. 583.	**1.** Use a variable to describe the quantity that other quantities are compared to. **2.** Write an expression in terms of that variable for each of the other quantities.	Write algebraic expressions for the size of each angle of a triangle. The second angle of a triangle is 7° less than the first angle. The third angle of a triangle is double the first angle. Use the letter x. Since other angles are compared to the first angle, we let the variable x represent that angle. Let x = the number of degrees in the first angle. Let $x - 7$ = the number of degrees in the second angle. Let $2x$ = the number of degrees in the third angle.
Solving applied problems using equations, p. 583.	**1.** *Understand the problem.* **(a)** Draw a sketch. **(b)** Choose a variable. **(c)** Represent other variables in terms of the first variable. **2.** *Write an equation.* **3.** *Solve the equation and state the answer.* **4.** *Check.*	The perimeter of a field is 128 meters. The length of this rectangular field is 4 meters less than triple the width. Find the dimensions of the field. **1.** *Understand the problem.* We need to find the dimensions of a rectangular field. Length Width Width Length We label the sides of our sketch. Let w = the width of the rectangle in meters. Let $3w - 4$ = the length of the rectangle in meters. **2.** *Write an equation.* Perimeter = 2(width) + 2(length) $128 = 2(w) + 2(3w - 4)$ **3.** *Solve the equation and state the answer.* $128 = 2w + 6w - 8$ $128 = 8w - 8$ $136 = 8w$ $17 = w$ The width is 17 meters. $3w - 4 = 3(17) - 4 = 47$ The length is 47 meters. **4.** *Check.* Is the perimeter 128 meters? Does $17 + 47 + 17 + 47 = 128$? $128 = 128$ ✓ Is the length 4 less than triple the width? Is $47 = 3(17) - 4$? $47 = 47$ ✓

Chapter 10 Review Problems

10.1 *Combine like terms.*

1. $x + 2y - 6x$
$-5x + 2y$

2. $5x + 6y - 7x$
$-2x + 6y$

3. $a + 8 - 2a + 4$
$-a + 12$

4. $7 + 2a + 6a - 5$
$8a + 2$

5. $5x + 2y - 7x - 9y$
$-2x - 7y$

6. $3x - 7y + 8x + 2y$
$11x - 5y$

7. $5x - 9y - 12 - 6x - 3y + 18$
$-x - 12y + 6$

8. $7x - 2y - 20 - 5x - 8y + 13$
$2x - 10y - 7$

9. $1.2a + 5.6b - 3 - 4a - 2.2b + 1$
$-2.8a + 3.4b - 2$

10. $-1.5a + 3.4b - 7 - 6a + 5.6b + 3$
$-7.5a + 9b - 4$

10.2 *Simplify.*

11. $(-3)(5x + y)$
$-15x - 3y$

12. $(-4)(2x + 3y)$
$-8x - 12y$

13. $(2)(x - 3y + 4)$
$2x - 6y + 8$

14. $(3)(2x - 6y - 1)$
$6x - 18y - 3$

15. $(-8)(3a - 5b - c)$
$-24a + 40b + 8c$

16. $(-9)(2a - 8b - c)$
$-18a + 72b + 9c$

17. $(5)(1.2x + 3y - 5.5)$
$6x + 15y - 27.5$

18. $(6)(1.4x - 2y + 3.4)$
$8.4x - 12y + 20.4$

Simplify.

19. $(2)(x + 3y) - (4)(x - 2y)$
$-2x + 14y$

20. $(2)(5x - y) - (3)(x + 2y)$
$7x - 8y$

21. $(-2)(a + b) - (3)(2a + 8)$
$-8a - 2b - 24$

22. $(-4)(a - 2b) + (3)(5 - a)$
$-7a + 8b + 15$

10.3 *Solve for the variable.*

23. $x - 3 = 9$
$x = 12$

24. $5 + x = 13$
$x = 8$

25. $-8 = x - 12$
$x = 4$

26. $x + 17 = 9$
$x = -8$

27. $x + 8.3 = 20$
$x = 11.7$

28. $2.4 = x - 5$
$x = 7.4$

29. $3.1 + x = -9$
$x = -12.1$

30. $x - 5 = 1.8$
$x = 6.8$

31. $x - \frac{3}{4} = 2$
$x = 2\frac{3}{4}$

32. $x + \frac{1}{2} = 3\frac{3}{4}$
$x = 3\frac{1}{4}$

33. $x + \frac{3}{8} = \frac{1}{2}$
$x = \frac{1}{8}$

34. $x - \frac{5}{6} = \frac{2}{3}$
$x = \frac{9}{6}$ or $1\frac{1}{2}$

35. $2x - 5 = x + 1$
$x = 6$

36. $3x + 9 = 2x + 5$
$x = -4$

37. $2x + 20 = 25 + x$
$x = 5$

38. $5x - 3 = 4x - 15$
$x = -12$

10.4 *Solve for the variable.*

39. $7x = 56$
$x = 8$

40. $75 = 5y$
$y = 15$

41. $8x = -96$
$x = -12$

42. $-12y = 72$
$y = -6$

43. $1.5x = 9$
$x = 6$

44. $1.8y = 12.6$
$y = 7$

45. $-7.2x = 36$
$x = -5$

46. $6x = 1.5$
$x = 0.25$

47. $\frac{3}{4}x = 6$
$x = 8$

48. $\frac{2}{3}x = \frac{5}{9}$
$x = \frac{5}{6}$

49. $\frac{1}{8}x = 2\frac{5}{8}$
$x = 21$

50. $1\frac{1}{3}x = 8$
$x = 6$

10.5 *Solve for the variable.*

51. $7x + 8 = 43$

$x = 5$

52. $9x + 5 = 50$

$x = 5$

53. $3 - 2x = 9 - 8x$

$x = 1$

54. $8 - 6x = -7 - 3x$

$x = 5$

55. $10 + x = 3x - 6$

$x = 8$

56. $8x - 7 = 5x + 8$

$x = 5$

57. $9x - 3x + 18 = 36$

$x = 3$

58. $4 + 3x - 8 = 12 + 5x + 4$

$x = -10$

59. $5x - 2 = 27$

$x = \frac{29}{5}$

60. $(2)(x + 3) = -3x + 10$

$x = \frac{4}{5}$

61. $2x - 3 - 5x = 13 + (2)(2x - 1)$

$x = -2$

62. $(2)(3x - 4) = 7 - 2x + 5x$

$x = 5$

63. $5 + 2y + (5)(y - 3) = (6)(y + 1)$

$y = 16$

64. $3 + (5)(y + 4) = (4)(y - 2) + 3$

$y = -28$

10.6 *Translate the English sentence into an equation using the variables indicated for each problem.*

65. The weight of the truck is 3000 lb more than the weight of the car. Use w for the weight of the truck and c for the weight of the car.

$w = c + 3000$

66. The afternoon class had 18 fewer students than the morning class. Use a for the number of students in the afternoon class and m for the number of students in the morning class.

$m = a + 18$

67. The number of degrees in angle A is triple the number of degrees in angle B. Use A for the number of degrees in angle A and B for the number of degrees in angle B.

$A = 3B$

68. The length of a rectangle is 3 in. shorter than double the width of the rectangle. Use w for the width of the rectangle in inches and l for the length of the rectangle in inches.

$l = 2w - 3$

Write algebraic expressions for each of the specified quantities using the same variable.

69. Michael's salary is $2050 more than Roberto's salary. Use the letter r.

r = Roberto's salary $r + 2050$ = Michael's salary

70. The length of the second side of a triangle is double the length of the first side of the triangle. Use the letter x.

x = length 1st side $2x$ = length 2nd side

71. Nancy has completed six fewer graduate courses than Connie has. Use the letter c.

c = number of Connie's courses

$c - 6$ = number of Nancy's courses

72. The number of books in the new library is 450 more than double the number of books in the old library. Use the letter b.

b = number of books in old library

$2b + 450$ = number of books in new library

10.7 *Solve each problem using an equation. Show what you let the variable equal.*

73. A 60-ft length of pipe is divided into two pieces. One piece is 6.5 ft longer than the other. Find the length of each piece.
26.75 ft 33.25 ft

74. Two clerks work in a store. The new employee earns \$28 less per week than a person hired six months ago. Together they earn \$412 per week. What is the weekly salary of each person?
\$192 \$220

75. A local fast-food restaurant had twice as many customers in March as in February. It had 3000 more customers in April than in February. Over the three months 45,200 customers came to the restaurant. How many came each month?
Feb. = 10,550 Mar. = 21,100 Apr. = 13,550

76. Alfredo drove 856 mi during three days of travel. He drove 106 more miles on Friday than on Thursday. He drove 39 fewer miles on Saturday than on Thursday. How many miles did he drive each day?
Thurs. = 263 mi Fri. = 369 mi Sat. = 224 mi

77. A rectangle has a perimeter of 72 in. The length is 3 in. less than double the width. Find the dimensions of the rectangle.
width = 13 in. length = 23 in.

78. A rectangle has a perimeter of 180 m. The length is 2 m more than triple the width. Find the dimensions of the rectangle.
width = 22 m length = 68 m

79. A triangle has a perimeter of 99 m. The second side is 7 m longer than the first side. The third side is 4 m shorter than the first side. How long is each side?
1st side = 32 m 2nd side = 39 m 3rd side = 28 m

80. A triangle has three angles labeled *A*, *B*, and *C*. Angle *C* is triple the measure of angle *B*. Angle *A* is 74 degrees larger than angle *B*. Find the measure of each angle.
$A = 95.2°$ $B = 21.2°$ $C = 63.6°$

Chapter 10 Test

Collect like terms.

1. $5a - 11a$

2. $7b - 12b + 3b$

3. $-3x + 7y - 8x - 5y$

4. $6a - 5b + 7 - 5a - 3b + 4$

5. $7x - 8y + 2z - 9z + 8y$

6. $x + 5y - 6 - 5x - 7y + 11$

Simplify.

7. $(5)(12x - 5y)$

8. $(-4)(2x - 3y + 7)$

9. $(-1.5)(3a - 2b + c - 8)$

10. $(2)(-3a + 2b) - (5)(a - 2b)$

Solve for the variable.

11. $-5 - 3x = 19$

12. $(-3)(4 - x) = (5)(6 + 2x)$

13. $-5x + 9 = -4x - 6$

14. $8x - 2 - x = 3x - 9 - 10x$

15. $2x - 5 + 7x = 4x - 1 + x$

16. $3 - (x + 2) = 5 + (3)(x + 2)$

Translate the English sentence into an equation using the variables indicated for each problem.

17. The second floor of Trabor Laboratory has 15 more classrooms than the first floor. Use s to represent the number of classrooms on the second floor and f to represent the number of classrooms on the first floor.

18. The north field yields 15,000 fewer bushels of wheat than the south field. Use n to represent the number of bushels of wheat in the north field and s to represent the number of bushels of wheat in the south field.

1. $-6a$

2. $-2b$

3. $-11x + 2y$

4. $a - 8b + 11$

5. $7x - 7z$

6. $-4x - 2y + 5$

7. $60x - 25y$

8. $-8x + 12y - 28$

9. $-4.5a + 3b - 1.5c + 12$

10. $-11a + 14b$

11. $x = -8$

12. $x = -6$

13. $x = 15$

14. $x = -\frac{1}{2}$

15. $x = 1$

16. $x = -\frac{5}{2}$ *or* $-2\frac{1}{2}$

17. $s = f + 15$

18. $n = s - 15,000$

19. $\frac{1}{2}s$ = first
s = second
$2s$ = third

20. w = width
$2w - 5$ = length

21. Prentice farm = 87 acres
Smithfield farm = 261 acres

22. Marcia = \$24,000
Sam = \$22,500

23. Monday = 311 mi
Tuesday = 367 mi
Wednesday = 297 mi

24. width = 25 ft
length = 34 ft

Write algebraic expressions for each quantity using the given variable.

19. The first angle of a triangle is half the second angle. The third angle of the triangle is twice the second angle. Use the variable s.

20. The length of a rectangle is 5 in. shorter than double the width. Use the letter w.

Solve each problem by using an equation.

21. The number of acres of land in the old Smithfield farm is three times the number of acres of land in the Prentice farm. Together the two farms have 348 acres. How many acres of land are there on each farm?

22. Sam earns \$1500 less per year than Marcia does. The combined income of the two people is \$46,500 per year. How much does each person earn?

23. Gina drove 975 mi in three days. She drove 56 more miles on Tuesday than on Monday. She drove 14 fewer miles on Wednesday than on Monday. How many miles did she drive each day?

24. A rectangular field has a perimeter of 118 ft. The width is 8 ft longer than half the length. Find the dimensions of the rectangle.

Cumulative Test for Chapters 1–10

Approximately one half of this test is based on Chapter 10 material. The remainder is based on material covered in Chapters 1–9.

Do each problem. Simplify your answers.

1. Add. $456 + 89 + 123 + 79$

2. Multiply. $\begin{array}{r} 309 \\ \times\ 35 \\ \hline \end{array}$

3. Round to the nearest hundred. 45,678,934

4. Divide. $\frac{1}{2} \div \frac{1}{4}$

5. Multiply. $3\frac{1}{4} \times 2\frac{1}{2}$

6. Multiply. 9.3×0.0078

7. Subtract. $34{,}007.090 - 3456.789$

8. Find n. $\frac{9}{n} = \frac{40.5}{72}$

9. What is 28.5% of \$5600?

10. 34% of what number is 1870?

11. Convert 345 mm to m.

12. Convert 10 ft to in.

13. Find the circumference of a circle with diameter of 12 yd. Round your answer to the nearest tenth.

14. Find the area of a triangle that has a base of 13 m and a height of 22 m.

Perform the following operations

15. $4 - 8 + 12 - 32 - 7$

16. $(5)(-2)(3)(-1)$

Collect like terms.

17. $3a - 5b - 12a - 6b$

18. $-4x + 5y - 9 - 2x - 3y + 12$

1. 747

2. 10,815

3. 45,678,900

4. 2

5. $\frac{65}{8}$ *or* $8\frac{1}{8}$

6. 0.07254

7. 30,550.301

8. $n = 16$

9. 1596

10. 5500

11. 0.345 m

12. 120 in.

13. 37.7 yd

14. 143 m^2

15. -31

16. 30

17. $-9a - 11b$

18. $-6x + 2y + 3$

Simplify.

19. $(-7)(-3x + y - 8)$

20. $(2)(3x - 4y) - (8)(x + 2y)$

Solve for the variable.

21. $5x - 5 = 7x - 13$

22. $7 - 9y - 12 = 3y + 5 - 8y$

23. $x - 2 + 5x + 3 = 183 - x$

24. $9(2x + 8) = 20 - (x + 5)$

Write algebraic expressions for each of the specified quantities, using the given variable.

25. The weight of the computer was 322 lb more than the weight of the printer. Use the letter p.

26. The summer enrollment in algebra was 87 students less than the fall enrollment. Use the letter f.

Solve each word problem by using an equation. Show what you set the variable equal to.

27. Barbara drove 1081 mi in three days. She drove 48 more miles on Friday than on Thursday. She drove 95 fewer miles on Saturday than on Thursday. How many miles did she drive each day?

28. A rectangle has a perimeter of 98 ft. The length is 8 ft longer than double the width. Find each dimension.

19. $21x - 7y + 56$

20. $-2x - 24y$

21. $x = 4$

22. $y = -\frac{5}{2}$ *or* $-2\frac{1}{2}$

23. $x = 26$

24. $x = -3$

25. p = weight of printer
$p + 322$ = weight of computer

26. f = students during fall
$f - 87$ = students during summer

27. Thursday = 376 miles
Friday = 424 miles
Saturday = 281 miles

28. width = $13\frac{2}{3}$ ft
length = $35\frac{1}{3}$ ft

Practice Final Examination

This examination is based on Chapters 1–10 of the book. There are 10 problems covering the content of each chapter.

Chapter 1

1. Write in words. 82,367

2. Add. $\begin{array}{r} 13{,}428 \\ +\ 16{,}905 \\ \hline \end{array}$

3. Add. $\begin{array}{r} 19 \\ 23 \\ 16 \\ 45 \\ +\ 70 \\ \hline \end{array}$

4. Subtract. $\begin{array}{r} 89{,}071 \\ -\ 54{,}968 \\ \hline \end{array}$

Multiply the following.

5. $\begin{array}{r} 78 \\ \times\ 54 \\ \hline \end{array}$

6. $\begin{array}{r} 2035 \\ \times\ 107 \\ \hline \end{array}$

Divide the following. (Be sure to indicate the remainder if one exists.)

7. $7\overline{)1106}$

8. $26\overline{)15{,}756}$

9. Evaluate. Perform operations in their proper order.
$3^4 + 20 \div 4 \times 2 + 5^2$

10. Melinda traveled 512 mi in her car. The car used 16 gal of gas on the entire trip. How many miles per gallon did the car achieve?

Chapter 2

11. Reduce the fraction. $\dfrac{14}{30}$

12. Change to an improper fraction. $3\dfrac{9}{11}$

Add the following fractions.

13. $\dfrac{1}{10} + \dfrac{3}{4} + \dfrac{4}{5}$

14. $2\dfrac{1}{3} + 3\dfrac{3}{5}$

15. Subtract. $4\dfrac{5}{7} - 2\dfrac{1}{2}$

16. Multiply. $1\dfrac{1}{4} \times 3\dfrac{1}{5}$

Divide the following.

17. $\dfrac{7}{9} \div \dfrac{5}{18}$

18. $\dfrac{5\frac{1}{2}}{3\frac{1}{4}}$

19. Lucinda jogged $1\frac{1}{2}$ mi on Monday, $3\frac{1}{4}$ mi on Tuesday, and $2\frac{1}{10}$ mi on Wednesday. How many miles in all did she jog over the three-day period?

20. A butcher has $11\frac{2}{3}$ lb of steak. She wishes to place them in several equal-sized packages of steak. Each package will hold $2\frac{1}{3}$ lb of steak. How many packages can be made?

Answers

1. eighty-two thousand three hundred sixty-seven
2. 30,333
3. 173
4. 34,103
5. 4212
6. 217,745
7. 158
8. 606
9. 116
10. 32 mi/gal
11. $\frac{7}{15}$
12. $\frac{42}{11}$
13. $\frac{33}{20}$ *or* $1\frac{13}{20}$
14. $\frac{89}{15}$ *or* $5\frac{14}{15}$
15. $\frac{31}{14}$ *or* $2\frac{3}{14}$
16. 4
17. $\frac{14}{5}$ *or* $2\frac{4}{5}$
18. $\frac{22}{13}$ *or* $1\frac{9}{13}$
19. $6\frac{17}{20}$ mi
20. 5 packages

21.	0.719
22.	$\frac{43}{50}$
23.	>
24.	506.38
25.	21.77
26.	0.757
27.	0.492
28.	3.69
29.	0.8125
30.	0.7056
31.	$\frac{1400 \text{ students}}{43 \text{ faculty}}$
32.	False
33.	$n \approx 9.4$
34.	$n \approx 7.7$
35.	$n = 15$
36.	$n = 9$
37.	\$3333.33
38.	9.75 in.
39.	\$294.12 was withheld
40.	1.6 lb of butter

Chapter 3

21. Express as a decimal. $\frac{719}{1000}$

22. Write in reduced fractional notation. 0.86

23. Fill in the blank with $<$, $=$, or $>$
0.315 ____ 0.309

24. Round to the nearest hundredth. 506.3782

25. Add.
$$\begin{array}{r} 9.6 \\ 3.82 \\ 1.05 \\ +\ 7.3 \\ \hline \end{array}$$

26. Subtract.
$$\begin{array}{r} 3.61 \\ -\ 2.853 \\ \hline \end{array}$$

27. Multiply.
$$\begin{array}{r} 1.23 \\ \times\ 0.4 \\ \hline \end{array}$$

28. Divide. $0.24\overline{)0.8856}$

29. Write as a decimal. $\frac{13}{16}$

30. Evaluate, by doing operations in the proper order.
$0.7 + (0.2)^3 - 0.08(0.03)$

Chapter 4

31. Write a rate in simplest form to compare 7000 students to 215 faculty.

32. Is this proportion true or false? $\frac{12}{15} = \frac{17}{21}$

Solve the following proportions. Round to the nearest tenth when necessary.

33. $\frac{5}{9} = \frac{n}{17}$

34. $\frac{3}{n} = \frac{7}{18}$

35. $\frac{n}{12} = \frac{5}{4}$

36. $\frac{n}{7} = \frac{36}{28}$

Solve each of the following problems by using a proportion. Round your answers to the nearest hundredth when necessary.

37. Bob earned \$2000 for painting 3 houses. How much would he earn for painting 5 houses?

38. Two cities that are actually 200 mi apart appear 6 in. apart on the map. Two other cities are 325 mi apart. How far apart will they appear on the same map?

39. Roberta earned \$68 last week on her part-time job. She had \$5 withheld for federal income tax. Last year she earned \$4000 on her part-time job. Assuming the same rate, how much was withheld for federal income tax last year?

40. Malaga's recipe feeds 18 people and calls for 1.2 lb of butter. If she wants to feed 24 people, how many pounds of butter does she need?

Chapter 5

41. Write as a percent. 0.0063

42. Change $\frac{17}{80}$ to a percent.

Round all answers to the nearest tenth when necessary.

43. Write as a decimal. 164%

44. What percent of 300 is 52?

45. Find 6.3% of 4800.

46. 145 is 58% of what number?

47. 126% of 3400 is what number?

48. Pauline bought a new car. She got an 8% discount. The car listed for $11,800. How much did she pay for the car?

49. A total of 1260 freshmen were admitted to Central College. This is 28% of the student body. How big is the student body?

50. There are 11.28 cm of water in the rain gauge this week. Last week the rain gauge held 8.40 cm of water. What is the percent of increase from last week to this week?

Chapter 6

Convert the following. Express your answers as a decimal rounded to the nearest hundredth when necessary.

51. 17 qt = ____ gal

52. 3.25 tons = ____ lb

53. 16 ft = ____ in.

54. 5.6 km = ____ m

55. 69.8 g = ____ kg

56. 2.48 ml = ____ L

Round to the nearest hundredth.

57. 12 mi = ____ km

Write in scientific notation.

58. 0.00063182

59. 126,400,000,000

60. Two metal sheets are 0.623 cm and 0.74 cm thick, respectively. An insulating foil is 0.0428 mm thick. When all three layers are placed tightly together, what is the total thickness? Express your answer in centimeters.

41. 0.63%

42. 21.25%

43. 1.64

44. 17.3%

45. 302.4

46. 250

47. 4284

48. $10,856

49. 4500 students

50. 34.3% increase

51. 4.25 gal

52. 6500 lb

53. 192 in.

54. 5600 m

55. 0.0698 kg

56. 0.00248 L

57. 19.32 km

58. 6.3182×10^{-4}

59. 1.264×10^{11}

60. 1.36728 cm thick

61.	14.4 m
62.	206 cm
63.	5.4 sq ft
64.	75 sq m
65.	113.04 sq m
66.	56.52 m
67.	167.47 cu cm
68.	205.2 cu ft
69.	32.5 sq m
70.	$n = 32.5$
71.	8 million dollars
72.	one million dollars

Chapter 7

Round answers to the nearest hundredth when necessary. Use $\pi = 3.14$ when necessary.

61. Find the perimeter of a rectangle that is 6 m long and 1.2 m wide.

62. Find the perimeter of a trapezoid with sides of 82 cm, 13 cm, 98 cm, and 13 cm.

63. Find the area of a triangle with base 6 ft and height 1.8 ft.

64. Find the area of a trapezoid with bases of 12 m and 8 m and a height of 7.5 m.

65. Find the area of a circle with radius 6 m.

66. Find the circumference of a circle with diameter 18 m.

67. Find the volume of a cone with a radius of 4 cm and a height of 10 cm.

68. Find the volume of a rectangular pyramid with a base of 12 ft by 19 ft and a height of 2.7 ft.

69. Find the area of this object, consisting of a square and a triangle.

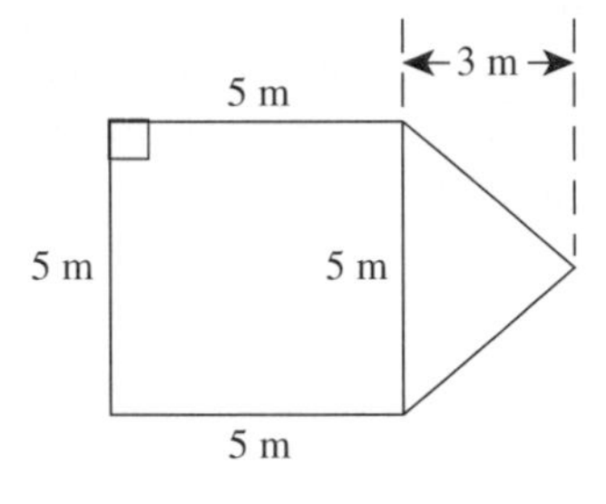

70. In the following pair of similar triangles find n.

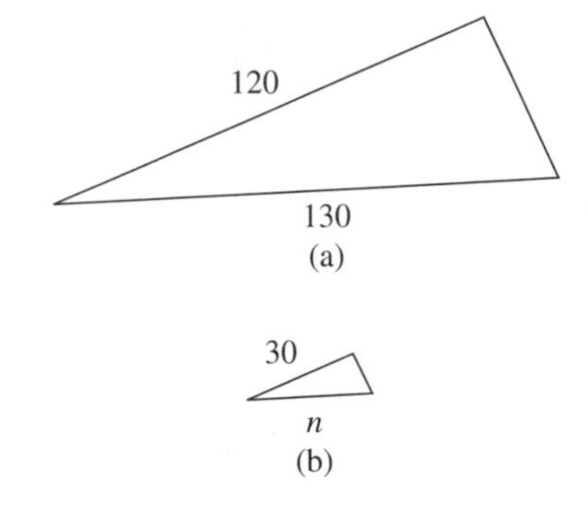

Chapter 8

The following double-bar graph indicates the quarterly profits for Westar Corporation in 1990 and 1991.

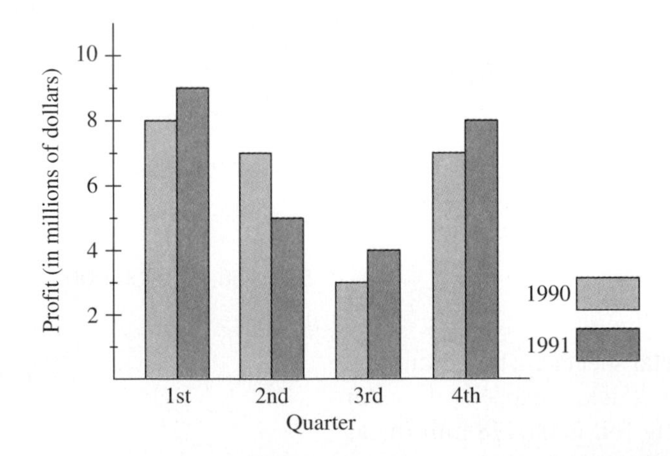

71. What were the profits in the fourth quarter of 1991?

72. How much greater were the profits in the first quarter of 1991 than the profits in the first quarter of 1990?

The following line graph depicts the average annual temperature at West Valley for the years 1950, 1960, 1970, 1980, 1990.

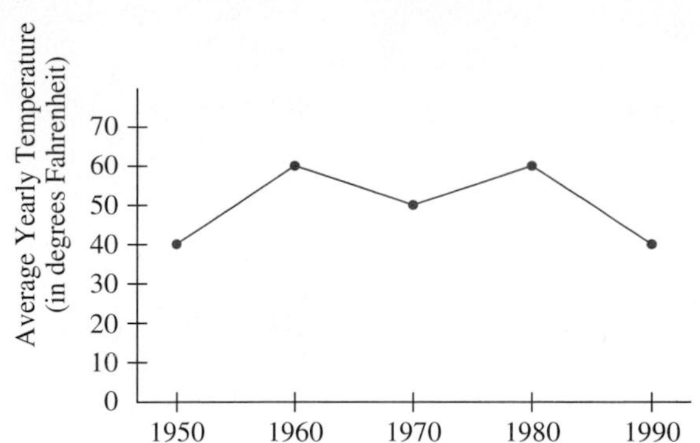

73. What was the average temperature in 1970?

74. In what 10-year period did the average temperature show the greatest decline?

The following histogram shows the number of students in each age category at Center City College.

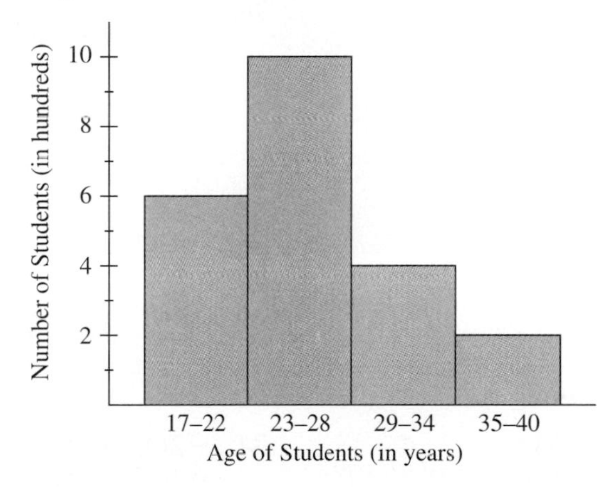

75. How many students are between 17 and 22 years old?

76. How many students are between 23 and 34 years old?

77. Find the *mean* and the *median* of the following. 8, 12, 16, 17, 20, 22

78. Evaluate exactly. $\sqrt{49} + \sqrt{81}$

79. Approximate to the nearest thousandth using a calculator or the square root table. $\sqrt{123}$

80. Find the unknown side of the right triangle.

?

12 ft

9 ft

73. 50°F

74. from 1980 to 1990

75. 600 students

76. 1400 students

77. mean ≈ 15.83
median = 16.5

78. 16

79. 11.091

80. 15 ft

81. -13

82. $\frac{1}{8}$

83. -3

84. -17

85. 24

86. $-\frac{8}{3}$ *or* $-2\frac{2}{3}$

87. 4

88. 27

89. 8

90. 0.5 *or* $\frac{1}{2}$

91. $-3x - 7y$

92. $-7 - 4a - 17b$

93. $-2x + 6y + 10$

94. $-11x - 9y - 4$

95. $x = 2$

96. $x = -2$

97. $x = -\frac{1}{2}$ *or* $x = -0.5$

98. $x = \frac{-2}{5}$ *or* $x = -0.4$

99. 122 students are taking history. 110 students are taking math.

100. The length is 37 m. The width is 16 m.

Chapter 9

Add the following.

81. $-8 + (-2) + (-3)$

82. $-\frac{1}{4} + \frac{3}{8}$

Subtract the following.

83. $9 - 12$

84. $-20 - (-3)$

85. Multiply. $(2)(-3)(4)(-1)$

86. Divide. $-\frac{2}{3} \div \frac{1}{4}$

Perform the following operations in the proper order.

87. $(-16) \div (-2) + (-4)$

88. $12 - 3(-5)$

89. $7 - (-3) + 12 \div (-6)$

90. $\frac{(-3)(-1) + (-4)(2)}{(0)(6) + (-5)(2)}$

Chapter 10

Collect like terms.

91. $5x - 3y - 8x - 4y$

92. $5 + 2a - 8b - 12 - 6a - 9b$

Simplify the following.

93. $-2(x - 3y - 5)$

94. $-2(4x + 2) - 3(x + 3y)$

Solve for the variable.

95. $5 - 4x = -3$

96. $5 - 2(x - 3) = 15$

97. $7 - 2x = 10 + 4x$

98. $-3(x + 4) = 2(x - 5)$

Solve the following word problems by using an equation.

99. There are 12 more students taking history than math. There are twice as many students taking psychology as there are students taking math. There are 452 students taking these 3 subjects. How many are taking history? How many are taking math?

100. A rectangle has a perimeter of 106 ft. The length is 5 m longer than double the width. Find the length and width of the rectangle.

Glossary

Absolute value of a number (9.1) The absolute value of a number is the distance between that number and zero on the number line. When we find the absolute value of a number, we use the $| \ |$ notation. To illustrate, $|-4| = 4$, $|6| = 6$, $|-20 - 3| = |-23| = 23$, $|0| = 0$.

Addends (1.2) When two or more numbers are added, the numbers being added are called addends. In the problem $3 + 4 = 7$, the numbers 3 and 4 are both addends.

Algebraic expression (10.6) An algebraic expression consists of variables, numerals, and operation signs.

Altitude of a triangle (7.3) The height of a triangle.

Amount of a percent equation (5.3A) The product we obtain when we multiply a percent times a number. In the equation $75 = 50\% \times 150$, the amount is 75.

Angle (7.3) An angle is formed whenever two lines meet.

Area (7.1) The measure of the surface inside a geometric figure. Area is measured in square units, such as square feet.

Associative property of addition (1.2) The property that tells us that when three numbers are added, it does not matter which two numbers are added first. An example of the associative property is $5 + (1 + 2) = (5 + 1) + 2$. Whether we add $1 + 2$ first and then add 5 to that, or add $5 + 1$ first and then add that result to 2, we will obtain the same result.

Associative property of multiplication (1.4) The property that tells us that when we multiply three numbers, it does not matter which two numbers we group together first to multiply; the result will be the same. An example of the associative property of multiplication is $2 \times (5 \times 3) = (2 \times 5) \times 3$.

Base (1.6) The number that is to be repeatedly multiplied in exponent form. When we write $16 = 2^4$, the number 2 is the base.

Base of a percent equation (5.3A) The quantity we take a percent of. In the equation $80 = 20\% \times 400$, the base is 400.

Billion (1.1) The number 1,000,000,000.

Borrowing (1.3) The renaming of a number in order to facilitate subtraction. When we subtract $42 - 28$, we rename 42 as 3 tens plus 12. This represents 3 tens and 12 ones. This renaming is called borrowing.

Box (7.7) A three-dimensional object whose every side is a rectangle. Another name for a box is a *rectangular solid.*

Building fraction property (2.6) For whole numbers a, b, c where neither b nor c equals zero

$$\frac{a}{b} = \frac{a}{b} \times 1 = \frac{a}{b} \times \frac{c}{c} = \frac{a \times c}{b \times c}$$

Building up a fraction (2.6) To make one fraction into an equivalent fraction by making the denominator and numerator larger numbers. For example, the fraction $\frac{3}{4}$ can be built up to the fraction $\frac{30}{40}$.

Caret (3.5) A symbol $\wedge$ used to indicate the new location of a decimal point when performing division of decimal fractions.

Celsius temperature (6.4) A temperature scale in which water boils at 100 degrees (100°C) and freezes at 0 degrees (0°C). To convert Celsius temperature to Fahrenheit, we use the formula $F = 1.8 \times C + 32$.

Center of a circle (7.6) The point in the middle of a circle from which all points on the circle are an equal distance.

Centimeter (6.2) A unit of length commonly used in the metric system to measure small distances. 1 centimeter = 0.01 meter.

Circle (7.6) A figure for which all points are at an equal distance from a given point.

Circumference of a circle (7.6) The distance around the rim of a circle.

Commission (5.4) The amount of money a salesperson is paid that is a percentage of the value of the sales made by that salesperson. The commission is obtained by multiplying the commission rate times the value of the sales. If a salesman sells $120,000 of insurance and his commission rate is 0.5%, then his commission is $0.5\% \times 120{,}000 = \$600.00$.

Common denominator (2.7) Two fractions have a common denominator if the same number appears in the denominator of each fraction. $\frac{3}{7}$ and $\frac{1}{7}$ have a common denominator of 7.

Commutative property of addition (1.2) The property that tells us that the order in which two numbers are added does not change the sum. An example of the commutative property of addition is $3 + 6 = 6 + 3$.

Commutative property of multiplication (1.4) The property that tells us that the order in which two numbers are multiplied does not change the value of the answer. An example of the commutative property of multiplication is $7 \times 3 = 3 \times 7$.

Composite number (2.2) A composite number is a whole number greater than 1 that can be divided by whole numbers other than itself. The number 6 is a composite number since it can be divided exactly by 2 and 3 (as well as by 1 and 6).

Cone (7.7) A three-dimensional object shaped like an ice-cream cone or the sharpened end of a pencil.

Cross-multiplying (4.3) If you have a proportion such as $\frac{n}{5} = \frac{12}{15}$, then to cross-multiply, you form products to have $n \times 15 = 5 \times 12$.

Cubic centimeter (6.3) A metric measurement of volume equal to 1 milliliter.

Cup (6.1) One of the smallest units of volume in the American system. 2 cups = 1 pint.

Cylinder (7.7) A three-dimensional object shaped like a tin can.

Debit (1.2) A debit in banking is the removing of money from an account. If you had a savings account and took $300 out of it on Wednesday, we would say that you had a debit of $300 from your account. Often a bank will add a service charge to your account and use the word *debit* to mean that it has removed money from your account to cover the charge.

Decimal fraction (3.1) A fraction whose denominator is a power of 10.

Decimal places (3.4) The number of digits to the right of the decimal point in a decimal fraction. The number 1.234 has three decimal places, while the number 0.129845 has six decimal places. A whole number such as 42 is considered to have zero decimal places.

Decimal point (3.1) The period that is used when writing a decimal fraction. In the number 5.346, the period between the 5 and the 3 is the decimal point. It separates the whole number from the fractional part that is less than 1.

Decimal system (1.1) Our number system is called the decimal system or base 10 system because the value of numbers written in our system is based on tens and ones.

Decimeter (6.2) A unit of length not commonly used in the metric system. 1 decimeter = 0.1 meter.

Degree (7.3) A unit used to measure an angle. A degree is $\frac{1}{360}$ of a complete revolution. An angle of 32 degrees is written as 32°.

Dekameter (6.2) A unit of length not commonly used in the metric system. 1 dekameter = 10 meters.

Denominator (2.1) The number on the bottom of a fraction. In the fraction $\frac{2}{9}$ the denominator is 9.

Deposit (1.2) A deposit in banking is the placing of money in an account. If you had a checking account and on Tuesday you placed $124 into that account, we would say that you made a deposit of $124.

Diameter of a circle (7.6) A line segment across the circle that passes through the center of the circle. The diameter of a circle is equal to twice the radius of the circle.

Difference (1.3) The result of performing a subtraction. In the problem $9 - 2 = 7$ the number 7 is the difference.

Digits (1.1) The symbols 0, 1, 2, 3, 4, 5, 6, 7, 8, and 9 are called digits.

Discount (5.4) The amount of reduction in a price. The discount is a product of the discount rate times the list price. If the list price of a television is $430.00 and it has a discount rate of 35%, then the amount of discount is 35% × $430.00 = $150.50. The price would be reduced by $150.50.

Distributive property of multiplication over addition (1.4) The property illustrated by $5 \times (4 + 3) = (5 \times 4) + (5 \times 3)$. In general, for any numbers a, b, c, it is true that $a(b + c) = a \times b + a \times c$.

Dividend (1.5) The number that is being divided by another. In the problem $14 \div 7 = 2$, the number 14 is the dividend.

Divisor (1.5) The number that you divide into another number. In the problem $30 \div 5 = 6$, the number 5 is the divisor.

Earned run average (4.4) A ratio formed by finding the number of runs a pitcher would give up in a

9-inning game. If a pitcher has an earned run average of 2, it means that, on the average, he gives up two runs for every nine innings he pitches.

Equal fractions (2.2) Fractions that represent the same number. The fractions $\frac{3}{4}$ and $\frac{6}{8}$ are equal fractions.

Equality test of fractions (2.2) Two fractions $\frac{a}{b}$ and $\frac{c}{d}$ are equal if the product $a \times d = b \times c$. In this case, A, B, C, D are whole numbers and $B, D \neq 0$.

Equations (10.3) Mathematical statements with variables that say that two expressions are equal, such as $x + 3 = -8$ and $2s + 5s = 34 - 4s$.

Equilateral triangle (7.3) A triangle with three equal sides.

Equivalent equations (10.3) Equations that have the same solution.

Equivalent fractions (2.2) Two fractions that are equal.

Expanded notation for a number (1.1) A number is written in expanded notation if it is written as a sum of hundreds, tens, ones, etc. The expanded notation for 763 is $700 + 60 + 3$.

Exponent (1.6) The number that indicates the number of times a factor occurs. When we write $8 = 2^3$, the number 3 is the exponent.

Factors (1.4) Each of the numbers that are multiplied. In the problem $8 \times 9 = 72$, the numbers 8 and 9 are factors.

Fahrenheit temperature (6.4) A temperature scale in which water boils at 212 degrees (212°F) and freezes at 32 degrees (32°F). To convert Fahrenheit temperature to Celsius, we use the formula $C = \frac{5 \times F - 160}{9}$.

Foot (6.1) American system unit of length. 3 feet = 1 yard.

Fundamental theorem of arithmetic (2.2) Every composite number has a unique product of prime numbers.

Gallon (6.1) A unit of volume in the American system. 4 quarts = 1 gallon.

Gigameter (6.2) A metric unit of length equal to 1,000,000,000 meters.

Gram (6.3) The basic unit of weight in the metric system. A gram is defined as the weight of the water in a box that is 1 centimeter on each side. 1 gram = 1000 milligrams. 1 gram = 0.001 kilograms.

Hectometer (6.2) A unit of length not commonly used in the metric system. 1 hectometer = 100 meters.

Height (7.2) The distance between two parallel sides in a four-sided figure such as a parallelogram or a trapezoid.

Height of a cone (7.7) The distance from the vertex of a cone to the base of the cone.

Height of a pyramid (7.7) The distance from the point on a pyramid to the base of the pyramid.

Height of a triangle (7.3) The distance of a line drawn from a vertex perpendicular to the other side, or an extension of the other side, of the triangle. This is sometimes called the *altitude of a triangle*.

Hexagon (7.2) A six-sided figure.

Hypotenuse (7.5) The side opposite the right angle in a right triangle. The hypotenuse is always the longest side of a right triangle.

Improper fraction (2.3) A fraction in which the numerator is greater than or equal to the denominator. The fractions $\frac{34}{29}$, $\frac{8}{7}$, and $\frac{6}{6}$ are all improper fractions.

Inch (6.1) The smallest unit of length in the American system. 12 inches = 1 foot.

Inequality symbol (3.2) The symbol that is used to indicate if a number is greater than another number or less than another number. Since 5 is greater than 3, we would write this with a "greater than" symbol as follows: $5 > 3$. The statement 7 is less than 12 would be written as follows: $7 < 12$.

Interest (5.4) The money that is paid for the use of money. If you deposit money in a bank, the bank uses that money and pays you interest. If you borrow money, you pay the bank interest for the use of that money. Simple interest is determined by the formula $I = P \times R \times T$. Compound interest is usually determined by a table, a calculator, or a computer.

Invert a fraction (2.5) To invert a fraction is to interchange the numerator and the denominator. If we invert $\frac{5}{9}$, we obtain the fraction $\frac{9}{5}$. To invert a fraction is sometimes referred to as *to take the reciprocal of a fraction*.

Irreducible (2.2) A fraction that cannot be reduced (simplified) is called irreducible.

Isosceles triangle (7.3) A triangle with two sides equal.

Kilogram (6.3) The most commonly used metric unit of weight. 1 kilogram = 1000 grams.

Kiloliter (6.3) The metric unit of volume normally used to measure large volumes. 1 kiloliter = 1000 liters.

Kilometer (6.2) The unit of length commonly used in the metric system to measure large distances. 1 kilometer = 1000 meters.

Least common denominator (LCD) (2.6) The least common denominator (LCD) of two or more fractions is the smallest number that can be divided without remainder by each fraction's denominator. The LCD of $\frac{1}{3}$ and $\frac{1}{4}$ is 12. The LCD of $\frac{5}{6}$ and $\frac{4}{15}$ is 30.

Legs of a right triangle (7.5) The two shortest sides of a right triangle.

Length of a rectangle (7.1) Each of the longer sides of a rectangle.

Like terms (10.1) Like terms have identical variables with identical exponents. $-5x$ and $3x$ are like terms. $-7xyz$ and $-12xyz$ are like terms.

Line segment (7.3) A portion of a straight line that has a beginning and an end.

Liter (6.3) The standard metric measurement of volume. 1 liter = 1000 milliliters. 1 liter = 0.001 kiloliter.

Mean (8.4) The mean of a set of values is the sum of the values divided by the number of values. The mean of the numbers 10, 11, 14, 15 is 12.5. In everyday language, when people use the word "average," they are usually referring to the mean.

Median (8.4) If a set of numbers is arranged in order from smallest to largest, the median is that value that has the same number of values above it as below it. The median of the numbers 3, 7, 8 is 7. If the list contains an even number of items, we obtain the median by finding the mean of the two middle numbers. The median of the numbers 5, 6, 10, 11 is 8.

Megameter (6.2) A metric unit of length equal to 1,000,000 meters.

Meter (6.2) The basic unit of length in the metric system. 1 meter = 1000 millimeters. 1 meter = 0.001 kilometer.

Metric ton (6.3) A metric unit of measurement for very heavy weights. 1 metric ton = 1,000,000 grams.

Microgram (6.3) A unit of weight equal to 0.000001 gram.

Micrometer (6.2) A metric unit of length equal to 0.000001 meter.

Mile (6.1) Largest unit of length in the American system. 5280 feet = 1 mile, 1760 yards = 1 mile.

Milligram (6.3) A metric unit of weight used for very, very small objects. 1 milligram = 0.001 gram.

Milliliter (6.3) The metric unit of volume normally used to measure small volumes. 1 milliliter = 0.001 liter.

Millimeter (6.2) A unit of length commonly used in the metric system to measure very small distances. 1 millimeter = 0.001 meter.

Million (1.1) The number 1,000,000.

Minuend (1.3) The number being subtracted from in a subtraction problem. In the problem $8 - 5 = 3$, the number 8 is the minuend.

Mixed number (2.3) A number created by the sum of a whole number greater than 1 and a proper fraction. The numbers $4\frac{5}{6}$ and $1\frac{1}{8}$ are both mixed numbers. Mixed numbers are sometimes referred to as *mixed fractions*.

Mode (8.4) The mode of a set of data is the number or numbers that occur most often.

Multiplicand (1.4) The first factor in a multiplication problem. In the problem $7 \times 2 = 14$, the number 7 is the multiplicand.

Multiplier (1.4) The second factor in a multiplication problem. In the problem $6 \times 3 = 18$, the number 3 is the multiplier.

Nanogram (6.3) A unit of weight equal to 0.000000001 gram.

Nanometer (6.2) A metric unit of length equal to 0.000000001 meter.

Negative numbers (9.1) All of the numbers to the left of zero on the number line. The numbers -1.5, -16, -200.5, -4500 are all negative numbers. All negative numbers are written with a negative sign in front of the digits.

Number line (1.7) A line on which numbers are placed in order from smallest to largest.

Numerator (2.1) The number on the top of a fraction. In the fraction $\frac{3}{7}$ the numerator is 3.

Numerical coefficients (10.1) The numbers in front of the variables in one or more terms. If we look at $-3xy + 12w$, we find that the numerical coefficient of the xy term is -3 while the numerical coefficient of the w term is 12.

Octagon (7.2) An eight-sided figure.

Odometer (1.8) A device on an automobile that displays how many miles the car has been driven since it was first put into operation.

Opposite of a signed number (9.2) The opposite of a signed number is a number that has the same absolute value. The opposite of -5 is 5. The opposite of 7 is -7.

Order of operations (1.6) An agreed-upon procedure to do a problem with several arithmetic operations in the proper order.

Ounce (6.1) Smallest unit of weight in the American system. 16 ounces = 1 pound.

Overtime (2.9) The pay earned by a person if he works more than a certain number of hours per week. In most jobs that pay by the hour, a person will earn $1\frac{1}{2}$ times as much per hour for every hour beyond 40 hours worked in one workweek. For example, Carlos earns $6.00 per hour for the first 40 hours in a week and overtime for each additional hour. He would earn $9.00 per hour for all hours he worked in that week beyond 40 hours.

Parallel lines (7.2) Two straight lines that are always the same distance apart.

Parallelogram (7.2) A four-sided figure with both pairs of opposite sides parallel.

Parentheses (1.4) One of several symbols used in mathematics to indicate multiplication. For example, (3)(5) means 3 multiplied by 5. A parentheses is also used as a grouping symbol.

Percent (5.1) The word *percent* means per one hundred. For example, 14 percent means $\frac{14}{100}$.

Percent of decrease (5.4) The percent that something decreases is determined by dividing the amount of decrease by the original amount. If a tape deck sold for $300 and its price was decreased by $60, the percent of decrease would be $\frac{60}{300} = 0.20 = 20\%$.

Percent of increase (5.4) The percent that something increases is determined by dividing the amount of increase by the original amount. If the population of a town was 5000 people and the population increased by 500 people, the percent of increase would be $\frac{500}{5000} = 0.10 = 10\%$.

Percent proportion (5.3B) The percent proportion is the equation $\frac{a}{b} = \frac{p}{100}$ where a is the amount, b is the base, and p is the percent number.

Percent symbol (5.1) A symbol that is used to indicate percent. To indicate 23 percent, we write 23%.

Perfect square (7.4) When a whole number is multiplied by itself, the number that is obtained is a perfect square. The numbers 1, 4, 9, 16, 25, 36, 49, 64, 81, and 100 are all perfect squares.

Perimeter (7.1) The distance around a figure.

Perpendicular lines (7.1) Lines that meet at an angle of 90 degrees.

Pi (7.6) Pi is an irrational number that we obtain if we divide the circumference of a circle by the diameter of a circle. It is represented by the symbol π. Accurate to eleven decimal places, the value of pi is given by 3.14159265359. For most work in this textbook, the value of 3.14 is used to approximate the value of pi.

Picogram (6.3) A unit of weight equal to 0.000000000001 gram.

Pint (6.1) Unit of volume in the American system. 2 pints = 1 quart.

Placeholder (1.1) The use of a digit to indicate a place. Zero is a placeholder in our number system. It holds a position and shows that there is no other digit in that place.

Place-value system (1.1) Our number system is called a place-value system because the placement of the digits tells the value of the number. If we use the digits 5 and 4 to write the number 54, the result is different than if we placed them in opposite order and wrote 45.

Positive numbers (9.1) All of the numbers to the right of zero on the number line. The numbers 5, 6.2, 124.186, 5000 are all positive numbers. A positive number such as +5 is usually written without the positive sign.

Pound (6.1) Basic unit of weight in the American system. 2000 pounds = 1 ton. 16 ounces = 1 pound.

Power of 10 (1.4) Whole numbers that begin with 1 and end in one or more zeros are called powers of 10. The numbers 10, 100, 1000, etc., are all powers of 10.

Prime factors (2.2) Factors that are prime numbers. If we write 15 as a product of prime factors, we have $15 = 5 \times 3$.

Prime number (2.2) A prime number is a whole number greater than 1 that can only be divided by 1 and itself. The first fifteen prime numbers are 2, 3, 5, 7, 11, 13, 17, 19, 23, 29, 31, 37, 41, 43, 47. The list of prime numbers goes on indefinitely.

Principal (5.4) The amount of money deposited or borrowed on which interest is computed. In the simple interest formula $I = P \times R \times T$, the P stands for the principal. (The other letters are I = interest, R = interest rate, and T = the amount of time.)

Product (1.4) The answer in a multiplication problem. In the problem $3 \times 4 = 12$ the number 12 is the product.

Proper fraction (2.3) A fraction in which the numerator is less than the denominator. The fractions $\frac{3}{4}$ and $\frac{15}{16}$ are proper fractions.

Proportion (4.2) A statement that two ratios or two rates are equal. The statement $\frac{3}{4} = \frac{15}{20}$ is a proportion. The statement $\frac{5}{7} = \frac{7}{9}$ is false, and is therefore not a proportion.

Pyramid (7.7) A three-dimensional object made up of a geometric figure for a base and triangular sides that meet at a point. Some pyramids are shaped like the great pyramids of Egypt.

Pythagorean Theorem (7.5) A statement that for any right triangle the square of the hypotenuse equals the sum of the squares of the two legs of the triangle.

Quadrilateral (7.2) A four-sided geometric figure.

Quadrillion (1.1) The number 1,000,000,000,000,000.

Quart (6.1) Unit of volume in the American system. 4 quarts = 1 gallon.

Quotient (1.5) The answer after performing a division problem. In the problem $60 \div 6 = 10$ the number 10 is the quotient.

Radius of a circle (7.6) A line segment from the center of a circle to any point on the circle. The radius of a circle is equal to one-half of the diameter of the circle.

Rate (4.1) A rate compares two quantities that have different units. Examples of rates are $5.00 an hour and 13 pounds for every 2 inches. In fraction form, these two rates would be written as $\frac{\$5.00}{1 \text{ hour}}$ and $\frac{13 \text{ pounds}}{2 \text{ inches}}$.

Ratio (4.1) A ratio is a comparison of two quantities that have the same units. To compare 2 to 3, we can express the ratio in three ways: the ratio of 2 to 3; 2:3; or the fraction $\frac{2}{3}$.

Ratio in simplest form (4.1) A ratio is in simplest form when the two numbers do not have a common factor.

Rectangle (7.1) A four-sided figure that has four right angles.

Reduced fraction (2.2) A fraction for which the numerator and denominator have no common factor other than 1. The fraction $\frac{5}{7}$ is a reduced fraction. The fraction $\frac{15}{21}$ is not a reduced fraction because both numerator and denominator have a common factor of 3.

Regular hexagon (7.2) A six-sided figure with all sides equal.

Regular octagon (7.2) An eight-sided figure with all sides equal.

Remainder (1.5) When two numbers do not divide exactly, a part is left over. This part is called the remainder. For example, $13 \div 2 = 6$ with 1 left over; the 1 is the remainder.

Repeating decimals (3.6) Decimals that have a digit or a group of digits that repeat. The decimals 0.33333333333 . . . and 1.234234234234 . . . are repeating decimals. The pattern of repeating continues indefinitely. Repeating decimals can be written in a form with a bar over the repeating digit(s). Thus the decimals above could be written as $0.\overline{3}$ and $1.\overline{234}$.

Right angle (7.1) and (7.3) An angle that measures 90 degrees.

Right triangle (7.3) A triangle with one 90-degree angle.

Rounding (1.7) The process of writing a number in an approximate form for convenience. The number 9756 rounded to the nearest hundred is 9800.

Sales tax (5.4) The amount of tax on a purchase. The sales tax for any item is a product of the sales tax rate times the purchase price. If an item is purchased for $12.00 and the sales tax rate is 5%, the sales tax is $5\% \times 12.00 = \$0.60$.

Scientific notation (9.5) A positive number is written in scientific notation if it is in the form $a \times 10^n$ where a is a number greater than or equal to 1, but less than 10, and n is an integer. If we write 5678 in scientific notation, we have 5.678×10^3. If we write 0.00825 in scientific notation, we have 8.25×10^{-3}.

Semicircle (7.6) One-half of a circle. The semicircle usually includes the diameter of a circle connected to one-half of the circumference of the circle.

Sides of an angle (7.3) The two line segments that meet to form an angle.

Signed numbers (9.1) All of the numbers on a number line. Numbers like −33, 2, 5, −4.2, 18.678, −8.432 are all signed numbers. A negative number always has a negative sign in front of the digits. A positive number such as +3 is usually written without the positive sign in front of it.

Similar triangles (7.8) Two triangles that have the same shape but are not necessarily the same size. The corresponding angles of similar triangles are equal. The corresponding sides of similar triangles have the same ratio.

Simple interest (5.4) The interest determined by the formula $I = P \times R \times T$ where I = the interest obtained, P = the principal or the amount borrowed or invested, R = the interest rate (usually on an annual basis), and T = the number of time periods (usually years).

Solution of an equation (10.3) A number is a solution of an equation if replacing the variable by the number makes the equation always true. The solution of $x - 5 = -20$ is the number −15.

Sphere (7.7) A three-dimensional object shaped like a perfectly round ball.

Square (7.1) A rectangle with all four sides equal.

Square root (7.4) The square root of a number is one of two identical factors of that number. The square root of 9 is 3. The square root of 121 is 11.

Square root sign (7.4) The symbol $\sqrt{}$. When we want to find the square root of 25, we write $\sqrt{25}$. The answer is 5.

Standard notation for a number (1.1) A number written in ordinary terms. For example, $70 + 2$ in standard notation is 72.

Subtrahend (1.3) The number being subtracted. In the problem $7 - 1 = 6$, the number 1 is the subtrahend.

Sum (1.2) The result of an addition of two or more numbers. In the problem $7 + 3 + 5 = 15$, the number 15 is the sum.

Trapezoid (7.2) A four-sided figure with at least two parallel sides.

Term (10.1) A number, a variable, or a product of a number and one or more variables. $5x$, $2ab$, $-43cdef$ are three examples of terms, separated in an expression by a + sign or a − sign.

Terminating decimals (3.6) Every fraction can be written as a decimal. If the division process of dividing denominator into numerator ends with a remainder of zero, the decimal is a terminating decimal. Decimals such as 1.28, 0.007856, and 5.123 are terminating decimals.

Triangle (7.3) A three-sided figure.

Trillion (1.1) The number 1,000,000,000,000.

Unit fraction (6.1) A fraction used to change one unit to another. For example, to change 180 inches to feet, we multiply by the unit fraction $\frac{1 \text{ foot}}{12 \text{ inches}}$. Thus we have

$$180 \text{ inches} \times \frac{1 \text{ foot}}{12 \text{ inches}} = 15 \text{ feet}$$

Variable (10.1) A letter that is used to represent a number.

Vertex of an angle (7.3) The point at which two line segments meet to form an angle.

Vertex of a cone (7.7) The sharp point of a cone.

Volume (7.7) The measure of the space inside a three-dimensional object. Volume is measured in cubic units such as cubic feet.

Width of a rectangle (7.1) Each of the shorter sides of a rectangle.

Whole numbers (1.1) The whole numbers are the set of numbers 0, 1, 2, 3, 4, 5, 6, 7, 8, 9, 10, 11, 12, ... The set goes on forever. There is no largest whole number.

Word names for whole numbers (1.1) The notation for a number in which each digit is expressed by a word. To write 389 with a word name, we would write three hundred eighty-nine.

Zero (1.1) The smallest whole number. It is normally written 0.

APPENDIX A
Tables

TABLE OF BASIC ADDITION FACTS

+	0	1	2	3	4	5	6	7	8	9
0	0	1	2	3	4	5	6	7	8	9
1	1	2	3	4	5	6	7	8	9	10
2	2	3	4	5	6	7	8	9	10	11
3	3	4	5	6	7	8	9	10	11	12
4	4	5	6	7	8	9	10	11	12	13
5	5	6	7	8	9	10	11	12	13	14
6	6	7	8	9	10	11	12	13	14	15
7	7	8	9	10	11	12	13	14	15	16
8	8	9	10	11	12	13	14	15	16	17
9	9	10	11	12	13	14	15	16	17	18

TABLE OF BASIC MULTIPLICATION FACTS

×	0	1	2	3	4	5	6	7	8	9
0	0	0	0	0	0	0	0	0	0	0
1	0	1	2	3	4	5	6	7	8	9
2	0	2	4	6	8	10	12	14	16	18
3	0	3	6	9	12	15	18	21	24	27
4	0	4	8	12	16	20	24	28	32	36
5	0	5	10	15	20	25	30	35	40	45
6	0	6	12	18	24	30	36	42	48	54
7	0	7	14	21	28	35	42	49	56	63
8	0	8	16	24	32	40	48	56	64	72
9	0	9	18	27	36	45	54	63	72	81

TABLE OF PRIME FACTORS

Number	Prime Factors	Number	Prime Factors	Number	Prime Factors	Number	Prime Factors
		51	3×17	**101**	prime	**151**	prime
2	prime	**52**	$2^2 \times 13$	**102**	$2 \times 3 \times 17$	**152**	$2^3 \times 19$
3	prime	**53**	prime	**103**	prime	**153**	$3^2 \times 17$
4	2^2	**54**	2×3^3	**104**	$2^3 \times 13$	**154**	$2 \times 7 \times 11$
5	prime	**55**	5×11	**105**	$3 \times 5 \times 7$	**155**	5×31
6	2×3	**56**	$2^3 \times 7$	**106**	2×53	**156**	$2^2 \times 3 \times 13$
7	prime	**57**	3×19	**107**	prime	**157**	prime
8	2^3	**58**	2×29	**108**	$2^2 \times 3^3$	**158**	2×79
9	3^2	**59**	prime	**109**	prime	**159**	3×53
10	2×5	**60**	$2^2 \times 3 \times 5$	**110**	$2 \times 5 \times 11$	**160**	$2^5 \times 5$
11	prime	**61**	prime	**111**	3×37	**161**	7×23
12	$2^2 \times 3$	**62**	2×31	**112**	$2^4 \times 7$	**162**	2×3^4
13	prime	**63**	$3^2 \times 7$	**113**	prime	**163**	prime
14	2×7	**64**	2^6	**114**	$2 \times 3 \times 19$	**164**	$2^2 \times 41$
15	3×5	**65**	5×13	**115**	5×23	**165**	$3 \times 5 \times 11$
16	2^4	**66**	$2 \times 3 \times 11$	**116**	$2^2 \times 29$	**166**	2×83
17	prime	**67**	prime	**117**	$3^2 \times 13$	**167**	prime
18	2×3^2	**68**	$2^2 \times 17$	**118**	2×59	**168**	$2^3 \times 3 \times 7$
19	prime	**69**	3×23	**119**	7×17	**169**	13^2
20	$2^2 \times 5$	**70**	$2 \times 5 \times 7$	**120**	$2^3 \times 3 \times 5$	**170**	$2 \times 5 \times 17$
21	3×7	**71**	prime	**121**	11^2	**171**	$3^2 \times 19$
22	2×11	**72**	$2^3 \times 3^2$	**122**	2×61	**172**	$2^2 \times 43$
23	prime	**73**	prime	**123**	3×41	**173**	prime
24	$2^3 \times 3$	**74**	2×37	**124**	$2^2 \times 31$	**174**	$2 \times 3 \times 29$
25	5^2	**75**	3×5^2	**125**	5^3	**175**	$5^2 \times 7$
26	2×13	**76**	$2^2 \times 19$	**126**	$2 \times 3^2 \times 7$	**176**	$2^4 \times 11$
27	3^3	**77**	7×11	**127**	prime	**177**	3×59
28	$2^2 \times 7$	**78**	$2 \times 3 \times 13$	**128**	2^7	**178**	2×89
29	prime	**79**	prime	**129**	3×43	**179**	prime
30	$2 \times 3 \times 5$	**80**	$2^4 \times 5$	**130**	$2 \times 5 \times 13$	**180**	$2^2 \times 3^2 \times 5$
31	prime	**81**	3^4	**131**	prime	**181**	prime
32	2^5	**82**	2×41	**132**	$2^2 \times 3 \times 11$	**182**	$2 \times 7 \times 13$
33	3×11	**83**	prime	**133**	7×19	**183**	3×61
34	2×17	**84**	$2^2 \times 3 \times 7$	**134**	2×67	**184**	$2^3 \times 23$
35	5×7	**85**	5×17	**135**	$3^3 \times 5$	**185**	5×37
36	$2^2 \times 3^2$	**86**	2×43	**136**	$2^3 \times 17$	**186**	$2 \times 3 \times 31$
37	prime	**87**	3×29	**137**	prime	**187**	11×17
38	2×19	**88**	$2^3 \times 11$	**138**	$2 \times 3 \times 23$	**188**	$2^2 \times 47$
39	3×13	**89**	prime	**139**	prime	**189**	$3^3 \times 7$
40	$2^3 \times 5$	**90**	$2 \times 3^2 \times 5$	**140**	$2^2 \times 5 \times 7$	**190**	$2 \times 5 \times 19$
41	prime	**91**	7×13	**141**	3×47	**191**	prime
42	$2 \times 3 \times 7$	**92**	$2^2 \times 23$	**142**	2×71	**192**	$2^6 \times 3$
43	prime	**93**	3×31	**143**	11×13	**193**	prime
44	$2^2 \times 11$	**94**	2×47	**144**	$2^4 \times 3^2$	**194**	2×97
45	$3^2 \times 5$	**95**	5×19	**145**	5×29	**195**	$3 \times 5 \times 13$
46	2×23	**96**	$2^5 \times 3$	**146**	2×73	**196**	$2^2 \times 7^2$
47	prime	**97**	prime	**147**	3×7^2	**197**	prime
48	$2^4 \times 3$	**98**	2×7^2	**148**	$2^2 \times 37$	**198**	$2 \times 3^2 \times 11$
49	7^2	**99**	$3^2 \times 11$	**149**	prime	**199**	prime
50	2×5^2	**100**	$2^2 \times 5^2$	**150**	$2 \times 3 \times 5^2$	**200**	$2^3 \times 5^2$

TABLE OF SQUARE ROOTS

Square Root Values Are Rounded to the Nearest Thousandth Unless the Answer Ends in .000

n	$\sqrt{n}$	n	$\sqrt{n}$	n	$\sqrt{n}$	n	$\sqrt{n}$	n	$\sqrt{n}$
1	1.000	**41**	6.403	**81**	9.000	**121**	11.000	**161**	12.689
2	1.414	**42**	6.481	**82**	9.055	**122**	11.045	**162**	12.728
3	1.732	**43**	6.557	**83**	9.110	**123**	11.091	**163**	12.767
4	2.000	**44**	6.633	**84**	9.165	**124**	11.136	**164**	12.806
5	2.236	**45**	6.708	**85**	9.220	**125**	11.180	**165**	12.845
6	2.449	**46**	6.782	**86**	9.274	**126**	11.225	**166**	12.884
7	2.646	**47**	6.856	**87**	9.327	**127**	11.269	**167**	12.923
8	2.828	**48**	6.928	**88**	9.381	**128**	11.314	**168**	12.961
9	3.000	**49**	7.000	**89**	9.434	**129**	11.358	**169**	13.000
10	3.162	**50**	7.071	**90**	9.487	**130**	11.402	**170**	13.038
11	3.317	**51**	7.141	**91**	9.539	**131**	11.446	**171**	13.077
12	3.464	**52**	7.211	**92**	9.592	**132**	11.489	**172**	13.115
13	3.606	**53**	7.280	**93**	9.644	**133**	11.533	**173**	13.153
14	3.742	**54**	7.348	**94**	9.695	**134**	11.576	**174**	13.191
15	3.873	**55**	7.416	**95**	9.747	**135**	11.619	**175**	13.229
16	4.000	**56**	7.483	**96**	9.798	**136**	11.662	**176**	13.266
17	4.123	**57**	7.550	**97**	9.849	**137**	11.705	**177**	13.304
18	4.243	**58**	7.616	**98**	9.899	**138**	11.747	**178**	13.342
19	4.359	**59**	7.681	**99**	9.950	**139**	11.790	**179**	13.379
20	4.472	**60**	7.746	**100**	10.000	**140**	11.832	**180**	13.416
21	4.583	**61**	7.810	**101**	10.050	**141**	11.874	**181**	13.454
22	4.690	**62**	7.874	**102**	10.100	**142**	11.916	**182**	13.491
23	4.796	**63**	7.937	**103**	10.149	**143**	11.958	**183**	13.528
24	4.899	**64**	8.000	**104**	10.198	**144**	12.000	**184**	13.565
25	5.000	**65**	8.062	**105**	10.247	**145**	12.042	**185**	13.601
26	5.099	**66**	8.124	**106**	10.296	**146**	12.083	**186**	13.638
27	5.196	**67**	8.185	**107**	10.344	**147**	12.124	**187**	13.675
28	5.292	**68**	8.246	**108**	10.392	**148**	12.166	**188**	13.711
29	5.385	**69**	8.307	**109**	10.440	**149**	12.207	**189**	13.748
30	5.477	**70**	8.367	**110**	10.488	**150**	12.247	**190**	13.784
31	5.568	**71**	8.426	**111**	10.536	**151**	12.288	**191**	13.820
32	5.657	**72**	8.485	**112**	10.583	**152**	12.329	**192**	13.856
33	5.745	**73**	8.544	**113**	10.630	**153**	12.369	**193**	13.892
34	5.831	**74**	8.602	**114**	10.677	**154**	12.410	**194**	13.928
35	5.916	**75**	8.660	**115**	10.724	**155**	12.450	**195**	13.964
36	6.000	**76**	8.718	**116**	10.770	**156**	12.490	**196**	14.000
37	6.083	**77**	8.775	**117**	10.817	**157**	12.530	**197**	14.036
38	6.164	**78**	8.832	**118**	10.863	**158**	12.570	**198**	14.071
39	6.245	**79**	8.888	**119**	10.909	**159**	12.610	**199**	14.107
40	6.325	**80**	8.944	**120**	10.954	**160**	12.649	**200**	14.142

APPENDIX B
Scientific Calculators

This text does *not require* the use of a calculator. However, you may want to consider the purchase of an inexpensive scientific calculator. It is wise to ask your instructor for advice before you purchase any calculator for this course. It should be stressed that students are asked to avoid using a calculator for any of the exercises in which the calculations can be readily done by hand. The only problems in the text that really demand the use of a scientific calculator are marked with the ▦ symbol. Dependence on the use of the scientific calculator for regular exercises in the text will only hurt the student in the long run.

The Two Types of Logic Used in Scientific Calculators

There are two major types of scientific calculators that are popular today. The most common type employs a type of logic known as **algebraic** logic. The calculators manufactured by Casio, Sharp, and Texas Instruments as well as many other companies employ this type of logic. An example of calculation on such a calculator would be the following. To add 14 + 26 on an algebraic logic calculator, the sequence of buttons would be:

14 [+] 26 [=]

The second type of scientific calculator requires the entry of data in **Reverse Polish Notation (RPN).** Calculators manufactured by Hewlett-Packard and a few other specialized calculators are made to use RPN. To add 14 + 26 on a RPN calculator, the sequence of buttons would be:

14 [enter] 26 [+]

Graphing scientific calculators such as the TI-82 and TI-83 have a large liquid display for viewing graphs. To perform the calculation on most graphing calculators, the sequence of buttons would be:

14 [+] 26 [enter]

Mathematicians and scientists do not agree on which type of scientific calculator is superior. However, the clear majority of college students own calculators that employ *algebraic* logic. Therefore this section of the text is explained with reference to the sequence of steps employed by an *algebraic* logic calculator. If you already own or intend to purchase a scientific calculator that uses RPN, you are encouraged to study the instruction booklet that comes with the calculator and practice the problems shown in the booklet. After this practice you will be able to solve the calculator problems discussed in this section.

Performing Simple Calculations

The following example will illustrate the use of the scientific calculator in doing basic arithmetic calculations.

EXAMPLE 1 Add. $156 + 298$

We first key in the number 156, then press the [+] key, then enter the number 298, and finally press the [=] key.

156 [+] 298 [=] 454

Practice Problem 1 Add. $3792 + 5896$ ■

EXAMPLE 2 Subtract. $1508 - 963$

We first enter the number 1508, then press the [−] key, then enter the number 963, and finally press the [=] key.

1508 [−] 963 [=] 545

Practice Problem 2 Subtract. $7930 - 5096$ ■

EXAMPLE 3 Multiply. 196×358

196 [×] 358 [=] 70168

Practice Problem 3 Multiply. 896×273 ■

EXAMPLE 4 Divide. $2054 \div 13$

2054 [÷] 13 [=] 158

Practice Problem 4 Divide. $2352 \div 16$ ■

Decimal Problems

Problems involving decimals can be readily done on the calculator. Entering numbers with a decimal point is done by pressing the [•] key at the appropriate time.

EXAMPLE 5 Calculate. 4.56×283

To enter 4.56, we press the [4] key, the decimal point key, then the [5] key, and finally the [6] key.

4.56 [×] 283 [=] 1290.48

The answer is 1290.48. Observe how your calculator displays the decimal point.

Practice Problem 5 Calculate. 72.8×197 ■

EXAMPLE 6 Add. $128.6 + 343.7 + 103.4 + 207.5$

128.6 [+] 343.7 [+] 103.4 [+] 207.5 [=] 783.2

The answer is 783.2. Observe how your calculator displays the answer.

Practice Problem 6 Add. $52.98 + 31.74 + 40.37 + 99.82$ ■

Combined Operations

You must use extra caution concerning the order of mathematical operations when you are doing a problem on the calculator that involves two or more different operations.

Any scientific calculator with algebraic logic uses a priority system that has a clearly defined order of operations. It is the same order we use in performing arithmetic operations by hand. In either situation, calculations are performed in the following order:

1. First calculations within a parentheses are completed.

2. Then numbers are raised to a power or a square root is calculated.

3. Then multiplication and division operations are performed from left to right.

4. Then addition and subtraction operations are performed from left to right.

This order is carefully followed on *scientific calculators*. Small inexpensive calculators that do not have scientific functions often do not follow this order of operations.

The number of digits displayed in the answer varies from calculator to calculator. In the following examples, your calculator may display more or fewer digits than the answer we have listed.

EXAMPLE 7 Evaluate. $5.3 \times 1.62 + 1.78 \div 3.51$

This problem requires that we multiply 5.3 by 1.62 and divide 1.78 by 3.51 first and then add the two results. If the numbers are entered directly into the claculator exactly as the problem is written, the calculator will perform the calculations in the correct order.

5.3 [×] 1.62 [+] 1.78 [÷] 3.51 [=] 9.09312251

Practice Problem 7 Evaluate. $0.0618 \times 19.22 - 59.38 \div 166.3$ ■

The Use of Parentheses

In order to perform some calculations on the calculator the use of parentheses is helpful. These parentheses may or may not appear in the original problem.

EXAMPLE 8 Evaluate. $5 \times (2.123 + 5.786 - 12.063)$

The problem requires that the numbers in the parentheses be combined first. By entering the parentheses on the calculator this will be accomplished.

5 [×] [(] 2.123 [+] 5.786 [−] 12.063 [)] [=] −20.77

Note: The result is a negative number.

Practice Problem 8 Evaluate. $3.152 \times (0.1628 + 3.715 - 4.985)$ ■

Negative Numbers

To enter a negative number, enter the number followed by the [+/−] button.

EXAMPLE 9 Evaluate. $(-8.634)(5.821) + (1.634)(-16.082)$

The products will be evaluated first by the calculator. Therefore parentheses are not needed as we enter the data.

8.634 [+/−] [×] 5.821 [+] 1.634 [×] 16.082 [+/−] [=] −76.536502

Note: The result is negative.

Practice Problem 9 Evaluate. $(0.5618)(-98.3) - (76.31)(-2.98)$ ■

Scientific Notation

If you wish to enter a number in scientific notation, you should use the special scientific notation button. On most calculators it is denoted as [EXP] or [EE].

EXAMPLE 10 Multiply. $(9.32 \times 10^6)(3.52 \times 10^8)$

9.32 [EXP] 6 [×] 3.52 [EXP] 8 [=] 3.28064 15

This notation means the answer is 3.28064×10^{15}.

Practice Problem 10 Divide. $(3.76 \times 10^{15}) \div (7.76 \times 10^7)$ ■

Raising a Number to a Power

All scientific calculators have a key for finding powers of numbers. It is usually labeled [y^x]. (On a few calculators the notation is [x^y] or sometimes [^].) To raise a number to a power, first you enter the base, then push the [y^x] key. Then you enter the exponent, then finally the [=] button.

EXAMPLE 11 Evaluate. $(2.16)^9$

2.16 [y^x] 9 [=] 1023.490369

Practice Problem 11 Evaluate. $(6.238)^6$ ■

There is a special key to square something. It is usually labeled [x^2].

EXAMPLE 12 Evaluate. $(76.04)^2$

76.04 [x^2] 5782.0816

Practice Problem 12 Evaluate. $(132.56)^2$ ■

Finding Square Roots of Numbers

There is usually a key to approximate square roots on a scientific calculator labeled [$\sqrt{\ }$]. In this example we will need to use a parentheses.

EXAMPLE 13 Evaluate. $\sqrt{5618 + 2734 + 3913}$

[(] 5618 [+] 2734 [+] 3913 [)] [$\sqrt{\ }$] 110.747605

Practice Problem 13 Evaluate. $\sqrt{0.0782 - 0.0132 + 0.1364}$ ■

Scientific Calculator Exercises

Use your calculator to complete each of the following. Your answers may vary slightly because of the characteristics of individual calculators.

Complete the following table.

To Do This Operation	Use These Keystrokes	Record Your Answer Here
1. 8963 + 2784	8963 [+] 2784 [=]	11,747
2. 15,308 − 7980	15308 [−] 7980 [=]	7328
3. 2631 × 134	2631 [×] 134 [=]	352,554
4. 70,221 ÷ 89	70221 [÷] 89 [=]	789
5. 5.325 − 4.031	5.325 [−] 4.031 [=]	1.294
6. 184.68 + 73.98	184.68 [+] 73.98 [=]	258.66
7. 2004.06 ÷ 7.89	2004.06 [÷] 7.89 [=]	254
8. 1.34 × 0.763	1.34 [×] 0.763 [=]	1.02242

Write down the answer for the following and then show what problem you have solved.

9. 123.45 [+] 45.9876 [+] 8765.3 [=] 8934.7376
123.45 + 45.9876 + 8765.3

10. 0.0897 [×] 234.56 [×] 2.5428 [=] 53.50059337
0.0897 × 234.56 × 2.5428

11. 34 [÷] 8 [+] 12.56 [=] 16.81
$\frac{34}{8} + 12.56$

12. 458 [÷] 4 [−] 16.897 [=] 97.603
$\frac{458}{4} - 16.897$

Perform the following calculations using your calculator.

13. 9.467 + 0.563
10.03

14. 0.347 + 23.457
23.804

15. 34.89 + 39.6 + 214.897
289.387

16. 12.567 + 48.31 + 189.38
250.257

17. 412,899 − 34,675
378,224

18. 87,456 − 2876
84,580

19. 3,567,089 − 2,876,805
690,284

20. 8,345,802 − 4,985,004
3,360,798

21. 234 × 4.567
1068.678

22. 1.9876 × 347
689.6972

23. 0.456 × 3.48
1.58688

24. 67,876 × 0.0946
6421.0696

25. 3458 ÷ 2.5
1383.2

26. 9764 ÷ 8
1220.5

27. 12.107524 ÷ 15.86
0.7634

28. 16.06513 ÷ 17.98
0.8935

Perform the following calculations using your calculator.

29. 1.98 + 6.34 + 7.71 = 16.03

30. 8.92 + 9.31 + 7.79 = 26.02

31. \$103.91 + \$2653.82 + \$9804.61 = \$12,562.34

32. \$3986.21 + \$4502.89 + \$989.30 = \$9478.40

33. 368,781.5 − 283,617.8 = 85,163.7

34. 571,809.6 − 539,376.8 = 32,432.8

35. \$1,393,271.86 − \$1,289,663.21 = \$103,608.65

36. \$8,571,300.76 − \$4,098,789.39 = \$4,472,511.37

37. 345.34 × 45.7 = 15,782.038

38. 8954.34 × 425.4 = 3,809,176.236

39. 0.6314 × 3.96 = 2.500344

40. 0.0789 × 12.38 = 0.976782

41. 40.36)36202.92 — 897

42. 52.98)172,608.84 — 3258

43. 0.7613)17.12925 — 22.5

44. 0.9854)3.59671 — 3.65

Perform the following operations in the proper order using your calculator.

45. $4.567 + 87.89 - 2.45 \times 3.3$
84.372

46. $4.891 + 234.5 - 0.98 \times 23.4$
216.459

47. $7 \div 8 + 3.56$
4.435

48. $9 \div 4.5 + 0.6754$
2.6754

49. $(9.34)(0.345) + 98.345$
101.5673

50. $(0.628)(398) + 34.4581$
284.4021

51. $\dfrac{(95.34)(0.9874)}{381.36}$
0.24685

52. $\dfrac{(0.8759)(45.87)}{183.48}$
0.218975

53. $2.56 + 8.98 \times 3.14$
30.7572

54. $1.62 + 3.81 - 5.23 \times 6.18$
−26.8914

55. $(-4.23)(1.863) - 5.998$
−13.87849

56. $12.34 - (26.314)(-1.856)$
61.178784

57. $5.62(5 \times 3.16 - 18.12)$
−13.0384

58. $9.356(4.8 - 7.2 - 15.94)$
−171.58904

59. $(3.42 \times 10^8)(0.97 \times 10^{10})$
3.3174×10^{18}

60. $(6.27 \times 10^{20})(1.35 \times 10^3)$
8.4645×10^{23}

61. $\dfrac{(2.16 \times 10^3)(1.37 \times 10^{14})}{6.39 \times 10^5}$
$4.630985915 \times 10^{11}$

62. $\dfrac{(3.84 \times 10^{12})(1.62 \times 10^5)}{7.78 \times 10^8}$
7.9958869×10^8

63. $\dfrac{2.3 + 5.8 - 2.6 - 3.9}{5.3 - 8.2}$
−0.5517241379

64. $\dfrac{(2.6)(-3.2) + (5.8)(-0.9)}{2.614 + 5.832}$
−1.60312574

65. $\sqrt{253.12}$
15.90974544

66. $\sqrt{0.0713}$
0.267020599

67. $\sqrt{5.6213 - 3.7214}$
1.378368601

68. $\sqrt{3417.2 - 2216.3}$
34.6540041

69. $(1.78)^3 + 6.342$
11.981752

70. $(2.26)^8 - 3.1413$
677.4204134

71. $\sqrt{(6.13)^2 + (5.28)^2}$
8.090444982

72. $\sqrt{(0.3614)^2 + (0.9217)^2}$
0.9900206311

73. $\sqrt{56 + 83} - \sqrt{12}$
8.325724507

74. $\sqrt{98 + 33} - \sqrt{17}$
7.322417517

Find an approximate value for each of the following. Round your answer to five decimal places.

75. $\dfrac{7}{18} + \dfrac{9}{13}$
1.08120

76. $\dfrac{5}{22} + \dfrac{1}{31}$
0.25953

77. $\dfrac{7}{8} + \dfrac{3}{11}$
1.14773

78. $\dfrac{9}{14} + \dfrac{5}{19}$
0.90602

Solutions to Practice Problems

Chapter 1

1.1 Practice Problems

1. **(a)** 3182 = 3000 + 100 + 80 + 2
   (b) 520,890 = 500,000 + 20,000 + 800 + 90
   (c) 709,680,059 = 700,000,000 + 9,000,000 + 600,000 + 80,000 + 50 + 9
2. **(a)** 492 **(b)** 80,427
3. **(a)** 7 **(b)** 9 **(c)** 4,000
   (d) 900,000 for the first 9, 9 for the last 9
4. **(a)** Two thousand, seven hundred thirty-six
   (b) Nine hundred eighty thousand, three hundred six
   (c) Twelve million, twenty-one
5. The national debt in 1989 was two trillion, eight hundred fifty-seven billion, four hundred six million, three hundred thousand dollars.
6. **(a)** 803 **(b)** 30,229
7. **(a)** 13,000
   (b) 88,000
   (c) 10,000

1.2 Practice Problems

1. **(a)** $\begin{array}{r} 7 \\ +5 \\ \hline 12 \end{array}$ **(b)** $\begin{array}{r} 9 \\ +4 \\ \hline 13 \end{array}$ **(c)** $\begin{array}{r} 3 \\ +0 \\ \hline 3 \end{array}$

2. 7 + 6 + 5 + 8 + 2: 7; 6 + 5 = 11; 8 + 2 = 10; total $\begin{array}{r} 7 \\ +6 \\ +5 \\ +8 \\ +2 \\ \hline 28 \end{array}$

3. 1 + 7 + 2 + 9 + 3: 1 + 9 = 10; 7 + 3 = 10; total $\begin{array}{r} 1 \\ +7 \\ +2 \\ +9 \\ +3 \\ \hline 22 \end{array}$

4. $\begin{array}{r} 8246 \\ +1702 \\ \hline 9948 \end{array}$

5. $\begin{array}{r} {\scriptstyle 1} \\ 56 \\ +36 \\ \hline 92 \end{array}$

6. $\begin{array}{r} {\scriptstyle 21} \\ 789 \\ 63 \\ +297 \\ \hline 1149 \end{array}$

7. **(a)** $\begin{array}{r} {\scriptstyle 111} \\ 127 \\ 9876 \\ +\ 342 \\ \hline 10,345 \end{array}$ **(b)** Check by adding in opposite order. $\begin{array}{r} {\scriptstyle 111} \\ 342 \\ 9876 \\ +\ 127 \\ \hline 10,345 \end{array}$ same

8. $\begin{array}{r} {\scriptstyle 12\ 12} \\ 18,316 \\ 24,789 \\ +22,965 \\ \hline 66,070 \end{array}$ total women

9. $\begin{array}{r} 1000 \\ 2000 \\ 1000 \\ +2000 \\ \hline 6000 \end{array}$ ft

1.3 Practice Problems

1. **(a)** $\begin{array}{r} 9 \\ -6 \\ \hline 3 \end{array}$ **(b)** $\begin{array}{r} 12 \\ -5 \\ \hline 7 \end{array}$ **(c)** $\begin{array}{r} 17 \\ -8 \\ \hline 9 \end{array}$ **(d)** $\begin{array}{r} 14 \\ -0 \\ \hline 14 \end{array}$ **(e)** $\begin{array}{r} 18 \\ -9 \\ \hline 9 \end{array}$

2. $\begin{array}{r} 7695 \\ -3481 \\ \hline 4214 \end{array}$

3. $\begin{array}{r} {\scriptstyle 2\,14} \\ \not{3}\not{4} \\ -16 \\ \hline 18 \end{array}$

4. $\begin{array}{r} {\scriptstyle 8\,13} \\ 6\not{9}\not{3} \\ -426 \\ \hline 267 \end{array}$

5. $\begin{array}{r} {\scriptstyle 16} \\ {\scriptstyle 8\,9\,\not{6}\,10} \\ \not{9}\not{0}\not{7}\not{0} \\ -5886 \\ \hline 3184 \end{array}$

6. **(a)** $\begin{array}{r} 8964 \\ -\ 985 \\ \hline 7979 \end{array}$ **(b)** $\begin{array}{r} 50,000 \\ -32,508 \\ \hline 17,492 \end{array}$

7. Subtraction $\begin{array}{r} 9763 \\ -5732 \\ \hline 4031 \end{array}$ Checking by Addition $\begin{array}{r} 5732 \\ +4031 \\ \hline 9763 \end{array}$ IT CHECKS

8. **(a)** $\begin{array}{r} 284,000 \\ -\ 96,327 \\ \hline 187,673 \end{array}$ Checking by Addition $\begin{array}{r} 96,327 \\ +187,673 \\ \hline 284,000 \end{array}$ IT CHECKS

   (b) $\begin{array}{r} 8,526,024 \\ -6,397,518 \\ \hline 2,128,506 \end{array}$ Checking by Addition $\begin{array}{r} 6,397,518 \\ +2,128,506 \\ \hline 8,526,024 \end{array}$ IT CHECKS

9. **(a)** $x = 5$ **(b)** $x = 12$

10. **(a)** $\begin{array}{r} 23,667,947 \\ -14,227,799 \\ \hline 9,440,148 \end{array}$ **(b)** $\begin{array}{r} 11,198,655 \\ -\ 9,579,677 \\ \hline 1,618,978 \end{array}$

11. **(a)** From the bar graph:
    1995 sales 114
    1994 sales − 78
    Sales increase 36

    (b) From the bar graph:
    Springfield 91
    Riverside − 78
    13 more homes

    (c) 1995 sales 271
    1994 sales − 240
    31

    1996 sales 284
    1995 sales − 271
    13

    Therefore, the greatest increase in sales occurred from 1994 to 1995.

1.4 Practice Problems

1. **(a)** 64 **(b)** 42 **(c)** 40 **(d)** 63 **(e)** 81

2.
$$\begin{array}{r} 3021 \\ \times\ \ \ 3 \\ \hline 9063 \end{array}$$

3.
$$\begin{array}{r} {}^{2} \\ 43 \\ \times\ 8 \\ \hline 344 \end{array}$$

4.
$$\begin{array}{r} {}^{5\,6} \\ 579 \\ \times\ \ 7 \\ \hline 4053 \end{array}$$

5. **(a)** $1267 \times 10 = 12{,}670$
(b) $1267 \times 1000 = 1{,}267{,}000$
(c) $1267 \times 10{,}000 = 12{,}670{,}000$
(d) $1267 \times 1{,}000{,}000 = 1{,}267{,}000{,}000$

6. **(a)** $9 \times 6 \times 10{,}000 = 54 \times 10{,}000 = 540{,}000$
(b) $15 \times 4 \times 100 = 60 \times 100 = 6{,}000$
(c) $21 \times 8 \times 1000 = 168 \times 1000 = 168{,}000$

7.
$$\begin{array}{r} 323 \\ \times\ 32 \\ \hline 646 \\ 969\ \ \\ \hline 10{,}336 \end{array}$$

8.
$$\begin{array}{r} 385 \\ \times\ 69 \\ \hline 3465 \\ 2310\ \ \\ \hline 26{,}565 \end{array}$$

9.
$$\begin{array}{r} 34 \\ \times 20 \\ \hline 680 \end{array}$$

10.
$$\begin{array}{r} 130 \\ \times\ 50 \\ \hline 6500 \end{array}$$

11.
$$\begin{array}{r} 923 \\ \times 675 \\ \hline 4615 \\ 6461\ \ \\ 5538\ \ \ \ \\ \hline 623{,}025 \end{array}$$

12. $25 \times 4 \times 17 = (25 \times 4) \times 17 = 100 \times 17 = 1700$

13. $8 \times 4 \times 3 \times 25 = 8 \times 3 \times (4 \times 25)$
$= 24 \times 100$
$= 2400$

14.
$$\begin{array}{r} 17348 \\ \times\ \ 378 \\ \hline 138784 \\ 121436\ \ \\ 52044\ \ \ \ \\ \hline 6{,}557{,}544 \end{array}$$

The total sales of cars is about $6,557,544.

15. Area = 5 yards × 7 yards = 35 square yards.

1.5 Practice Problems

1. **(a)** $4\overline{)36}$ = 9 **(b)** $5\overline{)25}$ = 5 **(c)** $9\overline{)72}$ = 8 **(d)** $6\overline{)30}$ = 5

2. **(a)** $\frac{7}{1} = 7$ **(b)** $\frac{8}{8} = 1$ **(c)** $\frac{0}{5} = 0$
(d) $\frac{12}{0}$ Cannot be done **(e)** $\frac{0}{0}$ Cannot be done

3.
$$\begin{array}{r} 7\text{ R3} \\ 6\overline{)45} \\ \underline{42} \\ 3 \end{array}$$
Check.
$$\begin{array}{r} 6 \\ \times\ 7 \\ \hline 42 \\ +\ 3 \\ \hline 45 \end{array}$$

4.
$$\begin{array}{r} 21\text{ R3} \\ 6\overline{)129} \\ \underline{12}\ \ \\ 9 \\ \underline{6} \\ 3 \end{array}$$
Check.
$$\begin{array}{r} 21 \\ \times\ \ 6 \\ \hline 126 \\ +\ \ 3\text{ R} \\ \hline 129 \end{array}$$

5.
$$\begin{array}{r} 529\text{ R5} \\ 8\overline{)4237} \\ \underline{40}\ \ \\ 23 \\ \underline{16} \\ 77 \\ \underline{72} \\ 5 \end{array}$$

6.
$$\begin{array}{r} 7\text{ R5} \\ 32\overline{)229} \\ \underline{224} \\ 5 \end{array}$$

7.
$$\begin{array}{r} 1278\text{ R9} \\ 33\overline{)42183} \\ \underline{33}\ \ \ \\ 91\ \ \\ \underline{66}\ \ \\ 258 \\ \underline{231} \\ 273 \\ \underline{264} \\ 9 \end{array}$$

8.
$$\begin{array}{r} 25\text{ R27} \\ 128\overline{)3227} \\ \underline{256}\ \ \\ 667 \\ \underline{640} \\ 27 \end{array}$$

9.
$$\begin{array}{r} 16{,}852 \\ 7\overline{)117{,}964} \end{array}$$
Check.
$$\begin{array}{r} 16{,}852 \\ \times\ \ \ \ \ \ 7 \\ \hline 117{,}964 \end{array}$$

The cost of one car is $16,852.

10.
$$\begin{array}{r} 367 \\ 14\overline{)5138} \end{array}$$
The average speed was 367 mph.

1.6 Practice Problems

1. **(a)** $12 \times 12 \times 12 \times 12 = 12^4$. 12 is the base, 4 is the exponent.
(b) $2 \times 2 \times 2 \times 2 \times 2 \times 2 = 2^6$. 2 is the base, 6 is the exponent.

2. **(a)** $12^2 = 12 \times 12 = 144$
(b) $6^3 = 6 \times 6 \times 6 = 216$
(c) $2^6 = 2 \times 2 \times 2 \times 2 \times 2 \times 2 = 64$
(d) $1^{10} = 1 \times 1 \times 1 \times 1 \times 1 \times 1 \times 1 \times 1 \times 1 \times 1 = 1$

3. **(a)** $(7)(7)(7) + (8)(8) = 343 + 64 = 407$
(b) $(9)(9) + 1 = 81 + 1 = 82$
(c) $5^4 + 5 = (5)(5)(5)(5) + 5 = 625 + 5 = 630$

4. $7 + 4^3 \times 3 = 7 + 64 \times 3$ Exponents
$= 7 + 192$ Multiply
$= 199$ Add

5. $7 - 3 \times 4 \div 2 + 5 = 7 - 12 \div 2 + 5$ Multiplying
$= 7 - 6 + 5$ Dividing
$= 1 + 5$ Subtracting
$= 6$ Adding

6. $4^3 - 2 + 3^2$

$= 4 \times 4 \times 4 - 2 + 3 \times 3$	Evaluating exponents.
$= 64 - 2 + 9$	$4^3 = 64$ and $3^2 = 9$.
$= 62 + 9$	Subtracting
$= 71$	Adding

7. $(17 + 7) \div 6 \times 2 + 7 \times 3 - 4$

$= 24 \div 6 \times 2 + 7 \times 3 - 4$	Combining inside parenthesis
$= 4 \times 2 + 7 \times 3 - 4$	Dividing
$= 8 + 7 \times 3 - 4$	Multiplying
$= 8 + 21 - 4$	Multiplying
$= 29 - 4$	Adding
$= 25$	Subtracting

8. $5^2 - 6 \div 2 + 3^4 + 7 \times (12 - 10)$

$= 5^2 - 6 \div 2 + 3^4 + 7 \times 2$	Parenthesis
$= 25 - 6 \div 2 + 81 + 7 \times 2$	Exponents
$= 25 - 3 + 81 + 7 \times 2$	Divide
$= 25 - 3 + 81 + 14$	Multiply
$= 22 + 81 + 14$	Subtract
$= 117$	Add

1.7 Practice Problems

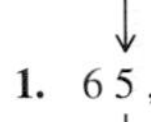

1. 6 5 , 5 2 8 — Locate the thousands round-off place.

   6 5 , (5) 2 8 — The first digit to the right is 5 or more. We will increase the thousands digit by 1.

   6 6 , 0 0 0 — All digits to the right of thousands are replaced by zero.

2. 1 7(2), 9 6 3 = 170,000 to the nearest ten thousand.

3. (a) 5 3 , 2 8 2 = 53,280 to nearest ten. The digit to the right of tens was less than 5.

   (b) 1 6 4 , 4 8 5 = 164,000 to nearest thousand. The digit to the right of thousands was less than 5.

   (c) 1 , 3 6 5 , 2 7 3 = 1,400,000 to nearest hundred thousand. The digit to the right of hundred thousands was greater than 5.

4. (a) 9 3 5 , 6 8 2 = 936,000. The digit to the right of thousands was greater than 5.

   (b) 9 3 5 , 6 8 2 = 900,000. The digit to the right of hundred thousands was less than 5.

   (c) 9 3 5 , 6 8 2 = 1,000,000. The digit to the right of millions was greater than 5.

5. 9,460,000,000,000,000 = 9,500,000,000,000,000 meters to the nearest hundred trillion.

6.

Actual Sum	Estimated Sum	
3456	3000	
9876	10000	
5421	5000	
+ 1278	+ 1000	
20,031	19,000	Close to the actual sum

7.

\$697	\$700
35	40
+ 19	+ 20
	\$760

We estimate that the total cost is \$760. (The exact answer is \$751, so we can see that our answer is quite close.

8. 10,000 + 10,000 + 20,000 + 60,000 = 100,000
   This is significantly different from 81,358, so we would suspect that an error has been made. In fact, Ming did make an error. The exact sum is actually 101,358!

9. 30,000,000 − 10,000,000 = 20,000,000
   We estimate that 20,000,000 more people lived in California than in Florida.

10. 9000 × 7000 = 63,000,000
    We estimate the product to be 63,000,000.

11. $40\overline{)80,000}$ = 2,000   Our estimate is 2,000.

12. $60\overline{)2,000,000}$ = 33,333   Our estimate is \$33,333 for one truck.

1.8 Practice Problems

Practice Problem 1

1. Understand the problem.

MATHEMATICS BLUEPRINT FOR PROBLEM SOLVING			
Gather the Facts	**What Am I Asked to Do?**	**How Do I Proceed?**	**Key Points to Remember**
The deductions are: \$135 \$28 \$13 \$34	Find out the total amount of deductions.	I must add the four deductions to obtain the total.	Watch out! Gross pay of \$1352 is not needed to solve the problem.

2. Solve and state the answer:
   135 + 28 + 13 + 34 = 210
   The total amount taken out of Diane's paycheck is \$210.
3. Check. Estimate the answer to see if it is reasonable.

Practice Problem 2

1. Understand the problem.

MATHEMATICS BLUEPRINT FOR PROBLEM SOLVING			
Gather the Facts	What Am I Asked to Do?	How Do I Proceed?	Key Points to Remember
Clinton had 1,964,842 votes. Bush had 1,876,495 votes.	Find out by how many votes Clinton beat Bush.	I must subtract the amounts.	Clinton is a Democrat and Bush is a Republican.

2. Solve and state the answer:

$$\begin{array}{r} 1{,}964{,}842 \\ -\ 1{,}876{,}495 \\ \hline 88{,}347 \end{array}$$

Clinton beat Bush by 88,347 votes.

3. Check. Estimate the answer to see if it is reasonable.

Practice Problem 3

1. Understand the problem.

MATHEMATICS BLUEPRINT FOR PROBLEM SOLVING			
Gather the Facts	What Am I Asked to Do?	How Do I Proceed?	Key Points to Remember
1 gallon is 1,024 fluid drams.	Find out how many fluid drams in 9 gallons.	I need to multiply 1024 by 9.	I must use fluid drams as my measure in my answer.

2. Solve and state the answer:

$$\frac{1024 \text{ fluid drams}}{1 \cancel{\text{gal}}} \times 9 \cancel{\text{gal}} = 9{,}216 \text{ fluid drams.}$$

There would be 9,216 fluid drams in 9 gallons.

3. Check. Estimate the answer to see if it is reasonable.

Practice Problem 4

1. Understand the problem.

MATHEMATICS BLUEPRINT FOR PROBLEM SOLVING			
Gather the Facts	What Am I Asked to Do?	How Do I Proceed?	Key Points to Remember
Donna bought 45 shares of stock. She paid $1620 for them.	Find out the cost per share of stock.	I need to divide $1620 by 45.	Use dollars as the unit in the answer.

2. Solve and state the answer:

$$\begin{array}{r} \$36 \\ 45\overline{)1620} \\ 135 \\ \hline 270 \\ 270 \\ \hline 0 \end{array}$$

Donna paid $36 per share for the stock.

3. Check. Estimate the answer to see if it is reasonable.

Practice Problem 5

1. Understand the problem.

MATHEMATICS BLUEPRINT FOR PROBLEM SOLVING			
Gather the Facts	**What Am I Asked to Do?**	**How Do I Proceed?**	**Key Points to Remember**
50 tables at \$200 each. 180 chairs at \$40 each. 6 carts at \$65 each.	Find the total cost.	1. Multiply the number of purchases by each price. 2. Add the total costs of each of the items.	There are three types of items involved in the total purchase.

2. Solve and state the answer:

$$\begin{array}{lrrl} \text{50 tables at \$200 each} = & 50 \times 200 = & \$10{,}000 & \text{cost of tables} \\ \text{180 chairs at \$40 each} = & 180 \times 40 = & 7{,}200 & \text{cost of chairs} \\ \text{6 carts at \$65 each} \quad = & 6 \times 65 = & \underline{390} & \text{cost of carts} \\ & & \$17{,}590 & \text{total cost} \end{array}$$

The total purchase was \$17,590

3. Check. Estimate the answer to see if it is reasonable.

Practice Problem 6

1. Understand the problem.

MATHEMATICS BLUEPRINT FOR PROBLEM SOLVING			
Gather the Facts	**What Am I Asked to Do?**	**How Do I Proceed?**	**Key Points to Remember**
Old Balance: \$498 New Deposits: \$607 \$163 Interest: \$36 Withdrawals: \$ 19 \$158 \$582 \$ 74	Find her new balance after the transactions.	(a) Add the new deposits and interest to the old balance. (b) Add the withdrawals. (c) Subtract the results from steps (a) and (b).	Deposits and interest are added and withdrawals are subtracted from savings accounts.

2. Solve and state the answer:

$$\textbf{(a)}\ \begin{array}{r} \$498 \\ 607 \\ 163 \\ \underline{+\ 36} \\ \$1304 \end{array} \qquad \textbf{(b)}\ \begin{array}{r} \$\ 19 \\ 158 \\ 582 \\ \underline{+\ 74} \\ \$833 \end{array} \qquad \textbf{(c)}\ \begin{array}{r} \$1304 \\ \underline{-\ 833} \\ \$471 \end{array}$$

Her balance this month is \$471.

3. Check. Estimate the answer to see if it is reasonable.

Practice Problem 7

1. Understand the problem.

MATHEMATICS BLUEPRINT FOR PROBLEM SOLVING			
Gather the Facts	**What Am I Asked to Do?**	**How Do I Proceed?**	**Key Points to Remember**
Odometer at end of trip: 51,118 miles Beginning of trip: 50,698 Used on trip: 12 gallons of gas.	Find the number of miles per gallon that the car obtained on the trip.	(a) Subtract the two odometer readings. (b) Divide that number by 12.	The gas tank was full at the beginning of the trip. 12 gallons fills the tank at the end of the trip.

2. Solve and state the answer:

$$\begin{array}{rl} 51{,}118 & \text{odometer at end of trip} \\ \underline{-\ 50{,}698} & \underline{\text{odometer at start of trip}} \\ 420 & \text{miles traveled on trip} \end{array}$$

$$\frac{420 \text{ miles}}{12 \text{ gallons of gas used}} = \begin{array}{r} 35 \\ 12\overline{)420} \\ \underline{36} \\ 60 \\ \underline{60} \end{array} = 35 \text{ miles per gallon on the trip}$$

3. Check. Estimate the answer to see if it is reasonable.

Chapter 2

2.1 Practice Problems

1. (a) Five parts out of twelve are shaded. The fraction is $\frac{5}{12}$.

(b) One part out of six is shaded. The fraction is $\frac{1}{6}$.

(c) Two parts out of three are shaded. The fraction is $\frac{2}{3}$.

2. (a) $\frac{4}{5}$ of the object is shaded.

(b) $\frac{3}{7}$ of the group is shaded.

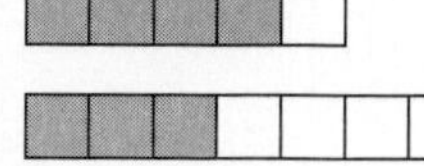

3. (a) $\frac{9}{17}$ represents 9 women out of 17.

(b) The total class is $382 + 351 = 733$.

The fractional part that is women is $\frac{351}{733}$.

(c) $\frac{7}{8}$ of a yard of material.

4. Total number of defective items $1 + 2 = 3$. Total number of items $7 + 9 = 16$. A fraction that represents the portion of the items that were defective is $\frac{3}{16}$.

2.2 Practice Problems

1. (a)
$$\begin{aligned} 18 &= 9 \times 2 \\ &= 3 \times 3 \times 2 \\ &= 3^2 \times 2 \end{aligned}$$

(b)
$$\begin{aligned} 72 &= 9 \times 8 \\ &= 3 \times 3 \times 2 \times 2 \times 2 \\ &= 3^2 \times 2^3 \end{aligned}$$

(c)
$$\begin{aligned} 400 &= 10 \times 40 \\ &= 5 \times 2 \times 5 \times 8 \\ &= 5 \times 2 \times 5 \times 2 \times 2 \times 2 \\ &= 5^2 \times 2^4 \end{aligned}$$

2. (a) $\frac{30 \div 6}{42 \div 6} = \frac{5}{7}$

(b) $\frac{60 \div 12}{132 \div 12} = \frac{5}{11}$

3. (a) $\frac{120}{135} = \frac{2 \times 2 \times 2 \times \cancel{3} \times \cancel{5}}{3 \times 3 \times \cancel{3} \times \cancel{5}} = \frac{8}{9}$

(b) $\frac{715}{880} = \frac{\cancel{5} \times \cancel{11} \times 13}{\cancel{5} \times \cancel{11} \times 2 \times 2 \times 2} = \frac{13}{16}$

4. (a) $\frac{84}{108} \stackrel{?}{=} \frac{7}{9}$

$84 \times 9 = 108 \times 7$

$756 = 756$ Yes

(b) $\frac{3}{7} \stackrel{?}{=} \frac{79}{182}$

$3 \times 182 = 7 \times 79$

$546 = 553$ No

2.3 Practice Problems

1. (a) $4\frac{3}{7} = \frac{7 \times 4 + 3}{7} = \frac{28 + 3}{7} = \frac{31}{7}$

(b) $6\frac{2}{3} = \frac{3 \times 6 + 2}{3} = \frac{18 + 2}{3} = \frac{20}{3}$

(c) $19\frac{4}{7} = \frac{7 \times 19 + 4}{7} = \frac{133 + 4}{7} = \frac{137}{7}$

2. (a) $17 \div 4 = 4$ remainder 1 ($4 \times 4 = 16$; $17 - 16 = 1$) so $\frac{17}{4} = 4\frac{1}{4}$

(b) $36 \div 5 = 7$ remainder 1 ($5 \times 7 = 35$; $36 - 35 = 1$) so $\frac{36}{5} = 7\frac{1}{5}$

(c) $116 \div 27 = 4$ remainder 8 ($27 \times 4 = 108$; $116 - 108 = 8$) so $\frac{116}{27} = 4\frac{8}{27}$

(d) $91 \div 13 = 7$ remainder 0 ($13 \times 7 = 91$; $91 - 91 = 0$) so $\frac{91}{13} = 7$

3. $\frac{51}{15} = \frac{3 \times 17}{3 \times 5} = \frac{\overset{1}{\cancel{3}} \times 17}{\underset{1}{\cancel{3}} \times 5} = \frac{17}{5}$

4. $\frac{16}{80} = \frac{1}{5}$

Therefore, $3\frac{16}{80} = 3\frac{1}{5}$

5. $\frac{1001}{572} = 1\frac{429}{572}$

Now the fraction $\frac{429}{572} = \frac{3 \times \overset{1}{\cancel{11}} \times \overset{1}{\cancel{13}}}{2 \times 2 \times \underset{1}{\cancel{11}} \times \underset{1}{\cancel{13}}} = \frac{3}{4}$.

Thus $\frac{1001}{572} = 1\frac{429}{572} = 1\frac{3}{4}$.

2.4 Practice Problems

1. (a) $\frac{6}{7} \times \frac{3}{13} = \frac{18}{91}$ **(b)** $\frac{1}{5} \times \frac{11}{12} = \frac{11}{60}$

2. $\frac{\overset{5}{\cancel{55}}}{\underset{9}{\cancel{72}}} \times \frac{\overset{2}{\cancel{16}}}{\underset{3}{\cancel{33}}} = \frac{5}{9} \times \frac{2}{3} = \frac{10}{27}$

3. (a) $7 \times \frac{5}{13} = \frac{7}{1} \times \frac{5}{13} = \frac{35}{13}$ or $2\frac{9}{13}$

(b) $\frac{13}{16} \times 8 = \frac{13}{\underset{2}{\cancel{16}}} \times \frac{\overset{1}{\cancel{8}}}{1} = \frac{13}{2}$ or $6\frac{1}{2}$

4. $\frac{3}{\underset{1}{\cancel{8}}} \times \overset{12,300}{\cancel{98,400}} = \frac{3}{1} \times 12,300 = 36,900$

There are 36,900 ft^2 in the wetland area.

5. (a) $2\frac{1}{6} \times \frac{4}{7} = \frac{13}{\underset{3}{\cancel{6}}} \times \frac{\overset{2}{\cancel{4}}}{7} = \frac{26}{21}$ or $1\frac{5}{21}$

(b) $10\frac{2}{3} \times 13\frac{1}{2} = \frac{\overset{16}{\cancel{32}}}{\underset{1}{\cancel{3}}} \times \frac{\overset{9}{\cancel{27}}}{\underset{1}{\cancel{2}}} = \frac{144}{1} = 144$

(c) $\frac{3}{5} \times 1\frac{1}{3} \times \frac{5}{8} = \frac{\overset{1}{\cancel{3}}}{\underset{1}{\cancel{5}}} \times \frac{\overset{1}{\cancel{4}}}{\underset{1}{\cancel{3}}} \times \frac{\overset{1}{\cancel{5}}}{\underset{2}{\cancel{8}}} = \frac{1}{2}$

(d) $3\frac{1}{5} \times 2\frac{1}{2} = \frac{\overset{8}{\cancel{16}}}{\underset{1}{\cancel{5}}} \times \frac{\overset{1}{\cancel{5}}}{\underset{1}{\cancel{2}}} = \frac{8}{1} = 8$

6. Area $= lw = 1\frac{1}{5} \times 4\frac{5}{6} = \frac{\overset{1}{\cancel{6}}}{5} \times \frac{29}{\underset{1}{\cancel{6}}} = \frac{29}{5} = 5\frac{4}{5}$ m^2

7. Since $8 \cdot 10 = 80$ and $9 \cdot 9 = 81$

We know that $\frac{8}{9} \cdot \frac{10}{9} = \frac{80}{81}$

Therefore, $x = \frac{10}{9}$

2.5 Practice Problems

1. **(a)** $\frac{7}{13} \div \frac{3}{4} = \frac{7}{13} \times \frac{4}{3} = \frac{28}{39}$

(b) $\frac{16}{35} \div \frac{24}{25} = \frac{\cancel{16}^{\,2}}{\cancel{35}_{\,7}} \times \frac{\cancel{25}^{\,5}}{\cancel{24}_{\,3}} = \frac{10}{21}$

2. **(a)** $\frac{3}{17} \div \frac{6}{1} = \frac{\cancel{3}^{\,1}}{17} \times \frac{1}{\cancel{6}_{\,2}} = \frac{1}{34}$

(b) $\frac{14}{1} \div \frac{7}{15} = \frac{\cancel{14}^{\,2}}{1} \times \frac{15}{\cancel{7}_{\,1}} = 30$

3. **(a)** $1 \div \frac{11}{13} = \frac{1}{1} \times \frac{13}{11} = \frac{13}{11}$ or $1\frac{2}{11}$

(b) $\frac{14}{17} \div 1 = \frac{14}{17} \times \frac{1}{1} = \frac{14}{17}$

(c) $\frac{3}{11} \div 0$ Division by zero cannot be done. This problem cannot be done.

(d) $0 \div \frac{9}{16} = \frac{0}{1} \times \frac{16}{9} = \frac{0}{9} = 0$

4. **(a)** $1\frac{1}{5} \div \frac{7}{10} = \frac{6}{\cancel{5}_{\,1}} \times \frac{\cancel{10}^{\,2}}{7} = \frac{12}{7}$ or $1\frac{5}{7}$

(b) $2\frac{1}{4} \div 1\frac{7}{8} = \frac{9}{4} \div \frac{15}{8} = \frac{\cancel{9}^{\,3}}{\cancel{4}_{\,1}} \times \frac{\cancel{8}^{\,2}}{\cancel{15}_{\,5}} = \frac{6}{5}$ or $1\frac{1}{5}$

5. **(a)** $\dfrac{5\frac{2}{3}}{7} = 5\frac{2}{3} \div 7 = \frac{17}{3} \times \frac{1}{7} = \frac{17}{21}$

(b) $\dfrac{1\frac{2}{5}}{2\frac{1}{3}} = 1\frac{2}{5} \div 2\frac{1}{3} = \frac{7}{5} \div \frac{7}{3} = \frac{\cancel{7}^{\,1}}{5} \times \frac{3}{\cancel{7}_{\,1}} = \frac{3}{5}$

6. $x \div \frac{3}{2} = \frac{22}{36}$

$x \cdot \frac{2}{3} = \frac{22}{36}$

$\frac{11}{12} \cdot \frac{2}{3} = \frac{22}{36}$ Thus $x = \frac{11}{12}$

7. $19\frac{1}{4} \div 14 = \frac{\cancel{77}^{\,11}}{4} \times \frac{1}{\cancel{14}_{\,2}} = \frac{11}{8}$ or $1\frac{3}{8}$

Each piece will be $1\frac{3}{8}$ inches long.

2.6 Practice Problems

1. **(a)** The LCD of $\frac{3}{4}$ and $\frac{11}{12}$ is 12.

12 can be divided by 4 and 12.

(b) The LCD of $\frac{1}{7}$ and $\frac{8}{35}$ is 35.

35 can be divided by 7 and 35.

2. The LCD of $\frac{3}{7}$ and $\frac{5}{6}$ is 42.

42 can be divided by 7 and 6.

3. **(a)** $14 = 2 \times 7$

$10 = 2 \times 5$

The LCD $= 2 \times 5 \times 7 = 70$.

(b) $15 = 3 \times 5$

$50 = 2 \times 5 \times 5$

The LCD $= 2 \times 3 \times 5 \times 5 = 150$

(c) $16 = 2 \times 2 \times 2 \times 2$

$12 = 2 \times 2 \times 3$

The LCD $= 2 \times 2 \times 2 \times 2 \times 3 = 48$

4. $49 = 7 \times 7$

$21 = 7 \times 3$

$7 = 7 \times 1$

LCD $= 7 \times 7 \times 3 = 147$.

5. **(a)** $\frac{3}{5} \times \frac{8}{8} = \frac{24}{40}$ **(c)** $\frac{2}{7} = \frac{2}{7} \times \frac{4}{4} = \frac{8}{28}$

(b) $\frac{7}{11} \times \frac{4}{4} = \frac{28}{44}$ $\frac{3}{4} = \frac{3}{4} \times \frac{7}{7} = \frac{21}{8}$

6. **(a)** The LCD is 60 because it can be divided by 20 and 15.

(b) $\frac{3}{20} \times \frac{3}{3} = \frac{9}{60}$

$\frac{11}{15} \times \frac{4}{4} = \frac{44}{60}$

2.7 Practice Problems

1. $\frac{3}{17} + \frac{12}{17} = \frac{15}{17}$

2. **(a)** $\frac{1}{12} + \frac{5}{12} = \frac{6}{12} = \frac{1}{2}$

(b) $\frac{13}{15} + \frac{7}{15} = \frac{20}{15} = \frac{4}{3}$ or $1\frac{1}{3}$

3. **(a)** $\frac{5}{19} - \frac{2}{19} = \frac{3}{19}$ **(b)** $\frac{21}{25} - \frac{6}{25} = \frac{15}{25} = \frac{3}{5}$

4.

$$\begin{array}{r} \frac{2}{15} = \frac{2}{15} \\ \frac{2}{5} \times \frac{3}{3} = \frac{6}{15} \\ \hline \frac{8}{15} \end{array}$$

5. LCD = 48 $\frac{5}{12} \times \frac{4}{4} = \frac{20}{48}$ $\frac{5}{16} \times \frac{3}{3} = \frac{15}{48}$

$\frac{5}{12} + \frac{5}{16} = \frac{20}{48} + \frac{15}{48} = \frac{35}{48}$

6. LCD = 48

$$\begin{array}{r} \frac{3}{16} \times \frac{3}{3} = \frac{9}{48} \\ \frac{1}{8} \times \frac{6}{6} = \frac{6}{48} \\ \frac{1}{12} \times \frac{4}{4} = \frac{4}{48} \\ \hline \frac{19}{48} \end{array}$$

7. LCD = 96 $\frac{9}{48} \times \frac{2}{2} = \frac{18}{96}$

$\frac{5}{32} \times \frac{3}{3} = \frac{15}{96}$ $\frac{9}{48} - \frac{5}{32} = \frac{18}{96} - \frac{15}{96} = \frac{3}{96} = \frac{1}{32}$

8.

$$\begin{array}{r} \frac{9}{10} \times \frac{2}{2} = \frac{18}{20} \\ -\frac{1}{4} \times \frac{5}{5} = -\frac{5}{20} \\ \hline \frac{13}{20} \end{array}$$

There is $\frac{13}{20}$ gallon left.

9. The LCD of $\frac{3}{10}$ and $\frac{23}{25}$ is 50.

$\frac{3}{10} \times \frac{5}{5} = \frac{15}{50}$ Now rewriting: $x + \frac{15}{50} = \frac{46}{50}$

$\frac{23}{25} \times \frac{2}{2} = \frac{46}{50}$ $\frac{31}{50} + \frac{15}{50} = \frac{46}{50}$

2.8 Practice Problems

1.
$$\begin{array}{r} 5\frac{1}{12} \\ +\,9\frac{5}{12} \\ \hline 14\frac{6}{12} = 14\frac{1}{2} \end{array}$$

2.
$$\begin{array}{rcr} 6\frac{1}{4} & = & 6\frac{5}{20} \\ +\,2\frac{2}{5} & = & +\,2\frac{8}{20} \\ \hline & & 8\frac{13}{20} \end{array}$$

3. LCD = 12
$$7\boxed{\frac{1}{4}\times\frac{3}{3}} = 7\frac{3}{12}$$
$$+3\boxed{\frac{5}{6}\times\frac{2}{2}} \quad +3\frac{10}{12}$$
$$10\frac{13}{12} = 10 + 1\frac{1}{12} = 11\frac{1}{12}$$

4. LCD = 12
$$\begin{array}{rcr} 12\frac{5}{6} & = & 12\frac{10}{12} \\ -\,7\frac{5}{12} & = & -\,7\frac{5}{12} \\ \hline & & 5\frac{5}{12} \end{array}$$

5a. LCD = 24
$$9\boxed{\frac{1}{8}\times\frac{3}{3}} = 9\frac{3}{24} = 8\frac{27}{24}$$
$$-3\boxed{\frac{2}{3}\times\frac{8}{8}} = -3\frac{16}{24} \quad -3\frac{16}{24}$$
$$5\frac{11}{24}$$

Borrow 1 from 9:
$$9\frac{3}{24} = 8 + 1 + \frac{3}{24} = 8\frac{27}{24}$$

b.
$$\begin{array}{rcr} 18 & = & 17\frac{18}{18} \\ -\,6\frac{7}{18} & = & -\,6\frac{7}{18} \\ \hline & & 11\frac{11}{18} \end{array}$$

6.
$$\begin{array}{rcrcr} 6\frac{1}{4} & = & 6\frac{3}{12} & = & 5\frac{15}{12} \\ -\,4\frac{2}{3} & = & -\,4\frac{8}{12} & = & -\,4\frac{8}{12} \\ \hline & & & & 1\frac{7}{12} \end{array}$$

They had $1\frac{7}{12}$ gallons left over.

2.9 Practice Problems

Practice Problem 1

1. Understand the problem.

MATHEMATICS BLUEPRINT FOR PROBLEM SOLVING			
Gather the Facts	**What Am I Asked to Do?**	**How Do I Proceed?**	**Key Points to Remember**
Gas amounts: $18\frac{7}{10}$ gal, $15\frac{2}{5}$ gal, $14\frac{1}{2}$ gal	Find out how many gallons of gas she bought altogether.	Add the three amounts.	When adding mixed numbers, the LCD is needed for the fractions.

2. Solve and state the answer:

$$\text{LCD} = 10 \quad 18\frac{7}{10} = 18\frac{7}{10}$$
$$15\frac{2}{5} = 15\frac{4}{10}$$
$$14\frac{1}{2} = 14\frac{5}{10}$$
$$47\frac{16}{10} = 48\frac{6}{10}$$
$$= 48\frac{3}{5} \text{ gallons}$$

Total is $48\frac{3}{5}$ gallons.

3. Check. Estimate the answer to see if it is reasonable.

Practice Problem 2

1. Understand the problem.

MATHEMATICS BLUEPRINT FOR PROBLEM SOLVING			
Gather the Facts	**What Am I Asked to Do?**	**How Do I Proceed?**	**Key Points to Remember**
Poster: $12\frac{1}{4}$ in. Top border: $1\frac{3}{8}$ in. Bottom border: 2 in.	Find the length of the inside portion of the poster.	(a) Add the two border lengths. (b) Subtract this total from the poster length.	When adding mixed numbers, the LCD is needed for the fractions.

2. Solve and state the answer:

(a) $1\frac{3}{8} + 2 = 3\frac{3}{8}$

(b) $12\frac{1}{4} - 3\frac{3}{8} = 12\frac{2}{8} - 3\frac{3}{8} = 11\frac{10}{8} - 3\frac{3}{8} = 8\frac{7}{8}$

The inside portion is $8\frac{7}{8}$ inches.

3. Check. Estimate the answer to see if it is reasonable.

Practice Problem 3

1. Understand the problem.

MATHEMATICS BLUEPRINT FOR PROBLEM SOLVING			
Gather the Facts	**What Am I Asked to Do?**	**How Do I Proceed?**	**Key Points to Remember**
Regular tent uses $8\frac{1}{4}$ yards. Large tent uses $1\frac{1}{2}$ times the regular. She makes 6 regular and 15 large tents.	Find out how many yards of cloth will be needed to make the tents.	Find the amount used for regular tents, and the amount used for large tents. Then add the two.	Large tents use $1\frac{1}{2}$ times the regular amount.

2. Solve and state the answer:

We multiply $6 \times 8\frac{1}{4}$ for regular tents and $16 \times 1\frac{1}{2} \times 8\frac{1}{4}$ for large tents. Then add total yardage.

$$6 \times 8\frac{1}{4} \text{ (regular tents)} = \overset{3}{\cancel{6}} \times \frac{33}{\underset{2}{\cancel{4}}} = \frac{99}{2} = 49\frac{1}{2} \text{ yards.}$$

$$\text{(Large tents)} \quad 16 \times 1\frac{1}{2} \times 8\frac{1}{4} = \overset{\overset{2}{\cancel{8}}}{\cancel{16}} \times \frac{3}{\underset{1}{\cancel{2}}} \times \frac{33}{\underset{1}{\cancel{4}}} = \frac{198}{1} = 198 \text{ yards.}$$

Total yards for all tents $198 + 49\frac{1}{2} = 247\frac{1}{2}$ yards.

3. Check. Estimate the answer to see if it is reasonable.

Practice Problem 4

1. Understand the problem.

MATHEMATICS BLUEPRINT FOR PROBLEM SOLVING			
Gather the Facts	**What Am I Asked to Do?**	**How Do I Proceed?**	**Key Points to Remember**
He purchases 12-foot boards. Each shelf is $2\frac{3}{4}$ ft. He needs four shelves for each bookcase and he is making two bookcases.	(a) Find out how many boards he needs to buy. (b) Find out how many feet of shelving are actually needed. (c) Find out how many feet will be left over.	Find out how many $2\frac{3}{4}$ shelves he can get from one board. Then see how many boards he needs to make all eight shelves.	There will be three answers to this problem. Don't forget to calculate the leftover wood.

2. Solve and state the answer:
We want to know how many $2\frac{3}{4}$ ft. shelves are in a 12-ft. board.
$12 \div 2\frac{3}{4} = \frac{12}{1} \div \frac{11}{4} = \frac{12}{1} \times \frac{4}{11} = \frac{48}{11} = 4\frac{4}{11}$ shelves
He will get 4 shelves from each board with some left over.
(a) For two bookcases, he needs 8 shelves. He gets four shelves out of each board. $8 \div 4 = 2$. He will need two 12-ft. boards.
(b) He needs 8 shelves at $2\frac{3}{4}$ feet.
$8 \times 2\frac{3}{4} = 8 \times \frac{11}{4} = 22$ feet.
He actually needs 22 feet of shelving.
(c) 24 feet of shelving bought
− 22 feet of shelving used
2 feet of shelving left over.
3. Check. Estimate the answer to see if it is reasonable.

Practice Problem 5

1. Understand the problem.

MATHEMATICS BLUEPRINT FOR PROBLEM SOLVING			
Gather the Facts	**What Am I Asked to Do?**	**How Do I Proceed?**	**Key Points to Remember**
Distance is $199\frac{3}{4}$ miles. He uses $8\frac{1}{2}$ gallons of gas	Find out how many miles per gallon he gets.	Divide the distance by the number of gallons.	Change mixed numbers to improper fractions before dividing.

2. Solve and state the answer:
$$199\frac{3}{4} \div 8\frac{1}{2} = \frac{799}{4} \div \frac{17}{2}$$
$$= \frac{\overset{47}{\cancel{799}}}{\underset{2}{\cancel{4}}} \times \frac{\overset{1}{\cancel{2}}}{\underset{1}{\cancel{17}}}$$
$$= \frac{47}{2} = 23\frac{1}{2}$$
He gets $23\frac{1}{2}$ miles per gallon.
3. Check. Estimate the answer to see if it is reasonable.

Chapter 3

3.1 Practice Problems

1. **(a)** 0.073 seventy-three thousandths
(b) 4.68 four and sixty-eight hundredths
(c) 0.0017 seventeen ten-thousandths
(d) 561.78 five hundred sixty-one and seventy-eight hundredths
2. Seven thousand eight hundred sixty-three and $\frac{4}{100}$

3. (a) $\frac{9}{10} = 0.9$ (b) $\frac{136}{1000} = 0.136$ (c) $2\frac{56}{100} = 2.56$

(d) $34\frac{86}{1000} = 34.086$

4. (a) $0.37 = \frac{37}{100}$

(b) $182.3 = 182\frac{3}{10}$

(c) $0.7131 = \frac{7131}{10{,}000}$

(d) $42.019 = 42\frac{19}{1000}$

5. (a) $8.5 = 8\frac{5}{10} = 8\frac{1}{2}$ (b) $0.58 = \frac{58}{100} = \frac{29}{50}$

(c) $36.25 = 36\frac{25}{100} = 36\frac{1}{4}$ (d) $106.013 = 106\frac{13}{1000}$

6. $\frac{2}{1{,}000{,}000{,}000} = \frac{1}{500{,}000{,}000}$ The concentration of PCBs is $\frac{1}{500{,}000{,}000}$

3.2 Practice Problems

1. Since 4 < 5

5.74 5.75 therefore, 5.74 < 5.75

2. 0.894 > 0.890, so 0.894 > 0.89
3. 2.45, 2.543, 2.46, 2.54, 2.5

It is helpful to add extra zeros and to place the decimals that begin with 2.4 in a group and the decimals that begin with 2.5 in the other.

2.450, 2.460 2.543, 2.540, 2.500

In order, we have from smallest to largest

2.450, 2.460, 2.500, 2.540, 2.543

It is OK to leave the extra terminal zeros in our answer.

4. 723.9
5. (a) 12.92 6 47

↑ Since digit to right of thousandth is less than 5, we drop the digits 4 and 7.

12.926

(b) 0.00 7 892

↑ Since digit to right of thousandths is greater than 5, we round up.

0.008

6. 15,700.0
7.

		Rounded to Nearest Dollar
Medical bills	375.50	376
Taxes	981.39	981
Retirement	980.49	980
Charity	817.65	818

3.3 Practice Problems

1. (a)

```
    1
   9.8
   3.6
 + 5.4
 -----
  18.8
```

(b)

```
   1 1 1
  300.72
  163.75
+ 291.08
--------
  755.55
```

(c)

```
   21 1
   8.9000
  37.0560
   0.0023
 + 9.4500
 --------
  55.4083
```

2.

```
   1  1
  93,521.8
+  1,634.8
----------
  95,156.6 miles
```

3.

```
  $ 80.95
   133.91
   256.47
    53.08
 + 381.32
 --------
  $905.73
```

4. (a)

```
    7 18
   38.8
 - 26.9
 ------
   11.9
```

(b)

```
   2034.908
 - 1986.325
 ----------
     48.583
```

5. **(a)** 19.000 − 12.579 = 6.421 **(b)** 283.076 − 96.380 = 186.696

6. 87,160.1 − 82,370.9 = 4,789.2 miles between oil changes

7. 15.3 − 10.8 = 4.5 x is 4.5

3.4 Practice Problems

1.

0.09	2 decimal places
× 0.6	+ 1 decimal place
0.054	3 decimal places in product

2. **(a)**

5.3	1 decimal place
× 0.9	+ 1 decimal place
4.77	2 decimal places in product

(b)

2.831	3 decimal places
× 0.07	+ 1 decimal place
1.9817	4 decimal places in product

(c)

0.47	2 decimal places
× 0.28	+ 2 decimal places
376	
94	
0.1316	4 decimal places in product

(d)

0.436	3 decimal places
× 18.39	+ 2 decimal places
8.01804	5 decimal places in product

3.

0.4264	4 decimal places
× 38	+ 0 decimal place
16.2032	4 decimal places in product

4. Area = length × width

1.26 × 2.3 = 2.898 square millimeters

5. **(a)** $0.0561 \times \underline{10} = 0.561$ Decimal point moved one place to the right.

(b) $1462.37 \times \underline{100} = 146{,}237$ Decimal point moved two places to the right.

6. **(a)** $0.26 \times \underline{1000} = 260$ Decimal point moved three places to the right. One extra zero needed.

(b) $5862.89 \times \underline{10{,}000} = 58{,}628{,}900$ Decimal point moved four places to the right. Two extra zeros needed.

7. **(a)** $7{,}684 \times 10^4 = 76{,}840$ Decimal point moved four places to the right. One extra zero needed.

(b) $0.00073 \times 10^2 = 0.073$ Decimal point moved two places to the right.

8. $156.2 \times 1000 = 156{,}200$ meters

3.5 Practice Problems

1. **(a)** 7)1.806 = 0.258
14
40
35
56
56
0

(b) 16)0.0928 = 0.0058
80
128
128
0

2. 46)23.820 = 0.517 = 0.52 to the nearest hundredth
23 0
82
46
360
322

3. 19)$3538.75 = $186.25 each month
19
163
152
118
114
4 7
3 8
95
95

4. **(a)** 0.09‸)0.10‸08 = 1.12
9
10
9
18
18
0

(b) 0.037‸)1.702‸ = 46.
1 48
222
222
0

5. **(a)** 1.8‸)0.0‸414 = 0.023
36
54
54

(b) 0.0036‸)8.3160‸ = 2310.
72
111
108
36
36
0

6. **(a)** 3.8‸)521.6‸00 = 137.26 The answer rounded to the nearest tenth is 137.3
38
141
114
27 6
26 6
1 00
76
24 0
22 8

(b) 8.05‸)0.17‸0000 = 0.0211 The answer rounded to the nearest thousandth is 0.021
16 10
900
805
950
805

7. 28.5‸)454.4‸0 = 15.9 The truck got approximately 16 miles per gallon.
285
169 4
142 5
269 0
256 5

8. 0.12‸)0.69‸6 = 5.8 n is 5.8
60
9 6
9 6

3.6 Practice Problems

1. **(a)**
$$\begin{array}{r} 0.3125 \\ 16\overline{)5.0000} \\ \underline{48} \\ 20 \\ \underline{16} \\ 40 \\ \underline{32} \\ 80 \\ \underline{80} \\ 0 \end{array}$$

(b)
$$\begin{array}{r} 0.1375 \\ 80\overline{)11.0000} \\ \underline{80} \\ 300 \\ \underline{240} \\ 600 \\ \underline{560} \\ 400 \\ \underline{400} \\ 0 \end{array}$$

2. **(a)** $0.6363 = 0.\overline{63}$
$$\begin{array}{r} 0.6363 \\ 11\overline{)7.0000} \\ \underline{66} \\ 40 \\ \underline{33} \\ 70 \\ \underline{66} \\ 40 \\ \underline{33} \\ 7 \end{array}$$

(b) $0.533 = 0.5\overline{3}$
$$\begin{array}{r} 0.533 \\ 15\overline{)8.000} \\ \underline{75} \\ 50 \\ \underline{45} \\ 50 \\ \underline{45} \\ 5 \end{array}$$

(c) $0.295454 = 0.29\overline{54}$
$$\begin{array}{r} 0.295454 \\ 44\overline{)13.000000} \\ \underline{88} \\ 420 \\ \underline{396} \\ 240 \\ \underline{220} \\ 200 \\ \underline{176} \\ 240 \\ \underline{220} \\ 200 \\ \underline{176} \\ 24 \end{array}$$

3. **(a)**
$$2\frac{11}{18} = 2 + \frac{11}{18} = 2 + .6\overline{1} = 2.6\overline{1}$$

$0.611 = 0.6\overline{1}$
$$\begin{array}{r} 0.611 \\ 18\overline{)11.000} \\ \underline{108} \\ 20 \\ \underline{18} \\ 20 \\ \underline{18} \\ 2 \end{array}$$

(b) $1.03703 = 1.\overline{037}$
$$\begin{array}{r} 1.03703 \\ 27\overline{)28.00000} \\ \underline{27} \\ 100 \\ \underline{81} \\ 190 \\ \underline{189} \\ 100 \\ \underline{81} \\ 19 \end{array}$$

4. $0.7916 = 0.792$ rounded to the nearest thousandth
$$\begin{array}{r} 0.7916 \\ 24\overline{)19.0000} \\ \underline{168} \\ 220 \\ \underline{216} \\ 40 \\ \underline{24} \\ 160 \\ \underline{144} \\ 16 \end{array}$$

5.
$$\begin{aligned} 0.9 - 0.3 + 0.8 \times 0.3 &= 0.9 - 0.3 + 0.24 \\ &= 0.6 + 0.24 \\ &= 0.84 \end{aligned}$$

6.
$$\begin{aligned} 0.3 \times 0.5 + (0.4)^3 - 0.036 &= 0.3 \times 0.5 + 0.064 - 0.036 \\ &= 0.15 + 0.064 - 0.036 \\ &= 0.214 - 0.036 \\ &= 0.178 \end{aligned}$$

7. $6.56 \div (2 - 0.36) + (8.5 - 8.3)^2$
$= 6.56 \div (1.64) + (0.2)^2$ Parenthesis
$= 6.56 \div (1.64) + 0.04$ Exponents
$= 4 + 0.04$ Divide
$= 4.04$ Add

3.7 Practice Problems

Practice Problem 1

1. Understand the problem.

MATHEMATICS BLUEPRINT FOR PROBLEM SOLVING			
Gather the Facts	**What Am I Asked to Do?**	**How Do I Proceed?**	**Key Points to Remember**
Coat Cost: \$198.98 Tax on Coat: \$9.95 Boot Cost: \$59.95 Tax on Boots: \$3.00 Renée has \$280 to spend.	Find out how much money Renée had left after her purchases.	Add the cost of each item and the tax on each item to obtain the total purchase price. Subtract that total from \$280.	Be sure to line up the decimal points when adding and subtracting the decimal values.

2. Solve and state the answer:

Add the four amounts.
$$\begin{array}{r} \$198.98 \\ 9.95 \\ 59.95 \\ \underline{+\quad 3.00} \\ \$271.88 \end{array}$$

Subtract all expenses from \$280.
$$\begin{array}{r} \$280.00 \\ \underline{-\ 271.88} \\ \$\quad 8.12 \end{array}$$

Renée has \$8.12 left after her purchases.

3. Check. $200 + $10 + $60 + $3 = $273.

$$\begin{array}{r} \$280 \\ -\ 273 \\ \hline \$\ \ 7 \end{array}$$

The answer is reasonable.

Practice Problem 2

1. Understand the problem.

MATHEMATICS BLUEPRINT FOR PROBLEM SOLVING			
Gather the Facts	**What Am I Asked to Do?**	**How Do I Proceed?**	**Key Points to Remember**
She works 51 hours. She gets paid $9.36 per hour for 40 hours. She gets paid time-and-a-half for 11 hours.	Find the amount Melinda earned working 51 hours last week.	Add the earnings of 40 hours at $9.36 per hour to the earnings of 11 hours at overtime pay.	Overtime pay is time-and-a-half, which is 1.5 × $9.36.

2. Solve and state the answer:

(a) Calculate regular earnings for 40 hours.

$$\begin{array}{r} \$9.36 \\ \times\ \ 40 \\ \hline \$374.40 \end{array}$$

(b) Calculate overtime pay rate.

$$\begin{array}{r} \$9.36 \\ \times\ 1.5 \\ \hline \$14.04 \end{array}$$

(c) Calculate overtime earnings for 11 hours.

$$\begin{array}{r} \$\ 14.04 \\ \times\ \ \ \ \ 11 \\ \hline \$154.44 \end{array}$$

(d) Add the two amounts.

$$\begin{array}{rl} \$374.40 & \text{Regular earnings} \\ +\ 154.44 & \text{Overtime earnings} \\ \hline \$528.84 & \text{Total earnings} \end{array}$$

Melinda earned $528.84 last week.

3. Check. 40 × $9 = $360; 10 × $14 = $140

$$\begin{array}{r} \$360 \\ +\ 140 \\ \hline \$500 \end{array}$$

The answer is reasonable.

Practice Problem 3

1. Understand the problem.

MATHEMATICS BLUEPRINT FOR PROBLEM SOLVING			
Gather the Facts	**What Am I Asked to Do?**	**How Do I Proceed?**	**Key Points to Remember**
The total amount of steak is 17.4 pounds. Each package contains 1.45 pounds. Each package costs $4.60 per pound.	(a) Find out how many packages of steak the butcher will have. (b) Find the cost of each package.	(a) Divide the total, 17.4, by the amount in each package, 1.45, to find the number of packages. (b) Multiply the cost of one pound, $4.60, by the amount in one package, 1.45.	There will be two answers to this problem.

2. Solve and state the answer:

(a)

$$\begin{array}{r} 12. \\ 1.45_{\wedge}\overline{)17.40_{\wedge}} \\ \underline{14\ 5} \\ 290 \\ \underline{290} \end{array}$$

The butcher will have 12 packages of steak.

(b)

$$\begin{array}{r} \$4.60 \\ \times\ 1.45 \\ \hline \$6.67 \end{array}$$

Each package will cost $6.67.

3. Check.

(a) $1.5_\wedge\overline{)17.0_\wedge}$ quotient 11.; 15; 20 (b) $\$5 \times 1 = \$5$

The answers are reasonable.

Chapter 4

4.1 Practice Problems

1. (a) $\frac{36}{40} = \frac{9}{10}$ (b) $\frac{18}{15} = \frac{6}{5}$ (c) $\frac{220}{270} = \frac{22}{27}$

2. (a) $\frac{18}{36} = \frac{1}{2}$ (b) $\frac{25}{90} = \frac{5}{18}$

3. $\frac{44 \text{ dollars}}{900 \text{ tons}} = \frac{11 \text{ dollars}}{225 \text{ tons}}$

4. $\frac{212 \text{ miles}}{4 \text{ hours}} = \frac{53 \text{ miles}}{1 \text{ hour}}$ 53 miles/hour

5. Selling price − purchase price = profit: $\$170.40 - 129.60 = \$40.80$. She made a profit of \$40.80 on 120 batteries.

$120\overline{)40.80}$ = .34; 360; 480; 480; 0. Her profit was \$0.34 per battery.

6. (a) $\frac{\$2.04}{12 \text{ ounces}} = \$0.17$ per ounce

$\frac{\$2.80}{20 \text{ ounces}} = \$0.14$ per ounce

(b) Fred saves \$0.03 per ounce by buying the larger size.

4.2 Practice Problems

1. 6 is to 8 as 9 is to 12.

$\frac{6}{8} = \frac{9}{12}$

2. $\frac{2 \text{ hours}}{72 \text{ miles}} = \frac{3 \text{ hours}}{108 \text{ miles}}$

3. (a) $\frac{10}{18} \times \frac{25}{45}$

$10 \times 45 \stackrel{?}{=} 18 \times 25$

$450 = 450$ yes

Thus $\frac{10}{18} = \frac{25}{45}$ is a proportion.

(b) $\frac{42}{100} \times \frac{22}{55}$

$42 \times 55 \stackrel{?}{=} 100 \times 22$

$2310 \neq 2200$ no

Thus $\frac{42}{100} \neq \frac{22}{55}$ is *not* a proportion.

4. (a) $\frac{2.4}{3} \stackrel{?}{=} \frac{12}{15}$

$\frac{2.4}{3} \times \frac{12}{15}$ $3 \times 12 = 36$ ↑ products equal ↓ $2.4 \times 15 = 36$

Thus $\frac{2.4}{3} = \frac{12}{15}$ *is* a proportion.

(b) $\frac{2\frac{1}{3}}{6} \stackrel{?}{=} \frac{14}{38}$

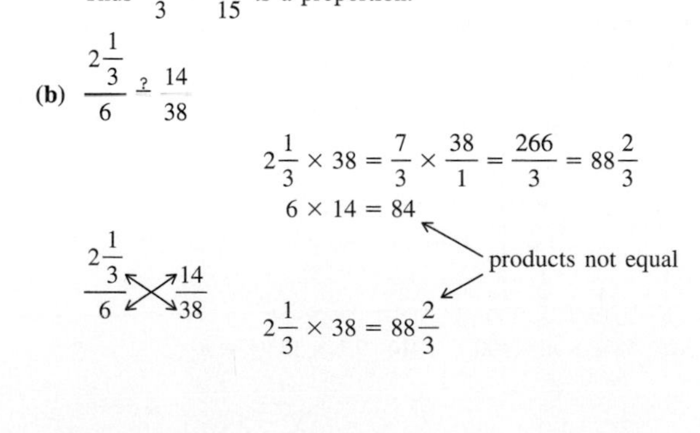

$2\frac{1}{3} \times 38 = \frac{7}{3} \times \frac{38}{1} = \frac{266}{3} = 88\frac{2}{3}$
$6 \times 14 = 84$
products not equal
$2\frac{1}{3} \times 38 = 88\frac{2}{3}$

Thus $\frac{2\frac{1}{3}}{6} \neq \frac{14}{38}$ is *not* a proportion.

5. (a) $\frac{1260}{7} \stackrel{?}{=} \frac{3530}{20}$

$1260 \times 20 \stackrel{?}{=} 7 \times 3530$

$25{,}200 \neq 24{,}710$ No

The rates are *not* equal.

(b) $\frac{2}{11} \stackrel{?}{=} \frac{16}{88}$

$2 \times 88 \stackrel{?}{=} 11 \times 16$

$176 = 176$ Yes

The rates *are* equal.

4.3 Practice Problems

1. (a) $5 \times n = 45$; $\frac{5 \times n}{5} = \frac{45}{5}$; $n = 9$ (b) $7 \times n = 84$; $\frac{7 \times n}{7} = \frac{84}{7}$; $n = 12$

2. (a) $108 = 9 \times n$; $\frac{108}{9} = \frac{9 \times n}{9}$; $12 = n$ (b) $210 = 14 \times n$; $\frac{210}{14} = \frac{14 \times n}{14}$; $15 = n$

3. (a) $15 \times n = 63$; $\frac{15 \times n}{15} = \frac{63}{15}$; $n = 4.2$ (b) $39.2 = 5.6 \times n$; $\frac{39.2}{5.6} = \frac{5.6 \times n}{5.6}$; $7 = n$

4. $\frac{24}{n} = \frac{3}{7}$

$24 \times 7 = n \times 3$

$168 = n \times 3$

$\frac{168}{3} = \frac{n \times 3}{3}$

$56 = n$

5. $\frac{176}{4} = \frac{286}{n}$

$176 \times n = 286 \times 4 = 1144$

$\frac{176 \times n}{176} = \frac{1144}{176}$

$n = 6.5$

6. $80 \times 6 = 5 \times n$

$480 = 5 \times n$

$\frac{480}{5} = \frac{5 \times n}{5}$

$96 = n$

n is \$96

7. $264.25 \times 2 = 3.5 \times n$

$528.5 = 3.5 \times n$

$\frac{528.5}{3.5} = \frac{3.5 \times n}{3.5}$

$151 = n$

n is 151 meters

4.4 Practice Problems

1. $\frac{27 \text{ defective engines}}{243 \text{ engines produced}} = \frac{n \text{ defective engines}}{4131 \text{ engines produced}}$

$27 \times 4131 = 243 \times n$

$111{,}537 = 243 \times n$

$\frac{111{,}537}{243} = \frac{243 \times n}{243}$

$459 = n$

Thus we estimate that 459 engines would be defective.

2. $\frac{5 \text{ students who quit}}{19 \text{ students}} = \frac{n \text{ students who quit}}{7410 \text{ total students}}$

$5 \times 7410 = 19 \times n$

$37{,}050 = 19 \times n$

$\frac{37{,}050}{19} = \frac{19 \times n}{19}$

$1950 = n$

Approximately 1,950 students quit smoking.

3. $\dfrac{9 \text{ gallons of gas}}{234 \text{ miles traveled}} = \dfrac{n \text{ gallons of gas}}{312 \text{ miles traveled}}$

$$9 \times 312 = 234 \times n$$
$$2808 = 234 \times n$$
$$\frac{2808}{234} = \frac{234 \times n}{234}$$
$$12 = n$$

She would need 12 gallons of gas.

4. $\dfrac{80 \text{ rpm}}{16 \text{ mph}} = \dfrac{90 \text{ rpm}}{n \text{ mph}}$

$$80 \times n = 16 \times 90$$
$$80 \times n = 1440$$
$$\frac{80 \times n}{80} = \frac{1440}{80}$$
$$n = 18$$

Alicia will be riding 18 miles per hour.

5. $\dfrac{4050 \text{ walk-in}}{729 \text{ purchase}} = \dfrac{5500 \text{ walk-in}}{n \text{ purchase}}$

$$4050 \times n = 729 \times 5500$$
$$4050 \times n = 4{,}009{,}500$$
$$\frac{4050 \times n}{4050} = \frac{4{,}009{,}500}{4050}$$
$$n = 990$$

Tom will expect 990 people to make a purchase in his store.

6. $\dfrac{50 \text{ bears tagged in 1st sample}}{n \text{ bears in forest}} = \dfrac{4 \text{ bears tagged in 2nd sample}}{50 \text{ bears caught in 2nd sample}}$

$$50 \times 50 = n \times 4$$
$$2500 = n \times 4$$
$$\frac{2500}{4} = \frac{n \times 4}{4}$$
$$625 = n$$

We estimate that there are 625 bears in the forest.

Chapter 5

5.1 Practice Problems

1. **(a)** $\dfrac{51}{100} = 51\%$ **(b)** $\dfrac{68}{100} = 68\%$
   (c) $\dfrac{7}{100} = 7\%$ **(d)** $\dfrac{26}{100} = 26\%$
2. **(a)** $\dfrac{238}{100} = 238\%$
   (b) $\dfrac{505}{100} = 505\%$
   (c) $\dfrac{121}{100} = 121\%$
3. **(a)** $\dfrac{0.5}{100} = 0.5\%$ **(b)** $\dfrac{0.06}{100} = 0.06\%$ **(c)** $\dfrac{0.003}{100} = 0.003\%$
4. **(a)** $47\% = .47$
   (b) $2\% = 0.02$
   (c) $90\% = 0.90$
5. **(a)** $80.6\% = 0.806$ **(b)** $2.5\% = 0.025$
   (c) $0.29\% = 0.0029$ **(d)** $231\% = 2.31$
6. **(a)** $0.78 = 78\%$
   (b) $0.02 = 2\%$
   (c) $5.07 = 507\%$
   (d) $0.029 = 2.9\%$
   (e) $0.006 = 0.6\%$

5.2 Practice Problems

1. **(a)** $71\% = \dfrac{71}{100}$ **(b)** $25\% = \dfrac{25}{100} = \dfrac{1}{4}$ **(c)** $8\% = \dfrac{8}{100} = \dfrac{2}{25}$
2. **(a)** $8.4\% = 0.084 = \dfrac{84}{1000} = \dfrac{21}{250}$
   (b) $55.25\% = .5525 = \dfrac{5525}{10{,}000} = \dfrac{221}{400}$
   (c) $28.5\% = .285 = \dfrac{285}{1000} = \dfrac{57}{200}$
3. **(a)** $170\% = 1.70 = 1\dfrac{7}{10}$
   (b) $288\% = 2.88 = 2\dfrac{88}{100} = 2\dfrac{22}{25}$
4. $7\dfrac{5}{8}\% = 7\dfrac{5}{8} \div 100$
   $= \dfrac{61}{8} \times \dfrac{1}{100}$
   $= \dfrac{61}{800}$
5. $21\dfrac{1}{4}\% = 21\dfrac{1}{4} \div 100 = \dfrac{85}{4} \times \dfrac{1}{100} = \dfrac{85}{400} = \dfrac{17}{80}$
6. $\dfrac{5}{8}$ $\quad 8\overline{)5.000}$ quotient 0.625 $\quad 62.5\%$
7. **(a)** $\dfrac{3}{5} = 0.60 = 60\%$
   (b) $\dfrac{21}{25} = 0.84 = 84\%$
   (c) $\dfrac{7}{16} = 0.4375 = 43.75\%$
8. **(a)** $\dfrac{7}{9} = 0.7777\overline{7} = 0.7778 = 77.78\%$
   (b) $\dfrac{19}{30} = 0.6333\overline{3} = 0.6333 = 63.33\%$
9. $\dfrac{7}{12}$ If we divide

```
     .58
 12)7.00
    60
    100
     96
      4
```

we see that $\dfrac{7}{12} = 0.58\dfrac{4}{12} = 0.58\dfrac{1}{3}$.

Thus $\dfrac{7}{12} = 58\dfrac{1}{3}\%$.

10.

Fraction	Decimal	Percent
$\dfrac{23}{99}$	0.2323	23.23%
$\dfrac{129}{250}$	0.516	51.6%
$\dfrac{97}{250}$	0.388	$38\dfrac{4}{5}\%$

5.3A Practice Problems

1. What is 26% of 35?
   ↓ ↓ ↓ ↓ ↓
   $n = 26\% \times 35$
2. 155% of 20 is what?
   ↓ ↓ ↓ ↓ ↓
   $155\% \times 20 = n$
3. Find 0.08% of 350
   ↓ ↓ ↓ ↓
   $n = 0.08\% \times 350$
4. **(a)** $58\% \times n = 400$ **(b)** $9.1 = 135\% \times n$

5. What percent of 250 is 36?

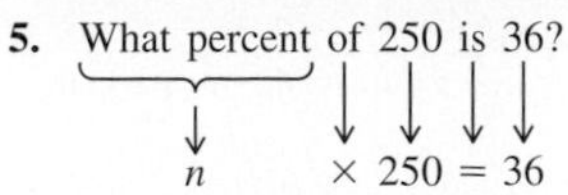

$n \quad \times 250 = 36$

6. (a) $50 = n \times 20$ (b) $n \times 2000 = 4.5$

7. What is 82% of 350?

$n = 82\% \times 350$

$n = 0.82(350)$

$n = 287$

8. (a) $5.2\% \times 800 = n$ (b) $n = 230\% \times 400$

$(.052)(800) = n$ $\quad n = (2.30)(400)$

$41.6 = n$ $\quad n = 920$

9. The problem asks: What is 8% of \$350?

$n = 8\% \times 350$

$n = \$28$ tax

10. $45\% \times n = 162$

$.45 \times n = 162$

$$\frac{.45 \times n}{.45} = \frac{162}{.45}$$

$n = 360$

11. $32 = 0.4\% \times n$

$32 = 0.004n$

$$\frac{32}{0.004} = \frac{0.004n}{0.004}$$

$8000 = n$

12. The problem asks: 30% of what is 6?

$30\% \times n = 6$

$.30n = 6$

$$\frac{.30n}{.30} = \frac{6}{.30}$$

$n = 20$

13. What percent of 9000 is 4.5?

$n \quad \times 9000 = 4.5$

$9000n = 4.5$

$$\frac{9000n}{9000} = \frac{4.5}{9000}$$

$n = 0.0005$

$n = 0.05\%$

14. $198 = n \times 33$

$$\frac{198}{33} = \frac{n \times 33}{33}$$

$6 = n$

Now express n as a percent: 600%

15. The problem asks: 5 is what percent of 16?

$5 = n \times 16$

$$\frac{5}{16} = \frac{n \times 16}{16}$$

$0.3125 = n$

Now express n as a percent rounded to tenths: 31.3%

5.3B Practice Problems

1. (a) Find 83% of 460.

$p = 83$

(b) 18% of what number is 90?

$p = 18$

(c) What percent of 64 is 8?

The percent is unknown. Use the variable p.

2. (a) $b = 52, a = 15.6$ (b) Base $= b, a = 170$

3. (a) What is 18% of 240?

Percent $p = 18$

Base $b = 240$

Amount is unknown; use the variable a.

(b) What percent of 64 is 4?

Percent is unknown; use the variable p.

Base $b = 64$

Amount $a = 4$

4. Percent $p = 32$

Base $b = 550$

Variable a

$$\frac{a}{b} = \frac{p}{100}$$

$$\frac{a}{550} = \frac{32}{100}$$

$100a = (550)(32)$

$a = 176$

Thus 32% of 550 is 176.

5. Percent $p = 340$

Base $b = 70$

Variable a

$$\frac{a}{70} = \frac{340}{100}$$

$100a = (70)(340)$

$a = 238$

Thus 340% of 70 is 238.

6. 68% of what is 476?

Percent $p = 68$

Base is unknown; use base $= b$.

Amount $a = 476$

$$\frac{a}{b} = \frac{p}{100} \quad \text{becomes} \quad \frac{476}{b} = \frac{68}{100}$$

If we reduce the right-hand fraction, we have

$$\frac{476}{b} = \frac{17}{25}$$

$(476)(25) = 17b$ Using cross multiplication.

$11{,}900 = 17b$ Simplifying.

$$\frac{11{,}900}{17} = \frac{17b}{17}$$ Dividing each side by 17.

$700 = b$ Result of $11{,}900 \div 17$.

Thus 68% of 700 is 476.

7. Percent $p = 0.3$

Variable b

Amount $a = 216$

$$\frac{216}{b} = \frac{0.3}{100}$$

$0.3b = (216)(100)$

$b = 72{,}000$

Thus 216 is 0.3% of 72,000

8. Variable p

Base $b = 3500$

Amount $a = 105$

$$\frac{105}{3500} = \frac{p}{100}$$

$$\frac{21}{700} = \frac{p}{100}$$

$700p = (21)(100)$

$p = 3$

Thus 3% of 3500 is 105.

9. Variable p

Base $b = 140$

Amount $a = 42$

$$\frac{42}{140} = \frac{p}{100}$$

$$\frac{21}{70} = \frac{p}{100}$$

$70p = (21)(100)$

$p = 30$

Thus 42 is 30% of 140

5.4 Practice Problems

1. Let n = number of people with reserved airline tickets.

12% of $n = 4800$

$0.12 \times n = 4800$

$$\frac{0.12 \times n}{0.12} = \frac{4800}{0.12} \quad n = 40{,}000$$

40,000 people had airline tickets that month.

2. The problem asks: What is 8% of \$62.30?

$n = .08 \times 62.30$

$n = 4.984$

The tax is \$4.98.

3. The problem asks: 105 is what percent of 130?

$105 = n \times 130$

$$\frac{105}{130} = n$$

$0.8077 = n$

Now express n as a percent rounded to tenths: 80.8%

Thus 80.8% of the flights were on time.

4. Commission = commission rate × value of sales
Commission = 6% × \$156,000
= 0.06 × 156,000
= 9360
His commission is \$9360.
5. 100% Cost of meal + tip of 15% = \$46.00
Let n = Cost of meal
100% of n + 15% of n = \$46.00
115% of n = 46.00
$1.15 \times n = 46.00$
$n = 40.00$
They can spend \$40.00 on the meal itself.
6. 15,000 − 10,500 = 4,500 the amount of decrease

$$\text{Percent of decrease} = \frac{\text{amount of decrease}}{\text{original amount}} = \frac{\$4500}{\$15,000} = 0.30 = 30\%$$

The percent of decrease is 30%.
7. **(a)** 7% of \$13,600 is the discount
0.07 × 13,600 = the discount
\$952 is the discount
(b) \$13,000 list price
− 952 discount
\$12,048 amount Betty paid for the car.
8. $I = P \times R \times T$
$P = \$5600 \quad R = 12\% \quad T = 1$ year
$I = 5600 \times 12\% \times 1$
$= 5600 \times 0.12$
$= 672$
The interest is \$672 for 1 year.
9. **(a)** $I = P \times R \times T$
$I = 1800 \times 0.11 \times 4$
$I = \$792$
(b) $I = P \times R \times T$
$I = 1800 \times 0.11 \times \frac{1}{2}$
$I = \$99$

Chapter 6

6.1 Practice Problems

1. **(a)** 3 feet in 1 yard **(b)** 5280 feet in 1 mile
(c) 60 minutes in 1 hour **(d)** 7 days in 1 week
(e) 16 ounces in 1 pound **(f)** 2 pints in 1 quart
(g) 4 quarts in 1 gallon
2. **(a)** $15,840 \cancel{\text{ft.}} \times \frac{1 \text{ mile}}{5280 \cancel{\text{ft.}}} = \frac{15,840}{5280}$ miles = 3 miles.
(b) $17 \cancel{\text{ft.}} \times \frac{12 \text{ inches}}{1 \cancel{\text{ft.}}} = 17 \times 12$ inches = 204 inches.
3. **(a)** $18.93 \cancel{\text{miles}} \times \frac{5280 \text{ feet}}{1 \cancel{\text{mile}}} = 99,950.4$ feet
(b) $16\frac{1}{2} \cancel{\text{inches}} \times \frac{1 \text{ yard}}{36 \cancel{\text{inches}}} = 16\frac{1}{2} \times \frac{1}{36}$ yards
$= \frac{\cancel{33}^{11}}{2} \times \frac{1}{\cancel{36}_{12}}$ yards $= \frac{11}{24}$ yards
4. $760.5 \cancel{\text{lbs.}} \times \frac{16 \text{ oz.}}{1 \cancel{\text{lb.}}} = 760.5 \times 16$ oz. = 12,168 oz.
5. $19 \cancel{\text{pints}} \times \frac{1 \text{ quart}}{2 \cancel{\text{pints}}} = \frac{19}{2}$ quarts = 9.5 quarts
6. $7.2 \cancel{\text{hrs.}} \times \frac{60 \text{ min.}}{1 \cancel{\text{hr.}}} = 7.2 \times 60$ min. = 432 minutes.
7. **Step 1:** $7 \cancel{\text{lbs.}} \times \frac{16 \text{ oz.}}{1 \cancel{\text{lb.}}} = 7 \times 16$ oz. = 112 oz.
Step 2: 112 oz. + 12 oz. = 124 ounces.
The potatoes weigh 124 ounces.
8. **Step 1:** $1\frac{3}{4} \cancel{\text{days}} \times \frac{24 \text{ hours}}{1 \cancel{\text{day}}} = \frac{7}{4} \times \frac{24}{1}$ hours = 42 hrs parked.
Step 2: $42 \text{ hrs} \times \frac{1.50 \text{ dollars}}{1 \text{ hour}} = 63.00$ dollars.
She paid \$63.00.

6.2 Practice Problems

1. **(a)** deka- **(b)** milli-
2. **(a)** 4 meters = 4.00‸ centimeters = 400 cm.
(b) 30 centimeters = 30.0‸ millimeters = 300 mm.
3. **(a)** 3 mm = ‸003. m = 0.003 m
(b) 47 cm = ‸00047. km. = 0.00047 km.
4. The car length would logically be choice (b) 3.8 meters. [A meter is close to a yard and 3.8 yards seems reasonable!]
5. **(a)** 1.527 km **(c)** 12.8 cm
(b) 3.75 m **(d)** 46,000 mm
6. **(a)** 389 mm = 0.0389 dekameter (four places to left)
(b) 0.48 hectometer = 4800 cm (four places to right)
7. 782 cm = 7.82 m
2 m = 2.00 m
537 m = 537.00 m
546.82 m

6.3 Practice Problems

1. **(a)** 5 L = 5000 mL
(b) 84 kL = 84,000 L
(c) 0.732 L = 732 mL
2. **(a)** 15.8 mL = 0.0158 L
(b) 12340 mL = 12.34 L
(c) 863.3 L = 0.0863 kL
3. **(a)** 396 mL = 396 cm^3
(because 1 milliliter = 1 cubic centimeter)
(b) 0.096 L = 96 cm^3
4. $\frac{\$110.00}{1000} = \$0.11$ per milliliter
5. **(a)** 3.2 t = 3200 kg
(b) 7.08 kg = 7080 g
(c) 0.526 g = 526 mg
6. **(a)** 59 kg = 0.059 t
(b) 6.152 g = 0.006152 kg
(c) 28.3 mg = 0.0283 g
7. A gram is $\frac{1}{1000}$ of a kilogram. If the coffee costs \$10.00 per kilogram, then 1 gram would be $\frac{1}{1000}$ of \$10.
$\frac{1}{1000} \times \$10 = \frac{\$10.00}{1000} = \$0.01$
The coffee costs 1¢ per gram.
8. The most reasonable answer is (a) 120 kg. [A kg is slightly more than 2 lbs.]

6.4 Practice Problems

1. **(a)** $7 \cancel{\text{ft}} \times \frac{0.305 \text{ m}}{1 \cancel{\text{ft}}} = 2.135$ m
(b) $4 \cancel{\text{in.}} \times \frac{2.54 \text{ cm}}{1 \cancel{\text{in.}}} = 10.16$ cm
2. **(a)** $17 \cancel{\text{m}} \times \frac{1.09 \text{ yd.}}{1 \cancel{\text{m}}} = 18.53$ yd.
(b) $29.6 \cancel{\text{km}} \times \frac{0.62 \text{ mi.}}{1 \cancel{\text{km}}} = 18.352$ mi.
(c) $26 \cancel{\text{gal}} \times \frac{3.79 \text{ L}}{1 \cancel{\text{gal}}} = 98.54$ L
(d) $6.2 \cancel{\text{L}} \times \frac{1.06 \text{ qt.}}{1 \cancel{\text{L}}} = 6.572$ qt.
(e) $16 \cancel{\text{lb.}} \times \frac{0.454 \text{ kg}}{1 \cancel{\text{lb}}} = 7.264$ kg.

(f) $280\,\cancel{g} \times \dfrac{0.0353\text{ oz.}}{1\,\cancel{g}} = 9.884\text{ oz.}$

3. $180\,\cancel{cm} \times \dfrac{0.394\,\cancel{in.}}{1\,\cancel{cm}} \times \dfrac{1\text{ ft}}{12\,\cancel{in.}} = 5.91\text{ ft}$

4. $\dfrac{88\,\cancel{km}}{\text{hr}} \times \dfrac{0.62\text{ mi}}{1\,\cancel{km}} = 54.56\text{ mi/hr}$

5. $9\text{ mm} = 0.9\text{ cm}$

$0.9\,\cancel{cm} \times \dfrac{0.394\text{ in.}}{\cancel{cm}} = 0.3546\text{ in.}$ rounded to 0.35 in.

6. $F = 1.8 \times C + 32$
$= 1.8 \times 20 + 32$
$= 36 + 32$
$= 68$

The temperature is 68°F.

7. $C = \dfrac{5 \times 181 - 160}{9} = \dfrac{905 - 160}{9}$

$= \dfrac{745}{9} = 82.87°\text{C}$ rounded to the nearest tenth

6.5 Practice Problems

Practice Problem 1

Step 1:

$2\frac{2}{3}$ yd
$8\frac{1}{3}$ yd
$2\frac{2}{3}$ yd
$+\ 8\frac{1}{3}$ yd
22yd

Step 2: $22\,\cancel{yd} \times \dfrac{3\text{ ft}}{1\,\cancel{yd}} = 66\text{ ft.}$

The perimeter is 66 ft.

Practice Problem 2

1. Understand the problem.

MATHEMATICS BLUEPRINT FOR PROBLEM SOLVING			
Gather the Facts	**What Am I Asked to Do?**	**How Do I Proceed?**	**Key Points to Remember**
He must use 18.06 liters of solution. He has 42 jars to fill.	Find out how many milliliters of a solvent will go into each jar.	We need to convert 18.06 liters to milliliters, and then divide that result by 42.	To convert 18.06 liters to milliliters we move the decimal point 3 places to the right.

2. Solve and state the answer:

(a) $18.06\text{ L} = 18{,}060\text{ mL}$

(b) $\dfrac{18{,}060\text{ mL}}{42\text{ jars}} = 430\text{ mL/jar}$

Thus, 430 mL of a solvent will go into each jar.

3. Check. 18.06 L is approximately 18 L, or 18,000 mL.

$\dfrac{18{,}000}{40} = 450$ and is reasonable.

Chapter 7

7.1 Practice Problems

1. $P = 2l + 2w$
$= (2)(6) + (2)(1.5)$
$= 12 + 3 = 15\text{ m}$

2. $P = 2l + 2w$
$= (2)(1200) + 2(860)$
$= 2400 + 1720$
$= 4120\text{ ft.}$

3. $P = 4s$
$= (4)(5.8) = 23.2\text{ cm}$

4. $P = 4 + 2 + 5 + 1 + 1 + 1 = 14\text{ cm.}$

5. $P = 4 + 4 + 5.5 + 2.5 + 1.5 + 1.5$
$= 19\text{ ft}$

$\text{Cost} = 19\,\cancel{ft} \times \dfrac{0.16\text{ dollar}}{1\,\cancel{ft}} = \$3.04$

6. $A = lw = (29)(17) = 493\text{ m}^2$

7. $A = (s)^2$
$= (11.8)^2$
$= 139.24\text{ mm}^2$

8. Area of rectangle $= (18)(20) = 360\text{ ft}^2$
Area of square $= 6^2 = 36\text{ ft}^2$
The total area $= 396\text{ ft}^2$

7.2 Practice Problems

1. $\text{Perimeter} = (2)(7.6) + (2)(3.5)$
$= 15.2 + 7.0 = 22.2\text{ cm}$

2. $A = bh$
$= (10.3)(1.5)$
$= 15.45\text{ km}^2$

3. $\text{Perimeter} = (2)(4.1) + (2)(11.6)$
$= 8.2 + 23.2 = 31.4\text{ km}$

$\text{Area} = bh$
$= (4.1)(9.8) = 40.18\text{ km}^2$

4. $P = 7 + 15 + 21 + 13 = 56\text{ yd.}$

5. $A = \dfrac{h(b + B)}{2} = \dfrac{(2.6)(1.3 + 1.9)}{2} = \dfrac{(2.6)(3.2)}{2} = \dfrac{8.32}{2} = 4.16\text{ mm}^2$

6. (a) $A = \dfrac{h(b + B)}{2}$

$= \dfrac{140(180 + 130)}{2}$

$= 21{,}700\text{ yd}^2$

(b) $21{,}700\,\cancel{yd^2} \times \dfrac{1\text{ gallon}}{100\,\cancel{yd^2}} = 217\text{ gallons}$

Thus 217 gallons of sealer are needed.

7. The area of the trapezoid

$A = \frac{(9.2)(12.6 + 19.8)}{2}$

$= \frac{(9.2)(32.4)}{2} = \frac{298.08}{2}$

$= 149.04 \text{ cm}^2$

The area of the rectangle is

$A = (8.3)(12.6) = 104.58 \text{ cm}^2$

Total area $= 253.62 \text{ cm}^2$

7.3 Practice Problems

1. The sum of the angles in a triangle is 180°. The two given angles total 125° + 15° = 140°. Thus 180° − 140° = 40°. Angle A must be 40°.
2. $P = 10.5 + 10.5 + 8.5 = 29.5$ m
3. $P = 30 \qquad 30 \div 3 = 10$
   The length of each side is 10 cm
4. $A = \frac{bh}{2} = \frac{(38)(13)}{2} = \frac{494}{2} = 247 \text{ m}^2$
5. $A = \frac{bh}{2} = \frac{(5)(16)}{2} = 40 \text{ cm}^2$
6. Area of rectangle = $(11)(24) = 264 \text{ cm}^2$
   Area of triangle $= \frac{(11)(7)}{2} = \frac{77}{2} = 38.5 \text{ cm}^2$
   Total area $= 302.5 \text{ cm}^2$

7.4 Practice Problems

1. **(a)** $\sqrt{49} = 7$ **(b)** $\sqrt{169} = 13$
2. $\sqrt{49} - \sqrt{4} = 7 - 2 = 5$
3. **(a)** Yes, because (12)(12) = 144 **(b)** $\sqrt{144} = 12$
4. **(a)** $\sqrt{3} \approx 1.732$ **(b)** $\sqrt{13} \approx 3.606$ **(c)** $\sqrt{5} \approx 2.236$
5. $\sqrt{100 \text{ in}^2} = 10$ in.
6. $\sqrt{22 \text{ m}^2} \approx 4.690$ m

7.5 Practice Problems

1. Hypotenuse $= \sqrt{(6)^2 + (8)^2}$
   $= \sqrt{36 + 64}$
   $= \sqrt{100}$
   $= 10$ m
2. Hypotenuse $= \sqrt{(12)^2 + (16)^2}$
   $= \sqrt{144 + 256}$
   $= \sqrt{400}$
   $= 20$ m
3. Hypotenuse $= \sqrt{(3)^2 + (7)^2}$
   $= \sqrt{9 + 49}$
   $= \sqrt{58}$
   Rounded to the nearest thousandth, hypotenuse ≈ 7.616 cm.
4. Leg $= \sqrt{(17)^2 - (15)^2}$
   $= \sqrt{289 - 225}$
   $= \sqrt{64}$
   $= 8$ m
5. Leg $= \sqrt{(10)^2 - (5)^2}$
   $= \sqrt{100 - 25}$
   $= \sqrt{75}$
   Rounded to the nearest thousandth,
   the leg ≈ 8.660 m.
6. The distance between the holes is the hypotenuse.
   Hypotenuse $= \sqrt{(5)^2 + (2)^2}$
   $= \sqrt{25 + 4}$
   $= \sqrt{29}$
   ≈ 5.385 cm
7. Leg $= \sqrt{(30)^2 - (27)^2}$
   $= \sqrt{900 - 729}$
   $= \sqrt{171}$
   The distance to the nearest tenth is 13.1 yd.
8. **(a)** $y = \frac{1}{2} \times 12 = 6$ ft.
   $x = \sqrt{(12)^2 - (6)^2}$
   $= \sqrt{144 - 36}$
   $= \sqrt{108}$
   ≈ 10.4 ft.
   (b) Hypotenuse $= \sqrt{2} \times \text{leg}$
   $\approx 1.414 \times 8$
   ≈ 11.3 m

7.6 Practice Problems

1. $C = \pi d$
   $= (3.14)(9)$
   $= 28.26$
   $= 28.3$ m to nearest tenth
2. $C = 2\pi r$
   $= (2)(3.14)(14)$
   $= 87.9$ in.
3. **(a)** $C = \pi d$
   $= (3.14)(30)$
   $= 94.2$ in
   (b) Change 94.2 in. to ft.
   $94.2 \cancel{\text{in}} \times \frac{1 \text{ ft}}{12 \cancel{\text{in}}} = 7.85$ ft
   (c) When the wheel makes 2 revolutions, the bicycle travels 15.7 ft. [7.85 × 2]
4. $A = \pi r^2$
   $= (3.14)(5^2)$
   $= (3.14)(25)$
   $= 78.5 \text{ km}^2$
5. $r = \frac{d}{2} = \frac{22}{2} = 11$ m
   $A = \pi r^2$
   $= (3.14)(11)^2$
   $= 379.9 \text{ m}^2$
   The area of the circle is 379.9 m^2
6. $r = \frac{d}{2} = \frac{10}{2} = 5$ ft.
   $A = \pi r^2$
   $= (3.14)(5)^2$
   $= 78.5 \text{ ft}^2$
   Change 78.5 ft^2 to yd^2
   $78.5 \cancel{\text{ft}^2} \times \frac{1 \text{ yd}^2}{9 \cancel{\text{ft}^2}} = 8.7 \text{ yd}^2$
   Find the cost: $\frac{\$12}{1 \cancel{\text{yd}^2}} \times 8.7 \cancel{\text{yd}^2} = \$104.40$
   The cost of the pool cover is \$104.40.
7. Area of square − area of circle = shared area
   $s^2 \quad - \quad \pi r^2 \quad =$ shaded area
   $5^2 \quad - \quad (3.14)(2^2)$
   $25 \quad - \quad (3.14)(4)$
   $25.00 \quad - \quad 12.56 \quad = 12.4 \text{ ft}^2$ shaded area (rounded to nearest tenth)
8. $r = \frac{d}{2} = \frac{8}{2} = 4$ ft.
   $A \text{ semicircle} = \frac{\pi r^2}{2}$
   $= \frac{(3.14)(4)^2}{2}$
   $= 25.12 \text{ ft}^2$
   $A \text{ rectangle} = lw = (12)(8) = 96 \text{ ft}^2$
   25.12 ft^2
   $+ 96.00 \text{ ft}^2$
   121.12 ft^2 The total area is 121.1 ft^2

7.7 Practice Problems

1. $V = lwh$
   $= (6)(5)(2)$
   $= (30)(2)$
   $= 60 \text{ m}^3$
2. $V = \pi r^2 h$
   $= (3.14)(2)^2(5)$
   $= (3.14)(4)(5)$
   $= 62.8 \text{ in}^3$
3. $V = \frac{4\pi r^3}{3} = \frac{(4)(3.14)(6^3)}{3} = \frac{(4)(3.14)(6)(6)(\cancel{6})^{2}}{\cancel{3}_{1}}$
   $= (12.56)(36)(2) = 904.32$
   $= 904.3 \text{ m}^3$ Rounded to nearest tenth

4. $V = \frac{\pi r^2 h}{3}$

$= \frac{(3.14)(5)^2(12)}{3}$

$= 314 \text{ m}^3$

5. (a) $V = \frac{Bh}{3}$

$= \frac{(6)(6)(10)}{3} = \frac{(36)(10)}{3}$

$= \frac{\cancel{36}^{12}(10)}{\cancel{3}_1} = 120 \text{ m}^3$

(b) $V = \frac{Bh}{3}$

$= \frac{(7)(8)(15)}{3} = \frac{(7)(8)(\cancel{15}^{5})}{\cancel{3}_1}$

$= (56)(5) = 280 \text{ m}^3$

7.8 Practice Problems

1. $\frac{11}{27} = \frac{15}{n}$

$11n = (27)(15)$

$11n = 405$

$\frac{11n}{11} = \frac{405}{11}$

$n = 36.\overline{81}$

$= 36.8$ meters measured to the nearest tenth.

2. a corresponds to p, b corresponds to m, c corresponds to n

3. Small triangle has a perimeter of 30 yards.
Let p = the unknown perimeter.

$\frac{12}{40} = \frac{30}{p}$

$12p = (40)(30)$

$12p = 1200$

$\frac{12p}{12} = \frac{1200}{12}$

$p = 100$

The larger triangle has a perimeter of 100 yards.

4. $\frac{h}{5} = \frac{20}{2}$

$2h = 100$

$h = 50$ ft.

5. $\frac{3}{29} = \frac{1.8}{n}$

$3n = (1.8)(29)$

$3n = 52.2$

$\frac{3n}{3} = \frac{52.2}{3}$

$n = 17.4$ The width is 17.4 meters.

7.9 Practice Problems

Practice Problem 1

1. Understand the problem.

MATHEMATICS BLUEPRINT FOR PROBLEM SOLVING			
Gather the Facts	**What Am I Asked to Do?**	**How Do I Proceed?**	**Key Points to Remember**
Mike needs to sand three rooms: 24 ft × 13 ft 12 ft × 9 ft 16 ft × 3 ft He can sand 80 ft² in 15 min.	Find out how long it will take him to sand all three rooms.	(a) Find the total area to be sanded. (b) Then find out how long it will take him to sand the total area.	Area = length × width. To get the total time, multiply the total area by the fraction: $\frac{15 \text{ min.}}{80 \text{ ft}^2}$

2. Solve and state the answer:

(a) $24 \times 13 = 312 \text{ ft}^2$ room 1
$12 \times 9 = 108 \text{ ft}^2$ room 2
$16 \times 3 = 48 \text{ ft}^2$ room 3
468 ft^2 total area

(b) $468 \cancel{\text{ft}^2} \times \frac{15 \text{ min}}{80 \cancel{\text{ft}^2}} = 87.75$ min. It takes Mike 88 min. to sand the rooms.

3. Check. Estimate the answer to see if it is reasonable.

Practice Problem 2

1. Understand the problem.

MATHEMATICS BLUEPRINT FOR PROBLEM SOLVING			
Gather the Facts	**What Am I Asked to Do?**	**How Do I Proceed?**	**Key Points to Remember**
The trapezoid has a height of 9 ft. The bases are 18 ft and 12 ft. The rectangular portion measures 24 ft × 15 ft. Roofing costs $2.75 per square yard.	(a) Find the area of the roof. (b) Find the cost to install new roofing.	(a) Find the area of the entire roof. Change square feet to square yards. (b) Multiply by $2.75.	9 square feet = 1 square yard.

2. Solve and state the answer:

(a) Area of trapezoid $= \frac{1}{2}h(b + B)$

$= \frac{1}{2}(9)(12 + 18)$

$= 135 \text{ ft}^2$

Area of rectangle $= lw$

$= (15)(24)$

$= 360 \text{ ft}^2$

Total Area $= 135 \text{ ft}^2 + 360 \text{ ft}^2 = 495 \text{ ft}^2$

Change square feet to square yards.

$495 \cancel{\text{ft}^2} \times \frac{1 \text{ yd}^2}{9 \cancel{\text{ft}^2}} = 55 \text{ yd}^2$

The area of the roof is 55 yd^2

(b) Cost $= 55 \cancel{\text{yd}^2} \times \frac{\$2.75}{1 \cancel{\text{yd}^2}} = \$151.25$

The cost to install new roofing would be \$151.25.

3. Check. Estimate the answers to see if they seem reasonable.

Chapter 8

8.1 Practice Problems

1. The smallest category of students is *special students.*
2. 3000 freshmen + 200 *special students* = 3200
   There are 3200 students who are either freshmen or *special students.*
3. There are 3000 freshmen but only 2600 sophomores. The ratio of freshmen to sophomores is $\frac{3000}{2600} = \frac{15}{13}$.
4. The ratio of freshmen to the total number of students is: $\frac{3000}{10000} = \frac{3}{10}$.
5. Lake Michigan occupies the third largest area with 23%.
6. The percent of the total area occupied by either Lake Superior or Lake Michigan is: 28% + 23% = 51%.
7. Lake Superior has 28% of the area. Now 28% of 290,000 mi^2 is $(0.28)(290{,}000) = 81{,}200 \text{ mi}^2$
8. (a) 10% + 8% = 18%
   (b) 20% of 690,000 = (0.20)(690,000) = 138,000

8.2 Practice Problems

1. The bar rises to 24. The approximate population was 24,000,000.
2. 16,000,000 − 7,000,000 = 9,000,000. The bar graph increases 9,000,000.
3. The bar rises to 250. The number of new cars sold in the fourth quarter of 1996 was 250.
4. 150 − 100 = 50. Thus, 50 fewer cars were sold.
5. The greatest number of customers came in July since the highest point of the graph occurs in July.
6. (a) 3500 (b) The number decreased.
7. The sharpest line of decrease is from July to August. The line slopes downward most steeply between those two months. Thus the biggest decrease in customers is between July and August.
8. Because the dot corresponding to 1992–93 is opposite 15 and the scale is in hundreds, we have 15 × 100 = 1500. Thus, 1500 degrees in visual and performing arts were awarded.
9. The computer science line goes above the visual and performing arts line first in 1990–91. Thus the first academic year with more degrees in computer science was 1990–91.

8.3 Practice Problems

1. The 60–69 bar rises to a height of 6. Thus six tests would have a D grade.
2. From the histogram, *16* tests were 70–79, *8* tests were 80–89, and *6* tests were 90–99. When we combine 16 + 8 + 6 = 30, we can see that 30 students scored greater than 69 on the test.
3. The 800–999 bar rises to a height of 20. Thus 20 light bulbs lasted between 800 and 999 hours.
4. From the histogram, *25* bulbs lasted 1200–1399 hours, *10* lasted 1400–1599 hours, and *5* lasted 1600–1799 hours. When we combine 25 + 10 + 5 = 40, we can see that 40 light bulbs lasted more than 1199 hours.
5. Weight in Pounds

Class Interval	Tally	Frequency
1600–1799	||	2
1800–1999	||||	4
2000–2199	||||	4
2200–2399	𝍸	5

6.

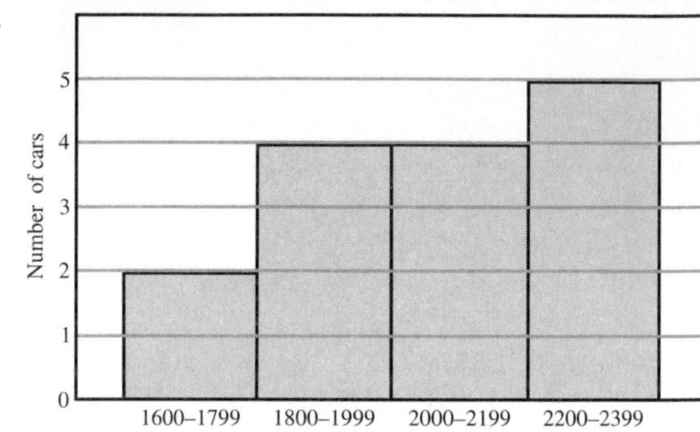

7. The greatest difference occurs between the under age 25 category and the 25–34 age category.

8.4 Practice Problems

1. $\frac{88 + 77 + 84 + 97 + 89}{5} = \frac{435}{5} = 87$ mean value
2. $\frac{\$39.20 + \$43.50 + \$81.90 + \$34.20 + \$51.70 + \$48.10}{6} = \$49.77$
   The mean monthly phone bill is \$49.77.
3. $\underbrace{2, 3, 7, 12,}_{\text{four values}}$ $\quad 13 \quad$ (↑ middle value) $\quad \underbrace{15, 18, 20, 21}_{\text{four values}}$
   The median value is 13 minutes.
4. $\underbrace{\$150, \$180, \$290,}_{\text{three numbers}}$ $\quad \$320, \quad$ (↑ middle number) $\quad \underbrace{\$400, \$450, \$600}_{\text{three numbers}}$
   Thus, \$320 is the median salary.
5. $\underbrace{88, 90,}_{\text{two values}}$ $\quad \underbrace{100, 105,}_{\text{two middle values}}$ $\quad \underbrace{118, 126}_{\text{two values}}$
   $\frac{100 + 105}{2} = \frac{205}{2} = 102.5$ median value

Chapter 9

9.1 Practice Problems

1. (a) $9 + 14 = 23$ (b) $-7 + (-3) = -10$ (c) $-4.5 + (-1.9) = -6.4$
2. (a) $-\frac{2}{15} + \left(-\frac{12}{15}\right) = -\frac{14}{15}$
   (b) $\frac{5}{12} = \frac{5}{12}$; $+\frac{1}{4} \times \frac{3}{3} = +\frac{3}{12}$; $\frac{8}{12} = \frac{2}{3}$
   (c) $-\frac{1}{6} = -\frac{7}{42}$; $+\left(-\frac{2}{7}\right) = +\left(-\frac{6}{42}\right)$; $= -\frac{13}{42}$

3. The value of $-\$4{,}200$ million or $-\$4.2$ billion. The debt for these two years is \$4,200,000,000.

4. **(a)** $\begin{array}{r} 7 \\ +\,-12 \\ \hline -5 \end{array}$ **(b)** $\begin{array}{r} -7 \\ +\,12 \\ \hline +5 \end{array}$

(c) $\begin{array}{r} -20.8 \\ +\;\;15.2 \\ \hline -5.6 \end{array}$ **(d)** $\begin{array}{r} 20.8 \\ +\,-15.2 \\ \hline +5.6 \end{array}$

5. $-19°\text{F} + 28°\text{F} = 9°\text{F}$.

6. $\begin{array}{r} 36 \\ +\,-21 \\ \hline 15 \end{array}$ Then we add $\begin{array}{r} 15 \\ +\,-18 \\ \hline -3 \end{array}$

7. $\begin{array}{r} -5 \\ -2 \\ +\,-7 \\ \hline -14 \end{array}$ $\begin{array}{r} 3 \\ 8 \\ +\,9 \\ \hline 20 \end{array}$ $\begin{array}{r} -14 \\ +\;\;20 \\ \hline +6 \end{array}$

8. $\begin{array}{r} \$30{,}000 \\ +\,\$40{,}000 \\ \hline \$70{,}000 \end{array}$ $\begin{array}{r} -\$20{,}000 \\ -\$5{,}000 \\ +\,-\$35{,}000 \\ \hline -\$60{,}000 \end{array}$ $\begin{array}{r} \$70{,}000 \\ +\,-\$60{,}000 \\ \hline \$10{,}000 \end{array}$

The company had an overall profit of \$10,000 in the five-month period.

9.2 Practice Problems

1. $-10 - (-5) = -10 + 5 = -5$

2. **(a)** $5 - 12 = 5 + (-12) = -7$

(b) $-11 - 17 = -11 + (-17) = -28$

(c) $4 - (-17) = 4 + 17 = 21$

3. **(a)** $3.6 - (-9.5) = 3.6 + 9.5 = 13.1$

(b) $-\frac{5}{8} = -\frac{5}{8} \times \frac{3}{3} = -\frac{15}{24}$

$\quad -\,-\frac{5}{24} = \quad +\frac{5}{24} = +\frac{5}{24}$

$\quad -\frac{10}{24} = -\frac{5}{12}$

4. **(a)** $8 - 14 = 8 + (-14) = -6$

(b) $-6 - 13 = -6 + (-13) = -19$

(c) $20 - (-5) = 20 + 5 = 25$

(d) $-9 - (-3) = -9 + 3 = -6$

(e) $-\frac{1}{5} - \left(-\frac{1}{2}\right) = -\frac{1}{5} + \frac{1}{2} = -\frac{2}{10} + \frac{5}{10} = \frac{3}{10}$

(f) $3.6 - (-5.5) = 3.6 + 5.5 = 9.1$

5. $-5 - (-9) + (-14) = -5 + 9 + (-14) = 4 + (-14) = -10$

6. $31 - (-37) = 31 + 37 = 68$ The difference is 68°F.

9.3 Practice Problems

1. **(a)** $(6)(9) = 54$ **(b)** $(7)(12) = 84$

(c) $3(5)(8) = 15(8) = 120$ **(d)** $(6)(2)(7) = (12)(7) = 84$

2. **(a)** $(-8)(-8) = 64$ **(b)** $-10(-6) = 60$

(c) $\left(-\frac{1}{3}\right)\left(-\frac{2}{7}\right) = \frac{2}{21}$ **(d)** $(1.2)(0.4) = 0.48$

3. **(a)** $\frac{-16}{-8} = 2$ **(b)** $-78 \div (-2) = 39$

(c) $\frac{-1.2}{-0.5} = 2.4$ **(d)** $\frac{3}{5} \div \frac{1}{10} = \frac{3}{\cancel{5}_1} \cdot \frac{\cancel{10}^2}{1} = \frac{6}{1} = 6$

4. **(a)** $(-8)(5) = -40$ **(b)** $3(-60) = -180$

(c) $\left(-\frac{1}{5}\right)\left(\frac{3}{7}\right) = -\frac{3}{35}$ **(d)** $0.5(-6.7) = -3.35$

5. **(a)** $-50 \div 25 = -\frac{50}{25} = -2$

(b) $49 \div (-7) = \frac{49}{-7} = -7$

(c) $\frac{21.12}{-3.2} = -6.6$ $\quad \begin{array}{r} 6.6 \\ 3.2_\wedge\overline{)21.1_\wedge2} \\ \underline{192} \\ 192 \\ \underline{192} \\ 0 \end{array}$

(d) $-\frac{2}{7} \div \frac{4}{13} = -\frac{\cancel{2}^1}{7} \cdot \frac{13}{\cancel{4}_2} = -\frac{13}{14}$

6. $(-6)(3)(-4) = (-18)(-4) = 72$

7. $4(-8)\left(-\frac{1}{4}\right)(-3) = (-32)\left(-\frac{1}{4}\right)(-3) = (+8)(-3) = -24$

8. $(-2)(6) = -12$

Thus the change in charge would be -12.

9.4 Practice Problems

1. $\underbrace{20 \div (-5)}(-3)$

$= (-4) \quad (-3)$

$= +12$

2. **(a)** $\underbrace{25 \div (-5)} + \underbrace{16 \div (-8)}$

$= (-5) \quad + \quad (-2)$

$= -7$

(b) $9 + \underbrace{20 \div (-4)}$

$= 9 + (-5)$

$= 4$

(c) $13 + 7(-6) - 8$

$= 13 + (-42) - 8$

$= -29 - 8$

$= -37$

3. **(a)** $\frac{12 - 2 - 3}{21 + 3 - 10} = \frac{7}{14} = \frac{1}{2}$

(b) $\frac{9(-3) - 5}{2(-4) \div (-2)} = \frac{-27 - 5}{-8 \div (-2)} = \frac{-32}{+4} = -8$

4. $\frac{-10{,}000 + 30{,}000 - 18{,}000 - 6{,}000}{4} = \frac{-4000}{4} = -\$1000$

Average loss is \$1000 per month.

9.5 Practice Problems

1. $896 = 8.96 \times 10^2$

2. **(a)** 4.8×10^1 **(b)** 3.729×10^3 **(c)** 5.06936×10^5

3. **(a)** $4600 = 4.6 \times 10^3$ **(b)** $900 = 9 \times 10^2$

(c) $3{,}800{,}000 = 3.8 \times 10^6$

4. **(a)** 7.6×10^{-2} **(b)** 9.82×10^{-1}

(c) 3.12×10^{-4} **(d)** 6×10^{-3}

5. **(a)** $3.08 \times 10^2 = 308$ **(b)** $6.543 \times 10^3 = 6543$

6. **(a)** 850 **(b)** 430,000 **(c)** 60,000

7. **(a)** $8.56 \times 10^{-1} = 0.856$ **(b)** $7.72 \times 10^{-3} = 0.00772$

(c) $2.6 \times 10^{-5} = 0.000026$

8. **(a)** $\begin{array}{r} 6.85 \times 10^{22}\text{ kg} \\ +\,2.09 \times 10^{22}\text{ kg} \\ \hline 8.94 \times 10^{22}\text{ kg} \end{array}$ **(b)** $\begin{array}{r} 8.04 \times 10^{30}\text{ tons} \\ -\,6.98 \times 10^{30}\text{ tons} \\ \hline 1.06 \times 10^{30}\text{ tons} \end{array}$

9. $3.1 \times 10^4 = 31{,}000$

But $31{,}000 = 0.31 \times 10^5$

$\begin{array}{r} 4.36 \times 10^5 \\ -\,0.31 \times 10^5 \\ \hline 4.05 \times 10^5 \end{array}$

Chapter 10

10.1 Practice Problems

1. **(a)** The variables are *A*, *b* and *h*.

(b) The variables are *d* and *C*.

(c) The variables are *V*, *l*, *w*, and *h*.

2. **(a)** $p = 2w + 2l$

(b) $A = \pi r^2$

3. Since $9 + 2 = 11$, therefore $9x + 2x = 11x$

4. (a) $26x - 18x = 26x + (-18x) = 8x$
(b) $8x - 22x + 5x = 8x + (-22x) + 5x = (-14x) + 5x = -9x$
(c) $19x - 7x - 12x = 19x + (-7x) + (-12x) = 12x + (-12x) = 0$

5. (a) $9x + (-12x) + 1x = -3x + 1x = -2x$
(b) $5.6x + (-8.0x) + 10.0x = -2.4x + 10.0x = 7.6x$

6. (a) $-8 + (-9) + (2x) + (-15x) = -17 - 13x$
(b) $17.5 + 10.5 + (-6.3x) + (-8.2x) = 28 - 14.5x$

7. (a) $2w + (-5w) + 3z + (-1z) + (-12) + (-16) = -3w + 2z - 28$
(b) $-3a + (-7a) + 2b + (-5b) + 1c + (-2c) = -10a - 3b - c$

10.2 Practice Problems

1. (a) $(7)(x + 5) = 7(x) + 7(5) = 7x + 35$
(b) $(-4)(x + 2y) = -4(x) + (-4)(2y) = -4x - 8y$

2. (a) $(x + 3y)(8) = (x)(8) + (3y)(8) = 8x + 24y$
(b) $(2x - 7)(2) = (2x)(2) + (-7)(2) = 4x - 14$

3. (a) $(8)(3x - y - 6) = (8)(3x) + (8)(-1y) + (8)(-6) = 24x - 8y - 48$
(b) $(-5)(x + 4y + 5) = (-5)(1x) + (-5)(4y) + (-5)(5) = -5x - 20y - 25$
(c) $(3)(2.2x + 5.5y + 6) = (3)(2.2x) + (3)(5.5y) + (3)(6) = 6.6x + 16.5y + 18$

4. (a) $8(1.2x + 2.4y - 3z - 1.5) = 9.6x + 19.2y - 24z - 12$
(b) $(-9)(x + 3y - 4z + 5) = -9x - 27y + 36z - 45$
(c) $\left(\frac{3}{2}\right)\left(\frac{1}{2}x - \frac{1}{3}y + 4z - \frac{1}{2}\right)$

$$= \left(\frac{3}{2}\right)\left(\frac{1}{2}\right)(x) + \left(\frac{3}{2}\right)\left(-\frac{1}{3}\right)(y) + \left(\frac{3}{2}\right)(4)(z) + \left(\frac{3}{2}\right)\left(-\frac{1}{2}\right)$$

$$= \frac{3}{4}x - \frac{1}{2}y + 6z - \frac{3}{4}$$

5. $3(2x + 4y) + 2(5x + y) = 6x + 12y + 10x + 2y = 16x + 14y$

6. (a) $-4(x - 5) + 3(-1 + 2x) = -4x + 20 - 3 + 6x = 2x + 17$
(b) $2(x - 3y) + 5(2x + 6) = 2x - 6y + 10x + 30 = 12x - 6y + 30$

10.3 Practice Problems

1.
$$x + 7 = -8$$
$$x + 7 + (-7) = -8 + (-7)$$
$$x = -15$$

2. (a)
$$x + 8 = 2$$
$$x + 8 + (-8) = 2 + (-8)$$
$$x = -6$$

Check.
$$(-6) + 8 \stackrel{?}{=} 2$$
$$2 = 2 \checkmark$$

(b)
$$y - 3.2 = 9$$
$$y - 3.2 + 3.2 = 9.0 + 3.2$$
$$y = 12.2$$

Check.
$$(12.2) - 3.2 \stackrel{?}{=} 9$$
$$9 = 9 \checkmark$$

(c)
$$x + \frac{1}{6} = \frac{2}{3}$$
$$x + \frac{1}{6} + \left(-\frac{1}{6}\right) = \frac{4}{6} + \left(-\frac{1}{6}\right)$$
$$x = \frac{3}{6}$$
$$x = \frac{1}{2}$$

Check.
$$\left(\frac{1}{2}\right) + \frac{1}{6} \stackrel{?}{=} \frac{2}{3}$$
$$\frac{3}{6} + \frac{1}{6} \stackrel{?}{=} \frac{2}{3}$$
$$\frac{4}{6} = \frac{2}{3} \checkmark$$

3.
$$3x - 5 = 2x + 1$$
$$3x - 5 + 5 = 2x + 1 + 5$$
$$3x = 2x + 6$$
$$3x + (-2x) = 2x + (-2x) + 6$$
$$x = 6$$

Check.
$$3(6) - 5 \stackrel{?}{=} 2(6) + 1$$
$$18 - 5 \stackrel{?}{=} 12 + 1$$
$$13 = 13 \checkmark$$

10.4 Practice Problems

1. (a)
$$8n = 104$$
$$\frac{8n}{8} = \frac{104}{8}$$
$$n = 13$$

(b)
$$5n = 36$$
$$\frac{5n}{5} = \frac{36}{5}$$
$$n = \frac{36}{5}$$

2. (a)
$$\frac{-7n}{-7} = \frac{35}{-7}$$
$$n = -5$$

(b)
$$\frac{-9n}{-9} = \frac{-108}{-9}$$
$$n = 12$$

3. (a)
$$\frac{3.2x}{3.2} = \frac{16}{3.2}$$
$$x = 5$$

Check.
$$3.2(5) \stackrel{?}{=} 16$$
$$16 = 16 \checkmark$$

(b)
$$\frac{0.5y}{0.5} = \frac{2.9}{0.5}$$
$$y = 5.8$$

Check.
$$0.5(5.8) \stackrel{?}{=} 2.9$$
$$2.9 = 2.9 \checkmark$$

4. (a)
$$\frac{3}{5}x = \frac{1}{4}$$
$$\left(\frac{5}{3}\right)\left(\frac{3}{5}\right)x = \frac{5}{3}\left(\frac{1}{4}\right)$$
$$1x = \frac{5}{12}$$
$$x = \frac{5}{12}$$

Check.
$$\left(\frac{3}{5}\right)\left(\frac{5}{12}\right) \stackrel{?}{=} \frac{1}{4}$$
$$\frac{1}{4} = \frac{1}{4} \checkmark$$

(b)
$$\left(\frac{1}{6}\right)y = \frac{8}{3}$$
$$\left(\frac{6}{1}\right)\left(\frac{1}{6}\right)y = \left(\frac{6}{1}\right)\left(\frac{8}{3}\right)$$
$$1y = 16$$
$$y = 16$$

Check.
$$\left(\frac{1}{6}\right)(16) \stackrel{?}{=} 2\frac{2}{3}$$
$$\frac{16}{6} \stackrel{?}{=} \frac{8}{3}$$
$$\frac{8}{3} = \frac{8}{3} \checkmark$$

(c)
$$\left(3\frac{1}{5}\right)z = 4$$
$$\left(\frac{16}{5}\right)z = 4$$
$$\left(\frac{5}{1}\right)\left(\frac{16}{5}\right)z = \left(\frac{5}{1}\right)(4)$$
$$16z = 20$$
$$\frac{16z}{16} = \frac{20}{16}$$
$$z = \frac{5}{4}$$

Check.
$$\left(3\frac{1}{5}\right)\left(\frac{5}{4}\right) \stackrel{?}{=} 4$$
$$\left(\frac{16}{5}\right)\left(\frac{5}{4}\right) \stackrel{?}{=} 4$$
$$4 = 4 \checkmark$$

10.5 Practice Problems

1.
$$5x + 13 = 33$$
$$5x + 13 + (-13) = 33 + (-13)$$
$$5x = 20$$
$$\frac{5x}{5} = \frac{20}{5}$$
$$x = 4$$

Check.
$$5(4) + 13 \stackrel{?}{=} 33$$
$$20 + 13 \stackrel{?}{=} 33$$
$$33 = 33 \checkmark$$

2.
$$7x - 8 = -50$$
$$7x - 8 + 8 = -50 + 8$$
$$7x = -42$$
$$\frac{7x}{7} = \frac{-42}{7}$$
$$x = -6$$

Check.
$$7(-6) - 8 \stackrel{?}{=} -50$$
$$-42 - 8 \stackrel{?}{=} -50$$
$$-50 = -50 \checkmark$$

3.
$$4x = -8x + 42$$
$$4x + 8x = -8x + 8x + 42$$
$$12x = 42$$
$$\frac{12x}{12} = \frac{42}{12}$$
$$x = \frac{7}{2}$$

4.
$$4x - 7 = 9x + 13$$
$$4x - 7 + 7 = 9x + 13 + 7$$
$$4x = 9x + 20$$
$$4x + (-9x) = 9x + (-9x) + 20$$
$$-5x = 20$$
$$\frac{-5x}{5} = \frac{20}{-5}$$
$$x = -4$$

5. (a)
$$4x - 23 = 3x + 7 - 2x$$
$$4x - 23 = x + 7$$
$$4x + (-1x) - 23 = (-1x) + x + 7$$
$$3x - 23 = 7$$
$$3x - 23 + 23 = 7 + 23$$
$$3x = 30$$
$$x = 10$$

(b)
$$7y - 5 + 2y = 2 + 3y + 7$$
$$9y - 5 = 9 + 3y$$
$$9y + (-3y) - 5 = 9 + 3y + (-3y)$$
$$6y - 5 = 9$$
$$6y - 5 + 5 = 9 + 5$$
$$6y = 14$$
$$\frac{6y}{6} = \frac{14}{6}$$
$$y = \frac{7}{3}$$

6.
$$(3)(x - 2) + 5x = 7x + 10$$
$$3x - 6 + 5x = 7x + 10$$
$$8x - 6 = 7x + 10$$
$$8x - 6 + 6 = 7x + 10 + 6$$
$$8x = 7x + 16$$
$$8x + (-7x) = 7x + (-7x) + 16$$
$$x = 16$$

10.6 Practice Problems

1. $t = 7 + a$

2. $n = m - 24$

3. $t = 5 + f$

4. $l = 7 + 3w$

5. Let s = length in miles of Sally's trip
$s + 380$ = length in miles of Melinda's trip

6. Let m = height in feet of McCormick Hall
$m - 126$ = height in feet of Larson Center

7. Let x = cost in dollars of a chair
$x + 215$ = cost in dollars of a sofa

8. Let x = length in inches of the first side of the triangle
$2x$ = length in inches of the second side of the triangle
$x + 6$ = length in inches of the third side of the triangle

10.7 Practice Problems

1. Let x = length in feet of shorter piece of board
$x + 4.5$ = length in feet of longer piece of board
$$x + x + 4.5 = 18$$
$$2x + 4.5 = 18$$
$$2x = 13.5$$
$$x = 6.75$$
The shorter piece is 6.75 feet long.
$x + 4.5 = 6.75 + 4.5 = 11.25$
The longer piece is 11.25 feet long.

2. Let x = cost in dollars of a sofa
$x - 186$ = cost in dollars of an easy chair
$$x + x - 186 = 649$$
$$2x - 186 = 649$$
$$2x = 835$$
$$x = 417.50$$
The sofa cost \$417.50.
$x - 186 = 417.50 - 186.00 = 231.50$
The easy chair cost \$231.50.

3. Let x = the number of departures on Monday
$x + 29$ = the number of departures on Tuesday
$x - 16$ = the number of departures on Wednesday
$$x + x + 29 + x - 16 = 349$$
$$3x + 13 = 349$$
$$3x = 336$$
$$x = 112$$
$$x + 29 = 112 + 29 = 141$$
$$x - 16 = 112 - 16 = 96$$
There were 112 departures on Monday, 141 departures on Tuesday, and 96 departures on Wednesday.

4. Let w = the width of the field measured in feet
$2w + 8$ = the length of the field measured in feet
$$2(\text{width}) + 2(\text{length}) = 772$$
$$2(w) + 2(2w + 8) = 772$$
$$2w + 4w + 16 = 772$$
$$6w + 16 = 772$$
$$6w = 756$$
$$w = 126$$
The width of the field is 126 feet.
$2w + 8 = 2(126) + 8 = 252 + 8 = 260$
The length of the field is 260 feet.

5. Let x = the length in meters of the first side of the triangle
$2x$ = the length in meters of the second side of the triangle
$x + 10$ = the length in meters of the third side of the triangle
$$x + 2x + x + 10 = 36$$
$$4x + 10 = 36$$
$$4x = 26$$
$$x = 6.5$$
The first side of the triangle is 6.5 meters in length.
$2x = 2(6.5) = 13,\ x + 10 = 6.5 + 10 = 16.5$
The second side of the triangle is 13 meters in length.
The third side of the triangle is 16.5 meters in length.

6. Let x = the measure of angle A in degrees
$3x$ = the measure of angle C in degrees
$x - 30$ = the measure of angle B in degrees
$$x + 3x + x - 30 = 180$$
$$5x - 30 = 180$$
$$5x = 210$$
$$x = 42$$
Angle A measures 42 degrees.
$3x = 3(42) = 126,\ x - 30 = 42 - 30 = 12$
Angle C measures 126 degrees. Angle B measures 12 degrees.

7. Let s = the total amount of sales of the boats in dollars
$0.03s$ = the amount of the commission earned on sales of s dollars
$$3250 = 1000 + 0.03s$$
$$2250 = 0.03s$$
$$\frac{2250}{0.03} = \frac{0.03s}{0.03}$$
$$75{,}000 = s$$
Therefore he sold \$75,000 worth of boats for the month.

Selected Answers

Chapter 1

Pretest Chapter 1

1. Seventy-eight million, three hundred ten thousand, four hundred thirty-six **2.** 30,000 + 8,000 + 200 + 40 + 7 **3.** 5,064,122 **4.** 41,217,000 **5.** 2,237,000 **6.** 244 **7.** 50,570 **8.** 14,729 **9.** 2111 **10.** 76,311 **11.** 1,981,652 **12.** 108 **13.** 18,606 **14.** 6734 **15.** 331,420 **16.** 10,605 **17.** 7376 R 1 **18.** 139 **19.** 6^4 **20.** 81 **21.** 29 **22.** 52 **23.** 23 **24.** 271,000 **25.** 26,500 **26.** 60,000,000 **27.** 2600 **28.** 30,000,000,000 **29.** 1166 miles **30.** $22.00 **31.** $299 **32.** **(a)** 148,192 **(b)** 684,933

1.1 Exercises

1. 6000 + 700 + 30 + 1 **3.** 100,000 + 8,000 + 200 + 70 + 6 **5.** 20,000,000 + 3,000,000 + 700,000 + 60,000 + 1000 + 300 + 40 + 5 **7.** 100,000,000 + 3,000,000 + 200,000 + 60,000 + 700 + 60 + 8 **9.** 671 **11.** 9863 **13.** 76,036 **15.** 706,200 **21.** Fifty-three **23.** Eight thousand, nine hundred thirty-six **25.** Thirty-six thousand, one hundred eighteen **27.** One hundred five thousand, two hundred sixty-one **29.** Twenty-three million, five hundred sixty-one thousand, two hundred forty-eight **31.** Four billion, three hundred two million, one hundred fifty-six thousand, two hundred **33.** 375 **35.** 56,281 **37.** 100,079,826 **39.** One thousand, nine hundred sixty-five **41.** 9 million or 9,000,000 **43.** 16 million or 16,000,000 **45.** $201 billion or $201,000,000,000 **47.** $380 billion or $380,000,000,000 **49.** **(a)** 5 **(b)** 2 **50.** **(a)** 1 **(b)** 3 **51.** **(a)** 3 **(b)** 4 **52.** **(a)** 9 **(b)** 7 **53.** 613,001,033,208,003 **54.** Three quintillion, six hundred eighty-two quadrillion, nine hundred sixty-eight trillion, nine billion, nine hundred thirty-one million, nine hundred sixty thousand, seven hundred forty-seven

1.2 Exercises

3.

+	3	5	4	8	0	6	7	2	9	1
2	5	7	6	10	2	8	9	4	11	3
7	10	12	11	15	7	13	14	9	16	8
5	8	10	9	13	5	11	12	7	14	6
3	6	8	7	11	3	9	10	5	12	4
0	3	5	4	8	0	6	7	2	9	1
4	7	9	8	12	4	10	11	6	13	5
1	4	6	5	9	1	7	8	3	10	2
8	11	13	12	16	8	14	15	10	17	9
6	9	11	10	14	6	12	13	8	15	7
9	12	14	13	17	9	15	16	11	18	10

5. 23 **7.** 26 **9.** 60 **11.** 99 **13.** 4579 **15.** 13,336 **17.** 13,861 **19.** 42,739 **21.** 121 **23.** 1132 **25.** 17,909 **27.** 11,579,426 **29.** 1,135,280,240 **31.** 2,303,820 **33.** 248 **35.** 20,909 **37.** $794 **39.** $7986 **41.** 855 feet **43.** 121,100,000 square miles **45.** 2,922,339 votes **47.** **(a)** 1134 students **(b)** 1392 students **49.** **(a)** $9553 **(b)** $7319 **(c)** $13,047 **50.** 12,117,289,695 **51.** 1161 **52.** Answers may vary. A sample is: You could not add the addends in reverse order to check the addition. **53.** Answers may vary. A sample is: You could not group the addends in groups that sum to 10 to make column addition easier. **54.** Seventy-six million, two hundred eight thousand, nine hundred forty-one **55.** One hundred twenty-one million, three hundred seventy-four **56.** 8,724,396 **57.** 9,051,719

1.3 Exercises

1. 6 **3.** 8 **5.** 1 **7.** 9 **9.** 16 **11.** 9 **13.** 7 **15.** 6 **17.** 3 **19.** 9 **21.** 61 $\begin{array}{r}37\\+\ 61\\\hline 98\end{array}$ **23.** 12 $\begin{array}{r}73\\+\ 12\\\hline 85\end{array}$

25. 31 $\begin{array}{r}95\\+\ 31\\\hline 126\end{array}$ **27.** 625 $\begin{array}{r}143\\+\ 625\\\hline 768\end{array}$ **29.** 1341 $\begin{array}{r}422\\+\ 1341\\\hline 1763\end{array}$ **31.** 11,191 $\begin{array}{r}13{,}205\\+11{,}191\\\hline 24{,}396\end{array}$ **33.** 553,101 $\begin{array}{r}433{,}201\\+\ 553{,}101\\\hline 986{,}302\end{array}$ **35.** $\begin{array}{r}19\\+\ 110\\\hline 129\end{array}$ Correct

37. $\begin{array}{r}1113\\+\ 5067\\\hline 6180\end{array}$ Correct **39.** $\begin{array}{r}6030\\-\ 5020\\\hline 1010\end{array}$ Incorrect **41.** $\begin{array}{r}98{,}763\\-\ 42{,}531\\\hline 56{,}232\end{array}$ Incorrect **43.** 46 **45.** 37 **47.** 75 **49.** 581 **51.** 3,296 **53.** 34,092 **55.** 2,314

57. 223,116 **59.** $x = 5$ **61.** $x = 21$ **63.** $x = 27$ **65.** He got 182 more votes. **67.** 6,984,914 **69.** He received $248. **71.** 961,235 people **73.** 3,686,434 people **75.** 320,317 people **77.** 29 homes **79.** 80 homes **81.** Between 1994 and 1995 **83.** Willow Creek and Irving **85.** If a and b represent the same number. For example, if $a = 10$ and $b = 10$. **86.** It is true for all a and b if $c = 0$. For example, if $a = 5$, $b = 3$, and $c = 0$. **87.** 21,222,415,864 **88.** 1,909,111,121,312 **89.** 8,466,084 **90.** Two hundred ninety-six thousand, three hundred eight **91.** 168 **92.** 1,119,534

1.4 Exercises

3.

×	6	2	3	8	0	5	7	9	1	4
5	30	10	15	40	0	25	35	45	5	20
7	42	14	21	56	0	35	49	63	7	28
1	6	2	3	8	0	5	7	9	1	4
0	0	0	0	0	0	0	0	0	0	0
6	36	12	18	48	0	30	42	54	6	24
2	12	4	6	16	0	10	14	18	2	8
3	18	6	9	24	0	15	21	27	3	12
8	48	16	24	64	0	40	56	72	8	32
4	24	8	12	32	0	20	28	36	4	16
9	54	18	27	72	0	45	63	81	9	36

5. 96 **7.** 208 **9.** 18,306 **11.** 303,612 **13.** 70 **15.** 522 **17.** 1630 **19.** 7609 **21.** 100,208 **23.** 942,808 **25.** 1560 **27.** 2,715,800 **29.** 482,000 **31.** 3,162,500,000 **33.** 8460 **35.** 63,600 **37.** 56,000,000 **39.** 31,682 **41.** 7884 **43.** 5696 **45.** 15,175 **47.** 41,537 **49.** 69,312 **51.** 148,567 **53.** 823,823 **55.** 41,830 **57.** 89,496 **59.** 217,980 **61.** 1,435,083 **63.** 4,097,115 **65.** 10,400 **67.** 9210 **69.** 70 **71.** 308 **73.** 700 **75.** 360 **77.** 0 **79.** 3216 square feet **81.** $12,075 **83.** $1200 **85.** $3192 **87.** 612 miles **89.** 1020 hamburgers **91.** $5,224,000,000 **93.** 54,000 **94.** 8 **95.** 8 **96.** 9 **97.** 7 **98.** No, it would not always be true. In our number system 62 = 60 + 2. But in roman numerals IV ≠ I + V. The digit system in roman numerals involves subtraction. Thus (XII) × (IV) ≠ (XII × I) + (XII × V). **99.** Yes. $5 \times (8 - 3) = 5 \times 8 - 5 \times 3$; $a \times (b - c) = (a \times b) - (a \times c)$ **100.** 6,756 **101.** 1249 **102.** $139 **103.** $59

1.5 Exercises

3. 7 **5.** 3 **7.** 5 **9.** 4 **11.** 3 **13.** 7 **15.** 9 **17.** 9 **19.** 7 **21.** 4 **23.** 7 **25.** 0 **27.** 9 **29.** 1 **31.** 5 R 2 **33.** 9 R 4 **35.** 25 R 3 **37.** 23 R 4 **39.** 32 **41.** 37 **43.** 322 R 1 **45.** 127 R 1 **47.** 753 **49.** 1357 R 4 **51.** 1757 R 5 **53.** 2478 R 3 **55.** 5 R 5 **57.** 5 R 7 **59.** 7 **61.** 160 R 10 **63.** 48 R 12 **65.** 615 R 11 **67.** 210 R 8 **69.** 202 R 7 **71.** 4 R 4 **73.** 7 R 26 **75.** 27 **77.** 730 pounds of feed per horse **79.** $245,192 per hour **81.** $46,179 **83.** 34 minutes **85.** a and b must represent the same number. For example, if $a = 12$, then $b = 12$. **87.** 5504 **88.** 1,038,490 **89.** 406,195 **90.** 66,844

Putting Your Skills to Work

1. $10,552 **3.** $18,760

1.6 Exercises

7. 6^4 **9.** 5^3 **11.** 8^4 **13.** 9^1 **15.** 16 **17.** 16 **19.** 216 **21.** 10,000 **23.** 1 **25.** 64 **27.** 81 **29.** 225 **31.** 343 **33.** 256 **35.** 1 **37.** 625 **39.** 1,000,000 **41.** 1024 **43.** 9 **45.** 2401 **47.** 82 **49.** 225 **51.** 520 **53.** $30 - 3 + 10$; $27 + 10 = 37$ **55.** $2 + 72 \div 2$; $2 + 36 = 38$ **57.** $100 \div 25 + 3$; $4 + 3 = 7$ **59.** $7 \times 9 + 4 - 8$; $63 + 4 - 8 = 59$ **61.** 304 **63.** $20 \div 20 = 1$ **65.** $950 \div 5 = 190$ **67.** $64 - 20 = 44$ **69.** $9 + 16 \div 4 = 9 + 4 = 13$ **71.** $42 - 4 \div 4 = 42 - 1 = 41$ **73.** $100 - 9 \times 4 = 100 - 36 = 64$ **75.** 56 **77.** $20 \times 3 \times 2 \div 2$; $120 \div 2 = 60$ **79.** $144 - 0$; 144 **81.** 49 **83.** $5 + 3 - 1$; 7 **85.** 61 **87.** $12 \div 2 \times 3 - 16$; $18 - 16 = 2$ **89.** $9 \times 6 \div 9 + 4 \times 3$; $6 + 12 = 18$ **91.** $4(4)^3 \div 8 = 4(64) \div 8 = 32$ **93.** $1200 - 8(3) \div 6 = 1200 - 4 = 1196$ **95.** $12 - 10 + 8 = 10$ **97.** $15 - 6 + 10$; 19 **99.** $15 + 21 - 20$; 16 **101.** 100,000 + 50,000 + 6000 + 300 + 10 + 2 **102.** 200,765,909 **103.** Two hundred sixty-one million, seven hundred sixty-three thousand, two **104.** 14,528 **105.** 1156 **106.** 90,000 **107.** 612 minutes; approximately 36,700 seconds

1.7 Exercises

3. 80 **5.** 70 **7.** 90 **9.** 530 **11.** 4240 **13.** 400 **15.** 2800 **17.** 7700 **19.** 8000 **21.** 1000 **23.** 28,000 **25.** 800,000 **27.** 15,000,000 stars **29.** **(a)** 143,000 **(b)** 143,400 **31.** **(a)** 3,700,000 square miles or 9,600,000 square kilometers **(b)** 3,710,000 square miles or 9,600,000 square kilometers

33. $\begin{array}{r}500\\200\\+\ 900\\\hline 1600\end{array}$ **35.** $\begin{array}{r}30\\80\\60\\+\ 30\\\hline 200\end{array}$ **37.** $\begin{array}{r}100{,}000\\50{,}000\\+\ 100{,}000\\\hline 250{,}000\end{array}$ **39.** $\begin{array}{r}600{,}000\\-\ 100{,}000\\\hline 500{,}000\end{array}$ **41.** $\begin{array}{r}80{,}000{,}000\\-\ 60{,}000{,}000\\\hline 20{,}000{,}000\end{array}$ **43.** $\begin{array}{r}30{,}000{,}000\\-20{,}000{,}000\\\hline 10{,}000{,}000\end{array}$ **45.** $\begin{array}{r}60\\\times\ 50\\\hline 3000\end{array}$ **47.** $\begin{array}{r}2000\\\times\ \ 7\\\hline 14{,}000\end{array}$

49. $\begin{array}{r}600{,}000\\\times\ \ 300\\\hline 180{,}000{,}000\end{array}$ **51.** $\begin{array}{r}1{,}000\\20\overline{)20{,}000}\end{array}$ **53.** $\begin{array}{r}5{,}000\\40\overline{)200{,}000}\end{array}$ **55.** $\begin{array}{r}14{,}000\\500\overline{)7{,}000{,}000}\end{array}$ **57.** Incorrect $\begin{array}{r}400\\500\\900\\+\ 200\\\hline 2000\end{array}$ **59.** Incorrect $\begin{array}{r}100{,}000\\50{,}000\\+\ \ 40{,}000\\\hline 190{,}000\end{array}$

61. Correct $\begin{array}{r}300{,}000\\-\ 100{,}000\\\hline 200{,}000\end{array}$ **63.** Incorrect $\begin{array}{r}80{,}000{,}000\\-\ 50{,}000{,}000\\\hline 30{,}000{,}000\end{array}$ **65.** Incorrect $\begin{array}{r}200\\\times\ \ 20\\\hline 4000\end{array}$ **67.** Correct $\begin{array}{r}6000\\\times\ \ \ 70\\\hline 420{,}000\end{array}$ **69.** $\begin{array}{r}5000\\80\overline{)400{,}000}\end{array}$ Correct **71.** $\begin{array}{r}500\\400\overline{)200{,}000}\end{array}$ Correct

73. 4000 square yards **75.** \$1500 **77.** 5400 sit-ups **79.** Estimate = \$170; exact amount = \$145 **81.** About \$16 per week **83.** **(a)** 400,000 hours **(b)** 20,000 days **84.** **(a)** 300,000 hours **(b)** 15,000 days **85.** 83 **86.** 27 **87.** 28 **88.** 66

1.8 Exercises

1. 25,231 **3.** \$4668 **5.** 3750 hors d'oeuvres **7.** 6¢ per ounce **9.** \$23 **11.** 9 hours. This is equivalent to 540 minutes. **13.** 800,000 people **15.** \$20,382 **17.** \$786 **19.** Her balance will be \$1482. **21.** \$16,405 **23.** 25 miles per gallon **25.** 343 **26.** 21 **27.** 4788 **28.** 258

Chapter 1 Review Problems

1. Three hundred seventy-six **2.** Five thousand, eighty-two **3.** One hundred nine thousand, two hundred seventy-six **4.** Four hundred twenty-three million, five hundred seventy-six thousand, fifty-five **5.** 4000 + 300 + 60 + 4 **6.** 20,000 + 7000 + 900 + 80 + 6 **7.** 1,000,000 + 300,000 + 5000 + 100 + 20 + 8 **8.** 40,000,000 + 2,000,000 + 100,000 + 60,000 + 6000 + 30 + 7 **9.** 924 **10.** 6095 **11.** 1,328,828 **12.** 45,092,651 **13.** 130 **14.** 115 **15.** 690 **16.** 150 **17.** 400 **18.** 1469 **19.** 1598 **20.** 125,423 **21.** 14,703 **22.** 10,582 **23.** 17 **24.** 6 **25.** 27 **26.** 171 **27.** 159 **28.** 4828 **29.** 3026 **30.** 63,146 **31.** 224,757 **32.** 1,230,691 **33.** 36 **34.** 114 **35.** 0 **36.** 24 **37.** 18 **38.** 64 **39.** 105 **40.** 120 **41.** 144 **42.** 0 **43.** 240 **44.** 168 **45.** 2,612,100 **46.** 84,312,000 **47.** 83,200,000 **48.** 563,000,000 **49.** 864 **50.** 1856 **51.** 4050 **52.** 13,680 **53.** 25,524 **54.** 24,096 **55.** 87,822 **56.** 268,513 **57.** 543,510 **58.** 255,068 **59.** 150,000 **60.** 240,000 **61.** 7,200,000 **62.** 7,500,000 **63.** 2,000,000,000 **64.** 12,000,000,000 **65.** 2 **66.** 5 **67.** 14 **68.** 4 **69.** 0 **70.** 12 **71.** 7 **72.** 0 **73.** 7 **74.** 7 **75.** not possible **76.** 4 **77.** 7 **78.** 6 **79.** 8 **80.** not possible **81.** 125 **82.** 125 **83.** 258 **84.** 369 **85.** 25,874 **86.** 3692 **87.** 36,958 **88.** 36,921 **88.** 15,986 R 2 **90.** 35,783 R 4 **91.** 7 R 21 **92.** 4 R 37 **93.** 31 R 15 **94.** 14 R 11 **95.** 38 R 30 **96.** 54 R 38 **97.** 258 **98.** 95 **99.** 54 **100.** 19 **101.** 13^2 **102.** 24^2 **103.** 8^5 **104.** 9^5 **105.** 64 **106.** 81 **107.** 125 **108.** 128 **109.** 49 **110.** 81 **111.** 216 **112.** 256 **113.** 11 **114.** 8 **115.** 22 **116.** 66 **117.** 107 **118.** 76 **119.** 17 **120.** 48 **121.** 44 **122.** 93 **123.** 5670 **124.** 1280 **125.** 15,310 **126.** 42,640 **127.** 12,000 **128.** 23,000 **129.** 676,000 **130.** 202,000 **131.** 5,700,000 **132.** 10,000,000

133. $\begin{array}{r}600\\600\\900\\+\ 900\\\hline 3000\end{array}$ **134.** $\begin{array}{r}30{,}000\\40{,}000\\+\ 70{,}000\\\hline 140{,}000\end{array}$ **135.** $\begin{array}{r}30{,}000\\-\ 20{,}000\\\hline 10{,}000\end{array}$ **136.** $\begin{array}{r}4{,}000{,}000\\-\ 3{,}000{,}000\\\hline 1{,}000{,}000\end{array}$ **137.** $\begin{array}{r}2000\\\times\ 6000\\\hline 12{,}000{,}000\end{array}$ **138.** $\begin{array}{r}3{,}000{,}000\\\times\ 900\\\hline 2{,}700{,}000{,}000\end{array}$ **139.** $\begin{array}{r}4{,}000\\20\overline{)80{,}000}\end{array}$

140. $\begin{array}{r}20{,}000\\400\overline{)8{,}000{,}000}\end{array}$ **141.** Correct $\begin{array}{r}90\\90\\40\\+\ 60\\\hline 280\end{array}$ **142.** Correct $\begin{array}{r}900{,}000\\-\ 400{,}000\\\hline 500{,}000\end{array}$ **143.** Incorrect $\begin{array}{r}200{,}000\\\times\ 5000\\\hline 1{,}000{,}000{,}000\end{array}$ **144.** Correct $\begin{array}{r}30{,}000\\30\overline{)900{,}000}\end{array}$

145. 175 words **146.** 204 cans **147.** \$59,470 **148.** 2397 km **149.** \$7028 **150.** 10,301 feet **151.** \$64 per share **152.** \$1356 **153.** \$334 **154.** \$278 **155.** 25 miles per gallon **156.** 24 miles per gallon **157.** \$5041 **158.** \$3456

Chapter 1 Test

1. forty-four million, seven thousand, six hundred thirty-five **2.** 20,000 + 6000 + 800 + 50 + 9 **3.** 3,581,076 **4.** 700 **5.** 1045 **6.** 318,977 **7.** 7419 **8.** 38,341 **9.** 5,225,768 **10.** 378 **11.** 2184 **12.** 91,875 **13.** 129,437 **14.** 3014 R 1 **15.** 2189 **16.** 352 **17.** 11^3 **18.** 64 **19.** 23 **20.** 78 **21.** 79 **22.** 26,450 **23.** 6,460,000 **24.** 3,600,000 **25.** 150,000,000,000 **26.** 18,000 **27.** \$2148 **28.** 467 feet **29.** \$127 **30.** \$292

Chapter 2

Pretest Chapter 2

1. $\frac{3}{8}$ **2.** Answers will vary **3.** $\frac{17}{124}$ **4.** $\frac{1}{6}$ **5.** $\frac{1}{3}$ **6.** $\frac{1}{7}$ **7** $\frac{5}{7}$ **8.** $\frac{4}{11}$ **9.** $\frac{9}{7}$ **10.** $\frac{55}{9}$ **11.** $24\frac{1}{4}$ **12.** $5\frac{4}{5}$
13. $2\frac{2}{17}$ **14.** $\frac{5}{44}$ **15.** $\frac{2}{3}$ **16.** $67\frac{5}{6}$ **17.** 1 **18.** $\frac{1}{2}$ **19.** $5\frac{4}{13}$ **20.** $4\frac{2}{3}$ **21.** 8 **22.** 45 **23.** 55 **24.** 72 **25.** $\frac{19}{72}$
26. $\frac{49}{72}$ **27.** $\frac{13}{3}$ or $4\frac{1}{3}$ **28.** $3\frac{41}{42}$ **29.** $\frac{77}{18}$ or $4\frac{5}{18}$ **30.** $7\frac{1}{6}$ miles **31.** $7\frac{11}{36}$ tons **32.** 18 students

2.1 Exercises

5. N: 3; D: 5 **7.** N: 2; D: 3 **9.** N: 1; D: 17 **11.** $\frac{1}{2}$ **13.** $\frac{5}{6}$ **15.** $\frac{2}{3}$ **17.** $\frac{5}{6}$ **19.** $\frac{1}{4}$ **21.** $\frac{3}{10}$ **23.** $\frac{5}{8}$ **25.** $\frac{4}{7}$
27. $\frac{7}{8}$ **29.** $\frac{2}{5}$ **31.**  **33.**

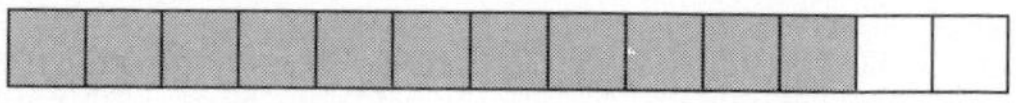

35. 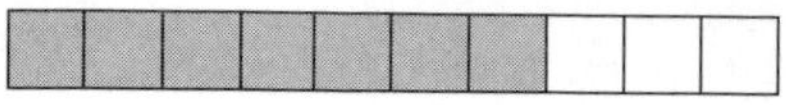 **37.** $\frac{29}{165}$ **39.** $\frac{209}{750}$ **41.** $\frac{89}{211}$ **43.** $\frac{17}{50}$ **45.** $\frac{9}{14}$ **47.** **(a)** $\frac{90}{195}$ **(b)** $\frac{22}{195}$

49. The amount of money each of 6 business owners gets if the business has a profit of \$0. **50.** We cannot do it. Division by zero is undefined.
51. 241 **52.** 11,106 **53.** 119,944 **54.** 177 R 6

2.2 Exercises

7. 3×5 **9.** 2×3 **11.** 7^2 **13.** 2^6 **15.** 5×11 **17.** 5×7 **19.** 3×5^2 **21.** 2×3^3 **23.** $2^2 \times 3 \times 7$ **25.** 2×7^2
27. Prime **29.** 3×19 **31.** Prime **33.** 2×31 **35.** Prime **37.** Prime **39.** 11×11 **41.** 7×23 **43.** $\frac{18 \div 9}{27 \div 9} = \frac{2}{3}$
45. $\frac{32 \div 16}{48 \div 16} = \frac{2}{3}$ **47.** $\frac{30 \div 6}{48 \div 6} = \frac{5}{8}$ **49.** $\frac{42 \div 6}{48 \div 6} = \frac{7}{8}$ **51.** $\frac{3 \times 1}{3 \times 5} = \frac{1}{5}$ **53.** $\frac{3 \times 13}{2 \times 2 \times 13} = \frac{3}{4}$ **55.** $\frac{2 \times 3 \times 3}{2 \times 2 \times 2 \times 3} = \frac{3}{4}$
57. $\frac{3 \times 3 \times 3}{3 \times 3 \times 5} = \frac{3}{5}$ **59.** $\frac{3 \times 11}{3 \times 12} = \frac{11}{12}$ **61.** $\frac{5 \times 13}{13 \times 13} = \frac{5}{13}$ **63.** $\frac{11 \times 8}{11 \times 11} = \frac{8}{11}$ **65.** $\frac{3 \times 50}{4 \times 50} = \frac{3}{4}$ **67.** $\frac{7 \times 17}{7 \times 30} = \frac{17}{30}$
69. Yes **71.** Yes **73.** No **75.** No **77.** Yes **79.** $\frac{2}{3}$ **81.** $\frac{6}{19}$ **83.** $\frac{1}{9}$ **85.** 164,050 **86.** 1296 **87.** 1,630,000
88. 25,920,000,000

2.3 Exercises

1. $\frac{7}{2}$ **3.** $\frac{14}{3}$ **5.** $\frac{17}{7}$ **7.** $\frac{53}{10}$ **9.** $\frac{29}{6}$ **11.** $\frac{65}{3}$ **13.** $\frac{55}{6}$ **15.** $\frac{169}{6}$ **17.** $\frac{131}{12}$ **19.** $\frac{79}{10}$ **21.** $\frac{201}{25}$ **23.** $\frac{211}{2}$
25. $\frac{494}{3}$ **27.** $\frac{119}{15}$ **29.** $\frac{138}{25}$ **31.** $1\frac{2}{5}$ **33.** $2\frac{3}{4}$ **35.** $2\frac{1}{2}$ **37.** $3\frac{3}{8}$ **39.** 5 **41.** $9\frac{5}{9}$ **43.** $2\frac{2}{13}$ **45.** $3\frac{3}{16}$
47. $9\frac{1}{3}$ **49.** $17\frac{1}{2}$ **51.** 13 **53.** $22\frac{2}{9}$ **55.** 6 **57.** $36\frac{7}{11}$ **59.** $2\frac{3}{4}$ **61.** $4\frac{1}{6}$ **63.** $12\frac{3}{8}$ **65.** 4 **67.** $\frac{12}{5}$ **69.** $\frac{26}{3}$
71. $2\frac{88}{126} = 2\frac{44}{63}$ **73.** $2\frac{138}{424} = 2\frac{69}{212}$ **75.** $2\frac{250}{350} = 2\frac{5}{7}$ **77.** 1,082 **79.** $50\frac{1}{3}$ acres **81.** No. 101 is prime and is not a factor of 5687. **82.** No. 157 is prime and not a factor of 9810. **83.** 28,567 **84.** 260,247 **85.** 16,000,000,000 **86.** 3000

2.4 Exercises

1. $\frac{21}{55}$ **3.** $\frac{15}{52}$ **5.** 1 **7.** $\frac{24}{35}$ **9.** $\frac{5}{12}$ **11.** $\frac{21}{8}$ *or* $2\frac{5}{8}$ **13.** $\frac{24}{7}$ *or* $3\frac{3}{7}$ **15.** $\frac{5}{2}$ *or* $2\frac{1}{2}$ **17.** $\frac{4}{15}$ **19.** $\frac{1}{2}$
21. $\frac{14}{5}$ *or* $2\frac{4}{5}$ **23.** $\frac{55}{12}$ *or* $4\frac{7}{12}$ **25.** 15 **27.** $\frac{69}{50}$ *or* $1\frac{19}{50}$ **29.** 0 **31.** $\frac{154}{3}$ *or* $51\frac{1}{3}$ **33.** $\frac{53}{15}$ *or* $3\frac{8}{15}$ **35.** $11\frac{5}{7}$
37. $x = \frac{9}{5}$ **39.** $x = \frac{11}{13}$ **41.** $37\frac{11}{12}$ square miles **43.** 134 miles **45.** 146 yards **47.** $9\frac{3}{16}$ ounces **49.** $\$6141\frac{1}{4}$
51. **(a)** $\frac{5}{8}$ of the students **(b)** 5375 students voted **53.** 529 cars **54.** 368 calls **55.** 1752 lines **56.** 173,040 gallons

2.5 Exercises

1. $\frac{4}{35}$ **3.** $\frac{2}{3}$ **5.** $\frac{9}{10}$ **7.** $\frac{3}{2}$ *or* $1\frac{1}{2}$ **9.** $\frac{4}{5}$ **11.** $\frac{16}{7}$ *or* $2\frac{2}{7}$ **13.** $\frac{4}{3}$ *or* $1\frac{1}{3}$ **15.** 1 **17.** $\frac{9}{49}$ **19.** $\frac{36}{7}$ *or* $5\frac{1}{7}$ **21.** 0
23. Cannot be done **25.** $\frac{3}{4}$ **27.** $\frac{4}{5}$ **29.** 1 **31.** 16 **33.** $\frac{7}{32}$ **35.** 5000 **37.** $\frac{2}{15}$ **39.** $\frac{7}{40}$ **41.** $\frac{32}{3}$ *or* $10\frac{2}{3}$
43. $\frac{7}{18}$ **45.** $\frac{2}{3}$ **47.** 4 **49.** $\frac{15}{7}$ *or* $2\frac{1}{7}$ **51.** 3 **53.** $\frac{63}{32}$ *or* $1\frac{31}{32}$ **55.** $\frac{1}{2}$ **57.** 12 **59.** $\frac{2}{3}$ **61.** $\frac{5}{21}$
63. $\frac{32}{15}$ *or* $2\frac{2}{15}$ **65.** $x = \frac{7}{5}$ **67.** $x = \frac{4}{7}$ **69.** $2\frac{1}{4}$ gallons **71.** $37\frac{1}{2}$ miles per hour **73.** 28 flags **75.** $a = 2, b = 3, c = 4, d = 6.$
$\frac{2}{3} \div \frac{4}{6} = \frac{4}{6} \div \frac{2}{3}$. In general, if $\frac{a}{b} = \frac{c}{d}$, then it is true. **76.** $3\frac{7}{12} \div \frac{1}{4} = \left(3 + \frac{7}{12}\right) \times 4 = 12 + \frac{28}{12} = 12 + \frac{7}{3} = 14\frac{1}{3}$
77. Thirty-nine million, five hundred seventy-six thousand, three hundred four. **78.** 400,000 + 50,000 + 9000 + 200 + 70 + 3 **79.** 1099
80. 87,595,631

2.6 Exercises

1. 10 **3.** 28 **5.** 35 **7.** 18 **9.** 48 **11.** 16 **13.** 90 **15.** 60 **17.** 105 **19.** 36 **21.** 6 **23.** 120
25. 144 **27.** 84 **29.** 120 **31.** 3 **33.** 45 **35.** 20 **37.** 14 **39.** 96 **41.** 39 **43.** $\frac{21}{36}$ and $\frac{20}{36}$
45. $\frac{24}{200}$ and $\frac{35}{200}$ **47.** $\frac{114}{150}$ and $\frac{52}{150}$ **49.** LCD = 42; $\frac{30}{42}$ and $\frac{7}{42}$ **51.** LCD = 48; $\frac{20}{48}$ and $\frac{3}{48}$ **53.** LCD = 80; $\frac{52}{80}$ and $\frac{55}{80}$
55. LCD = 60; $\frac{35}{60}$ and $\frac{46}{60}$ **57.** LCD = 72; $\frac{15}{72}, \frac{22}{72}, \frac{3}{72}$ **59.** LCD = 56; $\frac{3}{56}, \frac{49}{56}, \frac{40}{56}$ **61.** LCD = 63; $\frac{5}{63}, \frac{12}{63}, \frac{56}{63}$
63. (a) 16 **(b)** $\frac{3}{16}, \frac{12}{16}, \frac{6}{16}$ **65.** 178 R 3 **66.** 3624 **72.** 9963

2.7 Exercises

1. $\frac{7}{9}$ **3.** $\frac{11}{15}$ **5.** $\frac{20}{23}$ **7.** $\frac{17}{44}$ **9.** $\frac{7}{12}$ **11.** $\frac{9}{20}$ **13.** $\frac{7}{8}$ **15.** $\frac{3}{2}$ *or* $1\frac{1}{2}$ **17.** $\frac{37}{100}$ **19.** $\frac{26}{175}$ **21.** $\frac{31}{24}$ *or* $1\frac{7}{24}$
23. $\frac{27}{40}$ **25.** $\frac{1}{4}$ **27.** $\frac{8}{35}$ **29.** $\frac{9}{16}$ **31.** $\frac{11}{60}$ **33.** $\frac{1}{4}$ **35.** $\frac{1}{2}$ **37.** $\frac{5}{48}$ **39.** 0 **41.** 0 **43.** $\frac{8}{5}$ *or* $1\frac{3}{5}$ **45.** $\frac{11}{30}$
47. $1\frac{3}{5}$ **49.** $\frac{3}{14}$ **51.** $\frac{5}{33}$ **53.** $\frac{8}{15}$ **55.** $1\frac{7}{15}$ cup **57.** $1\frac{1}{2}$ pounds **59.** $1\frac{23}{36}$ inches **61.** $7\frac{7}{12}$ cups
63. $\frac{4}{21}$ of the membership **65.** $\frac{3}{17}$ **66.** $\frac{2}{3}$ **67.** $7\frac{11}{16}$ **68.** $\frac{101}{7}$

2.8 Exercises

1. $9\frac{3}{4}$ **3.** $4\frac{1}{7}$ **5.** $17\frac{1}{2}$ **7.** $16\frac{1}{10}$ **9.** $\frac{4}{7}$ **11.** $2\frac{17}{24}$ **13.** $16\frac{1}{4}$ **15.** $3\frac{1}{2}$ **17.** $7\frac{5}{12}$ **19.** $14\frac{4}{7}$ **21.** $41\frac{4}{5}$
23. $6\frac{7}{12}$ **25.** $16\frac{2}{3}$ **27.** $73\frac{37}{40}$ **29.** $5\frac{1}{2}$ **31.** 0 **33.** $4\frac{41}{60}$ **35.** $8\frac{8}{15}$ **37.** $102\frac{5}{8}$ **39.** $13\frac{5}{24}$ **41.** $43\frac{1}{8}$ miles
43. $2\frac{1}{24}$ pounds **45. (a)** $9\frac{19}{24}$ pounds **(b)** $6\frac{5}{24}$ pounds **47.** $\frac{2607}{40}$ *or* $65\frac{7}{40}$ **48.** $\frac{277}{42}$ *or* $6\frac{25}{42}$ **49.** $\frac{941}{21}$ *or* $44\frac{17}{21}$
50. $\frac{589}{9}$ *or* $65\frac{4}{9}$ **51.** 111,303 **52.** 38,912 **53.** 512,012 **54.** 8,529,300

Putting Your Skills to Work

1. $16,600 **2.** $5060

2.9 Exercises

1. $16\frac{7}{24}$ tons **3.** $25\frac{11}{24}$ tons **5.** $1\frac{9}{16}$ inches **7.** 17 pieces **9.** $\$93\frac{1}{3}$ **11.** $275\frac{5}{8}$ gallons **13.** $106\frac{7}{8}$ nautical miles
15. $451 per week is left **17. (a)** 42 bags **(b)** $147 **(c)** $145 profit **19. (a)** $14\frac{1}{8}$ ounces of bread **(b)** $\frac{5}{8}$ ounce
21. (a) $30\frac{1}{2}$ knots **(b)** 7 hours **23. (a)** 5485 bushels **(b)** $11{,}998\frac{7}{16}$ cubic feet **(c)** $9598\frac{3}{4}$ bushels **25.** 44,245 **26.** 27,814
27. 45,441 **28.** 278 R 3

Chapter 2 Review Problems

1. $\frac{5}{12}$ **2.** $\frac{3}{8}$ **3.** Answers will vary. **4.** Answers will vary. **5.** $\frac{6}{31}$ **6.** $\frac{87}{100}$ **7.** $2 \times 3 \times 7$ **8.** 2×3^3 **9.** $2^3 \times 3 \times 7$ **10.** Prime **11.** $2 \times 3 \times 13$ **12.** Prime **13.** $\frac{1}{4}$ **14.** $\frac{2}{7}$ **15.** $\frac{7}{12}$ **16.** $\frac{13}{17}$ **17.** $\frac{7}{8}$ **18.** $\frac{17}{35}$ **19.** $\frac{35}{8}$ **20.** $\frac{65}{23}$ **21.** $2\frac{5}{7}$ **22.** $4\frac{11}{13}$ **23.** $3\frac{3}{11}$ **24.** $\frac{117}{8}$ **25.** $1\frac{29}{48}$ **26.** $\frac{20}{77}$ **27.** $\frac{7}{15}$ **28.** 0 **29.** $\frac{4}{63}$ **30.** $\frac{492}{5}$ *or* $98\frac{2}{5}$ **31.** $\frac{817}{40}$ *or* $20\frac{17}{40}$ **32.** $\frac{51}{2}$ *or* $25\frac{1}{2}$ **33.** $24\frac{1}{2}$ **34.** $\$667\frac{1}{4}$ **35.** $486\frac{17}{20}$ square inches **36.** $\frac{15}{14}$ *or* $1\frac{1}{14}$ **37.** $\frac{5}{34}$ **38.** 1920 **39.** $\frac{27}{40}$ **40.** $\frac{25}{6}$ *or* $4\frac{1}{6}$ **41.** $\frac{17}{164}$ **42.** 0 **43.** $\frac{46}{33}$ *or* $1\frac{13}{33}$ **44.** $186\frac{2}{3}$ calories **45.** 7 dresses **46.** 98 **47.** 120 **48.** 90 **49.** $\frac{24}{56}$ **50.** $\frac{33}{72}$ **51.** $\frac{36}{172}$ **52.** $\frac{187}{198}$ **53.** $\frac{1}{14}$ **54.** $\frac{13}{12}$ *or* $1\frac{1}{12}$ **55.** $\frac{85}{63}$ *or* $1\frac{22}{63}$ **56.** $\frac{11}{40}$ **57.** $\frac{23}{70}$ **58.** $\frac{25}{36}$ **59.** $\frac{61}{75}$ **60.** $\frac{19}{48}$ **61.** $\frac{6}{23}$ **62.** $5\frac{4}{9}$ **63.** $8\frac{2}{3}$ **64.** $20\frac{5}{7}$ **65.** $\frac{39}{8}$ *or* $4\frac{7}{8}$ **66.** $\frac{209}{48}$ *or* $4\frac{17}{48}$ **67.** $\frac{310}{3}$ *or* $103\frac{1}{3}$ **68.** $\frac{857}{12}$ *or* $71\frac{5}{12}$ **69.** $8\frac{29}{40}$ miles **70.** $\$1\frac{5}{8}$ **71.** $1\frac{2}{3}$ cups sugar, $2\frac{1}{8}$ cups flour **72.** $206\frac{1}{8}$ miles **73.** 15 lengths **74.** $9\frac{5}{8}$ liters **75.** 30 words per minute **76.** $\$56\frac{1}{4}$ **77.** $\$8\frac{3}{4}$ **78.** $1\frac{1}{16}$ inch **79.** $242 **80.** **(a)** 25 miles per gallon **(b)** $\$22\frac{2}{25}$

Chapter 2 Test

1. $\frac{3}{5}$ **2.** $\frac{311}{388}$ **3.** $\frac{3}{14}$ **4.** $\frac{3}{7}$ **5.** $\frac{9}{2}$ **6.** $\frac{34}{5}$ **7.** $8\frac{1}{7}$ **8.** 12 **9.** $\frac{14}{45}$ **10.** $31\frac{1}{5}$ **11.** $\frac{77}{40}$ or $1\frac{37}{40}$ **12.** $\frac{12}{7}$ or $1\frac{5}{7}$ **13.** $6\frac{12}{13}$ **14.** $\frac{39}{62}$ **15.** 72 **16.** 48 **17.** 24 **18.** $\frac{30}{72}$ **19.** $\frac{13}{36}$ **20.** $\frac{11}{20}$ **21.** $\frac{25}{28}$ **22.** $14\frac{6}{35}$ **23.** $4\frac{13}{14}$ **24.** $2\frac{4}{9}$ **25.** $21\frac{1}{4}$ **26.** 42 square yards **27.** 8 packages **28.** $\frac{7}{10}$ mile **29.** $14\frac{1}{24}$ miles **30.** **(a)** 40 oranges **(b)** $9.60 or 960¢

Cumulative Test for Chapters 1–2

1. Eighty-four million, three hundred sixty-one thousand, two hundred eight **2.** 869 **3.** 719,220 **4.** 2075 **5.** 17,216 **6.** 4788 **7.** 202,896 **8.** 4307 R 1 **9.** 369 **10.** 49 **11.** 6,037,000 **12.** 38 **13.** $174 **14.** $306 **15.** $\frac{55}{84}$ **16.** $\frac{7}{13}$ **17.** $\frac{75}{4}$ **18.** $14\frac{2}{7}$ **19.** $\frac{527}{48}$ *or* $10\frac{47}{48}$ **20.** $\frac{12}{35}$ **21.** 39 **22.** $\frac{31}{54}$ **23.** $\frac{71}{8}$ *or* $8\frac{7}{8}$ **24.** $\frac{113}{15}$ *or* $7\frac{8}{15}$ **25.** $\frac{13}{28}$ **26.** $23\frac{1}{8}$ tons **27.** $24\frac{3}{5}$ miles per gallon **28.** $8\frac{1}{8}$ cups of sugar; $5\frac{5}{6}$ cups of flour **29.** 60,000,000 miles

Chapter 3

Pretest Chapter 3

1. Forty-seven and eight hundred thirteen thousandths **2.** 0.0567 **3.** $2\frac{11}{100}$ **4.** $\frac{21}{40}$ **5.** 1.59, 1.6, 1.601, 1.61 **6.** 10.5 **7.** 1.053 **8.** 19.45 **9.** 27.191 **10.** 10.59 **11.** 8.892 **12.** 0.3501 **13.** 4780.5 **14.** 5.129 **15.** 0.354 **16.** 0.128 **17.** 0.4375 **18.** $0.2\overline{27}$ or 0.22727 . . . **19.** 0.97 **20.** 17.9 miles per gallon **21.** $489.51 **22.** $8.53

3.1 Exercises

5. Fifty-seven hundredths **7.** Three and eight tenths **9.** One hundred twenty-four thousandths **11.** Twenty-eight and seven ten thousandths **13.** Thirty-six and $\frac{18}{100}$ dollars **15.** One thousand two hundred thirty-six and $\frac{8}{100}$ dollars **17.** Ten thousand and $\frac{76}{100}$ dollars **19.** 0.7 **21.** 0.039 **23.** 0.0065 **25.** 0.000286 **27.** 0.3 **29.** 0.76 **31.** 0.771 **33.** 0.053 **35.** 0.0026 **37.** 8.3 **39.** 84.13 **41.** 1.019 **43.** 126.0571 **45.** $\frac{1}{50}$ **47.** $3\frac{3}{5}$ **49.** $\frac{121}{1000}$ **51.** $12\frac{5}{8}$ **53.** $7\frac{3}{2000}$ **55.** $235\frac{627}{5000}$ **57.** $\frac{187}{10,000}$ **59.** $8\frac{27}{2500}$ **61.** $289\frac{47}{125}$ **63.** $\frac{9889}{10,000}$ **65.** $\frac{1}{50,000,000}$ **67.** 818 **68.** 938 **69.** 56,800 **70.** 8,069,000

3.2 Exercises

1. > **3.** < **5.** < **7.** > **9.** < **11.** > **13.** < **15.** > **17.** 12.6, 12.65, 12.8 **19.** 0.007, 0.0071, 0.05 **21.** 1.1, 1.79, 1.8, 1.81 **23.** 26.003, 26.033, 26.034, 26.04 **25.** 18.006, 18.060, 18.065, 18.066, 18.606 **27.** 5.7 **29.** 29.5 **31.** 578.1 **33.** 2176.8 **35.** 26.03 **37.** 5.77 **39.** 156.12 **41.** 2786.71 **43.** 1.061 **45.** 0.0474 **47.** 5.00761 **49.** 0.007537 **51.** 129 **53.** $2537 **55.** $10,098 **57.** $56.98 **59.** $5783.72 **61.** 0.10 kilogram, 0.07 kilogram **63.** 0.0059, 0.006, 0.0519, $\frac{6}{100}$, 0.0601, 0.0612, 0.062, $\frac{6}{10}$, 0.61 **64.** 0.049, 0.0515, 0.052, 1.05, 1.051, 1.0513, $\frac{15}{10}$, $\frac{151}{100}$, 1.512 **65.** You should consider only one digit to the right of the decimal place that you wish to round to. 86.23498 is closer to 86.23 than to 86.24. **66.** The bank rounds up for any fractional part of a cent. **67.** $12\frac{1}{8}$ **68.** $10\frac{9}{20}$ **69.** 692 miles **70.** $9658

3.3 Exercises

1. 72.8 **3.** 1215.55 **5.** 7.025 **7.** 136.844 **9.** 323.9 **11.** 23.00 **13.** 36.7287 **15.** 67.42 **17.** 1112.21 **19.** 19.86 m **21.** 60.1 gallons **23.** $78.12 **25.** $1,411.97 **27.** 3.9 **29.** 27.17 **31.** 64.18 **33.** 508.313 **35.** 0.02151 **37.** 4.6465 **39.** 6.737 **41.** 1189.07 **43.** 1.4635 **45.** $33.50 **47.** 17.661 kilograms **49.** $36,947.16 more **51.** $45.30 **53.** 1.46 cm **55.** $601,409.07 revenue shortage **57.** 0.008 milligrams, no **59.** $5.7 billion; 5,700,000,000 dollars **61.** $13.3 billion; 13,300,000,000 dollars **63.** $x = 8.4$ **65.** $x = 43.7$ **67.** $x = 2.109$ **69.** 20,288 **70.** 18,213 **71.** $\frac{77}{25}$ *or* $3\frac{2}{25}$ **72.** $\frac{35}{4}$ *or* $8\frac{3}{4}$

Putting Your Skills to Work

1. $83.24 **2.**

	Check Number	Date	Transaction	Amount of Withdrawal (−)		Amount of Deposit (+)		New Balance	
								83	24
(a)		3/20	Service fee	3	50			79	74
(b)		3/22	Deposit			301	64	381	38
(c)	158	3/30	Sears	159	95			221	43
(d)		3/31	Deposit			450	36	671	79
(e)	159	4/2	Gulf Oil	46	52			625	27
(f)	160	4/3	Western Bell	39	64			585	63
(g)	161	4/14	IRS	561	18			24	45
(h)		4/18	New Checks	6	95			17	50

3. $290.20

3.4 Exercises

1. 0.12 **3.** 0.06 **5.** 0.00288 **7.** 0.00711 **9.** 0.002025 **11.** 0.6582 **13.** 2555.52 **15.** 0.013244 **17.** 768.1517 **19.** 4911.3 **21.** 441 **23.** 43,358.6 **25.** 0.000238 **27.** 6.5237 **29.** $234.00 **31.** $382.00 **33.** 235.48 square feet **35.** $664.20 **37.** $0.77 **39.** 514.8 miles **41.** 28.6 **43.** 70.1 **45.** 128,650 **47.** 56,098.2 **49.** 28,056,020 **51.** 816,320 **53.** 671.8 **55.** 81.376 **57.** 593.2 centimeters **59.** 2980 meters **61.** $36,405,000.00 **63.** $62,279.00 **65.** 98 **66.** 86 **67.** 125 R 4 **68.** 129 R 4

3.5 Exercises

1. 2.1 **3.** 0.0369 **5.** 18.31 **7.** 0.0565 **9.** 0.0029 **11.** 12.2 **13.** 64.3 **15.** 8.01 **17.** 21 **19.** 230 **21.** 5.3 **23.** 2.3 **25.** 33.8 **27.** 65.96 **29.** 12.24 **31.** 24.92 **33.** 31.020 **35.** 12.246 **37.** 50 **39.** 13 **41.** 6.2 ounces in each portion **43.** 20 miles per gallon **45.** $376.50 **47.** 9 payments **49.** There are 38 blocks of chalk in the box. The error was made by placing 2 fewer blocks of chalk in the box than what is listed on the label. **51.** 0.123 **53.** 69 **55.** 45 **57.** 810 **59.** 74 **60.** 41 **61.** $1\frac{31}{40}$ **62.** $\frac{15}{16}$ **63.** $\frac{91}{12}$ or $7\frac{7}{12}$ **64.** $\frac{5}{3}$ or $1\frac{2}{3}$

3.6 Exercises

5. 0.25 **7.** 0.875 **9.** 0.4375 **11.** 0.35 **13.** 0.62 **15.** 2.25 **17.** 2.125 **19.** 1.4375 **21.** $0.\overline{6}$ **23.** $0.\overline{5}$ **25.** $0.8\overline{3}$ **27.** $0.\overline{90}$ **29.** $0.9\overline{3}$ **31.** $0.\overline{15}$ **33.** $1.\overline{037}$ **35.** $2.2\overline{7}$ **37.** 0.571 **39.** 0.905 **41.** 0.146 **43.** 1.296 **45.** 0.247 **47.** 0.944 **49.** 3.143 **51.** 0.474 **53.** $0.208\overline{3}$ inch thick **55.** It is too small by 0.125 inch. **57.** Too thick, 0.00125 inch **59.** 5.078 **61.** 114 **63.** 374.65 **65.** 156.0664 **67.** 21.414 **69.** 2.5 **71.** 0.325 **73.** 28.6 **75.** 20.836 **77.** 0.586930 **79. (a)** $\begin{array}{r} 0.1616\overline{16} \\ -\ 0.0016\overline{16} \\ \hline 0.16 \end{array}$ **(b)** $\begin{array}{r} 0.1616\overline{16} \\ -\ 0.0166\overline{6} \\ \hline 0.144949 \end{array}$ **(c)** The repeating patterns line up differently. **80. (a)** $\begin{array}{r} 1.8989\overline{89} \\ -\ 0.0189\overline{89} \\ \hline 1.88 \end{array}$ **(b)** $\begin{array}{r} 1.898989\overline{89} \\ -\ 0.1899999\overline{9} \\ \hline 1.70898989 \end{array}$ **(c)** The repeating patterns line up differently. **81.** 312 square feet **82.** 869 votes **83.** $377 was deposited **84.** $83 per employee

3.7 Exercises

1. 621.2 pounds **3.** $8868.88 **5.** 7889.42 square meters **7.** 46 packages **9.** Yes. After she bought the eggs she had $0.33. **11.** 11.59 meters of rainfall per year **13.** 12 days **15.** $1263.09 **17.** $743.60 **19.** The balance is $426.55 **21.** $17,319. He will pay back $5819 more than the loan. **23.** $1872 **25.** Yes, by 0.149 milligram **27.** 137 minutes **29.** 17.7 quadrillion Btu **31.** Approximately 47.8 quadrillion Btu. This is the same as 47,800,000,000,000,000 Btu. **33.** $\frac{79}{70}$ or $1\frac{9}{70}$ **34.** $\frac{11}{34}$ **35.** $\frac{5}{9}$ **36.** 22

Chapter 3 Review Problems

1. Thirteen and six hundred seventy-two thousandths **2.** Eighty-four hundred thousandths **3.** 0.7 **4.** 0.81 **5.** 1.523 **6.** 0.0079 **7.** $\frac{17}{100}$ **8.** $\frac{73}{200}$ **9.** $26\frac{22}{25}$ **10.** $1\frac{1}{4000}$ **11.** > **12.** > **13.** 0.901, 0.918, 0.98, 0.981 **14.** 5.2, 5.26, 5.59, 5.6, 5.62 **15.** 0.6 **16.** 19.21 **17.** 1.100 **18.** $156.00 **19.** 6.639 **20.** 77.6 **21.** **(a)** 8.5 gallons **(b)** 9 gallons **22.** 3.894 **23.** 90.739 **24.** 236.9 miles **25.** 0.0228 **26.** 0.90118 **27.** 0.000364 **28.** 887.81 **29.** 86.4 **30.** 2398.02 **31.** 78.6 **32.** 15,637.1 **33.** 0.613 **34.** 123,540 **35.** $0.90 **36.** 24,075 **37.** 0.00258 **38.** 36.8 **39.** 232.9 **40.** 574.4 **41.** 0.059 **42.** $66.75 **43.** $0.2\overline{7}$ **44.** 0.175 **45.** $1.8\overline{3}$ **46.** 0.9375 **47.** 0.786 **48.** 2.294 **49.** 3.6 **50.** 14.1 **51.** 1.152 **52.** 1.301208 **53.** 87.13 **54.** 6 **55.** 0.2128 **56.** 0.55 **57.** 0.6 **58.** 13.122 **59.** 439.19 **60.** 0.402216 **61.** 0.8403 **62.** 69.2 **63.** 20.004 **64.** 1.25 **65.** $9.13 **66.** $84.46 **67.** $368.08 **68.** 24.8 miles **69.** $2170.30 **70.** He will earn more at ABC Company. **71.** no; by 0.0005 mg **72.** 7 test tubes

Chapter 3 Test

1. One hundred fifty-seven thousandths **2.** 0.3977 **3.** $7\frac{3}{20}$ **4.** $\frac{261}{1000}$ **5.** 2.19, 2.9, 2.907, 2.91 **6.** 78.66 **7.** 0.0342 **8.** 43.989 **9.** 37.53 **10.** 0.0979 **11.** 71.155 **12.** 0.5817 **13.** 218.9 **14.** 25.7 **15.** 47 **16.** $1.\overline{2}$ **17.** 0.5625 **18.** 1.487 **19.** 6.1952 **20.** $33.15 **21.** 18.8 miles per gallon **22.** 3.43 centimeters less

Cumulative Test for Chapters 1–3

1. Thirty-eight million, fifty-six thousand, nine hundred fifty-four **2.** 479,587 **3.** 54,480 **4.** 39,463 **5.** 258 **6.** 16 **7.** $\frac{3}{8}$ **8.** $7\frac{1}{2}$ **9.** $\frac{9}{35}$ **10.** $\frac{7}{6}$ or $1\frac{1}{6}$ **11.** 16 **12.** $\frac{33}{10}$ or $3\frac{3}{10}$ **13.** 24,000,000,000 **14.** 0.571 **15.** 2.01, 2.1, 2.11, 2.12, 20.1 **16.** 26.080 **17.** 19.54 **18.** 8.639 **19.** 1.136 **20.** 36,512.3 **21.** 1.058 **22.** 0.8125 **23.** 13.597 **24.** 456 miles **25.** $195.57

Chapter 4

Pretest Chapter 4

1. $\frac{13}{18}$ **2.** $\frac{1}{5}$ **3.** $\frac{9}{2}$ **4.** $\frac{11}{12}$ **5.** **(a)** $\frac{7}{24}$ **(b)** $\frac{11}{120}$ **6.** $\frac{\$41}{12 \text{ cabinets}}$ **7.** $\frac{31 \text{ gallons}}{42 \text{ square feet}}$ **8.** $\frac{30.5 \text{ miles}}{1 \text{ hour}}$ or 30.5 miles/hour **9.** $\frac{\$37}{1 \text{ radio}}$ or $37/radio **10.** $\frac{13}{40} = \frac{39}{120}$ **11.** $\frac{42}{78} = \frac{21}{39}$ **12.** true **13.** false **14.** $n = 17$ **15.** $n = 18$ **16.** $n = 4$ **17.** $n = 7$ **18.** $n = 600$ **19.** 3.5 cups **20.** 364.5 miles **21.** 146 miles **22.** 54 defective bulbs

4.1 Exercises

5. $\frac{3}{4}$ **7.** $\frac{7}{6}$ **9.** $\frac{5}{11}$ **11.** $\frac{2}{3}$ **13.** $\frac{15}{16}$ **15.** $\frac{2}{3}$ **17.** $\frac{8}{5}$ **19.** $\frac{2}{3}$ **21.** $\frac{2}{3}$ **23.** $\frac{10}{17}$ **25.** $\frac{43}{60}$ **27.** $\frac{9}{1}$ **29.** $\frac{13}{19}$ **31.** $\frac{11}{19}$ **33.** $\frac{7}{33}$ **35.** $\frac{41}{245}$ **37.** $\frac{90}{41}$ **39.** $\frac{9}{16}$ **41.** $\frac{\$10}{3 \text{ plants}}$ **43.** $\frac{\$85}{6 \text{ bushels}}$ **45.** $\frac{62 \text{ gallons}}{125 \text{ sq. ft.}}$ **47.** $\frac{410 \text{ revolutions}}{1 \text{ mile}}$ *or* 410 rev/mile **49.** $\frac{9 \text{ miles}}{4 \text{ hours}}$ **51.** $13/hr **53.** 16 mi/gal **55.** 155 gal/hr **57.** 230 words/page **59.** 517 km/hr **61.** $30/share **63.** $14 profit/watch **65.** **(a)** $0.08/oz small box; $0.07/oz large box **(b)** 1¢/oz or $.01/oz **(c)** $0.48 **67.** **(a)** 13 moose **(b)** 12 moose **(c)** North Slope **69.** $24.53 **71.** Increased by Mach 0.2 **72.** Decreased by Mach 0.1 **73.** $2\frac{5}{8}$ **74.** 5 **75.** $\frac{5}{46}$ **76.** $1\frac{1}{48}$ **77.** $12.25/sq yd

4.2 Exercises

3. $\frac{48}{32} = \frac{3}{2}$ **5.** $\frac{8}{3} = \frac{32}{12}$ **7.** $\frac{20}{36} = \frac{5}{9}$ **9.** $\frac{27}{15} = \frac{9}{5}$ **11.** $\frac{22}{30} = \frac{11}{15}$ **13.** $\frac{45}{135} = \frac{9}{27}$ **15.** $\frac{5.5}{10} = \frac{11}{20}$ **17.** $\frac{12 \text{ pounds}}{\$4} = \frac{33 \text{ pounds}}{\$11}$ **19.** $\frac{10 \text{ runs}}{45 \text{ games}} = \frac{36 \text{ runs}}{162 \text{ games}}$ **21.** $\frac{20 \text{ pounds}}{\$75} = \frac{30 \text{ pounds}}{\$112.50}$ **23.** $\frac{2{,}200 \text{ people}}{100 \text{ benches}} = \frac{2{,}750 \text{ people}}{125 \text{ benches}}$ **25.** $\frac{16 \text{ pounds}}{1{,}520 \text{ sq ft}} = \frac{19 \text{ pounds}}{1{,}805 \text{ sq ft}}$ **27.** 525 = 525 True **29.** 504 = 504 True **31.** $2160 \neq 2240$ False **33.** 180 = 180 True **35.** $4950 \neq 4900$ False **37.** 31,500 = 31,500 True **39.** 63 = 63 True **41.** $110 \neq 114$ False **43.** 26 = 26 True **45.** 38 = 38 True **47.** 880 = 880 True **49.** 5148 = 5148 True **51.** $2160 \neq 2240$ False **53.** Yes **55.** $6570 \neq 6540$ No **57.** **(a)** $\frac{63}{161} = \frac{9}{23}$ $\frac{171}{437} = \frac{9}{23}$ True **(b)** $63 \times 437 \stackrel{?}{=} 161 \times 171$; 27,531 = 27,531 True **(c)** For most students it is faster to multiply than to reduce fractions. **58.** **(a)** $\frac{169}{221} = \frac{13}{17}$ $\frac{247}{323} = \frac{13}{17}$ True **(b)** $169 \times 323 \stackrel{?}{=} 221 \times 247$; 54,587 = 54,587 True **(c)** For most students it is faster to multiply than to reduce fractions. **59.** 23.1405 **60.** 14.15566 **61.** 402.408 **62.** 25.8

4.3 Exercises

1. $n = 8$ **3.** $n = 38$ **5.** $n = 16$ **7.** $n = 13$ **9.** $n = 26$ **11.** $n = 5.6$ **13.** $n = 5$ **15.** $n = 7$ **17.** $n = 5.5$ **19.** $n = 15$ **21.** $n = 16$ **23.** $n = 7.5$ **25.** $n = 32$ **27.** $n = 24$ **29.** $n = 7$ **31.** $n = 66$ **33.** $n \approx 8.8$ **35.** $n = 20$ **37.** $n = 25$ **39.** $n = 18$ **41.** $n = 14$ **43.** $n = 63$ **45.** $n = 50$ **47.** $n \approx 2.9$ **49.** $n \approx 9.4$ **51.** $n \approx 31.3$ **53.** $n = 2.8$ **55.** $n = 121.60$ **57.** $n = 8.64$ **59.** $n = 4.35$ **61.** $n = 1.36$ **63.** $n = 15.75$ **65.** $n = 3\frac{5}{8}$ **66.** $n = 3\frac{13}{15}$ **67.** $n = 10\frac{8}{9}$ **68.** $n = 4\frac{33}{34}$ **69.** $64 + 4 + 18 - 10 = 76$ **70.** $2.56 - 0.42 + 14.72 = 16.86$ **71.** five hundred sixty-three thousandths **72.** 0.0034

4.4 Exercises

1. 380 desserts **3.** 9 shots of coffee **5.** 916 drops of pigment **7.** $336 Canadian **9.** 16.9 miles per hour **11.** 197.6 feet **13.** 33 minutes **15.** 356 miles **17.** 42 millimeters **19.** $11\frac{2}{3}$ cups or approximately 11.7 cups **21.** 15 cars **23.** 35 errors **25.** 4898 people **27.** 78 giraffes **29.** $3570 **31.** 903 pounds **33.** 3410 people **35.** Yes. The units show that the parts of the proportion correspond correctly. **36.** Yes. The units show the parts of the proportion correspond correctly. **37.** No. Look at the denominators: 49 miles per hour does not correspond to 88 feet per second; 60 miles per hour does. This should be $\frac{n \text{ feet per second}}{88 \text{ feet per second}} = \frac{49 \text{ miles per hour}}{60 \text{ miles per hour}}$ **38.** No. 49 miles per hour should be in the numerator on the right. Then the parts of the proportion will correspond. $\frac{60 \text{ miles per hour}}{88 \text{ feet per second}} = \frac{49 \text{ miles per hour}}{n \text{ feet per second}}$ **39.** 56,200 **40.** 196,380,000 **41.** 56.1 **42.** 1.96 **43.** 0.0762 **44.** 598.321

Putting Your Skills to Work

1. There are 4668 people per square mile in Cincinnati and 4014 people per square mile in Austin. Cincinnati is more densely populated.
2. There are 12,209 people per square mile in Chicago and 12,484 people per square mile in Boston. Boston is more densely populated.

Chapter 4 Review Problems

1. $\frac{11}{5}$ **2.** $\frac{5}{3}$ **3.** $\frac{4}{5}$ **4.** $\frac{10}{19}$ **5.** $\frac{25}{62}$ **6.** $\frac{1}{3}$ **7.** $\frac{52}{147}$ **8.** $\frac{40}{93}$ **9.** $\frac{2}{5}$ **10.** $\frac{1}{5}$ **11.** $\frac{3}{4}$ **12.** $\frac{23}{26}$ **13.** $\frac{7}{43}$ **14.** $\frac{4}{43}$ **15.** $\frac{5 \text{ gallons}}{9 \text{ people}}$ **16.** $\frac{4 \text{ revolutions}}{11 \text{ minutes}}$ **17.** $\frac{47 \text{ vibrations}}{4 \text{ seconds}}$ **18.** $\frac{6 \text{ cups}}{19 \text{ people}}$ **19.** $17.00/share **20.** $17.00/chair **21.** $13.50/square yard **22.** $12.40/ticket **23.** **(a)** $0.74 **(b)** $0.58 **(c)** $0.16 **24.** **(a)** $0.22 **(b)** $0.25 **(c)** $0.03 **25.** $\frac{12}{48} = \frac{7}{28}$ **26.** $\frac{10}{21} = \frac{30}{63}$ **27.** $\frac{7.5}{45} = \frac{22.5}{135}$ **28.** $\frac{8.6}{43} = \frac{17.2}{86}$ **29.** $\frac{136}{17} = \frac{408}{51}$ **30.** $\frac{117}{61} = \frac{351}{183}$ **31.** $\frac{4.50 \text{ dollars}}{15 \text{ pounds}} = \frac{8.10 \text{ dollars}}{27 \text{ pounds}}$ **32.** $\frac{138 \text{ passengers}}{3 \text{ buses}} = \frac{230 \text{ passengers}}{5 \text{ buses}}$ **33.** False **34.** True **35.** True **36.** True **37.** False **38.** False **39.** True **40.** False **41.** True **42.** False **43.** $n = 23$ **44.** $n = 32$ **45.** $n = 31$ **46.** $n = 17$ **47.** $n = 33$ **48.** $n = 42$ **49.** $n = 7$ **50.** $n = 24$ **51.** $n \approx 3.9$ **52.** $n \approx 3.6$ **53.** $n = 3$ **54.** $n = 1$ **55.** $n = 87$ **56.** $n = 324$ **57.** $n = 16.8$ **58.** $n \approx 4.7$ **59.** $n = 12$ **60.** $n = 550$ **61.** $n \approx 12.6$ **62.** $n \approx 4.5$ **63.** 15 gallons **64.** $231.00 **65.** 9 nurses **66.** 2184 students **67.** 2016 francs **68.** 50 pounds **69.** 600 miles **70.** 16 miles per hour **71.** 67,200 feet **72.** $3045.69 **73.** 120 feet **74.** **(a)** 7.65 gallons **(b)** $8.80

Putting Your Skills to Work

1. No, the ratio is slightly less. **2.** $\frac{3}{25}$ The room expense would have been $2640. **3.** $21,862

Chapter 4 Test

1. $\frac{9}{26}$ **2.** $\frac{14}{37}$ **3.** $\frac{98 \text{ miles}}{3 \text{ gallons}}$ **4.** $\frac{140 \text{ sq. ft.}}{3 \text{ pounds}}$ **5.** 3.8 tons/day **6.** $8.28/hour **7.** 245.45 feet/pole **8.** $85.21/share **9.** $\frac{17}{29} = \frac{51}{87}$ **10.** $\frac{12}{19} = \frac{18}{28.5}$ **11.** $\frac{490 \text{ miles}}{21 \text{ gallons}} = \frac{280 \text{ miles}}{12 \text{ gallons}}$ **12.** $\frac{5 \text{ tablespoons}}{18 \text{ people}} = \frac{15 \text{ tablespoons}}{54 \text{ people}}$ **13.** False **14.** True **15.** True **16.** False **17.** $n = 16$ **18.** $n = 14$ **19.** $n = 12$ **20.** $n = 29.4$ **21.** $n = 120$ **22.** $n = 8.4$ **23.** $n = 120$ **24.** $n = 5.13$ **25.** 6 eggs **26.** 80.95 pounds **27.** 19 miles **28.** $360 **29.** 4 quarts **30.** 696.67 kilometers
30. 696.67 kilometers

Cumulative Test for Chapters 1–4

1. Twenty-six million, five hundred ninety-seven thousand, eighty-nine **2.** 411 **3.** 13,936 **4.** 68 **5.** $\frac{43}{40}$ *or* $1\frac{3}{40}$ **6.** $\frac{27}{35}$ **7.** $\frac{117}{8}$ *or* $14\frac{5}{8}$ **8.** $\frac{17}{6}$ *or* $2\frac{5}{6}$ **9.** 163.58 **10.** 8.2584 **11.** 0.179586 **12.** 14 **13.** $\frac{3}{1}$ **14.** $\frac{\$0.14}{1 \text{ banana}}$ **15.** $\frac{4 \text{ yen}}{1 \text{ peso}}$ **16.** True **17.** True **18.** $n = 3$ **19.** $n = 2$ **20.** $n = 128$ **21.** $n = 9$ **22.** $n \approx 3.4$ **23.** $n = 21$ **24.** 8.33 inches **25.** $750.00 **26.** 5 pounds

Chapter 5

Pretest Chapter 5

1. 13% **2.** 21% **3.** 18.5% **4.** 37.2% **5.** 134% **6.** 894% **7.** 0.2% **8.** 0.4% **9.** 17% **10.** 27% **11.** 13.4% **12.** 19.8% **13.** $6\frac{1}{2}$% **14.** $1\frac{3}{8}$% **15.** 60% **16.** 2.5% **17.** 115% **18.** 106.25% **19.** 71.43% **20.** 28.57% **21.** 95.65% **22.** 68.42% **23.** 440% **24.** 275% **25.** 0.33% **26.** 0.25% **27.** $\frac{11}{50}$ **28.** $\frac{19}{50}$ **29.** $\frac{53}{100}$ **30.** $\frac{41}{100}$ **31.** $\frac{3}{2}$ or $1\frac{1}{2}$ **32.** $\frac{8}{5}$ or $1\frac{3}{5}$ **33.** $\frac{19}{300}$ **34.** $\frac{7}{150}$ **35.** $\frac{41}{80}$ **36.** $\frac{7}{16}$ **37.** 55.2 **38.** 22.36 **39.** 94.44% **40.** 81.82% **41.** 3000 **42.** 1850 **43.** They won 63.16% of the games. **44.** $5400 commission would be paid. **45.** The dinner without the tax was $14.40. **46.** The percent decrease was 30%.

5.1 Exercises

5. 45% **7.** 7% **9.** 80% **11.** 245% **13.** 5.3% **15.** 0.6% **17.** 29% **19.** 78% **21.** 32% **23.** 0.51 **25.** 0.07 **27.** 0.2 **29.** 0.436 **31.** 0.003 **33.** 0.0072 **35.** 1.26 **37.** 3.66 **39.** 74% **41.** 50% **43.** 8% **45.** 56.3% **47.** 0.2% **49.** 0.57% **51.** 135% **53.** 272% **55.** 27% **57.** 36% **59.** 143% **61.** 30% **63.** 0.5% **65.** 0.62 **67.** 1.28 **69.** 0.005 **71.** 0.8 **73.** 49% **75.** 3.413% **77.** 36% = 36 percent = 36 "per one hundred" = $36 \times \frac{1}{100} = \frac{36}{100} = 0.36$. The rule is using the fact that 36% means 36 per one hundred. **78.** $10.65 = 1065 \times \frac{1}{100}$ = 1065 "per one hundred" = 1065 percent = 1065%. We change 10.65 to $1065 \times \frac{1}{100}$ and use the idea that percent means "per one hundred." **79. (a)** 555.62 **(b)** $\frac{55{,}562}{100}$ **(c)** $\frac{27{,}781}{50}$ **81.** $\frac{14}{25}$ **82.** $\frac{18}{25}$ **83.** 0.6875 **84.** 0.8125

5.2 Exercises

1. $\frac{22}{25}$ **3.** $\frac{7}{100}$ **5.** $\frac{33}{100}$ **7.** $\frac{11}{20}$ **9.** $\frac{3}{4}$ **11.** $\frac{1}{5}$ **13.** $\frac{29}{200}$ **15.** $\frac{22}{125}$ **17.** $\frac{81}{125}$ **19.** $\frac{57}{80}$ **21.** $1\frac{19}{25}$ **23.** $3\frac{2}{5}$ **25.** 12 **27.** $\frac{13}{600}$ **29.** $\frac{99}{800}$ **31.** $\frac{11}{125}$ **33.** $\frac{7}{325}$ **35.** $\frac{9}{220}$ **37.** 75% **39.** 33.33% **41.** 31.25% **43.** 28% **45.** 27.5% **47.** 41.67% **49.** 360% **51.** 283.33% **53.** 412.5% **55.** 42.86% **57.** 93.75% **59.** 52% **61.** 34.31% **63.** 34.84% **65.** 2.5% **67.** 1.23% **69.** $83\frac{1}{3}$% **71.** $68\frac{3}{4}$% **73.** $37\frac{1}{2}$% **75.** $7\frac{1}{2}$%

	Fraction	Decimal	Percent
77.	$\frac{5}{12}$	0.4167	41.67%
79.	$\frac{3}{50}$	0.06	6%
81.	$\frac{2}{5}$	0.4	40%
83.	$\frac{69}{200}$	0.345	34.5%
85.	$\frac{3}{200}$	0.015	1.5%
87.	$\frac{5}{9}$	0.5556	55.56%
89.	$\frac{1}{32}$	0.0313	$3\frac{1}{8}$%

	Fraction	Decimal	Percent
78.	$\frac{7}{12}$	0.5833	58.33%
80.	$\frac{2}{25}$	0.08	8%
82.	$\frac{3}{5}$	0.6	60%
84.	$\frac{5}{8}$	0.625	62.5%
86.	$\frac{9}{200}$	0.045	4.5%
88.	$\frac{7}{9}$	0.7778	77.78%
90.	$\frac{21}{800}$	0.0263	$2\frac{5}{8}$%

91. $\frac{463}{1600}$ **93.** 15.375% **95.** $n = 5.625$ **96.** $n = 4$ **97.** $n = 88$ **98.** $n = 96$

5.3A Exercises

1. $n = 38\% \times 500$ **3.** $50\% \times n = 7$ **5.** $17 = n \times 85$ **7.** $n = 128\% \times 4000$ **9.** $n \times 400 = 15$ **11.** $136 = 145\% \times n$ **13.** $n = 150$ **15.** $n = 912$ **17.** $6.39 **19.** $n = 1200$ **21.** $n = 1300$ **23.** $150 **25.** 60% **27.** 28% **29.** 5% **31.** 50.4 **33.** 68 **35.** 12% **37.** 3.28 **39.** 64% **41.** 445 **43.** 35% **45.** 39.6 **47.** 1.44 **49.** 902.825 **51.** 59.29% (rounded) **53.** 85% **55.** 554 students **57.** 40 years **59.** 57.6 **61.** 2.448 **62.** 14.892 **63.** 2834 **64.** 2917

5.3B Exercises

	p	b	a
1.	75	660	495
3.	42	400	a
5.	49	b	2450
7.	p	50	30
9.	p	25	10
11.	160	b	400

13. 72 **15.** 75 **17.** 56 **19.** 75 **21.** 60 **23.** 600,000 **25.** 25 **27.** 2.5 **29.** 1.8 **31.** 91 **33.** 300 **35.** 16.4% **37.** 3.64 **39.** 25% **41.** 170 **43.** 43% **45.** 47.5 **47.** 2.8 **49.** 15.82% **51.** 10.91 **53.** 153.62 **55.** $118.80 **57.** $1\frac{31}{45}$ **58.** $\frac{1}{26}$ **59.** $4\frac{1}{5}$ **60.** $1\frac{13}{15}$

5.4 Exercises

1. 180,000 pencils **3.** $94.25 **5.** 20% **7.** Approximately $42.26 **9.** $6.44 **11.** 9% **13.** 333 people **15.** $240 **17.** 25% **19.** $9,600,000 **21.** 1365 M&Ms are blue **23.** $75,000 **25.** 216 babies **27.** 4% **29.** $55 **31.** $12,210,000 for personnel, food, and decorations; $20,790,000 for security facility rental and all other expenses **33. (a)** $14.16 **(b)** $250.16 **35.** 36.94% **37.** 60% **39. (a)** $1279.20 **(b)** $14,710.80 **41. (a)** $255 **(b)** $1870 **43. (a)** $200 **(b)** $2700 **45. (a)** $118.40 **(b)** $3818.40 **47.** $849.01 **49.** $10.18, $708.62 **51.** $17,108.38 **53.** 1,698,000 **54.** 2,452,400 **55.** 1.63 **56.** 0.793 **57.** 0.0556 **58.** 0.0792

Putting Your Skills to Work

1. An increase of 153.8%; health benefits of reducing amount of fat consumed in milk **2.** Decrease of 21.6%; health benefits of reducing amount of caffeine consumed

Chapter 5 Review Problems

1. 87% **2.** 59% **3.** 27.6% **4.** 32.9% **5.** 7.13% **6.** 6.08% **7.** 252% **8.** 437% **9.** 103.6% **10.** 105.2% **11.** 0.6% **12.** 0.2% **13.** 0.29% **14.** 0.53% **15.** 72% **16.** 61% **17.** 19.5% **18.** 21.6% **19.** 0.24% **20.** 0.98% **21.** $4\frac{1}{12}\%$ **22.** $3\frac{5}{12}\%$ **23.** 317% **24.** 225% **25.** 64% **26.** 88% **27.** 90% **28.** 70% **29.** 45.45% **30.** 44.44% **31.** 225% **32.** 375% **33.** 442.86% **34.** 555.56% **35.** 190% **36.** 183.33% **37.** 0.38% **38.** 0.63% **39.** 0.002 **40.** 0.007 **41.** 0.219 **42.** 0.431 **43.** 1.66 **44.** 1.39 **45.** 0.32125 **46.** 0.26375 **47.** $\frac{41}{50}$ **48.** $\frac{9}{25}$ **49.** $\frac{37}{20}$ **50.** $\frac{9}{4}$ **51.** $\frac{41}{250}$ **52.** $\frac{61}{200}$ **53.** $\frac{5}{16}$ **54.** $\frac{7}{16}$ **55.** $\frac{1}{2000}$ **56.** $\frac{3}{5000}$

	Fraction	Decimal	Percent
57.		0.6	60%
58.		0.875	87.5%
59.	$\frac{3}{8}$	0.375	
60.	$\frac{9}{16}$	0.5625	
61.	$\frac{1}{125}$		0.8%
62.	$\frac{9}{20}$		45%

63. 288 **64.** 340 **65.** 15 **66.** 225 **67.** 40% **68.** 40% **69.** 57.5 **70.** 70.8 **71.** 160 **72.** 140 **73.** 20% **74.** 46.67% **75.** 13 **76.** 28 **77.** 464.29% **78.** 324.32% **79.** 51 students **80.** 96 trucks **81.** 4.17% **82.** 15.69% **83.** $11,200 **84.** $80,200 **85.** 15% **86.** 7.5% **87.** $1200 **88.** $558 **89.** 4.09% **90. (a)** $330 **(b)** $1320 **91. (a)** $319 **(b)** $1276 **92.** $1800

Chapter 5 Test

1. 42% **2.** 1% **3.** 0.6% **4.** 13.9% **5.** 218% **6.** 71% **7.** 2.7% **8.** $3\frac{1}{7}\%$ **9.** 47.5% **10.** 40% **11.** 140% **12.** 175% **13.** 17.13% **14.** 302.4% **15.** $1\frac{13}{25}$ **16.** $\frac{31}{400}$ **17.** 26.69 **18.** 130 **19.** 55.56% **20.** 200 **21.** 5000 **22.** 46% **23.** 699.6 **24.** 2.29% **25.** $6092 commission **26. (a)** $150.81 discount **(b)** They paid $306.19 **27.** 89.29% were not defective **28.** 8.93% is the percent of decrease **29.** 12,000 registered voters **30. (a)** $240 in interest in 6 months **(b)** $960 in interest in 2 years

Cumulative Test for Chapters 1–5

1. 2241 **2.** 8444 **3.** 5292 **4.** 89 **5.** $\frac{67}{12}$ *or* $5\frac{7}{12}$ **6.** $\frac{1}{12}$ **7.** $\frac{35}{12}$ *or* $2\frac{11}{12}$ **8.** $\frac{5}{21}$ **9.** 5731.7 **10.** 34.118 **11.** 1.686 **12.** 0.368 **13.** $\frac{3 \text{ pounds}}{5 \text{ square feet}}$ **14.** True **15.** $n = 24$ **16.** 673 faculty **17.** 2.3% **18.** 46.8% **19.** 198% **20.** 3.75% **21.** 2.43 **22.** 0.0675 **23.** 17.76% **24.** 114.58 **25.** 300 **26.** 718.2 **27.** $8370 **28.** 3200 students **29.** 11.31% increase **30.** $352

Chapter 6

Pretest Chapter 6

1. 204 **2.** 56 **3.** 3520 **4.** 6400 **5.** 1320 **6.** 24 **7.** 5320 **8.** 4680 **9.** 98.6 **10.** 0.027 **11.** 529.6 **12.** 0.123 **13.** 2376 m **14.** 94.262 m **15.** 3820 **16.** 3.162 **17.** 0.0563 **18.** 4800 **19.** 0.568 **20.** 8900 **21.** 4.73 **22.** 1.28 **23.** 59.52 **24.** 1826.78 **25.** 39.69 **26.** 103.4 **27.** 55.8 feet **28.** **(a)** 95°F **(b)** No **29.** 12.2 miles farther **30.** 22.5 gal/hr

6.1 Exercises

1. 12 **3.** 1760 **5.** 2000 **7.** 4 **9.** 2 **11.** 60 **13.** 5 **15.** 7 **17.** 2 **19.** 12,320 **21.** 144 **23.** 123 **25.** 6.25 **27.** 12 **29.** 26,000 **31.** 36 **33.** 28 **35.** 9 **37.** 62 **39.** 64 **41.** 84 **43.** 11 **45.** 264 **47.** 4200 **49.** 64,800 **51.** 3 **53.** 20,265 **55.** $135 **57.** $9.75 **59.** 199,056 feet **61.** $161\frac{3}{4}$ inches **63.** 133,953 tons **65.** 26,280 hours **67.** 14,739 land miles

6.2 Exercises

7. 370 **9.** 3600 **11.** 0.328 **13.** 0.563 **15.** 200,000 **17.** 0.078 **19.** 538,600 **21.** 3.5; 0.035 **23.** 3.582; 0.003582 **25.** 0.0032; 0.0000032 **27.** a **29.** c **31.** 30 **33.** 270 **35.** 1.98 **37.** 0.482 **39.** 5236 **41.** 3255 m **43.** 463 cm **45.** 406.71 m **47.** 2.5464 cm *or* 25.464 mm **49.** False **51.** True **53.** True **55.** False **57.** True **59.** **(a)** 11.92 km **(b)** 11,920 meters **61.** **(a)** 481,800 centimeters **(b)** 4.818 kilometers **63.** 0.000000004 kilometer

6.3 Exercises

7. 64,000 **9.** 4700 **11.** 0.0189 **13.** 0.752 **15.** 2,430,000 **17.** 82 **19.** 0.005261 **21.** 74,000 **23.** 0.162 **25.** 0.035 **27.** 6.328 **29.** 2920 **31.** 17,000 **33.** 0.00032 **35.** 0.007896 **37.** 5,900,000 **39.** 0.007; 0.000007 **41.** 0.128; 0.000128 **43.** 0.522; 0.000522 **45.** 3.607; 0.003607 **47.** b **49.** a **51.** 113.922 L **53.** 9.803 t **55.** 260,160 mg **57.** True **59.** False **61.** True **63.** False **65.** True **67.** $358 **69.** $573,000 **71.** $340,000 **73.** $10,102,500 **75.** 0.005632 **76.** 76,182 **77.** 20% **78.** 57.5 **79.** $321.30 **80.** $716.80

6.4 Exercises

1. 2.14 m **3.** 22.86 cm **5.** 15.26 yd **7.** 28.89 yd **9.** 9.3 mi **11.** 21.94 m **13.** 132.02 km **15.** 82 ft **17.** 6.90 in. **19.** 218 yd **21.** 18.95 L **23.** 1061.2 L **25.** 21.76 L **27.** 5.02 gal **29.** 4.77 qt **31.** 14.53 kg **33.** 198.45 g **35.** 35.2 lb **37.** 4.45 oz **39.** 5.45 ft **41.** 502.92 cm **43.** 31 mi/hr **45.** 96.6 km/hr **47.** 0.51 in. **49.** 104°F **51.** 185°F **53.** 53.6°F **55.** 20°C **57.** 75.56°C **59.** 30°C **61.** 143.87 km **63.** 18.85 L **65.** 1397 lb **67.** It is 66.2°F at 4:00 A.M. The temperature may reach 113°F after 7:00 A.M. **69.** 3420°C **71.** 21.773 g **73.** 180.6448 sq cm **75.** 32.7273 mi/hr **77.** 47 **78.** 49 **79.** 91 **80.** 74

Putting Your Skills to Work

1. 1,182,633,000 bytes of space

6.5 Exercises

1. 8 yd **3.** 11 yd **5.** $33.46 **7.** 775 mL per bottle **9.** 310 cm per piece **11.** 1340 cm **13.** 125 samples **15.** The discrepancy is 4.4°F. The sign in the store is 4.4°F less than it should be. **17.** The difference is 6°F. The temperature reading of 180°C is hotter. **19.** 195 miles further **21.** **(a)** About 105 kilometers per hour **(b)** Probably not. We cannot be sure, but we have no evidence to indicate that they broke the speed limit. **23.** 60 gal/hr **25.** $102 **27.** 85.05 g **29.** **(a)** She bought 5 qt extra. **(b)** $6.95 **31.** **(a)** $5.46 **(b)** 132 miles per gallon **33.** $n = 0.64$ **34.** $n = 0.2$ **35.** 15.5 miles **36.** 41.25 yards

Chapter 6 Review Problems

1. 9 **2.** 11 **3.** 5280 **4.** 7040 **5.** 7.5 **6.** 6.5 **7.** 3 **8.** 2 **9.** 8000 **10.** 14,000 **11.** 5.75 **12.** 6.25 **13.** 60 **14.** 84 **15.** 15.5 **16.** 13.5 **17.** 36 **18.** 23 **19.** 840 **20.** 660 **21.** 560 **22.** 290 **23.** 176.3 **24.** 259.8 **25.** 920 **26.** 740 **27.** 5000 **28.** 7000 **29.** 0.285 **30.** 0.473 **31.** 7.93 m **32.** 17.01 m **33.** 35.63 m **34.** 89.59 m **35.** 17,000 **36.** 23,000 **37.** 0.059 **38.** 0.077 **39.** 196,000 **40.** 721,000 **41.** 0.778 **42.** 0.459 **43.** 0.125 **44.** 0.705 **45.** 76,000 **46.** 41,000 **47.** 765 **48.** 423 **49.** 2430 **50.** 1930 **51.** 92.4 **52.** 72.6 **53.** 4.58 **54.** 2.75 **55.** 368.55 **56.** 481.95 **57.** 54.86 **58.** 39.62 **59.** 5.52 **60.** 7.09 **61.** 9.08 **62.** 13.62 **63.** 10.97 **64.** 12.80 **65.** 49.6 **66.** 43.4 **67.** 59 **68.** 77 **69.** 105 **70.** 85 **71.** 0 **72.** 3.12 cm **73. (a)** 17 ft **(b)** 204 in. **74. (a)** 200 m **(b)** 0.2 km **75.** $2.22 **76.** 32.6 miles farther **77.** Yes 1.67 feet extra **78.** 166.67 milliliters per jar **79.** $581.12 **80.** 380 centimeters **81.** 25°F too hot

Chapter 6 Test

1. 3200 **2.** 228 **3.** 84 **4.** 7 **5.** 0.75 **6.** 30 **7.** 0.273 **8.** 9200 **9.** 4.6 **10.** 0.0988 **11.** 1270 **12.** 9.36 **13.** 0.046 **14.** 127,000 **15.** 0.0289 **16.** 0.983 **17.** 920 **18.** 9420 **19.** 67.62 **20.** 1.63 **21.** 3.55 **22.** 10.03 **23.** 16.06 **24.** 85.05 **25. (a)** 20 meters **(b)** 21.8 yards **26. (a)** 15°F **(b)** Yes **27.** 82.5 gal/hr **28. (a)** 300 km **(b)** 14 miles

Cumulative Test for Chapters 1–6

1. 6028 **2.** 185,440 **3.** 69 **4.** $\frac{19}{42}$ **5.** $1\frac{3}{8}$ **6.** True **7.** $n = 6$ **8.** 209.23 grams **9.** 250% **10.** 64.8 **11.** 20,000 **12.** 9.5 **13.** 5000 **14.** 3.5 **15.** 300 **16.** 3700 **17.** 0.0628 **18.** 790 **19.** 0.05 **20.** 672 **21.** 106.12 **22.** 43.58 **23.** 25.81 **24.** 14.49 **25.** 11.88 meters **26.** 59°F The difference is 44°F. The 15°C temperature is higher. **27.** 7 miles **28.** 1.738 centimeters

Chapter 7

Pretest Chapter 7

1. 18 m **2.** 14 m **3.** 23 sq cm **4.** 2.4 sq cm **5.** 25.6 yd **6.** 52 ft **7.** 135 sq in. **8.** 171 sq in. **9.** 97 sq m **10.** 59° **11.** 15.3 m **12.** 30 sq m **13.** 8 **14.** 12 **15.** 6.782 **16.** 8 ft **17.** 12 ft **18.** 28 in. **19.** 94.2 cm **20.** 153.9 sq m **21.** 27.4 sq m **22.** 240 cu yd **23.** 113 cu ft **24.** 1846.3 cu in. **25.** 4375 cu m **26.** 1130.4 cu m **27.** $n = 120$ cm **28.** $n = 28$ m **29. (a)** 3706.5 sq yd **(b)** $555.98

7.1 Exercises

5. 15 mi **7.** 23.6 ft **9.** 49.2 ft **11.** 1.92 mm **13.** 25.46 in. **15.** 17.12 km **17.** 0.0246 cm **19.** 43.7 m **21.** 52 m **23.** 4.8 mi **25.** 0.0208 mm **27.** 31.84 cm **29.** 94 m **31.** 180 cm **33.** 72 in.^{2} **35.** 96.04 ft^2 **37.** 0.288 m^2 **39.** 14,976 yd^2 **41.** 294 m^2 **43.** $132,000 **45.** $100.70 **47.** $953.22 **49.** $598.22 **51.** 223.3 **52.** 7.18 **53.** 21,842.8 **54.** 1.58

7.2 Exercises

5. 40.2 m **7.** 29.8 in. **9.** 1057.5 m^2 **11.** 440.75 cm^2 **13.** 3528 yd^2 **15.** 82 m **17.** 380 cm **19.** 42 m^2 **21.** 86 cm^2 **23.** 344 yd^2 **25.** 550 km^2 **27.** 718 m^2 **29.** 162.5 cm^2 **31.** 345 ft^2 **33.** $80,960.00 **35.** 694.49 in.^{2} (rounded) **37.** Area = $5.2415 \times b^2$ sq units **39.** 30 **40.** 3 **41.** 1800 **42.** 2.6

7.3 Exercises

7. True **9.** True **11.** False **13.** 60° **15.** 30° **17.** 90° **19.** 128° **21.** 104 m **23.** 130 in. **25.** 10.5 mi **27.** 25.5 in. **29.** 22.5 ft^2 **31.** 15.75 in.^{2} **33.** 83.125 cm^2 **35.** 14.07 m^2 **37.** 12.25 yd^2 **39.** 16.4 cm **41.** 188 yd^2 **43.** 1740 ft^2 **45.** $21,060.00 **47.** $n = 12$ **48.** $n = 67.67$ **49.** 716 **50.** 96 magazines

7.4 Exercises

5. 1 **7.** 4 **9.** 5 **11.** 7 **13.** 10 **15.** 11 **17.** 13 **19.** 0 **21.** 10 **23.** 11 **25.** 14 **27.** 4 **29.** 10 **31.** 5 **33.** 22 **35. (a)** Yes **(b)** 7 **37. (a)** Yes **(b)** 16 **39.** 4.243 **41.** 5.568 **43.** 9.110 **45.** 10.954 **47.** 11.180 **49.** 11 m **51.** 5.099 m **53.** 8.660 m **55. (a)** 2 **(b)** 0.2 **(c)** 0.02 **(d)** Each answer is obtained from the previous answer by dividing by 10. **(e)** No, because 0.004 isn't a perfect square. **57.** 205.027 **59.** 39.299 **61.** 4800 in.^{2} **62.** 80,500 meters **63.** 0.92 meter **64.** 0.0989 kilogram

7.5 Exercises

1. 5 in. **3.** 8.544 yd **5.** 15.199 ft **7.** 3.606 **9.** 10.247 km **11.** 11.402 m **13.** 7.071 m **15.** 3 ft **17.** 15 in. **19.** 6.928 yd **21.** 17 ft **23.** 5 mi **25.** 8.7 ft **27.** 4 in., 6.9 in. **29.** 7 m, 12.1 m **31.** 8.5 m **33.** 15.6 cm **35.** 7.1 in. **36.** 10 m **37.** 52 yd **39.** 341 m^2 **40.** 168 m^2 **41.** 441 in.^{2} **42.** 4224 yd^2

7.6 Exercises

5. 58 in. **7.** 5 mm **9.** 22.5 yd **11.** 1.9 cm **13.** 75.4 cm **15.** 69.1 in. **17.** 81.6 in. **19.** 41.87 ft **21.** 78.5 yd^2 **23.** 907.5 m^2 **25.** 803.8 cm^2 **27.** 153.9 ft^2 **29.** 11,304 mi^2 **31.** 0.70057457 m **33.** 628 m^2 **35.** 163.3 m^2 **37.** 189.3 m^2 **39.** 31.0 m^2 **41.** $1211.20 **43.** 6.28 ft **45.** 256.43 ft **47.** 200 revolutions **49.** 3215.36 $in.^2$ **51.** $226.08 **53.** **(a)** $0.75 per slice; 22.1 $in.^2$ **(b)** $0.67 per slice; 18.8 $in.^2$ **(c)** The 12-in. pizza is $0.04 per $in.^2$; the 15-in. pizza is $0.03 per $in.^2$ The 15-in. pizza is a better value. **55.** 13.92 **56.** 0.3 **57.** 6000 **58.** 3000

7.7 Exercises

7. 24 m^3 **9.** 700 mm^3 **11.** 87.9 m^3 **13.** 2260.8 m^3 **15.** 3052.1 yd^3 **17.** 267.9 m^3 **19.** 718.0 m^3 **21.** 1017.4 cm^3 **23.** 261.7 ft^3 **25.** 21 m^3 **27.** 120 m^3 **29.** 170.832816 m^3 **31.** 592.3814 m^3 **33.** 95.6887288 ft^3 **35.** 40 yd^3 **37.** 1004.8 $in.^3$ **39.** 381,510,000,000,000 mi^3 **41.** $942 **43.** 263,900 yd^3 **45.** $9\frac{7}{12}$ **46.** $6\frac{3}{8}$ **47.** $\frac{135}{16}$ *or* $8\frac{7}{16}$ **48.** $\frac{25}{14}$ *or* $1\frac{11}{14}$

Putting Your Skills to Work

1. 224 pounds **2.** 30 $in.^2$

7.8 Exercises

5. 8 m **7.** 17.5 cm **9.** 1.9 yd **11.** *a* corresponds to *f*; *b* corresponds to *e*; *c* corresponds to *d* **13.** 2.2 in. **15.** 3.36 meters long **17.** 2.1 ft **19.** 36 ft **21.** 81.2 ft **23.** 1.4 km **25.** 16.3 cm **27.** 33.3 m **29.** 11.6 yd^2 **30.** 220.4 m^2 **31.** 12 **32.** 31 **33.** 42 **34.** 62

7.9 Exercises

1. $22.10 **3.** **(a)** 75 km/hr **(b)** 76 km/hr **(c)** Through Woodville and Palermo **5.** 118 minutes or 1 hour, 58 minutes **7.** $510.00 **9.** 953.8 gal **11.** $74.42 **13.** **(a)** 40,820 km **(b)** 20,410 km/hr **15.** $9880.34 **17.** 128 **18.** 915 **19.** 0.25 **20.** 4.87

Chapter 7 Review Problems

1. 19.8 m **2.** 24.8 cm **3.** 23.2 yd **4.** 9.6 yd **5.** 16.5 cm^2 **6.** 73.5 m^2 **7.** 18.5 $in.^2$ **8.** 51.8 $in.^2$ **9.** 38 ft **10.** 58 ft **11.** 68 m^2 **12.** 63.5 m^2 **13.** 100.4 m **14.** 76.4 cm **15.** 80 in. **16.** 62 mi **17.** 2700 m^2 **18.** 2050 m^2 **19.** 336 yd^2 **20.** 720 yd^2 **21.** 422 cm^2 **22.** 357 cm^2 **23.** 22 ft **24.** 14 ft **25.** 153° **26.** 40° **27.** 24.5 m^2 **28.** 59.4 m^2 **29.** 45.6 cm^2 **30.** 40.0 m^2 **31.** 450 m^2 **32.** 87 m^2 **33.** 9 **34.** 8 **35.** 10 **36.** 11 **37.** 12 **38.** 15 **39.** 6 **40.** 6 **41.** 5 **42.** 12 **43.** $\approx$ 5.916 **44.** $\approx$ 6.708 **45.** $\approx$ 9.381 **46.** $\approx$ 8.718 **47.** $\approx$ 13.077 **48.** $\approx$ 13.416 **49.** 5 m **50.** 5 m **51.** 8.25 cm **52.** 8.06 m **53.** 6.4 cm **54.** 9.2 ft **55.** 6.3 ft **56.** 3.6 ft **57.** 106 cm **58.** 24 cm **59.** 37.7 in. **60.** 44.0 in. **61.** 113.0 m^2 **62.** 254.3 m^2 **63.** 201.0 ft^2 **64.** 153.9 ft^2 **65.** 226.1 $in.^2$ **66.** 201.0 m^2 **67.** 318.5 ft^2 **68.** 126.1 m^2 **69.** 107.4 ft^2 **70.** 80.1 m^2 **71.** 45 ft^3 **72.** 450 m^3 **73.** 14,130 ft^3 **74.** 7.2 ft^3 **75.** 307.7 m^3 **76.** 4578.1 cm^3 **77.** 1728 m^3 **78.** 100 m^3 **79.** 3768 ft^3 **80.** 9074.6 yd^3 **81.** 30 m **82.** 3.3 m **83.** 33.8 cm **84.** 9.6 mi **85.** 324 cm **86.** 175 ft **87.** $92.40 **88.** **(a)** 210 cm **(b)** 1764 times larger **89.** $736.00 **90.** $V = 2034.7$ $in.^3$ $W = 32{,}555.2$ g **91.** **(a)** 21,873.2 ft^3 **(b)** 17,498.6 bushels **92.** **(a)** 50 km 100 km/h **(b)** 56 km 70 km/hr **(c)** Through Ipswich

Chapter 7 Test

1. 40 yd **2.** 34 ft **3.** 20 m **4.** 80 m **5.** 24.8 m **6.** 180 yd^2 **7.** 104.0 m^2 **8.** 78 m^2 **9.** 144 m^2 **10.** 12 cm^2 **11.** 52.5 m^2 **12.** 9 **13.** 8 **14.** 12 **15.** 7 **16.** $\approx$ 7.348 **17.** $\approx$ 10.954 **18.** $\approx$ 13.675 **19.** 9.22 **20.** 10 **21.** 5.83 cm **22.** 9 ft **23.** 37.7 in. **24.** 254.3 ft^2 **25.** 107.4 $in.^2$ **26.** 144.3 $in.^2$ **27.** 840 m^3 **28.** 803.8 m^3 **29.** 33.5 m^3 **30.** 508.7 ft^3 **31.** 56 m^3 **32.** 43.2 cm **33.** 46.7 ft **34.** 6456 yd^2 $2582.40

Cumulative Test for Chapters 1–7

1. 935,760 **2.** 33,415 **3.** $\frac{26}{45}$ **4.** $\frac{4}{21}$ **5.** 56.13 **6.** 7.2272 **7.** 83 **8.** $n = 27$ **9.** 800 students **10.** 75% **11.** 2000 **12.** 40.7 **13.** 5.86 m **14.** 1512 in. **15.** 54.56 mi **16.** 50 m **17.** 208 cm **18.** 56.5 yd **19.** 1.4 cm^2 **20.** 540 m^2 **21.** 192 m^2 **22.** 664 yd^2 **23.** 50.2 m^2 **24.** 2411.5 m^3 **25.** 3052.1 cm^3 **26.** 3136 cm^3 **27.** 2713.0 m^3 **28.** 33.4 m **29.** 4.1 ft **30.** 124 yd^2 $992.00 **31.** 11 **32.** $\approx$ 7.550 **33.** 10.440 in. **34.** 4.899 m **35.** 13.9 mi

Chapter 8

Pretest Chapter 8

1. Under age 18 **2.** 43% **3.** 17% **4.** 1650 students **5.** 350 students **6.** 300 people **7.** 600 people **8.** 3rd quarter of both 1993 and 1994 **9.** 4th quarter of 1993 **10.** 100 more people **11.** 100 more people **12.** August and December **13.** December **14.** November **15.** 20,000 sets **16.** 30,000 sets **17.** 55,000 cars **18.** 60,000 cars **19.** 25,000 cars **20.** 20,000 cars **21.** 37 pages per day **22.** 40 pages per day

8.1 Exercises

1. Rent **3.** \$200 **5.** \$500 **7.** $\frac{10}{7}$ **9.** $\frac{3}{10}$ **11.** Hit batters **13.** 294 **15.** 379 pitches **17.** $\frac{72}{325}$ **19.** $\frac{49}{24}$ **21.** 15% **23.** 72% **25.** 44,840 **27.** 1,850,000,000 **29.** 37% **31.** 81% **33.** 1.5% **35.** 42 in.^{2} **36.** 204 in.^{2} **37.** 16 gal **38.** About 3 g

8.2 Exercises

1. 10 million people **3.** 14 million people **5.** 1960–1970 **7.** 50,000 women **9.** 1990 **11.** 25,000 more **13.** 1930 to 1950 **15.** 3.5 million dollars **17.** 1992 **19.** one million dollars **21.** 2.5 in. **23.** October, November, and December **25.** 1.5 in. more

27.

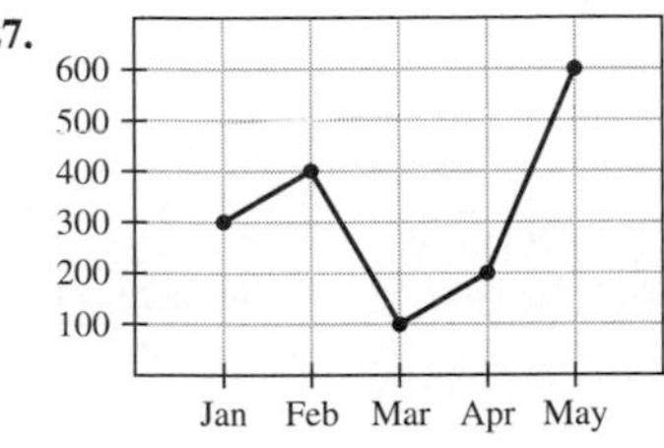

28. increase **29.** 35 **30.** 115

Putting Your Skills to Work

1. 1989 **2.** Trucks that are 12 years old or more

8.3 Exercises

1. 10 cars **3.** 35 cars **5.** 45 cars **7.** 145 cars **9.** 12 days **11.** 3 days **13.** 42 days **15.** 37 days **17.** 63.8%

	TALLY	FREQUENCY
19.	|||	3
21.	~~||||~~ |	6
23.	|||	3
25.	||	2

27.

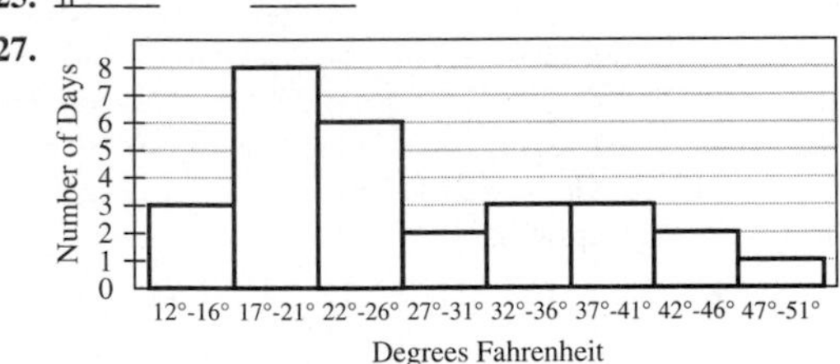

	TALLY	FREQUENCY
29.	|||	3
31.	||	2
33.	|	1
35.	||	2

37. $n = 59.5$ **38.** $n = 7$ **39.** 12 pounds **40.** 13 in.

8.4 Exercises

1. 91.8 **3.** 38 **5.** \$87,000 **7.** 5 hrs **9.** 98,500 **11.** 0.375 **13.** 23.7 mi/gal **15.** 47 **17.** 1011 **19.** 0.58 **21.** \$20,250 **23.** 35.5 min **25.** \$207 **27.** 46 actors **29.** \$69,161.88 **31.** 4850 **33.** **(a)** 2,340,400,000 barrels of oil per year **(b)** 98,296,800,000 gallons of oil per year **35.** **(a)** \$2157 **(b)** \$1615 **(c)** The median because the mean is affected by the high amount \$6300. **36.** **(a)** 12.65 sec **(b)** 11.95 sec **(c)** The median because the large time, 18 seconds, affects the mean. **37.** 60 **39.** 121 and 150 **41.** \$249

Chapter 8 Review Problems

1. 36 computers **2.** 43 computers **3.** 20 computers **4.** 21 computers **5.** $\frac{4}{1}$ **6.** $\frac{2}{1}$ **7.** 7.5% **8.** 30% **9.** 15% **10.** 4% **11.** 57% **12.** 27% **13.** \$192 **14.** \$360 **15.** \$816 **16.** \$1008 **17.** 6000 customers **18.** 4000 customers **19.** 4th quarter 1994 **20.** 3rd quarter 1993 **21.** 1000 customers **22.** 1500 customers **23.** Yes **24.** Yes **25.** 400 students **26.** 500 students **27.** 650 students **28.** 450 students **29.** 100 students more **30.** 50 students more **31.** 1992–1993 **32.** 1993–1994 **33.** 45,000 cones **34.** 30,000 cones **35.** 10,000 cones more **36.** 30,000 cones more **37.** 25,000 cones more **38.** 30,000 cones more **39.** The cooler the temperature, the fewer cones sold. **40.** The warmer the temperature, the more cones sold. **41.** 50 bridges **42.** 25 bridges **43.** 20 and 39 **44.** 25 bridges **45.** 150 bridges **46.** 5 bridges more

	TALLY	FREQUENCY
47.	卌 卌	10
48.	卌 |||	8
49.	|||	3
50.	卌	5
51.	||	2

52.

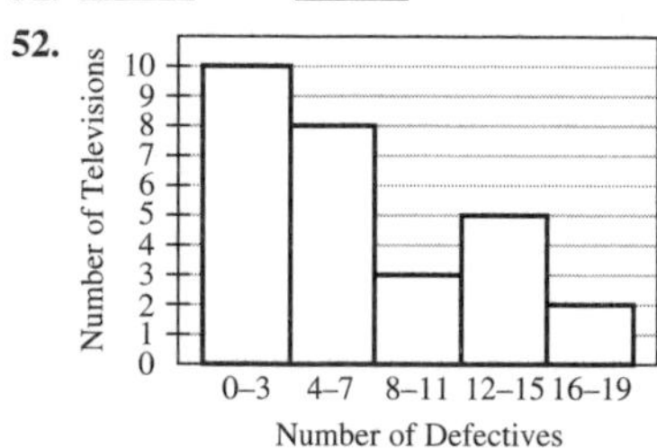

53. 18 times **54.** 90° **55.** \$114 **56.** 58 textbooks **57.** 188 women **58.** 1421 cars **59.** 1353 employees **60.** 82 **61.** 83 students **62.** \$36,000 **63.** \$141,500 **64.** 19.5 cups **65.** 18.5 deliveries **66.** The median, because of one low score, 31. **67.** The median, because of the one high data item, 39.

Chapter 8 Test

1. 37% **2.** 21% **3.** 12% **4.** 90,000 automobiles **5.** 81,000 automobiles **6.** 350 cars **7.** 500 cars **8.** 3rd quarter 1993 **9.** 1st quarter **10.** 50 cars more **11.** 150 cars more **12.** 20 yrs **13.** 26 yrs **14.** 12 yrs **15.** Age 35 **16.** Age 65 **17.** 60,000 televisions **18.** 25,000 televisions **19.** 20,000 televisions **20.** 60,000 televisions **21.** 15.125 **22.** 15.5

Cumulative Test for Chapters 1–8

1. 20,825 **2.** 78,104 **3.** $\frac{153}{40}$ *or* $3\frac{33}{40}$ **4.** $\frac{108}{35}$ *or* $3\frac{3}{35}$ **5.** 2864.37 **6.** 72.65 **7.** 72.23 **8.** $n = 0.6$ **9.** 39 cars **10.** 0.325 **11.** 350 **12.** 1.98 m **13.** 54 ft **14.** 28.3 in.^{2} **15.** 68 in. **16.** 34% **17.** 3840 students **18.** 3 million dollars **19.** 1 million dollars **20.** 16 in. **21.** 1950, 1960 **22.** 8 students **23.** 16 students **24.** \$6.50 **25.** \$4.95

Chapter 9

Pretest Chapter 9

1. -19 **2.** -9 **3.** 4.5 **4.** 0 **5.** $-\frac{4}{12}$ *or* $-\frac{1}{3}$ **6.** $-\frac{7}{6}$ **7.** -7 **8.** 1.7 **9.** -8 **10.** -31 **11.** $\frac{14}{17}$ **12.** -12 **13.** 1.4 **14.** -2.8 **15.** 42 **16.** $\frac{19}{15}$ *or* $1\frac{4}{15}$ **17.** 24 **18.** 4 **19.** -8 **20.** -10 **21.** -24 **22.** $\frac{15}{16}$ **23.** -64 **24.** -10 **25.** -10 **26.** 31 **27.** -12 **28.** 4.8 **29.** 5 **30.** -29 **31.** $-\frac{1}{3}$ **32.** $\frac{1}{5}$ **33.** 8×10^4 **34.** 5×10^{-4} **35.** 0.0067 **36.** 1,320,000

9.1 Exercises

1. 24 **3.** -17 **5.** -7 **7.** 16.5 **9.** $\frac{17}{35}$ **11.** $-\frac{11}{12}$ **13.** 9 **15.** -7 **17.** -5 **19.** 22 **21.** -2.8 **23.** $-\frac{2}{3}$ **25.** $\frac{5}{9}$ **27.** -22 **29.** -0.7 **31.** $\frac{1}{2}$ **33.** -38 **35.** -7.6 **37.** 4 **39.** $-\frac{1}{6}$ **41.** 18.1 **43.** $-5\frac{1}{2}$ **45.** -4 **47.** -11 **49.** -6 **51.** 10 **53.** $-\frac{184}{225}$ **55.** $-\$94,000$ **57.** \$9000 **59.** $-\$38,000$ **61.** $-18°F$ **63.** $-1°F$ **65.** 904.3 ft^3 **66.** 210 m^3

9.2 Exercises

1. -6 **3.** -14 **5.** -2 **7.** 24 **9.** -18 **11.** 3 **13.** -60 **15.** 556 **17.** -90 **19.** 40 **21.** -6.7 **23.** 7.1 **25.** -8.6 **27.** 6.9 **29.** 1 **31.** $-\frac{7}{6}$ *or* $-1\frac{1}{6}$ **33.** $-\frac{1}{6}$ **35.** $\frac{3}{22}$ **37.** 15 **39.** 0 **41.** 17 **43.** -18 **45.** -2 **47.** -13 **49.** $-\frac{1}{2}$ **51.** 6121 ft **53.** 42°F **55.** $-26{,}500$ **57.** 25,000

59.

Bal.	\$50.00
Credit	\$80.00
New Bal.	\$130.00

60.

Bal.	0.00
Ck #1	−\$100.00
Ck #2	−\$50.00
New Bal.	−\$150.00

61. 5 **62.** 45

9.3 Exercises

1. -27 **3.** 24 **5.** -63 **7.** 27 **9.** 60 **11.** -160 **13.** -1.5 **15.** 3.75 **17.** $-\frac{6}{35}$ **19.** $\frac{1}{23}$ **21.** -10 **23.** 11 **25.** 14 **27.** -2 **29.** -8 **31.** 9 **33.** $-\frac{5}{6}$ **35.** $1\frac{1}{7}$ *or* $\frac{8}{7}$ **37.** -8.2 **39.** 6.2 **41.** 72 **43.** 192 **45.** 16 **47.** -30 **49.** 0 **51.** $-\frac{21}{64}$ **53.** -9.656964 **55.** 15,436.61972 **57.** 34 **59.** 2 **61.** 34 **63.** $+8$ **65.** 0 **66.** 0 **67.** b is negative **68.** c is positive **69.** 90 in.^{2} **70.** 264 m^2

9.4 Exercises

1. -1 **3.** -14 **5.** 8 **7.** -12 **9.** -19 **11.** 0 **13.** -5 **15.** -6 **17.** 20 **19.** -27 **21.** -50 **23.** -18 **25.** -25 **27.** 1 **29.** 4 **31.** $-\frac{1}{2}$ **33.** -1 **35.** $\frac{2}{3}$ **37.** $-\frac{1}{28}$ **39.** -15.7°F **41.** -8.3°F **43.** 9°F **45.** 3.84 km **46.** 36,800 mg **47.** 420 in.^{2} **48.** 330 m^2

Putting Your Skills to Work

1. 10 P.M. **2.** 5 P.M. **3.** 1 A.M. **4.** No. It is 7 A.M. the next day in the Fiji Islands.

9.5 Exercises

1. 3.5×10^1 **3.** 1.37×10^2 **5.** 2.148×10^3 **7.** 1.2×10^2 **9.** 5×10^2 **11.** 2.63×10^4 **13.** 2.88×10^5 **15.** 4.632×10^6 **17.** 1.2×10^7 **19.** 6.7×10^{-1} **21.** 3.98×10^{-1} **23.** 2.79×10^{-3} **25.** 4×10^{-1} **27.** 1.5×10^{-3} **29.** 1.6×10^{-5} **31.** 5.31×10^{-6} **33.** 7×10^{-4} **35.** 1×10^{-7} **37.** 16 **39.** 53,600 **41.** 0.062 **43.** 0.000056 **45.** 0.00063 **47.** 900,000,000,000 **49.** 0.0000003 **51.** 0.00000003862 **53.** 4,600,000,000,000 **55.** 67,210,000,000 **57.** 5.878×10^{12} miles **59.** 5.9×10^{-7} meter **61.** 2.64×10^8 people **63.** 12,500,000,000,000 insects **65.** 0.000075 centimeter **67.** 14,000,000,000 tons **69.** 9.01×10^7 dollars **71.** 7.93×10^9 dollars **73.** 4.39×10^{15} miles **75.** 3.624×10^8 feet **77.** 16.6334 **78.** 11.7216 **79.** 0.258 **80.** 0.362

Chapter 9 Review Problems

1. 14 **2.** -20 **3.** -22 **4.** 8 **5.** -15 **6.** -14 **7.** -8.8 **8.** -8.8 **9.** $-\frac{8}{15}$ **10.** $\frac{1}{14}$ **11.** 6 **12.** -20 **13.** 11 **14.** 2 **15.** 4 **16.** -4 **17.** -4 **18.** -18 **19.** -9 **20.** -13 **21.** -15 **22.** 7 **23.** 19 **24.** 17 **25.** -1.6 **26.** -12.3 **27.** $-\frac{1}{15}$ **28.** $\frac{7}{12}$ **29.** 13 **30.** -3 **31.** -9 **32.** -12 **33.** -14 **34.** -56 **35.** 50 **36.** 48 **37.** $\frac{2}{35}$ **38.** $-\frac{1}{6}$ **39.** -7.8 **40.** 4.32 **41.** 3 **42.** 6 **43.** -9 **44.** -5 **45.** 6 **46.** -15 **47.** $-\frac{7}{10}$ **48.** $\frac{3}{7}$ **49.** 30 **50.** 48 **51.** -112 **52.** -18 **53.** -34 **54.** 1 **55.** 14 **56.** -54 **57.** 13 **58.** 19 **59.** -11 **60.** 1 **61.** -5 **62.** 23 **63.** -8 **64.** 12 **65.** -72 **66.** 1 **67.** $-\frac{9}{5}$ or $-1\frac{4}{5}$ **68.** $\frac{1}{10}$ **69.** 1 **70.** 4.16×10^3 **71.** 3.7×10^6 **72.** 2.18×10^5 **73.** 4.732×10^{10} **74.** 4.0×10^{-3} **75.** 7.0×10^{-3} **76.** 2.18×10^{-5} **77.** 7.63×10^{-6} **78.** 5.136×10^{-1} **79.** 2.173×10^{-2} **80.** 9,000,000 **81.** 700,000 **82.** 18,900 **83.** 3760 **84.** 0.0752 **85.** 0.00661 **86.** 0.0000009 **87.** 0.00000008 **88.** 0.000536 **89.** 0.0000198 **90.** 8.44×10^{11} **91.** 9.72×10^{15} **92.** 1.342×10^{31} **93.** 1.443×10^{27} **94.** 1.44×10^{14} **95.** 6.8×10^{25}

Chapter 9 Test

1. -5 **2.** -32 **3.** 3.8 **4.** -6 **5.** -3 **6.** $-\frac{7}{8}$ **7.** -38 **8.** 5 **9.** $\frac{17}{15}$ *or* $1\frac{2}{15}$ **10.** -43 **11.** 4 **12.** 2.1 **13.** $\frac{11}{12}$ **14.** 46 **15.** 120 **16.** -8 **17.** 10 **18.** -18 **19.** 3 **20.** $-\frac{7}{10}$ **21.** 56 **22.** -32 **23.** 17 **24.** 14 **25.** -8 **26.** -88 **27.** -32 **28.** 22 **29.** $-\frac{1}{7}$ **30.** $-\frac{1}{6}$ **31.** 8.054×10^4 **32.** 7×10^{-6} **33.** 0.0000936 **34.** 72,000

Cumulative Test for Chapters 1–9

1. 12,383 **2.** 127 **3.** $\frac{143}{12}$ *or* $11\frac{11}{12}$ **4.** $\frac{55}{12}$ *or* $4\frac{7}{12}$ **5.** 9.812 **6.** 63.46 **7.** 65.9968 **8.** $n = 64$ **9.** 126 defective parts **10.** 0.304 **11.** 4000 **12.** 94,000 m **13.** 5 yd **14.** 78.5 m^2 **15. (a)** 300 students **(b)** 1100 students **16.** 13 **17.** -4.7 **18.** $\frac{5}{12}$ **19.** -11 **20.** -5 **21.** -60 **22.** $\frac{6}{5}$ *or* $1\frac{1}{5}$ **23.** 18 **24.** 4 **25.** -2 **26.** $\frac{4}{15}$ **27.** 5.79863×10^5 **28.** 7.8×10^{-4} **29.** 38,500,000 **30.** 0.00007

Chapter 10

Pretest Chapter 10

1. $-6x$ **2.** y **3.** $-3a + 2b$ **4.** $-12x - 4y + 10$ **5.** $16x - 2y - 6$ **6.** $4a - 12b + 3c$ **7.** $42x - 18y$ **8.** $-3a - 15b + 3$ **9.** $-3a - 6b + 12c + 10$ **10.** $x - 8y$ **11.** $x = 37$ **12.** $x = 3.5$ **13.** $x = \frac{5}{4}$ *or* $1\frac{1}{4}$ **14.** $x = -8$ **15.** $x = 5$ **16.** $x = \frac{3}{2}$ *or* $1\frac{1}{2}$ **17.** $x = 7$ **18.** $x = 1.6$ **19.** $x = 3$ **20.** $x = 2.25$ **21.** $c = 9 + p$ **22.** $l = 2w + 5$ **23.** a = height of Mt. Ararat; $a - 1758$ = height of Mt. Hood **24.** width = 19 m; length = 35 m **25.** 7.75 ft; 10.25 ft

10.1 Exercises

1. G, x, y **3.** S, r **5.** p, a, b **7.** $p = 4s^2$ **9.** $A = \dfrac{bh}{2}$ **11.** $V = \dfrac{4\pi r^3}{3}$ **13.** $H = 2a - 3b$ **15.** $17x$ **17.** $10x$ **19.** $-x$ **21.** $-21x$ **23.** $4x + 1$ **25.** $-1.1x + 6.4$ **27.** $-2.1x - 1.3$ **29.** $-32x - 3$ **31.** $2x - 3y - 4$ **33.** $13a - 5b - 7c$ **35.** $-3r - 8s - 5$ **37.** $8n - 29p + 4q$ **39.** $-28.5x - 4.4$ **41.** 0 **43.** $-2x - 12.9y + 6.8$ **45.** 4.8 **46.** 8.1 **47.** 3 **48.** 8.5

10.2 Exercises

5. $5x + 35$ **7.** $8x + 28$ **9.** $-2x - 2y$ **11.** $-10.5x + 21y$ **13.** $30x - 70y$ **15.** $2x + 6y - 10$ **17.** $-3a - 3b - 12c$ **19.** $-16a + 40b - 8c$ **21.** $-180a + 33b + 100.5$ **23.** $4a - 12b + 20c + 28$ **25.** $-2.6x + 17y + 10z - 24$ **27.** $x - \dfrac{3}{2}y + 2z - \dfrac{1}{4}$ **29.** $A = \dfrac{hB + hb}{2}$ **31.** $8x + y$ **33.** $27x - 39$ **35.** $-5a + 18b$ **37.** $14x + 12y - 28$ **39.** $8.1x + 8.1y$ **41.** $-9a + 7b - 2c$ **43.** $A = ab + ac$; $A = a(b + c)$ **44.** $A = ab - ac$; $A = a(b - c)$ **45.** 37.68 in. **46.** 200.96 in.^2 **47.** 77 cm.^2 **48.** 301.4 in.^3

10.3 Exercises

5. $y = 14$ **7.** $x = 9$ **9.** $x = -18$ **11.** $x = -8$ **13.** $x = 9$ **15.** $x = 19$ **17.** $x = 3.2$ **19.** $x = 9.8$ **21.** $x = -8.7$ **23.** $x = -0.4$ **25.** $x = \dfrac{2}{4}$ *or* $\dfrac{1}{2}$ **27.** $x = \dfrac{5}{5}$ *or* 1 **29.** $x = -\dfrac{9}{6}$ *or* $-1\dfrac{1}{2}$ **31.** $x = -\dfrac{7}{20}$ **33.** $x = 14$ **35.** $x = -12$ **37.** $x = 2$ **39.** $x = -7$ **41.** $y = 6\dfrac{1}{2}$ **43.** $z = -8$ **45.** $y = -10.9$ **47.** $x = 6$ **49.** To solve the equation $3x = 12$, divide both sides of the equation by 3 so that x stands alone. **50.** $-x$ **51.** $6a - 4b$ **52.** $3x + 3y + 3$ **53.** $-x + 21y + 3$ **54.** 5.6%

10.4 Exercises

5. $x = 9$ **7.** $y = -4$ **9.** $x = -2$ **11.** $x = 8$ **13.** $9 = n$ **15.** $-9 = m$ **17.** $x = 6$ **19.** $x = 10$ **21.** $z = 1.8$ **23.** $x = -4.5$ **25.** $x = 9$ **27.** $y = 10$ **29.** $n = \dfrac{5}{4}$ *or* $1\dfrac{1}{4}$ **31.** $x = \dfrac{2}{3}$ **33.** $x = \dfrac{8}{3}$ *or* $2\dfrac{2}{3}$ **35.** $z = 8$ **37.** Undo the multiplication. $4.5 \div 0.5 = 9$; Now divide. $9 \div 0.5 = 18$; 18 is the correct answer. **38.** Undo the multiplication. $4\left(\dfrac{3}{2}\right) = 6$; Now multiply by the reciprocal. $6\left(\dfrac{3}{2}\right) = 9$; 9 is the correct answer. **39.** -20 **40.** 14 **41.** $4x - 7y + 6$ **42.** $-8a + 3b + 8$ **43.** 401.5% **44.** 10.4%

10.5 Exercises

1. Yes **3.** No **5.** $4x = -4 \quad x = -1$ **7.** $12x = 36 \quad x = 3$ **9.** $9x = -4 \quad x = -\frac{4}{9}$ **11.** $-5x = 25 \quad x = -5$
13. $-12x = -10 \quad x = \frac{5}{6}$ **15.** $-6x = -12 \quad x = 2$ **17.** $-2x = -14 \quad x = 7$ **19.** $-3x = -15 \quad x = 5$ **21.** $-3y = 2 \quad y = -\frac{2}{3}$
23. $2y = -14y = -7$ **25.** $3y = -18 \quad y = -6$ **27.** $-2y = -14 \quad y = 7$ **29.** $11z = 22 \quad z = 2$ **31.** $36 = 12y \quad y = 3$
33. $5x + 20 = 4x + 15 \quad x = -5$ **35.** $3x + 12 + 2x = 7 \quad x = -1$ **37.** $3x = 9 \quad x = 3$ **39.** $2y + 6 = 4y + 13 \quad y = -\frac{7}{2}$
41. $3x - 18 = 4x - 18 \quad x = 0$ **43.** $x + 0.1 = 2x + 0.1 \quad -x = 0 \quad x = 0$ **45. (a)** $-3x = -15 \quad x = 5$ **(b)** $15 = 3x \quad 5 = x$
46. (a) $8 + 5x = 3x - 6 \quad 2x = -14 \quad x = -7$ **(b)** $14 = -2x \quad -7 = x$ **47.** $407{,}513.4 \text{ cm}^3$ **48.** 23.4 in.^2

Putting Your Skills to Work

1. $N = (1.395)(c)$ **2.** $N = (1.186)(c)$ **3.** $N = (1.022)(c)$

10.6 Exercises

1. $e = 7 + m$ **3.** $h = 42 + f$ **5.** $b = n - 107$ **7.** $f = s - 111$ **9.** $d = e - 5$ **11.** $l = 2w + 7$ **13.** $s = 3f - 2$
15. $n = 3000 + 3s$ **17.** $j + s = 26$ **19.** $ht = 500$ **21.** c = Charlie's salary $c + 600$ = Jim's salary **23.** j = Julie's mileage
$j + 386$ = Barbara's mileage **25.** b = number of degrees in angle B $b - 46°$ = number of degrees in angle A
27. w = height of Mt. Whitney in meters $w + 4430$ = height of Mt. Everest in meters **29.** e = classes taken by Ernesto $2e$ = classes taken by Wally **31.** x = width $3x + 12$ = length **33.** x = 1st angle $2x$ = second angle $x - 14$ = third angle **35.** 8 **36.** 1 **37.** -1
38. -18

10.7 Exercises

1. x = length of shorter piece
$x + 5.5$ = length of longer piece
$x + 5.5 + x = 16$
$x = 5.25$ ft $\quad x + 5.5 = 10.75$ ft

3. x = no. of miles driven 2nd day
$x + 97$ = no. of miles driven 1st day
$x + 97 + x = 560$
$x = 231.5$ mi $\quad x + 97 = 328.5$ mi

5. Jake can leg press 405 lb.
Jodie can leg press 265 lb.

7. 119 cars in November; 203 cars in May; 76 cars in July

9. The shorter piece is 3.65 feet long. The longer piece is 8.35 feet long.
11. x = width of rectangle; $3x - 2$ = length of rectangle; $2x + 2(3x - 2) = 52$; $x = 7$ cm $3x - 2 = 19$ cm
13. The sides are 8 cm, 16 cm, and 20 cm.
15. The sides are 61 mm, 81 mm, and 57 mm.
17. x = no. of degrees in angle A $3x$ = no. of degrees in angle B; $x + 40°$ = no. of degrees in angle C; $x + 3x + x + 40 = 180$; $x = 28°$ $3x = 84°$ $x + 40 = 68°$
19. x = total sales $2000 + 0.02x = \$2800$ $x = \$40{,}000$ **21.** x = yearly rent $100 + 0.12x = 820$; $0.12x = 720$ $x = \$6000$
23. The population last year was 2,838,500. **25.** Width = 476.6628 yd Length = 961.5322 yd **27.** First program = \$70.37; second program = \$120.74; third program = \$379.22 **29.** 57.75 **30.** 60% **31.** 500 **32.** 20.8

Chapter 10 Review Problems

1. $-5x + 2y$ **2.** $-2x + 6y$ **3.** $-a + 12$ **4.** $8a + 2$ **5.** $-2x - 7y$ **6.** $11x - 5y$ **7.** $-x - 12y + 6$ **8.** $2x - 10y - 7$
9. $-2.8a + 3.4b - 2$ **10.** $-7.5a + 9b - 4$ **11.** $-15x - 3y$ **12.** $-8x - 12y$ **13.** $2x - 6y + 8$ **14.** $6x - 18y - 3$
15. $-24a + 40b + 8c$ **16.** $-18a + 72b + 9c$ **17.** $6x + 15y - 27.5$ **18.** $8.4x - 12y + 20.4$ **19.** $-2x + 14y$ **20.** $7x - 8y$
21. $-8a - 2b - 24$ **22.** $-7a + 8b + 15$ **23.** $x = 12$ **24.** $x = 8$ **25.** $x = 4$ **26.** $x = -8$ **27.** $x = 11.7$ **28.** $x = 7.4$
29. $x = -12.1$ **30.** $x = 6.8$ **31.** $x = 2\frac{3}{4}$ **32.** $x = 3\frac{1}{4}$ **33.** $x = \frac{1}{8}$ **34.** $x = \frac{9}{6}$ or $1\frac{1}{2}$ **35.** $x = 6$ **36.** $x = -4$
37. $x = 5$ **38.** $x = -12$ **39.** $x = 8$ **40.** $y = 15$ **41.** $x = -12$ **42.** $y = -6$ **43.** $x = 6$ **44.** $y = 7$ **45.** $x = -5$
46. $x = 0.25$ **47.** $x = 8$ **48.** $x = \frac{5}{6}$ **49.** $x = 21$ **50.** $x = 6$ **51.** $x = 5$ **52.** $x = 5$ **53.** $x = 1$ **54.** $x = 5$ **55.** $x = 8$
56. $x = 5$ **57.** $x = 3$ **58.** $x = -10$ **59.** $x = \frac{29}{5}$ **60.** $x = \frac{4}{5}$ **61.** $x = -2$ **62.** $x = 5$ **63.** $y = 16$ **64.** $y = -28$
65. $w = c + 3000$ **66.** $m = a + 18$ **67.** $A = 3B$ **68.** $l = 2w - 3$ **69.** r = Roberto's salary $r + 2050$ = Michael's salary
70. x = length 1st side $2x$ = length 2nd side **71.** c = number of Connie's courses; $c - 6$ = number of Nancy's courses
72. b = number of books in old library; $2b + 450$ = number of books in new library **73.** 26.75 ft 33.25 ft **74.** \$192 \$220
75. Feb. = 10,550 Mar. = 21,100 Apr. = 13,550 **76.** Thurs. = 263 mi Fri. = 369 mi Sat. = 224 mi **77.** width = 13 in. length = 23 in.
78. width = 22 m length = 68 m **79.** 1st side = 32 m 2nd side = 39 m 3rd side = 28 m **80.** $A = 95.2°$ $B = 21.2°$ $C = 63.6°$

Chapter 10 Test

1. $-6a$ **2.** $-2b$ **3.** $-11x + 2y$ **4.** $a - 8b + 11$ **5.** $7x - 7z$ **6.** $-4x - 2y + 5$ **7.** $60x - 25y$ **8.** $-8x + 12y - 28$ **9.** $-4.5a + 3b - 1.5c + 12$ **10.** $-11a + 14b$ **11.** $x = -8$ **12.** $x = -6$ **13.** $x = 15$ **14.** $x = -\frac{1}{2}$ **15.** $x = 1$ **16.** $x = -\frac{5}{2}$ *or* $-2\frac{1}{2}$ **17.** $s = f + 15$ **18.** $n = s - 15{,}000$ **19.** $\frac{1}{2}s$ = first; s = second; $2s$ = third **20.** w = width; $2w - 5$ = length **21.** Prentice farm = 87 acres; Smithfield farm = 261 acres **22.** Marcia = \$24,000; Sam = \$22,500 **23.** Monday = 311 mi; Tuesday = 367 mi; Wednesday = 297 mi **24.** width = 25 ft; length = 34 ft

Cumulative Test for Chapters 1–10

1. 747 **2.** 10,815 **3.** 45,678,900 **4.** 2 **5.** $\frac{65}{8}$ *or* $8\frac{1}{8}$ **6.** 0.07254 **7.** 30,550.301 **8.** $n = 16$ **9.** 1596 **10.** 5500 **11.** 0.345 m **12.** 120 in. **13.** 37.7 yd **14.** 143 m^2 **15.** -31 **16.** 30 **17.** $-9a - 11b$ **18.** $-6x + 2y + 3$ **19.** $21x - 7y + 56$ **20.** $-2x - 24y$ **21.** $x = 4$ **22.** $y = -\frac{5}{2}$ *or* $-2\frac{1}{2}$ **23.** $x = 26$ **24.** $x = -3$ **25.** p = weight of printer; $p + 322$ = weight of computer **26.** f = students during fall; $f - 87$ = students during summer **27.** Thursday = 376 miles; Friday = 424 miles; Saturday = 281 miles **28.** width = $13\frac{2}{3}$ ft; length = $35\frac{1}{3}$ ft

Practice Final Examination

1. eighty-two thousand, three hundred sixty-seven **2.** 30,333 **3.** 173 **4.** 34,103 **5.** 4212 **6.** 217,745 **7.** 158 **8.** 606 **9.** 116 **10.** 32 mi/gal **11.** $\frac{7}{15}$ **12.** $\frac{42}{11}$ **13.** $\frac{33}{20}$ *or* $1\frac{13}{20}$ **14.** $\frac{89}{15}$ *or* $5\frac{14}{15}$ **15.** $\frac{31}{14}$ *or* $2\frac{3}{14}$ **16.** 4 **17.** $\frac{14}{5}$ *or* $2\frac{4}{5}$ **18.** $\frac{22}{13}$ *or* $1\frac{9}{13}$ **19.** $6\frac{17}{20}$ mi **20.** 5 packages **21.** 0.719 **22.** $\frac{43}{50}$ **23.** $>$ **24.** 506.38 **25.** 21.77 **26.** 0.757 **27.** 0.492 **28.** 3.69 **29.** 0.8125 **30.** 0.7056 **31.** $\frac{1400 \text{ students}}{43 \text{ faculty}}$ **32.** False **33.** $n \approx 9.4$ **34.** $n \approx 7.7$ **35.** $n = 15$ **36.** $n = 9$ **37.** \$3333.33 **38.** 9.75 in. **39.** \$294.12 was withheld **40.** 1.6 lb of butter **41.** 0.63% **42.** 21.25% **43.** 1.64 **44.** 17.3% **45.** 302.4 **46.** 250 **47.** 4284 **48.** \$10,856 **49.** 4500 students **50.** 34.3% increase **51.** 4.25 gal **52.** 6500 lb **53.** 192 in. **54.** 5600 m **55.** 0.0698 kg **56.** 0.00248 L **57.** 19.32 km **58.** 6.3182×10^{-4} **59.** 1.264×10^{11} **60.** 1.36728 cm thick **61.** 14.4 m **62.** 206 cm **63.** 5.4 sq ft **64.** 75 sq m **65.** 113.04 sq m **66.** 56.52 m **67.** 167.47 cu cm **68.** 205.2 cu ft **69.** 32.5 sq m **70.** $n = 32.5$ **71.** 8 million dollars **72.** one million dollars **73.** 50°F **74.** From 1980 to 1990 **75.** 600 students **76.** 1400 students **77.** mean ≈ 15.83; median = 16.5 **78.** 16 **79.** 11.091 **80.** 15 ft **81.** -13 **82.** $\frac{1}{8}$ **83.** -3 **84.** -17 **85.** 24 **86.** $-\frac{8}{3}$ *or* $-2\frac{2}{3}$ **87.** 4 **88.** 27 **89.** 8 **90.** 0.5 *or* $\frac{1}{2}$ **91.** $-3x - 7y$ **92.** $-7 - 4a - 17b$ **93.** $-2x + 6y + 10$ **94.** $-11x - 9y - 4$ **95.** $x = 2$ **96.** $x = -2$ **97.** $x = -\frac{1}{2}$ *or* $x = -0.5$ **98.** $x = \frac{-2}{5}$ *or* $x = -0.4$ **99.** 122 students are taking history. 110 students are taking math. **100.** The length is 37 m. The width is 16 m.

Appendix B

1. 11,747 **3.** 352,554 **5.** 1.294 **7.** 254 **9.** 8934.7376; 123.45 + 45.9876 + 8765.3 **11.** 16.81; $\frac{34}{8} + 12.56$ **13.** 10.03 **15.** 289.387 **17.** 378,224 **19.** 690,284 **21.** 1068.678 **23.** 1.58688 **25.** 1383.2 **27.** 0.7634 **29.** 16.03 **31.** \$12,562.34 **33.** 85,163.7 **35.** \$103,608.65 **37.** 15,782.038 **39.** 2.500344 **41.** 897 **43.** 22.5 **45.** 84.372 **47.** 4.435 **49.** 101.5673 **51.** 0.24685 **53.** 30.7572 **55.** -13.87849 **57.** -13.0384 **59.** 3.3174×10^{18} **61.** $4.630985915 \times 10^{11}$ **63.** -0.5517241379 **65.** 15.90974544 **67.** 1.378368601 **69.** 11.981752 **71.** 8.090444982 **73.** 8.325724507 **75.** 1.08120 **77.** 1.14773 **79.** 18,307.52 **81.** 5890

Applications Index

Index

Photo Credits

CHAPTER 1 **CO** Don Mason/The Stock Market **p. 5** David Young-Wolff/PhotoEdit **p. 23** Crandall/The Image Works **p. 52** Michael Newman/PhotoEdit **p. 53** James Marshall/The Stock Market **p. 57** Robert Essel/The Stock Market **p. 60** Randy Duchaine/The Stock Market **p. 64** NASA **p. 67** Mugshots/The Stock Market **p. 73 (left), (right)** Peter Gisolfi Associates Architects

CHAPTER 2 **CO** Mark C. Burnett/Stock Boston **p. 99** Mugshots/The Stock Market **p. 111** Cameramann/The Image Works **p. 118** Jeff Greenberg/PhotoEdit **p. 120** Bob Daemmrich/The Image Works **p. 132** Siteman/Monkmeyer Press **p. 149** Michael Newman/PhotoEdit **p. 158** David Ulmer/Stock Boston **p. 160** Mary Kate Denny/PhotoEdit **p. 161** George Holz/The Image Works

CHAPTER 3 **CO** Renate Hiller/Monkmeyer Press **p. 179** M. Antman/The Image Works **p. 181** Bill Horsman/Stock Boston **p. 187** Tony Freeman/PhotoEdit **p. 192** Grantpix/Monkmeyer Press **p. 199** Robert Brenner/PhotoEdit **p. 207** Ariel Skelley/The Stock Market **p. 221** LeDuc/Monkmeyer Press **p. 225** Wolfgang Spunbarg/PhotoEdit

CHAPTER 4 **CO** Bopp/Monkmeyer Press **p. 243** Jack Novak/The Stock Market **p. 259** Elizabeth Zuckerman/PhotoEdit **p. 263** Daemmrich/Stock Boston **p. 265** Tony Freeman/PhotoEdit **p. 276** Deborah Davis/PhotoEdit

CHAPTER 5 **CO** Tony Arruza/Tony Stone Images **p. 285** Peter Chapman/Stock Boston **p. 292** Berenholtz/The Stock Market **p. 303** Robert Cerri/The Stock Market **p. 315** Robert Brenner/PhotoEdit **p. 324** Don Mason/The Stock Market

CHAPTER 6 **CO** Michael Keller/The Stock Market **p. 339** Bachmann/The Image Works **p. 341** Terry Vine/Tony Stone Images **p. 347** Lee Snider/The Image Works **p. 364** Ken Kerbs/Monkmeyer Press

CHAPTER 7 **p. 381** Chris Sorensen/The Stock Market **p. 382** Bill Luster/Matrix International **p. 392** Daniel MacDonald **p. 412** Tom McCarthy/PhotoEdit **p. 419** Matthew Borkoski/Stock Boston **p. 422** Richard V. Procopio/Stock Boston **p. 429** Bill Gallery/Stock Boston **p. 443** John Elk/Stock Boston **p. 444** Daemmrich/Stock Boston

CHAPTER 8 **CO** Myrleen Ferguson/PhotoEdit **p. 485** Bob Daemmrich/Stock Boston

CHAPTER 9 **CO** Ken Straiton/The Stock Market **p. 505** John Henley/The Stock Market **p. 517** Ken Straiton/The Stock Market **p. 520** David Schaefer/Monkmeyer Press **p. 532** Julian Baum/Science Photo Library/Photo Researchers, Inc.

CHAPTER 10 **CO** Daemmrich/The Image Works **p. 575** Wells/The Image Works **p. 589** Dean Abramson/Stock Boston